Rotifera X

Developments in Hydrobiology 181

Series editor

K. Martens

Rotifera X

Rotifer Research: Trends, New Tools and Recent Advances

Proceedings of the Xth International Rotifer Symposium, held in Illmitz, Austria, 7–13 June 2003

Edited by

Alois Herzig,[1] Ramesh D. Gulati,[2]
Christian D. Jersabek[3] & Linda May[4]

[1]*Biological Station Neusiedler See, Illmitz, Austria*
[2]*NIOO, Centre of Limnology, Nieuwersluis, The Netherlands*
[3]*University of Salzburg, Austria*
[4]*Centre for Ecology and Hydrology, Penicuik, Midlothian, Scotland*

Reprinted from Hydrobiologia, volume 546 (2005)

Springer

Library of Congress Cataloging-in-Publication Data

A C.I.P. Catalogue record for this book is available from the Library of Congress.

ISBN 1-4020-3493-8

Published by Springer,
P.O. Box 17, 3300 AA Dordrecht, The Netherlands

Cover illustration: *Hexarthra polyodonta*, drawing by Walter Koste

Printed on acid-free paper

Printed in the Netherlands

TABLE OF CONTENTS

PART V: MATING, RESTING EGGS, DIAPAUSE, ANHYDROBIOSIS, EMBRYONIC DEVELOPMENT

PART VI: POPULATION AND COMMUNITY ECOLOGY

PART IX: AQUACULTURE AND ECOTOXICOLOGY

Hydrobiologia (2005) 546:xi–xiv
A. Herzig, R.D. Gulati, C.D. Jersabek & L. May (eds.) Rotifera X: Rotifer Research: Trends, New Tools and Recent Advances
DOI 10.1007/s10750-005-4085-6

Preface

The Xth International Rotifer Symposium was held in Illmitz, Austria, 7–13 June 2003, at the Information Centre of the National Park Neusiedler See – Seewinkel. The Symposium was returning to Austria 27 years after the first rotifer meeting was organized there by Prof. Agnes Ruttner-Kolisko at the Biological Station Lunz in 1976. The Xth meeting was attended by 113 participants from 28 countries. It was organized by Alois Herzig with the assistance of Christian Jersabek, Institute of Zoology, University of Salzburg and Alois Lang, Information Centre of the Nationalpark Neusiedler See – Seewinkel. It was hosted by the Biological Station Neusiedler See (Provincial Administration of Burgenland) and the National Park Society. The symposium venue provided an excellent opportunity for the community of rotifer researchers to follow the scientific programme combined with enjoyable breaks and nice sundowners.

After the opening ceremony and a short appraisal by Alois Herzig of the contents and topics of the last nine meetings, Norbert Walz paid a tribute to the life-time works of Walter Koste. Subsequently, the scientific programme followed the traditions of the previous symposia with 6 invited main lectures, 56 oral contributions and 45 poster presentations. The papers were grouped into thematic sessions: Phylogeny, evolution and genetics; Molecular ecology; Biogeography and development; Diapause, anhydrobiosis and resting eggs; Morphology, ultrastructure and behaviour; Feeding; Population ecology; Culture of rotifers; Physiology and ecotoxicology. In addition, two late afternoon sessions were devoted to the careers in rotiferology of John Gilbert and Henri Dumont.

Special thanks to John J. Gilbert, Ramesh D. Gulati, Charles E. King, Linda May, Claudia Ricci, Terry W. Snell and Robert L. Wallace for their involvement with the arrangements for the scientific programme.

Social activities began with a Welcome Party that was held on Saturday evening at the Information Centre of the National Park. The occasion was made all the more enjoyable by the wonderful atmosphere created by a brass band playing music typical of Czech Republic, Slovakia and eastern Austria. A full-day excursion was organized on the Wednesday. The participants enjoyed a cruise around Neusiedler See, a visit and an introduction to the activities of the Wine Academy at Rust (which runs courses in the Science of Wine Making) combined with a short wine tasting, a guided tour through the old town of Rust (buildings dating back to the 16th century), and a visit to the Baroque Esterházy Palace at Eisenstadt. As an preamble to Joseph Haydn's music, a chamber concert was given in the famous Haydn Hall of the Esterházy Palace. Participants were invited to a lunch by the Head of the Government of the Province of Burgenland, which took place in representative rooms of the Palace. The excursion in the afternoon to Hungary included a visit to typical farm houses in a small village situated south of Neusiedler See (Fertöszéplak) and the baroque Esterházy Castle of Fertöd, the place where Haydn spent nearly two decades of his creative life. The Conference Dinner was hosted by the Government of the Province of Burgenland on Friday at Johanneszeche, a typical restaurant with Hungarian ambience and Croatian (Tamburizza) and gipsy music in the backdrop. Accompanying guests made several day trips to places of natural, cultural and historical interests in the Neusiedler See area and the adjacent Hungarian neighbourhood townships.

Kluwer Academic Publishers, now Springer Aquatic Sciences, and Prof. Dr. Koen Martens, *Editor-in-chief Hydrobiologia*, have accepted to publish the symposium proceedings as a special volume in the series *Developments in Hydrobiology*. The manuscripts accepted for publication have undergone a careful review and revision process and appropriate editorial amendments needed for clarity and conciseness. The final product is the result of the efforts of the authors, reviewers, editors and the Editor-in-chief.

The *Proceedings* are composed of nine parts (See Contents). The introductory paper by Funch et al. (Part I) offers a detailed discussion of the phylogenetic position of rotifers *vis-a-vis* gnathiferan groups. Originally, Gnathifera only comprised the hermaphroditic Gnathostomulida and the Syndermata. On the basis of the ultrastructure of the trophy, the rotifers belong to the Gnathifera; moreover, molecular evidence strongly suggests that they are closely related to the parasitic acanthocephalans and the two together form the clade Syndermata. In his paper Mark Welch provides evidence for a monophyletic Eurotatoria based on maximum likelihood and Bayesian analysis of the protein-coding gene *hsp82* and for the placement of Acanthocephala within the Phylum Rotifera as a sister clade to either Eurotatoria or Seisonidea.

Substantial differences in both life-table characteristics and reproductive patterns distinguish bdelloid rotifers from monogonont rotifers. King et al. explore some of the adaptive consequences of these life-history differences using a computer model to simulate the evolutionary acquisition of new beneficial mutations. Birky et al., isolated more than 100 females of the obligately asexual bdelloid rotifers from nature and sequenced their mitochondrial *cox*1 genes and conclude that in the absence of sexual reproduction the bdelloids have undergone substantial cladogenesis; bdelloid clades are adapted to different niches and have undergone substantial speciation. The authors failed to detect a decrease in the effectiveness of natural selection on bdelloid genes.

The development of cost-effective molecular tools that allow the amplification of minute amounts of DNA, effectively opened the field of molecular ecology of rotifers. In Part II (*Genetics and Molecular Ecology*), Gómez critically reviews: (1) methodological advances that have facilitated the application of molecular techniques to rotifers, (2) recent advances in the field of rotifer molecular ecology, and (3) future developments and areas which are likely to benefit further from the molecular ecological approach. It is now feasible to obtain representative DNA sequences from identified rotifer species for use in genomic-based surveys for determining rotifers in new sample collections, circumventing the difficulties that go with traditional surveys. D.B. Mark Welch and J.L. Mark Welch discuss in their paper the application of two genomic-based tools used in surveys of microbial communities to rotifer taxonomy: serial analysis of gene tags (SAGT) and microarray hybridization. They also report the construction and hybridization of a small microarray of rotifer sequences, thereby demonstrating that these techniques are most powerful if combined with traditional rotifer systematics.

Bdelloids show a rather uniform morphology of jaws (trophi), most recognizably feature is the presence of a series of teeth forming unci plates, each with one to ten major median teeth. Using SEM photomicrographs of trophi and literature data, Melone and Fontaneto deduce that few major teeth are common in species living in water bodies, where these species possibly eat unicellular algae, while more major teeth are more common in species inhabiting mosses and lichens, where they possibly consume bacteria.

Modern techniques can contribute significantly to our understanding of the rotifers anatomy, especially relating to the musculature and the nervous system. Very recently, immunostaining has been applied in combination with confocal laser scanning microscopy (CLSM). Santo et al., applied CLSM to describe the muscle arrangement of *Brachionus urceolaris*, *Floscularia ringens*, *Hexarthra mira* and *Notommata glyphura*. Sørensen describes the musculature of *Testudinella patina*. Kotikova et al. presents data on the immuno-reactivity patterns in the nervous system of *Platyas patulus*, *Euchlanis dilatata* and *Asplanchna herricki* using CLSM. That species considered to be cosmopolitan can be complexes of sibling species has been recently clearly demonstrated for the rotifer *Brachionus plicatilis*: within the Iberian Peninsula three species have been described and another three have been identified (see Gómez, Part II). Gilbert and Walsh who observed mating behaviour and genetic differentiation among four geographically isolated strains of *Brachionus calyciflorus* conclude that it is a species complex in which some geographically and genetically distinct strains are reproductively isolated from one another.

Bdelloid rotifers can withstand desiccation by entering a state of suspended animation: anhydrobiosis. Ricci and Covino describe and discuss costs and benefits of anhydrobiosis of a new species, *Adineta ricciae*, a new species described by Segers and Shiel (see part III). A comparison of this with *Macrotrachela quadricornifera* reveals both species suspend their life activities during desiccation as well do not "age" during anhydrobiosis. This meets the prediction of the "Sleeping Beauty" model. Anhydrobiosis is thought to require accumulation of the non-reducing disaccharides trehalose (in animals and fungi) or sucrose (in plant seeds and resurrection plants), which may protect proteins and membranes by acting as water replacement molecules and vitrifying agents. However, in clone cultures of the bdelloids *Philodina roseola* and *Adineta vaga* Tunnacliffe et al. were unable to detect trehalose or other disaccharides. Instead, on dehydration, *P. roseola* upregulates a hydrophilic protein related to the *late embryogenesis abundant* (LEA) proteins associated with desiccation tolerance in plants. It could well be that hydrophilic biosynthesis represents a common element of anhydrobiosis.

A number of studies examine spatial and temporal variability in rotifer abundance and community structure in relation to variations in abiotic and biotic factors along gradients in natural water bodies, and in laboratory. M. Gophen presents the results of a long-term study on seasonal rotifer dynamics in Lake Kinneret (Israel) and T.E. Andrew and J.A.M. Andrew in Lough Neagh (Northern Ireland). N. Lair focussed on rotifers in river plankton, especially the response of various rotifer taxa to hydraulic conditions, and the role of rotifers in the river food web.

As a special topic, Z. Brandl reviews the trophic relationship between freshwater copepods and rotifers. Most copepod species, at least in their later developmental stages, predate efficiently, preferably on rotifers. Generally, soft-bodied species of rotifers are more vulnerable to predation than those that possess spines or lorica. But also behavioural characteristics, e.g., movements and escape reactions, and temporal and spatial distribution of rotifers are important in these trophic interactions. The reported predation rates by freshwater copepods can effectively exercise a top-down control of rotifer populations.

During the preparation of the present proceedings, we heard about the sudden death of Andrzej Karabin, who had designed logos for previous Rotifer Symposia as well as contributed a paper to this volume. On behalf of the rotifer family, we extend our profound sympathy to Andrzej's wife, Jolanta Ejsmont-Karabin.

The editors wish to thank the "local organizers" for their invaluable help during the symposium. They also wish to acknowledge the Government of the Province of Burgenland for financial support.

The Editors

Alois Herzig
Ramesh D. Gulati
Christian D. Jersabek
Linda May

List of Reviewers

Ahlrichs, W.
Andrew, T.E.
Birky, Jr., C.W.
Carmona, M.J.
Carvalho, G.R.
De Meester, L.
De Paggi, J.
De Smet, W.H.
Dingmann, B.J.
Duggan, I.C.
Dumont, H.J.
Ejsmont-Karabin, J.
Ferrando, M.D.
Fussmann, G.
Gilbert, J.J.
Gómez, Á.
Gophen, M.
Green, J.
Gulati, R.D.
Hagiwara, A.
Hampton, S.E.
Herzig, A.
Hessen, D.
Jersabek, C.D.
Joaquim-Justo, C.
King, C.E.
Lair, N.
Lubzens, E.
Mark Welch, D.B.
May, L.

Melone, G.
Miracle, M.R.
Morales-Baquero, R.
Nandini, S.
Nogrady, T.
Ortells, R.
Raikova, O.I.
Rao, T.R.
Reckendorfer, W.
Ricci, C.
Rico-Martinez, R.
Rieger, R.
Sanoamuang, L.
Sarma, S.S.S.
Schmid-Araya, J.M.
Schröder, T.
Segers, H.
Serra, M.
Shiel, R.J.
Snell, T.W.
Starkweather, P.L.
Stemberger, R.
Tunnacliffe, A.
Van der Stap, I.
Viroux, L.
Wallace, R.L.
Walsh, E.J.
Walz, N.
Wilke, T.
Winkler, H.
Yúfera, M.
Zimmermann-Timm, H.

Hydrobiologia (2005) 546:1–8

A. Herzig, R.D. Gulati, C.D. Jersabek & L. May (eds.) Rotifera X: Rotifer Research: Trends, New Tools and Recent Advances
DOI 10.1007/s10750-005-4092-7

Walter Koste - a K-strategist? A laudatio

Norbert Walz
Leibniz Institute of Freshwater Ecology and Inland Fisheries, Mueggelseedamm 301, D-12555 Berlin, Germany
E-mail: walz@igb-berlin.de

Key words: Rotifers, taxonomy, publications, long life

Abstract

Walter Koste is one of the architects of modern rotiferology. Here I take a look at his scientific career using ecological concepts. Although he has more than 150 publications to his credit, he did not start publishing his works on rotifers before the age of 49, and then most of his publications appeared after his retirement. Long life and late reproduction are characteristics for K-strategists. Thus, it appears that Koste has followed K-strategy for his contributions to rotiferology.

Introduction

On July 19th, 2003 Walter Koste celebrated his 91th birthday (Fig. 1). At the Xth Rotifer Symposium in Illmitz in 2003 this laudatio was presented in his honour. For additional information on this remarkable scientist the reader is requested to consult Ruth Laxhuber's review of his curriculum vitae (Laxhuber, 1993) and Jürgen Schwoerbel's tribute presented on the occasion of his 75th birthday (Schwoerbel, 1987).

Although in retrospect Koste's contributions have been outstanding, the way to his success was hard and difficult. While he was interested in biology, especially in aspects of biodiversity since his school days, circumstances prevented him from attending university. This did not deter Koste's innate interest to become skilled at about rotifers. After World War II he became a teacher and a headmaster of a one-class country school in a village in North-Western Germany. After additional training in education at the college level he received a certificate to teach at the secondary school level for biology and geography. Later he moved to Quakenbrück, a small town in Lower Saxony, where until his retirement he was headmaster of the Artland Secondary School. He still resides in this small town.

Even though he did not start to publish his works until he became 49 years Koste has more than 150 publications to his credit, that is the envy of many a university professor. Moreover, a good deal of this work was accomplished during his leisure time, i.e., after his school duties. On his retirement from teaching he continued to publish thereby establishing himself as one of the grand dons of rotiferology. Long life and long and late reproduction of qualitatively good offspring are the life history characteristics termed K-strategy (MacArthur & Wilson, 1967). K-strategists are more competitive and use their energy more efficiently. In contrast, r-strategists have rather high reproduction rates early in their life, compensating higher mortality of offspring. Based on these criteria I pose the question: is Walter Koste a r- or K-strategist?

Results

Koste's publication career began in 1961 and continued in a modest way with a total of four papers by the mid 1960s. At that point his efforts picked up momentum, so much so that by the end of 1970 he had already quadrupled the pace of his output of his early years. Koste continued to be

Figure 1. Walter Koste at the occasion of his 91st birthday at July 19th, 2003.

highly productive until his retirement in 1974. Nevertheless, retirement and aging did not slow down Koste's productivity for nearly 75% of his works, including his revision of Voigt's taxonomy of the rotifers (see below), were published after he had stopped formal teaching. However, teaching never left Koste's blood, as was witnessed by many at the rotifer symposia he attended. It was here that a second type of contribution to the rotifer world could be noted: the way Koste helped his colleagues sort out the details of rotifer taxonomy.

Koste's early studies were on the rotifer fauna of the countryside of north-western Germany, near his residence. Here he found many species for the first time. His very successful series, called "Das Rädertier-Portrait" (the rotifer portrait), published in the monthly journal Mikrokosmos, is exemplary of the way Koste wedded his teaching interests with his scientific explorations of rotifers. In this series he examines different rotifers both on a scientific level as well as making them understandable for the general public. Through these early years he worked on a revision of the identification guide of M. Voigt "Rotatoria – the Rotifers of Middle Europe" which he published in two volumes in 1978 (Koste 1978). This enormous effort brought him a break-through in public perception. This book demonstrated for the first time to a wide public arena Koste's capacities to combine exact scientific observations of high accuracy with excellent artistic illustrations. Through this works we get a glimpse of how deeply impressed Koste was by the famous book of Ernst Haeckel (1899): "Art Forms in Nature", which he received as a present from a former teacher when he was about 16 years old.

Recognition of the importance of "Rotatoria" by the scientific community was a critical point in Koste's scientific career as it legitimised him in the field as a major force, but it also allowed him to expand his studies from Middle Europe to the whole world. Beginning about 1980 many scientists began to send him rotifer samples for identification from different countries and this contributed to his success as expert in the rotifer fauna from many geographical regions, including Australia and South America, especially the Amazon region, followed later by Africa, Canada, India, Sri Lanka, Malaysia, Indonesia, China, and others. Along with

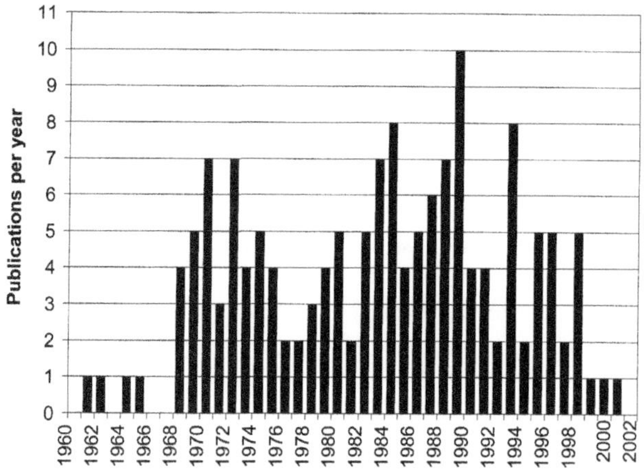

Figure 2. Number of scientific publications per year by Walter Koste.

this work he described many new species and subspecies (Schwoerbel, 1987). Altogether his publication record stands at 153 papers (Fig. 2).

One wonders if any other person will equal Koste's achievements, considering that he was equipped with only a good microscope and simple sampling equipment. Koste literally carried out all this work single handed and financed the costs from his pension as a retired teacher. I doubt very much if future generations will be able to exhibit such a patience and observe rotifers so precisely using a limiting equipment as Koste did. This is in sharp contrast with the present generation of taxonomists who are now using highly sophisticated techniques as scanning electron microscopy and molecular techniques to carry out their taxonomic studies. This, of course, is not an attempt to criticise those who aim at advancing the discipline. Clearly researchers must include other methods to reach other levels of understanding. So Koste – like a dinosaur – simultaneously marks the culmination of an ongoing development since the discovery of the animalcules by Antonio van Leeuwenhoek (1674). A few years after the publication of "Rotatoria" Koste began to win the honours appropriate to his accomplishments. In 1980 the University of Kiel promoted him to "Doctor honoris causa", a title very rarely conferred on scientists in Europe. The raison d'être given by the university for this honour makes it especially interesting: "For his achievements in the assessment of freshwater and marine life. For his exemplary and outstanding achievements in the taxonomy, biogeography, biology and ecology of rotifers," and: "For his efforts in imparting his knowledge and his enthusiasm for science to following generations". The latter has been Koste's hobby horse. For many years he gave courses at the University of Osnabrück (Germany) and other locations (e.g., Plön, Germany; Illmitz, Austria) in rotifer taxonomy, limnology, hydrobiology, and plankton biology as well as helping countless students in rotifer taxonomy from all over the world to identify their samples. In 1981 the Federal Order of Merit of the Federal Republic of Germany was conferred upon him. And in 1985 he was elected to a Corresponding Member of the Senckenberg Museum in Frankfurt and as an Honorary Member of the Queckett Microscope Club of the British Natural History Museum in London, and in 1986 he was established as an Associate Member of the Royal Society of South Australia of the South Australia Museum in Adelaide.

As Walter Koste had no external funding for his travels he was invited to many countries (Australia, Austria, Belgium, Brazil, China, England, France, Italy, Kenya, Romania, Scotland, Spain, Ukraine) to attend symposia or to introduce local scientists to the regional rotifer fauna at the expense of this hosting country.

Discussion

Koste's publication list of 153 papers is phenomenal! Between 1968 and 1998 he published at least 2 publications per year but often more so that he had on average 4.8 publications per year. It is however, most astonishing that he (1) did not start publishing his work until the age of 49, (2) published >75% of his work after his retirement in 1974 and (3) that he reached 77 years of age when he reached his output maximum 10 publications per annum. When Schwoerbel (1987) congratulated him to his 75th birthday he referred to Koste's publication list comprising over 100 publications, but he also reported that Koste was looking forward to go on with his scientific pursuits ... Jürgen Schwoerbel was quite right in his prediction because Koste added nearly 50% more to his already long list of publications.

Has Koste been following a K-strategy? The answer: it appears he is because long life with delayed but repeated reproduction is characteristic for this life history (Table 1). However, Koste had also numerous offspring (papers), a characteristic that is typical for r-strategists. R-strategists produce many offspring but they usually do not invest much energy in each. Clearly, this is not the case with Koste's publications. His papers have appeared in respected international journals such as Amazoniana, Annales de Limnologie, Archiv für Hydrobiologie, Australian Journal of Marine and Freshwater Research, Hydrobiologia, International Review of Hydrobiology, Limnologica, and Transactions of the Royal Society of South Australia. In fact, Koste has succeeded in combining the characteristics of both r- and K-strategist into a super-K-strategist.

Table 1. Characteristics of r-and K-strategy according to Pianka (1978) with data from Walter Koste

Characteristics	r-strategy	K-strategy
Length of life	Short	Long *Koste: 91 years*
Reproduction	Early	Delayed *Koste: started at age 49*
Type of reproduction	Often: single reproduction	Repeated reproduction *Koste: every year*
Number of offspring	Many	Fewer offspring *Koste: 153 papersc*
Quality of offspring	Low	High *Koste: offspring in excellent journals*
Intraspecific competition	Low	High *Koste: honours and decorations*
Use of resources	Maximum energy intake	Efficient energy intake *Koste:* in *retirement since* 1974
Body size	Small	Larger *Koste:* >190 *cm*
Environment	Variable	Constant *Koste: since 1962 in Quakenbrück*
Occurrence	Local	Ubiquitous *Koste: offspring in many libraries*

Not many are able to comprehend why Koste has been investing so much energy in scientific work at an age by which other senior people would have doing so already for long time. I think he simply could not have done otherwise. This way of life has been his hobby and his passion. I suppose his daily work observing rotifers is rather a contemplation for him from where he gets power for his life and for his contact with other people. This is the way how he could sustain such a steady output even long after his retirement. With this as basis Koste likes always to meet with friends and to share his humour. With this way of life Koste is a 'living fossil'; in the contemporary scientific environment where, unfortunately, r-strategy is the predominant paradigm and where people are urged to present them as young as possible and to burn up early like meteors. We, his friends and his students, living today with a longer life expectancy than ever before, have with Koste hopefully not a run-out model, which shows us how to combine scientific work and true life. He oversees his situation perfectly. Not for nothing is he often coquetting with his role as Methuselah. On the other side, nothing is farther away from him than to demonstrate his as a role as a super-K-strategist or as a king of rotiferologists. In fact, for all who like him, and these are many, he remains our "Uncle Walter".

A robust health has been a physical prerequisite for his success. This we, the members of the international family of rotiferologists, wish him many more years of healthy life – and, of course, we wish him to study and discover many new species of rotifers under his microscope.

Acknowledgements

I thank Walter Koste for many discussions where he also reported details of his life. It was always a true enrichment for me to visit him. Ruth Lax-Huber's documentation was also inspiring to me. I am very thankful to Bob Wallace who improved the English style.

References

Haeckel, E., 1899–1904. Kunstformen der Nature: Leipzig und Wien, Verlag des Bibliographischen Instituts.

Laxhuber, R., 1993. Das Rädertierportrait: Walter Koste. Hydrobiologia 255/256: xxiii–xxvi.

MacArthur, R. H. & E. O. Wilson, 1967. The Theory of Island Biogeography. Princeton University Press, Princeton NJ, 216 pp.

Pianka, E. R., 1978. Evolutionary Ecology (2nd edn). Harper & Row, Publishers, New York, 397 pp.

Schwoerbel. J., 1987. Dr. h. c. Walter Koste zum 75. Geburtstag. Archiv für Hydrobiologie 110: 631–638.

Bibliography of Walter Koste

Koste, W., 1961. *Paradicranophorus wockei* nov. spec., ein Rädertier aus dem Psammon eines norddeutschen Niederungsbaches. Zoologischer Anzeiger 167: 138–141.

Koste, W., 1962. Über die Rädertierfauna des Darnsees in Epe bei Bramsche, Kreis Bersenbrück. Veröffentlichungen Naturwissenschaftlicher Verein Osnabrück 30: 73–137.

Koste, W. & K. Wulfert, 1964. Rotatorien aus der Wüste Gobi. Limnologica 2: 483–490.

Koste, W., 1965. Rotatorien des Naturdenkmals "Engelbergs Moor" in Druchhorn, Kreis Bersenbrück. Veröffentlichungen Naturwissenschaftlicher Verein Osnabrück 31: 49–82.

Koste, W., 1968. Über moosbewohnende Rädertiere von Strohdächern alter Gebäude. Mitteilungen Kreisheimatbund Bersenbrück Nr. 15: 123–128.

Koste, W., 1968. Über *Proales sigmoida* (Skorikow) 1896 (eine für Mitteleuropa neue Rotatorienart) und *Proales daphnicola* (Thomson) 1892. Archiv für Hydrobiologie 65: 240–245.

Koste, W., 1968. Über die Rotatorienfauna des Naturschutzgebietes "Achmer Grasmoor", Kreis Bersenbrück. Veröffentlichungen Naturwissenschaftlicher Verein Osnabrück 32: 107–160.

Koste, W., 1968. Das moosbewohnende Rädertier *Tetrasiphon*. Mikrokosmos 57: 334–337.

Koste, W., 1969. Das parasitishe Räertier *Albertia naidis*. Mikrokosmos 58: 212–216.

Koste, W., 1969. Das sessile Rädertier *Ptygura velata*. Mikrokosmos 58: 1–6.

Koste, W., 1969. *Notommata copeus* und einige verwandte Arten. Mikrokosmos 58: 137–143.

Koste, W., 1969. *Filinia*, eine pelagisch lebende Rädertiergattung. Mikrokosmos 58: 298–301.

Koste, W., 1969. *Stephanoceros fimbriatus*. Mikrokosmos 58: 370–373.

Koste, W., 1970. Über eine parasitische Rotatorienart. *Albertia reichelti* nov. spec. Zoologischer Anzeiger 184: 428–434.

Koste, W., 1970. Das Putzer-Rädertier *Proales daphnicola*. Mikrokosmos 59: 49–51.

Koste, W., 1970. *Collotheca trilobata*, ein seltenes sessiles Rädertier. Mikrokosmos 59: 195–200.

Koste, W., 1970. Die Rädertiergattung *Lindia*. Mikrokosmos 59: 134–138.

Koste, W., 1970. *Macrotrachela quadricornifera*, ein moosbewohnendes bdelloides Rädertier. Mikrokosmos 59: 328–332.

Koste, W., 1970. Zur Rotatorienfauna Nordwestdeutschlands. Veröffentlichungen Naturwissenschaftlicher Verein Osnabrück 33: 139–163.

Koste, W., 1970. Über die sessilen Rotatorien einer Moorblänke in Nordwestdeutschland. Archiv für Hydrobiologie 68: 96–125.

Koste, W., 1971. Die Rädertiergattung *Collotheca*. Mitteleuropäische Arten mit besonders auffallenden Koronafortsätzen. Mikrokosmos 60: 161–167.

Koste, W., 1971. Kurt Wulfert, ein Nachruf. Archiv für Hydrobiologie 68: 457–461.

Koste, W., 1971. Kutikowa, L. A. Die Rädertiere der Fauna USSR (Unterkl. Eurotatoria, Ploimida, Monimotrochida, Paedotrochida). Best. Bücher zur Fauna USSR, Bd. 104: 1–744, Book Review. Archiv für Hydrobiologie 69: 142.

Koste, W., 1972. Ein seltener Außenparasit an Süßwasser-Oligochaeten: *Cephalodella parasitica*. Mikrokosmos 61: 10–12.

Koste, W., 1972. *Collotheca campanulata longicaudata*. Mikrokosmos 61: 97–99.

Koste, W., 1972. *Rhinoglena frontalis*, ein vivipares Rädertier. Mikrokosmos 61: 358–360.

Koste, W., 1972. Über zwei seltene parasitische Rotatorienarten, *Drilophaga bucephalus* Vejdovsky und *Proales giganthea* (Glascott). Osnabrücker Naturwissenschaftliche Mitteilungen 1: 149–218.

Koste, W., 1972. Über ein sessiles Rädertier aus Amazonien. *Floscularia noodti* n. sp. Archiv für Hydrobiologie 70: 534–540.

Koste, W., 1972. Rotatorien aus Gewässern Amazoniens. Amazoniana 3: 258–505.

Koste, W., 1972. Ein Rädertier des Hochmoores: *Monommata arndti*. Mikrokosmos 61: 269–273.

Koste, W., 1973. Über ein sessiles Rädertier aus Amazonien, *Ptygura elsteri* n. sp., mit Bemerkungen zur Taxonomie des Artkomplexes *Ptygura melicerta* (Ehrenberg) 1932. Internationale Revue der gesamten Hydrobiologie 57: 875–882.

Koste, W., 1973. *Cupelopagis vorax*, ein merkwürdiges festsitzendes Rädertier. Mikrokosmos 62: 101–106.

Koste, W., 1973. *Horaella thomassoni* n. sp., ein neues Rädertier aus Gewässern der Guiana-Brasilianischen Region der Neotropis. Archiv für Hydrobiologie 72: 375–383.

Koste, W., R. Chengalath & C. H. Fernando, 1973. Rotifera from Sri Lanka (Ceylon). 2. Further studies an the Eurotatoria including new records. Bulletin of the Fisheries Research Station, Sri Lanka (Ceylon) 24: 29–62 pp.

Koste, W., 1974. Die Gattung *Notholca*. Mikrokosmos 63: 48–52.

Koste, W., 1974. Zur Kenntnis der Rotatorienfauna der "schwimmenden Wiese" einer Uferlagune in der Varzea Amazoniens, Brasilien. Amazoniana 5: 25–60.

Koste, W., 1974. *Ptygura pectinifera*, die "Kammträgerin", eine Einwanderin in Warmwasseraquarien. Mikrokosmos 63: 182–187.

Koste, W., 1974. Rotatorien aus einem Ufersee des unteren Rio Tapajos, dem Lago Paroni. Gewässer and Abwässer 53/54: 43–68.

Koste, W., R. Chengalath & C. H. Fernando, 1974. Rotifera from Sri Lanka (Ceylon) 3. New species and records with a list of Rotifera recorded and their distribution in different habitats from Sri Lanka. Bulletin of the Fisheries Research Station, Sri Lanka (Ceylon) 25(1–2): 83–96.

Koste, W., 1975. Macrochaeten, die "Igel" unter den Rädertieren. Mikrokosmos 64: 143–147.

Koste, W., 1975. Über den Rotatorienbestand einer Mikrobiozönose in einem tropischen aquatischen Saumbiotop, der *Eichhornica-crassipes-Zone* im Litoral des Bung-Borapet, einem Stausee in Zentralthail and Gewässer Abwässer, 57/58: 43–58.

Koste, W., 1975. *Seison annulatus*, ein Ektoparasit des marinen Krebses *Nebalia*. Mikrokosmos 64: 341–347.

Koste, W., 1975. Ruttner-Kolisko, A., Plankton Rotifers, Biology and Taxonomy, In: Die Binnengewasser Bd. 16/1, Supplement. Stuttgart 1975, 146 pp., Book Review. Archiv für Hydrobiologie 76: 133–134.

Koste, W., 1976. *Trochosphaera aequatorialis*, das "Kugelrädertier". Mikrokosmos 65: 266–268.

Koste, W., 1976. Über die Rädertierbestände (Rotatoria) der oberen and mittleren Hase in den Jahren 1966–1969. Osnabrücker Naturwissenschaftliche Mitteilungen 4: 191–263.

Koste, W., 1977. *Ptygura pedunculata*. Mikrokosmos 66: 37–40.

Koste, W., 1977. Über drei neue Formen des Genus *Hexarthra* Schmarda 1854: *H. jenkinae f. nakuru* n. f., *H. brandorffi* n.

sp. and *H. polyodonta soaplakeiensis* n. sp. Gewässer und Abwässer 62/63: 7–16.

Koste, W, 1978. Über *Testudinella ohlei* Koste 1972, ein Rädertier der U. Ordnung Flosculariacea aus der Guiana-Brasilianischen Region der Neotropis. Archiv für Hydrobiologie 82: 359–363.

Koste, W., 1978. *Synchaeta grandis*, ein in Mitteleuropa vom Aussterben bedrohtes Planktonrädertier. Mikrokosmos 67: 331–336.

Koste, W., 1978, Die Rädertiere Mitteleuropas (Ü. Ordnung Monogononta), Begr. von M. Voigt, Textband: 1–8 + 1–673 pp., Tafelband: T. 1–234. Gebr. Borntraeger Verlagsbuchhandlung, Berlin-Stuttgart.

Koste, W., 1979. *Hexarthra mira*, ein sechsarmiges Planktonrädertier. Mikrokosmos 68: 134–139.

Koste, W., 1979. New Rotifera from the River Murray, Southeastern Australia, with a review of the Australian species of *Brachionus* and *Keratella*. Australian Journal of Marine and Freshwater Research 30: 237–253.

Koste, W., 1979. *Lindia deridderi* n. sp., ein Rädertier der Familie Lindiidae (Überordnung Monogononta) aus SE-Australien. Archiv für Hydrobiologie 87: 504–511.

Koste, W. & R. J. Shiel, 1979. Rotifera recorded from Australia. Transactions of the Royal Society of South Australia 103: 57–68.

Koste, W., 1980. *Brachionus plicatilis*, ein Salzwasserrädertier. Mikrokosmos 69: 148–155.

Koste, W., 1980. Zwei Plankton-Rädertiertaxa *Filinia australiensis* n. sp. und *Filinia hofmanni* n. sp., mit Bemerkungen zur Taxonomie der *longiseta-terminalis* Gruppe. Genus *Filinia* Bory de St. Vincent 1824, Familie Filiniidae Bartos 1959 (Überordnung Monogononta). Archiv für Hydrobiologie 90: 230–256.

Koste, W. & R. J Shiel, 1980. On *Brachionus dichotomus* Shephard 1911 (Rotatoria), Brachionidae from the Australian region. With a description of a new subspecies *Brachionus dichotomus reductus*. Transactions of the Royal Society of Victoria 91: 127–134.

Koste, W. & R. J. Shiel, 1980. Preliminary remarks on the characteristics of the rotifer fauna of Australia (Notogaea). Hydrobiologia 73: 221–222.

Koste, W. & R. J. Shiel, 1980. Rotifera from Australia. Transactions of the Royal Society of South Australia 104: 133–144.

Koste, W., 1981. Zur Morphologie, Systematik and Ökologie von neuen monogononten Rädertieren (Rotatoria) aus dem Überschwemmungsgebiet des Magela Creek in der Alligator-River-Region Australiens, N. T, Teil 1. Osnabrücker Naturwissenschaftliche Mitteilungen 8: 97–126.

Koste, W., 1981. Einige auffallende *Synchaeta*-Arten aus Küstengewässern. Mikrokosmos 71: 169–176.

Koste, W., 1982. *Ploesoma truncatum*, ein räuberisches Plankton-Rädertier. Mikrokosmos 71: 167–173.

Koste, W., 1982. Über *Dicranophorus robustus* Harring 1928, ein seltenes carnivores Rädertier aus der Familie Dicranophoridae (Überordnung Monongononta, Klasse Rotaroria). Osnabrücker Naturwissenschaftliche Mitteilungen 9: 65–84.

Koste, W., G.-O. Brandorff & N. N. Smirnov, 1982. The composition and structure of rotiferan and crustacean communities of the lower Rio Nhamunda, Amazonas, Brazil. Studies on Neotropical Fauna and Environment, Amazonas, Brazil Environment 17: 69–121.

Koste, W. & S. Jose de Paggi, 1982. Rotifera of the Superorder Monogononta recorded from Neotropis. Gewässer und Abwässer 68/69: 71–102.

Koste, W, & E. Wobbe, 1982. Kleiner Führer für den Moorlehrpfad durch das Hahlener Moor, Gemeinde Menslage, Ortsteil Hahnenmoor, Landkreis Osnabrück, 1. Aufl, 1–42, Berge Krs. Osnabrück.

Koste, W., 1983. *Lecane*, eine formen- und artenreiche Rädertiergattung. Mikrokosmos 72: 174–180.

Koste, W. & R. Chengalath, 1983. Rotifera from northeastern Quebec, Newfoundland and Labrador, Canada. Hydrobiologia 104: 49–56.

Koste, W & J. Poltz, 1983. Über einen Fund des seltenen schlammbewohnenden *Paradicranophorus hudsoni* (Glascott, 1893), einem schlammbewohnenden Rädertier im Dümmer, NW-Deutschland. Osnabrücker Naturwissenschaftliche Mitteilungen 10: 27–41.

Koste, W & B. Robertson, 1983. Taxonomic studies of the Rotifera (Phylum Aschelminthes) from a Central Amazonian varzea lake, Lago Camalleao, Ilha de Marchantaria, Rio Solimões, Amazonas, Brazil. Amazoniana 8: 225–254.

Koste, W. & R. J. Shiel, 1983. Rotifer communities of billabongs in northern and south-eastern Australia. Hydrobiologia 104: 41–47.

Koste, W. & R. J. Shiel, 1983. Morphology, systematics and ecology of new monogonont Rotifera (Rotatoria) from Alligator river region, Northern Territory. Transactions of the Royal Society of South Australia 107: 109–121.

Koste, W., R. J Shiel & M. A. Brock, 1983. Rotifera from Western Australian wetlands with descriptions of two new species. Hydrobiologia 104: 9–17.

Koste, W., 1984. *Trichotria tetractis* (Ehrenberg) und verwandte Arten. Mikrokosmos 73: 113–118.

Koste, W. & J. Poltz, 1984. Über die Rädertiere (Rotatoria, Phylum Aschelminthes) des Dümmers, NW-Deutschland. Osnabrücker Naturwissenschaftliche Mitteilungen 11: 91–125.

Koste, W., B. Robertson & E. R. Hardy, 1984. Further taxonomical studies of the Rotifera from Lago Camaleao, a Central Amazonia varzea lake (Ilha de Marchantaria, Rio Solimões, Amazonas Brazil). Amazoniana 8: 555–576.

Koste, W. & E. R. Hardy, 1984. Taxonomic studies and new distribution records of Rotifera (Phylum Aschelminthes) from Rio Jatapu and Uatuma, Amazonas, Brazil. Amazoniana 9: 17–29.

Hardy, E. R., B. Robertson & W. Koste, 1984. About the relationship between the zooplankton and fluctuating water levels of Lago Camaleão, a Central Amazonian varzea lake. Amazoniana 9: 43–52.

Jose de Paggi, S. & W. Koste, 1984. Checklist of the Rotifers recorded from Antarctic and Subantarctic areas. Senckenbergiana biologica 65: 169–178.

Lair, N. & W. Koste, 1984. The rotifer fauna and population dynamics of Lake Studer 2 (Kerguelen Archipelago) with description of *Filinia terminalis kerguelensis* n. sp. and a new record of *Keratella sancta* Russel 1944. Hydrobiologia 108: 57–64.

Tait, R. D., R. J. Shiel & W. Koste, 1984. Structure and dynamics of zooplankton communities, Alligator Region N.T., Australia. Archiv für Hydrobiologie 113: 1–13.

Koste, W., 1985. Zur Morphologie, Anatomie, Ökologie und Taxonomie von *Paradicranophorus wockei* Koste 1961 (Aschelminthes, Rotatoria, Dicranophoridae). Senckenbergiana biologica 66: 153–166.

Koste, W., 1985. Das Rädertier-Porträt. *Cephalodella gigantea*, ein seltenes Rädertier der Faulschlammzone. Mikrokosmos 74: 168–173.

Koste, W. & R. J. Shiel, 1985. New species and new records of Rotifera (Aschelminthes) from Australian Waters. Transactions of the Royal Society of South Australia 109: 1–15.

Shiel, R. J. & W. Koste, 1985. Records of rotifers epizoic on cladocerans from South Australia. Transactions of the Royal Society of South Australia 109: 179–180.

Koste, W., 1986. Über die Rotatorienfauna in Gewässer südöstlich von Concepcion, Paraguay, Südamerika. Osnabrücker Naturwissenschaftliche Mitteilungen 12: 129–155.

Koste, W., 1986. *Collotheca edentata*, ein sessiles Rädertier "ohne" Zähne? Mikrokosmos 75: 302–305.

Koste, W., & R. J. Shiel, 1986. Rotifera from Australian inland waters, 1. Bdelloidea (Rotifera: Digononta). Australian Journal of Marine and Freshwater Research 37: 765–792.

Koste, W. & R. J. Shiel, 1986. New Rotifera (Aschelminthes) from Tasmania. Transactions of the Royal Society of South Australia 109: 93–106.

Shiel, R. J., & W. Koste, 1986. Australian Rotifera: Ecology and Biogeography. In: De Deckker, P., & W. D. Williams (eds), Limnology in Australia CSIRO. Australia, Melbourne, Monographiae Biologicae 61: 141–150.

Koste, W., 1987. *Asplanchnopus multiceps*, ein Rädertier, das Krebse frißt. Mikrokosmos 76: 170–175.

Koste, W. & J. Poltz, 1987. Über Rädertiere (Rotatoria, Phylum Aschelminthes) des Alfsees, NW-Deutschland. Osnabrücker Naturwissenschaftliche Mitteilungen 13: 185–220.

Koste, W. & R. J. Shiel, 1987. Tasmanian Rotifera: Affinities with the Australian mainland fauna. Hydrobiologia 147: 31–43.

Koste, W. & R. J. Shiel, 1987. Rotifera from Australian inland waters, 2.Epiphanidae and Brachionidae, (Rotifera: Monogononta). Invertebrate Taxonomy 1: 949–1021.

Koste, W. & W. Tobias, 1987. Zur Rotatorienfauna des Sankarani-Stausees im Einzugsgebiet des Niger, Republik Mali, Westafrika. Archiv für Hydrobiologie 108: 499–515.

Chengalath, R. & W. Koste, 1987. Rotifera from Northwestern Canada. Hydrobiologia 147: 49–56.

Koste, W., 1988. Das Rädertierporträt. Die Gattung *Squatinella*. Mikrokosmos 77: 140–145.

Koste, W., 1988. Rotatorien aus Gewässern am Mittleren Sungai Makaham, einem Überschwemmungsgebiet in E-Kalimantan, Indonesian Borneo. Osnabrücker Naturwissenschaftliche Mitteilungen 14: 91–136.

Koste, W., 1988. Über die Rotatorien einiger Stillgewässer in der Umgebung der Biologischen Station Panguana im tropischen Regenwald in Peru. Amazoniana 10: 303–325.

Koste, W. & S. Jose Paggi, 1988. Rotifers from Saladillo River Basin (Santa Fe Province, Argentina). Hydrobiologia 157: 13–20.

Koste, W., R. J. Shiel & L. W. Tan, 1988. New rotifers (Rotifera) from Tasmania. Transactions of the Royal Society of South Australia 112: 119–131.

Koste, W. & E. Vasquez, 1988. Form variations of the rotifer *Brachionus variabilis* (Hempel, 1896) from the Orinoco River, Venezuela. Annales de Limnologie 24: 127–129.

Chengalath, R. & W. Koste, 1988. Composition of littoral rotifer communities of Cape Breton Island, Nova Scotia, Canada. Verhandlungen Internationale Vereinigung für Theoretische und Angewandte Limnologie 23: 2019–2027.

Koste, W., 1989. Über Rädertiere (Rotatoria) aus dem Lago do Macaco, einem Ufersee des mittleren Rio Trombetas, Amazonien. Osnabrücker Naturwissenschaftliche Mitteilungen 15: 199–214.

Koste, W., 1989. *Octotrocha speciosa*, eine sessile Art mit einem merkwürdigen Räderorgan. Mikrokosmos 78: 115–121.

Koste, W. & K. Böttger, 1989. Rotatorien aus Gewässern Ecuadors. Amazoniana 10: 407–438.

Koste, W. & R. J. Shiel, 1989. Rotifera from Australian inland waters. 3. Euchlanidae, Mytilinidae, Trichotriidae. Transactions of the Royal Society of South Australia 113: 85–114.

Koste, W., & R. J. Shiel, 1989. Rotifera from Australian inland waters. 4. Colurellidae. Transactions of the Royal Society of South Australia 113: 119–143.

Koste, W. & R. J. Shiel, 1990. Rotifers from Australian inland waters. Lecanidae. Transactions of the Royal Society of South Australia 114: 1–36.

Koste, W. & R. J. Shiel, 1989. Classical taxonomy and modern methodology. Hydrobiologia 186/187: 279–284.

Koste, W. & W. Tobias, 1989. Die Rädertierfauna der Sélingue-Talsperre in Mali, Westafrika (Aschelminthes: Rotato ria). Senekenbergiana biologica 69: 441–466.

Chengalath, R. & W. Koste, 1989. Composition and distributional patterns in arctic rotifers. Hydrobiologia 186/187: 191–200.

Herzig, A. & W. Koste, 1989. The development of *Hexarthra* spp. in a shallow alkaline lake. Hydrobiologia 186/187: 129–136.

Shiel, R. J., W. Koste & L. W. Tan, 1989. Tasmania revisited: Rotifer communities and habitat heterogeneity. Hydrobiologia 186/187: 239–245.

Koste, W. & B. Robertson, 1990. Taxonomic studies of the Rotifera from shallow waters on the Island of Maraca, Roraima, Brazil. Amazoniana 11: 185–200.

Koste, W. & R. J. Shiel, 1990. Rotifers from Australian inland waters. 6. Proalidae, Lindiidae. Transactions of the Royal Society of South Australia 114: 129–143.

Koste, W. & W. Tobias, 1990. Zur Kenntnis der Rädertierfauna des Kinda Stausees in Zentral-Burma (Aschelminthes: Rotatoria). Osnabrücker Naturwissenschaftliche Mitteilungen 16: 83–110.

Koste, W., 1991. *Anuraeopsis miraclei*, a new planktonic rotifer species in karstic lakes of Spain. Hydrobiologia 209: 169–173.

Koste, W., W. Janetzky & E. Vareschi, 1991. Über die Rotatorienfauna in Bromelien-Phytotelmata in Jamaika (Aschelminthes: Rotatoria). Osnabrücker Naturwissenschaftliche Mitteilungen 17: 143–170.

Koste, W. & R. J. Shiel, 1991. Rotifers from Australian inland waters. 7. Notommatidae (Rotifera), Monogononta. Transactions of the Royal Society of South Australia 115: 11–159.

Koste, W., E. Vasquez & M. L. Medina, 1991. Variaciones morfologicas del rotifero *Keratella americana* (Carlin, 1943) de una laguna de inundacion del Rio Orinoco, Venezuela. Revue d'Hydrobiologie Tropicale 24: 83–90.

Koste, W. & K. Böttger, 1992. Rotatorien aus den Gewässern Ecuadors. 2. Amazoniana 12: 263–303.

Shiel, R. J. & W. Koste, 1992. Rotifera from Australian inland waters.8.Trichocercidae (Monogononta). Transactions of the Royal Society of South Australia 116: 1–27.

Koste. W., et al. (eds.), 1993. Rotifera: Guides to the identification of the microinvertebrates of the continental waters of the world. The Hague, Vol. 4:1–7,1–137.

Koste, W. & E. D. Hollowday, 1993. A short history of western European rotifer research. Hydrobiologia 255/256: 557–572.

Brain, C. K. & W. Koste, 1993. Rotifers of the genus *Proales* from saline springs in the Namib desert, with the description of a new species. Hydrobiologia 255/256: 449–454.

Hirschfelder, A., W. Koste & H. Zucchi, 1993. Bdelloid rotifers in aerophytic mosses: influence of habitat structure and habitat age an species composition. Hydrobiologia 255/256: 343–344.

Jersabek, C. D. & W. Koste, 1993. Additional notes on taxonomy and ecology of *Anuraeopsis miraclei* Koste, 1991 (Rotatoria: Monogononta) from an Austrian alpine lake. Hydrobiologia 264: 55–60.

Kizito, Y. S., A. Nauwcrck, J. Chapman & W. Kostc, 1993. A limnological survey of some Western Uganda crater lakes. Limnologica 23: 335–347.

Peters, U., W. Koste & W. Westheide, 1993. A quantitative method to extract moss dwelling rotifers. Hydrobiologia 255/256: 339–341.

Shiel, R. J. & W. Koste, 1993. Rotifera from Australian inland waters. 9. Gastropodidae, Synchaetidae, Asplanchnidae (Rotifera: Monogononta). Transactions of the Royal Society of South Australia 117: 111–139.

Koste, W., W. Janetzky & E. Vareschi, 1994. Zur Kenntnis der limnischen Rotatorienfauna Jamaikas (Rotatoria: Aschelminthes).Teil I. Osnabrücker Naturwissenschaftliche Mitteilungen 19: 103–149.

Janetzky, W., W. Koste & E. Vareschi, 1994. Rotifers of Jamaica: ecology, diversity and biogeography. Jamaica Naturalist 4: 16–20.

Koste, W., W. Janetzky & E. Vareschi, 1995. Zur Kenntnis der limnischen Rotatorienfauna Jamaikas (Rotifera).Teil II. Osnabrücker Naturwissenschaftliche Mitteilungen 20/21: 399–433.

Koste, W. & Y. Zhuge, 1995. On *Paradicranophorus aculeatus* (Neistwestnova-Shadina 7, 1935), with remarks on all other species of the genus (Rotifera: Dicranophoridae). Internationale Revue der gesamten Hydrobiologie 80: 121–132.

Janetzky, W., W. Koste &, E. Vareschi, 1995. Rotatorien in limnischen Mikrohabitaten: Phytotelmata und Gastrotelmata. Deutsche Gesellschaft für Limnologie, Jahrestagung 1994 Hamburg. Erweiterte Zusammenfassungen. 933–937.

Janetzky, W., W. Koste & E. Varesci, 1995. Rotifers (Rotifera) of Jamaican inland waters. A synopsis. Ecotropica 1: 31–40.

Jose de Paggi, S. & W. Koste, 1995. Additions to the checklist of rotifers of the superorder Monogononta recorded from Neotropis. Internationale Revue der gesamten Hydrobiologie 80: 133–140.

Koste, W., 1996. On soil Rotatoria from a Lithotelma near Halali Lodge in Etosha National Park in N-Namibia, South Africa. Internationale Revue der gesamten Hydrobiologie 81: 353–365.

Koste, W., 1996. Über die moosbewohnende Rotatorienfauna Madagaskars. Osnabrücker Naturwissenschaftliche Mitteilungen 22: 235–253.

Koste, W. & Y. Zhuge, 1996. A preliminary report on the occurrence of Rotifera in Hainan. Quekett Journal of Microscopy 37: 666–883.

Segers, H., W. Koste & S. M. Yussuf, 1996. Contributions to the knowledge of the monogononta Rotifera of Zanzibar, with a note of *Filinia novaezealandiae* Shiel & Sanomuang, 1993. Internationale Revue der gesamten Hydrobiologie 81: 597–603.

Zhuge, Y. & W. Koste, 1996. Two new species of Rotifera from China. Internationale Revue der gesamten Hydrobiologie 81: 605–609.

Janetzky, W. J. & W. Koste, 1997. Rotifera in broemeliad phytotelma. http: //www. ifas. ufl. edu/~frank/robrom. htm.

Zhuge, Y. & W. Koste, 1997. Rotiferology in China. Rotifer News 1997: 9–10.

Koste, W. & W. Tobias, 1998. Rädertiere des Manantali-Stausees in der Republik Mali, Westafrika (Aschelminthes: Rotatoria). Senckenbergiana biologica 77: 257–267.

Koste, W. & Y. Zhuge, 1998. Zur Kenntnis der Rotatorienfauna (Rotifera) der Insel Hainan, China. Teil II. Osnabrücker Naturwissenschaftliche Mitteilungen 24: 183–222.

Holst, H., H. Zimmermann, H. Kausch & W. Koste, 1998. Temporal and spatial dynamics of planktonic rotifers in the Elbe Estuary during spring. Estuarine Coastal and Shelf Science 47: 261–273.

Leutbecher, C. & W. Koste, 1998. Die Rotatorienfauna des Dümmers unter besonderer Berücksichtigung der sessilen Arten. Teil 1. Osnabrücker Naturwissenschaftliche Mitteilungen 24: 223–255.

Zhuge, Y., H. Xiangfei & W. Koste, 1998. Rotifera recorded from China, 1893–1997, with remarks on their composition and distribution. International Review of Hydrobiology 83: 217–232.

Koste, W., 1999. Über Rädertiere (Rotifera) aus Gewässern des südlichen Pantanal (Brasilien). Osnabrücker Naturwissenschaftliche Mitteilungen 25: 179–209.

Koste, W., 2000. Study of the Rotatoria-Fauna of the littoral of the Rio Branco, South of Boa Vista, Northern Brazil. International Review of Hydrobiology 85: 433–469.

Koste, W. & H. Terlutter, 2001. Die Rotatorienfauna einiger Gewässer des Naturschutzgebietes "Heiliges Meer"im Kreis Steinfurt. Osnabrücker Naturwissenschaftliche Mitteilungen 27: 113–177.

Part I
Phylogeny and Evolution

Hydrobiologia (2005) 546:11–28
A. Herzig, R.D. Gulati, C.D. Jersabek & L. May (eds.) Rotifera X: Rotifer Research: Trends, New Tools and Recent Advances
DOI 10.1007/s10750-005-4093-6

On the phylogenetic position of Rotifera – have we come any further?

Peter Funch[1,*], Martin Vinther Sørensen[2] & Matthias Obst[1]
[1]*Department of Zoology, Institute of Biological Sciences, University of Aarhus, Universitetsparken, Building 135, 8000, Århus C, DK, Denmark*
[2]*Invertebrate Department, Zoological Museum, University of Copenhagen, Universitetsparken 15, 2100, Copenhagen Ø, DK, Denmark*
*(*Author for correspondence: E-mail: peter.funch@biology.au.dk; Tel.: +45-89422764; Fax: +45-86125175)*

Key words: Gnathifera, jaws, trophi, Syndermata, Micrognathozoa, Cycliophora

Abstract

Rotifers are bilateral symmetric animals belonging to Protostomia. The ultrastructure of the rotiferan trophi suggests that they belong to the Gnathifera, and ultrastructural similarities between the integuments and spermatozoa as well as molecular evidence strongly suggest that rotifers and the parasitic acanthocephalans are closely related and form the clade Syndermata. Here we discuss the phylogenetic position of rotifers with regard to the gnathiferan groups. Originally, Gnathifera only included the hermaphroditic Gnathostomulida and the Syndermata. The synapomorphy supporting Gnathifera is the presence of pharyngeal hard parts such as jaws and trophi with similar ultrastructure. The newly discovered Micrognathozoa possesses such jaws and is a strong candidate for inclusion in Gnathifera because their cellular integument also has an apical intracytoplasmic lamina as in Syndermata. But Gnathifera might include other taxa. Potential candidates include the commensalistic Myzostomida and Cycliophora. Traditionally, Myzostomida has been included in the annelids but recent studies regard them either as sister group to the Acanthocephala or Cycliophora. Whether Cycliophora belongs to Gnathifera is still uncertain. Some analyses based on molecular data or total evidence point towards a close relationship between Cycliophora and Syndermata. Other cladistic studies using molecular data, morphological characters or total evidence suggest a sister group relationship between Cycliophora and Entoprocta. More molecular and morphological data and an improved sampling of taxa are obviously needed to elucidate the phylogenetic position of the rotifers and identify which phyla belong to Gnathifera.

Introduction

Our knowledge about the phylogenetic affinities of the Rotifera has increased a lot within the latest decade. Throughout time they have been considered relatives to the Infusoria, Crustacea, Tardigrada, Nematoda, Annelida, Mollusca, Gastrotricha and Platyhelminthes (Remane, 1929–1933; Hyman, 1951; Clément, 1985) and more recently as a member of the obviously polyphyletic group named 'Aschelminthes' (Ruppert & Barnes, 1994; Wallace et al., 1996). A close relationship between the Rotifera and Acanthocephala has been broadly accepted since Storch & Welsch (1969, 1970) demonstrated the ultrastructural similarities in the integuments of the two groups, and today most taxonomists unite them in the taxon Syndermata Ahlrichs, 1995.

A possible homology between the jaws of rotifers and gnathostomulids was first suggested by Ax (1956) and Reisinger (1961). An increasing amount of data now supports a close relationship between Rotifera, Gnathostomulida, Micrognathozoa and Acanthocephala united in a group named Gnathifera (Ahlrichs, 1995a, b, 1997; Rieger & Tyler, 1995; Haszprunar, 1996a; Melone et al., 1998; Kristensen & Funch, 2000; Sørensen, 2000, 2003; Jondelius et al., 2002; Sørensen & Sterrer, 2002; Zrzavý, 2003). Several synapomorphies have been proposed for Gnathifera, and even though some of these may be questionable (see Jenner, 2004), the presence of jaws with a unique ultrastructure appears to be a strong support argument for gnathiferan monophyly.

Some problems, however, still remain and new questions appear. The phylogenetic position of Gnathifera in the Metazoa is still uncertain and recently, new studies have questioned the monophyly of Gnathifera (Giribet et al., 2004). Cladistic analyses based partly or solely on molecular data imply that Gnathifera might be polyphyletic (Littlewood et al., 1998; Zrzavý et al., 1998; Peterson & Eernisse, 2001) or paraphyletic, for example, in respect to Cycliophora, Gastrotricha or Myzostomida (Cavalier-Smith, 1998; Giribet et al., 2000; Zrzavý et al., 2001; Giribet, 2002). Many studies based on morphological as well as molecular data have proposed that the Acanthocephala should be considered highly advanced rotifers (Lorenzen, 1985; Garey et al., 1996; Ahlrichs, 1997; Garey et al., 1998; Mark Welch, 2000; Herlyn et al., 2003) and most recently it was suggested that the newly described taxon Micrognathozoa (Kristensen & Funch, 2000) is sister group to the monogonont rotifers (De Smet, 2002).

Here we will discuss some of the conflicting proposals about rotifer relationships. We focus primarily on two newly described taxa, Cycliophora and Micrognathozoa, but also make some comments to the position of the Acanthocephala, Myzostomida, and Gnathifera within the Metazoa.

Discussion

Cycliophora – newly discovered taxon with rotifer affinities?

Among metazoans the Cycliophora is the most recently described major taxon. So far, only one marine species *Symbion pandora* Funch & Kristensen 1995 is known. This microscopic animal lives as a commensal on the mouth parts of the Norway lobster *Nephrops norvegicus* (Linné). Throughout the metagenetic life cycle of *S. pandora* six different stages emerge of which the sessile feeding individual is the largest (approx. 350 μm) and most prominent one. It is the only stage with an alimentary tract and is permanently attached to the host with an adhesive disc (Fig. 1, ad). The body of the feeding individual has a bell-shaped buccal funnel (Fig. 1, bf) and an ovoid trunk. The buccal funnel carries a mouth ring (Fig. 1, mr_1) consisting of multiciliated cells with compound cilia. The feeding apparatus works as a downstream collecting system (Riisgård et al., 2000) that filters food particles generated from the feeding activities of the host. The mouth leads into a U-shaped gut that terminates in an anus that is situated close to the buccal funnel (Fig. 1, an). The entire alimentary apparatus is periodically replaced by internal budding (Fig. 1, ib). A fluid-filled body cavity is absent (Funch & Kristensen, 1997).

Three different stages develop in the brood chamber of the feeding individual (Fig. 1, bc). The asexual developed Pandora larva (Fig. 1, pl) is approximately 120 μm long and equipped with an antero-ventral ciliated disc, various frontal glands, long and stiff, sensory cilia, and an internal bud from which a new feeding individual will arise. The asexually developed male larva (Prometheus larva) is approximately 100 μm long and has an antero-ventral ciliated disc, various glands and several pairs of long, stiff cilia anteriorly. Several dwarf males arise from internal buds within this larva (Obst & Funch, 2003). The dwarf male, approximately 33 μm in length, has a complex morphology with two ciliated fields covering the anterior and ventral body. It has various sensory structures, gland complexes, a relatively large brain connected to a pair of ventral longitudinal nerves, and numerous muscle cells. The reproductive system of the male is located in the posterior part of the body and consists of an unpaired testis, several adjacent gland cells, and a ventral penis connected to some of the muscle cells. The female, which is also developed in the brood chamber of the feeding individual (Fig. 1, bc), grows to a size of about 150–190 μm, and is morphological like the Pandora larva. However, the female contains a single egg instead of a bud. After fertilization the embryo grows inside the female, nourished by the degenerating maternal tissue and develops into a chordoid larva with a size of about 200 μm and a

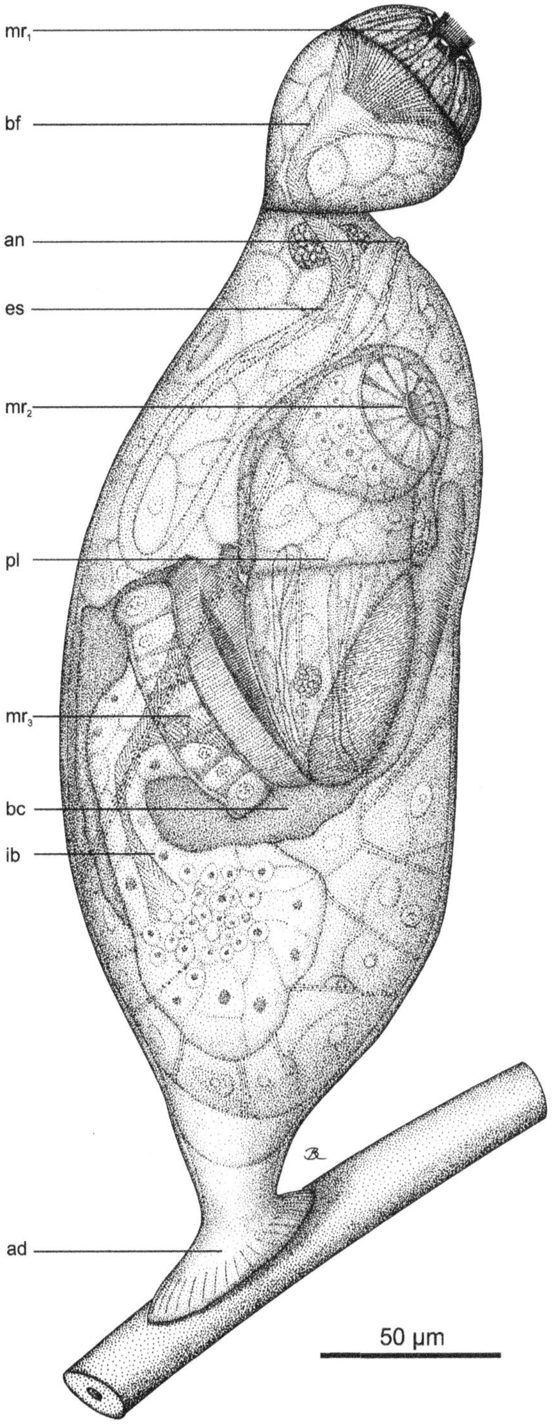

Figure 1. A young feeding stage of *Symbion pandora* (Cycliophora) on a seta from the host *Nephrops norvegicus*, lateral view. ad – adhesive disc; an – anus; bc – brood chamber; bf – buccal funnel; es – esophagus; ib – inner bud; mr_1 – functional mouth ring on feeding stage; mr_2 – mouth ring inside developing Pandora larva; mr_3 – developing mouth ring on inner bud; pl – Pandora larva. From Funch & Kristensen (1997).

complex morphology (Funch, 1996). The external ciliation of the chordoid larva consists of two anterior bands, two large ventral fields, and different sensory structures. The ciliated areas of the chordoid larva have been proposed to be homologous to those of a trochophore (Funch, 1996). Internally, the chordoid larva possesses a brain connected to a pair of ventral longitudinal nerves, a pair of protonephridia, a longitudinal rod of stacked muscle cells (chordoid organ), several gland and muscle complexes and one or two clusters of budding cells. Only a single host is known in the life cycle, and the chordoid larva is regarded as the dispersal stage between hosts.

Evaluation of the phylogenetic affinity between Cycliophora and Syndermata

In the original description of *Symbion pandora*, Funch & Kristensen (1995) proposed a close relationship between Cycliophora and Entoprocta and/or Bryozoa. Since then phylogenetic analyses have resulted in two competing hypotheses. Some analyses based on morphological data or total evidence support a Cycliophora–Entoprocta relationship (Zrzavý et al., 1998; Sørensen et al., 2000; Zrzavý, 2003), while others support a Cycliophora–Syndermata relationship (Peterson & Eernisse, 2001; Zrzavý et al., 2001). Analyses using molecular data or total evidence often favour a Cycliophora–Syndermata relationship (Winnepenninckx et al., 1998; Giribet et al., 2000; Peterson & Eernisse, 2001; Zrzavý et al., 2001). Based on both morphological and molecular data Zrzavý et al. (2001) proposed a monophyletic group including Cycliophora, Myzostomida, and Syndermata. This putative group was supported by three synapomorphies: (1) sperm with anteriorly inserted flagellum, (2) sperm without an accessory centriole, and possibly (3) the general tendency to live in association with crustaceans. However, it is not known if the flagellum in the cycliophoran sperm inserts anteriorly (Funch & Kristensen, 1997). An accessory centriole is apparently lacking in the sperm of *S. pandora*, but is also absent in several other taxa, i.e. some Gastrotricha and Platyhelminthes (Ferraguti & Balsamo, 1994; Ahlrichs, 1995b). Furthermore, the lack of an accessory centriole could be correlated with the evolution of the filiform sperm which probably occurred more than once (Jenner, 2004). The association with crustaceans as a 'possible support' for Cycliophora, Myzostomida, and Syndermata being monophyletic (Zrzavý et al., 2001) is of dubious character. Most rotifers, acanthocephalans, and all myzostomids are not associated with crustaceans, and this synapomorphy would require numerous independent losses of symbiosis with crustaceans. One also may speculate whether the cycliophoran affiliation to the lobster's mouth parts, the seisonid association with *Nebalia*, and the acanthocephalan endoparasitism in various crustaceans are so similar that they are products of one unique evolutionary event.

In a phylogenetic analysis using morphological data Peterson & Eernisse (2001) placed Cycliophora in a trichotomy with Rotifera and Gnathostomulida + Platyhelminthes, but Acanthocephala and *Seison* were not included. No unambiguous support for a Cycliophora–Rotifera relationship was found and the entire clade had a low Bremer and bootstrap support. Hence, none of the morphological analyses are able to produce consistent support for the suggested affinities between Cycliophora and Syndermata, even though they do share some superficial similarities.

Both the cycliophoran mouth ring and the rotiferan corona have bands of compound cilia that function as a downstream-collecting system, but this seems to be a general feature of ciliary suspension feeding protostomes (Riisgård et al., 2000; Nielsen, 2001). Also, the ability to retract the feeding structures differs, while the corona of rotifers can be retracted into the trunk, the mouth ring of cycliophorans cannot (Funch & Kristensen, 1997). The cycliophoran chordoid larva possesses one pair of multiciliated lateral pits and one paired dorsal ciliated organ, somewhat resembling the lateral and dorsal antennae in rotifers (Funch, 1996). In addition the sensory structures of rotifer and cycliophoran dwarf males have a somewhat similar morphology but their homology has yet to be assessed (see Obst & Funch, 2003).

Based on an ultrastructural study of the cycliophoran male and comparison with literature descriptions of rotiferan males, Obst & Funch (2003) argued that the presence of dwarf males in Rotifera and Cycliophora is a result of a convergent evolution. This is in agreement with the generally accepted idea that dwarf males have evolved within Syndermata (Wallace & Colburn, 1989; Ahlrichs, 1997; Melone et al., 1998; Sørensen, 2002), since *Seison* (Ricci et al., 1993) and some monogonont rotifers (Wesenberg-Lund, 1923; Hermes, 1932) possess fully developed males. The ontogeny differs as well, while the males of monogonont rotifers develop from haploid eggs produced by mictic females (Wallace, 1999); cycliophoran males develop from budding cells inside a male larva. Also, the copulatory organ in monogonont males consists of several cell types (Aloia & Moretti, 1973; Gilbert, 1983), and sometimes bears a ciliated crown around the genital pore (Clément et al., 1983). The cuticular penis of *S. pandora* is more simple and without cilia (Obst & Funch, 2003). The external ciliation of the dwarf male of *S. pandora* is more extensive than in rotiferan males and consists of two separated ciliated fields (Obst & Funch, 2003). The corona of monogonont dwarf males usually consists of a single anterior terminal disk of cilia or a girdle surrounding bundles of cilia (Hyman, 1951).

In contrast to syndermates *S. pandora* has a true cuticle that is formed from the cellular epidermis. The regenerative powers between rotifers and cycliophorans differ as well; while rotifers are poor in regeneration and apparently lack cell divisions in the adults (Hyman, 1951), *S. pandora* is able to replace the alimentary apparatus periodically or develop individual stages from internal buds. Also cuticular jaws are absent in *S. pandora*.

In summary, the morphological data supporting cycliophoran–syndermate affinity are weak and since all molecular analyses including cycliophorans mentioned above, used the same partial 18S rDNA sequence, the resulting relationships have to be treated with caution. For example Zrzavý et al. (2001) found support for the Syndermata relationship and Zrzavý (2003) for the Entoprocta relationship. Both studies used total evidence with 18S rDNA data. In a recent cladistic study Giribet et al. (2004) used four molecular loci (18S rDNA, a fragment of 28S rDNA, the nuclear protein-coding gene histone H3 and the mitochondrial gene cytochrome *c* oxidase subunit I) and a dense sampling of gnathiferan taxa, but the phylogenetic position of Cycliophora was still unstable. All four loci tended to place Cycliophora with Syndermata. Ribosomal and nuclear loci tended to place Cycliophora with Entoprocta. Alternatively, nuclear loci tended to place Cycliophora with Micrognathozoa and Syndermata. The support for closely related syndermates and cycliophorans is too weak (Fig. 6), and a relationship between Cycliophora and Entoprocta cannot be ruled out. Better comparative morphological studies of the possible homologies mentioned here are clearly needed to clarify a possible relationship between Cycliophora and Syndermata.

Micrognathozoa – sister group to Syndermata or aberrant rotifer?

Recently, a new microscopic taxon named Micrognathozoa was described from a cold spring on Disko Island, Greenland (Kristensen and Funch, 2000). At present it comprises a single species named *Limnognathia maerski* Kristensen and Funch and its possession of an intracytoplasmic lamina and complicated pharyngeal hard parts (Fig. 2, ja) suggest a close relationship with Rotifera.

With more than 15 paired or unpaired sclerites the micrognathozoan jaws are more complex than the pharyngeal hard parts found in any other microinvertebrate (Fig. 3). However, Kristensen & Funch (2000) demonstrated that the ultrastructure of the sclerites was very similar to that of the rotifer trophi. Based on a detailed comparison with the rotifer trophy, Kristensen & Funch (2000) proposed that the micrognathozoan main jaws and symphysis were homologous with the rotifer incus and that the micrognathozoan pseudophalangia and associated sclerites corresponded to the rotifer mallei, and these proposals were later supported by Sørensen (2003). Furthermore, Sørensen (2003) noted that the pharyngeal lamellae could be homologous with parts of the rotifer epipharynx, but it would be premature to draw this conclusion since the ground pattern of the rotifer epipharynx is not yet fully understood.

De Smet (2002) recently made a new attempt to understand the highly complicated jaws of Micrognathozoa. In a comprehensive description of the hard parts he pointed out several previously undescribed structures, such as the prominent brush on the main jaws and the circular platelet between the basal plates (Fig. 3, (cp), mj, vmj). Furthermore, he reinterpreted the already known

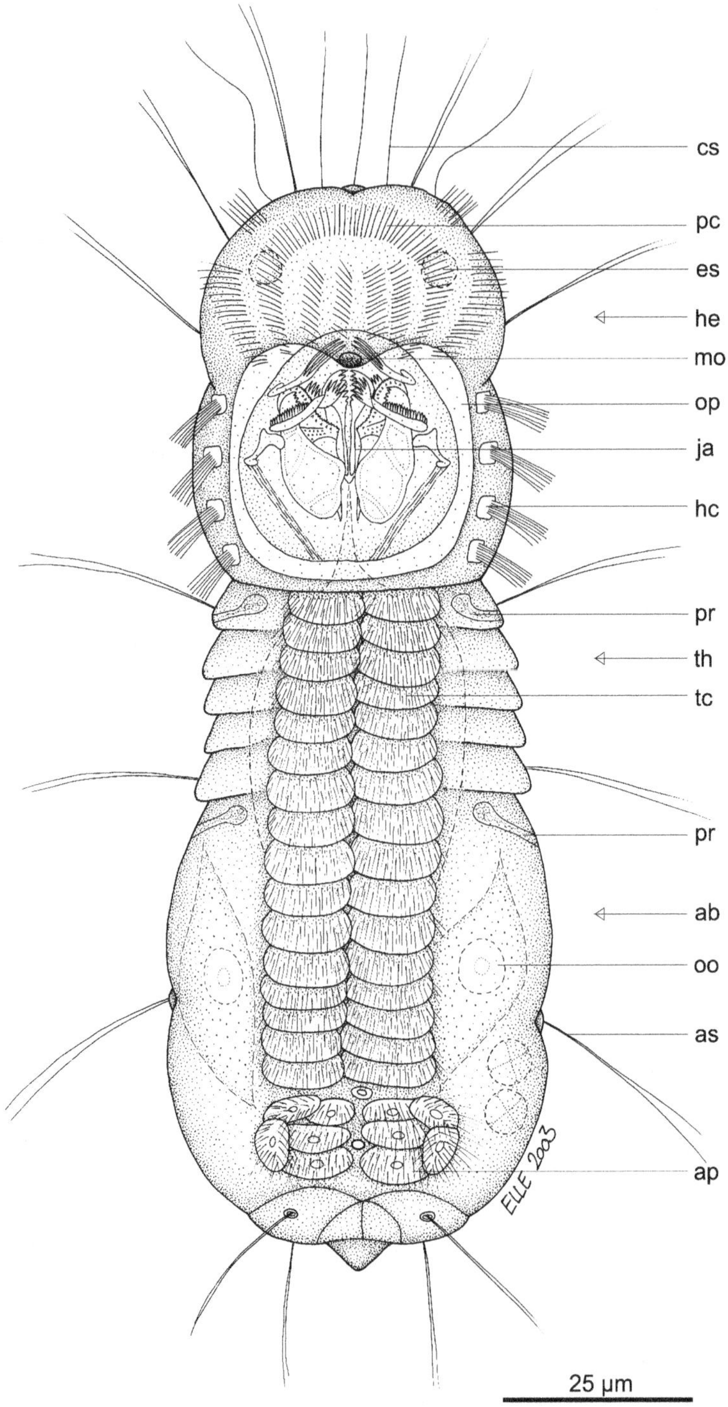

Figure 2. Drawing of *Limnognathia maerski* (Micrognathozoa), ventral view. ab – abdomen; as – abdominal sensory bristle; ap – adhesive pad; cs – cephalic sensory bristles; es – eyespot; hc – head ciliophore; he – head; ja – jaws; mo – mouth; oo – oocyte; op – oral plate; pc – preoral ciliary bands; pr – protonephridium; tc – trunk ciliophores; th – thorax. Courtesy of R. M. Kristensen, Zoological Museum, University of Copenhagen.

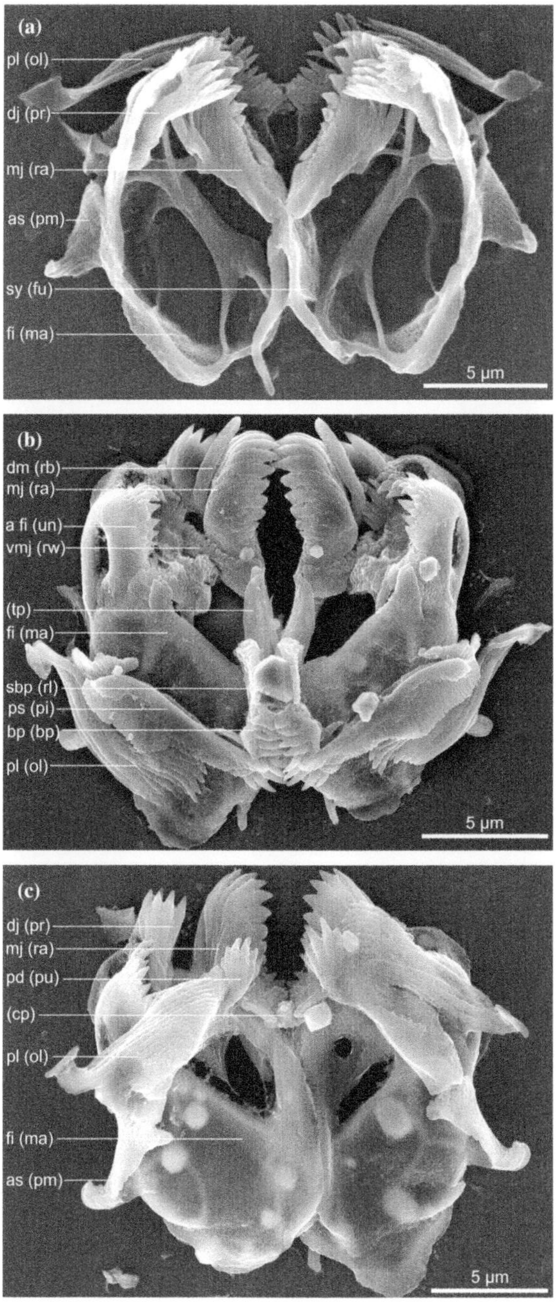

Figure 3. Limnognathia maerski (Micrognathozoa), SEM photographs of jaws. (a) Dorsal view. (b) Ventral view, note that basal plates, pharyngeal lamellae and pseudophalangia are tilted backwards. (c) Ventral view. Abbreviations are given *sensu* Kristensen & Funch (2000) and Sørensen (2003). Abbreviations in parenthesis are *sensu* De Smet (2002). a fi (un) – anterior part of fibularium (uncus); as (pm) – associate sclerite (pseudomanubrium); bp (bp) – basal plate (basal platelets); (cp) – (circular platelet); dj (pr) – dorsal jaw (pleural rod); dm (rb) – dentes medialis (rami brush); fi (ma) – fibularium (manubrium); mj (ra) – main jaw (ramus); ps (pi) – pseudophalangium (pseudintramalleus); pd (pu) – pseudodigits (pseuduncus); pl (ol) – pharyngeal lamella (oral lamella); sbp (rl) – shaft of basal plate (reinforced ligament); sy (fu) – symphysis (fulcrum); (tp) – triangular plate; vmj (rw) – ventral part of main jaw (reinforced web).

structures and made a detailed comparison with the rotiferan trophi. Based on these observations De Smet (2002) supports the previously proposed homology between the main jaws, plus symphysis (Fig. 3, mj, sy, vmj) and the rotifer incus, but rejects a possible homology between the pseudophalangia (Fig. 3, ps) including the associated sclerites and the rotifer mallei, and homologizes instead the mallei and the micrognathozoan fibularia (Fig. 3, fi). Moreover, he compares several of the remaining micrognathozoan sclerites with the rotifer trophi and homologizes them with different epipharyngeal sclerites, and concludes that these similarities support 'a sister-group relationship between Micrognathozoa and Rotifera Monogononta' (De Smet, 2002).

Evaluation of the proposed relationship between Micrognathozoa and Monogononta

De Smet's (2002) interpretation of the micrognathozoan jaws and suggested homologies with the rotiferan trophi are interesting in a phylogenetic as well as a comparative context and deserve some comment. The suggested homology between the fibularia (Fig. 3, fi) and the mallei are possible, but on the other hand it is difficult to find consistent morphological support for this assumption. De Smet (2002) interprets the anterior part of the fibularium as an uncus that is fused caudally with the manubrium, and notes that such an arrangement is not uncommon in rotifers. Fused unci and manubria are truly present in different taxa, for example, in all bdelloids and several monogonont taxa, such as *Birgea*, *Tylotrocha*, and Testudinellidae, but this character is nevertheless problematic. First, nothing indicates that fusion of the unci and manubria in *Birgea* and *Tylotrocha* is homologous with the arrangement found in Flosculariacea and Bdelloidea. Second, the general appearance of the micrognathozoan fibularium differs significantly from the mallei in Bdelloidea as well as in Flosculariacea and Ploima. In Micrognathozoa the fibularium has four chambers (exclusive the anterior-most one), whereas the rotifer manubrium has three or fewer. The fibularium is moreover connected with the main jaws via a unique structure named the reinforced web (Fig. 3, vmj (rw) (De Smet, 2002). Such an interconnection is not known in rotifers. Hence, from our point of view, nothing indicates that a homology between the malleus and fibularium is more likely than a homology between the malleus and the pseudophalangium with its associated sclerite. Nothing really favours the latter possibility, so at present this problem must be considered unresolved.

De Smet (2002) also homologizes the micrognathozoan dorsal jaws and pseudophalangia + associated sclerites (Fig. 3, dj, ps) with the rotiferan pleural rods and pseudomallei, respectively, but this assumption is questionable. First, the morphology of the epipharyngeal elements is very diverse, and our understanding of their basic patterns is still very limited. Thus, comparisons based on morphological similarities should be done with great care. The question can, however, be analyzed from a cladistic point of view. If *Limnognathia* is considered sister group to Rotifera, or even sister group to the Monogononta, and possesses pleural rods and pseudomallei that are homologous with those found in different ploimid taxa, it implies that these sclerites were present in the rotifer or monogonont ground plan. Pleural rods and pseudomallei are only present in some rotifers, and the structures do not necessarily co-occur in the same species. *Lindia* has pseudomallei, but lack pleural rods, while *Birgea* has pseudounci but lacks other pseudomallei sclerites. If all these elements should be present in the rotiferan ground plan it would require numerous secondary reductions, and if the presence of fused unci and manubria is added to this ground plan the number of necessary character transformations increases even more. Hence, we do not agree with the statement by De Smet (2002): 'that the jaws of *Limnognathia* can be homologized easily with the trophi of the monogonont Rotifera'. These uncertainties clearly demonstrate that it is premature to include Micrognathozoa as a subtaxon in Rotifera. It is furthermore noteworthy that Micrognathozoa deviate from monogonont rotifers at several points.

The integument is, as noted above, syncytial in both Acanthocephala and Rotifera, whereas the micrognathozoan epidermis is cellular. The structure of the integuments of Micrognathozoa and Syndermata could be explained by the following evolution: the ancestor of Micrognathozoa + Syndermata acquired a dense apical intracytoplasmic lamina that perhaps served as a cytoskeleton, and after the deviation of Micrognathozoa, the epidermis in the syndermate ancestor became a syncytium. Bender & Kleinow (1988) have shown that the intracytoplasmic lamina consists of keratin-like proteins in *Brachionus*. The integument of Acanthocephala also contains keratin (Dunagan & Miller, 1991), while the biochemical composition of the lamina in Micrognathozoa is unknown.

The ventral epidermis in *Limnognathia* is ciliated but posterior to the mouth the epidermis secretes a conspicuous cuticular oral plate (Fig. 2, op). This is in contrast to all known rotifers, where the ciliation of the epidermis is confined to the anterior corona and where no external cuticular oral plate is known. It is, however, interesting that Micrognathozoa has a ventral ciliation like in Gastrotricha. According to Rieger (1976) Gnathostomulida and Gastrotricha represent the most primitive bilaterians with respect to the construction of their integument, and Beauchamp (1909, 1965) actually suggested that both rotifers and gastrotrichs evolved from a ciliated crawling ancestor.

The gut epithelium of the micrognathozoans lacks cilia and has a brush border of microvilli as in Seisonidea (Ricci et al., 1993; Ahlrichs, 1995b) and Gnathostomulida (Lammert, 1991). The gut epithelium in monogonont rotifers is both ciliated and with microvilli, while a syncytial stomach epithelium seems to be autapomorphic for bdelloids (Clément & Wurdak, 1991; Melone et al., 1998). Of importance is also the fact that the micrognathozoan body plan is compact and has no fluid-filled extracellular compartment such as the often spacious pseudocoel in Syndermata (Wallace et al., 1996).

The Micrognathozoa possesses protonephridia (Fig. 2, pr) with two pairs of terminal cells. Interestingly, all the cells of these organs are monociliated as in Gnathostomulida and some Gastrotricha but contrary to all protonephridial cells investigated in Syndermata (Bartolomaeus & Ax, 1992).

The presence of paired gonads in both Bdelloidea and Seisonidea has been used as an argument to unite these two groups in the Digononta (Remane, 1929–1933; Pennak, 1989), but male Acanthocephala (Fig. 4) and Micrognathozoa (Fig. 2) have paired gonads as well (Dunagan & Miller, 1991; Kristensen & Funch, 2000). The characteristic vitellarium present in Monogononta and Bdelloidea are absent in *Seison* (Wallace & Colburn, 1989; Ricci et al., 1993), Micrognathozoa (Kristensen & Funch, 2000), and Acanthocephala (Monks, 2001) supporting monophyletic Eurotatoria. While bisexual reproduction is obligatory in Acanthocephala and *Seison*, males are unknown in Bdelloidea and Micrognathozoa. Hypodermic insemination is well known from monogonont rotifers (Wallace, 1999), but it is probably also occurring in gnathostomulids belonging to the Filospermoidea, a group that lack a vagina, have penises that seem unable to deliver the sperm, and have sperm with a spiralled head that could work as a drill when they rotate and thus actively go through the integument (Knauss & Rieger, 1979; Mainitz, 1989).

The dorsal and lateral antennae of Rotifera (Nogrady et al., 1993) are not present in Micrognathozoa (Kristensen & Funch, 2000). Also micrognathozoans seems to lack structures homologous to the retrocerebral and pedal glands of rotifers (Kristensen & Funch, 2000). The special adhesive pad situated at the ventral posterior trunk of Micrognathozoa (Fig. 2, ap) is very different in morphology compared to the pedal glands of rotifers and the cement glands of acanthocephalans (Fig. 4, ce).

Thus, Micrognathozoa differs from the Syndermata at several points, and their inclusion in Rotifera would require numerous reversals in the evolution of Micrognathozoa. Therefore, we still consider a sister group relationship between Micrognathozoa and Syndermata as the most likely (Fig. 6). However, it should be noted that Giribet et al. (2004) recently analyzed the phylogenetic position of Micrognathozoa using four molecular loci. They suggested that Micrognathozoa might constitute an independent lineage from those of Gnathostomulida and Syndermata.

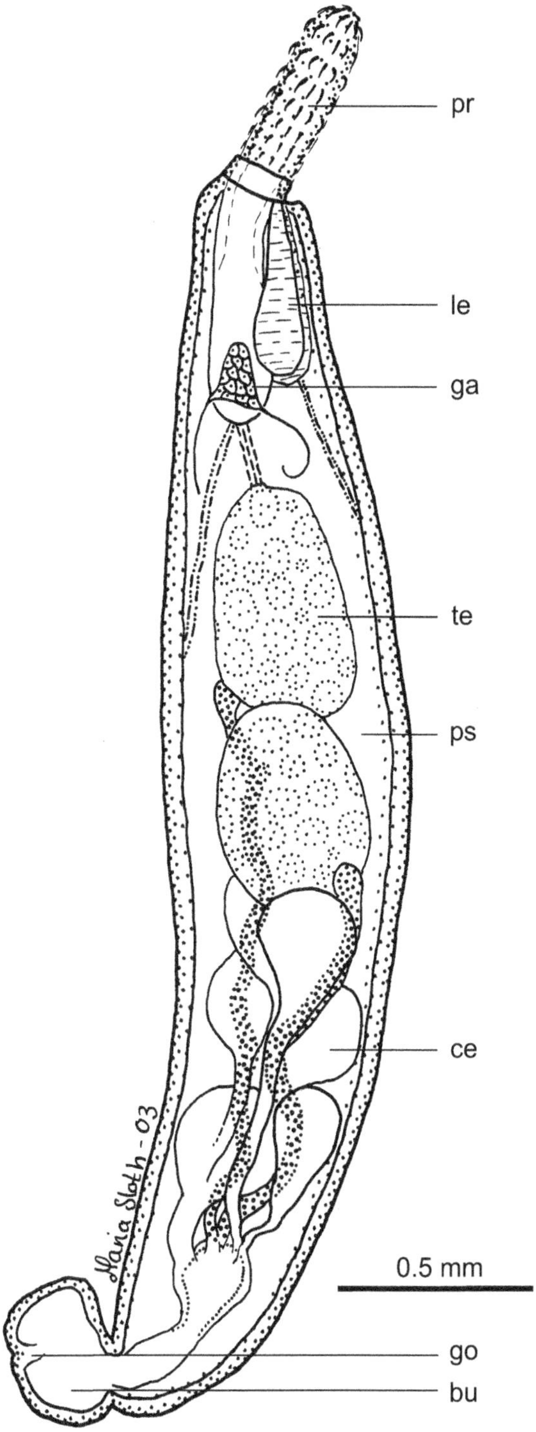

Figure 4. Drawing of male *Acanthocephalus dirus* (Acanthocephala). bu – bursa; ce – cement glands; ga – ganglion; go – gonopore; le – lemnisci; pr – proboscis; ps – pseudocoel; te – testes. Modified from Amin (1984).

Acanthocephala – gnathiferans without jaws and parasitic rotifers?

Adult acanthocephalans are endoparasites in the gut of vertebrates. They are unarticulated dioecious worms that attach themselves in the gut epithelium of the host with a retractile proboscis equipped with hooks (Fig. 4, pr). A general body cuticle is absent, collagen is present (Cain, 1970), and there is no trace of a digestive system. The epidermis is a syncytial tegument with a unique lacunar system. Males generally have paired testes and elaborate reproductive organs with cement glands, copulatory bursa, and penis (Fig. 4). Females develop free ovarian balls inside ligament sacs that sometimes rupture. The embryogenated eggs are sorted by the uterine bell. An acanthor larva develops from the fertilized egg and is capable of infecting an arthropod intermediate host. It invades the arthropod body cavity and develops into a larger acanthella stage in the right host. The growing acanthella develops the premature adult organs and inverts the proboscis. If the infected arthropod is eaten by a vertebrate the acanthella is capable of completing the life cycle by fixing itself to the gut epithelium of the vertebrate and grow into the adult (Hyman, 1951; Dunagan & Miller, 1991).

Evaluation of the phylogenetic position of Acanthocephala

Historically acanthocephalans were grouped with various parasitic worms such as Platyhelminthes, Nematoda, and Nematomorpha. Later they were often placed in the Aschelminthes together with Rotifera, Gastrotricha, Kinorhyncha, Nematoda, and Nematomorpha (Hyman, 1951; Ruppert & Barnes, 1994). More recent studies suggest that Aschelminthes is either paraphyletic or polyphyletic (Winnepenninckx et al., 1995; Ehlers et al., 1996; Ahlrichs, 1997). Although very different in morphology, acanthocephalans were early on suggested to be closely related to rotifers by Haffner (1950). This hypothesis gained further support from comparative ultrastructural studies on the syncytial integument (Storch & Welsch, 1969; Ahlrichs, 1997) and the sperm morphology (Ahlrichs, 1998; Ferraguti & Melone, 1999) and the clade Syndermata uniting Rotifera and Acanthocephala was suggested by Ahlrichs (1995b). A monophyletic Syndermata is also supported in most analyses using molecular data (Garey et al., 1996; Garcia-Varela et al., 2000; Mark Welch, 2000; Giribet et al., 2004).

Conclusively, there are currently four competing theories concerning the sister taxon of the Acanthocephala. One hypothesis suggests that Acanthocephala is a sister group to Bdelloidea (Lorenzen, 1985), based on supposed homology between the acanthocephalan proboscis and lemnisci (Fig. 4, le, pr), and the bdelloid rostrum and sac-like syncytial organs, respectively, but the presence of such bdelloid structures in Acanthocephala has been questioned (Melone et al., 1998; Ricci, 1998). Nevertheless, several molecular studies actually support the hypothesis (Garey et al., 1996; Near et al., 1998; Garey et al., 1998; Near, 2002), but the genes and taxa chosen in these studies were stated to be problematic due to long-branch attraction (Garey et al., 1998; Near, 2002). A sister group relationship between Acanthocephala and Bdelloidea was also recovered in a study of triploblastic animals using 18S rDNA combined with morphology (Giribet et al., 2000) and further supported in a recent study using four molecular loci (Giribet et al., 2004). Although *Seison* is a key taxon to understand the phylogeny of the Syndermata it was not included in any of the molecular studies mentioned above.

In the study by Near (2002) only the maximum-parsimony analysis placed bdelloid rotifers as the sister group to Acanthocephala. When the maximum-likelihood optimality criterion was used with the same 18S rDNA data; it resulted in another hypothesis, namely monophyletic Rotifera as sister group to Acanthocephala. In a cladistic study that assumed monophyletic Rotifera such a relationship was inferred from morphological characters (Nielsen et al., 1996) and supported in another molecular study using 18S rDNA (Garcia-Varela et al., 2000). However, Near (2002) criticized the latter study for using dendrogram-based similarity alignments that were adjusted by eye.

In a study that included *Seison*, Mark Welsch (2000) found strong support for a third possible sister group relationship between Acanthocephala and Eurotatoria using the nuclear coding gene for the heat shock protein hsp82. This also was supported by phylogenetic analyses using 18S rDNA (Miquelis et al., 2000). A cladistic study using morphological characters gave some additional support for this relationship (Sørensen et al., 2000), since spermatozoa of both monogononts and acanthocephalans lack an acrosome.

The fourth hypothesis suggests that Acanthocephala and *Seison* are sister groups supported by the presence of similar filament bundles in the epidermis and dense bodies in the spermatozoon (Ahlrichs, 1997, 1998). Though, Ahlrichs (1998) also showed the presence of an acrosome in *Seison*, a feature absent in the acanthocephalan spermatozoon (Dunagan & Miller, 1991). The Acanthocephala – *Seison* relationship got further support in a total evidence study by Zrzavý (2001) and a recent study based on 18S rDNA sequences with the inclusion of *Seison* (Herlyn et al., 2003).

In conclusion the support for monophyletic Syndermata is immense, while the sister group to the Acanthocephala is still not identified with certainty (Fig. 6).

Myzostomida – an annelid or a highly specialized gnathiferan?

Myzostomida is an enigmatic group of small marine worms that contains about 150 described species. Most myzostomids live in close association, as commensals or parasites, with asteroid and especially crinoid echinoderms. The animals can reach a size of several millimetres with a usually dorsoventral flattened body that can be elongated, oval, or irregular in shape (Fig. 5). Certain parts of the body are repeated along the anterior–posterior axis (Fig. 5, ci, lo, pa), which has traditionally been regarded as an indication of metamery. Five pairs of unarticulated appendages (parapodia) are present (Fig. 5, pa), normally containing a supportive rod and a protrusible hook. Usually myzostomids have pit-like lateral organs (Fig. 5, lo) that may vary in number and location, and have been interpreted as chemo- and mechanoreceptors (Eeckhaut & Jangoux, 1993). Most myzostomes have their mouth opening situated on a retractable proboscis (Fig. 5, mo, pr). The stomach extends into one or several

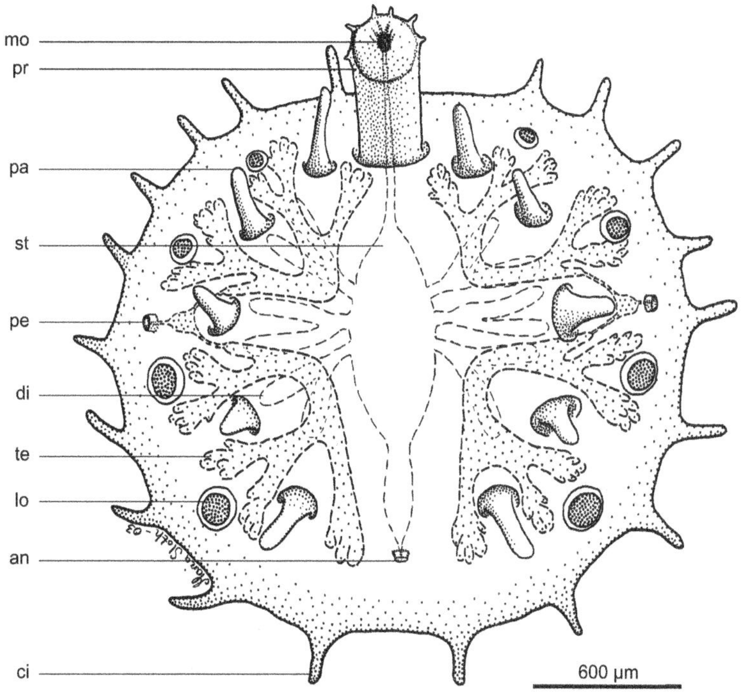

Figure 5. Drawing of *Myzostoma cirriferum* (Myzostomida), ventral view. an – anus; ci – cirrus; di – diverticulum; lo – lateral organ; mo – mouth opening; pa – parapodium; pe – penis; pr – proboscis; st – stomach; te – testis.

diverticula and a short intestine leads to a ventral anus (Fig. 5, an, di, st). Five pairs of protonephridia have been described from *Myzostoma cirriferum* (Pietsch & Westheide, 1987). The adult nervous system consists of a circumesophageal ring with two small cerebral ganglia, two nerve rings in the proboscis, and a ventral nerve cord. From here five pairs of nerves extend from the latter to the parapodia (Grygier, 2000; Müller & Westheide, 2000). The larval nervous system consists of five ventral longitudinal nerve cords (Eeckhaut et al., 2003). Most myzostomids are protandric hermaphrodites with paired testis diverticula that develop ventral to the digestive system (Fig. 5, te) and unpaired uterus diverticula that develop dorsally. The lumen of the ducts and branches of the reproductive system has been interpreted as the only remnants of a coelom (Jägersten, 1940). Some species have a pair of protrusible penises (Fig. 5, pe). In mature spermatozoa, the flagellum passes aside the main cell body and extends anteriorly (Afzelius, 1983; Grygier, 2000), a centriolar derivative is situated at their free end, and a nuclear membrane is absent. The chromatin is condensed into one or more rows of bead-like bodies. Sperms are transferred in a spermatophore during hypodermic impregnation (Eeckhaut & Jangoux, 1991). The development takes place via a free-swimming trochophore larva that temporarily develops chaetae as appendages.

Evaluation of phylogenetic affinity between Syndermata and Myzostomida

Myzostomida is one of the more problematic bilaterian taxa to place phylogenetically. The first scientist to describe a myzostomid regarded them as trematodes (Leuckart, 1827), but this idea is no longer considered valid. For a century Myzostomida have traditionally been placed within or close to Annelida (see Rouse & Fauchald, 1995), based on the presence of parapodia with chaetae, a trochophore larva and traces of segmentation (Fauchald & Rouse, 1997; Rouse & Fauchald, 1997). Recently, the group attracted special attention, and many authors have addressed the question of myzostomidan phylogenetic affinity.

Mattei & Marchand (1987, 1988) suggested a sister group relationship between Myzostomida and Acanthocephala based on similarities in spermatozoan ultrastructure and spermiogenesis. However, similar sperm morphology is also known from other taxa such as monogonont and seisonid rotifers (Melone & Ferraguti, 1994; Ahlrichs, 1998; Ferraguti & Melone, 1999). More support for a myzostomid affinity to Syndermata emerged when Zrzavý et al. (2001) analyzed combined morphological and molecular data (18S and 28S rDNA) and found evidence for a monophyletic clade including Myzostomida, Cycliophora, and Syndermata. Presumably the group was supported by synapomorphic presence of spermatozoa with anteriorly directed flagellum and no accessory centriole, although cycliophoran sperm is insufficiently known (Funch & Kristensen, 1997).

More evidence for a non-annelid relationship emerged in a cladistic analysis by Eeckhaut et al. (2000) who presented molecular evidence that myzostomids are closer related to Platyhelminthes than to Annelida. Zrzavý et al. (2001) argued that the characters supporting an annelid origin of myzostomids are usually either symplesiomorphic or convergent. Furthermore, the putative teloblastic growth of myzostomids needs to be confirmed, and the absence of an obvious coelom as well as the observed variation in the number of lateral organs suggests that 'segmented' patterns are only superficial.

On the contrary other studies have confirmed a strong annelid affinity. Müller & Westheide (2000) showed that the nervous system of *Myzostoma* is polychaete-like with obvious signs of segmentation and suggested to place them within Annelida. In a cladistic analysis by Rouse & Fauchald (1997) myzostomids are placed as a family nested within the polychaetes.

In conclusion, myzostomid relationships are unresolved (Zrzavý, 2003; Zrzavý & Hypša, 2003), but some evidence suggests that they are not within annelids (Haszprunar, 1996b; Eeckhaut et al., 2000). A recent study by Eeckhaut et al. (2003) showed several homologies between myzostomid and polychaete trochophores. However, the authors argued that these might be plesiomorphic and appeared early during the evolution of Spiralia.

The gnathiferan ground pattern and phylogenetic position

The phylogenetic position of Gnathifera is far from being resolved and different hypotheses about the metazoan phylogeny are being published instantly. The problem becomes even more puzzling because of the uncertainties about the gnathiferan ground pattern. Even though the possession of jaws appears to be a strong gnathiferan synapomorphy (Fig. 6), the basic body plan of, e.g., rotifers and gnathostomulids differs greatly. Rotifera possesses typical larval traits, such as the arrangement of the ciliary bands that resembles the trochophoran pattern with cilia forming a prototroch, metatroch, gastrotroch and telotroch (Nielsen, 1987), and this leads to the idea that the Gnathifera evolved by progenesis from an annelid-like ancestor, and hence were closely related with the trochozoans. However, it is difficult to explain how the gnathostomulids with their simple, monociliated epithelium and lack of distinct ciliary bands could be descendants of such an ancestor. Alternatively, the ancestor could have had a biphasic life cycle with a gnathostomulid-like, acoelomate adult and a trochophore-like larva. In this case one could imagine that the recent Gnathostomulida resemble the adult gnathiferan ancestor, but have undergone some modifications, such as, loss of larval stage and reversal to monociliated epidermis, whereas Micrognathozoa and Rotifera evolved from the ancestral larva by progenesis. However, this solution is highly speculative, and it has both advantages and disadvantages. If the gnathiferan evolution has followed this schedule it would support earlier proposed relationships with Platyhelminthes, Gastrotricha or both, and it would also explain the similarities in the platyhelminth and gnathostomulid body plans. The theory is, however, weakened by the fact that neither Platyhelminthes nor Gastrotricha appear to have a biphasic life cycle as a basal trait. Nielsen (1987) discussed some structural similarities between trochophores and the polyclad Müller's larva, but it is important to note that no platyhelminth phylogeny places Polycladia as a basal taxon, so it is most likely that Müller's larva is unique, and has evolved within the Platyhelminthes.

Gnathifera has been placed as a sister group to the Platyhelminthes (Ahlrichs, 1995b; Garey et al., 1998; Melone et al., 1998) or as a member of a clade Platyzoa also containing Platyhelminthes and Gastrotricha (see Giribet et al., 2000). An alternative position of Gnathifera as the most basal spiralian group was suggested by Sørensen et al. (2000) and later supported by some of the analyses presented by Zrzavý (2003).

The position of Gnathifera as a basal spiralian group is not unlikely, but it should be noted that to this point it is poorly supported. In the analysis of Sørensen et al. (2000) the Spiralia (*viz* Euspiralia + Gnathifera) are supported by the presence of egg cleavage with a spiralian cleavage pattern and ciliary bands formed by multiciliate cells that use a downstream system for suspension feeding (for further explanation see Nielsen, 1987, 2001; Riisgård et al., 2000). However, both characters may be problematic for Gnathifera. The cleavage pattern for Syndermata is far from being unambiguous spiralian and the assumed spiral cleavage in Gnathostomulida is based on a single observation (see Riedl, 1969) and needs to be confirmed with modern methods. Downstream collecting ciliary bands are certainly present in Rotifera (Nielsen, 1987), but if this is a basal trait for Gnathifera, it would require that radical modifications have happened on the branch that leads to Gnathostomulida. Also, the splits that take place in Euspiralia after deviation of Gnathifera are weakly supported. Euspiralia are solely supported by the shift from monociliated to multiciliated protonephridial terminal cells, and the following clade that comprises Teloblastica and Nemertea + Platyhelminthes simply lacks unambiguous support.

A sister group relationship between Platyhelminthes and Gnathifera has been supported by many authors and most recently by Ahlrichs (1995a, b, 1997) and Melone et al. (1998). However, Jenner (2004) recently reviewed the proposed synapomorphies for this clade and concluded that all of them were highly homoplastic characters that only could support Platyhelminthes + Gnathifera in pruned trees with a highly selective taxon choice.

The clade Platyzoa has been supported by molecular data (Winnepenninckx et al., 1995; Giribet, 2002), but the morphological support for this clade is scarce. Rieger (1976) showed that the epidermis of some macrodasyid gastrotrichs shares some striking similarities with the monociliated epidermis found in Gnathostomulida. He considered this as a plesiomorphic condition

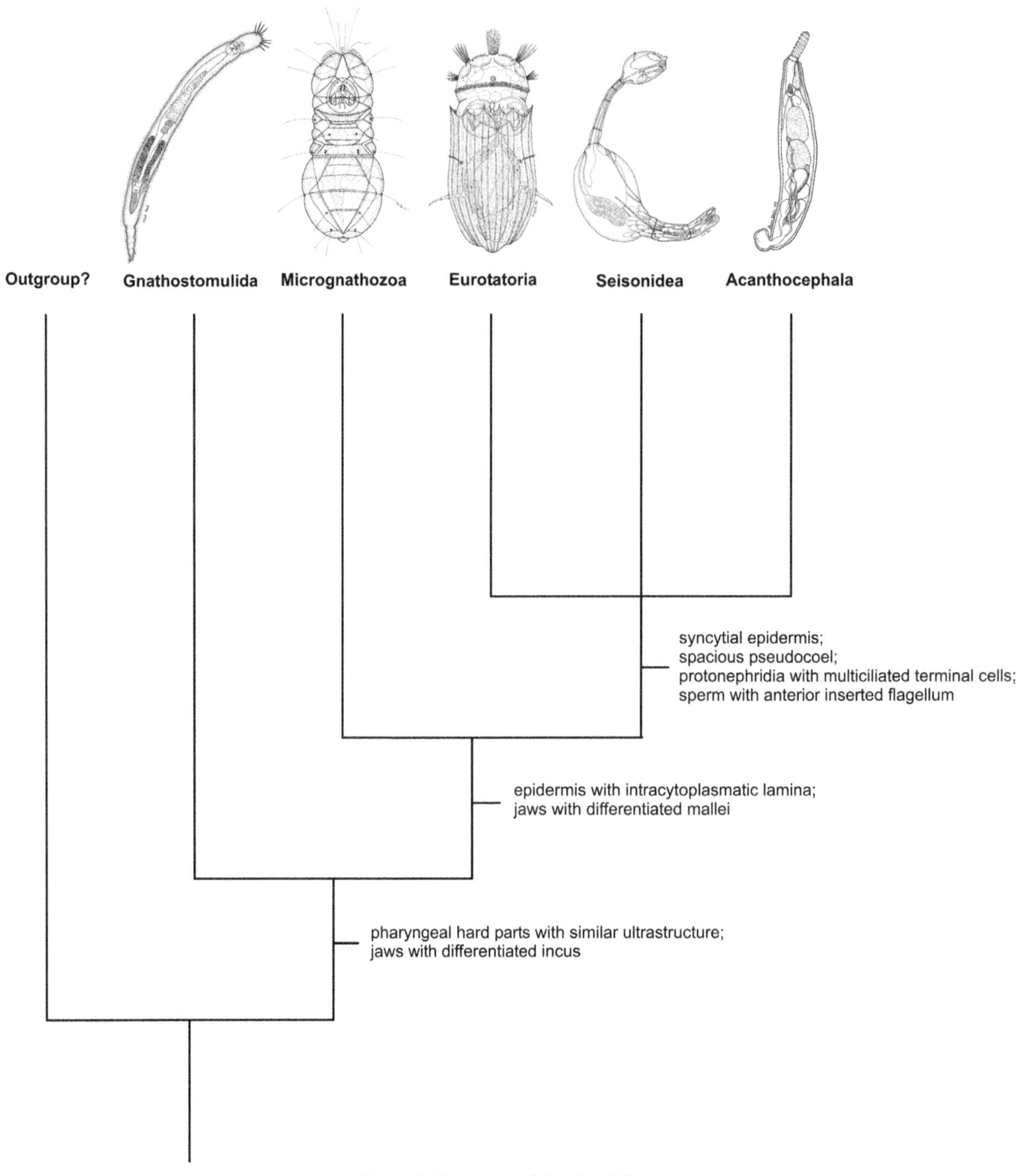

Figure 6. Phylogeny of the Gnathifera.

that indicated a basal position in Bilateria for both taxa. However, there are still conflicting opinions about the basal ciliary pattern in Gastrotricha (see Hochberg and Litvaitis, 2000; Zrzavý, 2003) and the question is at present unresolved (M. A. Todaro personal communication). Platyzoan monophyly also is challenged by the strong affinities between Gastrotricha and

Ecdysozoa (see Schmidt-Rhaesa et al., 1998; Zrzavý, 2003).

Thus the phylogenetic position of Gnathifera remains uncertain. Further data are needed to determine if any of the three possibilities discussed above should be favoured. New information and interpretations of the cleavage pattern in the gnathiferan groups could turn out to be extremely valuable, and a re-evaluation of the morphology in the platyzoan taxa could be interesting in light of the results that have been generated from molecular evidence.

Acknowledgements

We would like to thank Reinhardt Møbjerg Kristensen, Stine Elle, Birgitte Rubæk, and Maria Sloth Nielsen for the line drawings. Support from the Danish Natural Science Research Council (no. 21-02-0455 and 51-00-0278) and Zoologisk Rejse- og Ekskursionsfond, The Danish Natural History Society is greatly acknowledged. We appreciate the constructive comments of Reinhard Rieger and two anonymous referees.

References

Afzelius, B. A., 1983. The spermatozoon of *Myzostomum cirriferum* (Annelida, Myzostomida). Journal of Ultrastructure Research 83: 58–68.

Ahlrichs, W. H., 1995a. *Seison annulatus* und *Seison nebaliae*. Ultrastruktur und Phylogenie. Verhandlungen der Deutschen Zoologischen Gesellschaft 88: 155.

Ahlrichs, W. H., 1995b. Ultrastruktur und Phylogenie von *Seison nebaliae* (Grube 1859) und *Seison annulatus* (Claus 1876). Cuvillier Verlag, Göttingen, 310 pp.

Ahlrichs, W. H., 1997. Epidermal ultrastructure of *Seison nebaliae* and *Seison annulatus*, and a comparison of epidermal structures within the Gnathifera. Zoomorphology 117: 41–48.

Ahlrichs, W. H., 1998. Spermatogenesis and ultrastructure of the spermatozoa of *Seison nebaliae* (Syndermata). Zoomorphology 118: 255–261.

Aloia, R. C. & R. L. Moretti, 1973. Ultrastructural analysis of the functional copulatory organ of the male rotifer, *Asplanchna brightwelli*. Journal of Morphology 140: 285–306.

Amin O. M., 1984. Variability and redescription of *Acanthocephalus dirus* (Acanthocephala: Echinorhynchidae) from freshwater fishes in North America. Proceedings of the Helminthological Society of Washington 51: 225–237.

Ax, P., 1956. Die Gnathostomulida, eine rätselhafte Wurmgruppe aus dem Meeressand. Abhandlungen der Akademie der Wissenschaften und der Literatur in Mainz, mathematisch-naturwissenschaftliche Klasse 8: 1–32.

Bartolomaeus, T. & P. Ax, 1992. Protonephridia and metanephridia - their relation within the Bilateria. Zeitschrift für Zellforschung und Mikroskopische Anatomie 30: 21–45.

Beauchamp, P. d., 1909. Recherches sur les Rotifères: Les formations tégumentaires et l'appareil digestif. Archives De Zoologie Experimentale Et Generale 10: 1–410.

Beauchamp, P. d., 1965. Classe des Rotifère. In Grassé, P. P. (ed.), Traité de Zoologie, Anatomie, Systématique, Biologie. Masson, Paris: 1225–1379.

Bender, K. & W. Kleinow, 1988. Chemical properties of the lorica and related parts from the integument of *Brachionus plicatilis*. Comparative Biochemistry and Physiology B-Biochemistry and Molecular Biology 89: 483–487.

Cain, G. D., 1970. Collagen from the giant acanthocephalan *Macracanthorhynchus hirudinaceus*. Archives of Biochemistry and Biophysics 141: 264–270.

Cavalier-Smith, T., 1998. A revised six-kingdom system of life. Biological Reviews 73: 203–266.

Clément, P., 1985. The relationships of rotifers. In Conway-Morris, S., J. D. George, R. Gibson, & H. M. Platt (eds), The Origins and Relationships of Lower Invertebrates. Systematic Association, Clarendon Press, Oxford: 224–247.

Clément, P. & E. Wurdak, 1991. Rotifera. In Harrison, F. W. & E. Ruppert (eds), Microscopic Anatomy of Invertebrates, 4 Wiley-Liss, New York: 219–298.

Clément, P., E. Wurdak & J. Amsellem, 1983. Behavior and ultrastructure of sensory organs in rotifers. Hydrobiologia 104: 89–129.

De Smet, W. H., 2002. A new record of *Limnognathia maerski* Kristensen and Funch, 2000 (Micrognathozoa) from the subantarctic Crozet Islands, with redescription of the trophi. Journal of Zoology 258: 381–393.

Dunagan, T. & D. M. Miller, 1991. Acanthocephala. In Harrison, F. W. & E. Ruppert (eds), Microscopic Anatomy of Invertebrates 4. Wiley-Liss, New York: 299–332.

Eeckhaut, I., L. Fievez & M. C. Müller, 2003. Larval development of *Myzostoma cirriferum* (Myzostomida). Journal of Morphology 258: 269–283.

Eeckhaut, I. & M. Jangoux, 1991. Fine-structure of the spermatophore and intradermic penetration of sperm cells in *Myzostoma cirriferum* (Annelida, Myzostomida). Zoomorphology 111: 49–58.

Eeckhaut, I. & M. Jangoux, 1993. Integument and epidermal sensory structures of *Myzostoma cirriferum* (Myzostomida). Zoomorphology 113: 33–45.

Eeckhaut I., D. McHugh, P. Mardulyn, R. Tiedemann, D. Monteyne, M. Jangoux & M. C. Milinkovitch, 2000. Myzostomida: a link between trochozoans and flatworms? Proceedings of the Royal Society of London Series B-Biological Sciences 267: 1383–1392.

Ehlers, U., W. H. Ahlrichs, C. Lemburg & A. Schmidt-Rhaesa, 1996. Phylogenetic systematization of the Nemathelminthes (Aschelminthes). Verhandlungen der Deutschen Zoologischen Gesellschaft 89: 8.

Fauchald, K. & G. Rouse, 1997. Polychaete systematics: Past and present. Zoologica Scripta 26: 71–138.

Ferraguti, M. & M. Balsamo, 1994. Sperm morphology and anatomy of the genital organs in *Mesodasys laticaudatus* Remane, 1951 (Gastrotricha, Macrodasyida). Journal of Submicroscopic Cytology and Pathology 26: 21–28.

Ferraguti, M. & G. Melone, 1999. Spermiogenesis in *Seison nebaliae* (Rotifera, Seisonidea): further evidence of a rotifer-acanthocephalan relationship. Tissue and Cell 31: 428–440.

Funch, P., 1996. The chordoid larva of *Symbion pandora* (Cycliophora) is a modified trochophore. Journal of Morphology 230: 231–263.

Funch, P. & R. M. Kristensen, 1995. Cycliophora is a new phylum with affinities to Entoprocta and Ectoprocta. Nature 378: 711–714.

Funch, P. & R. M. Kristensen, 1997. Cycliophora. In Harrison, F. W. & R. M. Woollacott (eds), Microscopic Anatomy of Invertebrates 13 Wiley-Liss, New York: 409–474.

Garcia-Varela, M., G. P. P. Leon, P. la Torre, M. P. Cummings, S. Sarma & J. P. Laclette, 2000. Phylogenetic relationships of Acanthocephala based on analysis of 18S ribosomal RNA gene sequences. Journal of Molecular Evolution 50: 532–540.

Garey, J. R., T. J. Near, M. R. Nonnemacher & S. A. Nadler, 1996. Molecular evidence for Acanthocephala as a subtaxon of Rotifera. Journal of Molecular Evolution 43: 287–292.

Garey, J. R., A. Schmidt-Rhaesa, T. J. Near & S. A. Nadler, 1998. The evolutionary relationships of rotifers and acanthocephalans. Hydrobiologia 387/388: 83–91.

Gilbert, J. J., 1983. Rotifera. In Adiyodi, K. G. & R. G. Adiyodi (eds), Reproductive Biology of Invertebrates 2 John Wiley and Sons, New York: 181–193.

Giribet, G., 2002. Current advances in the phylogenetic reconstruction of metazoan evolution. A new paradigm for the Cambrian explosion? Molecular Phylogenetics and Evolution 24: 345–357.

Giribet, G., D. L. Distel, M. Polz, W. Sterrer & W. C. Wheeler, 2000. Triploblastic relationships with emphasis on the acoelomates and the position of Gnathostomulida, Cycliophora, Plathelminthes, and Chaetognatha: a combined approach of 18S rDNA sequences and morphology. Systematic Biology 49: 539–562.

Giribet, G., M. V. Sørensen, P. Funch, R. M. Kristensen & W. Sterrer, 2004. Investigations into the phylogenetic position of Micrognathozoa using four molecular loci. Cladistics 20: 1–13.

Grygier, M. J., 2000. Class Myzostomida. In Beesley, P. L., G. J. B. Ross, & C. J. Glasby (eds), Polychaetes and Allies: The Southern Synthesis 4A. CSIRO, Melbourne: 297–329.

Haffner, K. v., 1950. Organisation und systematische Stellung der Acanthocephalen. Zoologischer Anzeiger (Supplement) 145: 243–274.

Haszprunar, G., 1996a. Platyhelminthes and Platyhelminthomorpha - paraphyletic taxa. Journal of Zoological Systematics and Evolutionary Research 34: 41–48.

Haszprunar, G., 1996b. The Mollusca: coelomate turbellarians or mesenchymate annelids? In Taylor, J. (ed.), Origin and Evolutionary Radiation of the Mollusca. Oxford University Press, Oxford: 3–28.

Herlyn, H., O. Piskurek, J. Schmitz, U. Ehlers & H. Zischler, 2003. The syndermatan phylogeny and the evolution of acanthocephalan endoparasitism as inferred from 18S rDNA sequences. Molecular Phylogenetics and Evolution 26: 155–164.

Hermes, G., 1932. Studien über die Konstanz histologischer Elemente. IV. Die Männchen von *Hydatina senta* Ehrenberg, *Rhinops vitrea* Hudson und *Asplanchna priodonta* Gosse. Zeitschrift für Wissenschaftliche Zoologie 141: 581–725.

Hochberg, R. & M. K. Litvaitis, 2000. Phylogeny of Gastrotricha: a morphology-based framework of gastrotrich relationships. Biological Bulletin 198: 299–305.

Hyman, L. H, 1951. The Invertebrates. Acanthocephala, Aschelminthes and Entoprocta. The Pseudocoelomate Bilateria, McGraw-Hill, New York, 572 pp.

Jägersten, G., 1940. Zur Kenntnis der Morphologie, Entwicklung und Taxonomie der Myzostomida. Nova Acta Regiae Societatis Scientiarum Upsaliensis 11: 1–84.

Jenner, R. A., 2004. Towards a phylogeny of the Metazoa: evaluating alternative phylogenetic positions of Platyhelminthes, Nemertea, and Gnathostomulida, with a critical reappraisal of cladistic characters. Contributions to Zoology 73: 3–163.

Jondelius, U., I. Ruiz-Trillo, J. Baguna & M. Riutort, 2002. The Nemertodermatida are basal bilaterians and not members of the Platyhelminthes. Zoologica Scripta 31: 201–215.

Knauss, E. B. & R. M Rieger, 1979. Fine structure of the male reproductive system in two species of *Haplognathia* Sterrer (Gnathostomulida, Filospermoidea). Zoomorphologie 94: 33–48.

Kristensen, R. M. & P. Funch, 2000. Micrognathozoa: a new class with complicated jaws like those of Rotifera and Gnathostomulida. Journal of Morphology 246: 1–49.

Lammert, V., 1991. Gnathostomulida. In Harrison, F. W. & E. Ruppert (eds) Microscopic Anatomy of Invertebrates. 4. Wiley-Liss, New York: 19–39.

Leuckart, F. S., 1827. Versuch einer naturgemässen Einteilung der Helminthen nebst den Entwurf einer Verwandschafts- und Stuferfolge der Thiere ueberhaupt. Neue Akademische Buchhandlung von Karl Gross, Heidelberg und Leipzig, 88 pp.

Littlewood, D. T. J., M. J. Telford, K. A. Clough & K. Rohde, 1998. Gnathostomulida – An enigmatic metazoan phylum from both morphological and molecular perspectives. Molecular Phylogenetics and Evolution 9: 72–79.

Lorenzen, S., 1985. Phylogenetic aspects of pseudocoelomate evolution. In Conway-Morris, S., J. D. George, R. Gibson & H. M. Platt (eds), The Origins and Relationships of Lower Invertebrates. Systematic Association, Clarendon Press, Oxford: 210–223.

Mainitz, M., 1989. Gnathostomulida. In Adiyodi, K. G. & R. G. Adiyodi (eds), Reproductive Biology of Invertebrates John Wiley and Sons, Chicester: 167–177.

Mark Welch, D. B., 2000. Evidence from a protein-coding gene that acanthocephalans are rotifers. Invertebrate Biology 119: 17–26.

Mattei, X. & B. Marchand, 1987. Les spermatozoïdes des Acanthocéphales et des Myzostomides. Ressemblances et conséquences phylétiques. Comptes Rendus de l'Academie des Sciences Serie III Sciences de la Vie 305: 525–529.

Mattei, X. & B. Marchand, 1988. La spermiogenese de *Myzostomum* sp. (Procoelomata, Myzostomida). Journal of Ultrastructure and Molecular Structure Research 100: 75–85.

Melone, G. & M. Ferraguti, 1994. The spermatozoon of *Brachionus plicatilis* (Rotifera, Monogononta) with some notes on sperm ultrastructure in rotifera. Acta Zoologica 75: 81–88.

Melone, G., C. Ricci, H. Segers & R. L. Wallace, 1998. Phylogenetic relationships of phylum Rotifera with emphasis on the families of Bdelloidea. Hydrobiologia 387/388: 101–107.

Miquelis, A., J. F. Martin, E. W. Carson, G. Brun & A. Gilles, 2000. Performance of 18S rDNA helix E23 for phylogenetic relationships within and between the Rotifera–Acanthocephala clades. Comptes Rendus de l'Academie Des Sciences. Serie III, Sciences de la Vie 323: 925–941.

Monks, S., 2001. Phylogeny of the Acanthocephala based on morphological characters. Systematic Parasitology 48: 81–116.

Müller, M. C. & W. Westheide, 2000. Structure of the nervous system of *Myzostoma cirriferum* (Annelida) as revealed by immunohistochemistry and cLSM analyses. Journal of Morphology 245: 87–98.

Near, T. J., 2002. Acanthocephalan phylogeny and the evolution of parasitism. Integrative and Comparative Biology 42: 668–677.

Near, T. J., J. R. Garey & S. A. Nadler, 1998. Phylogenetic relationships of the Acanthocephala inferred from 18S ribosomal DNA sequences. Molecular Phylogenetics and Evolution 10: 287–298.

Nielsen, C., 1987. Structure and function of metazoan ciliary bands and their phylogenetic significance. Acta Zoologica 68: 205–262.

Nielsen, C, 2001. Animal Evolution: Interrelationships of the Living Phyla. Oxford University Press, Oxford, 563 pp.

Nielsen, C., N. Scharff & D. Eibye-Jacobsen, 1996. Cladistic analyses of the animal kingdom. Biological Journal of the Linnean Society 57: 385–410.

Nogrady, T., R. L. Wallace & T. W. Snell, 1993. Rotifera: Biology, Ecology and Systematics. SPB Academic Publishing, Amsterdam, 142 pp.

Obst, M. & P. Funch, 2003. The dwarf male of *Symbion pandora* (Cycliophora). Journal of Morphology 255: 261–278.

Pennak, R. W, 1989. Fresh water invertebrates of the United States. John Wiley, New York.

Peterson, K. J. & D. J. Eernisse, 2001. Animal phylogeny and the ancestry of bilaterians: inferences from morphology and 18S rDNA gene sequences. Evolution and Development 3: 170–205.

Pietsch, A. & W. Westheide, 1987. Protonephridial organs in *Myzostoma cirriferum* (Myzostomida). Acta Zoologica 68: 195–203.

Remane, A., 1929–1933. Rotatoria. Akademische Verlagsgesellschaft mbH, Leipzig, 576 pp.

Reisinger, E., 1961. Morphologie der Coelenteraten, acoelomaten und pseudocoelomaten Würmer. Fortschritte der Zoologie 13: 1–82.

Ricci, C., 1998. Are lemnisci and proboscis present in the Bdelloidea? Hydrobiologia 387/388: 93–96.

Ricci, C., G. Melone & C. Sotgia, 1993. Old and new data on Seisonidea (Rotifera). Hydrobiologia 255/256: 495–511.

Riedl, R. J., 1969. Gnathostomulida from America. Science 163: 445–452.

Rieger, R. M., 1976. Monociliated epidermal cells in Gastrotricha: significance for concepts of early metazoan evolution. Zeitschrift für Zoologische Systematik und Evolutionsforschung 14: 198–226.

Rieger, R. M. & S. Tyler, 1995. Sister-group relationship of Gnathostomulida and Rotifera-Acanthocephala. Invertebrate Biology 114: 186–188.

Riisgård, H. U., C. Nielsen & P. S. Larsen, 2000. Downstream collecting in ciliary suspension feeders: the catch-up principle. Marine Ecology-Progress Series 207: 33–51.

Rouse, G. W. & K. Fauchald, 1995. The articulation of annelids. Zoologica Scripta 24: 269–301.

Rouse, G. W. & K. Fauchald, 1997. Cladistics and polychaetes. Zoologica Scripta 26: 139–204.

Ruppert E. E. & R. D. Barnes, 1994. Invertebrate Zoology. Saunders College Publishing, New York, 1056 pp.

Schmidt-Rhaesa, A., T. Bartolomaeus, C. Lemburg, U. Ehlers & J. R. Garey, 1998. The position of the Arthropoda in the phylogenetic system. Journal of Morphology 238: 263–285.

Storch, V. & U. Welsch, 1969. Über den Aufbau des Rotatorienintegumentes. Zeitschrift für Zellforschung und Mikroskopische Anatomie 95: 405–414.

Storch, V. & U. Welsch, 1970. Über den Aufbau resorbierender Epithelien darmloser Endoparasiten. Zoologischer Anzeiger (Supplement) 33: 617–621.

Sørensen, M. V., 2000. An SEM study of the jaws of *Haplognathia rosea* and *Rastrognathia macrostoma* (Gnathostomulida), with a preliminary comparison with rotiferan trophi. Acta Zoologica 81: 9–16.

Sørensen, M. V., 2002. On the evolution and morphology of the rotiferan trophi, with a cladistic analysis of Rotifera. Journal of Zoological Systematics and Evolutionary Research 40: 129–154.

Sørensen, M. V., 2003. Further structures in the jaw apparatus of *Limnognathia maerski* (Micrognathozoa), with notes on the phylogeny of the Gnathifera. Journal of Morphology 255: 131–145.

Sørensen, M. V., P. Funch, E. Willerslev, A. J. Hansen & J. Olesen, 2000. On the phylogeny of the metazoa in the light of Cycliophora and Micrognathozoa. Zoologischer Anzeiger 239: 297–318.

Sørensen, M. V. & W. Sterrer, 2002. New characters in the gnathostomulid mouth parts revealed by scanning electron microscopy. Journal of Morphology 253: 310–334.

Wallace, R. L., 1999. Rotifera. In Knobil, E. & J. D. Neil (eds), Encyclopaedia of Reproduction. Academic Press, New York: 290–301.

Wallace, R. L. & R. A. Colburn, 1989. Phylogenetic relationships within phylum Rotifera - orders and genus *Notholca*. Hydrobiologia 186: 311–318.

Wallace, R. L., C. Ricci & G. Melone, 1996. A cladistic analysis of pseudocoelomate (aschelminth) morphology. Invertebrate Biology 115: 104–112.

Wesenberg-Lund, C., 1923. Contributions to the biology of the Rotifera. I. The males of the Rotifera. Det Kongelige

Danske Videnskabernes Selskabs Skrifter, Naturvidenskabelig og Mathematisk Afdeling. 8. Række 4: 189–345.

Winnepenninckx, B. M. H., T. Backeljau & R. M. Kristensen, 1998. Relations of the new phylum Cycliophora. Nature 393: 636–638.

Winnepenninckx, B. M. H., T. Backeljau, L. Y. Mackey, J. M. Brooks, R. DeWachter, S. Kumar & J. R. Garey, 1995. 18S rRNA data indicate that Aschelminthes are polyphyletic in origin and consist of at least three distinct clades. Molecular Biology and Evolution 12: 1132–1137.

Zrzavý, J., 2001. The interrelationships of metazoan parasites: a review of phylum- and higher-level hypotheses from recent morphological and molecular phylogenetic analyses. Folia Parasitologica 48: 81–103.

Zrzavý, J., 2003. Gastrotricha and metazoan phylogeny. Zoologica Scripta 32: 61–81.

Zrzavý, J. & V. Hypša, 2003. Myxozoa, *Polypodium*, and the origin of the Bilateria: The phylogenetic position of "Endocnidozoa" in the light of the rediscovery of *Buddenbrockia*. Cladistics 19: 164–169.

Zrzavý, J., V. Hypša & D. F. Tietz, 2001. Myzostomida are not annelids: Molecular and morphological support for a clade of animals with anterior sperm flagella. Cladistics 17: 170–198.

Zrzavý, J., S. Mihulka, P. Kepka, A. Bezdek & D. Tietz, 1998. Phylogeny of the Metazoa based on morphological and 18S ribosomal DNA evidence. Cladistics 14: 249–285.

Hydrobiologia (2005) 546:29–45
A. Herzig, R.D. Gulati, C.D. Jersabek & L. May (eds.) Rotifera X: Rotifer Research: Trends, New Tools and Recent Advances
DOI 10.1007/s10750-005-4097-2

Speciation and selection without sex

C. William Birky Jr.[1,*], Cynthia Wolf[2,3], Heather Maughan[4,5], Linnea Herbertson[6,7]
Elena Henry[8,9]
[1]*Department of Ecology and Evolutionary Biology and Graduate Interdisciplinary Program in Genetics, The University of Arizona, Biological Sciences West, Tucson, AZ, 85745 USA*
[2]*Department of Molecular and Cellular Biology, The University of Arizona, Biological Sciences West, Tucson, AZ, 85745 USA*
[3]*Present address: Program in Genetic Counseling, The University of Texas Graduate School of Biomedical Sciences, Houston, TX, 77225 USA*
[4]*Graduate Interdisciplinary Program in Genetics, The University of Arizona, Biological Sciences West, Tucson, AZ, 85745 USA*
[5]*Present address: Department of Ecology and Evolutionary Biology, The University of Arizona, Biological Sciences West, Tucson, AZ, 85745*
[6]*Department of Ecology and Evolutionary Biology, The University of Arizona, Biological Sciences West, Tucson, AZ 85745*
[7]*Present address: Aqauria, Inc., 6100 Condor Drive, Moorpark, CA, 93021*
[8]*Department of Molecular and Cellular Biology, The University of Arizona, Biological Sciences West, Tucson, AZ85745*
[9]*4645 Pueblo Ave., Sierra Vista, AZ 85650 USA*
*(*Author for correspondence: E-mail: birky@u.arizona.edu)*

Key words: cladogenesis, speciation, asexual reproduction, bdelloid rotifer, natural selection

Abstract

More than 100 females of the obligately asexual bdelloid rotifers were isolated from nature and their mitochondrial *cox1* genes (encoding cytochrome oxidase subunit 1) were sequenced. Phylogenetic analysis of the sequences showed that most of the isolates fall into 21 clades that show two characteristics of species: they are reciprocally monophyletic and have sequence diversities similar to that of species in other organisms. These clades have been evolving independently in spite of being effectively sympatric, indicating that they are adapted to different ecological niches. In support of this, at least some of the clades differ in morphology, food utilization, and temperature tolerance. We conclude that the bdelloid rotifers have undergone substantial speciation in the absence of sexual reproduction. We also used these sequences to test the prediction that asexual organisms should be subject to relaxed natural selection and hence will accumulate detrimental mutations. In contrast to this prediction, several estimates of the ratio K_a/K_s for the *cox1* gene showed that this gene is subject to strong selection in the bdelloid rotifers.

Introduction

"Sex is the queen of problems in evolutionary biology. Perhaps no other natural phenomenon has aroused so much interest; certainly none has sowed as much confusion." Graham Bell 1982 *The Masterpiece of Nature*

"These facts ... seemed to throw some light on the origin of species – that mystery of mysteries, as it has been called by one of our greatest philosophers." Charles Darwin 1859 *The Origin of Species* (p. 11 in The Modern Library edition).

The phylum Rotifera includes groups in which sexual reproduction is obligatory, interspersed with asexual reproduction, or lacking entirely and replaced by obligate asexual reproduction. Rotifers are thus ideal for studying two of the most important questions in biology: What is the evolutionary advantage of sexual reproduction? and What are species and how do they arise?

The evolutionary advantage of sex can be illustrated with the monogonont rotifers. Monogononts can lose facultative sexual reproduction as a consequence of a single mutation in any one of the genes required for sex. The loss quickly becomes irreversible as additional mutations accumulate in these genes. Moreover, asexual mutants have as much as a two-fold advantage and so should be quickly fixed in the population and species. This begs the question, why haven't all the sexual monogononts been replaced by obligately asexual lineages? Evidently sexual organisms enjoy compensating advantages which give them a higher speciation rate or lower extinction rate than asexual organisms. There are many theories about the selective advantages that favor sexual organisms and species over asexuals (Bell, 1982; Kondrashov, 1993; Barton & Charlesworth, 1998). Most of these can be summarized in one general statement: natural selection works better with sex. Asexual clones and species should accumulate more detrimental mutations than sexual lineages and species (Muller's ratchet). The increased genetic load should lead to extinction of the mutant clone or species (the meltdown). Moreover, asexual clones and species will be less able to fix advantageous mutations and adapt to different habitats. To the extent that speciation depends on adaptation to different niches, asexual species will be less able to speciate. If asexual lineages have a high rate of extinction and a low rate of speciation, then the net rate of speciation (speciation minus extinction) will be low in asexual organisms. Clearly, the question of the evolutionary advantage of sex and the question of the nature of species and speciation come to a common focus in asexual organisms.

In fact, the very possibility of speciation in asexuals is controversial for several reasons. One is the focus on the biological species concept, in which species boundaries are defined by the absence of sexual reproduction. As a result of this focus, most studies of speciation have dealt with the establishment of reproductive isolation in sexual organisms. The existence of species in asexual organisms is also controversial because of a misunderstanding of basic population genetics, leading to the misconception that the descendants of an asexual organism must form a continuum of genetic variation, or that sexual reproduction is necessary to hold a species together.

These misconceptions have been addressed theoretically by Barraclough et al. (2003). They used well-established results from coalescent theory to show that an asexual lineage can split into two or more independently evolving clades as a result of divergent selection for adaptation to different niches, geographic isolation, or both. This is illustrated in Figure 1. A common model of asexual reproduction is shown in Figure 1a in which an asexual organism divides repeatedly to produce a clone. If mutations occur randomly in the population, then individuals that separated in the sixth cell cycle will be very similar, while those that separated in the fifth cycle will have on average more mutational differences, and so on. The result is a continuum of genetic differences between individuals, with no gaps separating clades or species. However, this model assumes completely synchronous reproduction with all individuals having exactly the same number of offspring. In real life, reproduction is not synchronous; some individuals die without reproducing, while others leave variable numbers of offspring (Fig. 1b). At the level of the gene, this is random genetic drift. Clusters of similar organisms are produced but these clusters are transient and would not be considered different species. Divergent selection due to adaptation to different ecological niches (Fig. 1c) produces long-lasting clades that evolve independently of each other. These are likely to differ in morphology, physiology, or behavior, and would reasonably be called species.

Although it is clear that asexual organisms can speciate in theory at least, theory does not make a clear prediction about whether the rate of speciation will be less than it is in similar sexual organisms. As discussed above, natural selection is predicted to be less effective in asexual organisms, making it more difficult to fix the mutations required for adaptation to different niches. Also asexuals cannot speciate in allopatry without adapting to different niches, while sexual organisms can, at least in principle. On the other hand, asexuals do not have to evolve reproductive isolation in order to speciate, since they are already reproductively isolated from each other. Asexuals are generally more effective colonizers than sexuals and hence are more likely to be exposed to different selective pressures. Finally, genetic hitchhiking in asexuals facilitates cladogenesis during divergent selection. Barraclough et al. (2003) concluded that the extent to which asexual organisms speciate is an empirical question. We undertook to determine whether the obligately asexual bdelloid rotifers have undergone speciation.

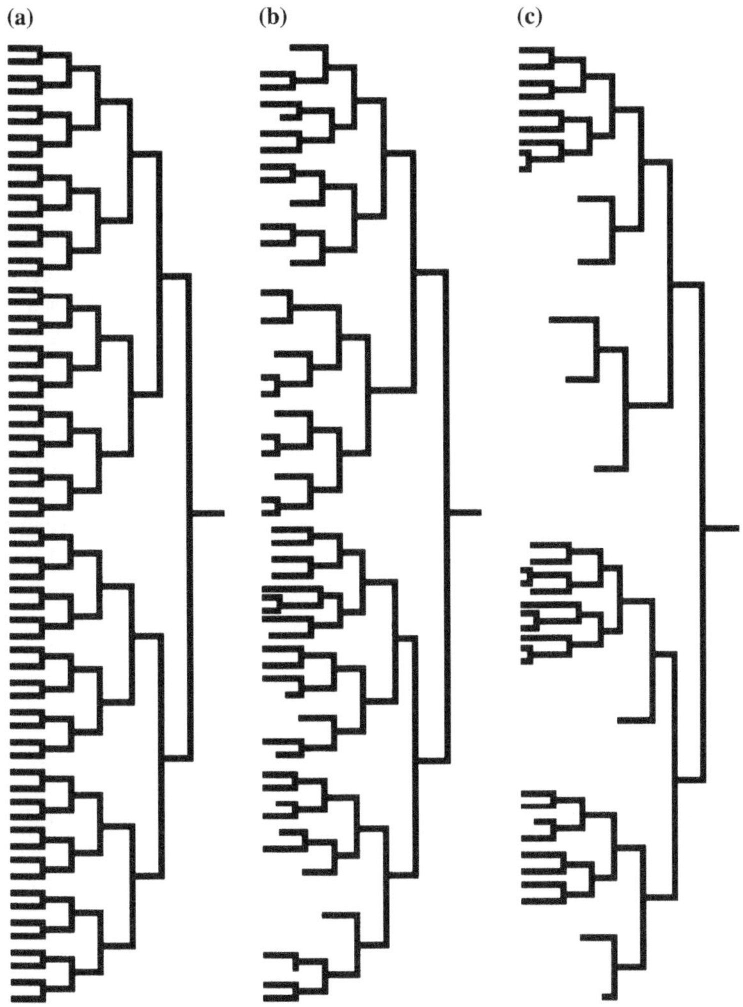

Figure 1. Diagrams illustrating asexual reproduction that is (a) synchronous with all individuals having the same number of offspring; (b) asynchronous with variable numbers of offspring and extinction (zero offspring), i.e. random drift at the gene level; and (c) drift plus divergent selection for adaptation to three different ecological niches.

Although the bdelloids have been divided into 4 families, 18 genera, and more than 374 species (Donner, 1965; Segers, 2002), this has been done largely on morphological grounds. While the species so described are as stable as species in the facultatively sexual monogononts (Holman, 1987), we believe that the definition and identification of species needs to be corroborated by phylogenetic analyses of DNA sequence data. Papers in this symposium and elsewhere show that traditional taxonomy of the monogononts does not always agree with molecular phylogenies. In particular, the identification of sibling or cryptic species seems to be common in rotifers as it is with many other invertebrates and protists (Derry et al., 2003; Gómez, 2004, Part II. Genetics and Molecular Ecology). We sequenced the mitochondrial *cox1* gene from a large collection of individual bdelloids isolated from nature. Phylogenetic analysis of these sequences revealed 21 independently evolving clades with diversity similar to that found in species of sexual organisms. Identification of these clades as species was supported by evidence for adaptive differences between some of the clades in morphology, food utilization, and temperature tolerance. Our results provide the first molecular phylogenetic evidence for substantial speciation in an asexual lineage.

The same sequence data used for phylogenetic analysis can also be used to test theories about the evolutionary advantage of sex. In particular, it can be used to test the prediction that asexual lineages should accumulate detrimental mutations to a greater extent than do sexual lineages. This phenomenon, called Muller's ratchet, is a special case of the Hill–Robertson effect (Barton & Charlesworth, 1998), in which selection acting on one site interferes with selection acting on all linked sites. Although the Hill–Robertson effect is based on robust theory and is almost certainly applicable to all asexual organisms, the strength of its effect depends on a number of factors, notably selection intensity and population size. Consequently, the accumulation of detrimental mutations (and likewise the reduction in speciation) may not be significant. The extent to which detrimental mutations are fixed in a lineage can be estimated by K_a/K_s, the ratio of amino acid substitutions (mostly detrimental or neutral) to synonymous substitutions (mostly neutral) along the lineage. Using this measure, we were unable to detect any greater accumulation of detrimental mutations in the bdelloids than in the monogononts or other sexual organisms. Thus, it seems unlikely that bdelloid species have an unusually high rate of extinction due to genetic load.

Materials and methods

We collected 102 individual female bdelloids from nature; 8 more were obtained from other laboratories or commercial sources. They were reared in the lab on algae (*Chlamydomonas reinhardtii* or *Chlorella vulgaris*) or bacteria (*Escherichia coli* or *Klebsiella pneumoniae pneumoniae*), or a combination of these, to produce clones of females. All were identified to genus and some to species. The descendants of a single bdelloid constitute a clone. Table 1 gives the classification and origin of each clone and the accession number of the corresponding *cox1* sequence.

Approximately 100 females from each clone were washed free of food organisms by exposure to dilute detergent (SDS), rinsed in sterile distilled water, starved for 5–7 days to eliminate food from their digestive tract, and washed again. *Adineta* and *Habrotrocha* were simply rinsed repeatedly in sterile distilled water before and after starvation, because they are sensitive to SDS. After the final wash or rinse, animals were collected in a minimal volume of distilled water in a microfuge tube and stored at −20 °C.

QIAmp or DNeasy spin columns (QiaGen) were used to isolate DNA from frozen rotifers. In a few cases, rotifers were lysed with SDS and proteinase K, then DNA was purified by a phenol–chloroform–isoamyl alcohol procedure. The mitochondrial *cox1* gene encoding cytochrome oxidase subunit 1 was amplified from the DNA by the PCR using primers HCO 5′ TAAACTTCAGGGTGACCAAAAAATCA, and LCO 5′ GGTCAACAAATCATAAAGATATTG G (Hoeh et al., 1998). The amplification product was purified with QIAquick columns (QiaGen) and sequenced in both directions by the Genomic Analysis and Technology Core at the University of Arizona using the amplification primers. From each clone, we obtained 591 bases of unambiguous sequence for *cox1*, corresponding to sites 77–666 of the human sequence, with the insertion of one codon in the bdelloid sequence.

The *cox1* sequences of monogonont rotifers used as the outgroup were taken from GenBank; accession numbers are in Table 1. Each *Brachionus plicatilis* sequence was obtained from a single fertilized egg (Gómez et al., 2000) which would have hatched to produce a clone of asexual monogonont females. Sequences from *B. quadridentatus*, *B. calyciflorus* 6ALM, and *Hexarthra* sp. 1ERA were provided by Africa Gómez and are unpublished. We will refer to each of the monogonont sequences as a clone.

Initially, *cox1* sequences were aligned with Clustal W implemented in the SeqApp or SeqPup programs (Gilbert, 1992) and the alignment was verified by examining the inferred amino acid sequences. Thereafter, additional bdelloid sequences could be unambiguously aligned by eye. There were some gaps and differences in sequence length; we used only regions present in all sequences. Phylogenetic trees were made in PAUP* (Swofford, 1998) using parsimony or neighbor-joining. The latter used pairwise sequence differences corrected for multiple hits

Table 1. List of clones used in this study, with their origins and identification, and GenBank accession numbers of their *cox1* sequences. Bdelloid clones collected by the authors are identified by 3– to 5–letter names indicating the location, followed by collection number/clone number. Exact locations of collecting sites are available from the author.

Species	Clone	Origin	GenBank
Adineta vaga	WAv1/1	Italy [a]	DQ078512
Adineta sp.	Rou1/6	Round Valley, Chniricahua Mts, AZ	DQ078513
Adineta oculata	War1/1	unknown [b]	DQ078515
Adineta sp.	Rou1/8	Round Valley, Chniricahua Mts, AZ	DQ078514
Habrotrocha sp.	HuH1-4a	Arizona	DQ078516
Habrotrocha sp.	Rob2/10d	Arizona	DQ078517
Habrotrocha sp.	Bir2/3	Tucson, AZ	DQ078518
Adineta sp.	BCB1/2b	Arizona	DQ078519
Habrotrocha constricta	WHc1/1	Sandwich, MA [c]	DQ078520
Habrotrocha sp.	Smok1/1	Great Smoky Mountain NP, TN	DQ078521
Habrotrocha sp.	Smok1/2	Great Smoky Mountain NP, TN	DQ078522
Habrotrocha sp.	Pim1/1	Pima Canyon, Santa Catalina Mts, AZ	DQ078523
Habrotrocha sp.	Glen1/10	Glenwood, NM	DQ078524
Habrotrocha sp.	Glen1/5	Glenwood, NM	DQ078525
Habrotrocha sp.	Rob2/8	Robinsom Spring, Santa Rita Mts, AZ	DQ078526
Habrotrocha sp.	Rob2/7	Robinsom Spring, Santa Rita Mts, AZ	DQ078527
Habrotrocha sp.	Rob2/9	Robinsom Spring, Santa Rita Mts, AZ	DQ078528
Habrotrocha sp.	HuH1/5b	Hex pool, Hueco Tanks State Park, TX	DQ078529
Habrotrocha sp.	Rob2/10a	Arizona	DQ078530
Habrotrocha sp.	HuH1/5a	Hex pool, Hueco Tanks State Park, TX	DQ078531
Habrotrocha sp.	Rob2/10c	Arizona	DQ078532
Habrotrocha sp.	Wad1/7	Tucson, AZ	DQ078533
Habrotrocha sp.	Wad1/9	Tucson, AZ	DQ078534
Abrochtha sp.	Angl1/1	Angelfish pool, Virginia Dale, CO	DQ078535
Macrotrachela quadricornifera	WMq	Italy [a]	DQ078536
Macrotrachela sp.	Trap1/2	Trap Lake, Front Range, CO	DQ078537
Macrotrachela sp.	HuH1/1	Hex pool, Hueco Tanks State Park, TX	DQ078538
Macrotrachela sp.	HuJu1/10	Julie?s pool, Hueco Tanks State Park, TX	DQ078540
Macrotrachela sp.	HuJu1/12	Julie?s pool, Hueco Tanks State Park, TX	DQ078539
Macrotrachela sp.	SnoB1/3	Snowy Range, WY	DQ078541
Macrotrachela sp.	SnoB1/7	Snowy Range, WY	DQ078542
Rotaria sp.	Trap1/1	Trap Lake, Front Range, CO	DQ078543
Philodina roseola	WPr1/1	unknown [d]	DQ078544
Philodina sp.	Car1/1	unknown [e]	DQ078545
Philodina sp.	Car/Pr2	unknown [e]	DQ078546
Philodina sp.	Car/Pr1	unknown [e]	DQ078547
Philodina sp.	HuJ1/1	Jenn?s pool, Hueco Tanks State Park, TX	DQ078548
Philodina sp.	Huf1/1	Pool f, Hueco Tanks State Park, TX	DQ078549
Philodina sp.	Huf1/2	Pool f, Hueco Tanks State Park, TX	DQ078550
Philodina sp.	Huf1/3	Pool f, Hueco Tanks State Park, TX	DQ078551
Philodina sp.	FlT2/1	Florida Canyon, Santa Rita Mts, AZ	DQ078552
Philodina sp.	HuN1/1	North pool, Huech Tanks State Park, TX	DQ078553
Philodina sp.	HuN1/2	North pool, Huech Tanks State Park, TX	DQ078554
Philodina sp.	ScM1/1	Scotia Canyon, Huachucha Mts, AZ	DQ078555

Continued on p. 34

Table 1. (Continued)

Species	Clone	Origin	GenBank
Philodina sp.	Duc1/1	Sierra Vista, AZ	DQ078556
Philodina sp.	Duc1/2	Sierra Vista, AZ	DQ078557
Philodina sp.	PaP1/3	Paton's pond, Patagonia, AZ	DQ078558
Philodina sp.	FRP1/3	Unnamed canyon, Santa Catalina Mts, AZ	DQ078559
Philodina sp.	FlT2/2	Florida Canyon, Santa Rita Mts, AZ	DQ078560
Philodina sp.	FlT2/3c	Florida Canyon, Santa Rita Mts, AZ	DQ078561
Philodina sp.	HuK1/1	Kettle Pool, Hueco Tanks State Park, TX	DQ078562
Philodina sp.	HuK1/2	Kettle Pool, Hueco Tanks State Park, TX	DQ078563
Philodina sp.	Chi1/1	Chino Canyon, Santa Rita Mts, AZ	DQ078564
Philodina sp.	Wil1/2	Wilderness of Rocks, Santa Catalina Mts, AZ	DQ078565
Philodina sp.	Wil1/3	Wilderness of Rocks, Santa Catalina Mts, AZ	DQ078566
Philodina sp.	SwT1/2	Sweetwater Wetlands, Tucson, AZ	DQ078567
Philodina sp.	Yet1/1	Yetman trail, Tucson Mts, AZ	DQ078568
Philodina sp.	Yet1/3	Yetman trail, Tucson Mts, AZ	DQ078569
Philodina sp.	Yet2/1	Yetman trail, Tucson Mts, AZ	DQ078570
Philodina sp.	Yet2/2	Yetman trail, Tucson Mts, AZ	DQ078571
Philodina sp.	Yet2/4	Yetman trail, Tucson Mts, AZ	DQ078572
Philodina sp.	Yet2/6	Yetman trail, Tucson Mts, AZ	DQ078573
Philodina sp.	Yet2/7	Yetman trail, Tucson Mts, AZ	DQ078574
Philodina sp.	Yet2/8	Yetman trail, Tucson Mts, AZ	DQ078575
Philodina sp.	Yet2/9	Yetman trail, Tucson Mts, AZ	DQ078576
Philodina sp.	Yet2/10	Yetman trail, Tucson Mts, AZ	DQ078577
Philodina sp.	Ven1/3	Ventana Canyon, Santa Catalina Mts, AZ	DQ078578
Philodina sp.	Yet1/2	Yetman trail, Tucson Mts, AZ	DQ078579
Philodina sp.	Amp1/1	Pontatoc Canyon, Santa Catalina Mts, AZ	DQ078580
Philodina sp.	Amp1/2	Pontatoc Canyon, Santa Catalina Mts, AZ	DQ078581
Philodina sp.	Amp1/3	Pontatoc Canyon, Santa Catalina Mts, AZ	DQ078582
Philodina sp.	Ven1/2	Ventana Canyon, Santa Catalina Mts, AZ	DQ078583
Philodina sp.	Kof1/4	Palm Canyon, Kofa NWR, AZ	DQ078584
Philodina sp.	Kof1/5	Palm Canyon, Kofa NWR, AZ	DQ078585
Philodina sp.	Kof1/2	Palm Canyon, Kofa NWR, AZ	DQ078587
Philodina sp.	Kof1/1	Palm Canyon, Kofa NWR, AZ	DQ078586
Philodina sp.	Kof1/3	Palm Canyon, Kofa NWR, AZ	DQ078588
Philodina sp.	Kof1/6	Palm Canyon, Kofa NWR, AZ	DQ078589
Philodina sp.	Hel1/3	Helvetia, Santa Rita Mts, AZ	DQ078590
Philodina sp.	Yet2/5	Yetman trail, Tucson Mts, AZ	DQ078591
Philodina sp.	Bird2/2	Tucson, AZ	DQ078592
Philodina sp.	BirdE1/1	Tucson, AZ	DQ078593
Philodina sp.	BirdT1/3	Tucson, AZ	DQ078594
Philodina sp.	Yet2/3	Yetman trail, Tucson Mts, AZ	DQ078595
Philodina sp.	Bir1/1	Tucson, AZ	DQ078596
Philodina sp.	Gut1b1b	Columbus, OH	DQ078597
Philodina sp.	Gut1/1c	Columbus, OH	DQ078598
Philodina sp.	Rou1/3	Round Valley, Chniricahua Mts, AZ	DQ078599
Philodina sp.	Hel1/2	Helvetia, Santa Rita Mts, AZ	DQ078600
Philodina sp.	Rou1/10	Round Valley, Chniricahua Mts, AZ	DQ078601

Continued on p. 35

Table 1. (Continued)

Species	Clone	Origin	GenBank
Philodina sp.	Bel1/1	Bellows Spring, Santa Rita Mts, AZ	DQ078606
Philodina sp.	Bel1/2	Bellows Spring, Santa Rita Mts, AZ	DQ078602
Philodina sp.	Bel1/4	Bellows Spring, Santa Rita Mts, AZ	DQ078607
Philodina sp.	Bel1/5	Bellows Spring, Santa Rita Mts, AZ	DQ078603
Philodina sp.	Bel1/8	Bellows Spring, Santa Rita Mts, AZ	DQ078604
Philodina sp.	Bel1/10	Bellows Spring, Santa Rita Mts, AZ	DQ078605
Philodina sp.	Bel2/1	Bellows Spring, Santa Rita Mts, AZ	DQ078609
Philodina sp.	Bel2/5	Bellows Spring, Santa Rita Mts, AZ	DQ078610
Philodina sp.	Bel2/6	Bellows Spring, Santa Rita Mts, AZ	DQ078611
Philodina sp.	Bel2/8	Bellows Spring, Santa Rita Mts, AZ	DQ078608
Philodina sp.	Fin1/1	Finger Rock Canyon, Santa Catalina Mts, AZ	DQ078612
Philodina sp.	Rou1/2	Round Valley, Chniricahua Mts, AZ	DQ078613
Philodina sp.	Rou1/4	Round Valley, Chniricahua Mts, AZ	DQ078614
Philodina sp.	Rat1/1	Rat Cave pool, Virginia Dale, CO	DQ078615
Philodina sp.	Rat1/2	Rat Cave pool, Virginia Dale, CO	DQ078616
Philodina sp.	Rat1/3	Rat Cave pool, Virginia Dale, CO	DQ078617
Philodina sp.	Angl1/3	Angelfish pool, Virginia Dale, CO	DQ078618
Philodina sp.	Bear1/2	Bear Canyon, Santa Catalina Mts, AZ	DQ078619
Philodina sp.	Bear1/1	Bear Canyon, Santa Catalina Mts, AZ	DQ078620
Philodina sp.	FIT3/3	Florida Canyon, Santa Rita Mts, AZ	DQ078621
Brachionus plicatilis	B.plicat(6TUR1)	Spain [f]	AF266859
Brachionus plicatilis	B.plicat(2SA21)	Spain [f]	AF266896
Brachionus quadridentatus	B.quadrid	Spain [g]	
Brachionus calyciflorus	B.calycif	Spain [g]	
Hexarthra sp.	Hexarthra_sp.	Spain [g]	

[a] From Claudia Ricci and Giulio Melone via David Mark Welch. [b] Purchased from Ward? Natural Science Establishment, Inc. [c] From David Mark Welch. [d] From Carolina Biological Supply Company via David Mark Welch. [e] Purchased from Carolina Biological Supply Company. [f] (Gómez et al. 2000). [g] Sequences provided by Africa Gómez.

with the GTR + I + G model and parameters selected with the maximum likelihood algorithm in ModelTest (Posada & Crandall, 1998).

For testing food utilization, each clone was reared for ≥7 days in parallel cultures on each of the experimental food sources (*E. coli* or *C. reinhardtii*). Then the growth of each clone on each food was assayed by isolating 10 randomly chosen females in each of 10 depression slides in approximately 1 ml of distilled water at 25 °C. Excess food organisms (*E. coli* or *C. reinhardtii*) were added. After approximately 2, 4, 6, and 8 days, the fluid was replaced with fresh water and food organisms and animals were counted. After an initial lag, the number of animals increased logarithmically. Log growth rates were compared using ANOVA.

The temperature tolerance of a clone was measured by isolating 12 females aged < 24 h in depression slides with excess *C. reinhardtii* at each temperature (16, 27, 36 °C). Animals were counted and the fluid was replaced with fresh food and water at intervals that varied with the growth rate of the cultures. After an initial lag, the number of animals increased logarithmically. Log growth rates were compared using ANOVA.

Results

Cladogenesis in bdelloid rotifers

Bdelloids were collected from sites in Arizona, Massachusetts, New Mexico, Colorado, Illinois, Ohio, and Tennessee in the U.S. and from Italy (Table 1). Collection sites included temporary and permanent lakes, ponds, streams, and springs, dirt, and moss, at elevations ranging from sea level to 3660 m. In the laboratory, individual females were isolated, a clone was reared from each female, and the *cox1* gene was sequenced from each clone. Phylogenetic trees were made from these sequences plus those of four monogonont rotifers as an outgroup. Details of this analysis, and its application to a larger dataset, will be presented elsewhere; here we give only a brief summary.

Trees made with either parsimony or neighbor-joining showed the striking pattern illustrated in Figure 2, in which 21 shallow clades with ≥70% bootstrap support are separated from each other and from the remaining clones by deep branches. The deep branches are joined at a polytomy, which can be partially resolved with the aid of additional sequence data from the mitochondrial *cob* gene (not shown). There are very few intermediate branches.

The well-supported shallow clades meet two criteria for potential species:

1. The within-clade nucleotide diversity is ≤2%, similar to that of well-established species in many other organisms (e.g., Avise, 1994; Moriyama & Powell, 1996).
2. The sequence differences between the clades are at least four times greater than the maximum diversity within the clades. This makes it likely that the clades represent samples from populations that are reciprocally monophyletic, based on the following reasoning: The neutral expectation of the pairwise differences within a population (nucleotide diversity π) is approximately $2N_eu$ for mitochondrial genes (Birky et al., 1983), or less due to hitchhiking (Maruyama & Birky, 1991). Also it takes $4N_e$ generations for 95% of pairs of sister species to become reciprocally monophyletic in an asexual genome (Avise & Ball, 1990; Avise, 1994), at which point the mean pairwise difference between the species is $d = 8N_eu$. Moreover, Rosenberg (2003) showed that when $d \geq 4N_eu$, one can infer reciprocal monophyly of the populations from which the clades are samples with 95% probability. This theory was developed for uniparentally inherited mitochondrial or chloroplast genes. Although it would not apply to nuclear genes in a sexual organism (Hudson & Coyne, 2002), it does apply to all genes in the asexual bdelloids because all nuclear and organelle genes behave as a single non-recombining unit

Niche adaptation in bdelloid clades

Two or more clades will be reciprocally monophyletic if the populations of bdelloids that they represent have been evolving independently of each other for a long time. Two or more lineages are most likely to evolve independently if they are permanently geographically isolated or if they are adapted to different niches.

Several lines of evidence make it very unlikely that our clades have been geographically isolated during their evolution. The evidence is described in detail elsewhere and will only be summarized here.

1. In several cases, members of different clades or singlets were isolated from the same sample. These cases are:

   Abr1, Pha10 from temporary pool, Virginia Dale, CO (sample Angl1)

   Adi singlet Bird2/3, Pha singlet Bird2/4, Pha6, from bird bath, Tucson, AZ (Bird2)

   Pha6, Pha8, from temporary stream, Helvetia, AZ (Hel1)

   Adi1, Pha8, Pha10, from temporary stream, Round Valley, AZ (Rou1)

   Pha5a, Pha5b, from temporary pool, Kofa National Wildlife Refuge, AZ (Kofa1)

   Pha4, Pha6, Pha7, from dust, Yetman trail, Tucson, AZ (Yet2)
2. Bdelloids survive dessication and disperse in the wind, and colonize habitats quickly and effectively (Cáceres & Soluk, 2002).
3. We were able to estimate the dispersal rate of animals in clade Pha8. The clones in this clade were collected in Columbus, Ohio and from Helvetia and Round Valley in Arizona. We divided the geographic distance between each site in Arizona and Ohio by the sequence difference between each pair of Arizona and Ohio rotifers to obtain a dispersal rate of 613 km/% sequence difference. These clones are separated from clones in other clades and from singlets by a median sequence difference of about 50%, so they could have dispersed by

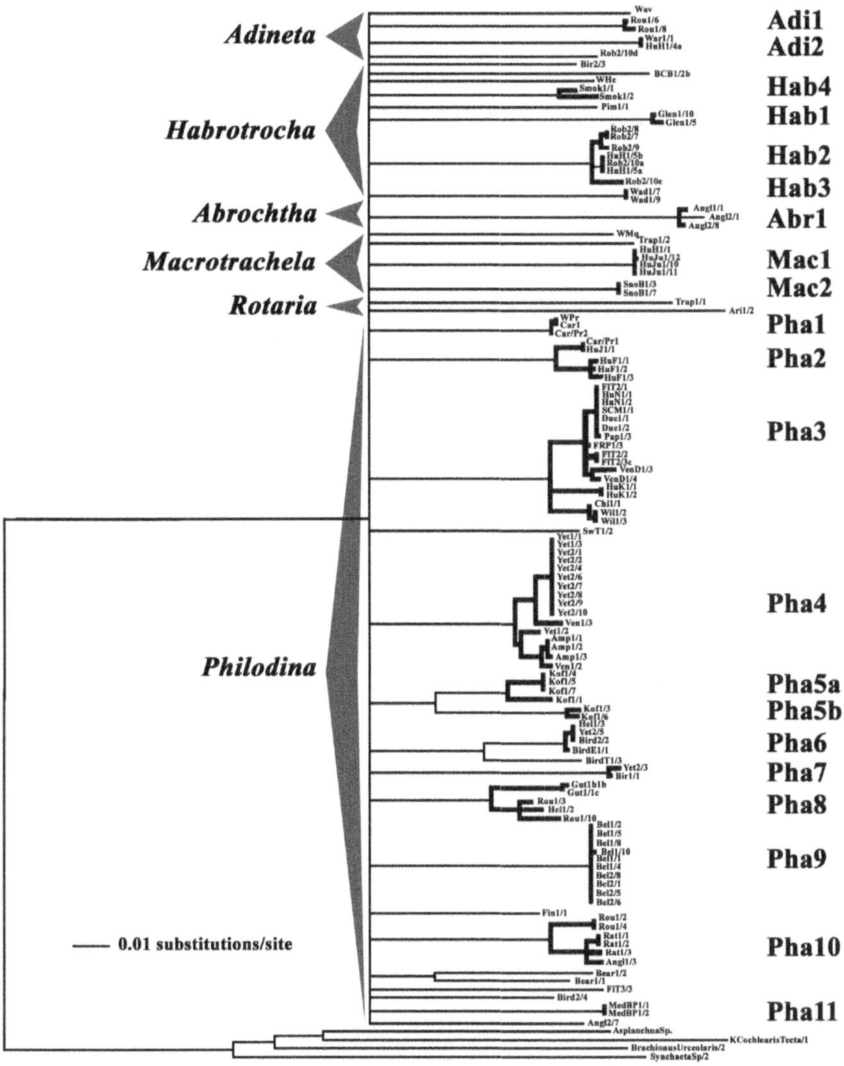

Figure 2. Bootstrapped neighbor-joining tree using pairwise distances corrected for multiple hits with the general time reversible plus invariant sites and gamma-distributed variable sites. Clades that are putative species are shown with bold lines and named on the right (e.g., Adi1 for *Adineta* clade 1, Pha6 for *Philodina* clade 6, etc. Genera are indicated left of the basal polytomy.

over $613 \times 50 = 30{,}646$ km since they diverged from the other clades. This is about 3/4 of the circumference of the earth.

We plotted the geographic distance versus the sequence difference between each pair of clades, using amino acid sequences to avoid possible problems with saturation. The Mantel test, implemented in IBD (Bohonak, 2002), showed a significant correlation between geographic distance and sequence difference but the regression line has a very small slope, indicating rapid dispersal. Using the rate of dispersal estimated by this slope, we found that the bdelloid clades could have dispersed around the world many times since their divergence.

It thus appears that most or all of our clades have repeatedly colonized the same bodies of water. Their continued independent evolution can then be most easily explained if they are adapted to different niches. We obtained direct evidence that at least some of our clades are adapted to different niches.

First, at least some of the clades utilize different, albeit overlapping, food sources. We identified all clones to genus based on discrete morphological differences, and some of these differences almost certainly result from adaptation to different feeding methods. For example, *Adineta* lack a ciliated corona and cannot swim or feed on suspended particles; instead they glide over the substrate and presumably scrape up food with the comb-like structure just behind the ventral mouth. In contrast, *Philodina*, *Macrotrachela*, and *Rotaria* have ciliated coronas which they can use while swimming or to create a vortex that brings food to their mouths while they are attached to the substrate. Besides feeding on suspended material, they can also pick up food from the substrate in their immediate perimeter while attached by the toes.

We hypothesized that *Philodina* and *Adineta* are adapted to feed on suspended organisms or organisms on the substrate, respectively. In preliminary experiments to test this, we compared the food utilization of a clone of *Adineta*, Bird2/3, and one of *Philodina*, FlT2/1. The population growth of these clones was compared on two different food sources, the motile green alga *C. reinhardtii* and the nonmotile bacterium *K. pneumoniae pneumoniae*. For each clone and each food source, 10 replicate cultures were initiated with 1 randomly chosen female in distilled water with excess food and counted every one to two days. Population growth became loglinear after an initial lag; the slope of the loglinear phase is an estimate of the intrinsic rate of natural increase *r*. The results (Table 2) show a significant difference ($p < 0.001$) in food utilization between these two clones. The difference might be related to swimming versus nonmotile food as hypothesized. However, because *Klebsiella* is much smaller than *Chlamydomonas*, there could also be a preference for particle size (Vadstein et al., 1993; Ronneberger, 1998), or there might be a difference in nutritional quality. More extensive experiments by Ricci (1984, 1991) demonstrated differences in food utilization between genus and also between species of the same genus and between isolates of one species.

Table 2. Log growth rates (in females $\times$ day^{-1} $\times$ female^{-1}) of *Adineta* and *Philodina* fed bacteria and algae

Bdelloid clone (clade)	Bacteria	Algae
Adineta (Bird2/3)	0.0138	0.0102
Philodina (FlT2/1)	0.0113	0.0173

Differences between food sources within clones and between clones are statistically significant ($p < 0.001$).

Another likely kind of adaptation in bdelloids is to temperature. We compared the temperature tolerances of clones from two *Philodina* clades, Pha2 and Pha3. From each clone, 12 replicate cultures were initiated with 1 immature female at 16, 27, and 36 °C with *Chlamydomonas* as food. Each of these is referred to as a line. Animals were counted daily and the number of animals plotted against time, with the following results:

1. At 36 °C, only lines from clade Pha2 (clones Huf 1/1, Hufl/2, and Hufl/3) survived. The rate of reproduction varied greatly among the 12 lines of each clone; some never achieved log growth; and the number of animals increased erratically in some. The percentage of females that survived and reproduced was 92% for Hufl/1, 67% for Hufl/2, and 50% for Hufl/3. All of the females from clade Pha3 (clones Ducl/2, FlT2/1, and HuN1/1) died without reproducing. Some of the death may have been due to damage during isolation, but this should affect all the clones about equally.
2. At 27 °C, clones from the two clades differed in the length of a lag period before achieving exponential growth of numbers and in log growth rate. Pha3 clones (Ducl/1, Ducl/2, FlT2/1, HuN1/1) consistently have a shorter lag (ca. 45 h versus 75 h) and higher log growth rates than the clones from Pha2 (Hufl/1, Hufl/3). Some females died without reproducing, possibly due to damage during isolation. These lines were not included in the analysis, reducing the sample size below 12. In contrast to the situation at 36 °C, the number of non-reproducing animals was not significantly different between the two clades.
3. At 16 °C, clones from clade Pha2 (Hufl/1, Hufl/2, Hufl/3) again had a longer lag time of ca. 250–400 h compared to those from Pha3 (Ducl/1, Ducl/2, FlT2/1, HuN1/1, HuN1/2, ScM1/1; ≤ 110 h), but had similar log growth rates. Some animals died without reproducing, reducing the sample size. The numbers of non-reproducing animals was not significantly different between the two clades.
4. Two clones from Pha2 (Hufl/1, Hufl/3) and three clones from Pha3 (Ducl/2, FlT2/1,

HuN1/1) were tested at all temperatures; Figure 3 compares their mean log growth rates. The growth rates of Pha2 clones were lower at both 16 and 27 °C but the 95% confidence intervals overlapped.

We conclude that clade Pha2 is adapted to a higher temperature range than is clade Pha3.

Because bdelloids disperse and colonize readily, a female isolated from a particular site may not be well adapted to it; it may belong to a clone that is a recent arrival rather than a long-term resident. Consequently, when animals from different clades are found in the same sample of water, this is not by itself definitive evidence that those clades are adapted to different niches. However, collectively the data strongly suggest that the bdelloid clades identified as putative species have occupied the same bodies of water repeatedly if not continuously, and geographic isolation cannot explain their independent evolutionary paths.

We conclude that these clades fit most or all of the species definitions that are applicable to asexual as well as sexual organisms. In particular, they fit the phylogenetic and genealogical definitions that focus on reciprocal monophyly. They satisfy the evolutionary definition, which requires that they be following independent evolutionary paths. Finally, they fit the cohesion definition, which explains independent evolution in terms of demographic non-interchangeability due to adaptation to different niches. Contrary to some predictions, this ancient asexual lineage has undergone substantial speciation, driven at least in part by divergent selection.

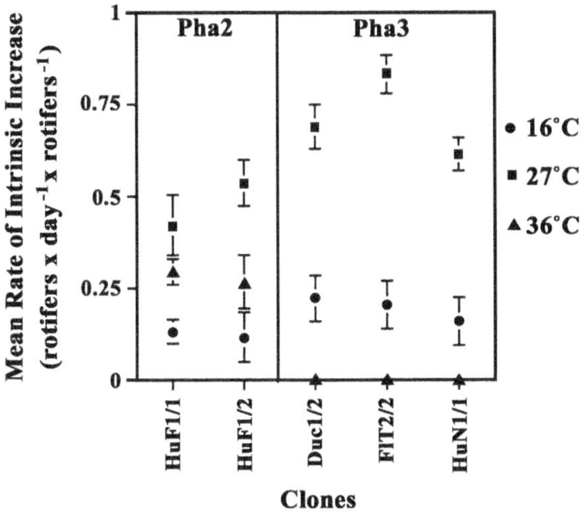

Figure 3. Means and 95% confidence intervals of the log growth rates of all surviving lines from each of two clones of clade Pha2 and three clones of clade Pha3.

Finally, we note that our results agree with the suggestion that *cox1* gene sequences could be used as a 'DNA barcode' to identify species (Hebert et al., 2003).

Effectiveness of selection on a bdelloid mitochondrial gene

The ability of bdelloids to speciate and survive for a very long time might indicate that natural selection continues to operate effectively on bdelloids in spite of their loss of sexual reproduction. We tested this possibility by estimating the proportion of mutations that change amino acids (nonsynonymous mutations) and are not eliminated by natural selection. This proportion is commonly estimated by K_a/K_s, the ratio of amino acid substitutions to synonymous substitutions. The rationale is that most synonymous mutations are neutral so K_s estimates the mutation rate. If selection is weaker in asexual than in sexual lineages due to the Hill–Robertson effect, a larger proportion of detrimental mutations will be fixed in the asexuals, and most of the detrimental mutations will be nonsynonymous.

Mitochondrial genes are inherited uniparentally, from the maternal parent only, in most animals (Birky, 1996). Therefore, it is likely that they are also inherited uniparentally in monogononts and in the common ancestor of monogononts and bdelloids. Nevertheless, the loss of sexual reproduction in bdelloids should result in a marked reduction in the effectiveness of selection for both mitochondrial and nuclear genes. The effectiveness of selection at any site in a genome is reduced by selection on other loci. This reduction is greatest when the other segregating loci are completely linked to the site (Hill & Robertson, 1966). The larger the number of such linked loci, the greater is this Hill–Robertson effect (Birky & Walsh, 1988). In sexual organisms, the nuclear and mitochondrial genomes are unlinked, and background selection on the nucleus has little effect on mitochondrial genes. In the bdelloids, loss of sex caused all the nuclear genes to become completely linked to each other and to the mitochondrial genes. Since nuclear genes outnumber mitochondrial genes by a factor of 500 or more, number of loci linked to a mitochondrial gene increases by the same factor with a resulting large increase in the Hill–Robertson effect.

We used two partial datasets to estimate K_a/K_s in the mitochondrial *cox1* gene. The 'large' data set consisted of 51 bdelloids selected to include a single member of most clades and some singlets, 5 monogononts, and the cephalopod *Gonatus onyx* as an outgroup. Figure 4 is a bootstrapped parsimony tree to illustrate the cladistic relationships of the clones in the large dataset. The 'small' dataset included 22 bdelloids, 3 monogononts, and the cephalopod. This dataset was used because some analyses required too much computation time to be applied to the larger dataset. The phylogenetic tree in Figure 5 was used for the maximum likelihood analysis of the large dataset.

We first estimated K_a/K_s for *cox1* among the clones in the large dataset. Maximum likelihood analysis of this dataset was not computationally feasible, so we used K_{12}/K_3 as an estimator of K_a/K_s (most substitutions in positions 1 and 2, K_{12}, are nonsynonymous, while most substitutions in position 3, K_3, are synonymous). We made neighbor-joining trees with the Jukes–Cantor model for the first and second codon positions, then separately for the third codon position. We then calculated K_{12}/K_3 for each pair of taxa in the

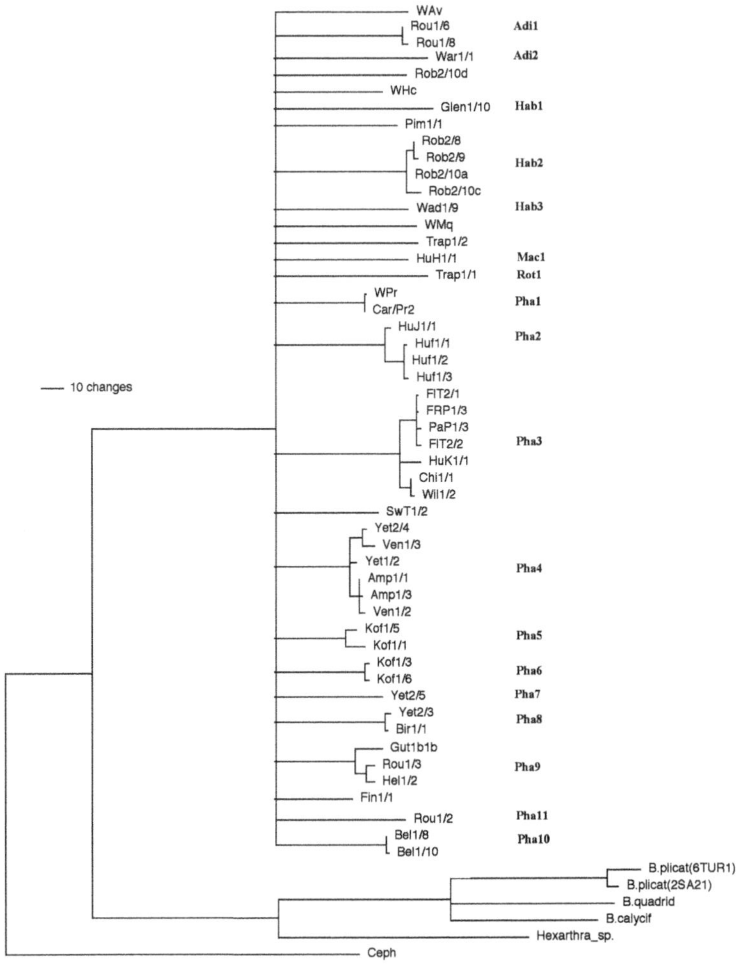

Figure 4. Phylogenetic tree of 56 rotifer clones (large dataset) plus a cephalopod, *Gonatus onyx*. The tree was made with the parsimony algorithm in PAUP* and rooted with the cephalopod as outgroup. All clades are supported by ≥70% of 1000 heuristic bootstraps. Clades are named on the right; in some cases only one member of a clade is represented in this dataset.

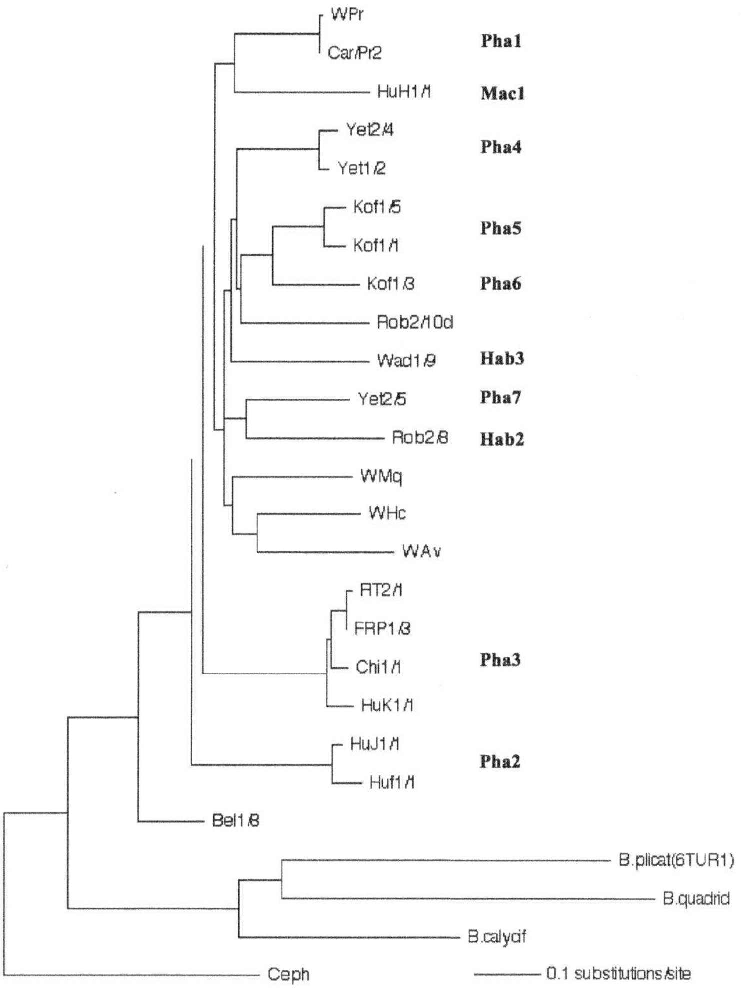

Figure 5. Phylogenetic tree of 25 rotifer clones (small dataset) plus a cephalopod, *Gonatus onyx*. The tree was made with the neighbor-joining algorithm in PAUP*, using the Jukes–Cantor model, and rooted with the cephalopod as outgroup. All clades are supported by ≥70% of 1000 bootstraps in a separate analysis. Clades are named on the right; in some cases only one member of a clade is represented in this dataset.

dataset. We did the same analysis with the small dataset to look for effects of taxon sampling. Because some substitutions in the first and second codon positions are synonymous and some substitutions in the third codon position are not, we used two other methods to estimate K_a and K_s. We used the method of Nei & Gojobori (1986), implemented in PAML (Yang, 2000), to estimate K_a and K_s for each pair of bdelloids and monogononts; then we calculated the pairwise values of K_a/K_s. Finally, we used the maximum likelihood model of Goldman & Yang (1994), also implemented in PAML, to estimate K_a/K_s on each branch in the phylogenetic tree of Figure 5. These methods were applied only to the small dataset because they are computationally intensive; the results are summarized in Table 3. All of our estimates of the mean K_a/K_s for the bdelloids are well below 1 (a value of 1 indicates complete relaxation of selection), and most are less than 0.1, indicating strong selection on the *cox1* gene. It is important to note that selection was strong not only on average but also in most of the individual branches of the bdelloid phylogenetic tree.

The Jukes–Cantor and Goldman and Yang models do not allow variation in substitution rates among sites. This is problematic *a priori*, because

Table 3. Effectiveness of selection on bdelloids and monogononts, indicated by various estimates of K_a/K_s for the *cox1* gene

Dataset	Estimate	Model	Organisms	Mean	Minimum	Maximum	Std. Dev.	Mann–Whitney probability
Large	Pairwise K_{12}/K_3	JC	Bdelloids	0.0624	0	1	0.0495	<0.02
			Monogononts	0.1006	0.0488	0.1617	0.0464	
Small	Pairwise K_{12}/K_3	JC	Bdelloids	0.0732	0.0037	0.178	0.0413	0.36
			Monogononts	0.0802	0.048	0.1434	0.0547	
Small	Pairwise K_{12}/K_3	NG	Bdelloids	0.1703	0.0172	0.3331	0.1446	<0.01
			Monogononts	0.0341	0	0.1018	0.0243	
Small	K_a/K_s on branches	ML	Bdelloids	0.0047	0.0006	0.02	0.0059	0.2
			Monogononts	0.004	0.0018	0.0067	0.0025	

Evolutionary models used to correct for multiple hits are those of Jukes and Cantor (JC), Nei and Gojobori (NG), and the maximum likelihood model of Goldman and Yang (ML). The Mann–Whitney probability is the probability that the monogononts and bdelloids have the same mean, determined with the Mann–Whitney *U* test.

we expect many sites at the first and second codon position and some amino acids to be invariant or nearly so when selection is strong. This would lead to an underestimate of K_a/K_s. Therefore, we also applied two of the variable sites models implemented in PAML. These models allow K_a/K_s to vary among sites (codons) in each branch, but the ratio K_a/K_s is constrained to be the same on all branches. Because the great majority of the branches are among the bdelloids, the overall K_a/K_s estimate is probably close to what would be obtained with bdelloids alone. The model of Yang (1994) with three discrete classes of substitutions (model M3 in PAML) estimated K_a/K_s as 0.0329 in the bdelloids, similar to the estimates in Table 3. In contrast, the Nielsen & Yang (1998) model with a continuous beta distribution (model M7 in PAML) gave $K_a/K_s = 0.2168$, slightly higher than the NG value.

We used several different methods to estimate K_a/K_s because there is some disagreement about which method is the most accurate; the differences among these estimates reflect the methodological differences. The important point is that all methods show strong selection.

There is some evidence that the synonymous substitutions on the deep branches are saturated in our data set (David Mark Welch, Matthew Meselson, and Michael Cummings, personal communication). If so, we have underestimated K_s and overestimated K_a/K_s and selection on *cox1* is even stronger than these numbers show.

We compared these K_a/K_s values for bdelloids to those of sexual organisms. For the small sample of monogononts in our analyses, K_a/K_s was not significantly different in two methods, was significantly higher in the bdelloids in one comparison, and was significantly lower in one (Table 3). Lynch & Blanchard (1998) calculated the pairwise K_a/K_s for a large sample of protein-coding genes from a broad spectrum of invertebrates; the mean was 0.13, higher than that for either the bdelloids or monogononts in our sample. Giribet et al. (2001) obtained *cox1* sequences from a large sample of arthropod taxa. We calculated pairwise values of K_{12}/K_3 with the Jukes–Cantor model for 651 bp of their sequences from 23 arthropods, obtaining a mean value of 0.15, again higher than in our rotifers. We calculated the mean pairwise $K_{12}/K_3 = 0.089$, again with the Jukes–Cantor model, for 957-bp *cox1* sequences from 75 leaf beetles (Funk, 1999). Most or all of the organisms in these datasets are obligately sexual. These comparisons lead us to conclude that the loss of sexual reproduction in the bdelloids resulted in little or no reduction in the effectiveness of selection on the *cox1* gene.

Discussion

Bdelloids have undergone substantial cladogenesis

Our data show that the bdelloid rotifers have undergone substantial cladogenesis. This is the first demonstration of substantial cladogenesis in an obligately asexual lineage using sequence data. Of these clades, 21 show two important characteristics of species: modest sequence diversity, characteristic of species in sexual animals, and reciprocal monophyly. The latter is a key feature of the cladistic and phylogenetic concepts. It also shows that the clades, and the natural populations they represent, have been evolving independently of each other for a long time, as required by the evolutionary species concept. Each is thus an independent arena for the basic evolutionary processes of mutation, drift, and selection. This is the basis of the cohesion species concept.

Bdelloid clades are adapted to different niches

At least some of the clades differ in morphology and behavior, showing that they are adapted to different niches. In further support of this, we demonstrated differences in food utilization between two clones, and in temperature tolerance between clones of two clades. We also found no evidence for isolation by distance in our bdelloids. If the different clades are not isolated from each other by distance, then they must be adapted to different niches in order to be evolving independently.

Bdelloids have undergone substantial speciation

We conclude that the clades we identified can be considered different species by most of the species definitions that are applicable to asexual organisms. If this is correct, then the bdelloid rotifers have undergone substantial speciation in the absence of sexual reproduction. This is the first demonstration of substantial speciation in an asexual lineage, based on phylogenetic analysis of DNA sequences. The extent to which these species correspond to the species already identified on the basis of morphology remains to be seen. We also note that our data support the suggestion of Hebert et al. (2003) that sequences of the *cox1* gene can be used as 'DNA barcodes' for the identification of species.

No detectable decrease in the effectiveness of natural selection on bdelloid genes

We failed to detect a greater accumulation of detrimental mutations in bdelloids, relative to monogononts and other sexual invertebrates, in the mitochondrial *cox1* gene. Mark Welch & Meselson (2001) previously showed that the nuclear *hsp82* gene of clones WPr, WMq, WAv, and WHc are under strong selection in both bdelloids and monogononts. There are several possible explanations:

1. The genes used in these studies may be too strongly selected to be useful for this purpose. The ratio K_a/K_s depends on the fixation probability of a new mutation. This in turn is a function of $N_e s$, where N_e is the effective population size and s is the selection coefficient of the mutation. The reduced effectiveness of selection in asexuals can be viewed as a reduction in N_e, but when s is very large, it takes a very large reduction in N_e to produce a detectable change in $N_e s$ and the fixation probability (theoretical results not shown). In other words, when selection is sufficiently intense, genetic drift is not very important in either sexual or asexual species. It is possible that an analysis of other genes that are subject to less stringent selection in monogononts, and hence more sensitive to a reduction in selection intensity, would show that some have accumulated dangerously large loads of detrimental mutations. Recently, David Mark Welch, Matthew Meselson, and Michael Cummings (personal communication) estimated $K_a/K_s = 0.05$ for our combined sequences of *cox1* and another mitochondrial gene, *cob*, for four bdelloid clones (WPr, WMq, WAv, and WHc). We also found a low value of K_a/K_s for *cob* alone in a larger sample of bdelloids (data not shown).
2. The effective population size of rotifers may be so large that the decrease due to loss of sexual reproduction isn't readily detectable. This is unlikely, because the independently evolving clades show sequence diversity similar to that of sexual organisms. This sequence diversity is a function of $N_e u$ where u is the mutation rate. If N_e is unusually large in bdelloids, then the mutation rate must be unusually small. Even if bdelloids may very large census population sizes, we suspect that the number of offspring per female is extremely variable due to their opportunistic life style. If so, their variance effective population size would still be modest.
3. The phenotypic effects of new mutations may have changed in bdelloids. As discussed above, the effect of the loss of sex on the ratio K_a/K_s

depends on the product $N_e s$. The reduction in N_e due to the loss of sex could be compensated by a decrease in the selection coefficient s for nonsynonymous substitutions (i.e., they became less detrimental), or to an increased effectiveness of selection for synonymous substitutions (selection for codon bias increased). Another possibility is that detrimental mutations may be more strongly detrimental in bdelloids than in otherwise-similar sexual lineages. Strongly detrimental mutations are rarely fixed by random drift in asexual or sexual lineages and are unlikely to contribute to K_a/K_s (Gabriel et al., 1993). A change in fitness effects of new mutations is plausible because the fitness of a new mutation is influenced by the background genotype and can be modified by selection (Rutherford, 2000; Hartman et al., 2001).

4. There may be an ascertainment bias. We are seeing only the lucky survivors who survived because they haven't accumulated many detrimental mutations by chance. We will use our larger sample of bdelloids to estimate the strength of selection along branches at varying distances from the tip of the tree. The newest lineages have had less time in which to become extinct and should not be subject to this bias.

We do not know how large a load of detrimental mutations is needed to push a population of bdelloids over the brink into extinction, but they do not appear to be any closer to extinction than their sexual relatives.

We said at the outset that two basic problems of biology intersect in asexual organisms like the bdelloids: the evolutionary advantage of sex, and the nature of species. Theory predicts that natural selection should be weaker in asexual organisms, compared to sexual organisms. Consequently, asexuals should be less able to retain and fix the mutations needed adapt to different niches, and have a lower rate of speciation. Theory also predicts that asexual organisms should accumulate more detrimental mutations than sexual organisms, leading to a higher rate of extinction of species. The net effect would be a lower rate of speciation in asexuals than in sexuals, *all else being equal*. Our results show that the bdelloid rotifers can speciate, possibly because all else is *not* equal. Future studies will look for features of the bdelloids that might compensate for the loss of sex.

Acknowledgements

Many people provided helpful comments and constructive criticism, including Timothy Barraclough, Douglas Futuyma, Africa Gómez, David Hillis, Charles King, Wayne Maddison, Brian McGill, Claudia Ricci, David Mark Welch, Robert Wallace, Elizabeth Walsh, and an anonymous reviewer. Michael Dellinger, Marlea Gemmel, Dee Haefner, Catarina Kazcurkin, Julia Perry, and Michael Nodine assisted in the lab. Our research was supported by startup funds provided to CWB by the University of Arizona and a Grant-in-Aid to LH from the Society of Sigma Xi; salary from the Undergraduate Biology Research Program for CW and EH; a MacNair Scholarship for LH; and a research associateship from the Genetics Program for HM.

References

Avise, J. C., 1994. Molecular Markers, Natural History and Evolution. Chapman & Hall, Inc., New York.

Avise, J. C. & R. M. Ball Jr., 1990. Principles of genealogical concordance in species concepts and biological taxonomy. In Futuyama, D. & J. Antonovics (eds), Oxford Surveys in Evolutionary Biology. Oxford University Press, Oxford: 45–67.

Barraclough, T. G., C. W. Birky, Jr. & A. Burt, 2003. Diversification in sexual and asexual organisms. Evolution 57: 2166–2172.

Barton, N. H. & B. Charlesworth, 1998. Why sex and recombination? Science 281: 1987–1990.

Bell, G., 1982. The Masterpiece of Nature. Croom Helm, London.

Birky, C. W. Jr., 1996. Uniparental inheritance of mitochondrial and chloroplast genes: mechanisms and evolution. Proceedings of the National Academy of Sciences of the United States of America 92: 11331–11338.

Birky, C. W. Jr. & J. B. Walsh, 1988. Effects of linkage on rates of molecular evolution. Proceedings of the National Academy of Sciences of the United States of America 85: 6414–6418.

Birky, C. W., T. Maruyama & P. Fuerst, 1983. An approach to population and evolutionary genetic theory for genes in mitochondria and chloroplasts and some results. Genetics 103: 513–527.

Bohonak, A. J., 2002. IBD (Isolation By Distance): a program for analysis of isolation by distance. Journal of Heredity 93: 153–154.

Cáceres, C. E. & D. A. Soluk, 2002. Blowing in the wind: a field test of overland dispersal and colonization by aquatic invertebrates. Oecologia 131: 402–408.

Derry, A. M., P. D. N. Hebert & E. E. Prepas, 2003. Evolution of rotifers in saline and subsaline lakes: a molecular phylogenetic approach. Limnology and Oceanography 48: 675–685.

Donner, J., 1965. Ordnung Bdelloidea. Akademie Verlag, Berlin, 297 pp.

Funk, D. J., 1999. Molecular systematics of cytochrome oxidase I and 16S from *Neochlamisus* leaf beetles and the importance of sampling. Molecular Biology and Evolution 16: 67–82.

Gabriel, W., M. Lynch & R. Bürger, 1993. Muller's ratchet and mutational meltdown. Evolution 47: 1744–1757.

Gilbert, D. G., 1992. SeqApp, a biological sequence editor and analysis program for Macintosh computers. Published electronically on the Internet, available via gopher or anonymous ftp to ftp.biol.indiana.edu.

Giribet, G., G. D. Edgecombe & W. C. Wheeler, 2001. Arthropod phylogeny based on eight molecular loci and morphology. Nature 413: 157–161.

Goldman, N. & Z. Yang, 1994. A codon-based model of nucleotide substitution for protein-coding DNA sequences. Molecular Biology and Evolution 11: 725–736.

Gómez, A., 2005. Molecular ecology of rotifers: from population differentiation to speciation. Hydrobiologia 546: 83–99.

Gómez A., G. R. Carvalho & D. H. Lunt, 2000. Phylogeography and regional endemism of a passively dispersing zooplankter: mitochondrial DNA variation in rotifer resting egg banks. Proceedings of the Royal Society of London. Series B, Biological Sciences 267: 2189–2197.

Hartman, J. L. I., B. Garvik & L. Hartwell, 2001. Principles for the buffering of genetic variation. Science 291: 1001–1004.

Hebert, P. D. N., A. Cywinska, S. L. Ball & J. R. deWaard, 2003. Biological identification through DNA barcodes. Proceedings of the Royal Society of London. Series B, Biological Sciences 270: 313–321.

Hill, W. G. & A. Robertson, 1966. The effect of linkage on limits to artificial selection. Genetical Research 38: 226–231.

Hoeh, W., M. Black, R. Gustafson, A. Bogan, R. Lutz & R. Vrijenhoek, 1998. Testing alternative hypotheses of Neotrigonia (Bivalvia:Trigonioida) relationships using cytochrome C oxidase subunit I DNA sequences. Malacalogia 40: 267–278.

Holman, E. W., 1987. Recognizability of sexual and asexual species of rotifers. Systematic Zoology 36: 381–386.

Hudson, R. R. & J. A. Coyne, 2002. Mathematical consequences of the genealogical species concept. Evolution 56: 1557–1565.

Kondrashov, A. S., 1993. Classification of hypotheses on the advantage of amphimixis. Journal of Heredity 84: 372–387.

Lynch, M. & J. L. Blanchard, 1998. Deleterious mutation accumulation in organelle genomes. Genetica 102/103: 29–39.

Mark Welch, D. B. & M. Meselson, 2001. Rates of nucleotide substitution in sexual and anciently asexual rotifers. Proceedings of the National Academy of Sciences of the United States of America 98: 6720–6724.

Maruyama, T. & C. W. Birky Jr., 1991. Effects of periodic selection on gene diversity in organelle genomes and other systems without recombination. Genetics 127: 449–451.

Moriyama, E. N. & J. R. Powell, 1996. Intraspecific nuclear DNA variation in *Drosophila*. Molecular Biology and Evolution 13: 261–277.

Nei, M. & T. Gojobori, 1986. Simple methods for estimating the numbers of synonymous and nonsynonymous nucleotide substitutions. Molecular Biology and Evolution 3: 418–426.

Nielsen, R. & Z. Yang, 1998. Likelihood models for detecting positively selected amino acid sites and applications to the HIV–1 envelope gene. Genetics 148: 929–936.

Posada, D. & K. A. Crandall, 1998. MODELTEST: testing the model of DNA substitution. Bioinformatics 14: 817–818.

Ricci, C., 1984. Culturing of some bdelloid rotifers. Hydrobiologia 112: 45–51.

Ricci, C., 1991. Comparison of five strains of a parthenogenetic species, *Macrotrachela quadricornifera* (Rotifera, Bdelloidea). Hydrobiologia 211: 147–155.

Ronneberger, D., 1998. Uptake of latex beads as size-model for food of planktonic rotifers. Hydrobiologia 387/388: 445–449.

Rosenberg, N. A., 2003. The shapes of neutral gene genealogies in two species: probabilities of monophyly, paraphyly, and polyphyly in a coalescent model. Evolution 57: 1465–1477.

Rutherford, S. L., 2000. From genotype to phenotype: buffering mechanisms and the storage of genetic information. BioEssays 22: 1095–1105.

Segers, H., 2002. The nomenclature of the Rotifera: annotated checklist of valid family- and genus-group names. Journal of Natural History 36: 631–640.

Swofford, D. L., 1998. PAUP*: Phylogenetic Analysis Using Parsimony (*and Other Methods). Version 4. Sinauer Associates, Sunderland, Massachusetts.

Vadstein, O., G. Oie & Y. Olsen, 1993. Particle size dependent feeding by the rotifer *Brachionus plicatilis*. Hydrobiologia 255/256: 261–267.

Yang, Z., 1994. Maximum likelihood estimation from DNA sequences with variable rates over sites: approximate methods. Journal of Molecular Evolution 39: 306–314.

Yang, Z., 2000. Phylogenetic Analysis by Maximum Likelihood (PAML). London, England, University College London.

Hydrobiologia (2005) 546:47–54

A. Herzig, R.D. Gulati, C.D. Jersabek & L. May (eds.) Rotifera X: Rotifer Research: Trends, New Tools and Recent Advances
DOI 10.1007/s10750-005-4100-y

Bayesian and maximum likelihood analyses of rotifer–acanthocephalan relationships

David B. Mark Welch
Josephine Bay Paul Center for Comparative Molecular Biology and Evolution, Marine Biological Laboratory, 7 MBL Street, Woods Hole, MA, 02543, USA
E-mail: dmarkwelch@mbl.edu

Key words: phylogenetics, 18S, Eurotatoria, Syndermata, Gnathifera

Abstract

Rotifera is composed of groups with unusual ultrastructural, physiological, and reproductive characters. Our ability to understand the evolution of these features is complicated by the fact that the phylogenetic relationships among the three traditional rotifer groups (Seisonidea, Monogononta, and Bdelloidea) and Acanthocephala remain unresolved. Here, I present maximum likelihood and Bayesian analyses of rotifer–acanthocephalan relationships using both the protein-coding gene *hsp82* and a combined data set of *hsp82* and ribosomal small subunit (SSU) DNA sequences, using nucleotide and codon based models of evolution. Statistical analysis of the phylogenetic support for any of the likely relationships among rotifer groups suggests that more than a combined *hsp82* + SSU data set will be needed to resolve rotifer–acanthocephalan phylogeny with any degree of certainty.

Introduction

The phylum Rotifera has traditionally included three monophyletic groups, the bdelloids, monogononts, and seisonids (here designated as the classes Bdelloidea, Monogononta, and Seisonidea). In addition to numerous morphological and physiological differences, each group has evolved different reproductive modes over tens of millions years: seisonids are obligately sexual, monogononts are generally facultative parthenogens producing haploid males (arrhenous thelytoky), and bdelloids are obligately parthenogenetic (amictic thelytoky) (Gilbert, 1983; Wallace, 1999; Mark Welch & Meselson, 2000). Recent ultrastructural and molecular phylogenetic analyses have suggested that Acanthocephala, a group of more than 1000 species of obligate parasites, long regarded as a phylum closely related to rotifers, may actually be a fourth group within Rotifera (reviewed in Mark Welch, 2001).

Rotifers present an opportunity to study a number of important issues in evolutionary biology, such as the role of sexual reproduction and recombination, the dynamic adaptations of parasitism, and the evolution of parasitic, colonial, and other symbiotic relationships (Near, 2002; Wallace, 2002). To take proper advantage of rotifers for evolutionary studies, it is necessary to understand the phylogenetic relationship between rotifer groups and between these groups and related taxa (such as micrognathozoans, gnathostomulids, and others). However, the phylogeny of Rotifera remains unresolved.

Preliminary molecular phylogenies of rotifer taxa based on nuclear 18S and mitochondrial 16S gene sequences were the first molecular analyses to suggest a close association between acanthocephalans and particular rotifer groups (Garey et al., 1996; 1998). These studies found acanthocephalans to be more closely related to bdelloids than they were to monogononts, but included only single bdelloid and monogonont species, one or three acanthocephalans, and no seisonids. A phylogeny of the heat shock protein gene *hsp82* (Mark Welch, 2000) was the first to include multiple monogonont and bdelloid species, as well as a member of the basal seisonids, and to explicitly test alternate hypotheses. This analysis supported a clade of Bdelloidea and Monogononta (Eurotatoria) and placed the Acanthocephala as a sister-group to Eurotatoria within Rotifera, but significance tests were not able to exclude alternate topologies.

Due in part to the economic importance of acanthocephalans as parasites, the 18S region of a large number of acanthocephalan species have since been examined to better understand acanthocephalan phylogeny (García-Varela et al., 2000; Near, 2002; Herlyn et al., 2003). These analyses supported Eurotatoria with a closely related Acanthocephala, and tested and statistically rejected the alternative hypothesis of a clade of Bdelloidea and Acanthocephala. Only the study of Herlyn and co-workers included a representative of Seisonidea, and their analysis significantly supported a clade of Acanthocephala and Seisonidea as a sister-clade to Eurotatoria, a relationship not excluded by the *hsp82* analysis (Mark Welch, 2000, 2001) and consistent with certain morphological similarities (Ahlrichs, 1997). It should be noted that the above 18S analyses included only 1–4 monogononts, all of the family Ploima.

The data set of rotifer *hsp82* sequences represents the broadest sampling of rotifer taxa now available. Broad species sampling can greatly increase phylogenetic resolution (Lecointre et al., 1993; Hillis, 1996; 1998; Graybeal, 1998). In addition, the use of a coding sequence gene circumvents many of the difficulties inherent in phylogenetic analysis of 18S sequence data, such as alignment, covariation, and rate ambiguity (Abouheif, et al. 1998; Maley & Marshall, 1998; Philippe & Germot, 2000; Mark Welch, 2001). Furthermore, models of evolution for coding sequences now available in maximum likelihood and Bayesian frameworks allow very sophisticated representation of known biological criteria. Because of their different but complementary approaches to arriving at optimal trees (discussed in Cummings et al., 2003), the two methods provide the best available means of investigating gene genealogies within a framework of explicit evolutionary modeling. Here, I present an examination of the phylogeny of Rotifera based on maximum likelihood and Bayesian analyses of *hsp82* and of a combined *hsp82* + 18S data set.

Materials and methods

Taxa and alignments

Taxa and Genbank Accession numbers are listed in Table 1. In addition to sequences previously described (Mark Welch, 2000; Mark Welch & Meselson, 2001) or acquired from Genbank, DNA was extracted from a clonal culture of *Adineta ricciae* Segers 2004 (Segers & Shiel, 2005) and from a mixed population of *Lepidodermella* sp (Gastrotricha) obtained from Carolina Biological Supply, and a region of the *hsp82* gene was amplified by PCR, cloned and sequenced as previously described (Mark Welch, 2000). The approximately 870 bp region of *hsp82* examined contains the invariant hsp82–90 signature region, the ATP-geldanamycin binding site, and a variable highly charged region which consists largely of aspartate, glutamate, and lysine residues, and is described more thoroughly in Mark Welch (2000). The *hsp82* sequences were translated and aligned to a template of a eukaryotic *hsp82* amino acid alignment as previously described (Mark Welch, 2000). The two *Adineta* species possess an intron in this region which was not used in subsequent phylogenetic analyses as its presence in only two sister-taxa is non-informative. The *hsp82* sequence from the monogonont *Brachionus plicatilis* was not included because previous analyses have shown that this gene is evolving in a non-representative manner in this species (Mark Welch, 2000, 2001) which can confound phylogenetic reconstruction even under likelihood (and, by extension, Bayesian) methods (Omilian & Taylor, 2001).

The 18S alignment was that of the European Ribosomal RNA Database (http://oberon.rug.ac.be:8080/rRNA/index.html) and is based on a proposed model of secondary structure; sequences not already aligned in the database were aligned using ClustalW (Thompson et al., 1994) with the

Table 1. Taxa and accession numbers

Species	Hsp82	18S
Acanthocephala		
Moniliformis moniliformis (Bremser, 1811)	AF143853	Z19562
Oligacanthorhynchus tortuosa (Leidy, 1850)	AF375825	AF064817
Oncicola sp.	AF375826	AF064818
Bdelloidea		
Adineta vaga (Davis, 1873)	AF143849	–
Adineta ricciae (Segers & Shiel 2005)	AY394701	–
Habrotrocha constricta (Dujardin, 1841)	AF143850	–
Macrotrachela quadricornifera (Milne, 1886)	AF143852	–
Philodina acuticornis (Murray, 1906)	–	U41281
Philodina roseola Ehrenberg, 1832	AF143851	AF154567
Monogononta		
Asplanchna sieboldi Leydig	–	AF092434
Brachionus calyciflorus (Pallas, 1766)	AF143855	–
Brachionus patulus (O.F. Muller, 1786)	–	AF154568
Brachionus plicatilis (Mueller, 1786)	AF143856	U49911
Eosphora ehrenbergi (Weber, 1918)	AF143858	–
Sinantherina socialis (Linnaeus, 1758)	AF143854	–
Seisonidea		
Seison nebaliae (Grube, 1859)	AF143857	AF469411
Seison sp.	–	AF053612
Gastrotricha		
Lepidodermalla squammatum (Dujardin, 1841)	–	U29198
Lepidodermella sp.	AY394702	–

gap creation and extension penalties both lowered to three. The *hsp82* + 18S data set was created by: (1) combining the *hsp82* and 18S sequences of those species for which both were available; (2) creating operational taxonomic units by combining the *hsp82* and 18S sequences of *B. calyciflorus* and *B. patulus*, of *E. ehrenbergi* and *Asplanchna* sp, and of *H. constricta* and *P. acuticornis*; and (3) adding the partial 18S sequence of *Seison* sp. to the *Seison nebaliae hsp82*+18S sequence. Sequences and alignments are available from WheelBase, http://jbpc.mbl.edu/wheelbase.

Phylogenetic analysis by maximum likelihood

The program ModelTest (Posada & Crandall, 1998) was used to determine appropriate nucleotide-based evolutionary models for three partitions of the *hsp82* dataset: all codon positions, codon 1st and 2nd positions, and codon 3rd positions. The model for all positions was that of Tamura & Nei (1993) with rate heterogeneity estimated by a gamma distribution with a shape parameter (alpha) of 1.56 and a fraction of invariant sites of 0.328; for codon 1st and 2nd positions the same model was used with an alpha of 0.326 and no invariant sites; for codon 3rd positions the General Time Reversible model (Rodríguez et al., 1990) with an alpha of 2.87 was used. PAUP* 4.0b10 (Swofford, 2002) was used to find the best ML trees, using a heuristic search with tree bisection-reconnection and 1000 random-addition-sequence replications. Bootstrap values were generated in heuristic searches with 1000 bootstrapped datasets and 10 random-addition sequence replications for each bootstrap replicate. Various codon-based models of evolution were explored using the codeml program of PAMLv3.13a (Yang, 1997), and likelihood ratio tests (LRT) and the Akaike Information Criterion (AIC) were used to evaluate the appropriateness of each model. The codon model favored by both criteria estimates codon equilibrium frequencies from existing codon frequencies and estimates synonymous–nonsynonymous ratios independently for each of 25 branches; these parameters may be inadequately specified by the available data. The second-best model, favored over all others by both criteria, does not estimate the synonymous–nonsynonymous ratio independently for each branch and produced an identical topology. Tree searching under codon-based models using maximum likelihood is extremely computationally intensive and bootstrap analysis is not currently practical.

Alternate tree topologies were evaluated with Kishino-Hasegawa (KH) and Shimodaira-Hasegawa (SH) tests (Kishino & Hasegawa, 1989; Shimodaira & Hasegawa, 1999). Topologies were tested with nucleotide-based models in PAUP with 1000 bootstrap resamplings of log-likelihood scores estimated with full optimization. For the combined *hsp82* + 18S data set, separate rates were estimated for *hsp82* codon 1st and 2nd positions, *hsp82* codon 3rd positions, and 18S independently using the site specific rate model option in PAUP (which does not allow for rate heterogeneity or independent estimates of nucleotide frequency or state change). For *hsp82*, the likelihood of alternate tree topologies was also evaluated with codon-based models using SH and KH tests implemented in CODEML.

Phylogenetic analysis by Bayesian inference

Bayesian analyses were performed with the program MRBAYES 3.0B4 (Huelsenbeck & Ronquist, 2001). The nucleotide substitution model was GTR + I + G. This program allows independent estimations of parameters for different partitions of the data; nucleotide frequencies, substitution rates, and rate heterogeneity parameters were estimated for codon 1st and 2nd positions and for codon 3rd positions independently using the 'unlink' option. The prior probability distribution of alpha was set to uniform distributions from 0.1–1.0 for codon 1st and 2nd positions and from 1.0–5.0 for codon 3rd positions, respectively. For the *hsp82* – 18S analysis, parameters for 18S were estimated as an additional unlinked partition. The chain length for all analyses was 2×10^6 generations with trees sampled every 100 generations. Chain parameters appeared to be stationary after several thousand sampled trees; the first 10^4 trees (10^6 generations) were discarded and the second 10^4 trees examined. Additional runs with the same conditions produced the same topology with insignificant differences in posterior probability of any node.

Results

Gene trees of *hsp82* with the best likelihood or prior probability scores are shown in Figure 1 with their associated bootstrap or posterior probability support. Accurate reconstruction of phylogenies from coding sequences can require a very different model of evolutionary change for codon 1st and 2nd positions than for codon 3rd positions, which cannot be performed with current maximum likelihood programs. Therefore, codon 1st and 2nd positions, codon 3rd positions, and all codon positions of *hsp82* were examined by maximum likelihood independently under different models of nucleotide evolution. Each analysis supports the monophyly of each rotifer group. When all codon positions are included, and when only 3rd positions are considered, the best ML tree shows Acanthocephala as a sister-taxon to Eurotatoria, with a basal Seisonidea; these nodes are significantly supported in the codon 3rd position analysis. When only codon 1st and 2nd positions are considered, the best ML tree found shows an unusual arrangement of taxa with no significant bootstrap support. The best tree obtained under a codon-based model, which accounts for the different rate of synonymous and non-synonymous change, shows the same arrangement of rotifer groups as the trees based on codon 3rd positions and on all codon positions using nucleotide-based models. The likelihood score of the best tree was compared to six alternate topologies for each partition. The likelihood scores and results of KH and SH tests are summarized in Table 2. Most topologies under most models cannot be rejected at the 0.05 level.

The program MRBAYES 3.0B4 allows independent evolutionary models for different portions of data, and thus a single Bayesian analysis could be performed using separate models for codon 1st and 2nd positions and for codon 3rd positions. The tree topology with the highest posterior probability has the same arrangement of rotifer groups as the maximum likelihood analysis of all codon positions and codon 3rd positions under a nucleotide model and as the maximum likelihood analysis of all codons under a codon model (Fig. 1, Table 2). This topology and the topology with the second highest posterior probability, which differs only within Monogononta, account for 69% of the cumulative posterior probability. The third best topology, which has Eurotatoria as a sister-taxon to a clade of Seisonidea + Acanthocephala, accounts for only 8.6% of the cumulative posterior probability. Trees with a clade of Bdelloidea + Acanthocephala account for less than 1% of the cumulative posterior probability (see Table 2).

Analysis of the combined *hsp82* + 18S data set also strongly supports Eurotatoria. Bayesian analysis allowing evolutionary parameters to be

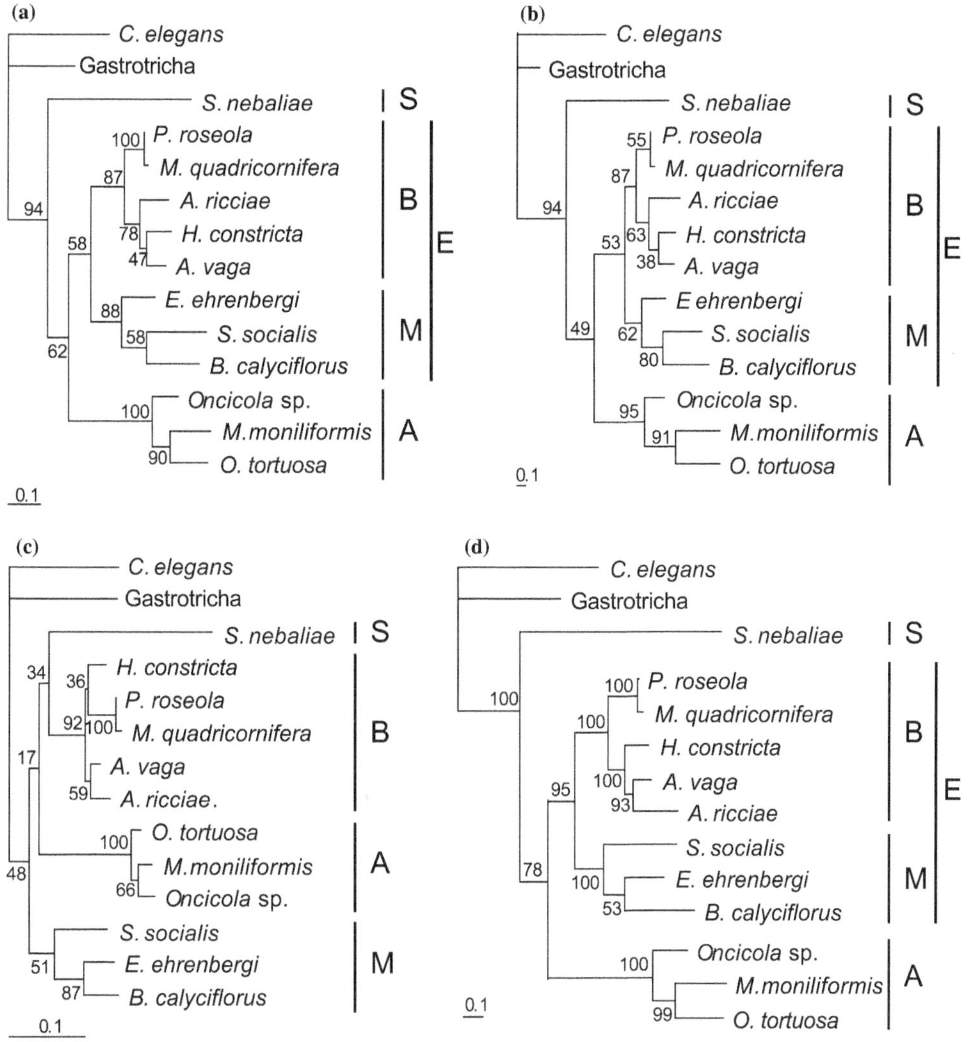

Figure 1. Best trees from analyses of *hsp82*. (a–c) ML gene trees with percent bootstrap support for each node for (a) all codon positions, (b) codon 3rd positions, and (c) codon 1st and 2nd positions. The paraphyly of Adinetidae in (a) and (b) is not significantly supported. (d) Bayesian gene tree, with percent posterior probability of each node. Trees are drawn proportional to the inferred number of changes per nucleotide per branch (see scale bars); for (d) this is the mean of the posterior probability density of branch lengths. Horizontal bars indicate higher-order clades: A, Acanthocephala; B, Bdelloidea; E, Eurotatoria; M, Monogononta; S, Seisonidea.

estimated independently for *hsp82* codon positions and 18S supports Eurotatoria + Acanthocephala with a basal Seisonidea, with most alternative trees having zero posterior probability (Fig. 2, Table 2). Maximum likelihood using a site-specific model prefers a tree in which Acanthocephala and Seisonidea are sister taxa, but this topology is not significantly better than one in which Seisonidea is basal (Table 2).

Discussion

All maximum likelihood and Bayesian analyses placed Acanthocephala within Rotifera. Maximum likelihood analyses of all codon positions and of codon 3rd positions of *hsp82* and all Bayesian analyses support a clade of Bdelloidea + Monogononta (Eurotatoria) and a basal Seisonidea.

Table 2. Statistical support for alternative topologies of Rotifera

Topology	ML score (-ln)						Bayesian posterior probability	
	hsp82 under four models				*hsp82* + *18S*	*18S*		
	all *pos*	1st + 2nd	3rd	Codo n			*hsp82*	*hsp82* + 18S
(S(A(B,M)))	**7313**	(6.1)	**3877**	**6547**	(3.6)	(0.4)	0.75	0.82
(S(M(A,B)))	(7.5)	(5.2)	(6.0)	(3.2)	(74)**	(33)**	<.01	0
((S,A)(M,B))	(2.1)	(5.3)	(2.1)	(2.3)	**14 946**	**8427**	0.16	0.14
((S,B)A)M))	(11)	(0)	(12.3)	(19)*	(112)**	(45)**	<.01	0
((B,S)(A,M))	(10)	**3091**	(12.3)	(51)**	(118)**	(45)**	<.01	0
(S(B(A,M)))	(5.2)	(2.7)	(6.0)	(1.5)	(76)**	(33)**	<.01	0
(A(M(B,S)))	(7.4)	(2.9)	(7.7)	(5.6)	(75)**	(27)**	<.01	0
(A(S(B,M)))	(2.3)	(6.4)	(2.3)	(60)**	(10)	(0.4)	0.06	0.04

Tree topologies are given in Newick format; A, Acanthocephala; B, Bdelloidea; M, Monogononta; S, Seisonidea. For each model under ML, the best -ln likelihood is shown in bold with the difference in likelihood for other topologies indicated in parentheses. *rejected at $p < 0.05$ by KH; ** rejected at $p < 0.01$ by KH and SH.

The best trees obtained by maximum likelihood analysis of all *hsp82* codon positions and of codon 3rd positions show a clade of Bdelloidea + Monogononta with a sister-clade of Acanthocephala and a basal Seisonidea (Fig. 1). The unusual tree found when only codon 1st and 2nd positions are used, with its very low bootstrap support, is probably a result of unknown factors disproportionately favoring an inappropriate model of evolution (Sullivan & Swofford, 1997). Bootstrap support for Eurotatoria and Eurotatoria + Acanthocephala is less than 70% and therefore probably not significant (Hillis & Hill, 1993), nor can alternate topologies be effectively excluded by hypothesis testing (Table 2). The same topology is obtained with a codon based model, which more accurately reflects the evolutionary history of the gene sequences. This model can exclude certain alternative topologies, particularly the traditional arrangement of Eurotatoria + Seisonidea with a basal Acanthocephala.

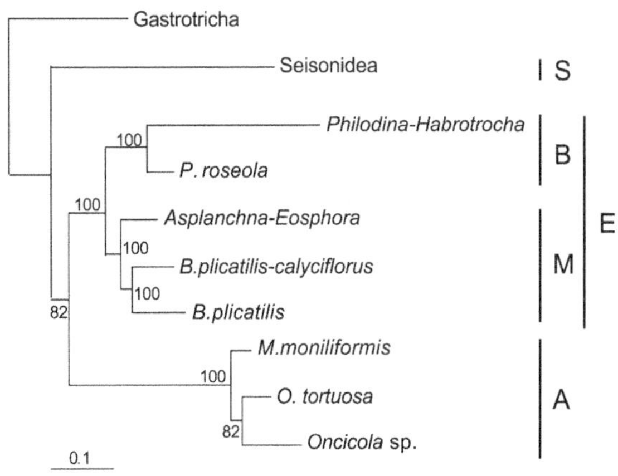

Figure 2. Bayesian tree with the highest posterior probability from an analysis of the *hsp82* + 18S data set. The percent posterior probability of each node is shown; branch lengths are the mean of the posterior probability density of branch lengths drawn proportional to the inferred number of changes per nucleotide per branch (see scale bar). Higher-order clades are indicated as in Figure 1.

A similar topology is obtained with Bayesian analysis of all *hsp82* codon positions, using a model that estimates evolutionary parameters of codon 1st and 2nd positions independently from 3rd positions. Bayesian analysis significantly supports Eurotatoria and can exclude most other topologies (Table 2). However, as there is some suggestion that current implementations of Bayesian inference in phylogenetic applications may produce inflated posterior probability estimates (Suzuki et al., 2002; Alfaro et al., 2003; Cummings et al., 2003; Douady et al., 2003; but see Wilcox et al., 2002), these results must be viewed with some caution.

Increasing the amount of sequence data in a phylogenetic data set can improve the resolution of phylogenies (Cummings et al., 1995, 1999; Otto et al., 1996; Graybeal, 1998; Poe, 1998; Poe & Swofford, 1999). Unfortunately, very few rotifer sequences have been examined and fewer still are appropriate for phylogenetic analysis at the phylum level. In an attempt to provide sufficient data to significantly discriminate between alternate topologies, a combined *hsp82* + 18S data set was created using all available data. Bayesian analysis of this data set strongly supports Eurotatoria and the inclusion of Acanthocephala within the Rotifera, and the tree with the highest posterior probability indicates that Acanthocephala is a sister-clade to Eurotatoria. However, the combined data set cannot exclude any topologies not excluded by Bayesian analysis of *hsp82* alone. Similarly, significance tests based on maximum likelihood scores using the combined data set do not exclude any topologies not excluded by the 18S data set.

In summary, maximum likelihood and Bayesian analysis of the protein-coding gene *hsp82* support a monophyletic Eurotatoria and the placement of Acanthocephala within the phylum Rotifera as a sister clade to either Eurotatoria or Seisonidea. As has been argued previously (Mark Welch, 2000, 2001; Segers, 2002), the absence of traditional morphological rotifer synapomorphies (ciliated corona, mastax, etc.) in Acanthocephala is undoubtably a result of the group's adaptation to obligate parasitism. As these characters are present in Seisonidea (Ricci et al., 1993) the term Rotifera remains a useful designation for the assemblage and there is no need to consider Rotifera paraphyletic. Molecular phylogenetic resolution of the exact relationship between Seisonidea and Acanthocephala will require additional sequence data from sampled species and sampling of additional species, particularly seisonids and multiple species from related phyla such as micrognathozoans and gnathostomulids.

Acknowledgements

I thank Michael Cummings for discussions about Bayesian phylogenetics and Jessica Mark Welch and two anonymous reviewers for careful reading of the manuscript. This work was supported by a grant from the United States National Science Foundation.

References

Abouheif, E., R. Zardoya & A. Meyer, 1998. Limitations of metazoan 18S rRNA sequence data: implications for reconstructing a phylogeny of the animal kingdom and inferring the reality of the Cambrian explosion. Journal of Molecular Evolution 47: 394–405.

Alfaro, M. E., S. Zoller & F. Lutzoni, 2003. Bayes or Bootstrap? A simulation study comparing the performance of Bayesian Markov Chain Monte Carlo sampling and bootstrapping in assessing phylogenetic confidence. Molecular Biology and Evolution 20: 255–266.

Ahlrichs, W. H., 1997. Epidermal ultrastructure of *Seison nebaliae* and *Seison annulatus*, and a comparison of epidermal structures within the Gnathifera. Zoomorphology 117: 41–48.

Cummings, M. P., S. P. Otto & J. Wakeley, 1995. Sampling properties of DNA sequence data in phylogenetic analysis. Molecular Biology and Evolution 12: 814–822.

Cummings, M. P., S. A. Handley, D. S. Myers, D. L. Reed, A. Rokas & K. Winka, 2003. Comparing bootstrap and posterior probability values in the four-taxon case. Systematic Biology 52: 477–487.

Cummings, M. P., S. P. Otto & J. Wakeley, 1999. Genes and other samples of DNA sequence data for phylogenetic inference. Biological Bulletin 196: 532–540.

Douady, C. J., F. Delsuc, Y. Baoucher, W. F. Doolittle & E. J. P. Douzery, 2003. Comparison of Bayesian and maximum likelihood bootstrap measures of phylogenetic reliability. Molecular Biology and Evolution 20: 248–254.

García-Varela, M., G. Pérez-Ponce de León, P. de la Torre, M. P. Cummings, S. S. S. Sarma & J. P. Laclette, 2000. Phylogenetic relationships of Acanthocephala based on analysis of 18S ribosomal RNA gene sequences. Journal of Molecular Evolution 50: 532–554.

Garey, J. R., T. J. Near, M. R. Nonnemacher & S. A. Nadler, 1996. Molecular evidence for Acanthocephala as a subtaxon of Rotifera. Journal of Molecular Evolution 43: 287–292.

Garey, J. R., A. Schmidt-Rhaesa, T. J. Near & S. A. Nadler, 1998. The evolutionary relationships of rotifers and acanthocephalans. Hydrobiologia 387/388: 83–91.

Gilbert, J. J., 1983. Rotifera. In Adiyodi, K. G. & R. G. Adiyodi (eds), Reproductive Biology of Invertebrates, Vol. 1. John Wiley & Sons, Chichester, NY: 181–209.

Graybeal, A., 1998. Is it better to add taxa or characters to a difficult phylogenetic problem?. Systematic Biology 47: 9–17.

Hancock, J. M. & A. P. Vogler, 2000. How slippage-derived sequences are incorporated into rRNA variable-region secondary structure: implications for phylogeny reconstruction. Molecular Phylogenetics and Evolution 14: 366–374.

Herlyn, H., O. Piskurek, J. Schmitz, U. Ehlers & H. Zischler, 2003. The syndermatan phylogeny and the evolution of acanthocephalan endoparasitism as inferred from 18S rDNA sequences. Molecular Phylogenetics and Evolution 26: 155–164.

Hillis, D. M., 1996. Inferring complex phylogenies. Nature 388: 130.

Hillis, D. M., 1998. Taxonomic sampling, phylogenetic accuracy, and investigator bias. Systematic Biology 47: 3–8.

Hillis, D. M. & J. J. Hill, 1993. An empirical test of bootstrapping as a method for assessing confidence in phylogenetic analysis. Systematic Biology 42: 182–192.

Huelsenbeck, J. P. & F. Ronquist, 2001. MrBayes: Bayesian inference of phylogenetic trees. Bioinformatics 17: 754–755.

Kishino, H. & M Hasegawa, 1989. Evaluation of the maximum likelihood estimate of the evolutionary tree topologies from DNA sequence data, and the branching order in Hominoidea. Journal of Molecular Evolution 29: 170–179.

Lecointre, G., H. Philippe, H. L. V. Lê & H. Le Guyader, 1993. Species sampling has a major impact on phylogenetic inference. Molecular Phylogenetics and Evolution 2: 205–224.

Maley, L. E. & C. R Marshall, 1998. The coming of age of molecular systematics. Science 279: 505–506.

Mark Welch, D. B., 2000. Evidence from a protein-coding gene that acanthocephalans are rotifers. Invertebrate Biology 111: 17–23.

Mark Welch, D. B., 2001. Early contributions of molecular phylogenetics to understanding the evolution of Rotifera. Hydrobiologia 446/447: 315–322.

Mark Welch, D. B. & M. Meselson, 2000. Evidence for the evolution of bdelloid rotifers without sexual reproduction or genetic exchange. Science 288: 2111–2115.

Mark Welch, D. B. & M. Meselson, 2001. Rates of nucleotide evolution in sexual and anciently asexual rotifers. Proceedings of the National Academy of Sciences USA 98: 6720–6724.

Near, T., 2002. Acanthocephalan phylogeny and the evolution of parasitism. Integrative and Comparative Biology 42: 668–677.

Omilian, A. R. & D. J. Taylor, 2001. Rate acceleration and long-branch attraction in a conserved gene of cryptic Daphniid (Crustacea) species. Molecular Biology and Evolution 18: 2201–2212.

Otto S. P., M. P. Cummings & J. Wakeley, 1996. Inferring phylogenies from DNA sequence data: the effects of sampling. In Harvey, P. H., A. J. Leigh Brown, J. Maynard Smith & S. Nee (eds), New Uses for New Phylogenies. Oxford University Press, Oxford: 103–115.

Philippe, H. & A. Germot, 2000. Phylogeny of eukaryotes based on ribosomal RNA: long-branch attraction and models of sequence evolution. Molecular Biology and Evolution 17: 830–834.

Poe, S., 1998. Sensitivity of phylogeny estimation to taxonomic sampling. Systematic Biology 47: 18–31.

Poe, S. & D. L. Swofford, 1999. Taxon sampling revisited. Nature 398: 299–300.

Posada, D. & K. A. Crandall, 1998. ModelTest: testing the model of DNA substitution. Bioinformatics 14: 917–818.

Ricci, C., G. Melone & C. Sotgia, 1993. Old and new data on Seisonidea (Rotifera). Hydrobiologia 255/256: 495–511.

Rodríguez, F., J. F. Oliver, A. Marín & J. R. Medina, 1990. The general stochastic model of nucleotide substitution. Journal of Theoretical Biology 142: 485–501.

Segers, H., 2002. The nomenclature of the Rotifera: annotated checklist of valid family- and genus-group names. Journal of Natural History 36: 631–640.

Segers, H. & R. J. Shiel, 2005. Tale of a sleeping beauty: a new and easily cultured model organism for experimental studies on bdelloid rotifers. Hydrobiologia 546: 141–145.

Shimodaira, H. & M. Hasegawa, 1999. Multiple comparisons of log-likelihoods with applications to phylogenetic inference. Molecular Biology and Evolution 16: 1114–1116.

Sullivan, J. & D. L Swofford, 1997. Are guinea pigs rodents? The importance of adequate models in molecular evolution. Journal of Mammalian Evolution 2: 77–86.

Suzuki, Y., G. V. Glazko & M. Nei, 2002. Overcredibility of molecular phylogenies obtained by Bayesian phylogenetics. Proceedings of the National Academy of Science USA 99: 16138–16143.

Swofford, D. L., 2002. PAUP*: Phylogenetic Analysis Using Parsimony (*and Other Methods). Version 4. Sinauer Associates, Sunderland, MA.

Tamura, K. & M. Nei, 1993. Estimation of the number of nucleotide substitutions in the control region of mitochondrial DNA in humans and chimpanzees. Molecular Biology and Evolution 10: 512–526.

Thompson, J. D., D. G. Higgins & T. J. Gibson, 1994. ClustalW: improving the sensitivity of progressive multiple sequence alignment through sequence weighting, position specific gap penalties and weight matrix choice. Nucleic Acids Research 22: 4673–4680.

Wallace, R. L., 1999. Phylum Rotifera. In Knobil E. & J. D. Neill (eds), Encyclopedia of Reproduction, Vol. 4. Academic Press, San Diego, CA: 118–129.

Wallace, R. L., 2002. Rotifers: exquisite metazoans. Integrative and Comparative Biology 42: 660–667.

Wilcox, T. P., D. J. Zwickl, T. A. Heath & D. M. Hillis, 2002. Phylogenetic relationships of dwarf boas and a comparison of Bayesian and bootstrap measures of phylogenetic support. Molecular Phylogenetics Evolution 25: 361–371.

Yang, Z., 1997. PAML: a program package for phylogenetic analysis by maximum likelihood. Computer Applications in the Biosciences 15: 555–556.

Hydrobiologia (2005) 546:55–70
A. Herzig, R.D. Gulati, C.D. Jersabek & L. May (eds.) Rotifera X: Rotifer Research: Trends, New Tools and Recent Advances
DOI 10.1007/s10750-005-4102-9

Evolutionary dynamics of 'the' bdelloid and monogonont rotifer life-history patterns

Charles E. King[1,*], Claudia Ricci[2], Justin Schonfeld[3] & Manuel Serra[4]
[1]*Department of Zoology, Oregon State University, Corvallis, OR, 97331, USA*
[2]*Dipartimento di Biologia, Università degli Studi di Milano, via Celoria 26, 20133, Milano, Italy*
[3]*Department of Computational Biology and Bioinformatics, Iowa State University, Ames, IA, 50011, USA*
[4]*Institut Cavanilles de Biodiversitat i Biologia Evolutiva, Universitat de València, Apartat 22085, E46071, Valencia, Spain*
*(*Author for correspondence: E-mail: kingc@science.oregonstate.edu)*

Key words: rotifers, life-history evolution, parthenogenesis, advantage of sex, random genetic drift

Abstract

Substantial differences in both life-table characteristics and reproductive patterns distinguish bdelloid from monogonont rotifers. Bdelloids reproduce only asexually, whereas most monogononts are cyclical parthenogens. We explore some of the adaptive consequences of these life-history differences using a computer model to simulate the evolutionary acquisition of new beneficial mutations. A one-locus mutation-selection regime based on the life-history characteristics of bdelloids indicates that asexuals can maintain higher levels of both allelic and genotypic diversity over a longer time period than obligate sexuals. These results are produced by differences in the magnitude of random genetic drift (RGD) associated with the different types of reproduction. Cyclical parthenogens have significantly higher evolutionary rates than sexual forms in a single-locus model, but incorporate beneficial mutations more slowly than sexuals in a two-locus simulation. Our results are therefore strongly influenced by the number of loci being evaluated as well as the pattern of reproduction. The asexual life history was found to maintain higher levels of allelic diversity than any pattern including sexual reproduction. This intriguing finding is amplified as the number of loci undergoing selection is increased. We end by considering the adaptive consequences of the remarkably divergent life histories found in typical bdelloid and monogonont rotifers.

Introduction

The Phylum Rotifera contains one class of obligate parthenogens, Bdelloidea, a second class of obligate sexual species, Seisonidea, and a third class composed predominantly of cyclical parthenogens, Monogononta. Virtually nothing is known of the life-history of the two species of seisonids, both of which are epizoic on leptostracan crustaceans. Numerous species from the other two classes have received extensive study. Although many other phyla have an occasional asexual species, rotifers are unique in having reproductively divergent, major phyletic branches that extend deeply into the taxonomic structure of the group.

The most perplexing taxon from an evolutionary viewpoint is the Bdelloidea. Asexuality, although it may be beneficial in the short term, is commonly thought to lead to extinction because it reduces evolutionary potential relative to life histories that incorporate sexual recombination (Maynard Smith, 1978, 1989). However, the apparent long-term success of the bdelloids is not only an exception to this pattern, it led Maynard Smith (1986) to refer to the bdelloids as an 'evolutionary scandal.' Recent molecular evidence convincingly demonstrates that there is no source of cryptic sexual recombination occurring in this group; furthermore, apomictic parthenogenesis, in which meiosis is suppressed, must be an ancient condition (at least 90 My) in the bdelloids (Mark Welch & Meselson, 2000; Birky et al., this volume, Part I).

Here we explore the potential evolutionary significance of life-history patterns that approximate those of typical bdelloids and monogononts. Although bdelloids and monogononts may be close relatives, they differ extensively in many of their life-table characteristics and niche parameters. Since they have taken quite divergent adaptive directions, relatively little insight can be obtained by making direct comparisons between species in the two classes. Instead we adopt a more informative approach. We first select either a bdelloid or monogonont life-history pattern and then, using a simulation model, expose the population to beneficial mutations. The fate of these mutations and the resultant changes in the population's genetic structure are followed separately under the assumptions of obligate sexual reproduction, obligate asexual reproduction and cyclical parthenogenesis in which a number of apomictic generations alternate with a single generation of sexual reproduction. Because of time and space constraints, we have not included all of the desirable permutations of life-history parameters and genetic structure. However, the included comparisons help to isolate the effect of reproductive pattern and provide insight into the genetic and ecological features that have permitted bdelloid and monogonont rotifers to flourish in the absence or de-emphasis of sexual reproduction.

Materials and Methods

The C++ model used in this paper is a modification of the 'CPA model' described and depicted by King & Schonfeld (2001); the reader is referred to that paper for a more complete description of the model's assumptions and structure. Before the start of a simulation, the genetic structure of the population to be modeled is determined by specifying the number of alleles at each of the loci. Selective values and mutation rates also are specified for each locus. Each locus is represented by a pool of its initiating genotypes in Hardy–Weinberg frequencies. A single individual is initially formed by random sampling from each of the genotype pools. If there are two or more loci present, the individual is specified by its multilocus genotype (MLG). This process of random sampling from the genotype pools is repeated until all N individuals needed to initiate the population have been formed (Fig. 1). The population is then subjected to random forward and reverse mutation at each locus; here we will use a mutation rate that produces an average of one new mutation per locus per generation. If reproduction is sexual, the population follows the top loop in Figure 1. After selection, the new sets of allele frequencies form Hardy–Weinberg genotype pools, and the random sampling process is repeated to construct the next breeding population. If the infinitely large genotype pools contain a low-frequency genotype, as might occur when there has been a new mutation in the previous breeding interval, there is a significant probability that the rare allele will not be included in one of the N individuals used to start the next generation. This step constitutes a potentially important source of RGD. Rare alleles, even if they are highly beneficial, are likely to be lost by chance events.

If reproduction is asexual, an alternative path is taken following mutation. Recombination is absent from the bottom loop and the entire genotype constitutes a clone that is the unit of selection. In King & Schonfeld (2001) the asexual cycle was assumed to produce only the N individuals needed to start the next breeding interval. RGD therefore operated in the sexual cycle but was absent in the asexual cycle. To remove this disparity we have applied genotype-specific finite rates of increase to reproduction during the asexual cycle. We define F_{max} as the number of offspring contributed to the zygote pool per allele of the most fit homozygote at each locus. Under the additive fitness assignments used in this paper, the fitness of the heterozygote at each locus is the average of those of the two homozygotes. We treat F_{max} as a constant for all loci in a given simulation in this paper; however, each locus can be associated with a unique coefficient of selection against the homozygote with lowest fitness. Thus variation among the resultant reproductive rates during asexual reproduction produces clonal

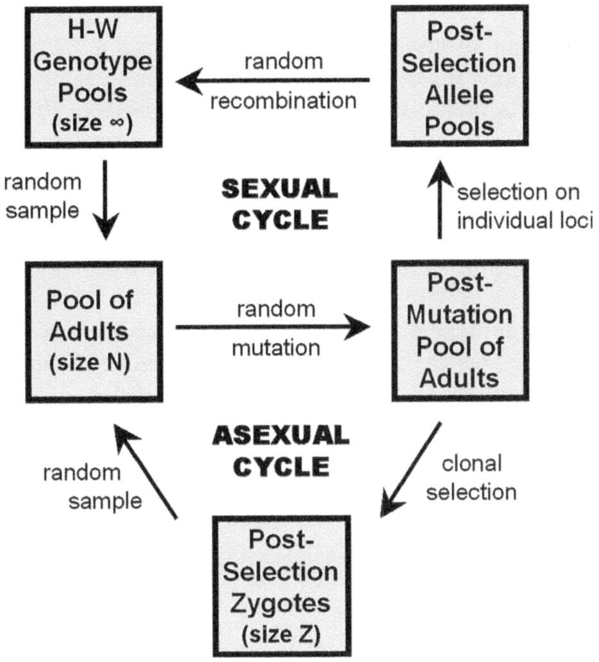

Figure 1. Flow chart of the CPA model used to determine evolutionary rates and population structure under variable frequencies of sexual and asexual reproduction. Note that we use the term 'zygote' herein to refer to diploid individuals that have been newly formed by asexual reproduction.

selection and alters both the number of zygotes and MLG frequencies in the pool that is formed. We assume a constant population size of N individuals at the start of each breeding interval and prevent decreases by requiring that the average fitness in each generation be ≥1.0. Stated differently, the number of zygotes (Z) must be at least as large as the assumed equilibrium number of individuals (N) at the start of each breeding interval; but depending on the finite rates of increase, it may be considerably larger. Specifically, the number of zygotes will increase as F_{max} increases and the intensity of selection decreases. If $Z >$ N, the pool of zygotes is randomly sampled N times without replacement to select the individuals that will start the next generation. Rare genotypes in the zygote pool are therefore subject to loss from RGD.

We also assume the absence of seasonal variation and other complicating environmental effects. Because of the model's structure, rates of reproduction can be chosen to simulate either semelparous or iteroparous reproduction. We have used the values in Table 2, which were calculated as averages from the cited sources in Table 1 as discussed below, to compress the iteroparous life history of rotifers into a discrete analog. Accordingly the 'breeding intervals' referred to in this paper are equivalent to generations and a single breeding interval (BI) can be precisely defined as one pass through either the top sexual or the bottom asexual loop in Figure 1.

All simulations were started with homozygotes for the B allele. Beneficial mutations occurred at rate u to produce allele A. The AA homozygote, whose fitness is $2F_{max}$, will always have a higher fitness than the BB homozygote whose fitness at locus x is $2F_{min,x} = 2(1-s_x)(F_{max})$, where $s_x =$ the coefficient of selection against the BB homozygote at locus x. For multiple loci, the fitness of the least successful (MLG) will be the arithmetic average of the $F_{min,x}$ values.

The criterion for fixation of the AA genotype in the one-locus case, or the AAAA MLG in the two-locus case, is arbitrarily set at a frequency of 95%. A higher number is not used because both forward and reverse mutation take place continuously at the same rate in our model. Under the selection parameters we have employed, the mutation-selection equilibrium occurs at a frequency that is greater than 0.95, but less than 1.00. In some cases, particularly in the two-locus simulations as discussed later, the population failed to convert to the AAAA genotype within the limits of the simulation (generally 500 breeding intervals for the single locus simulations, 1000 for those modeling two loci). To maintain homogeneity of sample size for the variables being considered here, we used the first 25 runs for each set of conditions that completed the evolutionary change.

Under asexual reproduction, genotype diversity is equivalent to clonal diversity, which can be measured using Simpson's index, $C = 1 - \Sigma\, p_i^2$, where p_i is the frequency of the ith genotype. The derivation and properties of this index are discussed by Pielou (1969). Allele diversity is calculated using Nei's index (1975), $h = 1 - \Sigma\, x_i^2$, where x_i is the frequency of the ith allele at a single locus and h is averaged over all loci.

Life-history parameters used in the simulations were taken from the sources cited in Table 1. While this list is not exhaustive, it does cover a variety of species in both of the rotifer classes considered in this paper. The eight monogonont (MG) species in Table 1 had a mean net reproduction rate (R_o) of 14.1 ± 9.6 and a mean life span (LS) of 10.4 ± 8.5 days for an average lifetime fecundity of 1.4 young per day. The 16 bdelloids (BD) had a mean R_o of 22.3 ± 9.9 and a mean LS of 37.3 ± 15.0 days. Note that seven of the eight MG species produced more than one young per day, whereas only one of the 16 BD produced more than one young per day. Statistical comparisons of the life-table characteristics of the above species would be meaningless since the data were obtained under non-comparable conditions. Moreover, while the precise values are irrelevant to this investigation, it is the general experience of rotifer workers that

Table 1. Net reproduction (R_o) and mean lifespan (LS) values for a variety of monogonont (MG) and bdelloid (BD) species. Moss-dwelling bdelloids (m) are distinguished from those collected in aquatic (a) habitats

	Class	R_0 (young /♀)	LS (days)	References
Euchlanis dilatata	MG	13–14	3–4	King (1967)
Euchlanis dilatata	MG	6.4–12.6	2.4–4.4	King (1970)
Asplanchna brightwelli	MG	3.5–5.2	2.5–4.7	Snell & King (1977)
Asplanchna girodi	MG	3.2–7.7	4.5–6.6	Snell (1979)
Lecane tenuiseta	MG	6.8	26	Hummon & Bevelhymer, (1980)
Notommata copeus	MG	26–36	15–25	Pourriot et al. (1986)
Brachionus calyciflorus	MG	17–29	7–18	Pourriot & Rougier (1986)
Brachionus plicatilis	MG	17–22	8–10	Korstad et al. (1989)
Adineta grandis (m)	BD	3.5–5.6	40–48	Dartnall (1992)
Philodina gregaria (m)	BD	15	60	Dartnall (1992)
Philodina vorax (m)	BD	13	22	Ricci & Fascio (1995)
Habrotrocha elusa vegeta (m)	BD	22	32	Ricci, 1983
Habrotrocha sylvestris (m)	BD	25.6	40	Ricci (1983)
Otostephanos torquatus (m)	BD	10	45	Ricci (1983)
Macrotrachela insolita (m)	BD	22	76	Ricci (1983)
Macrotrachela vanoyei (m)	BD	15–26	29–40	Ricci, (2001a)
Adineta vaga (m)	BD	14	17	Ricci (1983)
Adineta ricciae (m)	BD	32	38	Ricci & Covino (this volume)
Philodina roseola (a)	BD	45	48	Lebedeva & Gerasimova (1987)
Philodina roseola (a)	BD	30	25	Ricci (1983)
Habrotrocha constricta (a)	BD	21	38	Ricci (1983)
Macrotrachela inermis (a)	BD	24	31	Ricci. (1983)
Macrotrachela quadricornifera (a)	BD	27–29	24–33	Ricci (2001a)
Embata laticeps (a)	BD	20	27	Ricci (1983)

monogononts have shorter life spans but higher rates of egg production than bdelloids.

To limit the number of variables considered in this paper, we have chosen to evaluate population responses in the least complex genetic system, a single locus, and also in a two-locus system, which is the simplest case that provides the opportunity for gene interactions. We further simplify our presentation of the one-locus results by confining our analysis in this paper to that of the BD life history. The two-locus simulations are based on the contrasting MG life-history values.

Results

One-locus case

The bdelloid (BD) life table and run parameters in Table 2 were used for the one-locus simulations. Twenty-five replicate simulations were run for each of seven cases: obligate asexual reproduction, five different patterns of cyclical parthenogenesis, and obligate sexual reproduction. Since the population was started with low fitness homozygous BB individuals, mutations producing AB heterozygotes increased fitness, and the appearance of homozygous AA individuals raised fitness once again.

Results from typical one-locus runs using the bdelloid life-history characteristics are presented in Figure 2. The top pair of panels is for obligate parthenogenesis; there is an extended period of allele frequency pseudoequilibrium in the left panel of this No Sex simulation. During this period the population is dominated by the heterozygote (right panel). Pseudoequilibrium starts as the BB homozygote is eliminated and comes to an end as the heterozygote is in turn replaced by the AA homozygote.

The example of cyclical parthenogenesis plotted in Figure 2 has a pattern of 49 asexual (A) generations intervening between sexual (S) generations and is denoted S1-A49. The length of time required to complete the elimination of the B allele is shorter for the life histories with sexual reproduction than for the asexual case primarily because recombination eliminates the need for a second mutation. Note that cyclical parthenogenesis appears to lead to a more rapid fixation of the A allele than the All Sex case (83 ± 3.3 vs. 107 ± 4.6 BI). The explanation for this difference is that sexual individuals are exposed to a higher rate of RGD during recombination because new mutations are at a low frequency and are likely to be lost through sampling effects. By contrast, RGD is relatively unimportant in the asexual phase of the one locus bdelloid simulations in Figure 2 since 72% of the zygotes survive to join the breeding population. The two major distinctions between the No Sex and All Sex cases are therefore (1) the degree of exposure of the sexual population to RGD during recombination, and (2) the relative duration of the heterozygote.

Figure 3 shows the pattern of change of allele and genotype diversity for the simulations presented in Figure 2. Both measures of diversity peak when there is an allelic or genotypic substitution since diversity is maximized as the number of categories (i.e., number of alleles or genotypes) increases and as the categories attain equal frequencies. In the absence of sexual reproduction, the heterozygote has an extended representation in the population, producing a situation in which allele diversity exceeds genotype diversity (as measured by the area under their curves). With additive fitnesses, the heterozygote is a transient stage that is replaced by the higher fitness AA homozygotes.

Table 2. Base data assumed for the simulations of the bdelloid (BD) and monogonont (MG) life histories

Group	R_o	LS	T_{est}	r_{est}	R_{est}	F_{max}	u	S_1	S_2	N
BD	22	37	9	0.33	1.40	0.7	0.0001	0.25	–	10,000
MG	14	10	2.5	1.05	2.85	1.4	0.0001	0.33	0.26	10,000

R_o = net reproductive rate; LS = mean life span; T_{est} = estimated generation time = ¼ LS; r_{est} = estimated instantaneous rate of increase = ln R_o/T_{est}; R_{est} = estimated finite rate of increase = $e^{r_{est}}$; F_{max} = contribution to fitness of a beneficial mutation; u = mutation rate; s_1 and s_2 = coefficients of selection against the less fit homozygote (BB) at locus 1 and locus 2; N = number of adults exposed to mutation and reproduction in each breeding interval (Calculation of the fitness contribution of an A allele: ln R_{est} = r_{est} = ln R_o/T_{est}, where T_{est} = 0.25 LS; F_{max} = $R_{est}/2$).

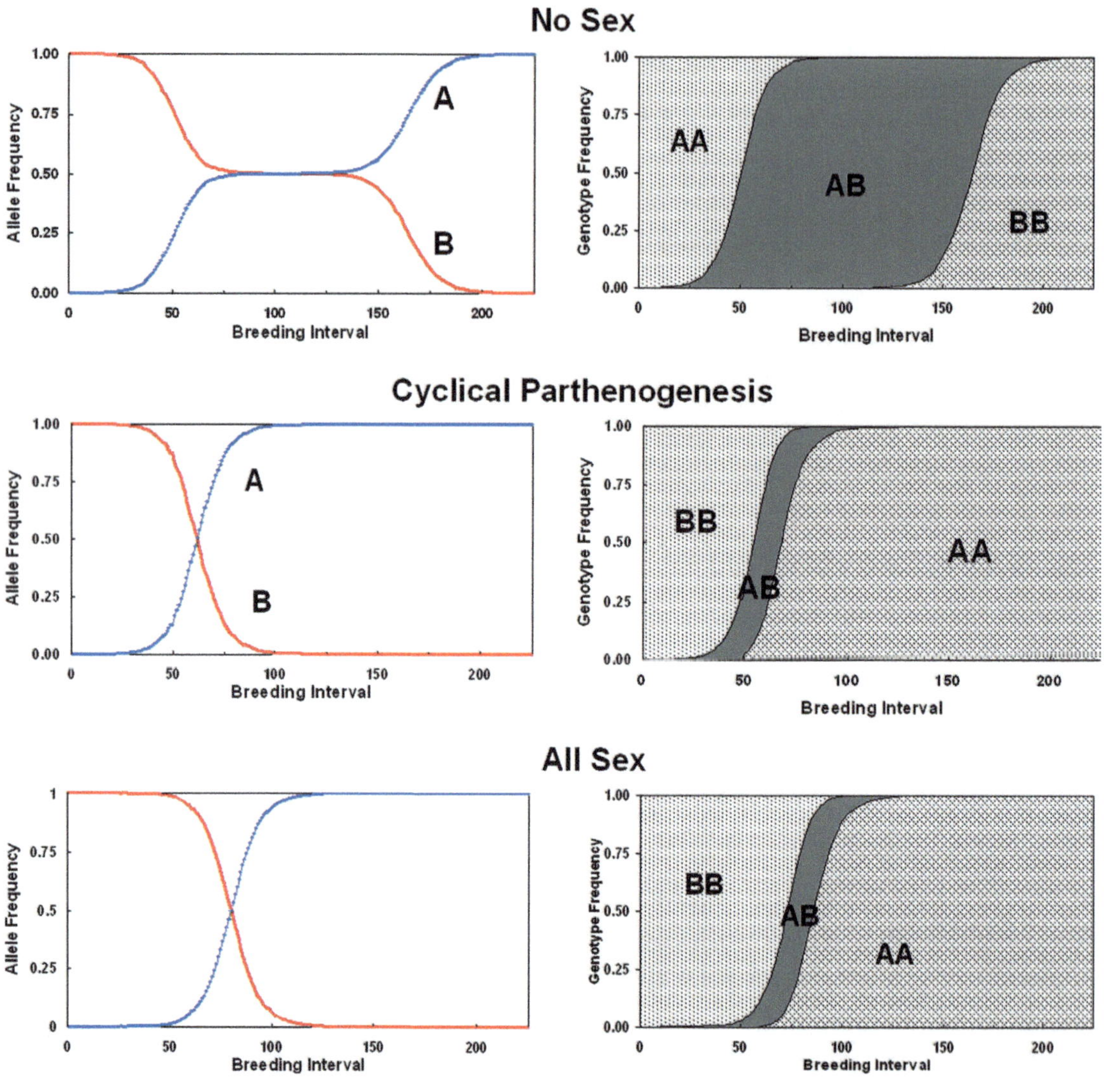

Figure 2. Change in allele and genotype frequencies as selection is applied to obligate asexual, cyclical parthenogenetic (S1-A49), and sexual reproductive patterns of the BD life history characteristics in Table 2. These examples were selected from the sets of 25 replicates on the basis of having rates of evolution closest to the observed mean observed for each group.

To explore the influence of varying schedules of cyclical parthenogenesis, we also ran 25 replicate series with sexual reproduction occurring once every 5, 10, 17, and 25 BI (Fig. 4). There were no significant differences in rates of evolution among the five series of cyclical parthenogens (Kruskal–Wallis one-way ANOVA on ranks, $p = 0.17$); however, this group differed from both the All Sex and No Sex groups as determined by a Tukey multiple comparisons test ($p < 0.05$).

In the S1-A49 treatment (Fig. 4), only 2% of the generations are sexual, yet the rate of evolution of this group is significantly faster than that of the obligate asexuals. This observation forces the question: What is so special about sexual reproduction that, even when rare, it leads to much more rapid evolutionary change? In other words: What is the advantage of sex to a cyclical parthenogen? At the same time we note that there is also a disparity between the evolutionary rates of cyclical parthenogens and obligate sexuals (Fig. 4). This raises a second important question: What is the advantage of asexual reproduction to a cyclical parthenogen?

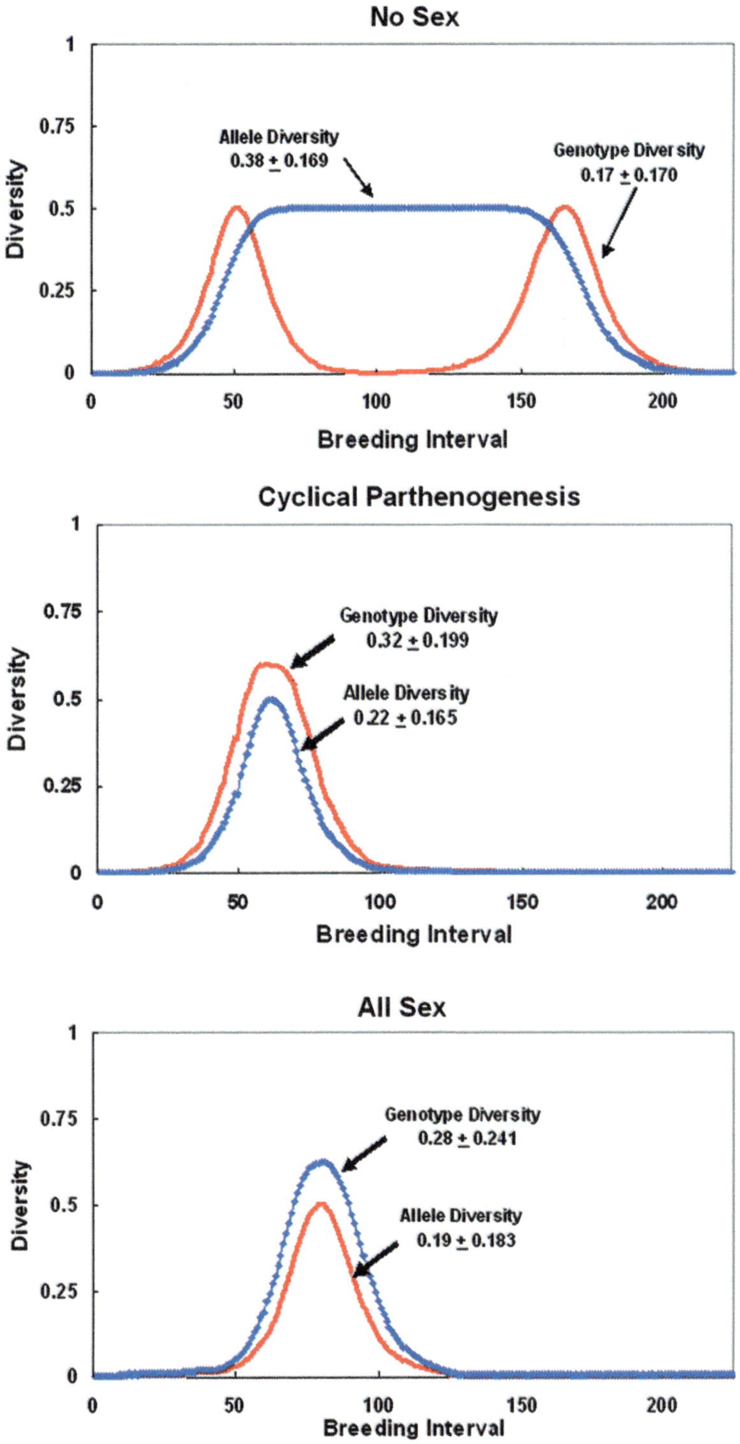

Figure 3. Change in allele and genotype diversity as selection is applied to the populations presented in Figure 2. Means and standard deviations of the diversity measures are taken over the breeding-interval period during which genotype diversity first becomes greater than 0.05 and is last greater than 0.05.

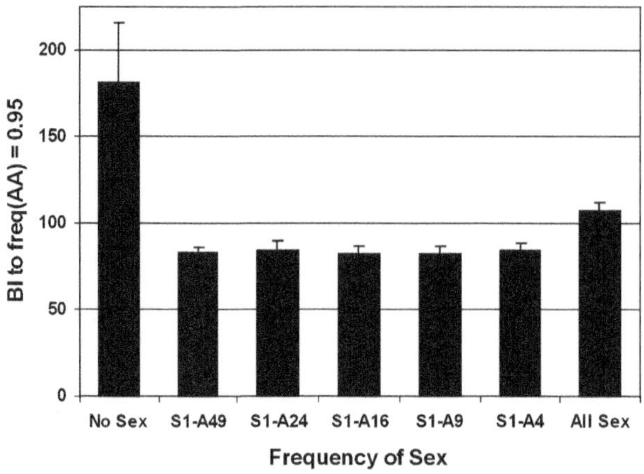

Figure 4. Generations required to change the genotype frequency from 1.0 BB to 0.95 AA under the BD selection parameters given in Table 2. Each bar presents the mean and standard deviation of 25 replicates.

What is the advantage of sex to a cyclical parthenogen?

Under asexual reproduction at least two mutations are required to produce an AA homozygote in a population that is homozygous for the B allele. The first mutation produces an AB heterozygote, and typically this genotype must expand its numbers and become common before a subsequent mutation converts a heterozygote to an AA homozygote. That is, the second mutation must occur in a descendant of the individual that acquired the first mutation. Under asexual reproduction the frequency of heterozygotes is a major determinant of the probability of acquiring the second mutation.

In contrast, there are two ways to form an AA homozygote under sexual reproduction. One of these is by the double mutation process described for asexuals. Alternatively, once a mutation from B to A has occurred and the resultant heterozygote has started to increase in frequency, it becomes more and more likely that two heterozygotes will mate. When that happens, homozygous AA individuals are likely to be produced by sexual recombination. In this latter scenario, only one mutation and one mictic event are required.

What is the advantage of asexual reproduction to a cyclical parthenogen?

To answer this question, let us compare and contrast the details of sexual and asexual reproduction. We assume that population replacement from one generation to the next in the sexual cycle is not constrained by the size of the gamete pool. When a mutation occurs, say from B to A in a population that was homozygous for allele B, the heterozygote that is formed and the remaining BB homozygotes undergo differential reproduction according to their assigned fitnesses. The end result of this process is a gamete pool in which random recombination produces genotypes in Hardy–Weinberg frequencies. This pool, which is infinitely large in our model, is randomly sampled to form the N individuals needed to initiate the next generation. Notice that rare mutations, even if they are highly beneficial, are likely to be lost during the sampling process due to RGD.

Contrast the above pattern to the events occurring in an asexual reproductive cycle. Instead of gametes, asexual individuals produce diploid eggs that will be referred to as here as zygotes. The size of the zygote pool (Z) is determined solely by the population's average reproductive rate in the current breeding interval. Average reproductive rates are determined by the rates of offspring production, by the properties of the most fit allele, $F_{\max}$, and by the coefficients of selection, s_x. As stated earlier, we constrain Z to be equal to or larger than N, the number of individuals needed to start the next generation. The zygote pool is randomly sampled N times to obtain the individuals needed to start the next generation. Notice that if Z = N, all individuals in the zygote pool become breeding adults and there is no RGD operating on

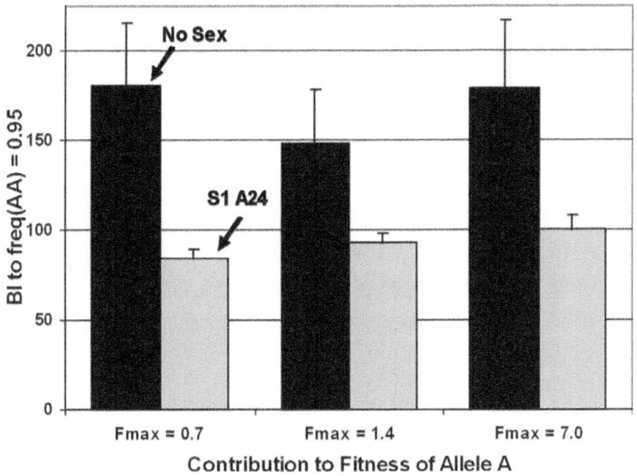

Figure 5. Effect of F_{max} on the rate of fixation of a beneficial allele under obligate asexual reproduction and under cyclical parthenogenesis (S1-A24) using the BD life-history characteristics presented in Table 2.

the asexual cycle. As F_{max} increases, $Z - N$ also increases. Zygote mortality rate is therefore directly related to the magnitude of F_{max}. As the rate of zygote mortality increases, the potential for RGD also increases, and one of the major distinctions between the sexual and asexual life histories is simultaneously reduced.

The above distinctions between asexual and sexual reproduction have important implications for understanding variations in life-history patterns between bdelloid and monogonont rotifers. In particular the bdelloid life history has two features that act in harmony to reduce the effects of RGD: (1) the absence of recombination, and (2) low values of F_{max}. We shall return to this point in the discussion.

Effect of F_{max} on rate of fixation

Figure 5 presents the results of a series of measures, with 25 replicates per treatment, of the number of breeding intervals required to change the frequency of the AA genotype from 0 to 0.95 for a range of F_{max} values. A Kruskal–Wallis ranks test indicates significant differences between the No Sex and S1-A24 groups. A Tukey multiple comparisons test found no significant differences among the three No Sex groups ($P > 0.05$). However, a significant difference was found between the two extreme S1-A24 groups; the $F_{max} = 0.7$ group had a significantly higher rate of evolution than the $F_{max} = 7.0$ group (84 ± 5.6 vs. 100 ± 7.8 BI). Notice that as F_{max} increases, the size of the zygote pool increases and the potential for RGD also increases. This constraint, however, applies only to constant environments in which the size of the breeding population is fixed. A case not considered by our model is an ameliorating environment which would favor higher F_{max} and the ability to genetically track environmental change.

Two-locus Case

The monogonont (MG) life-history features from Table 2 were used in the simulations for this series. The two loci had the same F_{max} values but differed in the coefficients of selection against their BB homozygotes. At locus one, the fitness of genotype BB was 34% lower than that of the AA homozygote. At the second locus, the coefficient of selection against genotype BB was 0.26. By convention, when the complete two-locus MLG is specified, the first pair of letters refers to locus one, the second pair to locus two.

Allele-replacement patterns in the two-locus case are much more complex than in the single-locus case. Instead of three genotypes with quite distinct fitness values, there are nine MLGs with a much more gradual series of fitness steps separating the extremes (Fig. 6). In the one-locus case there was only one path to increase fitness. Here there are five unique asexual paths, each producing an increase of fitness at each step, leading from the initial population of BBBB homozygotes to the

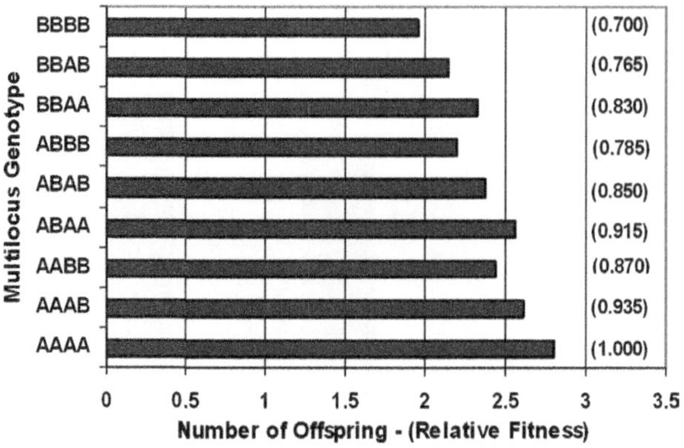

Figure 6. Fitness values for the two-locus runs based on the MG life-history characteristics in Table 2. Bars indicate the number of offspring produced by single individuals of the designated genotype; numbers in parentheses indicate the genotype's relative fitness. The first pair of letters indicates the genotype at locus one, the second pair at locus two.

final post-selection population of AAAA. It is therefore not surprising that this complexity is carried over to the typical selection simulations shown in Figure 7. Further complicating the pattern of change are the effects of stochastic forces leading to both the generation and loss of new mutations.

The top panels in Figure 7 indicate that the retarded response of obligate asexuals observed in the one-locus runs are not only carried over to the two-locus case, they are amplified. The time required to convert the population from BBBB to AAAA is more than twice as long under obligate asexual reproduction as under either cyclical parthenogenesis (S1-A24) or obligate sexual reproduction. Moreover, interactions among alleles and loci are obvious in both the asexual (top) and cyclical parthenogenetic (center) panels of Figure 7.

As demonstrated in Figure 8, sexual reproduction has a major advantage when two or more loci are undergoing simultaneous selection for beneficial mutations. This advantage is derived from the role of sexual recombination in uniting independent mutations occurring in different individuals at different loci. That is, in the presence of sex only one mutation from B to A is required at each of the two loci to produce a N individual that is homozygous for all four A alleles. By contrast, four independent mutations must take place in a single asexual lineage to obtain an AAAA descendant from a BBBB ancestor.

Of particular interest is the observation that the advantage of recombination is not confined to the one-locus case. A Kruskal–Wallis OW-ANOVA on ranks test indicated a significant ($p < 0.001$) difference among the six treatment groups containing sexual reproduction. The three central groups (S1-A32, S1-A24, and S1-A9) showed no significant difference in mean rates of fixation (Tukey test), but differed from the two groups on each end. There was no significant difference between the No Sex and the S1-A49 series, or between the All Sex and the S1-A4 series (Tukey test). In contrast to the single locus results, obligate sexual reproduction leads to more rapid fixation of beneficial mutations than cyclical parthenogenesis.

In the two-locus case it is again apparent that both allele and genotype diversities are maximized under obligate asexual reproduction during selection for beneficial mutations (Fig. 9) because the time course of selection is so extended. There is, however, a striking difference in the relative magnitude of these diversity measures in the obligate asexuals compared with either the sexual or cyclical parthenogenetic runs. Increasing the number of loci undergoing selection increases the relative diversity of the asexuals because as more loci are added there are more intermediate stages, each of which requires a new mutation that must be incorporated before the highest fitness homozygote is acquired. For two loci each having two alleles, there are nine genotypes; for three two-allele loci there are 27, and for five loci with

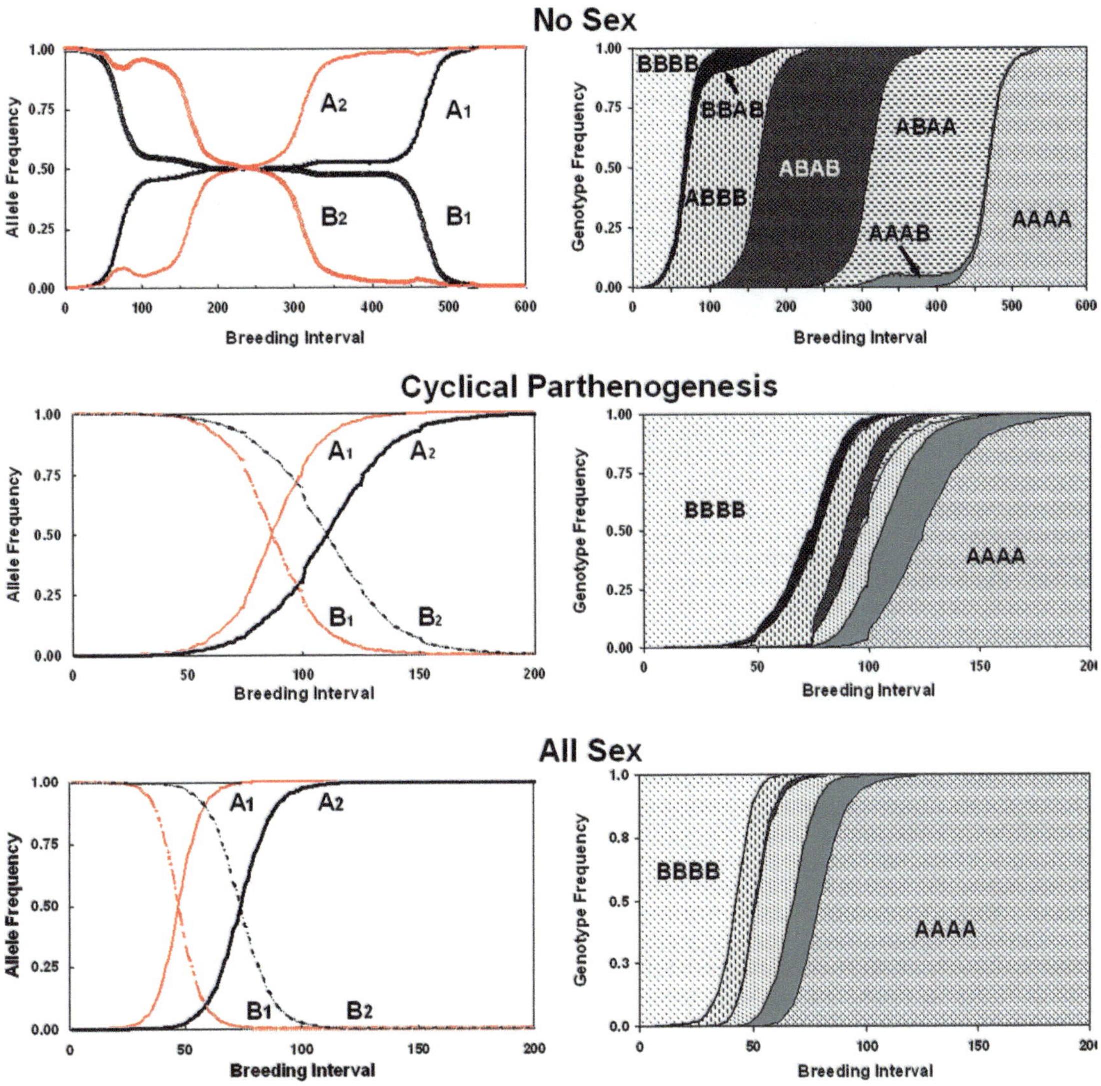

Figure 7. Change in allele and genotype frequencies as selection is applied to obligate asexual, cyclical parthenogenetic (S1-A24), and sexual reproductive patterns of the MG life history characteristics in Table 2. Note the abscissa scale differences between the two upper and the four lower panels. These examples were selected from groups of 25 replicates on the basis of having rates of evolution closest to the observed mean for each group.

two alleles each there are $3^5 = 243$ MLGs. Similarly, increasing the number of alleles at a single locus from two to five increases the number of genotypes at that locus from three to 15. As the number of MLGs increases, the magnitude of the selective differences among genotypes will decrease and fitness will in effect be continuously distributed thereby slowing the process of selection in obligate asexuals. For this reason we suggest that asexual reproduction will tend to maximize both allele and genotype diversity across time.

The number of loci simultaneously undergoing selection also influences the pattern of diversity in the sexual cycles. Both genotype and allele diversities can be seen to have bimodal distributions in the All Sex panel of Figure 9. This distribution is caused by the sequential fixation of the A allele at the two loci. In contrast, these separate events are in closer synchrony for the

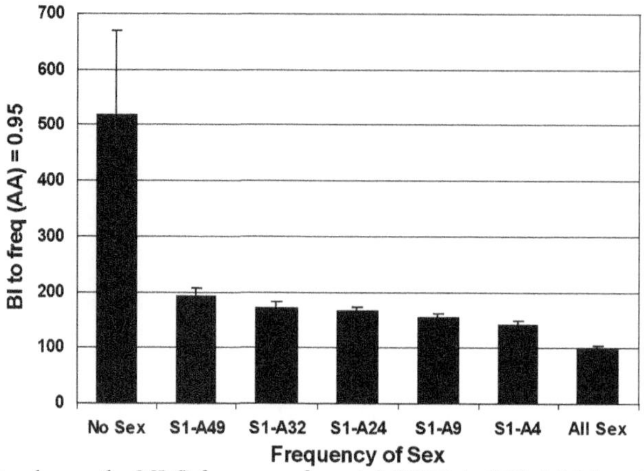

Figure 8. Generations required to change the MLG frequency from 1.0 BBBB to 0.95 AAAA under various mixes of asexual and sexual reproduction and the MG selection parameters in Table 2. Each bar presents the mean and standard deviation of 25 replicates.

cyclical parthenogens; thus their genotype and allele diversities appear to be unimodal (center panel).

Discussion

'Asexual reproduction,' in the words of Ridley (1996), 'is mainly confined to small twigs in the phylogenetic tree.' These twigs are spread broadly throughout the plant and animal kingdoms, but typically extend only to the level of a species or genus within a larger taxonomic group. The irregular taxonomic distribution of asexuality has been used by some investigators to suggest that parthenogenesis is inextricably linked to extinction (Maynard Smith, 1989). It has become part of the conventional wisdom that parthenogenesis is a dead end because it leads to the accumulation of deleterious mutations (Muller's Ratchet) and the loss of genetic variation needed to adapt to a changing environment.

An important prediction of our model is that both allele and genotype diversity will be larger under asexual reproduction than under any reproductive pattern involving sexual recombination. Ignoring loss through RGD, a diploid asexual system typically requires at least twice as many beneficial mutations to achieve a given gene substitution as the number required by a sexual system. Rates of beneficial mutation, while unknown, are generally regarded to be much lower than those of neutral and deleterious mutations (Lewin, 2000). Thus if we consider the numerous alternative and possibly long-lived MLG heterozygotes as well as the potential for reverse mutation, it becomes clear that obligate asexual reproduction under low F_{max} and weak selection produces an optimal system for the retention of genetic diversity. Under these circumstances high diversity is likely to take the form of a polymorphism in which a sizable fraction of the population at any given time departs from the fitness optimum. This departure may have long-term, beneficial fitness consequences since the slow response to selection maintains allelic variation which, in turn, may promote survival through future environmental changes. A counter argument is that slow response to selection may prevent adaptation to changing environments and thus increase the probability of local extinction.

Three major factors interact with each other and the environment to determine the fitness consequences of any particular life-history pattern: (1) type of reproduction, (2) life-table characteristics, and (3) response to environmental change. Our model is based on the first two of these factors; it does not consider the third. The physiological mechanisms involved in dormancy differ greatly between bdelloids and monogononts (Gilbert, 1974; Ricci, 2001b). Typically, in response to loss of water in their environment, bdelloids undergo a process of desiccation and enter an anhydrobiotic dormant state. When desiccated individuals are exposed to water, they break dormancy and typically resume normal

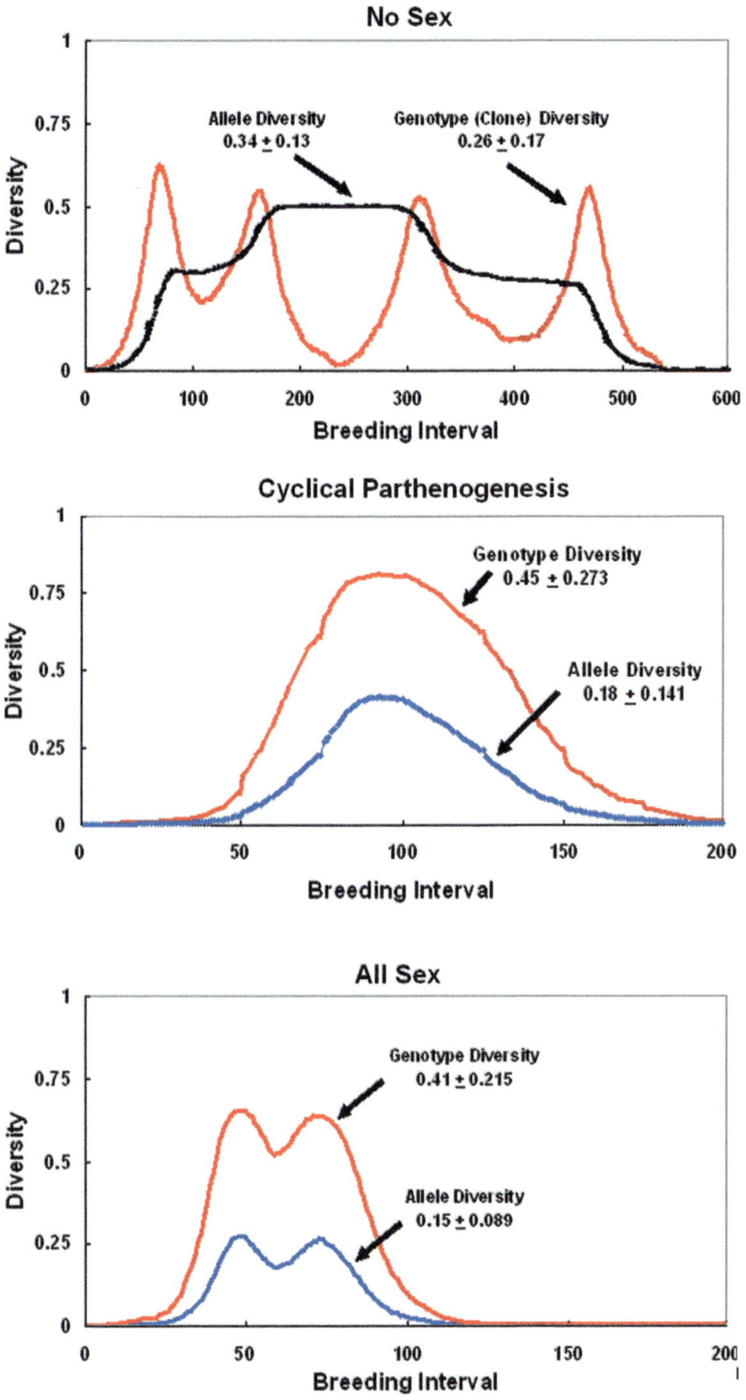

Figure 9. Change in allele and genotype diversity as selection is applied to the populations presented in Figure 7. Means and standard deviations of the diversity measures are taken over the breeding-interval period during which genotype diversity first becomes greater than 0.05 and is last greater than 0.05. Note the abscissa scale differences between the two upper and the four lower panels.

activity in a matter of a few hours. Although anhydrobiosis may exert a selective force because it reduces survival, it also shields genetic diversity from the effects of harsh environments. The key to this protection is the rapidity with which bdelloids can enter and recover from the anhydrobiotic state (Ricci & Covino this volume Part V).

By contrast the dormant stage of monogonont rotifers is the recombinant resting egg, and the process of their formation is anything but rapid. Before resting eggs can be formed, males must be produced, and before males can be formed mictic females must be produced in response to a mixis-inducing factor. The production of resting eggs requires a complex series of cross-generational, behavioral adaptations (Serra & King, 1999). Evidence from both continuous flow-through cultures and a model of mictic female production in chemostat systems suggests that genetic variation for the mictic response is reduced by the action of natural selection (Fussmann et al., 2003). This result is not surprising since to be successful in the production of resting eggs, a large proportion of the individuals in a monogonont population must respond to the mixis-inducing factor in a coordinated manner. Whereas anhydrobiosis in bdelloids is a reactive phenomenon because of the multiple generations required for resting egg production, dormancy in monogononts is best characterized as a predictive process based on the expectation of future environmental degradation. Low genetic diversities are therefore favored in monogononts both by selection against deleterious variants taking place during resting egg production, and by the relatively strong selection associated with cyclical parthenogenesis. In contrast, high genetic diversities are expected in bdelloids due to their low rates of evolution and the diversity-shielding effect of anhydrobiosis.

Maynard Smith and Szathmáry (1995) cite the conventional view that discrete species boundaries are maintained by sexual reproduction. In their view, Holman's (1987) finding that synonymous species names are less common in the bdelloids than in the monogononts is potential support for the proposition that bdelloid species are easier to recognize than monogononts. Holman looked for alternate explanations of his results, but was unsuccessful in identifying obvious sources of bias such as differences in time since first publication of species, effort or competence of systematists, size of genera, or taxonomic complexity of the species. Maynard Smith and Szathmáry suggested that if Holman's main finding could be confirmed by a direct morphometric study of the two groups, it 'might lead to a profound change in our view on the nature of species.' Our results take us in a different direction. Selection in bdelloids is a relatively slow process, so the typical population is likely to be genetically diverse. By contrast, cyclical parthenogenesis has the potential to lead to both rapid evolution and adaptation to a comparatively narrow range of environmental conditions. There is an obvious and important distinction in the time periods considered by Maynard Smith and Szathmáry on one hand, and by the CPA model on the other. One of the nascent themes in rotifer research over the past few years has been the study of temporal and spatial habitat partitioning that is particularly apparent among coexisting sibling species of monogononts (Serra et al., 1997, 1998; Gómez, A. et al., 2002). Based on our results we suggest that sibling species will be found in many genera of monogononts.

Cyclical parthenogens receive major benefits from both the asexual and sexual phases of their life histories. The asexual phase reduces effects of RGD. As shown by Fisher (1930), Muller (1932), and Crow & Kimura (1965), the sexual phase provides a means to combine multiple independent beneficial mutations without necessitating the passage of a long intermediate series of generations. One potentially important observation is that it takes relatively little sex to gain the full advantage of sexual reproduction. In the one-locus case, the S1-A49 life history with its 2% frequency of sexual generations had the same rate of evolution as the S1-A4 life history with its 20% frequency. A wide range of frequencies of sex also produced a relatively small effect in the two-locus study.

In the typical bdelloid life history, low values of F_{max} and the absence of recombination lead to exceedingly slow replacement of alleles during selection for new beneficial mutations. Note, however, that whether a given mutation is beneficial, neutral, or deleterious is often dependent on the present environmental state. If the environment were to stay in one state for an extended period, the genetic variation of bdelloids would tend to be reduced and their relatively small population sizes would make the acquisition of new variation difficult. But environments do not stay the same and though local populations are frequently very small, the number of populations may be very large. Finally, in addition to facilitating survival during harsh local conditions, anhydrobiosis appears to endow bdelloids with an ideal dispersal stage. These factors act in concert with the slow but deliberate response to beneficial mutations that is mandated by the bdelloid life history to protect the genetic diversity of this fascinating group of ancient asexuals.

Acknowledgements

We thank Paul Murtaugh of the Oregon State University Department of Statistics for discussion and advice. The manuscript was improved by comments from Robert L. Wallace and an anonymous referee. Travel funds were provided by a FIRST (University of Milan) Grant to CR. JS was supported in part by the IGERT training program funded through NSF Grant DGE-9982653. MS was supported by Grant BOS200-1451 from the Spanish Ministry of Science and Technology, and by a travel grant from the University of Valencia.

References

Birky, C. W., C. Wolf, H. Maughan, L. Herbertson & E. Henry, 2005. Speciation and selection without sex. Hydrobiologia 546: 29–45.

Crow, J. F. & M. Kimura, 1965. Evolution in sexual and asexual populations. The American Naturalist 99: 439–450.

Dartnall, H. J. G., 1992. The reproductive strategies of two Antarctic rotifers. Journal of Zoology 227: 145–162.

Fisher, R. A., 1930. The Genetical Theory of Natural Selection. Dover Reprint (1958), New York, 291 pp.

Fussmann, G. F., S. P. Ellner & N. G. Hairston Jr., 2003. Evolution as a critical component of plankton dynamics. Proceedings of the Royal Society of London Series B. 270: 1015–1022.

Gilbert, J.J., 1974. Dormancy in rotifers. Transactions of the American Microscopical Society 93: 490–513.

Gómez, A., M. Serra, G. R. Carvalho & D. H. Lunt, 2002. Speciation in ancient cryptic species complexes: evidence from the molecular phylogeny of *Brachionus plicatilis* (Rotifera). Evolution 56: 1431–1444.

Holman, E. W., 1987. Recognizability of sexual and asexual species of rotifers. Systematic Zoology 36: 381–386.

Hummon, W. D. & D. P. Bevelhymer, 1980. Life-table demography of the rotifer *Lecane tenuiseta* under culture conditions, and various age distributions. Hydrobiologia 70: 25–28.

King, C. E., 1967. Food, age, and the dynamics of a laboratory population of rotifers. Ecology 48: 111–128.

King, C. E., 1970. Comparative survivorship and fecundity of mictic and amictic female rotifers. Physiological Zoology 43: 206–212.

King, C. E. & J. Schonfeld, 2001. The approach to equilibrium of multilocus genotype diversity under clonal selection and cyclical parthenogenesis. Hydrobiologia 446/447: 323–331.

Korstad, J., Y. Olsen & O. Vadstein, 1989. Life-history characteristics of *Brachionus plicatilis* (Rotifera) fed different algae. Hydrobiologia 186/187: 43–50.

Lebedeva, L. I. & T. N. Gerasimova, 1987. Survival and reproduction potential of *Philodina roseola* (Ehrenberg) under various temperature conditions. Internationale Revue der gesamten Hydrobiologie 72: 695–707.

Lewin, B., 2000. Genes VII. Oxford University Press, Oxford, 990 pp.

Maynard Smith, J. M., 1978. The Evolution of Sex. Cambridge University Press, Cambridge, 222 pp.

Maynard Smith, J. M. & J. M. Maynard Smith, 1986. Contemplating life without sex. Nature 324: 300–301.

Maynard Smith, J. M., 1989. Evolutionary Genetics. Oxford University Press, Oxford, 325 pp.

Maynard Smith, J. M. & E. Száthmary, 1995. The Major Transitions in Evolution. W. H. Freeman, Oxford, 346 pp.

Mark Welch, D. B. & M. Meselson, 2000. Evidence for the evolution of bdelloid rotifers without sexual recombination or genetic exchange. Science 288: 1211–1215.

Muller, H. J., 1932. Some genetic aspects of sex. The American Naturalist 66: 118–138.

Nei, M., 1975. Molecular Population Genetics and Evolution. North-Holland, Amsterdam and New York, 288 pp.

Pielou, E. C., 1969. An Introduction to Mathematical Ecology. Wiley-Interscience, New York, 286 pp.

Pourriot, R. & C. Rougier, 1986. Rythmes de production de femelles sexuées chez le rotifère *Brachionus calyciflorous* en élevage à température constante. Bulletin de la Société Zoologique de France 111: 203–207.

Pourriot, R., C. Rougier & D. Benest, 1986. Influence de la température sur la production et la résponse mictique a la photopériode chez le rotifère *Notommata copeus* Ehrb. Vie et Milieu 36: 37–43.

Ricci, C., 1983. Life histories of some species of Rotifera Bdelloidea. Hydrobiologia 104: 175–180.

Ricci, C., 2001a. A reconsideration of the taxonomic status of *Macrotrachela quadricornifera* (Rotifera, Bdelloidea). Journal of Zoology 255: 273–277.

Ricci, C., 2001b. Dormancy patterns in rotifers. Hydrobiologia 446/447: 1–11.

Ricci, C. & C. Covino, 2005. Anhydrobiosis of *Adineta ricciae:* costs and benefits. Hydrobiologia 546: 307–314.

Ricci, C. & U. Fascio, 1995. Life-history consequences of resource allocation of two bdelloid rotifer species. Hydrobiologia 299: 231–239.

Ridley, M., 1996. Evolution (2nd ed.). Blackwell Science Inc., Cambridge, MA, 719 pp.

Serra, M., A. Galiana & A. Gómez, 1997. Speciation in monogonont rotifers. Hydrobiologia 358: 63–70.

Serra, M., A. Gómez & M. J. Carmona, 1998. Ecological genetics of *Brachionus* sibling species. Hydrobiologia 387/388: 373–384.

Serra, M. & C. E. King, 1999. Optimal rates of bisexual reproduction in cyclical parthenogens with density-dependent growth. Journal of Evolutionary Biology 12: 263–271.

Snell, T. W., 1979. Intraspecific competition and population structure in rotifers. Ecology 60: 494–502.

Snell, T. W. & C. E. King, 1977. Lifespan and fecundity patterns in rotifers. The cost of reproduction. Evolution 31: 882–890.

Hydrobiologia (2005) 546:71–80
A. Herzig, R.D. Gulati, C.D. Jersabek & L. May (eds.) Rotifera X: Rotifer Research: Trends, New Tools and Recent Advances
DOI 10.1007/s10750-005-4103-8

Toward a better understanding of the phylogeny of the Asplanchnidae (Rotifera)

Elizabeth J. Walsh[1,*], Robert L. Wallace[2] & Russell J. Shiel[3]
[1]*Department of Biological Sciences, University of Texas at El Paso, El Paso, TX 79968, USA*
[2]*Department of Biology, Ripon College, Ripon, WI 54971, USA*
[3]*The University of Adelaide, Adelaide, Australia*
*(*Author for correspondence: E-mail: ewalsh@utep.edu)*

Key words: cladistics, evolution, nrDNA, mtDNA, morphology, Monogononta, Synchaetidae

Abstract

We investigated the phylogenetic relationships of Family Asplanchnidae using both morphological and molecular data. The morphological database, comprising 23 characters from 19 taxa (15 Asplanchnidae and 4 outgroups), was compiled from a survey of the literature and our own observations; the molecular data (ITS and V4 region nuclear regions and mitochondrial *cox*1) was sequenced from specimens that we collected. Our analysis of the morphological data set (maximum parsimony) yielded 12 most-parsimonious trees with a tree length of 27 steps. From this analysis we conclude (1) Asplanchnidae is a monophyletic group as are the three genera comprising it, (2) there is no compelling support for the argument that *Asplanchna* should be separated into two discrete genera, and (3) there is some support for the proposal that Asplanchnidae and Synchaetidae are sister groups. Our analysis of the molecular data set supports the first two of these conclusions while the sister group of the family varied depending on the gene region analyzed and families and genera included. Current understanding of the phylogeny of Asplanchnidae is hampered by the need for additional informative morphological characters and a lack of molecular data for the genus *Harringia* and several other members of the Asplanchnidae.

Introduction

Family Asplanchnidae (Eurotatoria; Monogononta; Ploima) comprises 15 species of omnivorous rotifers, assigned to three genera (Ruttner-Kolisko, 1974; Koste, 1978; Jose de Paggi, 2002; Segers, 2002). These genera have specific habitat preferences that vary along a gradient from benthic to strictly planktonic. The benthic genus *Harringia* has a well-developed foot and is usually associated with plants; the semi-pelagic genus *Asplanchnopus* has a reduced foot and can be found with plants; the planktonic genus *Asplanchna* lacks a foot and is typically limited to open water. Whereas *Harringia* has a complete digestive system, the other two genera lack a gut; the etymon of both genera (G. *a*, lacking + G., *splanchnum*, the inward parts) refers to this feature. Although much is known about the general biology of Asplanchnidae (e.g., Salt et al., 1978; Gilbert et al., 1979; Gilbert 1980, 1999; Joanidopoulos & Marwan, 1998, 1999; Kappes et al., 2000), we are aware of only two specific hypotheses regarding its phylogeny and neither has been explored (Sudzuki, 1964; Kutikova, 1983). These hypotheses were formulated using the principles of evolutionary taxonomy (Ridley, 1986).

Sudzuki's (1964: Fig. 2 & 75–78) genus-level phylogeny separates *Asplanchna* into *Asplanchna* and *Asplanchnella* based on vitellarium morphology: *Asplanchna* = globular ovarium (sic) [vitellarium] (i.e., *priodonta*, *herrickii*); *Asplanchnella*

= elongated vitellarium (i.e., *brightwellii, girodi, intermedia, sieboldii*). He also emphasized the number of nuclei of the vitellarium in his key (see also Gilbert et al., 1979). Since Sudzuki's contribution, a third vitellarium shape has been described (sacciform, not spherical; Koste & Tobias, 1989), but no one has wedded this observation into his scheme. Both Ruttner-Kolisko (1974: 100) and Koste (1978: 449) support Sudzuki's position by subdividing *Asplanchna* based on vitellarium shape, but neither examined the suitability of this separation. However, in her taxonomic treatise Jose de Paggi (2002) retained *Asplanchna* (*sensu stricto*) without accepting Sudzuki's partition. Her assessment was based on the fact that vitellarium shape varies in other Asplanchnidae (i.e., *Asplanchnopus*), thus suggesting that vitellarium shape is too unreliable to warrant its use as a critical character for taxonomy within this family. This same point can be made for the nuclear number in the gastric and yolk glands and the number of flame cells in the protonephridia. The second phylogenetic hypothesis published on the Asplanchnidae was that of Kutikova (1983: Fig. 2). She argued for a sister relationship with Synchaetidae based on the details of coronal morphology and function.

Thus, our purpose was to investigate the phylogeny of the Asplanchnidae and specifically to examine the hypotheses of Sudzuki and Kutikova. To do this, we performed cladistical analyses on morphological and molecular databases.

Materials and methods

Morphological data

Based on a survey of the literature and our own observations, a data matrix of 23 characters on 19 monogonont rotifers comprising the 15 recognized species of Asplanchnidae and four outgroups *Brachionus* (Brachionidae), *Polyarthra* and *Synchaeta* (Synchaetidae), and *Trichocerca* (Trichocercidae) was constructed (Table 1). For the present analysis, we followed the general protocol used by Melone et al. (1998) in their study of rotiferan morphology.

The program PAUP* (4.0b10; Swofford, 2002) was used for all searches on the data matrix. A heuristic search strategy was employed in which the ancestral states were considered to be unrooted and the transformation types unordered. We ran searches first without and then with the characters vitellarium shape, cerebral eyespot, and ramus dentition. We also applied increased weight to all combinations of the following five characters (Foot, Pedal gland, Gut, Trophi type, Female polymorphism). Tree statistics reported here are (1) Tree Length (TL), (2) Consistency Index (CI) (excluding uninformative characters), and (3) Retention Index (RI). The reconstructions reported for the analysis include the two character optimization algorithms available in PAUP for rooted trees (ACCTRAN, DELTRAN). As a test of robustness we ran a bootstrap simulation ($n = 10^3$ replications) with and without including the characters vitellarium shape, cerebral eyespot, and ramus dentition. We also employed the permutation tail probability (PTP) test (Faith & Cranston, 1991) to examine cladistic structure of the data set again without the three characters noted above. Standard constraint methodologies were used to generate tree statistics for the alternative tree topologies reported as optimized on our data matrix.

Molecular data: sampling and processing

Species used for the present study were collected and analyzed from multiple sites as shown in Table 2. When possible, rotifers were starved before preservation and subsequent extraction of DNA. All rotifers were identified to species using keys of Koste (1978) and Jose de Paggi (2002). Voucher specimens of all species were preserved in 70% EtOH and are stored at Laboratory for Environmental Biology (UTEP). DNA was isolated and amplified from axenized fresh or ethanol preserved animals using Chelex-100 (Bio-Rad) as described in Walsh & De La Riva (unpublished). Primers used to amplify the 690 bp mitochondrial *cox*1 gene were LCO1490 (5′-GGTCAACAAATCATAAAGATATTGG-3′) and HCO2198 (5′-TAAACTTCAGGGTGACCAAAAAATCA-3′) (Folmer et al., 1994), and those for the nuclear ribosomal gene regions were those for the 18S variable region V4 (V4A: 5′-TGCAGTTAAAAAGCTCGTAGT-3′, V4B: 5′-CCCTTCCGTCAATTCCTTTAAG-3′) and internal transcriber spacer region (ITS4:

Table 1. Data matrix of 23 characters used in the morphological analysis. Outgroups, *Brachionus*, *Polyarthra*, *Synchaeta*, and *Trichocerca*; ingroups = 9 *Asplanchna* (*As*), 4 *Asplanchnopus* (*An*), and 2 *Harringia* (*H*)[1]

Taxa	Characters					
	1–4	5–8	9–12	13–16	17–20	–23
Brachionus	0000	0010	0000	0000	0000	110
Polyarthra	1000	0000	0010	0000	0001	020
Synchaeta	0000	0110	0011	0000	0000	020
Trichocerca	0000	1000	0010	0000	1010	110
As. brightwellii	1111	0002	0120	1101	0100	021
As. girodi	1111	0001	0120	1100	0000	020
As. priodonta	1111	0000	0120	1100	0000	020
As. herrickii	1111	0000	0120	1100	0000	020
As. intermedia	1111	0001	0120	1111	0000	021
As. sieboldii	1111	0002	0120	1101	0000	021
As. silvestrii	1111	0002	0120	1100	0000	021
As. tropica[2]	1111	0000	0120	1100	0000	02?
As. asymmetrica[3]	1111	0002	0120	1101	1000	02?
An. dahlgreni	0011	0000	1020	1100	0000	010
An. hyalinus	0011	0001	0020	1111	0000	010
An. multiceps	0011	0002	0020	1110	0000	010
An. bhimavaramensis	0011	0002	?020	1100	0000	010
H. eupoda	0001	0001	1?20	0000	0000	000
H. rousseleti	0001	0001	0?20	0000	0000	000

[1]This data set was derived from the reviews of Koste (1978), Jose de Paggi (2002), Hollowday (2002), original literature as available to us (Hudson, 1886, in Hudson & Gosse, 1886; Harring, 1913; Myers, 1934; Dhanapathi, 1975; Gilbert et al., 1979; Shiel & Koste, 1985; Koste & Tobias, 1989), and our own observations. Character states correspond to the descriptions provided in the Appendix (missing data = ?). NB: all *Asplanchnopus* species are oviparous (cf. Jose de Paggi, 2002); spellings of certain *Asplanchna* species have been corrected according to Jose de Paggi (2002). [2]*As. tropica* is known only from preserved material, thus the status of character #23 (Female polymorphism; i.e., the production of body wall outgrowths) is unknown. [3]*As. asymmetrica* has been observed alive by one of us (RJS) on several occasions (Ryan's Billabong, Australia). While none of these observations indicated the presence of body wall outgrowths, we coded the status of character #23 as unknown (=?) on the basis of its morphological similarities to *As. brightwellii*. However, additional cladistic analyses were run with a code of 0 for this character.

5′-TCCTCCGCTTATTGATATGC-3′, ITS5: 5′-GGAAGTAAAAGTCGTAACAAGG-3′; White et al., 1990). The ITS primers amplify ITS1, 5.8S, and ITS2 regions of the 18S nuclear ribosomal gene complex. All amplifications included a negative control. Amplification products were examined by electrophoresis to verify their size and cleaned with GeneClean kits (Bio101) before sequencing. Amplified products were sequenced directly using SequiTherm Excel LI kits (EpiCentre Technologies) and run on a LI-COR 4200 Series automated sequencer. All gene regions or genes were sequenced at least twice in both directions. Sequences were aligned using Clustal W (Thompson et al., 1994)

Phylogenetic trees of morphological characters were constructed using maximum parsimony (MP), and maximum likelihood (ML) algorithms, implemented in PAUP* (Swofford, 2002). All searches were heuristic, with TBR branch swapping and the MULPARS option in effect. For MP and ML searches, taxa were added randomly, with 10 addition-sequence replicates. To assess support for nodes in the MP and ML trees, we used non-parametric bootstrapping (Felsenstein, 1985). Neighbor-joining was used to construct all molecular phylogenies using uncorrected 'p' distances.

In the molecular analysis, in contrast to the morphological analysis, we do not include some species due to the lack of availability of samples.

Table 2. Species list and localities in which they were collected for molecular analysis

Species	Location
Ingroup	
Asplanchna brightwellii	Lake Carnegie, Princeton, NJ, USA
Asplanchna brightwellii	Lake Eyre Basin, AUS
Asplanchna girodi	Ascarate Pond, El Paso, TX, USA
Asplanchna priodonta	Lake Naomi, Pocono Pines, PA, USA
Asplanchna priodonta	Diamantina River, Lake Eyre Basin, AUS
Asplanchna herrickii	Lake Sagatan, Collegeville, MN, USA
Asplanchna intermedia	Hueco Tanks State Historic Site, El Paso, TX, USA
Asplanchna sieboldii	Several sites near El Paso, TX, USA
Asplanchna sp.	Kangeroo Island, AUS
Asplanchnopus hyalinus	Hueco Tanks State Historic Site, El Paso, TX, USA
Asplanchnopus multiceps	Hueco Tanks State Historic Site, El Paso, TX, USA
Asplanchnopus multiceps	Pelican Pond, UT, USA
Outgroup	
Trichocerca rattus	Hueco Tanks State Historic Site, El Paso, TX, USA
Synchaeta pectinata	Lake Naomi, Pocono Pines, PA, USA
Synchaeta littoralis	Elephant Butte Reservoir, NM, USA
Polyarthra dolichoptera	Elephant Butte Reservoir, NM, USA

This includes the genus *Harringia*, which has only been reported 3 times in the Zoological Records dating back to 1970. Other species that need to be included in future analyses are *A. intermedia*, *A. asymmetrica*, *A. sylvestrii* and *A. tropica*.

Results

Morphological analysis

Naturally because we ran our searches with the variations of excluding some characters and reweighting others, our study yielded several trees. Of all the analyses the ones in which the characteristics of vitellarium shape, cerebral eyespot, and ramus dentition were eliminated yielded the most interesting results; here we present an interpretation of that analysis. This search strategy yielded 12, equally most-parsimonious trees (TL = 27, CI = 0.746, RI = 0.915) (Fig. 1). However, with inclusion of any combination of the three characters noted above the topology of the trees generated remained fundamentally the same, but with many additional steps (TL ≥29) as well as additional polytomies.

Percentages of bootstrap replicates ($n = 10^3$) supporting nodes of the most-parsimonious tree are reported in Figure 1. None of the PTP tests yielded a tree with a TL equal to or with fewer steps than were otherwise obtained by any of the search strategies. Thus the null hypothesis, that the data defining these trees show no cladistic structure, is rejected ($p \approx 0.01$).

We used constraint analysis to examine the phylogenies of Sudzuki and Kutikova fixed on our data matrix, without the characters vitellarium shape, cerebral eyespot, and ramus dentition. Kutikova's hypothesis proved to be congruent to our tree with the same number of steps but yielding only six trees. We found similar results when we added the three characters noted above in all possible combinations, but again sometimes with additional steps and polytomies. However, Sudzuki's hypothesis possessed two more steps than our phylogeny. Again we had similar results when we added the three characters noted above in all possible combinations. Based on our morphological analysis we conform with Jose de Paggi's (2002) conclusion that Sudzuki's (1964) proposal to separate *Asplanchna* into two genera is not warranted. In fact, it makes more sense to group

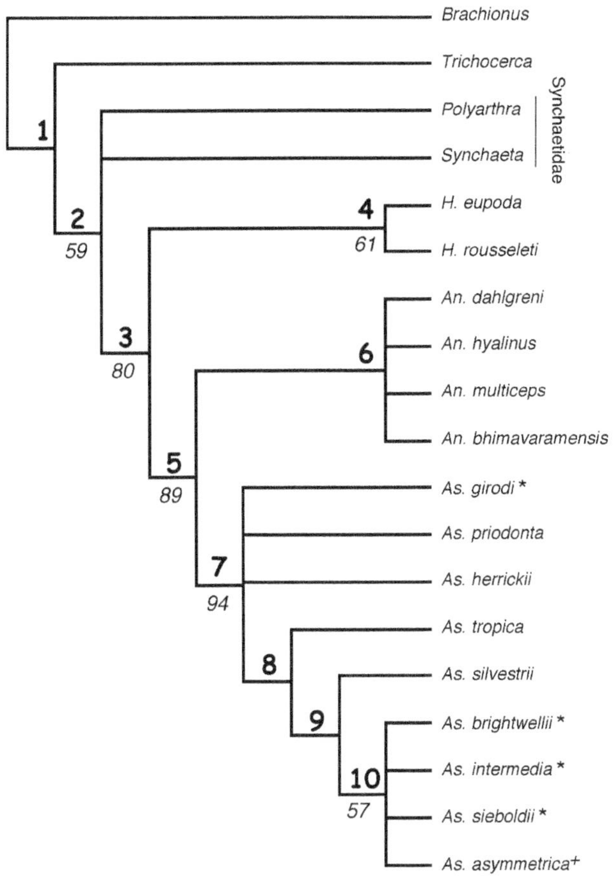

Figure 1. A phylogeny of Family Asplanchnidae based on morphological characters. This 50% majority-rule, consensus tree was generated using an unrooted and unordered search strategy that excluded the following three characters from our data matrix (Table 1): Vitellarium (#8), Cerebral eyespot (#9), and Ramus dentition (#15). Because some characters were uninformative, not all of the characters contributed to this tree. Numbers in bold print (**1**–**10**) refer to internal nodes. Numbers in italics below each node are the percentages of 1000 bootstrap replicates supporting that portion of the tree. Membership in Sudzuki's (1964) *Asplanchnella* (= elongated vitellarium) indicated by a * or a + (described after 1964). Changes in character state for the tree that are supported by both character optimization algorithms (ACCTRAN, DELTRAN) are reported below. These changes are annotated by node number or terminal taxon (in bold print). ***Brachionus***. Coronal bristles: present (#7); Trophus: malleate (#11). **(1)** ***Trichocerca***. Coronal palps: present (#5); Trophus: asymmetrical (#17); Body: asymmetrical (#19). **(2)** Lorica: illoricate (#21). **Synchaetidae** – ***Polyarthra***. Foot: absent (#1); Paddles: present (#20). ***Synchaeta***. Coronal auricles: present (#6); Coronal bristles: present (#7); Hypopharynx: present (#12). **(3)** **Asplanchnidae**: Coronal ciliation: single wreath (#4); Trophus: incudate (#11). **(4)** ***Harringia***: Cerebral eyespot (#9) (dropped in our standard analysis). **(5)** Gut: incomplete (#3); Manubrium function: reduced/vestigal (#13); Unci function: reduced/vestigal (#14). **(6)** ***Asplanchnopus***: Vitellarium Shape (#8) (dropped in our standard analysis); Cerebral eyespot (#9) (dropped in our standard analysis); Ramus dentition (#15) (dropped in our standard analysis); Apophysis-subapophysis (#16); other autapomorphies. **(7)** ***Asplanchna***. Foot: absent (#1); Pedal gland: absent (#2); Amictic egg development: oviviviparous (#10). **(8 & 9)** Apophysis-subapophysis absent (#16). **(10)** Apophysis-subapophysis present on trophus (#16).

members of this genus based on their propensity for female polymorphism (see Gilbert et al., 1979).

Molecular analysis

Analyses of all gene regions (except *cox*1) support the monophyly of the Asplanchnidae (Figs. 2–5). Analysis of the V4 region shows the sister group to Asplanchnidae as a clade represented by Brachionidae and *Polyartha* and *Trichocerca.* Since all taxa are not represented in phylogenies derived from the other gene regions, the molecular data are inconclusive at this time. Analyses of the ITS regions and *cox*1 resolve relationships at the species level. In the ITS1 analysis (Fig. 3), the epipelagic *Asplanchnopus* grouped with planktonic

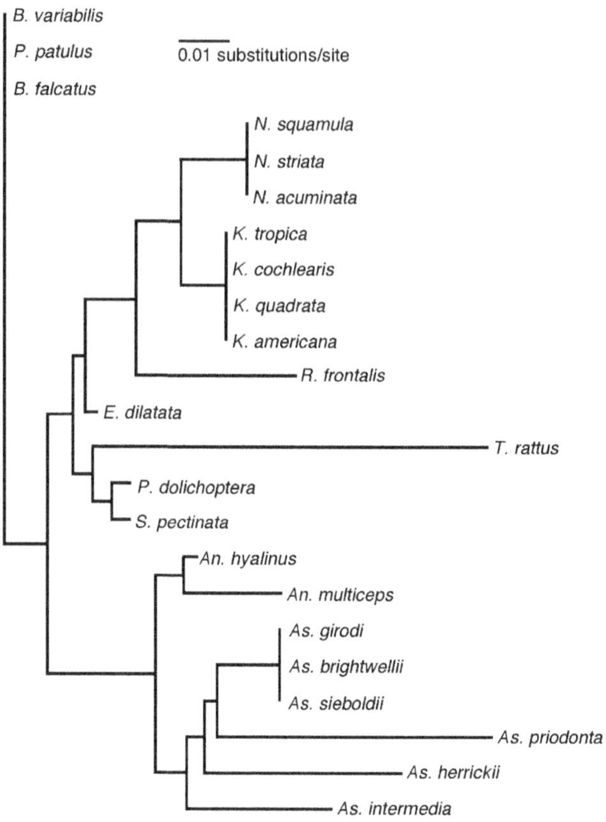

Figure 2. A Neighbor-Joining analysis of Family Asplanchnidae based on the V4 variable region of the 18S nuclear ribosomal gene. Brachionidae sequences were extracted from Walsh & De La Riva (unpublished). *Rhinoglena* sequence was extracted from Segers & Walsh (in prep.).

Asplanchna girodi, this is most likely a problem of incomplete taxon sampling since in all other analyses the *Asplanchnopus* are the sister group to *Asplanchna*. The ITS2 analysis reflects the ecological niches of rotifers as does the *cox*1 analysis (Figs. 4, 5).

Discussion

Based on both the morphological analysis and preliminary molecular analyses, we conclude the following: (1) Asplanchnidae is a monophyletic group, as are the three genera that comprise it; (2) there is no compelling support for the argument that *Asplanchna* should be separated into two discrete genera; (3) there is some support for the proposal that Asplanchnidae and Synchaetidae are sister groups. Unfortunately, using morphological characters, species relationships within genera were not well resolved. To aid in that resolution, improvements in the morphological dataset should be made. These refinements might include determining whether *A. asymmetrica* and *A. tropica* are capable of female polymorphism (i.e., producing body wall outgrowths) and establishing categories that distinguish among the extent to which body wall outgrowths develop.

The molecular analysis is promising in that for those taxa represented, species relationships were resolved. The utility of these gene regions in resolving phylogenies has been demonstrated for many organisms and recently in the brachionid rotifers (Gomez et al., 2002; Derry et al., 2003) (see papers in this volume Part I, Phylogeny and Evolution + II, Genetics and Molecular Ecology). We have found that a combination of gene regions works best to resolve family, generic and species relationships. The V4 nuclear region works well at the family level but poorly at the species

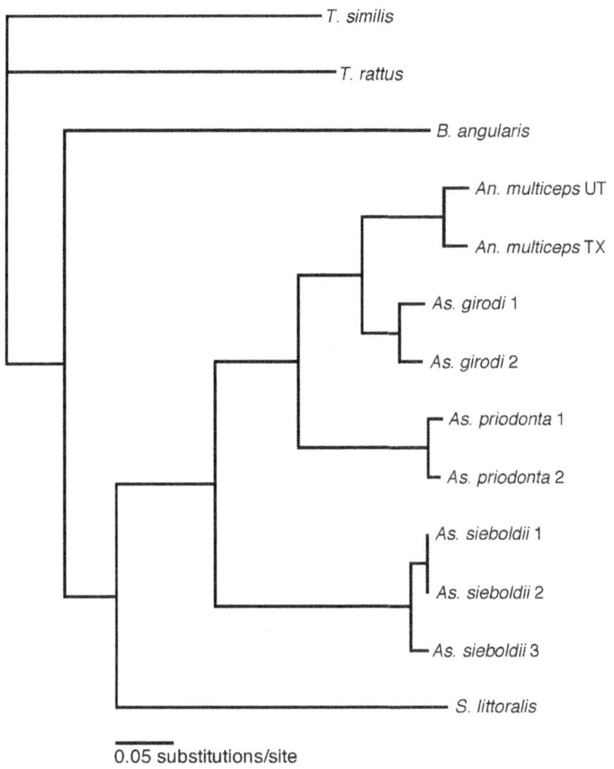

Figure 3. A Neighbor-Joining analysis of Family Asplanchnidae based on the internal transcribed spacer region 1 of the 18S nuclear ribosomal gene complex.

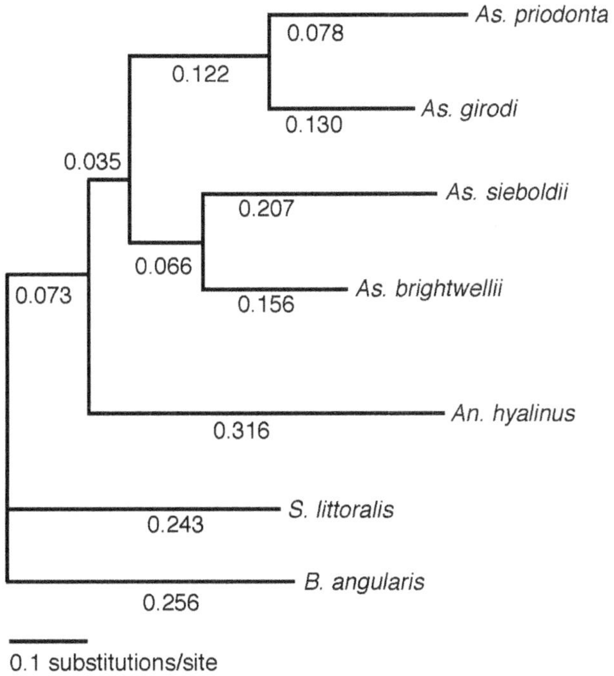

Figure 4. A Neighbor-Joining analysis of Family Asplanchnidae based on the internal transcribed spacer region 2 of the 18S nuclear ribosomal gene complex.

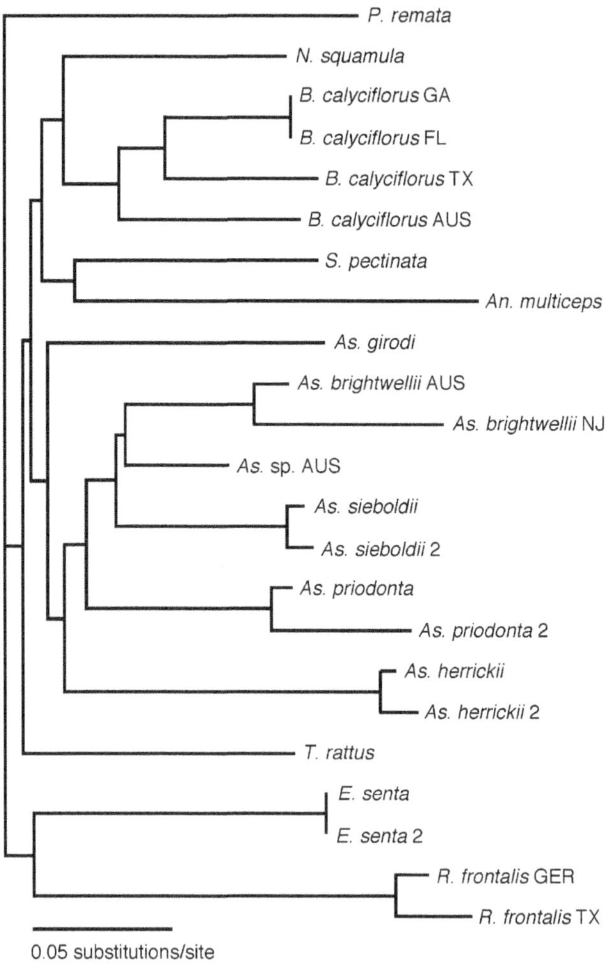

Figure 5. A Neighbor-Joining analysis of Family Asplanchnidae based on the mitochondrial cox1 gene. *Brachionus* sequences are from Gilbert & Walsh, this volume. *Rhinoglena* and *Epiphanes* sequences are from Segers & Walsh (in prep.).

level (Walsh & De La Riva, submitted) while the reverse is true for the *cox*1 and ITS regions (Gomez et al., 2004). Future work will focus on obtaining complete sequence data for the remaining members of the Asplanchnidae.

To our knowledge this is the first time that both morphological and molecular data have been employed in a cladistic analysis of a rotiferan family. In this regard we have taken up the challenge posed by Melone et al. (1998) who argued that should morphological and molecular studies remain firmly compartmentalized our understanding of rotifer phylogeny would be unobtainable. Unfortunately, significant gaps in our knowledge of the Asplanchnidae remain and this deficiency hampers our comprehension of their evolution.

Acknowledgements

The following researchers aided us in this study: B. J. Dingmann (MNSP Gallagher Fellow), A. Armendariz, M. Fout, and M. Ortega (Asplanchnidae); V. De La Riva (Brachionidae), and A. Frias (*Synchaeta, Polyarthra*). Elizabeth Wurdak kindly provided *Cisplanchna* for analyses. We thank J. J. Gilbert, C. D. Jersabek, T. Schröder, and an anonymous reviewer who provided valuable comments on our analysis. We also thank the staff at Hueco Tanks State Historic Site (collecting permits #66–99; 07–02). This research was supported by NSF HRD-9628568 and NIH 5G12RR008124. We also thank the Ripon College

(RLW) and University of Texas at El Paso (EJW) for additional funding.

References

Derry, A. M., P. D. N. Hebert & E. E. Prepas, 2003. Evolution of rotifers in saline and subsaline lakes: A molecular phylogenetic approach. Limnology and Oceanography 48: 675–685.

Dhanapathi, M. V. S. S., 1975. Rotifers from Andhra Pradesh, India. Zoological Journal of the Linnean Society 57: 85–94.

Faith, D. P. & P. S. Cranston, 1991. Could a cladogram this short have arisen by chance alone? On permutation tests for cladistic structure. Cladistics 7: 1–28.

Felsenstein, J., 1985. Confidence limits on phylogenies: An approach using the bootstrap. Evolution 39: 783–791.

Folmer, O., M. Black, W. Hoeh, R. Lutz & R. Vrijenhoek, 1994. DNA primers for amplification of mitochondrial cytochrome c oxidase subunit 1 from diverse metazoan invertebrates. Molecular Marine Biology and Biotechnology 3: 294–299.

Gilbert, J. J., 1980. Feeding in the rotifer *Asplanchna*: behavior, cannibalism, selectivity, prey defenses, and impact on rotifer communities. In Kerfoot, W. C. (ed.), Evolution and Ecology of Zooplankton Communities. Univ. Press New England, Hanover, NH, 158–172.

Gilbert, J. J., 1999. Kairomone-induced morphological defenses in rotifers. In Tollrian, R. & C. D. Harvell (eds), The Ecology and Evolution of Inducible Defenses. Princeton University Press, Princeton, NJ, 127–141.

Gilbert, J. J., C. W. Birky & E. S. Wurdak, 1979. Taxonomic relationships of *Asplanchna brightwelli*, *A. intermedia*, and *A. sieboldi*. Archiv für Hydrobiologie 87: 224–242.

Gómez, A., M. S. Serra, G. R. Carvalho & D. L. Hunt, 2002. Speciation in ancient cryptic species complexes. Evidence from molecular phylogeny of *Brachionus plicatilis* (Rotifera). Evolution 56: 1431–1444.

Harring, H. K., 1913. A list of the Rotatoria of Washington and vicinity with descriptions of a new genus and ten new species. Proceedings of the U.S. Natural Museum 46: 387–405.

Hollowday, E. D., 2002. Rotifera, Vol. 6. Family Synchaetidae. In Nogrady, T. & H. Segers (eds), Guides to the Identification of the Microinvertebrates of the Continental Waters of the World. Backhuys Publishers, The Hague, 87–211.

Hudson, C. T. & P. H. Gosse, 1886. The Rotifera or Wheel Animacules. Vol. I. Longmans, Green, and Co. London, 128 pp.

Joanidopoulos, K. D. & W. Marwan, 1998. Specific behavioural responses triggered by identified mechanosensory receptor cells in the apical field of the giant rotifer *Asplanchna sieboldi*. Journal of Experimental Biology 201: 169–177.

Joanidopoulos, K. D. & W. Marwan, 1999. A combination of chemosensory and mechanosensory stimuli triggers the male mating response in the giant rotifer *Asplanchna sieboldi*. Ethology 105: 465–475.

Jose de Paggi, S., 2002. Rotifera, Vol. 6. Family Asplanchnidae. In Nogrady, T. & H. Segers (eds), Guides to the Identification of the Microinvertebrates of the Continental Waters of the World. Backhuys Publishers, The Hague, 1–27.

Kappes, H., C. Mechenich & U. Sinsch, 2000. Long-term dynamics of *Asplanchna priodonta* in Lake Windsborn with comments on the diet. Hydrobiologia 432: 91–100.

Koste, W., 1978. Rotatoria. Die Rädertiere Mitteleuropas. 2 volumes. Gebrüder Borntraeger, Berlin, Stuttgart, Germany, Textband 673 pp., Tafelband 234 Tafeln.

Koste, W. & W. Tobias, 1989. Rotatorien der Sélingúe-Talsperre in Mali, Westafrika (Aschelminthes). Senckenbergiana biologica 69: 441–466.

Kutikova, L. A., 1983. Parallelism in the evolution of rotifers. Hydrobiologia 104: 3–7.

Melone, G., C. Ricci, H. Segers & R. L. Wallace, 1998. Phylogenetic relationships of phylum Rotifera with emphasis on the families of Bdelloidea. Hydrobiologia 387/388: 101–107.

Myers, F. J., 1934. The distribution of Rotifera on Mount Desert Island, Part V. A new species of Synchaetidae and new species of Asplanchnidae,Trichocercidae and Brachionidae. American Museum Novitates 700: 1–16.

Ridley, M., 1986. Evolution and classification: the reformation of cladism. Longman, NY, 201 pp.

Ruttner-Kolisko, A., 1974. Planktonic Rotifers: biology and taxonomy. Die Binnengewässer (Supplement) 26: 1–146.

Salt, G. W., G. F. Sabbadini & M. L. Commins, 1978. Trophi morphology relative to food habits in six species of rotifers (*Asplanchnidae*). Transactions of the American Microscopical Society 97: 469–485.

Segers, H., 2002. The nomenclature of the Rotifera: annotated checklist of valid family- and genus-group names. Journal of Natural History 36: 631–640.

Shiel, R. J. & W. Koste, 1985. New species and new records of Rotifera (Aschelminthes) from Australian waters. Transaction of the Royal Society of Australia 109: 1–15.

Sudzuki, M., 1964. New systematical approach to the Japanese planktonic Rotatoria. Hydrobiologia 23: 1–124.

Swofford, D. L., 2002. PAUP* - Phylogenetic Analysis Using Parsimony (* and Other Methods). Ver. 4 [Computer software]. Sinauer Associates, Sunderland, MA [with periodic on-line updates].

Thompson, J. D., D. G. Higgins & T. J. Gibson, 1994. ClustalW: Improving the sensitivity of progressive multiple sequence alignment through sequence weighting, position-specific gap penalties and weight matrix choice. Nucleic Acids Research 22: 4676–4680.

White, T. J., J. Bruns, S. Lee, & J. Taylor, 1990. Amplification and direct sequencing of fungal ribosomal RNA genes for phylogenetics. In Innis, M., D. Gelfand, J. Sninsky & T. White (eds), PCR Protocols: A Guide to Methods and Applications. Academic Press, Inc, San Diego.

Appendix

Character descriptions (unless otherwise defined, p, present, a, absent).

1. *Foot.* [p = 0, a = 1]. **2.** *Pedal Gland* [p = 0, reduced/absent = 1] (NB: *As. herrickii* rudimentary pedal gland = 1). **3.** *Gut.* [complete = 0, incomplete = 1]. **4**. *Coronal Ciliation* [(0 = circumapical bands, 1 = single wreath)]. **5**. *Coronal Palps* [a = 0, p = 1]. **6**. *Coronal Auricles* [a = 0, p = 1]. **7**. *Coronal Bristles* [a = 0, p = 1]. **8**. *Vitellarium Shape*. [spherical/sacciform = 0, elongate = 1] (dropped in our standard analysis). **9**. *Cerebral Eyespot*. [p = 0, a = 1] (dropped in our standard analysis). **10**. *Amictic Egg Development*. [oviparous = 0, ovovivparous = 1]. **11**. *Trophi*. [malleate = 0, virgate = 1, incudate = 2]. **12**. *Hypopharynx* [a = 0, p = 1]. **13**. *Manubria Function*. [functional = 0, reduced/vestigal = 1]. **14**. *Unci Function*. [functional = 0, reduced/vestigial = 1]. **15**. *Ramus denticulation*. [p = 0, a = 1] (dropped in our standard analysis). **16**. *Apophysis-subapophysis (*on trophus). [a = 0, p = 1]. **17**. *Trophus Symmetry*. [symmetrical = 0, asymmetrical = 1]. **18**. *Lamella* (behind rami apices). [a = 0, p = 1]. **19**. *Body Symmetry* [0 = symmetrical, 1 = asymmetrical]. **20**. *Paddles*. [a = 0, p = 1]. **21**. Lorica [a = 0, p = 1]. **22**. *Habitat*. [benthic = 0, semipelagic/littoral = 1, planktonic = 2]. **23**. *Female polymorphism* (body wall outgrowths). [a = 0, p = 1].

Part II
Genetics and Molecular Ecology

Hydrobiologia (2005) 546:83–99

A. Herzig, R.D. Gulati, C.D. Jersabek & L. May (eds.) Rotifera X: Rotifer Research: Trends, New Tools and Recent Advances
DOI 10.1007/s10750-005-4104-7

Molecular ecology of rotifers: from population differentiation to speciation

Africa Gómez
Department of Biological Sciences, University of Hull, Hull, HU6 7RX, United Kingdom
E-mail: a.gomez@hull.ac.uk

Key words: rotifera, cryptic species, *Brachionus plicatilis*, clonal structure, monopolisation, resting eggs

Abstract

The development of cost-effective molecular tools allowing the amplification of minute amounts of DNA effectively opened the field of molecular ecology for rotifers. Here I review these techniques and the advances they have provided in the understanding of sibling species complexes, clonal structure, resting egg banks, population structure, phylogeographic patterns and phylogenetic relationships in rotifers. Most of the research to date has focused on the rotifer species complex *Brachionus plicatilis*. The use of DNA sequence and microsatellite variation, in the context of the background knowledge of life history, mating behaviour, and temporal population dynamics in these organisms have revolutionised our views into the processes shaping the genetic diversity in aquatic invertebrates. Rotifers have populations with a very high number of clones in genetic equilibrium. In temporary populations clonal selection is effective in eroding the number of clones. Rotifer populations are strongly differentiated genetically for neutral markers, even at small geographical scales, and exhibit deep phylogeographic structure which might reflect the impact of Pleistocene glaciations. Despite the high potential for dispersal afforded by resting eggs, rotifers display persistent historical colonisation effects, with gene flow effective only at a local scale and with marked isolation by distance. Instances of long-distance transcontinental migration resulting in successful colonisation have also been revealed. *B. plicatilis* is composed of a group of several ancient species and sympatry is common. Despite this, the presence of cosmopolitan species in this species complex cannot be discounted. I discuss future priorities and point out the main areas where our knowledge is still insufficient.

Introduction

Molecular ecology concerns the application of molecular techniques to questions in ecology, evolution, behaviour and conservation biology (Carvalho, 1998). Several technical and methodological developments in molecular biology in the 1980s facilitated such application, among these, refinement in DNA extraction protocols, polymerase chain reaction (PCR), design of conserved primers which allowed amplification and sequencing of genes of virtually any organism, and, the publication of protocols for the development of species-specific microsatellite primers. The use of molecular tools has been revolutionary in diverse fields of ecology, evolution and behaviour and has yielded numerous insights into population structure, mating strategies, among many others.

The field of rotifer biology has much to gain from molecular ecology, as several themes have aroused much theoretical debate with little support from empirical data. Some of these long formulated questions include: how many clones are there in rotifer populations? Is clonal selection important in eroding genetic diversity? Are resting egg banks repositories of past and present genetic variation? And in a related question, is sex serving a function other than the way rotifers make resting eggs? How important are dispersal and gene flow in structuring rotifer populations? Are rotifer taxonomic species complexes of sibling species? Is cosmopolitanism an artefact of poor taxonomy, or are there true cosmopolitans? Is cyclomorphosis mainly determined by genetic replacement of clones or species, or by phenotypic plasticity? As discussed here, although still in its infancy, the molecular ecological approach in rotifers has provided novel insights into these questions and revitalised many areas of rotifer research.

The small body size of rotifers, and the difficulty of laboratory culture for many species, has hindered the application of allozyme electrophoresis to their study (Gómez, 1998). In spite of this, labour-intensive studies using allozyme markers in rotifers have yielded valuable data (King, 1977; King, 1980; Gómez et al., 1995; Ortells et al., 2000). The results based on allozyme markers were often limited due to the reduced amount of polymorphism detected. Therefore, and unlike in cladocerans, molecular ecology did not become a strong field in rotifers immediately. Although rotifers have long been a well-established model system in limnology, ecology and life history, the lack of basic genetic knowledge substantially hampered advances in some of these fields. For example, studies involving the rearing and the comparison of the ecological characteristics (e.g., optimal growth rates, mixis inducing cues) of wild-caught or domesticated clones (for some examples see King, 1980; Snell & Carrillo, 1984; Serra et al., 1994) revealed a strong genetic component in life history trait variation. However, we now know that these studies almost certainly involved comparison of strains belonging to sympatric cryptic species.

Here I will review and discuss (i) methodological advances that have facilitated the application of molecular techniques to rotifers; (ii) recent advances in the field of rotifer molecular ecology; and (iii) future developments and areas which are likely to benefit further from the molecular ecological approach. It must be emphasised that most of the research to date has been performed on the species complex *Brachionus plicatilis*, therefore the results obtained with this taxon will not necessarily apply to other rotifers. For the sake of simplicity however, the name 'rotifers' will be employed throughout the review mostly to refer to this taxon.

Technical and methodological advances in molecular ecology

The number of techniques being applied to molecular ecology is quite substantial and new methods are being added to the molecular ecologist's toolkit virtually every year. For the present purposes I will focus on those techniques crucial for the investigation of basic questions in rotifer ecology and evolution, and which have already provided some progress. In doing this, I have favoured reviews and monographs instead of the primary literature. The interested reader can trace back the relevant literature cited therein. Although allozyme electrophoresis can be considered a molecular technique, it will not be treated here as the methods and its applications have been reviewed elsewhere (Gómez, 1998).

DNA extraction procedures

The small body size of rotifers means that only procedures that involve PCR (polymerase chain reaction) amplification of extracted DNA can be applied to population problems cost-effectively. Techniques originally developed to recover DNA from forensic material, usually small dried blood or semen samples containing little or degraded DNA (Walsh et al., 1991) were adapted to extract DNA from single rotifer females, males and resting eggs (Gómez et al., 1998; Leutbecher, 2000). The advantages of such techniques were immense, as rotifer genetic variation could be examined without the need to culture them in the laboratory. Morevover, the genetic variation in sexual females, males and sediment borne resting eggs could be analysed, allowing wide-scale analysis of population structure, as well as long-term genetic temporal variation.

Polymerase Chain Reaction (PCR) amplification using conserved primers

The minute amounts of DNA recovered from single rotifers must be amplified prior to analysis using PCR (Palumbi, 1996; Birt & Baker, 2000), and this amplification necessitates a pair of oligonuclotides (the so called 'primers') of sufficient sequence similarity to regions flanking the target organism DNA sequence. Several sets of 'universal' primers, have been designed from conserved gene regions in mitochondrial DNA (mtDNA) and nuclear DNA (nDNA), and used to amplify rotifer DNA successfully. To amplify mtDNA regions, these include cytochrome *c* oxidase I primers developed by Folmer et al. (1994) used by Gómez et al. (2000; 2002b) and Derry et al. (2003), and 16S ribosomal genes developed by Palumbi (1996) used by Derry et al. (2003). Among nuclear genes Gómez et al. (Gómez et al., 2002b) amplified and sequenced the ITS1 ribosomal DNA using primers developed by Palumbi (1996), and Mark Welch & Meselson (2000) described and used several coding gene primers in bdelloids. Sufficient variation at a local or regional scale can often be found in a species when relatively fast-evolving genes are examined, and this is one reason why mtDNA is favoured for phylogeographic investigations and phylogenetics at low taxonomic ranks. Other reasons for the choice of mtDNA are (1) its haploidy, clonality and uniparental mode of inheritance (usually mother to offspring in animals), which reduces to ¼ its effective population size relative to nuclear markers making it more sensitive to demographic and evolutionary relevant events such as bottlenecks and population subdivision (Birky et al., 1989; Birky, 2001), (2) the fact that it occurs in multiple copies per cell, which favours its preservation and retrieval from ancient, poorly preserved or small tissue samples (Wayne et al., 1999). A cautionary note on using mtDNA used on its own is that, being a maternally inherited molecule, introgression and hybridisation may not be detected, therefore calling for the use of nuclear markers to support it. Although hybridisation not been detected in *B. plicatilis*, it might well be present in other monogononts.

Microsatellite loci

Population genetic studies in rotifers have been hampered by the scarcity of known polymorphic genetic markers. Although allozyme loci proved to be useful tools to detect sibling species complexes, little or no genetic variation has been reported within populations (Gómez et al., 1995; Ortells et al., 2000; but see Ortells, 2002). Furthermore, allozyme loci do not allow for the exploration of genetic diversity stored in rotifer dormant egg banks in lake sediments, or in the sexual individuals of the population, and individuals collected in the field need to be cultured in the laboratory to obtain enough biomass, which is often limiting sampling sizes. Analysis of microsatellite loci can help circumvent these problems. Microsatellites are DNA sequences made of short nucleotide motifs (up to six bases long) repeated in tandem, and can reach sizes of 200 bp (Goldstein & Schlotterer, 1999). Microsatellite loci are abundant and ubiquitous in the genome of eukaryotes and so they may provide a nearly unlimited set of markers for the study of clonal and population structure (Jarne & Lagoda, 1996; Li et al., 2002).

Microsatellites are used by molecular ecologists to address questions of population structure, migration and gene flow, mating patterns, parentage, and individual and clonal identification (Jarne & Lagoda, 1996). Several characteristics make microsatellites good markers for these ends, including high mutation rates, a large number of alleles per locus, codominant Mendelian inheritance and selective neutrality.

Microsatellite loci are amplified using PCR primers designed from unique flanking sequences. The main limitation for microsatellite analysis is in fact the availability of such primers, as they tend to be species-specific and have to be developed following a time-consuming protocol. Some degree of conservation across species can be present and cross-amplification of microsatellites has been reported for different animal species of the same genera or even the same family, but this has to be determined empirically on a case-by-case basis (Primmer et al., 1996; Primmer & Merila, 2002). In rotifers, microsatellite markers have been developed only for *Brachionus plicatilis* sensu stricto (Gómez et al., 1998) and, unfortunately, they have failed to cross-amplify even in other species of the complex (Gómez et al., 1998). Therefore, microsatellites might not be the markers of choice for future population studies of rotifers. Although, more promising recently developed methods, often with high-throughput, including AFLP (amplified fragment length polymorphism) (Vos et al., 1995) and SSCP (single stranded conformation polymorphism) (Sunnucks et al., 2000) or SNPs (single nucleotide polymorphisms) (Brumfield et al., 2003) need yet to be tested on these organisms.

Applications to the understanding of rotifer ecology and evolution

Cryptic species complexes and biogeography

Rotifera is considered a relatively minor metazoan phylum with less than 2000 described species (Segers, 2002). Due to their assumed considerable abilities for passive dispersal, rotifers were long considered to comprise mostly cosmopolitan species. However, in his review on the biogeography of rotifers, Dumont (1983) argued that rotifers show some evidence for vicariance and illustrated the levels of endemicity of the group in several continents. However, Segers (1996) noticed that in comparison with other animal groups, rotifer morphospecies display large distribution ranges, which could at least partly be a consequence to widespread dispersal, with vicariance playing a subordinate role. Both Dumont (1983) and Segers (1996) concur in attributing the apparently high proportion of widely distributed taxa to insufficient taxonomic resolution, that is, to the presence of cryptic taxa within the described morphospecies. In addition to increased sampling of poorly known habitats or regions of the world, the description of cryptic or sibling species could contribute substantially to increase our knowledge of rotifer biodiversity. King (cited in Dumont, 1983) suggested that, in addition to plankton nets, rotifer researchers should carry electrophoretic equipment with them. Although allozyme electrophoresis has indeed been used to identify cryptic species (Gómez & Snell, 1996; Ortells et al., 2000; Ortells, 2002), King's advice has not been followed widely and the confusion between cryptic species (compounded by cyclomorphosis, see section *The Proximate Causes of Cyclomorphosis* below) has crippled much of rotifer basic research.

The wealth of ecological information on many rotifer taxa has very little value unless the species used are 'real' biological entities. For example, variation among isolates of the same taxonomic species attributed to cyclomorphosis *sensu stricto*, or to intraspecific ecological variation, led to proposals of models of temporal adaptation of populations which had little connection with the reality of rotifer populations. As King (1980) put it the "population" investigated in many limnological studies may be an artifact with closer affinities to griffins, unicorns and mermaids than to the population as a biological unit'. Without a proper analysis of genetic differentiation or mating compatibility, many species have been lumped together and labelled 'generalists' or euryoic, polymorphic and cosmopolitan. In addition to seriously underestimate rotifer biodiversity, the lack of knowledge of cryptic species is preventing rotifer researchers from studying niche partitioning, population dynamics and many other ecological and evolutionary questions.

The taxonomic uncertainty surrounding cryptic species complexes has traditionally been resolved using lengthy and costly experimental approaches. For example, after decades of experimental work in *Brachionus plicatilis*, which suggested hidden species diversity, three species in this taxon were described or redescribed (Ciros-Pérez et al., 2001). In order to discriminate these species morphologically in a consistent way, different genetically characterised clones were grown in the laboratory in the same controlled conditions and a biometric study was performed on scanning electron microscopy photographs of females of the same age (Ciros-Pérez et al., 2001). It is doubtful that the same approach can be employed widely for the whole of the Rotifera. First, not all rotifers can be cultured readily, and, second, the workload involved would be insurmountable, given the human and economic limitations attached to rotifer studies. A second more straightforward and promising approach is to screen populations for one or a few genes in order to identify such cryptic species complexes (see Hebert et al., 2003; Tautz et al., 2003). Sequences obtained from a few genes can often yield the information necessary to conclude that two taxa are good species (see for example Baum & Shaw, 1995). Since the advent of PCR based techniques the number of cryptic species described in a variety of taxa is increasing steadily (Fig. 1) reflecting a tradition of lumping by taxonomists, but also morphological conservatism (for a recent review see Knowlton, 2000). The sequences obtained in molecular assessments can be annotated and deposited in public DNA databases such as GenBank/EMBL and subsequently retrieved by any interested researcher through the internet. Although these ideas are still controversial, it should be possible to base species descriptions on sequence information, while still

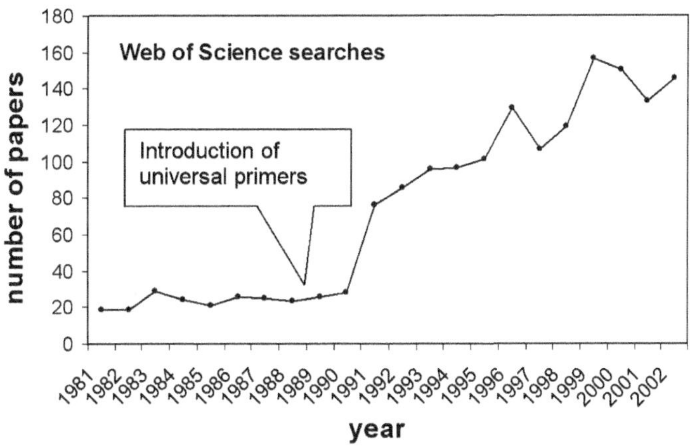

Figure 1. Effect of molecular tools on the recognition of cryptic species. Plot showing *Web of Science* searches on the keywords 'sibling species' or 'cryptic species'. The introduction of universal PCR primers is shown.

maintaining the importance of morphologically based descriptions (Tautz et al., 2003). Voucher specimens and details of collection localities would allow morphological or ecological appraisals *a posteriori.*

B. plicatilis remains by far the best studied monogonont species complex from a genetic point of view. Allozyme studies had previously indicated the occurrence of several species in this taxon, with no evidence of introgression in sympatry (Gómez et al., 1995; Ortells et al., 2000; Ortells, 2002). Using a collection of 57 specimens, including laboratory grown clones of known allozyme profiles, and field-collected resting eggs from 27 Iberian salt lakes and other worldwide locations, including North Africa, North America, Europe and Australia, Gómez et al. (2002b) obtained sequence information from two genes, a mtDNA gene, cytochrome *c* oxidase I (COI) and a nuclear gene, ribosomal internal transcribed spacer (ITS1). Phylogenetic analysis of both genes revealed nine concordant genetically divergent lineages (Fig. 2), six of them present in the Iberian Peninsula. COI evolves faster and therefore was more informative for the shallower branches of the phylogeny, whereas ITS1 was able to resolve deeper splits in the phylogeny. The three main branches of the phylogeny were strikingly concordant with the three described morphologies of the *B. plicatilis* complex, L, M and S (Fig. 2). The level of sequence divergence was well over that commonly found between different species and indicated that cladogenesis had not been a recent event.

Tentative dating of the radiation of the complex using molecular clocks for each gene goes back to 10–27 mya (Miocene) (Gómez et al., 2002b). Several additional lines of evidence support that these genetic lineages are different species or groups of species. First, the previously mentioned lack of hybridisation of these lineages when in sympatry (Gómez et al., 1995; Ortells et al., 2000; Ortells, 2002); second, cross-mating experiments performed between strains belonging to different lineages indicate behavioural reproductive isolation between them (Gómez & Serra, 1995; Gómez & Snell, 1996; Ortells et al., 2000; Berrieman et al., 2004); and third, ecological differences have been found when clones belonging to different lineages have been tested in the laboratory for optimal growth rates and mixis patterns (Gómez et al., 1997). The molecular phylogenetic assessment of Gómez et al. (2002b) was performed on a very patchy sampling of the geographical distribution of the *B. plicatilis* complex – Sub-Saharan Africa, South and Central America and most of Asia were not sampled – therefore it is very likely that several other species are present in this taxon, especially considering its thermophilic character. In addition, information on the geographic, genetic and ecological diversity and mating behaviour of some of these lineages is scarce or absent and therefore the detection of additional species in the already sampled lineages is likely.

A further consequence of this study concerns the issue of cosmopolitanism. Rotifer species show a high propensity towards local and regional

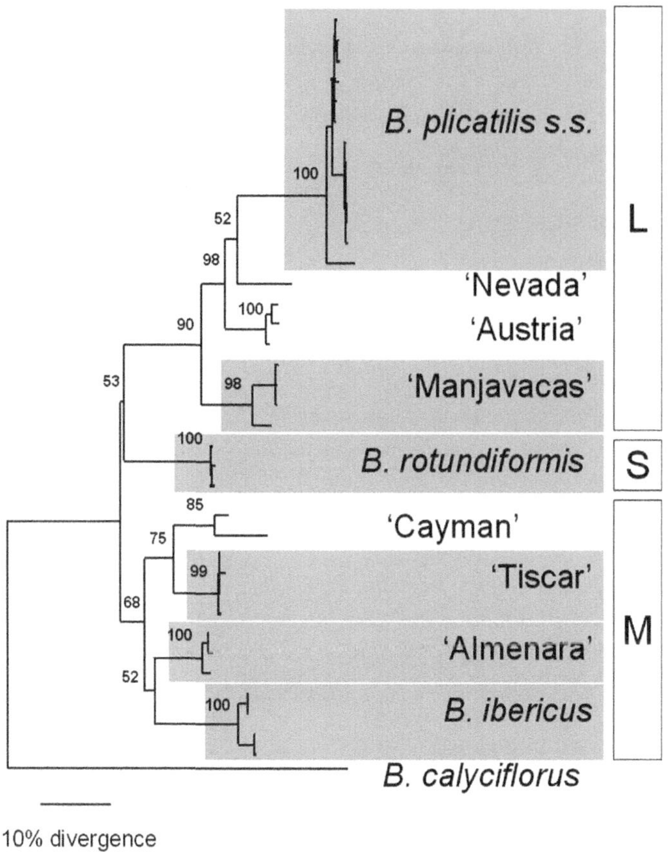

Figure 2. Phylogeny of the *Brachionus plicatilis* complex using *B. calyciflorus* as outgroup. Neighbour Joining tree based on a combined dataset using ITS1 and COI using a matrix of ML distances (see Gómez et al., 2002b for details). Values above branches indicate bootstrap support for nodes. The boxes indicate the size associated with the three main branches of the species complex (small, medium, large). Names boxed in grey indicate species present in the Iberian Peninsula.

genetic differentiation in spite of their capabilities for dispersal. Does the resolution of sibling species complexes in rotifers mean that good species will be regional endemics instead of cosmopolitan species? The answer seems to be: not completely. Undoubtedly, better geographic sampling is needed before strong conclusions can be drawn, but even with the relatively small effort undertaken so far, several of the lineages found are distributed in several continents (see Table 1). Whereas some of these lineages display a pattern of regional differentiation, others show very little genetic differences when clones retrieved from very distant geographical locations are compared, indicating that long-distance (even transcontinental) migration and colonisation was relatively recent. Human induced transportation is an issue to be considered here. Humans have been held responsible directly or indirectly for exotic introductions and range expansion of many organisms in aquatic habitats (Rahel, 2002). However, in contrast to other aquatic habitats (see for example Bailey et al., 2003 for data on several rotifer species transported through ballast water), salt lakes and coastal lagoons are usually remote, isolated from each other and from watercourses, little visited and devoid of any commercial use – other than the extraction of salt in some instances – and therefore it is unlikely that humans are responsible for such cases of species long-distance migrations. Therefore, rotifers from this sibling species complex undergo long-distance, intercontinental migrations and such events seem to be common enough to be detectable even with restricted sampling. A tentative conclusion regarding cosmopolitanism is that due to their high dispersal capabilities and occasional long-distance

Table 1. Geographic distribution of species in the *B. plicatilis* complex

Species	Europe	Asia	North America	North Africa	Australia
B. plicatilis s.s.	x		x		x
B. 'Manjavacas'	x	x		x	
B. 'Austria'	x	x	x		
B. 'Nevada'			x		
B. 'Cayman'		x	x		
B. 'Tiscar'	x				
B. 'Almenara'	x		x		
B. ibericus	x				
B. rotundiformis	x			x	

migration, rotifer species are always in the process of becoming cosmopolitan. A role for some dispersal limitation cannot be discarded, specially for very small and isolated lakes or for very large distances, although evidence for dispersal limitation in zooplankton is still controversial (Jenkins & Buikema Jr., 1998; Shurin, 2000).

A surprising finding of this study which supported previous data was that many species in the complex were often found in sympatry (Gómez et al., 2002b) (Fig. 3). Lakes containing two and three rotifer species were not rare and a lake has been found where four of the species coexist (Ortells, 2002). Due to the strong seasonality of salt lakes and coastal lagoons, the high level of sympatry in rotifers could be due to seasonal succession and temporal niche partitioning and/or different susceptibilities to predators and parasites. In fact, species in the *B. plicatilis* complex can be involved in seasonal succession (Gómez et al., 1995), and this has been attributed to their ecological specialisation to different salinities or temperatures (Serra et al., 1998) and also to their different food preferences (Ciros-Perez et al., 2001). If this is a common pattern, then it should be possible to predict the number of species from the complex likely to be found in a lake, based on the degree of temporal variability of that lake (ideally estimated across several years to account for the interannual variation of the habitats). However, Ortells et al. (2003) found that in some cases two species from the *B. plicatilis* complex coexist throughout most of their presence in the pond, which seems to suggest that factors mediating coexistence (disturbance, predators, etc.) must play a role in facilitating sympatry (Ciros-Perez et al., 2001).

A pattern of common sympatry reflects several processes: each species should (i) reach the lake or pond, (ii) establish populations with positive growth rates, and (iii) persist through time in the face of environmental variation. I have already mentioned that long-distance dispersal seems to be a common phenomenon in this species complex. A better understanding of ecological preferences (including pre-competitive and post-competitive niches) in these species would allow us to further understand the factors contributing to the establishment success of rotifer species when they reach new habitats. It is clear that the fact that these organisms maintain resting egg banks will contribute to the long-term persistence of species in a lake, even if its conditions are unsuitable for a few years (what has been termed a 'storage effect' (Cáceres, 1997). Thus, a combination of effective dispersal and colonisation, successful niche partitioning and occurrence of factors mediating coexistence, and storage-effect allowed by the resting egg bank, seems to be responsible for the high degree of sympatry observed in rotifer species from this complex.

Gómez et al. (2002b) study illustrates how different processes govern species and populations in rotifers. Populations show evidence of very low gene flow due to the monopolisation hypothesis (see the *Population Structure* section below), and as a consequence, a population in a lake contains a reduced proportion of the neutral genetic diversity of its species. In contrast, that very same lake can contain a significant proportion of sibling species,

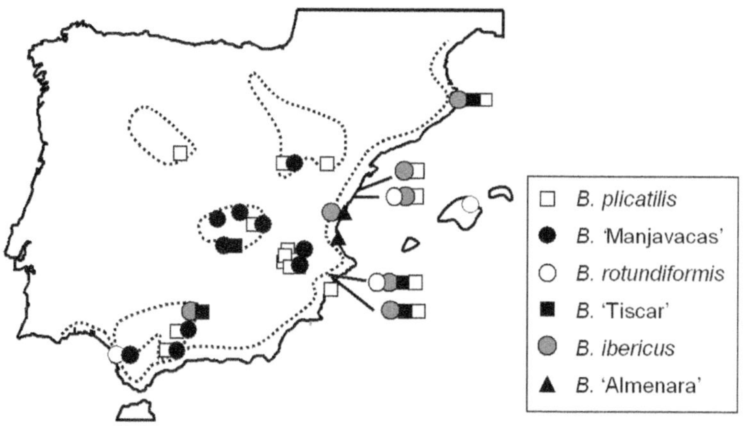

Figure 3. Sympatry in the *B. plicatilis* complex. Map of the Iberian Peninsula showing the distribution of the six species detected so far.

reflecting the colonisation abilities of the different species.

Finally, phenotypic plasticity, but also morphological conservatism, and a pattern of extensive sympatry among cryptic species are factors that might have contributed to the difficulties of recognising good rotifer species and led the average rotifer taxonomist to become a 'lumper'. If the patterns found in *B. plicatilis* are found to be common in other rotifer morphospecies the amount of hidden biodiversity in rotifers could be substantial. Indeed, the presence of cryptic species has been reported in *B. calyciflorus* using mating behaviour and sequence variation data (Gilbert & Walsh, this volume, Part V) and *Keratella* (Derry et al., 2003). The day might come when rotifers are not considered a minor phylum.

Routine species identification

Once the members of a species complex have been described using molecular techniques, a simple method available to a large number of rotifer researchers must be developed so that species can be identified from field samples routinely, or contamination detected in aquaculture facilities. In the case of *B. plicatilis*, several techniques are available, and they could be applied to other rotifer species. First, allozyme electrophoresis, which can be set up relatively cheaply (Gómez, 1998), could provide 'diagnostic loci' (Ortells et al., 2000; Ortells et al., 2003) which could be genotyped in laboratory grown clones. As mentioned before, the application of this technique is restricted to those species of rotifers that reproduce rapidly and can be cultured readily. For those species for which DNA sequence information is available, an approach based on PCR-RFLP could be used. This procedure consists of using PCR to amplify a fragment of DNA of known sequence and then digesting the amplification product with those restriction enzymes yielding a diagnostic restriction profile (restriction fragment length polymorphism, RLFP) when samples are run on a gel. Although this procedure requires the use of a thermocycler and basic electrophoresis tools, such equipment is basic to genetic and evolutionary laboratories of many universities, and the cost of materials could be relatively cheap, even cheaper than allozyme electrophoresis.

A PCR-RFLP method was used by (Berrieman et al., 2005, this volume) to discriminate between wild caught clones of the sympatric and morphologically very similar *Brachionus plicatilis* s.s. and *B.* 'Manjavacas', and between 'northern' and 'southern' *B. plicatilis* s.s. lineages in order to perform mating behaviour experiments.

Rotifer clonal structure and resting egg banks

Monogonont rotifer planktonic populations are made of numerous clones produced by parthenogenetic females that hatched from sexually produced resting eggs in the resting egg bank. In lakes and ponds that undergo periodic drying or freezing over, these planktonic populations are necessarily reconstituted every year from the resting egg bank. In mild years, parthenogenetic populations might survive several growing seasons with a variable input from the resting egg bank. In at least some cases, species are present in the water column during part of the year, partly reflecting their tolerance ranges or competitive abilities. The clonal composition and dynamics of planktonic rotifer populations and the interplay with the resting egg bank was virtually unknown until recently. Meanwhile, our understanding of clonal structure had advanced to a mature state in the other group of aquatic cyclical parthenogens, cladocerans. Cladoceran researchers working with *Daphnia* and mainly using allozyme markers had produced two models that seemingly accounted for the clonal and genetic structure of populations. In 'intermittent' ponds, where populations were founded every year from sediment banks, investment in sexual reproduction was important and genetic analysis revealed a large number of clones in genetic equilibrium (both Hardy–Weinberg and linkage) (see review in De Meester, 1996). No evidence for clonal selection was found in these ponds. In contrast, in permanent ponds some *Daphnia* populations persisted among years, sexual investment was reduced and there were often a low number of clones which underwent rapid changes in frequencies, indicating clonal selection. Although these models are simplifying and exceptions have been found, they can be used as a framework to help us understand what takes place in rotifer populations.

Gómez & Carvalho (2000) used a set of seven polymorphic microsatellite markers to screen an intermittent population of *Brachionus plicatilis* (Poza Sur, in Prat de Cabanes–Torreblanca Marsh). There were three consecutive planktonic samples along a parthenogenetic phase, a sample from the resting egg bank, and a sample after the re-establishment of populations after the summer drought. The set of seven polymorphic microsatellite loci previously developed for *Brachionus plicatilis* (Gómez et al., 1998) proved to be a useful tool for clonal identification because the probability that two clones produced by separate sexual recombination events (hatching from two different resting eggs) have the same multilocus genotype is very low. Overall, 349 different genotypes were found in the 390 individuals screened. A graph of the number of genotypes found versus sample size did not plateau, indicating that rotifer populations are made of a very large number of clones. Unexpectedly, most samples, including the resting egg bank, were in genetic equilibrium. However, evidence for linkage disequilibrium due to replicate genotypes was found in the planktonic sample at the end of the growth cycle (March), when the effects of clonal selection are expected to be noticeable. Indeed, in this sample 11 genotypes were found more than once, and a simulation revealed evidence of significantly small expected genotypic diversity, probably due to clonal selection. A different analysis on the same dataset (Stenberg et al., 2003) showed that at least four of the repeated multilocus genotypes are likely to be members of the same clone.

This study revealed that, although clonal selection is significant, and actually reduced the genotypic diversity along the parthenogenetic sample, its effects are weaker than might be predicted (King, 1980), as populations at the end of the parthenogenetic phase are still made of a very high number of clones, partly due to the very large number of initial clones. If the sample sizes in the study had been halved, the effects of clonal selection would not have been detected at all. Therefore, clonal selection might be effective in reducing genotypic diversity, but allelic diversity (at least in neutral alleles at significant frequencies) remains virtually the same. At least in the set of loci investigated, the observed genetic diversity is generated every year by recombination (input from the resting egg bank), rather than by mutational input, even for loci with relatively high mutation rates.

The clonal structure of this population resembles the 'incomplete genetic discontinuity model' of King (1977), in which clones might coexist during long periods, their frequencies fluctuating depending on the seasonal conditions of the lake. To further understand to what extent clones are ecological generalists or specialists, a joint investigation of genetic diversity and ecological characteristics should be undertaken. In spite of being a temporary population, the short generation times of rotifers facilitate the detection of clonal selection. In contrast, clonal selection in *Daphnia* has only been detected in permanent populations, in which selection has much longer time to take effect (Gómez & Carvalho, 2000). Recently, further evidence for the importance of directional clonal selection in rotifer populations was found

Figure 4. Phylogeography of species from the *B. plicatilis* complex. (a) Neighbour Joining phylogenetic tree representing the phylogeography of *B. plicatilis* s.s. colleted from Iberian lakes. (b) phylogeography of *B.* 'Manjavacas' lineage 1 collected from Iberian salt lakes. (c) phylogeography of rotifers tentatively classified as *B. plicatilis* s.s. collected in Wood Buffalo National Park (Canada) (tree produced from sequences deposited in GenBank, AF499054–AF499069, and published in Derry et al., 2003). Different symbols on each map and corresponding tree indicate geographically concordant lineages.

using allozyme analysis in several permanent and temporary Mediterranean ponds in a set of sibling species (Ortells, 2002).

The studies to date emphasize the importance of applying molecular tools to the understanding of rotifer clonal structure and the interplay with the resting egg bank. However, the body of research is still limited and, unfortunately, restricted to the *Brachionus plicatilis* complex. Comparisons with other rotifer species, particularly in large lakes, freshwater ponds, and riverine habitats are badly needed for a better understanding of rotifer clonal structure.

Phylogeography

Phylogeography is the study of the patterns and processes governing the geographic distribution of genetic lineages (Avise, 2000). Such analysis allows distinguishing between recurrent processes such as gene flow, and historical processes such as population subdivisions, long distance migration events or range expansions (Templeton, 1998). Phylogeographic information to date comes overwhelmingly from RFLPs, or from sequence variation of mtDNA (see review in Avise, 2000). At the time of Avise's influential book (Avise, 2000) phylogeographic research of freshwater zooplanktonic organisms was just beginning and the treatment they received was rather scant and biased: "This highly dispersive phase of the life cycle (the ephippium) probably accounts for near ubiquity of (*Daphnia*) mtDNA lineages across vast areas such as Northern Eurasia." Avise was reviewing the first mDNA assessments of Holarctic *Daphnia* (Taylor et al., 1996; Weider et al., 1996; Weider & Hobaek, 1997). These pioneering studies had been performed in areas strongly affected by the Pleistocene glaciations, and the results described the colonisation of very recently formed ponds and lake systems of the Arctic and Subarctic. The high dispersal and colonisation abilities of zooplankton (see review in De Meester et al., 2002) do indeed explain the rapid colonisation of such geographic areas. Since then, additional *Daphnia* studies in more temperate areas have supported strong geographic structure and regional endemism of lineages incompatible with high rates of gene flow (see review in De Meester et al., 2002).

The phylogeographic structure of *Brachionus plicatilis* (sensu Ciros-Pérez et al., 2001), in the Iberian Peninsula was investigated by sequencing 653 bp of the mitochondrial gene cytochrome *c* oxidase subunit 1 (COI) (Gómez et al., 2000). DNA was extracted from individual resting eggs retrieved from sediments of salt lakes in the Iberian Peninsula. Sampling resting eggs reduces biases due to stochastic variation in clonal populations due to selection or drift in a given parthenogenetic growth period. *B. plicatilis* s.s. was found in the resting egg banks of 18 of the 47 lakes sampled. A total of 98 individuals were sequenced for the mtDNA gene, yielding 21 different mtDNA haplotypes. Phylogenetic analysis revealed the occurrence of two mtDNA lineages (Fig. 4a). These lineages were strongly structured geographically, with one being present in the southern ponds, and the other in the northern ones. Both lineages were found to coexist in two ponds which formed a contact zone. The northern lineage was further divided in three subgroups (see Fig. 4a). Individual lakes had relatively low genetic diversity, and most of the haplotypes were restricted to single lakes. Examination of the data using Templeton's (1998) Nested Clade Analysis suggested a low level of gene flow, with isolation by distance and some episodes of long distance colonisation. The main process that structured genetic diversity was historical population fragmentation in allopatry. Given the degree of genetic divergence of the two groups of haplotypes, a hypothesis was proposed to explain the observed phylogeographic structure in Iberian rotifer populations. Such a pattern could have arisen from the climatic changes accompanying Pleistocene glaciations, which probably reduced and fragmented the area occupied by salt marshes and lakes in the Iberian Peninsula. Because rotifer resting eggs seem so well

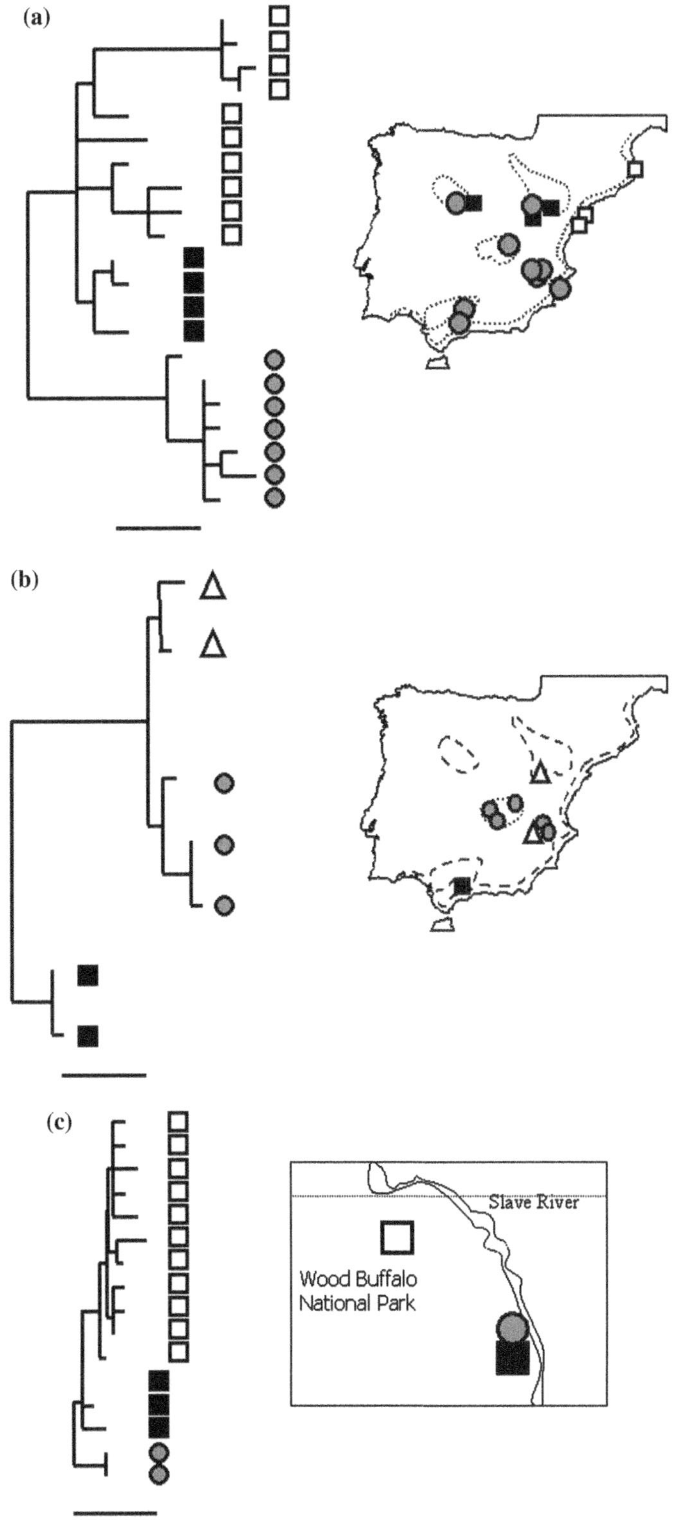
(a)
(b)
(c)
Slave River
Wood Buffalo
National Park

suited for dispersal (King, 1980), and the Iberian Peninsula is one of the main corridors of European waterfowl migration, the persistence of signatures of population fragmentation after several thousand years is surprising.

Recently, Gómez et al. (unpublished results) have examined the phylogeography of another rotifer from the *B. plicatilis* complex in the Iberian Peninsula, the *B.* 'Manjavacas' (see Gómez et al., 2002b). The mtDNA COI gene was sequenced in resting eggs collected from salt lake sediments. In agreement with the findings in *B. plicatilis* s.s., a strong phylogeographic structure was found in one of the lineages (Fig. 4b), but the two most divergent lineages of this species seem to overlap to a large extent (data not shown).

Derry et al. (2003) have recently investigated the sequence variation of two mtDNA genes of *B. plicatilis* in three Canadian salt lakes (Fig. 4c). Their results support the patterns found in the Iberian Peninsula. The species they worked with seems to be *B. plicatilis* s.s. (a 4.1% sequence divergence was found between the COI gene of Canadian and Iberian rotifers) and they found strong geographic structure, with haplotypes restricted to single lakes, and related haplotype lineages found in the same lake (Fig. 4c).

The pattern of intraspecific geographic differentiation found suggests that speciation could happen in allopatry, and that sympatry is secondary, although more detailed analyses are needed to reach conclusions regarding the tempo and mode of speciation in rotifers.

Population structure

The reasons underlying the high levels of inter-population differentiation, even at local scales, found in *Daphnia* and other zooplanktonic organisms despite their dispersal abilities have been much debated (see reviews in De Meester, 1996; De Meester et al., 2002). Rapid local adaptation might prevent survival of migrants due to inferior competitive abilities in the new habitat compared with the locals; intragenomic interactions, such as outbreeding depression with hybrid breakdown might arise due to genomic incompatibilities established during the historical colonisation process. In addition, habitats can be colonised by a few propagules which will reproduce rapidly, as they would grow unchecked by competitors giving rise to a 'persistent founding effect' (Boileau et al., 1992), by which the allelic frequencies established by the first colonists will be resistant to change due to migration. In order to investigate these processes in rotifers, Gómez et al. (2002a) typed between 20 and 50 rotifer resting eggs retrieved from sediment samples from the same group of salt lakes and coastal lagoons sampled for the mtDNA phylogeographic study for a set of 7 unlinked microsatellite loci. Of the 63 alleles found in the 440 eggs typed, 23 were private alleles, that is, they were alleles found in a single population. In accordance with the mtDNA findings, results show a strikingly high level of population genetic differentiation (global F_{st} estimate 0.43). Thirteen out of the fourteen populations for which more than 9 resting eggs were typed were in genetic equilibrium suggesting that ongoing inbreeding (due for example to a low number of clones in rotifer populations) is not a cause of population differentiation. A Principal Component Analysis revealed some differences with the phylogeographic structure of the species, as the microsatellite differentiation was not correlated with the mtDNA differentiation. Some populations seemed to be responsible for this discrepancy patterns as they were likely part of the contact zone where both historical mtDNA lineages had come into contact (see Berrieman et al., 2004). A strong pattern of isolation by distance was found, independently of the mtDNA constitution of the involved populations. This pattern indicates that populations harbouring different mtDNA lineages, in spite of their differentiation belonged to the same species and that gene flow might play some role at a local scale. It will be interesting to investigate if such local gene flow reflects local colonisation-extinction dynamics or genetic exchange between established populations.

The accumulation of data in several passively dispersed aquatic organisms has led to the proposal of an integrated hypothesis based on several ecological and evolutionary processes (De Meester et al., 2002). This monopolisation hypothesis explains the paradox of strong population structure (low gene flow) despite good colonisation abilities (high dispersal). Both neutral processes (persistent founder effects) and selective processes (local adaptation) have been shown to be particularly effective in several aquatic organisms and are hypothesised to act synergistically to diminish the genetic consequences of dispersal. The former is due to a 'dilution' effect: the migrant alleles form a much reduced proportion of the local gene pool. The latter acts by reducing the chances of migrants of surviving or leaving descendants in the population. The third component of the monopolisation hypothesis is the presence of large resting egg banks containing past and presently adapted genotypes (an effective archive of the cumulative time-space adaptive spectrum of the population, see for example Cousyn et al., 2001), against which any migrant has to compete.

Thus, in a similar manner to other zooplanktonic organisms, rotifers display marked population differentiation in neutral markers which can largely be explained by historical colonisation events. The importance of the dilution effect against migration afforfed by the resting egg bank can also be safely assumed.

The proximate causes of cyclomorphosis

Cyclomorphosis defines the temporal cyclic morphological changes that occur within a population (Black & Slobodkin, 1987). This widespread phenomenon (investigated particularly in *Brachionus*, *Keratella* and *Asplanchna*) has complicated rotifer taxonomy (Serra et al., 1997). The possible proximate causes determining cyclomorphosis include phenotypic plasticity, genetic replacement of clones, and, using a wider definition, seasonal succession of sibling species. Laboratory culture and exposure of clones to the same or different growing conditions can help disentangle the causes of morphological variation (see Gilbert, 2001 for a recent example). Molecular techniques might afford a more cost-effective and rapid demonstration of the genetic basis of cyclomorphosis, and the discrimination between species and clonal replacement can be more finely resolved. A recent contribution to the understanding of morphological changes in the genus *Keratella* was made by Derry et al. (2003). Samples from three morphs of *K. cochlearis* (*tecta*, *robusta* and *faluta*), 2 morphs of *K. hiemalis* (single spined and two-spined) and *K. quadrata* were collected in several lakes in Wood Buffalo National Park (Alberta, Canada). Fragments of two mtDNA genes, COI and 16S rRNA were sequenced in a total of 42 and 9 individuals respectively. In agreement with previous biometric analysis of seasonal variation (Hofmann, 1983), large sequence differences in their mtDNA suggest that *K. cochlearis* is a species complex (Fig. 5). In contrast, morphological variation in *K. hiemalis* in these lakes seemed to be due to phenotypic plasticity, as the single-spined and two-spined morphs were very similar genetically. Little genetic variation was found in *K. quadrata* in these Canadian lakes. This study illustrates that molecular techniques can be a powerful tool to investigate the nature of phenotypic variation in rotifers. More studies, and particularly, genetic analysis of well known systems, will be required for a more global assessment of the importance of clonal or species succession in cyclomorphosis.

Perspectives and future directions

The application of molecular techniques has given a new impetus to the fields of ecology and evolution of rotifers. There are two main causes for concern regarding general conclusions, though. First, the number of studies is still small, and second, the body of research is largely restricted to the species complex *B. plicatilis*. *B. plicatilis* s.l. inhabits salt lakes and lagoons – habitats that are not considered typical among rotifers, and therefore, the results obtained with this taxon might not be extendable to other monogononts. On the other hand, *B. plicatilis* has proven to be a good model organism and the conclusions attained so far have been largely in agreement with findings in freshwater zooplankters such as cladocerans and copepods. However, it must be emphasised that is important to test the generality of the patterns discovered in other rotifer models, for example inhabiting different habitat types (interstitial, riverine, lacustrine and other freshwater rotifers), or habitats with less seasonality than the ones investigated (tropical lakes, for example).

The future of rotifer Molecular Ecology looks very promising. Many exciting fields that have remained largely untouched in the past can now be tackled. For example, rotifer resting egg banks have not been investigated for historical temporal variation in ecologically relevant traits. This research has indeed provided surprising results in

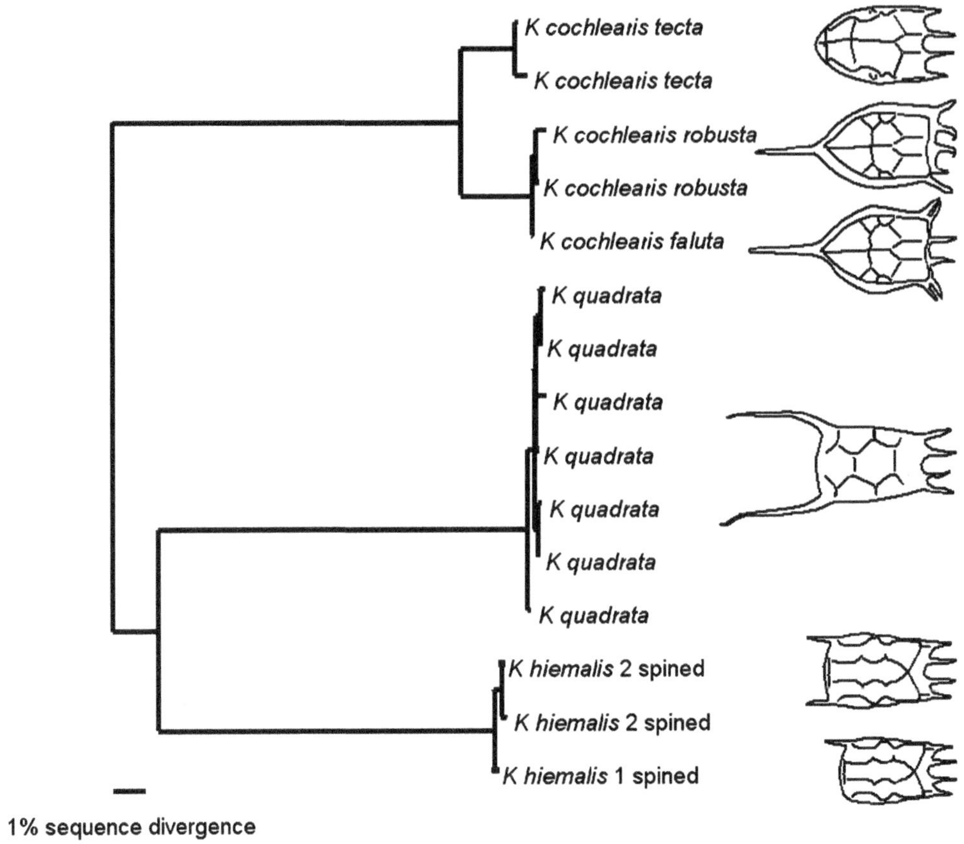

Figure 5. Cyclomorphosis in *Keratella*. Neighbour Joining phylogenetic tree (Log-Det distances) showing the phylogeny of species and morphs of three *Keratella* species collected in Canadian lakes. The tree was built from sequences downloaded from GenBank (AF499073–AF499087) originally published in Derry et al. (2003).

Daphnia (Cousyn et al., 2001). To further understand the structure of rotifer populations, studies must be undertaken on the importance of local adaptation, measuring the level of interpopulation differentiation regarding ecologically relevant traits relative to neutral genetic differentiation.

In addition, the application of metapopulation theory can be quite productive, especially because clusters of lakes or ponds can have very different demographic properties in terms of population extinction and colonisation. Some larger or more stable lakes could act as 'sources' and other smaller lakes – where temporal stochastic variation in resting egg production could lead to population extinctions – could be regarded as 'sinks'. In fact, sets of coastal pools were demonstrated to fit to a metapopulation structure in *Daphnia magna* (Ebert et al., 2002; Haag et al., 2002). The effect that this metapopulation structure, so different in its effects to the traditional island model, could have in explaining the genetic divergence of rotifer populations has not been explored at all.

Intraspecific clonal variation in ecologically relevant traits have been little investigated in rotifers (Zhao & King, 1989). Ecologically relevant traits are often determined by several quantitative trait loci and therefore, populations could harbour large variability for these traits allowing for rapid responses to selection (Lynch et al., 1999; Morgan et al., 2001). Indeed, some rotifer populations, especially those inhabiting temporary environments, have shown few temporal changes in their neutral population structure. However, important temporal population changes in ecologically relevant traits may be accompanied by few or no detectable changes in neutral genetic markers, as it has been shown with antipredator phototactic behaviour in *Daphnia magna* (Cousyn et al., 2001).

There is little information on the occurrence of local adaptation in rotifers and how it is achieved. For example, local adaptation in a population could be a property facilitated by its resting egg bank if adapted genotypes from previously selective environments are present and hatch in the appropriate environment or randomly (resting egg banks would constitute an archive of selections past), or it could be a property of the clonal population due to its rapid response to selection. If the former is important, the consequences of human-induced pollution, climate change or species invasions (or any other new environmental challenge) could have unforeseen effects as rotifer species could well lack the genetic diversity necessary to respond to such changes. On the contrary, if rotifers are capable to respond rapidly to selection, then their populations could be quite resilient to the aforementioned environmental challenges. Indeed microevolution in a rotifer planktonic population without access to a resting egg bank, due to selection against sexual reproduction, has been reported in chemostat cultures of *B. calyciflorus* (Fussmann et al., 2003).

Many interesting questions remain: What is driving speciation in rotifers? Is speciation mostly allopatric? Does reproductive isolation arise as a side-effect of population divergence or is reinforcement important? Is speciation ecological as proposed for *Daphnia* (Pfrender et al., 2000)? Will a global biogeography of the *B. plicatilis* complex reveal true cosmopolitans? What is the importance of inbreeding during population colonisation and its interaction with early dispersal?

Acknowledgements

Many people have helped me through the years to develop the ideas presented here, in particular I wish to thank Manuel Serra, Luc de Meester, Gary Carvalho, Raquel Ortells, and David Lunt. Bill W. Birky, Gary Carvalho, David Lunt, David Mark Welch, Raquel Ortells and Manuel Serra, read previous versions of this manuscript and provided constructive criticism and advice which greatly improved the manuscript. Any errors or omissions remain my own. Part of the research on which this review is based has been funded by the National Environmental Research Council (NERC) (U.K. grant No. GR9/04482), the Molecular Ecology and Fisheries Genetics Laboratory in Hull, and the University of Valencia.

References

Avise, J. C., 2000. Phylogeography: The History and Formation of Species. Harvard University Press, Cambridge, Massachusetts, 447 pp.

Bailey, S. A., I. C. Duggan, C. D. A. Overdijk, P. T. Jenkins & H. J. MacIsaac, 2003. Viability of invertebrate diapausing eggs collected from residual ballast water. Limnology and Oceanography 48: 1701–1710.

Baum, D. A. & K. L. Shaw, 1995. Genealogical perspectives on the species problem. In Hoch, P. C. & A. G. Stephenson (eds), Experimental and Molecular Approaches to Plant Biosystematics. Missouri Botanical Garden, Columbia, Missouri, 289–303.

Berrieman, H. K., D. H. Lunt & A. Gómez, (2005). Behavioural reproductive isolation in a rotifer hybrid zone. Hydrobiologia 546: 125–134.

Birky, C. W., P. Fuerst & T. Maruyama, 1989. Organelle gene diversity under migration, mutation, and drift – equilibrium expectations, approach to equilibrium, effects of heteroplasmic cells, and comparison to nuclear genes. Genetics 121: 613–627.

Birky, C. W., 2001. The inheritance of genes in mitochondria and chloroplasts: Laws, mechanisms, and models. Annual Review of Genetics 35: 125–148.

Birt, T. B. & A. J. Baker, 2000. Polymerase Chain Reaction. In Baker, A. J. (ed.), Molecular Methods in Ecology. Blackwell Science, Oxford, 50–64.

Black, R. W. & L. B. Slobodkin, 1987. What is cyclomorphosis. Freshwater Biology 18: 373–378.

Boileau, M. G., P. D. N. Hebert & S. S. Schwartz, 1992. Nonequilibrium gene frequency divergence – persistent founder effects in natural populations. Journal of Evolutionary Biology 5: 25–39.

Brumfield, R. T., P. Beerli, D. A. Nickerson & S. V. Edwards, 2003. The utility of single nucleotide polymorphisms in inferences of population history. Trends in Ecology & Evolution 18: 249–256.

Cáceres, C. E., 1997. Temporal variation, dormancy, and coexistence: a field test of the storage effect. Proceedings of the National Academy of Sciences of the United States of America 94: 9171–9175.

Carvalho, G. R., 1998. Advances in Molecular Ecology. IOS Press, Amsterdam, 313 pp.

Ciros-Pérez, J., M. J. Carmona & M. Serra, 2001. Resource competition between sympatric sibling rotifer species. Limnology and Oceanography 46: 1511–1523.

Ciros-Pérez, J., A. Gómez & M. Serra, 2001. On the taxonomy of three sympatric sibling species of the *Brachionus plicatilis* (Rotifera) complex from Spain, with the description of *B. ibericus* n. sp. Journal of Plankton Research 23: 1311–1328.

Cousyn, C., L. Meester, J. K. Colbourne, L. Brendonck, D. Verschuren & F. Volckaert, 2001. Rapid, local adaptation of zooplankton behavior to changes in predation pressure in the absence of neutral genetic changes. Proceedings of the National Academy of Sciences of the United States of America 98: 6256–6260.

De Meester, L., 1996. Local genetic differentiation and adaptation in freshwater zooplankton: patterns and processes. Ecoscience 3: 385–399.

De Meester, L., A. Gómez, B. Okamura & K. Schwenk, 2002. The Monopolization hypothesis and the dispersal-gene flow paradox in aquatic organisms. Acta Oecologica 23: 121–135.

Derry, A. M., P. D. N. Hebert & E. E. Prepas, 2003. Evolution of rotifers in saline and subsaline lakes: A molecular phylogenetic approach. Limnology and Oceanography 48: 675–685.

Dumont, H. J., 1983. Biogeography of rotifers. Hydrobiologia 104: 19–30.

Ebert, D., C. Haag, M. Kirkpatrick, M. Riek, J. W. Hottinger & V. I. Pajunen, 2002. A selective advantage to immigrant genes in a *Daphnia* metapopulation. Science 295: 485–488.

Folmer, O., M. Black, W. Hoeh, R. Lutz & R. Vrijenhoek, 1994. DNA primers for amplification of mitochondrial cytochrome *c* oxidase subunit I from diverse metazoan invertebrates. Molecular Marine Biology and Biotechnology 3: 294–299.

Fussmann, G., S. Ellner & N. De Stasio Jr, 2003. Microevolution as a critical component of plankton dynamics. Proceedings of the Royal Society of London Series B-Biological Sciences 270: 1015–1022.

Gilbert, J. J., 2001. Spine development in *Brachionus quadridentatus* from an Australian billabong: genetic variation and induction by *Asplanchna*. Hydrobiologia 446: 19–28.

Gilbert, J. J. and E. J. Walsh, 2005. *Brachionus calyciflorus* is a species complex: Mating behaviour and genetic differentation among four geographically isolated strains. Hydrobiologia 546: 257–265.

Goldstein, D. B. & C. Schlotterer, 1999. Microsatellites: Evolution and Applications. Oxford University Press, Oxford, 352 pp.

Gómez, A. & M. Serra, 1995. Behavioral reproductive isolation among sympatric strains of *Brachionus plicatilis* Muller 1786: Insights into the status of this taxonomic species. Hydrobiologia 313: 111–119.

Gómez, A., M. Temprano & M. Serra, 1995. Ecological genetics of a cyclical parthenogen in temporary habitats. Journal of Evolutionary Biology 8: 601–622.

Gómez, A. & T. W. Snell, 1996. Sibling species and cryptic speciation in the *Brachionus plicatilis* species complex (Rotifera). Journal of Evolutionary Biology 9: 953–964.

Gómez, A., M. J. Carmona & M. Serra, 1997. Ecological factors affecting gene flow in the *Brachionus plicatilis* complex (Rotifera). Oecologia 111: 350–356.

Gómez, A., 1998. Allozyme electrophoresis: its application to rotifers. Hydrobiologia 387/388: 385–393.

Gómez, A., C. Clabby & G. R. Carvalho, 1998. Isolation and characterization of microsatellite loci in a cyclically parthenogenetic rotifer, *Brachionus plicatilis*. Molecular Ecology 7: 1619–1621.

Gómez, A. & G. R. Carvalho, 2000. Sex, parthenogenesis and genetic structure of rotifers: microsatellite analysis of contemporary and resting egg bank populations. Molecular Ecology 9: 203–214.

Gómez, A., G. R. Carvalho & D. H. Lunt, 2000. Phylogeography and regional endemism of a passively dispersing zooplankter: mitochondrial DNA variation in rotifer resting egg banks. Proceedings of the Royal Society of London Series B-Biological Sciences 267: 2189–2197.

Gómez, A., G. J. Adcock, D. H. Lunt & G. R. Carvalho, 2002a. The interplay between colonisation history and gene flow in passively dispersing zooplankton: microsatellite analysis of rotifer resting egg banks. Journal of Evolutionary Biology 15: 158–171.

Gómez, A., M. Serra, G. R. Carvalho & D. H. Lunt, 2002b. Speciation in ancient cryptic species complexes: evidence from the molecular phylogeny of *Brachionus plicatilis* (Rotifera). Evolution 56: 1431–1444.

Haag, C. R., J. W. Hottinger, M. Riek & D. Ebert, 2002. Strong inbreeding depression in a *Daphnia* metapopulation. Evolution 56: 518–526.

Hebert, P. D. N., A. Cywinska, S. L. Ball & J. R. DeWaard, 2003. Biological identifications through DNA barcodes. Proceedings of the Royal Society of London Series B-Biological Sciences 270: 313–321.

Hofmann, W., 1983. On temporal variation in the rotifer *Keratella cochlearis* (Gosse) – the question of Lauterborn-cycles. Hydrobiologia 101: 247–254.

Jarne, P. & P. J. L. Lagoda, 1996. Microsatellites, from molecules to populations and back. Trends in Ecology & Evolution 11: 424–429.

Jenkins, D. G. & A. L. Buikema Jr., 1998. Do similar communities develop in similar sites? a test with zooplankton structure and function. Ecological Monographs 68: 421–443.

King, C. E., 1977. Genetics of reproduction, variation and adaptation in rotifers. Archiv für Hydrobiologie. Ergebnisse der Limnologie 8: 187–201.

King, C. E., 1980. The genetic structure of zooplankton communities. In Kerfoot, W. C. (ed.), Evolution and Ecology of Zooplankton Communities. The University Press of New England, Hanover (N.H.), 315–328.

Knowlton, N., 2000. Molecular genetic analyses of species boundaries in the sea. Hydrobiologia 420: 73–90.

Leutbecher, C., 2000. A routine method of DNA-extraction from extremely small metazoans, e.g. single rotifer specimens for RAPD-PCR analyses. Hydrobiologia 437: 133–137.

Li, Y. C., A. B. Korol, T. Fahima, A. Beiles & E. Nevo, 2002. Microsatellites: genomic distribution, putative functions and mutational mechanisms: a review. Molecular Ecology 11: 2453–2465.

Lynch, M., M. Pfrender, K. Spitze, N. Lehman, J. Hicks, D. Allen, L. Latta, M. Ottene, F. Bogue & J. Colbourne, 1999. The quantitative and molecular genetic architecture of a subdivided species. Evolution 53: 100–110.

Mark Welch, D. B. & M. Meselson, 2000. Evidence for the evolution of bdelloid rotifers without sexual reproduction or genetic exchange. Science 288: 1211–1215.

Morgan, K. K., J. Hicks, K. Spitze, L. Latta, M. E. Pfrender, C. S. Weaver, M. Ottone & M. Lynch, 2001. Patterns of

genetic architecture for life-history traits and molecular markers in a subdivided species. Evolution 55: 1753–1761.
Ortells, R., T. W. Snell, A. Gómez & M. Serra, 2000. Patterns of genetic differentiation in resting egg banks of a rotifer species complex in Spain. Archiv für Hydrobiologie 149: 529–551.
Ortells, R., 2002. Diversidad Genética y Ecológica en Especies Crípticas de Rotíferos. Patrones y Procesos, Universitat de Valencia, Valencia, Spain, 193 pp.
Ortells, R., A. Gómez & M. Serra, 2003. Coexistence of rotifer cryptic species: ecological and genetic characterisation of *Brachionus plicatilis*. Freshwater Biology 48: 2194–2202.
Palumbi, S. R., 1996. The Polymerase Chain Reaction. In Hillis, D. M. & C. B. K. Moritz Marble (eds), Molecular Systematics. Sinauer, Sunderland, MA., 205–247.
Pfrender, M. E., K. Spitze & N. Lehman, 2000. Multi-locus genetic evidence for rapid ecologically based speciation in *Daphnia*. Molecular Ecology 9: 1717–1735.
Primmer, C. R., A. P. Moller & H. Ellegren, 1996. A wide-range survey of cros-species microsatellite amplification in birds. Molecular Ecology 5: 365–378.
Primmer, C. R. & J. Merila, 2002. A low rate of microsatellite amplification success in Ranid frogs. Conservation Genetics 3: 445–449.
Rahel, F. J., 2002. Homogenisation of freshwater faunas. Annual Review of Ecology and Systematics 33: 291–315.
Segers, H., 1996. Biogeography of littoral *Lecane* Rotifera. Hydrobiologia 323: 169–197.
Segers, H., 2002. The nomenclature of the Rotifera: annotated checklist of valid family- and genus-group names. Journal of Natural History 36: 631–640.
Serra, M., M. J. Carmona & M. R. Miracle, 1994. Survival analysis of 3 clones of *Brachionus plicatilis* (Rotifera). Hydrobiologia 277: 97–105.
Serra, M., A. Galiana & A. Gómez, 1997. Speciation in monogonont rotifers. Hydrobiologia 358: 63–70.
Serra, M., A. Gómez & M. J. Carmona, 1998. Ecological genetics of *Brachionus* sympatric sibling species. Hydrobiologia 388: 373–384.
Shurin, J. B., 2000. Dispersal limitation, invasion resistance and the structure of pond zooplankton populations. Ecology 81: 3074–3086.
Snell, T. W. & K. Carrillo, 1984. Body size variation among strains of the rotifer *Brachionus plicatilis*. Aquaculture 37: 359–367.
Stenberg, P., M. Lundmark & A. Saura, 2003. MLGsim: a program for detecting clones using a simulation approach. Molecular Ecology Notes 3: 329–331.
Sunnucks, P., A. C. C. Wilson, L. B. Behegaray, L. B. Zenger, J. French & A. C. Taylor, 2000. SSCP is not that difficult: the application and utility of single-stranded conformation polymorphism in evolutionary biology and molecular ecology. Molecular Ecology 9: 1699–1710.
Tautz, D., P. Arctander, A. Minelli, R. H. Thomas & A. P. Vogler, 2003. A plea for DNA taxonomy. Trends in Ecology & Evolution 18: 70–74.
Taylor, D. J., P. D. N. Hebert & J. K. Colbourne, 1996. Phylogenetics and evolution of the *Daphnia longispina* group (Crustacea) based on 12S rDNA sequence and allozyme variation. Molecular Phylogenetics and Evolution 5: 495–510.
Templeton, A. R., 1998. Nested clade analyses of phylogeographic data: testing hypotheses about gene flow and population history. Molecular Ecology 7: 381–397.
Vos, P., R. Hogers, M. Bleeker, M. Reijans, T. Lee, M. Hornes, A. Fritjers, J. Pot, J. Peleman, M. Kuiper & M. Zabeau, 1995. AFLP: a new technique for DNA fingerprinting. Nucleic Acids Research 23: 4407–4414.
Walsh, P. S., D. A. Metzger & R. Higuchi, 1991. Chelex-100 as a Medium for Simple Extraction of DNA for Pcr- Based Typing from Forensic Material. Biotechniques 10: 506–513.
Wayne, R. K., J. A. Leonard & A. Cooper, 1999. Full of sound and fury: the recent history of ancient DNA. Annual Review of Ecology and Systematics 30: 457–477.
Weider, L. J., A. Hobaek, T. J. Crease & H. Stibor, 1996. Molecular characterization of clonal population structure and biogeography of arctic apomictic *Daphnia* from Greenland and Iceland. Molecular Ecology 5: 107–118.
Weider, L. J. & A. Hobaek, 1997. Postglacial dispersal, glacial refugia, and clonal structure in Russian/Siberian populations of the Arctic *Daphnia pulex* complex. Heredity 78: 363–372.
Zhao, Y. & C. E. King, 1989. Ecological genetics of the rotifer *Brachionus plicatilis* in Soda Lake, Nevada, USA. Hydrobiologia 185: 175–181.

Hydrobiologia (2005) 546:101–108
A. Herzig, R.D. Gulati, C.D. Jersabek & L. May (eds.) Rotifera X: Rotifer Research: Trends, New Tools and Recent Advances
DOI 10.1007/s10750-005-4105-6

The potential of genomic approaches to rotifer ecology

David B. Mark Welch* & Jessica L. Mark Welch
Josephine Bay Paul Center for Comparative Molecular Biology and Evolution, Marine Biological Laboratory, Woods Hole, MA, 02543, USA
*(*Author for correspondence: E-mail: dmarkwelch@mbl.edu)*

Key words: biological diversity, ecological genomics, genetic variation

Abstract

Rotifers are a key component of many freshwater ecosystems, but surveys of the composition of rotifer communities are limited by the labor-intensiveness of sample processing, particularly of non-planktonic taxa, and by the shortage of investigators qualified to identify a broad range of rotifer species. Additional problems are posed by species that must be identified from living specimens, and by members of cryptic species complexes. As DNA sequencing becomes easier and cheaper, it has become practical to obtain representative DNA sequences from identified rotifer species for use in genome-based surveys to determine which rotifers are present in a new sample, avoiding the difficulties of traditional surveys. Here we discuss two genome-based tools used in surveys of microbial communities: serial analysis of gene tags (SAGT) and microarray hybridization. SAGT is a method for inexpensively obtaining characteristic short DNA sequences from a sample that can both identify taxa for which the tag sequence is known and signal the presence of additional uncharacterized species. Microarray hybridization allows detection of DNA sequences in the sample that are identical or similar to sequences present on the microarray. We also report the construction and hybridization of a small microarray of rotifer sequences, demonstrating that this method can discriminate among bdelloid families, and is likely to make much finer discriminations if appropriate sequences are present on the microarray. These techniques are most powerful when combined with traditional systematics in collaborative efforts, which may be fostered through the data base of rotifer biology, WheelBase (http://jbpc.mbl.edu/wheelbase).

Introduction

Rotifers are found in large numbers in a wide variety of environments and play an important role in aquatic food webs (Hutchinson, 1967; Williamson, 1983; Arndt, 1993; Rublee, 1998; Ricci & Balsamo, 2000; Wallace & Ricci, 2002). However, quantitative examination of rotifer abundance and species diversity is hampered by the dearth of investigators qualified to make species-level identification of all but a few taxa. Furthermore, for the many non-planktonic taxa, particularly those that adhere to the substrate, quantitative sampling is extremely laborious (Ricci & Balsamo, 2000; Wallace & Ricci, 2002). Additional difficulties are posed by species that must be identified from living specimens, particularly bdelloids, whose nondescript morphology when fixed has precluded virtually all studies of bdelloids in the environment.

Interpretation of ecological phenomena may also be confounded by the presence of cryptic species. Recent molecular investigations of the *B. plicatilis* species complex show that genetically disparate rotifers can be difficult to discriminate morphologically, and suggest that cryptic speciation may be a common phenomenon among rotifers (Gomez et al., 2002a; Gomez, 2005 this volume Part II).

Many of these same difficulties have presented daunting challenges to microbial ecologists, who have responded with a variety of genome-based methods to study the ecology of Bacteria and Protozoa (see, for example, Liu & Stahl, 2001; Amaral Zettler et al., 2002; Polz et al., 2003; Spear et al., 2003). Two of these techniques, serial analysis of gene tags and microarray hybridization, would seem also to be appropriate to the study of rotifers and other micro-invertebrates.

Serial analysis of gene tags (SAGT)

A number of techniques have been developed to indirectly survey DNA sequence variation in ecological samples, such as rapid amplification of polymorphic DNA (RAPD), single stranded conformational polymorphism (SSCP), denaturing gradient gel electrophoresis (DGGE), and PCR amplification of microsatellites. Although these methods have been used to great benefit in the examination of rotifer populations (Gomez et al., 2002b), they can be difficult to reproduce, lack resolution, and provide a limited amount of information. In contrast, DNA sequences can be determined with high reproducibility and are specific, easily compared across studies, and provide much greater resolution of underlying biological processes. However, while sequencing provides enormous benefits, its cost can still be prohibitive for large ecological surveys. Ironically, sequencing an entire PCR product of ~1000 bp often provides more information than is necessary for many ecological studies. While phylogenetic analyses or tests for natural selection require as much information as possible, and thus long sequences, species identification is generally possible with much shorter sequences. Serial analysis of gene tags (SAGT) is a new sequencing approach that minimizes the cost of using sequencing to identify taxa in an ecological sample by increasing the information about the diversity of a sample that can be obtained from a single sequencing reaction. This is achieved by concatemerizing short stretches of DNA from multiple representatives in a sample and sequencing the concatemer, so that a single sequencing reaction provides information about more than one individual (Kysela et al., 2005).

The basic SAGT procedure is outlined in Figure 1. Short PCR products are generated from genomic regions that have sufficient sequence information to identify taxa of interest. These amplification products are ligated to form concatemers 2–3 kb in length and the concatemers are then cloned and sequenced. In order to maximize the amount of information obtained from each sequencing reaction, most of the PCR primer sequence is removed prior to ligation by digestion with a type IIS restriction enzyme, which recognizes a sequence incorporated into the 5′ end of each primer and cuts ~14 bp upstream of this sequence. This step increases the information in each clone by as much as 30%, and provides cohesive ends for ligation. To maximize the efficiency of ligation, the 5′ end of each primer is biotinylated so that undigested amplicons and 5′ digest fragments can be removed.

The utility of SAGT has been demonstrated in a survey of microbial diversity at the hydrothermically active Guaymas Basin in the Gulf of California (Kysela et al., 2005). Primers spanning a ~67 bp highly variable region of the bacterial SSU rDNA gene were used to amplify DNA extracted from a sediment sample. Twenty-one of the 35 primer bases were removed by digestion with *BsgI*, which recognized a restriction site incorporated into the primers. The resulting ~80 bp fragments were concatemerized, cloned, and sequenced. Four to six fragments were read from each sequencing reaction, and more than 90% of the fragments could be identified based on BLAST scores. The bacterial diversity detected was similar to a previous study (Teske et al., 2002) of the same sample in which each clone sequenced contained a single PCR product from a single organism, but sampled five times as many PCR products per sequencing reaction.

Microarray hybridization

Microarrays are typically used to examine the level of transcription of thousands of genes from a single organism simultaneously, often using competitive hybridization to compare two different metabolic states or tissues (Schena et al., 1996; Bowtell, 1999). However, microarray technology has also been used for the identification of organisms in an ecological sample by hybridizing DNA extracted from the sample to a set of known DNA specimens arrayed on a solid support,

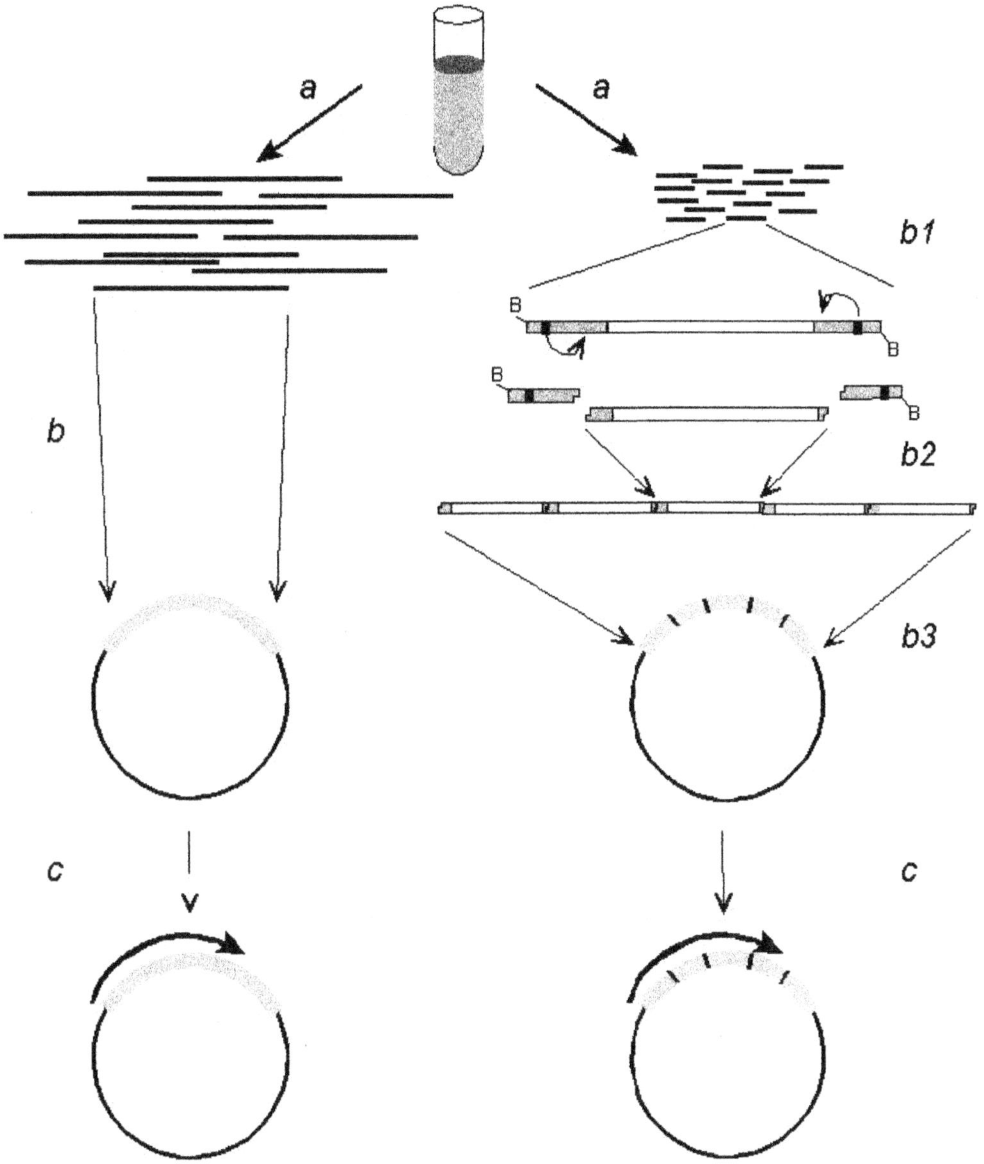

Figure 1. Outline of SAGT. In traditional sequencing surveys (left), DNA extracted from a biological sample is amplified by PCR (a), cloned (b), and sequenced (c). In SAGT (right) shorter amplification products (a) are created; each product consists of primers (grey bars) that are 5′ biotinylated (indicated as B) and contain a recognition sequence for a type IIS endonuclease (small black bars), which digests the primers 10–20 bases 3′ of the recognition sequence to remove most of the primer sequence and create 3′ overhangs (b1). The DNA is passed through a streptavidin column to remove the biotinylated primer fragments and undigested PCR product, and the remaining fragments are ligated together (b2). These concatemers are cloned (b3) and sequenced (c); thus a single sequencing reaction of a single clone yields information about multiple original PCR products.

usually a microscope slide (reviewed in Call et al., 2003). The sample DNA, or more usually a PCR product derived from it, is labeled with a fluorophore so that the specific sites of hybridization on the microarray can be visualized and the intensity of the hybridization signal quantified. The method is conceptually identical to Southern hybridization of a dot blot, but technological innovations allow thousands of known DNA samples to be arrayed on a single microscope slide and many such slides to be produced in a single batch at low cost (Fig. 2).

Microarray hybridization has been used to assess microbial diversity and to assay for specific

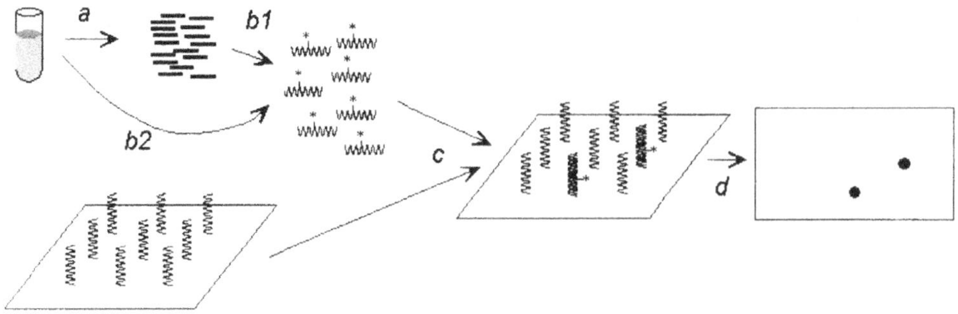

Figure 2. Outline of microarray hybridization. DNA extracted from an ecological sample is PCR amplified (a) and the amplification products labeled with a fluorophore (b1); alternatively, total DNA may be labeled (b2). Labeled DNA is used to probe previously characterized DNA immobilized in spots on a microscope slide (c). Spots of known DNA to which the fluorescently labeled probe DNA has hybridized can then be visualized (d).

pathogenic bacteria in numerous studies (reviewed in Ye et al., 2001; Call et al., 2003; Polz et al., 2003). Here we describe the construction of a pilot microarray of rotifer sequences, and the hybridization of the array to a PCR-generated probe to demonstrate the applicability of this method to the identification of rotifers.

Materials and methods

Construction of a rotifer microarray

The rotifer species represented on the microarray are listed in Table 1. DNA for spotting onto the microarray was generated by polymerase chain reaction (PCR) amplification of previously cloned and characterized fragments of the gene encoding the 82-kD heat shock protein, *hsp82*, as previously described (Mark Welch, 2000; Mark Welch & Meselson, 2000). Sequences, primers, and protocols are available at WheelBase (http://jbpc.mbl.edu/wheelbase). The ~930 bp amplification products were gel-purified and their concentrations estimated by spectrophotometry. Aliquots containing 0.5×10^{-6} and 1.0×10^{-6} g of the amplification products were put into the wells of a microtitre plate, the liquid evaporated under vacuum with moderate heat, and the DNA resuspended at 50 or 100 mg/l in 10 μl of a 50% solution of dimethyl sulfoxide. The DNA solution was spotted onto Corning GAPS II slides using a GeneMachines OmniGrid machine. Slides were UV-crosslinked with 0.3 Joules in a UV Stratalinker 2400 (Stratagene) and were stored over desiccant at room temperature.

Table 1. Taxa, sequences, and accession codes

Rotifer *hsp82* sequences	GenBank accession	% difference from *P. roseola* copy 1
Bdelloidea		
Adineta vaga (Davis, 1873) copy 1	AF143849	13.5
Adineta ricciae Segers & Shiel 2005 copy 1	AY394701	15.2
Habrotrocha constricta (Dujardin, 1841) copy 1	AF143850	15.1
Philodina roseola Ehrenberg, 1832 copy 1	AF143851	–
Philodina roseola Ehrenberg, 1832 copy 2	AF249997	0.6
Philodina roseola Ehrenberg, 1832 copy 3	AF250002	13.1
Philodina roseola Ehrenberg, 1832 copy 4	AF250004	13.4
Monogononta		
Brachionus calyciflorus Pallas, 1766	AF143855	25.0
Eosphora ehrenbergi Weber, 1918	AF143858	22.3
Sinantherina socialis (Linnaeus, 1758)	AF143854	25.0

Microarray hybridization and visualization

Nick-translation was used to incorporate Alexa 647-12-OBEA-dCTP (Molecular Probes) into 2 μg of the purified *P. roseola hsp82*-1 amplification product following standard protocols (Sambrook et al., 1989), using Alexa 647-12-OBEA-dCTP and unlabeled dCTP in a ratio of 1:1. The average length of labeled probe fragments was 200–400 nucleotides. Labeled DNA was separated from unincorporated dNTPs by passage through a G50-150 Sephadex (Sigma) spin column, precipitated with 10×10^{-6} g of blocking DNA (sheared salmon sperm DNA, Eppendorf) and resuspended in 0.2 ml hybridization solution (4 × SSC pH 7.0, 0.1% SDS, 50% deionized formamide, 0.14 g/l sheared salmon sperm DNA and 0.2 g/l tRNA).

Microarrays were denatured in deionized water at 95 °C for 2 min, dehydrated in 95% ethanol for 2 min at room temperature, and dried using compressed air. Microarrays were rehydrated by soaking in prehybridization solution (5 × SSC pH 7, 0.1% SDS, 1% fraction V BSA) for 45–90 min at 42 °C, then were washed in deionized water twice for 4–5 s each, in isopropanol once for 3–4 s, and dried by briefly spinning in a centrifuge (Beckman-Coulter TS-5.1-500 rotor at 45 rcf for 2 min). Labeled DNA was denatured at 95 °C for 3 min, cooled on ice, and 20 μl was applied to each microarray slide. Microarrays and labeled DNA were then covered with Hybri-Slip cover slips (Molecular Probes), placed in hybridization chambers (Corning) and hybridized for 18–23 h at 42 °C.

Microarray slides were washed in 2 × SSC, 0.1% SDS at 42 °C for 5 min, then in 0.1 × SSC, 0.1% SDS once for 10 min at room temperature (low stringency), or four times for 5 min each at 60 °C (high stringency). Slides were then washed three times for 1 min each at room temperature in 0.1 × SSC, immediately dried by briefly spinning in a centrifuge, and stored in a dark box. Hybridization of labeled DNA to the microarray was visualized using a Gene-Pix 4000 B scanner (Axon Instruments) and GenePix Pro 4.0 software.

Results and discussion

We printed a microarray of *hsp82* sequences from diverse bdelloid and monogonont rotifers onto each of twelve slides. Two slides were hybridized with fluorescently labeled DNA from *P. roseola hsp82* copy 1 (Pr1; see Mark Welch & Meselson, 2000); one was washed under low stringency conditions which allow some degree of hybridization between mismatched sequences to remain; the other was washed at a higher stringency which should denature all but the best-paired DNA duplexes. Scanned images of the slides are shown in Figure 3. All sequences other than Pr2 differ from Pr1 by >13%, and under the hybridization conditions used here Pr1 did not hybridize to them at significant levels (1–10% of the hybridization of Pr1 to itself). Pr1 and Pr2 differ by 0.6%, and the average hybridization of Pr1 to Pr2, under conditions of low and high stringency, respectively, was 92 and 78% that of Pr1 to itself.

These results demonstrate that hybridization of an exact match between labeled DNA and DNA on the microarray is measurably more intense than hybridization of two sequences that differ by as little as 0.6%. In the case of *hsp82*, this would allow discrimination within the *Brachionus* plicatilis species complex and within the genus *Adineta*. The use of sequences shown to be evolving more rapidly in rotifers, such as the ITS region or the mitochondrial cytochrome oxidase (COI) gene (E. Walsh, pers. comm.; Birky et al., 2005 this volume Part I), would allow even greater sensitivity. For example, cryptic species of the plicatilis group could easily be discriminated using ITS1 or COI sequences, which differ by 3–20% within sampled members of the complex (Gomez et al., 2002a). The cross-hybridization of related sequences, particularly under lower stringency conditions, may also be useful, as it can be used to detect the presence of sequences in the ecological sample that are not present on the microarray but are closely related to those that are.

Ecological samples will likely contain DNA from a mixture of species, at different concentrations depending on relative species abundance. We mimicked such a sample by mixing Pr1 and *H. constricta* copy 1 PCR products in a ratio of 1:5, then amplifying and labeling the mixture as described above. We performed five hybridizations using decreasing concentrations of labeled DNA (from 200 to 0.1 ng) and high stringency washes, which should have produced conditions in which the labeled DNA was limiting in at least some of

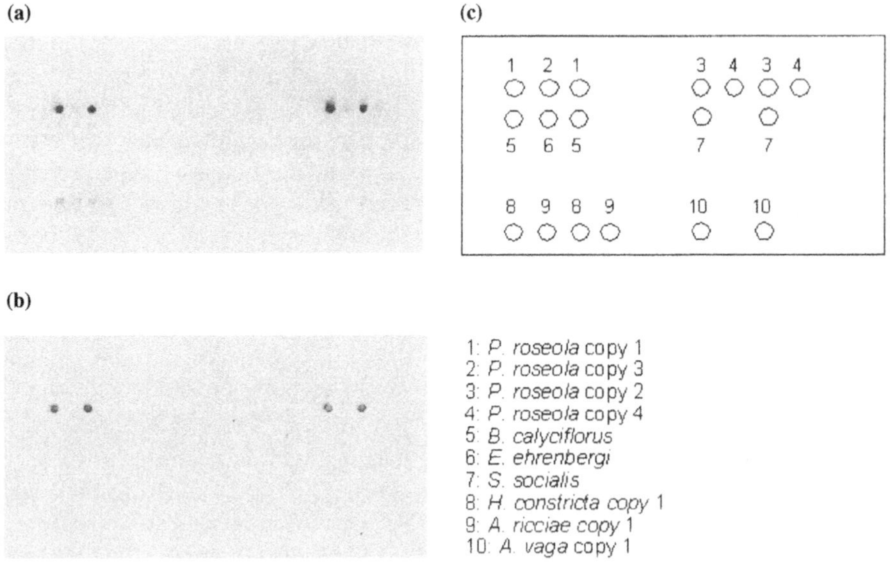

Figure 3. Hybridization of *P. roseola* hsp82-1 to a rotifer microarray. Fluorescence after low stringency washing (a) and after high stringency washing (b). Only portions of the microarray slides are shown. A schematic (c), enlarged and condensed for clarity, shows the location of rotifer DNA on the array, with numbers corresponding to each type of DNA listed below. Bdelloid copy numbers are as in Mark Welch & Meselson (2000).

the hybridizations. While hybridization signal to both Pr and Hc sequences on the microarray decreased with decreasing concentration of labeled DNA, the ratio of the hybridization signal of Pr1 to that of Hc1 never significantly differed from 1:1 (not shown). This suggests that the hybridization reactions in which the labeled DNA was limiting did not reach equilibrium, or that the amplification of the Pr1–Hc1 mixture was significantly biased (Schmalenberger et al., 2001). The use of unamplified total ecological sample DNA in the hybridization reaction avoids this latter problem, and has been used with some success with the smaller, less complex genomes of eubacteria (Small et al., 2001; Wu et al., 2001; Chandler et al., 2003). In preliminary experiments, hybridization of labeled rotifer genomic DNA was detectable only to very long (~40 kb) DNA fragments on the microarray (not shown), which were available only because of the existence of previously isolated cosmid clones (Mark Welch et al., 2004), which is not practical for most rotifer species. However, signal amplification and other improvements in technique may allow for quantitative microarray hybridization with total ecological sample DNA in the future (Belosludtsev et al., 2001; Small et al., 2001; Karsten et al., 2002).

A standard application of microarray hybridization in studies of genomic transcription levels involves competitive hybridization of two differently labeled samples of cDNA to the same microarray slide (Schena et al., 1996; Bowtell,1999). This allows quantitative differentiation between transcription levels in the two samples when the relative difference in abundance is greater than about 2-fold (Quackenbush, 2002). Using the same technique, it should be possible to quantitate 2-fold differences in species abundance between two ecological samples. This would allow rapid and sensitive examination of differences in species abundance in diurnal cycling, for example.

A method that does appear to be reasonably quantitative, and extremely sensitive, is SAGT. Unlike microarray hybridization, SAGT can be implemented by any research group currently using traditional sequencing. While SAGT was developed to allow high-throughput sequencing facilities to generate even more data, the technique may be even more relevant to researchers who are limited in the number of sequencing reactions they are able to run. Currently, SAGT can produce more than ten sequence tags per sequencing reaction (C. Palacios, pers. comm.), providing most of the power of full-length sequencing of ecological samples at a fraction of the cost. The use of primers specific for rotifers or specific sub-taxa would allow rapid quantitative surveys of new localities, or the same locality over time. Among the useful applications of this technique are the analysis of sediment core samples and resting egg banks, and assessing genetic variation within populations or between cryptic species. The positive identification of sequence tags generated from a rotifer SAGT experiment is currently limited by the small amount of sequence information available. However, SAGT surveys of little-known microbial flora have been quite successful using bioinformatic and phylogenetic analyses such as BLAST and maximum likelihood to classify tags into known clades (Kysela et al., 2004). A rotifer-specific example of the success of such a bioinformatically driven approach can be seen in a survey of the 18S sequences obtained from a sample taken from a cryoconite hole in Antarctica (Christner et al., 2003). Christner and colleagues sequenced full-length PCR clones and were able to identify a sequence as likely having come from a Philodina based on BLAST scores alone, despite the fact that only two bdelloid 18S sequences were available in their data base search (the sequence is much more similar to one sequence in the database than the two sequences in the database are to each other).

Both SAGT and microarray hybridization are powerful techniques that can greatly increase the scope and resolution of ecological studies by expanding the range and breadth of traditional ecological sampling. While both approaches are simple enough that they can be pursued by researchers working independently, their real power will come from collaborative efforts between biochemists, ecologists, geneticists, systematists, and others, as well as between groups working on different issues but united in the use of a common study organism. The traditional geniality among those who study rotifer biology makes this field well-suited to benefit from such possibilities.

To this end, detailed protocols for microarray hybridization and a list of specific genes currently available for production of a rotifer chip are available on WheelBase (http://jbpc.mbl.edu/wheelbase), the online data base of rotifer biology. Once DNA is available, a microarray can be produced in large quantities at relatively little cost, and the use of indirect detection with streptavidin or antibodies can reduce the cost of the hybridization reaction (Alexandre et al., 2001). Microarrays may be made for specialized applications, such as discriminating between members of the *plicatilis* group, or for general surveys of rotifers; however, as a single microarray can easily contain a sequence from every known rotifer species (or even multiple sequences from each species), a reasonable goal would be a microarray containing all known rotifer sequences, produced in quantities sufficient for use throughout the community. As new sequences became available, they could be added to new versions of the rotifer microarray chip at regular intervals.

Acknowledgements

We are grateful to Carmen Palacios for discussions of SAGT protocols, to CP, David Kysela, and Mitch Sogin for sharing unpublished SAGT results, and to Steve Biller and Jillian Ward for help in creation of the microarray. The helpful comments of Charles King and two anonymous reviewers greatly improved this manuscript. This work was supported by the United States National Science Foundation.

Note added in proof

Since submission of this manuscript a second protocol for serial analysis of gene tags has been published: Neufeld, J.D., Z. Yu, W. Larn & W.W. Mohn, 2004. Serial analysis of ribosomal sequence tags (SARST): a high-throughput method for profiling complex microbial communities. Environmental Microbiology 6: 131–144.

References

Alexandre, I., S. Hamels, S. Dufour, J. Collet, N. Zammatteo, F. Longueville, J. L. Gala & J. Remacle, 2001. Colorimetric silver detection of DNA microarrays. Analytical Biochemistry 295: 1–8.

Amaral Zettler, L. A., F. Gomez, E. Zettler, B. G. Keenan, R. Amils & M. L. Sogin, 2002. Eukaryotic diversity in Spain's River of Fire. Nature 417: 137.

Arndt, H., 1993. Rotifers as predators on components of the microbial web (bacteria, heterotrophic flagellates, ciliates) – a review. Hydrobiologia 255/256: 231–246.

Belosludtsev, Y., B. Iverson, S. Lemeshko, R. Eggers, R. Wiese, S. Lee, T. Powdrill & M. Hogan, 2001. DNA microarrays based on noncovalent oligonucleotide attachment and hybridization in two dimensions. Analytical Biochemistry 292: 250–256.

Birky, C. W., C. Wolf, H. Maugham, L. Herbertson & E. Henry, 2005. Speciation and selection without sex. Hydrobiologia 546: 29–45.

Bowtell, D. D., 1999. Options available – from start to finish – for obtaining expression data by microarray. Nature Genetics 21: 25–32.

Call, D. R., M. K. Borucki & F. J. Loge, 2003. Detection of bacterial pathogens in environmental samples using DNA microarrays. Journal of Microbiological Methods 53: 235–243.

Chandler, D. P., G. J. Newton, J. A. Small & D. S. Daly, 2003. Sequence versus structure for the direct detection of 16S rRNA on planar oligonucleotide microarrays. Applied and Environmental Microbiology 69: 2950–2958.

Christner, B. C., B. H. Kvitko 2nd & J. N. Reeve, 2003. Molecular identification of Bacteria and Eukarya inhabiting an Antarctic cryoconite hole. Extremophiles 7: 177–83.

Gómez, A., 2005. Molecular ecology of rotifers: from population differentiation to speciation. Hydrobiologia 546: 83–99.

Gomez, A., M. Serra, G. R. Carvalho & D. H. Lunt, 2002a. Speciation in ancient cryptic species complexes: evidence from the molecular phylogeny of *Brachionus plicatilis* (Rotifera). Evolution 56: 1431–1444.

Gomez, A., G. J. Adcock, D. H. Lunt & G. R. Carvalho, 2002b. The interplay between colonization history and gene flow in passively dispersing zooplankton: microsatellite analysis of rotifer resting egg banks. Journal of Evolutionary Biology 15: 158–171.

Hutchinson, G. E., 1967. A Treatise on Limnology, (vol. 2). John Wiley and Sons, New York, 115 pp.

Karsten, S. L., V. M. Deerlin, C. Sabatti, L. H. Gill & D. H. Geschwind, 2002. An evaluation of tyramide signal amplification and archived fixed and frozen tissue in microarray gene expression analysis. Nucleic Acids Research 30: E4 .

Kysela, D. T., C. Palacios & M. L. Sogin, 2005. Serial analysis of V6/ribosomal sequence tags (SARST-V6): A novel method for efficient, high-throughput analysis of microbial community composition. Environmental Microbiology 7: 356–364.

Liu, W-T., & D. A. Stahl, 2001. Molecular approaches for the measurement of density, diversity, and phylogeny. In Manual of Environmental Microbiology, 2nd edn. ASM Press, Washington, DC, 114–134.

Mark Welch, D. B., 2000. Evidence from a protein-coding gene that acanthocephalans are rotifers. Invertebrate Biology 119: 17–26.

Mark Welch, D. B. & M. Meselson, 2000. Evidence for the evolution of bdelloid rotifers without sexual recombination or genetic exchange. Science 288: 1211–1215.

Mark Welch, J. L., D. B. Mark Welch & M. Meselson, 2004. Cytogenetic evidence for asexual evolution of bdelloid rotifers. Proceedings of the National Academy of Sciences (USA) 101: 1618–1621.

Polz, M. F., S. Bertilsson, S. G. Acinas & D. Hunt, 2003. A(r)ray of hope in analysis of the function and diversity of microbial communities. Biological Bulletin 204: 196–199.

Quackenbush, J., 2002. Microarray data normalization and transformation. Nature Genetics 32 (supplement): 496–501.

Ricci, C. & M. Balsamo, 2000. The biology and ecology of lotic rotifers and gastrotrichs. Freshwater Biology 44: 15–28.

Rublee, P. A., 1998. Rotifers in arctic North America with particular reference to their role in microplankton community structure and response to ecosystem perturbations in Alaskan Arctic LTER lakes. Hydrobiologia 387/388: 153–160.

Sambrook, J., E. F. Fritsch & T. Maniatis, 1989. Molecular Cloning: A Laboratory Manual, 2nd edn. Cold Spring Harbor Press, New York.

Schena, M., D. Shalon, R. Heller, A. Chai, P. O. Brown & R. W. Davis, 1996. Parallel human genome analysis: microarray-based expression monitoring of 1000 genes. Proceedings of the National Academy of Sciences USA 93: 10614–10619.

Schmalenberger, A., F. Schwieger & C. C. Tebbe, 2001. Effect of primers hybridizing to different evolutionarily conserved regions of the small-subunit rRNA gene in PCR-based microbial community analyses and genetic profiling. Applied and Environmental Microbiology 617: 3557–3563.

Segers, H. & R. J. Shiel, 2005. Tale of a sleeping beauty: a new and easily cultured organism for experimental studies on bdelloid rotifers. Hydrobiologia 546: 141–145.

Small, J., D. R. Call, F. J. Brockman, T. M. Straub & D. P. Chandler, 2001. Direct detection of 16S rRNA in soil extracts by using oligonucleotide microarrays. Applied and Environmental Microbiology 67: 4708–4716.

Spear, J. R., R. E. Ley, A. B. Berger & N. R. Pace, 2003. Complexity in natural microbial ecosystems: the Guerrero Negro experience. Biological Bulletin 204: 168–173.

Teske, A., K-U. Hinrichs, V. Edgcomb, A. Vera Gomez, D. Kysela, S. P. Sylva, M. L. Sogin & H. W. Jannasch, 2002. Microbial diversity of hydrothermal sediments in the Guaymas basin: evidence for anaerobic methanotrophic communities. Applied and Environmental Microbiology 68: 1994–2007.

Wallace, R. L. & C. Ricci, 2002. Rotifera. In Rundle, S. D., A. L. Robertson & J. M. Schmid-Araya, (eds), Freshwater Meiofauna: Biology and Ecology. Backhuys Publishers, Leiden (The Netherlands), 15–44.

Williamson, C. E., 1983. Invertebrate predation on planktonic rotifers. Hydrobiologia 104: 385–396.

Wu, L., D. K. Thompson, G. Li, R. A. Hurt, J. M. Tiedje & J. Zhou, 2001. Development and evaluation of functional gene arrays for detection of selected genes in the environment. Applied and Environmental Microbiology 67: 5780–5790.

Ye, R. W., T. Wang, L. Bedzyk & K. M. Croker, 2001. Applications of DNA microarrays in microbial systems. Journal of Microbiological Methods 47: 257–272.

Hydrobiologia (2005) 546:109–115
A. Herzig, R.D. Gulati, C.D. Jersabek & L. May (eds.) Rotifera X: Rotifer Research: Trends, New Tools and Recent Advances
DOI 10.1007/s10750-005-4106-5

Using amplified fragment length polymorphisms (AFLP) to study genetic variability in several freshwater rotifer species

Sanjuana Hernández-Delgado[1], Netzahualcoyotl Mayek-Pérez[2], Gustavo Emilio Santos-Medrano[3] & Roberto Rico-Martínez[3,]*
[1]*Instituto Tecnológico Agropecuario de Aguascalientes (ITA 20), Km 18 Carretera Aguascalientes-San Luis Potosí, 20330, El Llano, Aguascalientes, México*
[2]*Centro de Biotecnología Genómica, Instituto Politécnico Nacional. Blvd. Del Maestro s/n esq. Elías Piña, Col. Narciso Mendoza, C.P. 88710, Reynosa, Tamaulipas, México*
[3]*Universidad Autónoma de Aguascalientes, Centro Básico. Departamento de Química, Avenida Universidad 940, Ags. C.P. 20100, Aguascalientes, México*
*(*Author for correspondence: E-mail: rrico@correo.uaa.mx)*

Key words: DNA, genetic diversity, AFLP, phylogeny, Rotifera

Abstract

We have used amplified fragment length polymorphisms (AFLP) to investigate the potential of this technique as a tool to measure genetic variability in eight species of freshwater rotifers: *Brachionus calyciflorus*, *Lecane bulla*, *L. luna*, *L. quadridentata*, *Plationus patulus*, *Philodina acuticornis odiosa*, *Rotaria neptunia*, and *R. rotatoria*. We used nine combinations of oligonucleotides. We observed a total of 806 amplified bands, 798 polymorphic and 8 monomorphic. The data were analyzed using cluster analysis with UPGMA, first within each set of oligonucleotide combination and finally using all nine combinations. Our best dendrogram clearly separated monogononts from digononts, and grouped the species of monogononts in the two genera. However, it grouped *R. neptunia* with *P. acuticornis odiosa* rather than with *R. rotatoria*. These results are discussed in view of recent works in the literature measuring genetic variability and discussing the phylogeny of the Rotifera.

Introduction

The amplified fragment length polymorphism (AFLP) technique is a powerful DNA marker methodology that was originally conceived for the construction of very high density DNA marker maps for application in genome research and positional cloning of genes (Vos & Kuiper, 1998). The AFLP is a random fingerprinting technique that may be applied to DNA of any origin or complexity and differs importantly from other random fingerprinting techniques such as random amplified polymorphic DNA (RAPDs) (Williams et al., 1990) by its robustness and reproducibility (Vos et al., 1995; Vos & Kuiper, 1998).

AFLP is based on the detection of DNA restriction fragments by PCR amplification. Amplification of restriction fragments is accomplished by the ligation of double-stranded adapter sequences to the ends of the restriction sites, which can subsequently serve as 'universal' binding sites for primer annealing in PCR. Restriction fragments of any DNA can be amplified with universal AFLP primers corresponding to the restriction site and adapter sequence. Since in most DNAs the number of fragments that will be simultaneously detected will be too high to be resolved in any fragment analysis system as gels, AFLP primers have at their 3′ end a number of selective bases that extend into the restriction fragments. The AFLP technique uses generic AFLP primers, which consists of two parts; one 'common' part corresponding to the adapter and restriction site sequence and a unique part corresponding to the selective bases. AFLP primers named '+0' when having no selective bases (only the common part), '+1' when having a single nucleotide base, '+2' for having two selective bases, etc. This results in selective amplification of those fragments in which the primer extensions match the nucleotides flanking the restriction site. In simple genomes the number of selective nucleotides required will be low because of the low number of restriction fragments available for amplification; while complex genomes will require more selective nucleotides to reduce the number of amplified fragments to a number suitable for resolution on sequence gels. For example, less of two selective nucleotides have been used in bacteria (Janssen et al., 1996), one or two selective nucleotides in fungi (Gonzalez et al., 1998; Mayek-Pérez et al., 2001), and three or more selective nucleotides in plants (Angiolillo et al., 1999; Eiadthong et al., 2000; Goulao et al., 2001). The number of fragments to be amplified can be tuned by the selection of the number of selective bases in the AFLP primers, and this number will be limited to 50–100 in order to allow an appropriate detection on sequence gels. Restriction fragment patterns generated by the AFLP technique are called AFLP fingerprints and these fingerprints are a rich source for restriction fragment polymorphisms, called AFLP markers. The frequency with which AFLP markers are found is dependent on the sequence polymorphisms between the tested DNA samples (Vos et al., 1995).

Briefly, AFLP consist on three steps (Vos et al., 1995). The first step is restriction of the DNA with two different restriction enzymes, a rare-cutter and a frequent-cutter, and the ligation of double-stranded adapters to the ends of the restriction fragments. The second step is the amplification by PCR of subsets of restriction fragments using selective AFLP primers. The amplification is carried out in two consecutive steps, a first step called pre-amplification and a second step called selective AFLP amplification. Detection of the AFLP fragments is achieved by radioactive (^{33}P) or chemical (digoxigenin) labeling of one from the two AFLP primers used in the selective amplification reactions or by using silver nitrate staining. The final step is the analysis of the fingerprints. Then, the reaction products are separated on denaturing polyacrylamide gels.

Phylogenetic relationships among rotifers have been the subject of several works that have resulted in controversial results, which sometimes contradict the accepted phylogeny (Wallace et al., 1996; see this volume Part I). Garey et al. (1996) analyzed PCR-amplified sequences of the 18S rRNA of the rotifers *Brachionus plicatilis* and *Philodina acuticornis* with three species of acanthocephalans, and at least one member of each of the major phyla of animals, and concluded that the Acanthocephala is a taxon within the Rotifera. More recently, Herlyn et al. (2003) using 18S rDNA proposed a classification that included the Eurotatoria (Monogononta + Bdelloidea) as part of a phylum called Syndermata, suggesting a monophyletic origin for the Eurotatoria and that *Seison* is an acanthocephalan sister group. However, Miquelis et al. (2000) analyzing the helix E23 of the 18S rDNA found 'vastly different' 18S E23 rDNA sequences among Ploimida species. Sørensen (2002) using cladistic analysis that included morphological characters (mainly trophy morphology) confirmed the monophyly of Eurotatoria. Later, Sørensen (2003) by a scanning electron microscopy study of the trophi concluded that Micrognathozoa is more closely related to Rotifera than to Gnathostomulida (see Funch et al., this volume, Part I).

Wallace et al. (1996) analyzed the possible relationships among pseudocelomates (including rotifers) to construct a phylogenetic tree congruent with those generated by molecular analysis (mainly of one molecule the 18S rDNA). They concluded that pseudocoelomates are probably polyphyletic. Aguinaldo et al. (1997) analyzed sequences of 18S rDNA of 22 taxa and grouped arthropods next to tardigrades and nematodes, and also grouped rotifers (using only one rotifer species *B. plicatilis*) in a sister clade with oligochaetes, brachiopods, polychaetes, and molluscs. Winnepenninckx et al. (1998) analyzing 18S rDNA and using only one rotifer species (*P. acuticornis*) grouped the recently described phylum

Cycliophora in a lophophorate-aschelminth-protostome clade and a sister relationship with a Rotifera-Acanthocephala clade.

Wallace & Colburn (1989) used computer-generated cladograms to study the troubled relationship among taxa within the genus *Notholca.* They argued that in general distinctions among Rotifera taxa are clear, but there are particular cases where cladistics can be used to resolve controversies. Fu et al. (1991) used allozyme variation to separate 67 strains of *B. plicatilis* 'L' and 'S' types. Their dendrogram clearly separated the strains in two groups corresponding to these two types. Later these types where described as different species (Segers, 1995). More recently, Gómez et al. (2002) using nuclear (ribosomal internal transcribed spacer 1) and mitochondrial (cytochrome *c* oxidase subunit 1) sequences concluded that the *B. plicatilis* complex worldwide is in fact a combination of nine specific lineages.

Derry et al. (2003) working with two mitochondrial gene sequences (16S rDNA and cytochrome oxidase I) found that spined and unspined forms of *Keratella cochlearis* are so different as to be considered different species. However, they also concluded that the morphological variation of *Keratella hiemalis* was environmentally induced.

Few works have been performed where the combination of molecular techniques and cladistics is employed to study phylogenetic relationships within the Rotifera, employing several rotifer species at once. Therefore, the aim of the present contribution was to assess the use of the AFLP technique as a potential tool for phylogenetic analysis within the Rotifera.

Materials and methods

Collection of rotifer material

Lecane bulla, *Plationus patulus*, *Rotaria neptunia*, and *Rotaria rotatoria* were collected in a pond at the Water Treatment Plant of the Universidad Autónoma de Aguascalientes. The approximate geographical coordinates of this sampling site are 21° 53′ 10″ N and 102° 28′ 54″ W (geopositioner GPS 4000 XL Satellite Navigator, Magellan Inc., 1997). *Brachionus calyciflorus* Gainesville strain was originally collected in Gainesville, Florida, USA (Rico-Martínez & Snell, 1997), and cysts were recently hatched in the lab with white fluorescent lamps at 20 °C. *Lecane luna* (O. F. Müller, 1776) was collected at Los Arquitos dam, in Aguascalientes, México. *Lecane quadridentata* (Ehrenberg, 1832) was collected at Lake Chapala (Mexico's biggest natural lake). For location of these two reservoirs see Rico-Martínez & Silva-Briano (1993). *P. acuticornis odiosa* was collected at the Moctezuma River near the city of Tamazunchale, S. L. P., Mexico. The approximate geographical coordinates of this sampling location are 21° 15′ 06″ N, 98° 50′ 53″ W. All species were cultured in EPA medium (192 mg $NaHCO_3$, 120 mg $CaSO_4 \cdot 2H_2O$, 246 mg $MgSO_4 \cdot 7H_2O$, and 8 mg KCl in 2 l) prepared with deionized water (16–18 megaohms) from a Water Pro PS deionizer (Labconco Co.), and fed the green algae *Nannochloropsis oculata* (strain UTEX LB2164) grown in Bold's Basal Medium (Nichols, 1973). At least 50 females were collected from clonal cultures of each species and place in a 1.5 ml Eppendorf tube with EPA medium to start the DNA extraction.

AFLP analysis

Total genomic DNAs were extracted from 50–200 individuals from each rotifer species by the method of Walsh & Starkweather (1993). DNA concentrations were visually estimated on agarose gels by comparison with standard λ phage digested by *Hin*dIII. The AFLP analysis was performed following the protocol described by Vos et al. (1995). Briefly, 150 ng genomic DNA was digested with 25 units of *Eco*RI and 25 units of *Tru*91 at 37 °C for 4 h and incubated at 70 °C for 15 min. The DNA fragments were ligated to *Eco*RI and *Mse*I adapters at 15 °C over night. After pre-selective amplification by PCR using the nucleotide A, a second selective amplification by PCR was performed with nine *Eco*RI/*Mse*I (three +2/+2, three +1/+1, and three +1/+2; see Table 1) primer combinations. AFLP reactions were denatured by boiling with formamide buffer (98% formamide, 10 mM EDTA, bromophenol blue, xylene cyanol). All samples were electrophoresed on 6% denaturing polyacrylamide gels (35 × 45 cm) for 3 h at 2000 V and then revealed by using the manufacturer's instructions in the Silver Sequence Staining Reagents kit (Promega^R) manual.

Table 1. Amplified Fragment Length Polymorphisms found in the eight rotifer species investigated

AFLP primer combination	Amplified products		
	Monomorphic	Polymorphic	Total
+1/+1			
A/A	0	115	115
A/C	3	122	125
G/A	1	82	83
+1/+2			
A/AC	1	119	120
A/AG	2	118	120
A/AT	1	106	107
+2/+2			
AG/AA	0	46	46
AC/AA	0	38	38
AC/AC	0	52	52
Total	8	798	806

Data analysis

A binary matrix reflecting the presence (1) or absence (0) of each AFLP band was generated for each species. The genetic distance among rotifer species was estimated using the simple matching coefficient (Skroch et al., 1992). The distance matrix generated was used to produce a dendrogram by the UPGMA method using software Statistica ver. 5.0.

Results

The bands generated by the AFLP technique are shown in Figure 1. The analysis of the bands generated with the different AFLP primer combinations is shown in Table 1. With this data, we generated four different dendrograms (three with three combinations, and one with nine combinations) that grouped the taxa in very different ways. The first dendrogram included data from 136 bands (data not shown) and was clearly inefficient separating the different taxa of rotifers in a conventional way (putting together members of the same genus or at the very least separating Bdelloidea from the Monogononta). The only correct inference obtained from this dendrogram was to

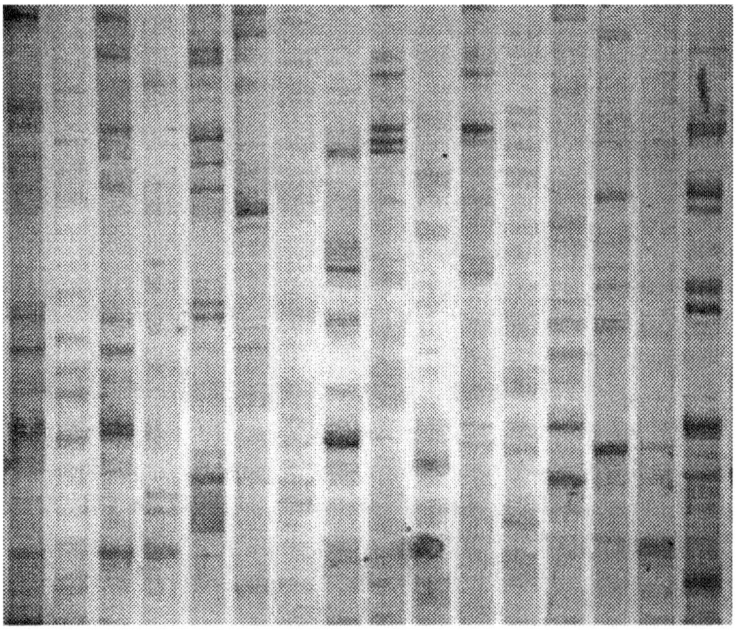

1 2 3 4 5 6 7 8 1 2 3 4 5 6 7 8

Figure 1. Acrylamide gel showing the results of the AFLP technique with eight freshwater rotifer species. Numbers below the bands correspond to: (1) *B. calyciflorus*. (2) *P. patulus*, (3) *L. bulla*, (4) *L. luna*, (5) *L. quadridentata*, (6) *P. acuticornis odiosa*, (7) *R. neptunia*, (8) *R. rotatoria*. There are two lines for each species. Each line represents a primer combination. Not all 806 bands are shown in the figure.

place in the same clade the two species of *Brachionus* used in the analysis. The second dendrogram (data not shown), included 323 bands and was as inefficient as the first one. The third dendrogram (data not shown), included 347 bands and also failed to separate both classes of rotifers. Finally, the last dendogram (Fig. 2) that included all the bands obtained (806) and 9 combinations represented, as expected, a better alternative. This dendrogram grouped the *Brachionus* and *Plationus* species together, they belong to the same family and some authors (Koste et al. 1993; Sarma 2003, pers. comm.) consider that *Plationus* is not really a valid genus name and that the two species of the genus should be considered species of *Brachionus*. It also grouped the *Lecane* species together and clearly separated monogononts from bdelloids. The only controversy with the accepted taxonomic status of the species investigated occurred within the Bdelloidea. The dendrogram grouped together *P. acuticornis odiosa* with *R. neptunia* rather than grouping together the two *Rotaria* species. Although the difference is of only two bands and may even represent a polytomy (see Birky Jr., 1996).

Discussion

The AFLP technique resulted in a promising tool that may allow rotifer taxonomists tackle important issues related to phylogenetic controversies. For the first time AFLP has been used in rotifers to analyze phylogenetic relationship between rotifer clades, and in general worked well. However, caution must be employed in using a small number of bands and/or combinations to construct the dendograms. Also, the selection of the species to be analyzed is an important process that should be carefully planned. In this particular exercise we used species available in the laboratory, that could be cultured with ease and that belonged to the two classes of rotifers (care was taken to include members of the same genus). However, in those cases where this technique could be useful, for example to study relationships within the genus *Lecane* (with more than 100 species) or *Trichocerca*, then the selection of the species and the culture to obtain at least one microgram of DNA could be a difficult process that can consume several years.

In our best dendrogram employing the data collected of all 806 bands the AFLP analysis clearly grouped the *Lecane* species together, but also grouped *L. bulla* next to *L. quadridentata* (with 8% of genetic dissimilarity), in a clade different of *L. luna*. This grouping supports the view that these species are closer since its foot is only slightly separated, a criterion that Koste (1978) considered to establish the subgenus *Monostyla*. *L. luna* is a species with the foot completely separated, a criterion that Koste (1978) used to

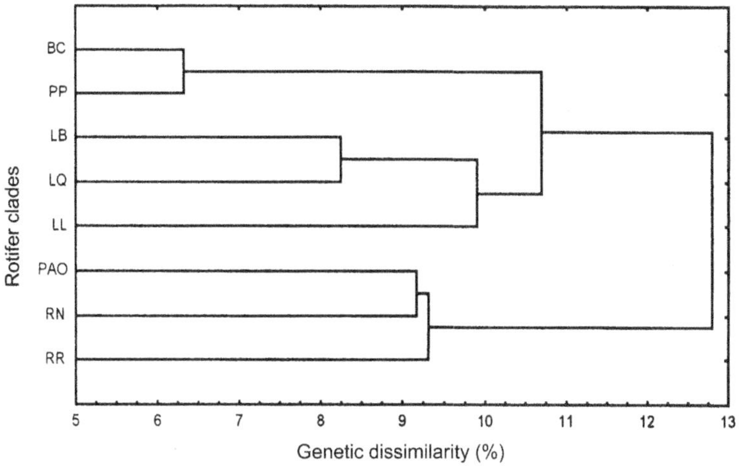

Figure 2. Dendrogram showing the relationship among the eight species of rotifers based on nine combinations of AFLP +2/+2, +1/+1, and +2/+1 (806 bands). BC – *B. calyciflorus*, PP – *P. patulus*, LB – *L. bulla*, LQ – *L. quadridentata*, LL – *L. luna*, PAO – *P. acuticornis odiosa*, RN – *R. neptunia*, RR – *R. rotatoria*.

establish the subgenus *Lecane* sensu strictu. Whether or not the genetic variability is directly associated with this phenotypic characteristic is a matter to be decided with further analysis that should include many more *Lecane* species. The AFLP technique will work nicely for such analysis.

What is the meaning of the *R. neptunia*–*P. acuticornis odiosa* pairing rather than the *R. neptunia*–*R. rotatoria* pairing that we might have expected? It is important to note for this case that the difference between *R. rotatoria* and the other two bdelloid species is of less than 0.25% in genetic dissimilarity (see Fig. 2). This means that out of the 806 bands analyzed there were 2 bands different between *R. rotatoria* and the other two bdelloids species. Perhaps, this small difference may also be a result of the lack of recombination due to sexual reproduction absent in bdelloids for millions of years (Welch & Meselson, 2000), although asexual reproduction does not slow evolution (C. William Birky Jr., 2003, pers. com.). Another possible explanation is that since most AFLP bands come from nuclear chromosomes, then asexual organisms might have ploidy cycles in which one chromosome of a pair is lost and the other duplicated. If the chromosomes were heteromorphic, as it is likely, then one can lose phylogenetic information (Birky Jr., 1996). Even so, the AFLP technique clearly separated bdelloids from monogononts with more than 4% difference (see Fig. 2).

The use of molecular analysis to construct phylogenetic trees among Rotifera and other taxa is highly recommended as a way to check the accepted taxonomic status. In our analysis we were able to obtain very different trees with phylogenetic controversies arising due to the lack of information. However in our best tree with more information (806 bands) we were able to obtain a dendrogram that in general agrees with the recent phylogenetic status of the species employed. However, many recent articles in the literature (Garey et al., 1996; Aguinaldo et al., 1997; Winnepenninckx et al., 1998; Miquelis et al., 2000; Herlyn et al., 2003) are based in comparisons of just one molecule (the 18S rDNA) and usually only one member of each taxa (at least in the case of rotifers). The conclusions obtained by these authors, although important, should be analyzed with extreme care, and perhaps be confirmed or denied by future works that ideally should combine the analysis of molecular and phenotypic traits. In that sense, the AFLP technique may represent an useful tool for such works.

Acknowledgements

We thank Aydin Örstan for identifying *Philodina acuticornis odiosa* for us.

References

Aguinaldo, A. M. A., J. M. Turbeville, L. S. Linford, M. C. Rivera, J. R. Garey, R. A. Raff & J. A. Lake, 1997. Evidence for a clade of nematodes, arthropods, and other moulting animals. Nature 387: 489–493.

Angiolillo, A., M. Mencuccini & L. Baldón, 1999. Olive genetic diversity assessed using amplified fragment length polymorphisms. Theoretical Applied Genetics 98: 411–421.

Birky Jr., C. W., 1996. Heterozygosity, heteromorphy, and phylogenetic trees in asexual eukaryotes. Genetics 144: 427–437.

Derry, A. M., P. D. N. Hebert & E. E. Prepas, 2003. Evolution of rotifers in saline and subsaline lakes: A molecular phylogenetic approach. Limnology and Oceanography 48: 675–685.

Eiadthong, W., K. Yonemori, S. Kanzaki, A. Sugiura N. Utsunomiya & S. Subhandrabandhu, 2000. Amplified fragment length polymorphism analysis for studying genetic relationships among *Mangifera* species in Thailand. Journal of the American Society of Horticultural Science 125: 160–164.

Fu, Y., K. Hirayama & Y. Natsukari, 1991. Genetic divergence between S and L type strains of the rotifer *Brachionus plicatilis* O F. Muller. Journal of Experimental Marine Biology and Ecology 151: 43–56.

Garey, J. R., T. J. Near, M. R. Nonnemacher & S. A. Nadler, 1996. Molecular evidence for Acanthocephala as a subtaxon of Rotifera. Journal of Molecular Evolution 43: 287–292.

Gómez, A., M. Serra, G. R. Carvalho & D. H. Luna, 2002. Speciation in ancient cryptic species complexes: evidence from the molecular phylogeny of *Brachionus plicatilis* (Rotifera). Evolution 56: 1431–1444.

González, M., R. Rodríguez, M. E. Zavala, J. L. Jacobo, F. Hernández, J. Acosta, O. Martínez & J. Simpson, 1998. Characterization of Mexican isolates of *Colletotrichum lindemuthianum* by using differential cultivars and molecular markers. Phytopathology 88: 292–299.

Goulao, L., L. Cabrita, C. M. Oliveira & J. M. Leitao, 2001. Comparing RAPD and AFLP analysis in discrimination and estimation of genetic similarities among apple (*Malus domestica* Borkh.) cultivars. Euphytica 119: 259–270.

Herlyn, H., O. Piskurek, J. Schmitz, U. Ehlers & H. Zischler, 2003. The syndermatan phylogeny and the evolution of acanthocephalan endoparasitim as inferred from 18S rDNA sequences. Molecular Phylogenetics and Evolution 26: 155–164.

Janssen, P., R. Coopman, G. Huys, J. Swings, M. Bleeker, P. Vos, M. Zabeau & K. Keraters, 1996. Evaluation of the DNA fingerprinting as a new tool in bacterial taxonomy. Microbiology 142: 1881–1893.

Koste, W., 1978. Rotatoria. Die Rädertiere Mitteleuropas. 2 vols, Gebrüder Borntraeger, Berlin, Stuttgart, Germany, Textband 673 pp., Tafelband 234 Tafelin.

Koste, W., W. Jametzky & E. Vareschi, 1993. Zur Kenntnis der limnischen Rotatorienfauna Jamaikas (Rotatoria: Aschelminthes). Teil I. Osnabrücker Naturwissenschaftliche Mitteilungen 19: 103–149.

Mark Welch, D. M. & M. Meselson, 2000. Evidence for the evolution of bdelloid rotifers without sexual reproduction or genetic exchange. Science 288: 1211–1215.

Mayek-Pérez, N., C. López-Castañeda, M. González-Chavira, R. García-Espinosa, J. A. Acosta-Gallegos, O. Martínez-De la Vega & J. Simpson, 2001. Variability of Mexican isolates of *Macrophomina phaseolina* based on pathogenesis and AFLP genotype. Physiology and Molecular Plant Pathology 59: 257–264.

Miquelis, A., J. -F. Martin, E. W. Carson, G. Brun & A. Gilles., 2000. Performance of 18S rDNA helix E23 for phylogenetic relationship within and between the Rotifera-Acanthocephala clades. C.R. Academia Scientifica Paris, Scienes de la vie/Life Sciences 323: 925–941.

Nichols, H. W., 1973. Growth media - freshwater. In Stein, J. R. (ed.), Handbook of Phycological Methods. Cambridge University Press, 7–24.

Rico-Martínez, R. & M. Silva-Briano, 1993. Contribution to the knowledge of the Rotifera of Mexico. Hydrobiologia 255/256: 467–474.

Rico-Martínez, R. & T. W. Snell, 1997. Mating behavior in eight rotifer species: Using cross-mating tests to study species boundaries. Hydrobiologia 356: 165–173.

Segers, H., 1995. Guides to the Identification of the Microinvertebrates of the Continental Waters of the World 6: Rotifera, Volume 2: The Lecanidae (Monogonta). SPB Academic Publishing, The Hague, 226 pp.

Skroch, P., J. Tivag, J. Nienhaus, 1992. Analysis of genetic relationships using RAPD marker data. In: Applications of RAPD technology to Plant Breeding. The American Society for Horticultural Science and American Genetic Association. Joint Plant Breeding Symposia Series. Crop Science Society of America. Madison, WI, USA, 26–32.

Sørensen, M. V., 2002. On the evolution and morphology of the rotiferan trophy, with a cladistic analysis of Rotifera. Journal of Zoological Systematics and Evolutionary Research 40: 129–154.

Sørensen, M. V., 2003. Further structures in the jaw apparatus of Limnognathia maerski (Micrognathozoa), with notes on the phylogeny of the Gnathifera. Journal of Morphology 255: 131–145.

Vos, P., R. Hogers, M. Bleeker, M. Reijans, T. Lee, M. Hornes, A. Frijters, J. Pot, J. Peleman, M. Kuiper & M Zabeau, 1995. AFLP: A new technique for DNA fingerprinting. Nucleic Acids Research 23: 4407–4414.

Vos, P. & M. Kuiper, 1998. AFLP analysis. In Caetano-Anollés, G. & P. M. Gresshoff (eds), DNA Markers. Protocols, Applications, and Overviews. Wiley-VCH, New York: 115–131.

Wallace, R. L. & R. A. Colburn, 1989. Phylogenetic relationships within phylum Rotifera: orders and genus *Notholca*. Hydrobiologia 186/187: 311–318.

Wallace, R. L., C. Ricci & G. Melone, 1996. A cladistic analysis of pseudocoelomate (aschelminth) morphology. Invertebrate Biology 115: 104–112.

Walsh, E. J. & P. L. Starkweather, 1993. Analysis of rotifer ribosomal gene structure using the Polymerase Chain Reaction (PCR). Hydrobiologia 255/256: 219–224.

Williams, J. G. K., A. R. Kubelik, K. J. Livak, J. A. Rafalski & S. Tingey, 1990. DNA polymorphisms amplified by arbitrary primers are useful as genetic markers. Nucleic Acids Research 18: 6531–6535.

Winnepenninckx, B. M. H., T. Backeljau & R. M. Kristensen, 1998. Relations of the new phylum Cycliophora. Nature 393 (6686): 636–638.

Hydrobiologia (2005) 546:117–123

A. Herzig, R.D. Gulati, C.D. Jersabek & L. May (eds.) Rotifera X: Rotifer Research: Trends, New Tools and Recent Advances
DOI 10.1007/s10750-005-4107-4

Molecular characterization of Mn-superoxide dismutase and gene expression studies in dietary restricted *Brachionus plicatilis* rotifers

Gen Kaneko[1], Tatsuki Yoshinaga[2], Yoshiko Yanagawa[1], Shigeharu Kinoshita[1], Katsumi Tsukamoto[2] & Shugo Watabe[1,*]
[1]*Graduate School of Agricultural and Life Sciences, The University of Tokyo, Bunkyo Tokyo, 113-8657, Japan*
[2]*Ocean Research Institute, The University of Tokyo, Nakano Tokyo, 164-8639, Japan*
*(*Author for correspondence: E-mail: awatabe@mail.ecc.u-tokyo.ac.jp)*

Key words: *Brachionus plicatilis*, caloric restriction, superoxide dismutase, lifespan, reactive oxygen species, rotifer

Abstract

Superoxide dismutases (SODs) promote a conversion of harmful reactive oxygen species (ROS) to relatively moderate forms, resulting in the extension of lifespan in the nematode *Caenorhabditis elegans* under caloric restriction. The lifespan of the rotifer *Brachionus plicatilis* is also markedly extended by caloric restriction. We, therefore, cloned cDNA encoding SOD activated with Mn (Mn SOD) from *B. plicatilis* and examined its expression pattern in rotifers raised with energy restricted diet. The full length deduced amino acid sequence of the rotifer Mn SOD showed 61% identity with the *C. elegans* ortholog. Four amino acid residues that are essential to the binding of this enzyme to Mn were conserved in the rotifer Mn SOD. Subsequently we examined the mRNA expression patterns of Mn SOD using highly sensitive quantitative real-time PCR for various rotifer populations that are likely to differ in their lifespans in experiments on calorie restricted diets. The accumulated mRNA levels of Mn SOD were found to increase in supposedly long-lived rotifers. These results suggest that Mn SOD is possibly related to the aging of *B. plicatilis*.

Introduction

Significant extension of lifespan under caloric restriction (CR) has been reported for *Asplanchna brightwelli* (Verdone-Smith & Enesco, 1982) and *Brachionus plicatilis* (Yoshinaga et al., 2000). Yoshinaga et al. (2003) suggested that the lifespan of rotifer *B. plicatilis* is shorter when it is well fed and produces a large number of offsprings. However, Enesco (1993) claimed that the reproduction of rotifers is not directly related to their lifespan. Thus, the relationship of trade-off between extension of lifespan and reproduction in rotifers under CR is still controversial and the mechanisms determining longevity have remained obscure.

It is widely accepted that lifespan is regulated by an interaction between oxidative stress and an enzymatic antioxidation system. Reactive oxygen species (ROS) mainly produced as metabolic by-products provoke massive damages to DNA, proteins and lipids (Finkel & Holbrook, 2000). Antioxidant enzymes catalyze decomposition of ROS, thereby moderating oxidative stresses and resulting in longevity. An enhancement of expression of the major antioxidant enzyme, superoxide dismutase (SOD), was observed in the nematode *C. elegans* (Honda & Honda, 1999) and the fruitfly *Drosophila melanogaster* (Parkes et al., 1998) with an extended lifespan. Since CR restriction leads to the extension of lifespan probably due to decreasing oxidative stress (Finkel & Holbrook, 2000), the mechanisms involved are likely to account for the longevity of rotifers.

The objective of this study was: (1) to clone and also obtain a sequence of rotifer manganese-SOD (Mn SOD) cDNA, and (2) to examine the mRNA level by highly sensitive real-time PCR, between the two groups of rotifers, under CR and well fed rotifers. As expected, the mRNA levels were higher in the long-lived, calorie-restricted rotifers than those of the well fed groups. These results suggest that the expression of Mn SOD is possibly related to the aging in rotifer.

Materials and methods

Materials

We used a genetically identical *B. plicatilis* population (Ishikawa strain) established from a single amictic female by parthenogenesis (Yoshinaga et al., 1999). Rotifers used for RNA extraction were cultured as described in Kaneko et al. (2002) except that *Chlorella* (Nikkai Center) was used as dietary algae. The algal cells were rinsed twice with a sterilized culture medium (Yoshinaga et al., 1999).

cDNA cloning

RNA extraction from rotifer and first strand cDNA synthesis were performed as described previously (Kaneko et al., 2002). Primers rMnSOD-f1 and rMnSOD-rl for cloning Mn SOD were designed with reference to DNA nucleotide sequences from various species of the Mn SOD genes reported for yeast *Saccharomyces cerevisiae* (Marres et al., 1985; GenBank accession number X02156), nematode *C. elegans* (GenBank accession number D85499), mouse (DiSilvestre et al., 1995; GenBank accession number S78832) and human (Wispe et al., 1989; GenBank accession number X14322) (Fig. 1, Table 1). PCR was conducted for 3 min at 94 °C followed by 30 cycles of 94 °C for 30 s, 55 °C for 30 s and 72 °C for 10 s. The final extension step was performed at 72 °C for 1 min. Each of the 20 μl reaction mixtures contained 20 pmol of forward and reverse primers, approximately 1 μg of first strand cDNA synthesized from total RNA of rotifer, 20 nmol of dNTP mixtures, 10 μl of 10 × PCR buffer (100 mM Tris–HCl, pH 8.3, 500 mM KCl, 15 mM $MgCl_2$, 0.01% (w/v) gelatin) and 1 U of Ex *Taq* DNA polymerase (Takara). The amplified products were diluted 100 times with sterilized water and used as a template for reamplification. PCR for the reamplification was the same as described above except that the total reaction volume was 100 μl. The final products were subcloned into the pGEM-T Easy vector (Promega).

Degenerated primer ActinF (Leblanc et al., 1999) and adapter primer AUAP were used for amplification of a DNA fragment encoding rotifer actin, which was used as the internal control for quantitative real-time PCR. PCR and subcloning were carried out by the same method as described in cDNA cloning of rotifer Mn SOD.

Rapid amplification of cDNA ends (RACE)

3′ RACE was carried out with first strand cDNA as a template, using AUAP and gene-specific primer rMn-3RACE1, that were designed from a partial nucleotide sequence obtained by a previous PCR (Fig. 1, Table 1). PCR was performed for 3 min at 94 °C followed by 30 cycles of 94 °C for 30 s, 58 °C for 30 s and 72 °C for 30 s. The final extension step was performed at 72 °C for 5 min. Nested PCR was performed using primers rMn-3RACE2 and AUAP with the same method except that the total reaction volume was 100 μl. The following procedures were the same as those for cDNA cloning of Mn SOD.

5′ RACE was performed using the 5′ RACE system for rapid amplification on cDNA ends version 2.0 (Invitrogen) according to the manufacture's protocol. Gene-specific primers rMn-5RACE1, rMn-5RACE2 and rMn-5RACE3 were designed from the sequence of cDNA amplified by the previous conventional PCR (Fig. 1, Table 1).

DNA nucleotide sequencing

DNA nucleotide sequencing was performed for both 5′ and 3′ strands of subclones labeled with BigDye terminator cycle sequencing kit (Applied Biosystems) using DNA sequencers model 310 and 3100 (Applied Biosystems). Sequence homology was examined using BLAST search (http://www.ncbi.nlm.nih.gov/BLAST/).

Culture conditions

Rotifer used for quantitative real-time PCR were cultured according to the previous study with

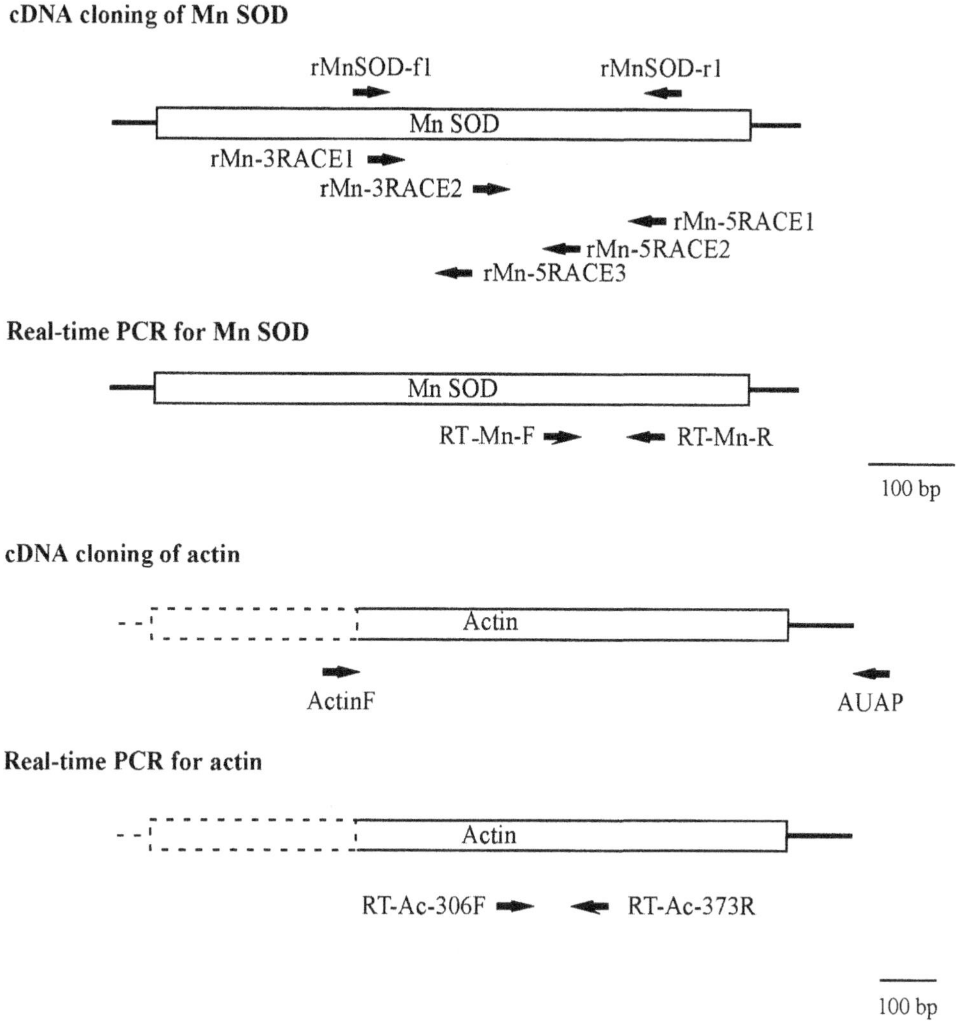

Figure 1. Locations of primers used for PCR amplification of cDNAs encoding Mn SOD and actin from the rotifer *Brachionus plicatilis* and those used for their quantitative analysis by real-time PCR of Mn SOD mRNA. The boxes and solid bars indicate the coding and non-coding regions, respectively.

minor modification (Yoshinaga et al., 2003). Eighty neonates (age < 1 h) were randomly divided into 4 groups (n = 20 each) and each cultured in 5 ml of medium (Yoshinaga et al., 1999). The rotifer cohorts were daily transferred to newly prepared culture medium. Each group was cultured with food alga until the age of 1 to 4 days (termed Fed-1d, Fed-2d, Fed-3d and Fed-4d, respectively), and maintained in the medium containing no algae until the age of 5 days. Subsequently, a small aliquot (ca. 10 μl) containing 20 individuals was suspended in 500 μl of ISOGEN (Nippon Gene), frozen immediately in liquid nitrogen and stored at −80 °C until use.

Quantitative real-time PCR

RNA extraction from rotifers and first strand cDNA synthesis were performed as described previously (Kaneko et al., 2002). The synthesized cDNAs were diluted 50 times with sterilized water and used as a template for quantitative real-time PCR. Primers for the amplification of rotifer Mn SOD and actin were designed from sequences obtained by PCR and RACE using Primer Express software (Applied Biosystems) (Table 1). Real-time PCR was carried out using SYBR Green I PCR Master Mix and ABI PRISM 7900 HT (Applied Biosystems) according

Table 1. Nucleotide sequences of primers used in PCR amplifications and quantitative real-time PCR for rotifer Mn SOD and actin cDNA

Primer	Sequence	Nucleotide position
rMnSOD-f1	5′-MTSAAGTTCMATGGYGGWGG-3′	308–324
rMnSOD-r1	5′-CTGCARGTAGTARGCGTGYTC-3′	598–618
rMn-3RACE1	5′-CCACATAAACCATTCAATTT-3′	325–344
rMn-3RACE2	5′-TAGTTGCCACATCCACCG-3′	447–464
rMn-5RACE1	5′-CAAACATCAATGCCAAACA-3′	579–597
rMn-5RACE2	5′-CTTGTTGTATCCTAACCATCC-3′	485–505
rMn-5RACE3	5′-TTCGGAGCTACCATTTTGG-3′	361–379
ActinF*	5′-ATGTTYGARACBTTCAACGT-3′	
AUAP	5′-GGCCACGCGTCGACTAGTAC-3′	
rMn-5NTR	5′-AATGTCTATCTAGAATTATTT-3′	1–21
rMn-3NTR	5′-AAAAATTGGGGTCTTTTTA-3′	713–731
RT-Mn-F	5′-CCGCGTGTGCTAACCAAGA-3′	529–546
RT-Mn-R	5′-TTCCCAAACATCAATGCCAA-3′	582–601
RT-Ac-306F	5′-CGTCCAGTTCCTCGTTGGAA-3′	306–325
RT-Ac-373R	5′-CTCGTTGCCAATGGTGATCA-3′	354–373

*The sequence could not be ascertained from the cDNA fragment cloned.

to the manufacturer's instructions. PCR consisted of 50 °C for 2 min and 95 °C for 10 min, followed by 40 cycles of 95 °C for 15 s and 50 °C for 1 min. Each of the primer pairs was optimized to ensure amplification of the specific gene product and absence of primer dimers, which were ascertained by agarose gel electrophoresis (data not shown). The real-time PCR results were analyzed at two threshold levels by the comparative Ct method using the Sequence Detection System software v2.0 release 4 (Applied Biosystems). The accumulated mRNA levels of rotifer β-actin, which were obtained by the same method as those for Mn SOD, were used as the internal control. The expression levels of Mn SOD mRNA in rotifer starved at various ages were statistically analyzed by one-way analysis of variance (ANOVA) followed by post-hoc Bonferroni test.

Results

cDNA cloning of rotifer SOD

A cDNA fragment amplified by PCR using primers rMnSOD-f1 and rMnSOD-r1 consisted of 274 nucleotides (nt), whereas 3′ and 5′ RACE yielded cDNA fragments consisting of 326 and 379 nt, respectively. The complete sequence of rotifer Mn SOD cDNA, which was ensured by a single PCR using primers designed from 5′ and 3′ non-coding regions (rMn-5NTR and rMn-3NTR, Table 1), consisted of 772 nt encoding 221 residues. The deduced amino acid sequence showed 36, 61, 59 and 60% identity with orthologs of yeast *S. cerevisiae* (Marres et al., 1985), nematode *C. elegans* (GenBank accession number D85499), mouse (DiSilvestre et al., 1995) and human (Wispe et al., 1989), respectively. Four amino acid residues that are essential for Mn SOD binding to Mn at the catalytic site were conserved in rotifer (Fig. 2).

PCR using primers ActinF and AUAP resulted in the amplification of a cDNA fragment consisting of 852 nt encoding a C-terminal part of the actin molecule of 246 amino acid residues. The deduced amino acid sequence showed approximately 95% identity with the sequence for the nematode *C. elegans* (Piano et al., 2000) and mouse (Chu & Paul, 1998). DNA nucleotide sequences of our rotifer Mn SOD and β-actin have been registered with the DDBJ/EMBL/GenBank databases with accession numbers AB111351 and AB111352, respectively.

```
(a) AATGTCTATCTAGAATTATTTACAACAAAGTTTCAAATTTTTGATCAAAATGTTTGCAAAT   61
                                                    M  F  A  N     4

    ATTGCCAGAGTTTTAGCTCAAAGCAGCAACAAATGCTTGCCACTAATCCGCGCCAAGCAT  121
     I  A  R  V  L  A  Q  S  S  N  K  C  L  P  L  I  R  A  K  H    24

    TCTCTTCCAGACTTGCCCTATGACTTTAATGCATTGGAGCCTGTCATATCAGCCGAAATT  181
     S  L  P  D  L  P  Y  D  F  N  A  L  E  P  V  I  S  A  E  I    44

    ATGCAAGTGCACCACCAAAAGCATCATGCCACTTATGTCAACAATCTAAATGTAGCAGAA  241
     M  Q  V  H  H  Q  K  H  H  A  T  Y  V  N  N  L  N  V  A  E    64

    GAGAAACTTTTCGAGGCAAAACAAAGAGGATCAGTATCCGACCAAATAGCTTTAACATCA  301
     E  K  L  F  E  A  K  Q  R  G  S  V  S  D  Q  I  A  L  T  S    84

    GCCTTGAAGTTTAATGGTGGTGGCCACATAAACCATTCAATTTTTTGGACTATTTTATGC  361
     A  L  K  F  N  G  G  G  H  I  N  H  S  I  F  W  T  I  L  C   104

    CAAAATGGTAGCTCCGAACCGGGGGGAGAACTAGCCGAAGCTATCAAAAGAGATTTTGGT  421
     Q  N  G  S  S  E  P  G  G  E  L  A  E  A  I  K  R  D  F  G   124

    TCTTATCAAAGGATGGTAGATTTGTTAGTTGCCACATCCACCGGTATTCAGGGGTCCGGA  481
     S  Y  Q  R  M  V  D  L  L  V  A  T  S  T  G  I  Q  G  S  G   144

    TGGGGATGGTTAGGATACAACAAGCAGCTGAATAGACTAGAGATTGCCGCGTGTGCTAAC  541
     W  G  W  L  G  Y  N  K  Q  L  N  R  L  E  I  A  A  C  A  N   164

    CAAGACCCTCTGCAAGCGACAACTGGCTTGCATCCATTGTTTGGCATTGATGTTTGGGAA  601
     Q  D  P  L  Q  A  T  T  G  L  H  P  L  F  G  I  D  V  W  E   184

    CATGCTTATTATCTGCAATACAAAAATGCTCGGCCCGACTATGTCAAGAGCATTTTCAAA  661
     H  A  Y  Y  L  Q  Y  K  N  A  R  P  D  Y  V  K  S  I  F  K   204

    ATTGCCAACTGGAGAAAAGTCGAAGAGCTTTTTGGAAAAGCTAGACACGACTAAAAAGAC  721
     I  A  N  W  R  K  V  E  E  L  F  G  K  A  R  H  D            221

    CCCAATTTTTTTTTCCAAATAAAAACTAATTCTAAAAAAAAAAAAAAAAAA           772
```

```
(b) Rotifer   MFANIARVLAQSSNKC--LPLI--RAKHSLPDLPYDFNALEPVISAEIMQ   46
    Yeast     ...KT.AANLTKKGGLSL.STTAR.T.VT....KW..G....Y..GQ.NE   50
    Nematode  .LQ.TV.CVSKLVQPI--TGVAAV.S..........YAD......H....   48
    Mouse     .LCRA.CSTGRRLGPV--AGAAGS.H..........YG....H.N.Q...   48
    Human     .LSRAVCGTSRQLAPA--.GYLGS.Q..........YG....H.N.Q...   48

    Rotifer   VHHQKHHATYVNNLNVAEEKLFEAK-------QRGSVSDQIALTSALKFN   89
    Yeast     L.YT...Q....GF.T.VDQFQ.LSDLLAKEPSPANARKM..IQQNI..H  100
    Nematode  L.............QI....H..V-------SK.N.KEA...QP.....   91
    Mouse     L..S....A......AT...YH..L-------AK.D.TT.V..QP.....   91
    Human     L..S....A.......T...YQ..L-------AK.D.TA.T..QP.....   91

    Rotifer   GGGHINHSIFWTILCQ---NGSSEPGGELAEAIKRDFGSYQRMVDLLVAT  136
    Yeast     ...FT..CL..EN.APESQG.GEP.T.A..K..DEQ...LDELIK.TNTK  150
    Nematode  ...........N.AK---D.G-..SA..LT...S....LDNLQKQ.S.S  137
    Mouse     .......T....N.SP---K.GG..K...L.........FEKFKEK.T.V  138
    Human     ...........N.SP---..GG..K...L.........FDKFKEK.T.A  138

    Rotifer   STGIQGSGWGWLGYNK-QLNRLEIAACANQDPLQATTGLHPLFGIDVWEH  185
    Yeast     LA.V.....AFIVK.LSNGGK.DVVQTY...TVTGP--.V..VA..A...  198
    Nematode  TVAV..........CP-KGKI.KV.T.......E.....V..........  186
    Mouse     .V.V.........F..-EQG..Q....S......G....I..L.......  187
    Human     .V.V.........F..-ERGH.Q....P......G....I..L.......  187

    Rotifer   AYYLQYKNARPDYVKSIFKIANWRKVEELFGKARHD                221
    Yeast     ......Q.KKA..F.A.WNVV..KEASRR.DAGKI-                233
    Nematode  ........V.....NA.W.....KN.S.R.A..QQ-                221
    Mouse     ........V....L.A.WNVI..EN.T.RYTACKK-                222
    Human     ........V....L.A.WNVI..EN.T.RYMACKK-                222
```

Figure 2. Nucleotide and deduced amino acid sequences of rotifer Mn SOD cDNA (a) and a comparison of the amino acid sequence with those of various organisms (b). The nucleotide sequence determined by RT-PCR is boxed, and unboxed sequences determined by 5′ and 3′ RACE. Stop codon and polyadenylation signal are shown with bold letters and an underline, respectively. Compared were yeast *Saccharomyces cerevisiae* (Johnston et al., 1994), nematode *Caenorhabditis elegans* (Genbank accession number D85499), mouse (DiSilvestre et al., 1995) and human (Wispe et al., 1989). The amino acid residues essential for Mn SOD to bind to Mn are meshed. Amino acid residues identical to those of rotifer Mn SOD are indicated with dots, and dashes denote gaps introduced to maximize homology. The numbers on the right margin of the sequences represent those from the N-terminal methionine.

Quantitative real-time PCR

The mRNA accumulation levels of Mn SOD in the Fed-2d group were 1.5-fold higher than those of other groups (ANOVA, $p < 0.0001$; post-hoc Bonferroni test, $p < 0.01$ in all cases; Fig. 3). Rotifers in the Fed-1d group also accumulated more Mn SOD mRNA than those of the short-lived Fed-3d and Fed-4d groups, but the differences were not statistically significant. The results from two independent experiments with two threshold values of 0.2 and 0.4 were consistent.

Discussion

Since a high expression of SODs can extend lifespans in several organisms (Parkes et al., 1998; Guarente & Kenyon, 2000), we hypothesized that the expression of SODs is enhanced in calorie-restricted rotifers, which have a longer lifespan. To test this hypothesis, we cloned the Mn SOD gene from *B. plicatilis* using both the conventional PCR and RACE, and subsequently examined for its mRNA expression pattern. According to our previous study (Yoshinaga et al., 2003), Fed-1d and Fed-2d groups are expected to show extended lifespan, whereas Fed-3d and Fed-4d groups are not. As expected, the accumulated mRNA levels of Mn SOD were significantly higher in long-lived Fed-2d group than those of the short-lived Fed-3d and Fed-4d groups (Fig. 3). These results support our hypothesis that Mn SOD expression is related to the longevity of rotifer. To our knowledge, this is the first report describing a candidate gene that determines the lifespan of rotifer.

On the other hand, the accumulated mRNA levels of Mn SOD were not significantly higher in long-lived Fed-1d group than those in short-lived groups. These results suggest that other genes are involved in the regulation of rotifer lifespan. In this study, we examined only the expression pattern of the Mn SOD gene out of the three types of SOD genes that show different subcellular localization and expression patterns (Zelko et al., 2002). Although the Mn SOD gene seems more important for longevity than other SODs, because it is induced by CR in certain species such as nematode *C. elegans* (Honda & Honda, 1999) and mouse (Furuyama et al., 2000), Cu/Zn and extracellular SODs may have also some effects on rotifer lifespan. Alternatively, it is possible that the Mn SOD gene was induced by CR at shorter ages in Fed-1d group, although it was induced in 5 days for Fed-2d group. To examine these possibilities, it is necessary to investigate time course of the expression patterns for all types of the SOD genes throughout the rotifer lifespan.

It has been difficult to determine the functions of the cellular components which may affect aging of rotifer, using one individual. Such molecular biological approaches as northern blot analysis have been accomplished with rotifer population consisting of over 10^4 individuals. The combined use of small scale cultures and the molecular

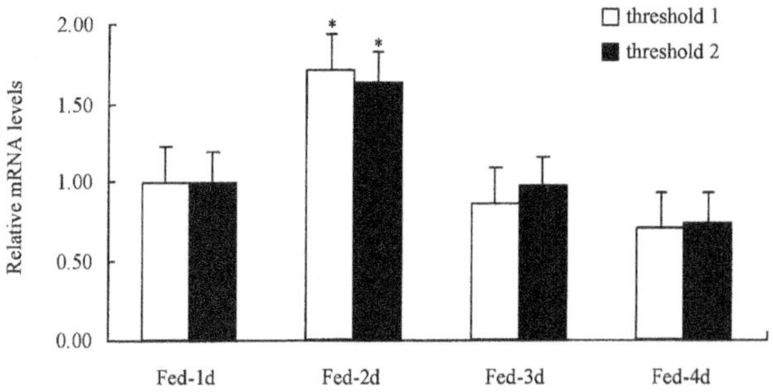

Figure 3. Relative mRNA levels of Mn SOD from the rotifer *Brachionus plicatilis* treated under various caloric restriction conditions. Fed-1d, Fed-2d, Fed-3d and Fed-4d represent rotifer groups fed until the age of 1, 2, 3 and 4 days, respectively. Each group contained 20 individuals. Bars represent standard errors. The accumulated mRNA levels were analyzed twice with two arbitrary threshold levels. The accumulated mRNA levels for the Fed-1d group were taken as 1.00, using those of β-actin as the internal control. The expression levels of Mn SOD mRNA were statistically analyzed by one-way analysis of variance (ANOVA) followed by post-hoc Bonferroni test. Asterisks indicate significant difference between the accumulated mRNA levels of the Fed-2d group and those of other groups of both threshold levels, which are the optional values in comparative Ct method ($p < 0.01$).

biological approaches as used in the present study will give new insights to the better understanding of the molecular mechanisms involved in the regulation of lifespan of rotifers.

Acknowledgements

This study was partly supported by Grants-in-Aids for Creative Scientific Research No. 12NP0201 and Exploratory Research No. 14656080 from the Ministry of Education, Culture, Sports, Science and Technology of Japan. GK, and TY were supported by Research Fellowships of the Japan Society for the Promotion of Science for Young Scientists.

References

Chu, C. C. & W. E. Paul, 1998. Expressed genes in interleukin-4 treated B cells identified by cDNA representational difference analysis. Molecular Immunology 35: 487–502.

DiSilvestre, D., S. R. Kleeberger, J. Johns & R. C. Levitt, 1995. Structure and DNA sequence of the mouse MnSOD gene. Mammalian Genome 6: 281–284.

Enesco, H. E., 1993. Rotifers in aging research: use of rotifers to test various theories of aging. Hydrobiologia 255/256: 59–70.

Finkel, T. & N. J. Holbrook, 2000. Oxidants, oxidative stress and the biology of ageing. Nature 408: 239–247.

Furuyama, T., T. Nakazawa, I. Nakano & N. Mori, 2000. Identification of the differential distribution patterns of mRNAs and consensus binding sequences for mouse DAF-16 homologues. Biochemical Journal 349: 629–634.

Guarente, L. & C. Kenyon, 2000. Genetic pathways that regulate ageing in model organisms. Nature 408: 255–262.

Honda, Y. & S. Honda, 1999. The *daf-2* gene network for longevity regulates oxidative stress resistance and Mn-superoxide dismutase gene expression in *Caenorhabditis elegans*. FASEB Journal 13: 1385–1393.

Johnston, M., S. Andrews, R. Brinkman, J. Cooper, H. Ding, J. Dover, Z. Du, A. Favello, L. Fulton, S. Gattung, et al., 1994. Complete nucleotide sequence of *Saccharomyces cerevisiae* chromosome VIII. Science 265: 2077–2082.

Kaneko, G., S. Kinoshita, T. Yoshinaga, K. Tsukamoto & S. Watabe, 2002. Changes in expression patterns of stress protein genes during population growth of the rotifer *Brachionus plicatilis*. Fisheries Science 68: 1317–1323.

Leblanc, C., A. Falciatore, M. Watanabe & C. Bowler, 1999. Semi-quantitative RT-PCR analysis of photoregulated gene expression in marine diatom. Plant Molecular Biology 40: 1031–1044.

Marres, C. A., A. P. Van Loon, P. Oudshoorn, H. Van Steeg, L. A. Grivell & E. C. Slater, 1985. Sequence analysis of the nuclear gene coding for manganese superoxide dismutase of yeast mitochondria, a gene previously assumed to code for the Rieske iron-sulphur protein. European Journal of Biochemistry 147: 153–161.

Parkes, T. L., A. J. Elia, D. Dickinson, A. J. Hilliker, J. P. Phillips & G. L. Boulianne, 1998. Extension of *Drosophila* lifespan by overexpression of human SOD1 in motorneurons. Nature Genetics 19: 171–174.

Piano, F., A. J. Schetter, M. Mangone, L. Stein & K. J. Kemphues, 2000. RNAi analysis of genes expressed in the ovary of *Caenorhabditis elegans*. Current Biology 10: 1619–1622.

Verdone-Smith, C. & H. E. Enesco, 1982. The effect of temperature and of dietary restriction on lifespan and reproduction in the rotifer *Asplanchna brightwelli*. Experimental Gerontology 17: 255–262.

Wispe, J. R., J. C. Clark , M. S. Burhans, K. E. Kropp, T. R. Korfhagen & J. A. Whitsett, 1989. Synthesis and processing of the precursor for human manganese-superoxide dismutase. Biochimica et Biophysica Acta 994: 30–36.

Yoshinaga, T., A. Hagiwara & K. Tsukamoto, 1999. Effect of conditioned media on the asexual reproduction of the monogonont rotifer *Brachionus plicatilis* O. F. Müller. Hydrobiologia 412: 103–110.

Yoshinaga, T., A. Hagiwara & K. Tsukamoto, 2000. Effect of periodical starvation on the life history of *Brachionus plicatilis* O. F. Müller (Rotifera): a possible strategy for population stability. Journal of Experimental Marine Biology and Ecology 253: 253–260.

Yoshinaga, T., A. Hagiwara & K. Tsukamoto, 2003. Life history response and age-specific tolerance to starvation in *Brachionus plicatilis* O. F. Müller. Journal of Experimental Marine Biology and Ecology 287: 261–271.

Zelko, I. N., T. J. Mariani & R. J. Folz, 2002. Superoxide dismutase multigene family; a comparison of the Cu/Zn-SOD (SOD1), Mn-SOD (SOD2), and EC-SOD (SOD3) gene structures, evolution, and expression. Free Radical Biology and Medicine 33: 337–349.

Hydrobiologia (2005) 546:125–134 © Springer 2005
A. Herzig, R.D. Gulati, C.D. Jersabek & L. May (eds.) Rotifera X: Rotifer Research: Trends, New Tools and Recent Advances
DOI 10.1007/s10750-005-4109-2

Behavioural reproductive isolation in a rotifer hybrid zone

Helen K. Berrieman, David H. Lunt & Africa Gómez*
Department of Biological Sciences, University of Hull, HU6 7RX Hull, UK
*(*Author for correspondence: E-mail: a.gomez@hull.ac.uk)*

Key words: Rotifera, *Brachionus plicatilis*, reproductive isolation, RFLP, mtDNA, hybrid zone

Abstract

A hybrid zone between two *Brachionus plicatilis* rotifer mitochondrial DNA (mtDNA) lineages was recently described in the Iberian Peninsula between a pond (Santed 2) and a lake (Gallocanta). The patterns of mitochondrial and nuclear genetic variation observed suggested that gene flow is mainly male-mediated from the lake to the pond. Here we test two hypotheses: (a) that male-mediated gene flow occurs through assortative mating between individuals from these ponds, (b) that behavioural isolation occurs between the two mtDNA lineages. We isolated, reared and genotyped rotifer clones from resting eggs collected in the sediments of these and two other distant ponds. We devised a quick, inexpensive RFLP method to discriminate between *B. plicatilis* and its sibling species *B.* 'Manjavaeas' and between both mtDNA *B. plicatilis* lineages. Behavioural no-choice tests using new-born, virgin males and females were performed between five clones. *B.* 'Manjavacas' and *B. plicatilis* were reproductively isolated. *B. plicatilis* clones did not show evidence of reproductive isolation, regardless of their mtDNA lineage, except Santed 2 males, which discriminated strongly against Gallocanta females. These results could help to explain the discrepancies between mitochondrial and nuclear genetic variation reported in the two populations.

Introduction

The application of molecular methods to the analysis of zooplanktonic populations has recently challenged widely held views on the biology of these organisms. First, molecular methods have uncovered a plethora of sibling species complexes (Gómez et al., 2002b) that previously contributed to confounding biodiversity and ecological assessments. Second, important aspects of the population biology of these organisms have been discovered (De Meester et al., 2002), which emphasise the surprisingly low impact of dispersal and gene flow in continental zooplanktonic populations. Third, phylogeographic assessments of continental zooplanktonic organisms, although relatively recent, have revealed an important amount of geographic diversification, leading to important insights into their microevolution and speciation processes (see review in De Meester et al., 2002).

To fully exploit the advantages of using molecular methods in zooplanktonic organisms, a multidisciplinary approach is necessary. Thus, although some of the aforementioned molecular assessments rely on mitochondrial DNA, none investigates reproductive isolation between the taxa involved and, therefore, it is not known if the molecular signatures detected reveal gene patterns or population level phenomena. A recent study of the phylogeography of *Brachionus plicatilis* Müller 1786 in Iberian saline lakes (Gómez et al., 2002a) revealed two distinct monophyletic mtDNA lineages with a geographic structure corresponding to 'northern' and 'southern' lineages. There was also evidence of separate evolutionary histories since the onset of the Pleistocene. A survey of seven unlinked microsatellite loci revealed strong population structure, but surprisingly little correlation with mtDNA differentiation, population structure showing isolation by distance (Gómez et al., 2002a). Therefore, unexpectedly, populations from two nearby locations – a lake, Laguna de Gallocanta, and a pond, Balsa de Santed 2 – dominated by mtDNA haplotypes from northern and southern lineages, respectively, proved very similar in their nuclear loci allele frequencies. This suggests that gene flow occurs between the two populations, most prominently from the lake to the pond, with some mechanism restricting the amount of female-mediated gene flow. Although hybrid swarms have been described in *Daphnia* (see for example Schwenk et al., 2000), to our knowledge, Gallocanta and Balsa de Santed 2 form the first classic hybrid zone described for a continental zooplanktonic organism.

Claims of reproductive isolation are rarely substantiated with direct evidence from mating tests in zooplanktonic organisms due to the impracticalities of carrying out such experiments. However, mating tests have been used to great success in rotifers to demonstrate assortative mating and behavioural reproductive isolation between different species (e.g., Snell & Hawkinson, 1983; Gómez & Serra, 1995; Rico-Martinez & Snell, 1995a,b). Many characteristics of *B. plicatilis* make it an ideal organism for this type of study: (a) ease of laboratory culture, (b) short generation time, (c) easy handling, (d) well known mating behaviour (Snell & Hawkinson, 1983).

Variation in assortative mating among domesticated and wild strains in *B plicatilis s.l.* has been reported several times (Snell & Hawkinson, 1983; Fu et al., 1993; Rico-Martinez & Snell, 1995a, b). However, few studies dealt with genetically well characterised wild strains (Gómez & Serra, 1996; Gómez & Snell, 1996), and therefore their conclusions have limited value.

In this paper we combine the use of molecular tools to identify rotifer species and genetic lineages from clones isolated from field sediments, and mating behaviour experiments in the laboratory to test two hypotheses: (a) mating behaviour can be responsible for the discrepancies between mitochondrial and nuclear DNA phylogeography in *B. plicatilis* observed in this hybrid zone, and (b) the occurrence of behavioural reproductive isolation between southern and northern mtDNA lineages. Using pair-wise mating tests, we investigate the existence of reproductive isolation between the northern and southern mtDNA lineages of Iberian *B. plicatilis s.s.* and the patterns of mating behaviour from rotifers from Laguna de Gallocanta, and Balsa de Santed 2. The results will help us understand the microevolutionary processes involved in rotifer speciation.

Methods

Experimental design

Individuals of the rotifer *B. plicatilis* sensu stricto (see Ciros-Pérez et al., 2001) from the lakes Balsa de Santed 2 (SA2) and Laguna de Gallocanta (GAL) were used in the mating experiments. Populations from these lakes are dominated by southern and northern mtDNA haplotypes (see Fig. 1), respectively, and are geographically close together (less than 10 km apart). Therefore, both populations are likely to exchange migrants, as suggested by microsatellite results (Gómez et al., 2002a). Individuals from Laguna de las Eras (ERA) (predominantly northern mtDNA) and Laguna de Mojón Blanco (MOJ) (southern) were also included in the experimental design; these lakes are equidistant from the other two. In this way, our experimental design examined reproductive isolation between populations with divergent mitochondrial and nuclear DNA (distant populations) and also between populations with different mitochondrial but similar nuclear DNA (nearby populations in the contact zone between mtDNA lineages). Finally, a strain from the lineage *B.* 'Manjavacas' (Gómez et al., 2002b), isolated from Laguna de Gallocanta, was also included. This lineage is highly divergent from *B. plicatilis* for both nuclear and mitochondrial sequenced genes (Gómez et al., 2002b) and allozyme loci (Ortells et al., 2000). The lack of hybrid genotypes between *B. plicatilis* and *B.* 'Manjavacas' despite co-occurrence in their natural habitat, and the observation that *B. plicatilis* males discriminate against 'Manjavacas' females (Ortells et al., 2000) has led to the suggestion of species status for this lineage (Gómez et al., 2002b).

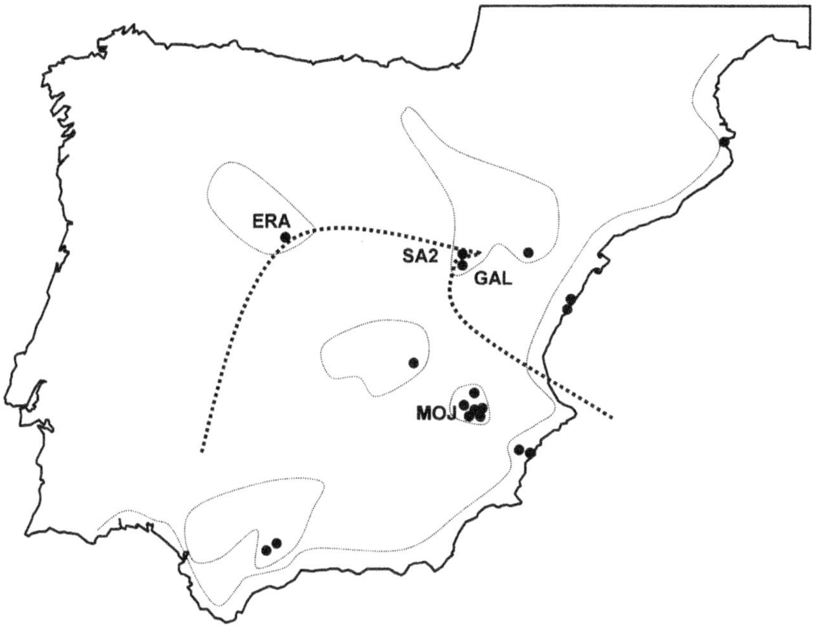

Figure 1. Map of the Iberian Peninsula showing the sampling locations indicated as follows: ERA: Laguna de las Eras; SA2: Balsa de Santed II; GAL: Laguna de Gallocanta; MOJ: Laguna de Mojón Blanco. Unlabelled points are sampling locations used by Gómez et al. (2000). Thin dashed lines indicate drainage basins and the thicker dashed line represents the boundary between northern and southern mitochondrial DNA lineages. Lakes are not drawn to scale.

Resting egg isolation, hatching and DNA extraction

Resting eggs were isolated from sediment samples collected from Laguna de Gallocanta (GAL), Balsa de Santed 2 (SA2), Laguna de Mojón Blanco (MOJ) and Laguna de Las Eras (ERA) (Fig. 1) using a sucrose flotation method (see Gómez et al., 2000 for details of the sampling locations). Approximately 5 ml of sediment was dispersed in 35–40 ml of 1:1 weight:volume sucrose solution and centrifuged at 700 rpm for 5 min. The supernatant was filtered through a 30 μm Nytal mesh and the filtrate rinsed with 6 g l^{-1} saltwater (Instant Ocean®; Aquarium Systems) and then resuspended in fresh saltwater for examination under a stereoscope at ×40 magnification. *B. plicatilis* eggs were identified by their morphology, and kept in a Petri dish at 20 °C under constant illumination (see below) for approximately 48 h to allow hatching. New-born rotifer females were transferred to individual wells in 24-well tissue culture plates (Nalge Nunc International) containing 1 ml of growth medium (see below) using a pipette and disposable tips. Rotifers were allowed to reproduce and some of the clones were used for the mating tests. Each isolated clone, offspring of a single hatched female from the sediment sample, was named with the pond initials and a letter.

For DNA extractions, individual live female rotifers from each clone were isolated and washed in fresh 12 g l^{-1} saltwater before being released into a 0.2 ml tube containing 40 μl of Chelex (6% InstaGene™ Matrix; Bio-Rad). Individuals were handled using sterile pipette tips to avoid cross-contamination. The tubes were heated for 20 min at 56 °C followed by 10 min at 99.9 °C and then cooled for 30 min at 4 °C before spinning down the resulting solution. Samples were kept frozen (−20 °C) until needed.

Culture conditions

Growth medium was a monoculture of the unicellular algae *Tetraselmis chuii* Butcher in 12 g l^{-1} saltwater enriched with f/2 medium (Guillard & Ryther, 1962) and aerated using an aquarium pump. Rotifer clones were transferred from culture wells to glass test tubes with fresh growth medium when several individuals could be observed in the growth wells. Algae and rotifer cultures were kept in a temperature-controlled chamber at 20 °C under constant illumination (ca. 47 Em^{-2} s^{-2}). Rotifers clones were maintained in duplicate and fed once a week by replacing half of the culture with fresh growth medium.

PCR amplification and RFLP analysis of mitochondrial DNA

A 712 bp (base pair) fragment of the mitochondria) gene cytochrome *c* oxidase I (COI) DNA was amplified for each clone using the polymerase chain reaction (PCR). Reactions were carried out in 0.2 ml tubes in a final volume of 10 μl containing 2 μl template DNA, PCR buffer (16 mM $(NH_4)_2SO_4$, 67 mM Tris–HCl (pH 8.8 at 25 °C), 0.01% Tween-20), 1.5 mM $MgCl_2$, 200 μM of each nucleotide, 2.5 pmol of each primer and 0.125 U *Taq* polymerase. The primers LCO1490 (5′-GGTCAACAAAT-CATAAAGATATTGG-3′) and HCO2198 (5′-TAAACTTCAGGGTGACC AAAAA-ATCA-3′) (Folmer et al., 1994) were used, with the following cycling conditions: 3 min at 93 °C followed by 40 cycles of 15 s at 92 °C, 20 s at 50 °C, 1 min at 70 °C, and a final 3 min at 72 °C.

As detailed sequence information is available for the populations and species used here (Gómez et al., 2000, 2002b), the species and COI haplotype group of each clone was determined by carrying out Restriction Fragment Length Polymorphism (RFLP) analysis on the amplified COI fragment. To find out which restriction enzymes could be used to discriminate between species and haplotype groups, a survey of expected digests for a range of restriction enzymes was carried out using DNASTAR ver. 2.88 (Lasergene Inc.) using consensus sequences of *B.* 'Manjavacas' vs. *B. plicatilis* and between consensus sequences of the northern and southern lineages of *B. plicatilis* (obtained from Gómez et al., 2000, 2002b). Only enzymes cutting non-polymorphic sites within species and lineages were selected. Endonuclease digestions were carried out in 10 μl final volume containing 2 μl DNA, 1 μl buffer (Kpn1$^+$ for *Kpn*I, R$^+$ for *Mbo*I, O$^+$ for *Bst*X1; MBI Fermentas) and 2 U restriction endonuclease. Reactions were incubated overnight at 37 °C (55 °C for *Bst*X1). The species or haplotype group was determined by examining the pattern of DNA bands produced on a 2% agarose gel stained with ethidium bromide.

Mating tests

The mating tests carried out here were pair-wise no-choice experiments in which males produced by a given clone were exposed to either homogamic (same clone) or heterogamic (different clone) females. This experimental design is a cost-effective way of performing behavioural reproductive isolation tests in rotifers for two reasons. First, no way of visually discriminating between males or females from different clones has been devised and, second, as mating behaviour in rotifers occurs only after contact between a male and female, there is no detection at a distance and thus choice can only be effected when a male meets a female. Procedures for the mating tests followed Gómez & Serra (1996). Rotifer cultures for mating tests were initiated in 250 ml conical flasks with 200 ml growth medium and fed daily or every other day to ensure high population densities and the presence of many females carrying multiple mictic (male) and amictic (female) eggs. Before each experiment, female rotifers carrying male or female eggs were isolated from the appropriate cultures and placed in tissue culture wells in 12 g l^{-1} saltwater. Male eggs were recognised as being smaller than female eggs. After approximately 2 h, new-born virgin males and females were isolated for use in the experiments. Females were less than 4 h old and males less than 8 h old when the experiments were carried out, because very young females are most attractive to males (Gómez & Serra, 1996) and sexual activity and fertilisation ability in males decreases as they age (Snell & Childress, 1987; Snell & Hoff, 1987).

For each experiment, 25 new-born females of a given clone were isolated in 45 μl saltwater in a small well (96 round-bottom wells per plate; Nalge Nunc International). A male of the same or a different clone was then introduced in a further 5 μl of saltwater and its behaviour monitored for 5 min under a binocular stereoscope at ×20 magnification. Experiments were conducted within the temperature range 21–25 °C. Three types of male behaviour were recognised – contact, circling and copulation. Contact was defined as a head-on collision between a male and female, circling involved the male completing one or more revolutions around the female's body while maintaining coronal contact, and copulation was recorded when the male stopped circling, attached his penis to the female and coronal contact was lost. Six replicate males from each clone were tested successively on each group of females. The mating tests were performed blind, so that the observer (H.K.B.) was unaware of the identity of the clones until all the experiments had been completed.

Data analyses

The percentages of male-female encounters resulting in circling and copulation per individual male for a given combination of male and female clone were used as raw data. This removed biases due to the fact that males spend different amounts of time copulating and circling females. Normal qantile plots (Sokal & Rohif, 1995) indicated that these data were not normal; therefore, we performed non-parametric analysis to test our hypothesis. Our null hypothesis was that male behaviour was the same to homogamic females as to heterogamic females. To test this hypothesis we used Mann–Whitney tests in which the behaviour (circling or copulation percentages) of the males towards the homogamic females was compared with their behaviour towards the heterogamic females (Zar, 1984) using the six male replicates of each mating experiment. All statistical analyses were carried out using SPSS v. 8.0.2. A p-value of 0.05 was used for all tests.

Results

Genetic characterisation of the clones

Thirty-two hatchlings were obtained from the sediment samples: 8 from ERA, 7 from GAL, 6 from SA2 and 11 from MOJ. Once the genotype of the resulting clones was known, the clones having the desired genotypes from each pond were selected for the experiments. The clones used in the experiments and some representatives of their ponds were kept in culture in the laboratory.

The restriction enzymes *Kpn*I and *Mbo*I were diagnostic to discriminate between amplified COI fragments from *B. plicatilis* and its sibling species 'Manjavacas'. A restriction site for *Kpn*I (5′-GGTAC$^{\downarrow}$C-3′) in *B. plicatilis* DNA generates fragments of 349 and 363 bp when digested, but no restriction site for this enzyme exists in 'Manjavacas' DNA, leaving the 712 bp COI DNA fragment intact. *Mbo*I (5′-$^{\downarrow}$GATC-3′) generates fragments of different size on each species (614, 27 and 71 bp in *B. plicatilis*, and 407, 108, 99, 27 and 71 bp in 'Manjavacas'). *Bst*X1(5′-CCAN$_5$$^{\downarrow}$NTGG-3′) was used to determine whether mtDNA of the *B. plicatilis* clones belonged to northern or southern haplotype groups (see Gómez et al., 2000 for details). A restriction site for this enzyme in southern haplotype COI DNA generates fragments of 435 and 277 bp, but there is no *Bst*X1 site in northern haplotype DNA. All clones showed the RFLP haplotype expected for the lake of origin, based on the results of Gómez et al. (2000). SA2 and MOJ clones were all southern, ERA clones were all northern and GAL clones were either northern or belonged to the 'Manjavacas' lineage. Examples of the band patterns produced by agarose gel electrophoresis of the RFLP fragments are presented in Figure 2. This RFLP technique was found to be entirely reliable when digestion of previously sequenced clones was performed (data not shown).

Mating experiments

Five clones were selected for the mating experiments: ERA-B and GAL-B, *B. plicatilis* clones with northern haplotypes, SA2-A and MOJ-C, *B. plicatilis* clones with southern haplotypes and MAN, a *B.* 'Manjavacas' clone. All male-female clone combinations were tested, including males and females from the same clone. The mean percentage of encounters ending in circling and copulation of the six replicate experiments (males) for each cross are shown in Figure 3. All male strains displayed a very low percentage of circling with *B.* 'Manjavacas' females (<5%, except for homogamic males, which show a higher value of 5.73%). Significant differences between the circling percentages to homogamic females and 'Manjavacas' females were found for male clones GAL-B, SA2-A and MOJ-C (Fig. 3). With the exception of one male (of strain GAL-B) no copulations were observed with *B.* 'Manjavacas' females but, due to the low number of copulations observed in the

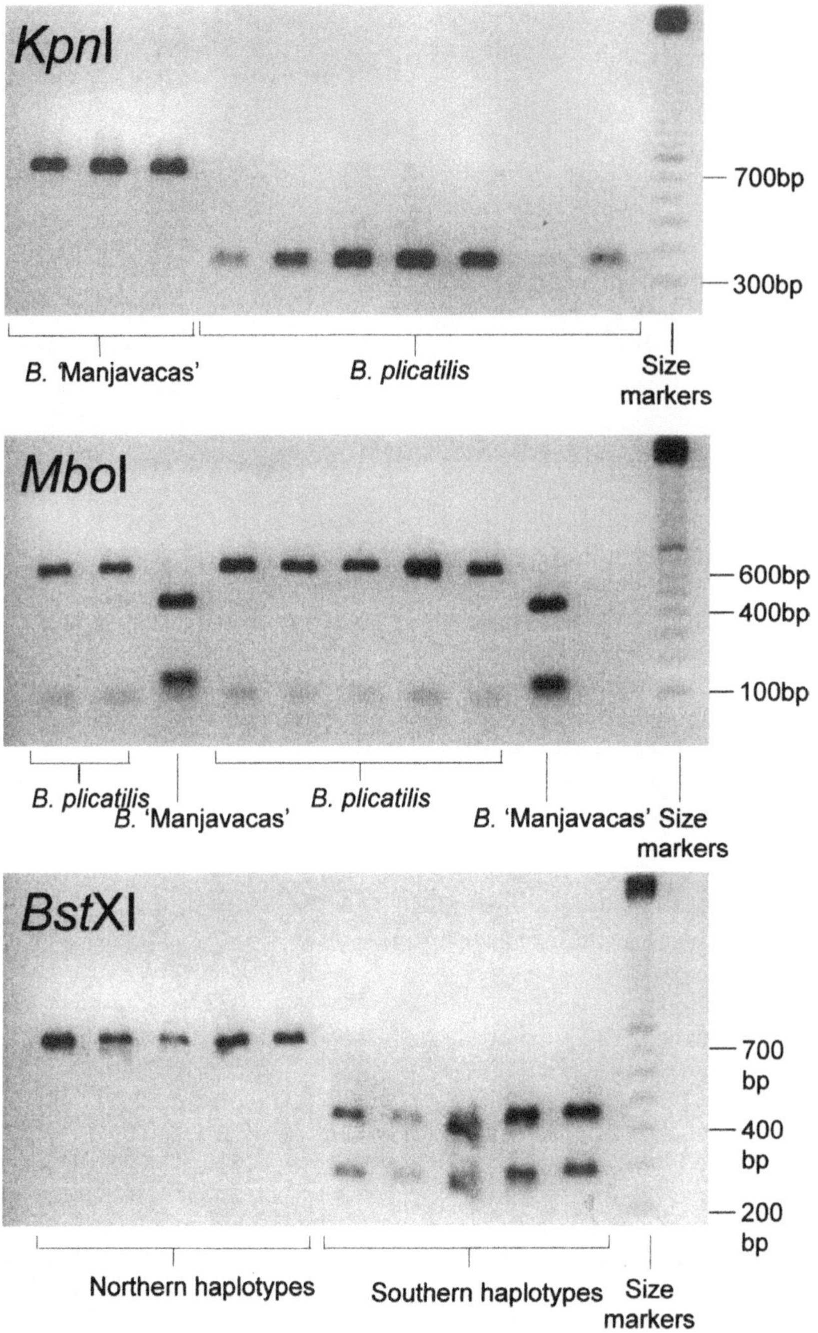

Figure 2. RFLP banding patterns produced when restriction enzyme digests of *B. plicatilis* and *B.* 'Manjavacas' DNA are run on agarose gels. Profiles generated by *Kpn*I (top gel) and *Mbo*I (middle gel) differentiate between *B. plicatilis* and *B.* 'Manjavacas'. *Bst*XI (bottom gel) differentiates northern and southern haplotypes of *B. plicatilis*.

intraspecific cross, these differences were not significant. *B.* 'Manjavacas' males showed significantly increased circling percentages towards females of the clones ERA-B, GAL-B, and SA2-A, and a consistent low percentage of copulations with all female strains (including their own) despite high percentages of circling to other female strains. Within the *B. plicatilis* male strains, the percentage

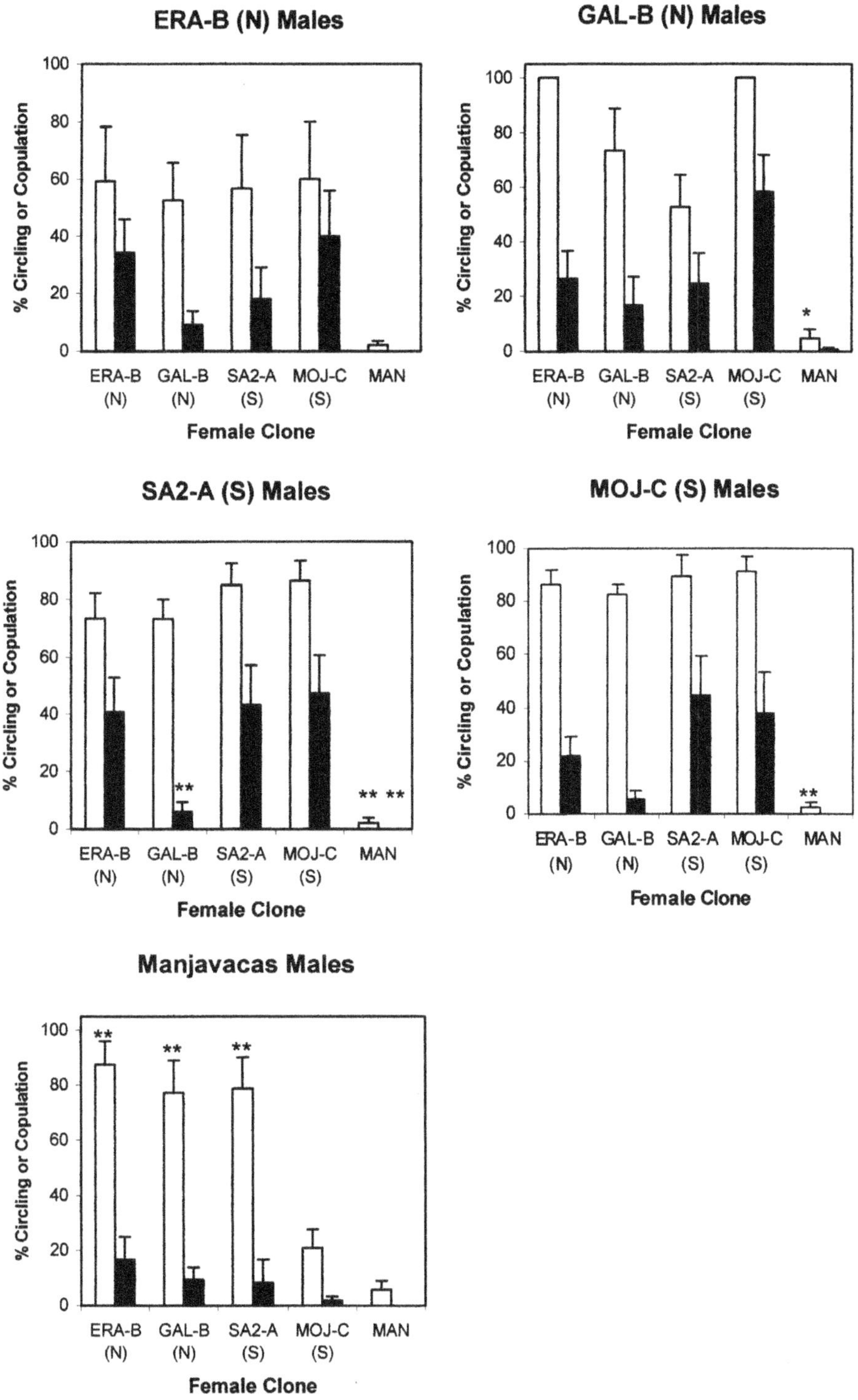

Figure 3. Mating behaviour of males of each clone towards each clone of females. Bars show the mean percentage of encounters ending in circling (clear bars) and copulation (black bars) of six replicate experiments for each cross. Vertical lines are the standard errors. Significant results of the Mann–Whitney tests used to compare the six replicate values of male behaviour (circling or copulation percentages) between homogamic and heterogamic females are also shown (*, $p < 0.05$ level, **, $p < 0.005$). The N and S in parentheses indicate the mtDNA lineage of each clone.

of encounters ending in circling was generally similar for all strains of females. The percentage of encounters ending in copulation was more variable than circling percentages. The least attractive female strain in terms of copulations seems to be GAL-B; this difference is more marked in southern strains than northern strains, but these differences were non-significant.

Discussion

The haplotype results from this study can be combined with those of Gómez et al. (2000), who found that only one out of the 10 individuals tested from ERA was of southern haplotype and only one of the 20 individuals tested from SA2 was of northern haplotype. Combined with the results presented above, a total of 1 southern and 16 northern haplotype individuals have been identified from ERA, and 1 northern and 25 southern from SA2, remaining consistent with the view that ERA is inhabited predominantly by northern haplotypes and SA2 by southern haplotypes. GAL and MOJ were respectively inhabited by northern and southern haplotypes exclusively. Therefore, our results contribute to highlighting the discrepancies between mitochondrial and nuclear DNA in the two nearby locations GAL and SA2. The RFLP technique presented here, given the absence of described diagnostic morphological features between the *B.* 'Manjavacas' lineage and *B. plicatilis*, and between both mtDNA lineages in *B. plicatilis*, is a rapid and cost-effective method of characterising clones isolated from the field.

Pair-wise mating experiments

Interactions between B. plicatilis *and* B. 'Manjavacas'

The mating behaviour data presented here does not lend support to the hypothesis that the northern and southern lineages are reproductively isolated. As for our first hypothesis, the data, although limited, is consistent with the discrepancies found between nuclear and mitochondrial DNA in Gallocanta and Balsa de Santed 2, as SA2-A males showed a significantly lower percentage of copulations to GAL-B females, which, if extended to the population of the pond (Balsa de Santed 2) as a whole, would result in a reduced level of mitochondrial introgression.

Furthermore, the results of the mating tests suggest the existence of behavioural reproductive isolation between *B. plicatilis* and *B.* 'Manjavacas' strains. 'Manjavacas' females consistently elicited fewer mating responses than *B. plicatilis* females with all strains of males, including their own. This result is consistent with that of Ortells et al. (2000), who found that *B. plicatilis* males discriminate against 'Manjavacas' females (the strain GAL-A6 in their paper). *B.* 'Manjavacas' males discriminate against *B. plicatilis* females mostly during the copulation step. The low percentage of homogamic circlings and the absence of homogamic matings between *B.* 'Manjavacas' males and females further supports the view that this lineage differs significantly from *B. plicatilis*, as this result may be due to different optimum culture conditions regarding salinity, temperature or food source. As the experiments involving *B.* 'Manjavacas' rotifers were not all executed at the same time this result cannot be attributed to different conditions on the day of the experiment. However, more work is needed regarding optimal conditions for mating behaviour tests for *B.* (Manjavacas), as previous work has shown that temperature in particular could importantly affect male mating behaviour (Kotani & Hagiwara, 1999, 2003), therefore, our results regarding the behaviour of 'Manjavacas' males must be taken with caution.

Interactions between northern and southern B. plicatilis

The percent of circlings in crosses involving *B. plicatilis* strains are in agreement with published data, with 50–100% of encounters leading to circling in cases where there has not been shown to be discrimination against females (Snell & Hawkinson, 1983; Gómez & Serra, 1995; Rico-Martinez & Snell, 1995a, b). Copulations were observed to occur less frequently than circlings, but often more frequently than has been reported in other studies; many crosses resulted in copulation in around 40% of encounters (maximum 58.33%), compared to published results of up to 12 or 14% (Gómez & Serra, 1995; Rico-Martinez & Snell,1995a, b), although Ortells et al. (2000) have reported copulation percentages up to 94.4% for homogamic matings. These results are most likely a consequence of performing the crosses with new-born females, as males show a marked preference for them (Gómez & Serra, 1996). In general, there was no clear discrimination for circling or copulations between clones of northern and southern haplotype, with the exception of SA2-A males discriminating against GAL-B females. These results indicate that the northern and southern lineages have not evolved significant behavioural reproductive isolation during the long period of time (since the beginning of the Pleistocene) that they have been in geographic isolation (Gómez et al., 2000), lending support to a hypothesis of slow evolution of mate recognition in allopatry.

The fact that nuclear introgression might explain the discrepancies between the nuclear and mitochondrial genomes in *B. plicatilis* SA2 and GAL populations could seem counter-intuitive given that the dispersing propagules of rotifers (resting eggs) hatch into parthenogenetic females. However, if the clones originating from hatchlings fail to produce resting eggs in their first season, only nuclear genes would be transmitted to the following generation, *via* the males produced by sexual daughters (Gómez et al., 2000), and mitochondrial genes will fail to introgress. According to the results presented here, if SA2-A males discriminate against GAL-B females, but the reverse is not true, GAL females migrating into SA2 will be rare in the population and will most probably not secure matings. They will consequently produce mostly males, not resting eggs, resulting in the biased transmission of nuclear genes. Therefore this pattern of asymmetric mating behaviour could explain the discrepancies between the mtDNA and microsatellite results (Gómez et al., 2000). As GAL-B males do not appear to discriminate against SA2-A females this may, indeed, be the case.

An additional factor affecting the direction of gene flow is the relative size of the two lakes. SA2 is a small temporary pond (0.8 ha), whereas GAL is a large lake (1200 ha). This could affect the net direction of dispersal (from GAL to SA2) and also gene flow, and helps explain why there is no observed nuclear introgression in the direction from SA2 to GAL.

A question still to be answered is: When did this discrimination evolve? An interesting possibility that fits our results is that a process of recent reinforcement of reproductive isolation has taken place in SA2. This requires the existence of decreased hybrid fitness in the offspring generated by crosses between both ponds. As, presumably, many fewer SA2 individuals migrate into GAL this would explain why no similar discrimination has developed between GAL-B males and SA2-A females and therefore there has been less selective pressure for GAL males to discriminate against SA2 females as inter-strain matings in this direction would be relatively rare events. If decreased fitness of hybrid offspring was the selective pressure causing mating discrimination (i.e., reinforcement), it would fit with the observation that behavioural isolation does not happen between allopatric populations, even those that have been isolated for more than 1 million years. The hypothesis of decreased fitness of hybrid offspring will need to be investigated in the future. An alternative to the reinforcement scenario would be that behavioural isolation evolved in allopatry in the SA2 pond, and then favoured male-mediated gene flow when the populations came into contact. The demonstration of reinforcement would imply that pre-mating isolation can evolve in sympatry in rotifers, suggesting that eventually complete reproductive isolation could come to exist via the process of sympatric speciation.

Conclusions

This work has presented a rapid and cost-effective method to discriminate between sympatric rotifer species. Behavioural cross-mating experiments showed that reproductive isolation appears to exist between *B. plicatilis* and *B.* 'Manjavacas' strains. Although there was no conclusive evidence of behavioural reproductive isolation between northern and southern *B. plicatilis* lineages, some reproductive isolation was found between Gallocanta and Santed 2 strains. Our results indicate that the pattern of reproductive isolation may be responsible for the population genetic structure reported for these ponds. These results suggest that further investigating the role reinforcement might have played in the evolution of reproductive isolation in these rotifers is warranted.

Acknowledgements

We thank Chris Mitchell for her assistance in the laboratory and her help with culturing rotifers. P.J. Wright (University of Bangor) provided us with the *Tetraselmis* culture. Raquel Ortells made helpful suggestions on a previous version of this manuscript. This work was part of the B.Sc. degree of H.K.B.

References

Ciros-Pérez, J., A. Gómez & M. Serra, 2001. On the taxonomy of three sympatric sibling species of the *Brachionus plicatilis* (Rotifera) complex from Spain, with the description of *B. ibericus* n. sp. Journal of Plankton Research 23: 1311–1328.

De Meester, L., A. Gómez, B. Okamura & K. Schwenk, 2002. The monopolization hypothesis and the dispersal-gene flow paradox in aquatic organisms. Acta Oecologica 23: 121–135.

Folmer, O., M. Black, W. Hoeh, R. Lutz & R. Vrijenhoek, 1994. DNA primers for amplification of mitochondrial cytochrome *c* oxidase subunit I from diverse metazoan invertebrates. Molecular Marine Biology and Biotechnology 3: 294–299.

Fu, Y., A. Hagiwara & K. Hirayama, 1993. Crossing between seven strains of the rotifer *Brachionus plicatilis*. Nippon Suisan Gakkaishi 59: 2009–2016.

Gómez, A., G. J. Adcock, D. H. Lunt & G. R. Carvalho, 2002a. The interplay between colonisation history and gene flow in passively dispersing zooplankton: microsatellite analysis of rotifer resting egg banks. Journal of Evolutionary Biology 15: 158–171.

Gómez, A., G. R. Carvalho & D. H. Lunt, 2000. Phylogeography and regional endemism of a passively dispersing zooplankter: mitochondrial DNA variation in rotifer resting egg banks. Proceedings of the Royal Society of London Series B-Biological Sciences 267: 2189–2197.

Gómez, A. & M. Serra, 1995. Behavioral reproductive isolation among sympatric strains of *Brachionus plicatilis* Muller 1786: insights into the status of this taxonomic species. Hydrobiologia 313: 111–119.

Gómez, A. & M. Serra, 1996. Mate choice in male *Brachionus plicatilis* rotifers. Functional Ecology 10: 681–687.

Gómez, A., M. Serra, G. R. Carvalho & D. H. Lunt, 2002b. Speciation in ancient cryptic species complexes: Evidence from the molecular phylogeny of *Brachionus plicatilis* (Rotifera). Evolution 56: 1431–1444.

Gómez, A. & T. W. Snell, 1996. Sibling species and cryptic speciation in the *Brachionus plicatilis* species complex (Rotifera). Journal of Evolutionary Biology 9: 953–964.

Guillard, R. R. L. & J. H. Ryther, 1962. Studies of marine planktonic diatoms I. *Cyclotella nana* Hustedt and *Detonula confervacea* (Cleve). Canadian Journal of Microbiology 8: 229–239.

Kotani, T. & A. Hagiwara, 1999. Mate recognition of the rotifer *Brachionus plicatilis* Muller at different temperatures. Bulletin of the Faculty of Fisheries of Nagasaki university 80: 55–59.

Kotani, T. & A. Hagiwara, 2003. Fertilisation between the rotifer *Brachionus plicatilis* strains at different temperatures. Fisheries Science 69: 1076–1078.

Ortells, R., T. W. Snell, A. Gómez & M. Serra, 2000. Patterns of genetic differentiation in resting egg banks of a rotifer species complex in Spain. Archiv für Hydrobiologie 149: 529–551.

Rico-Martinez, R. & T. W. Snell, 1995a. Mating behavior and mate recognition pheromone blocking of male receptors in *Brachionus plicatilis* Muller (Rotifera). Hydrobiologia 313: 105–110.

Rico-Martinez, R. & T. W. Snell, 1995b. Male discrimination of female *Brachionus plicatilis* Muller and *Brachionus rotundiformis* Tschugunoff (Rotifera). Journal of Experimental Marine Biology and Ecology 190: 39–49.

Schwenk, K., D. Posada & P. D. N. Hebert, 2000. Molecular systematics of European *Hyalodaphnia*: the role of contemporary hybridization in ancient species. Proceedings of the Royal Society of London Series B-Biological Sciences 267: 1833–1842.

Snell, T. W & M. Childress, 1987. Aging and loss of fertility in male and female *Brachionus plicatilis* (Rotifera). International Journal of Invertebrate Reproduction and Development 12: 103–110.

Snell, T. W. & C. A. Hawkinson, 1983. Behavioral reproductive isolation among populations of the rotifer *Brachionus plicatilis*. Evolution 37: 1294–1305.

Snell, T. W. & F. H. Hoff, 1987. Fertilization and male-fertility in the rotifer *Brachionus plicatilis*. Hydrobiologia 147: 329–334.

Sokal, R. R. & F. J. Rohlf, 1995. Biometry. W.H. Freeman and Co., New York, 776 pp.

Zar, J. H., 1984. Biostatistical Analyses. Prentice-Hall International, 718 pp.

Part III
Taxonomy and Biogeography

Hydrobiologia (2005) 546:137–140 © Springer 2005
A. Herzig, R.D. Gulati, C.D. Jersabek & L. May (eds.) Rotifera X: Rotifer Research: Trends, New Tools and Recent Advances
DOI 10.1007/s10750-005-4110-9

The 'Frank J. Myers Rotifera collection' at the Academy of Natural Sciences of Philadelphia

Christian D. Jersabek[1,2]
[1]*Department of Organismic Biology, University of Salzburg, A-5020, Salzburg, Austria*
[2]*Academy of Natural Sciences, Center for Systematic Biology and Evolution, Philadelphia, PA, 19103, USA*
E-mail: Christian.Jersabek@sbg.ac.at

Key words: natural history collections, glycerine mounts, curation, online catalog, specimen images

Abstract

The Academy of Natural Sciences of Philadelphia (ANSP) houses the world's most comprehensive collection of Rotifera on microscope slides. A particular strength of this major reference collection and type repository lies in the excellent preparation of specimens in life-like extended state and its balanced coverage of the phylum. The collection is almost worldwide in scope, and currently comprises 774 valid taxa, equalling a taxon coverage of 88.2% (families), 75.4% (genera), and 39.1% (species) compared to taxon numbers currently known to occur worldwide. A searchable database plus illustrated catalog has now been published on CD-ROM and on the Academy's website. The poor representation of Rotifera in natural history collections and their generally poor curation and conservation is commented upon.

Notwithstanding the ecological significance of rotifers in aquatic and semiaquatic environments, this phylum is only poorly represented in natural history collections. The few existing museum collections mostly consist of personal microscope slide collections, or are 'orphaned', uncurated and uncataloged slide assemblages dumped on a museum after the death of a researcher or when another museum closed down. Consequently, such taxonomic resources are not readily accessible to interested individuals, although the inspection of historic voucher and type material should form an integral part of systematic and taxonomic research.

As unveiled only recently (Segers, 1997), the Academy of Natural Sciences of Philadelphia (ANSP) houses a significant collection of Rotifera, permanently mounted on microscope slides. Since then, another 4400+ specimens were shipped to Philadelphia in 2000, after the death of W.T. Edmondson. It turned out that these preparations belong to the same collection, and were purchased by J.J. Gilbert in the 1970s. He then put them at the disposal of W.T. Edmondson at the University of Washington, when both researchers agreed the material should later go to the ANSP (J.J. Gilbert, personal communication). The large majority (96%) of the preparations is still in good to excellent condition, with even notoriously intractable taxa prepared in life-like extended state following narcotization, and mounted in pure glycerine or glycerine jelly. This is in contrast to other large rotifer collections, such as, e.g., the Natural History Museum's collection in London, U.K., which includes C.F. Rousselet's and D. Bryce's collections. Approximately 50% of these (mostly formalin) preparations are already deteriorated to an extent that they are of no future benefit, and many more are in jeopardy and need to be remounted (C. Hussey, personal communication). Ongoing losses are extinguishing complex information preserved in biological specimens of many collections, but this crisis has not yet been addressed.

Today, the Philadelphia collection preserves 8600+ specimens, which are catalogued both taxonomically and geographically. They are accessible to visiting researchers, and slides are available on loan to qualified individuals and institutions for research purposes. Preparations include complete specimens of more than 700 morphospecies, and isolated trophi (jaws) of about 250 species. These are mounted on approximately 2000 microscope slides, 104 of which contain type specimens. Accordingly, this collection is the largest of its kind in North America and in terms of species diversity the most comprehensive such collection in the world. Currently, 774 valid taxa equal a taxon coverage of 88.2% (families), 75.4% (genera), and 39.1% (species) compared to the taxon numbers currently known to occur worldwide (cf. Segers, 2002) (Fig. 1).

The collection was basically built up by the amateur rotiferologist and ANSP research fellow Frank J. Myers (1874–1954) and his student Leonard M. Bennetch in the 1920s–1950s. It further benefitted from Myer's collaboration with North American colleagues and overseas researchers, who also contributed slide material and/or specimens to the collection. Amongst these, E.H. Ahlstrom, L.M. Dorsey, W.T. Edmondson, H.S. Jennings, H.K. Harring, R.W. Pennak, A.W. Zahniser (all USA), G.M. Neal (Canada), B. Carlin (Sweden), J. Hauer (Germany), and C.F. Rousselet (United Kingdom) ought to be mentioned. More recently, additional material has been added by research fellows (H. Segers, C.D.J.) during their post-doctoral stays on the Academy's John J. & Anna H. Gallagher Fellowship for the study of Rotifera. The geographic scope of this collection is thus almost worldwide, with a strong focus on the northeastern United States (Fig. 2).

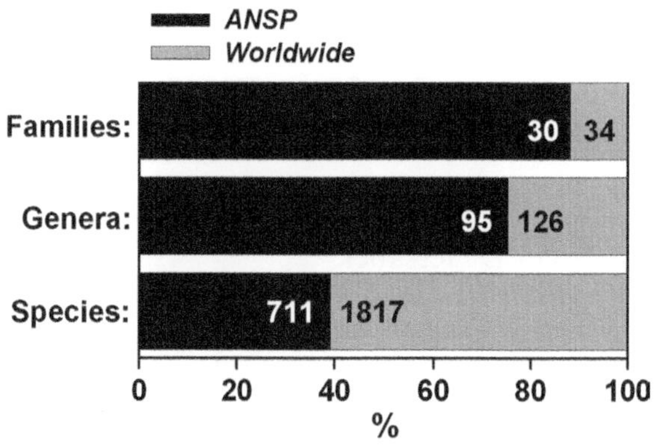

Figure 1. Taxon coverage of the Frank J. Myers Rotifera collection as compared to the taxon numbers currently known to occur worldwide.

The core collection reflects Myers' principal research on rotifers and can thus be used to revisit scientific findings of one of the foremost rotifer taxonomists. Notably, many voucher and type specimens originate from the Pocono Plateau (Pennsylvania) (Myers, 1940; 1942), the Pine Barrens, coastal freshwater and brackish habitats in New Jersey (Myers, 1936a), Mount Desert Island (Maine) (Myers, 1931; 1933a,1933b; 1934a, 1934b, 1934c, 1934d; 1936b), Wisconsin (Harring & Myers, 1922; 1924; 1926; 1928; Myers, 1930), and from the Adirondack Region (New York) (Myers, 1937).

Myers donated further rotifer slides to other museums, viz. the American Museum of Natural History in New York, and the Smithsonian

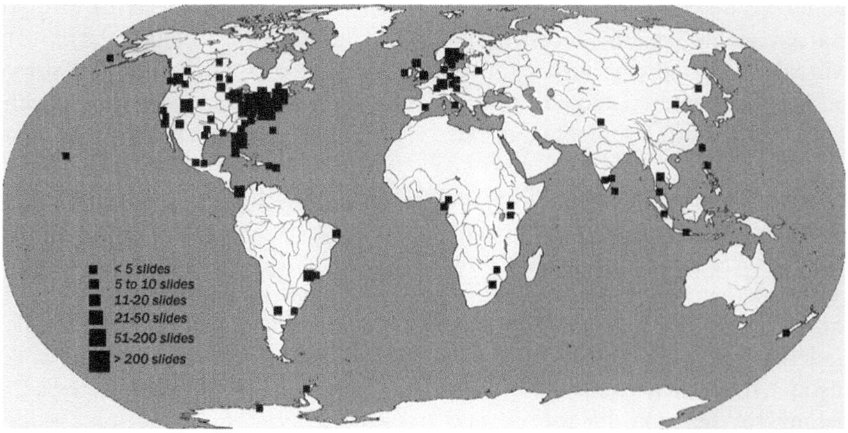

Figure 2. Geographic origin of slide material deposited at the Academy of Natural Sciences of Philadelphia.

Institution in Washington, D.C. The 'Frank J. Myers collection *s.str.*' is thus actually spread across three institutions. Of more than 1000 slides he deposited in the AMNH 'as a safe depository' (Myers, 1938), however, the large majority disappeared sometimes before the 1960s (H.L. Taylor, personal communication), including many type specimens (Boyko, 1994). As an alarming, albeit ironic example, this loss highlights the need for better curation of microscope slide collections, which are furthermore at constant risk of losing preparations because of deterioration from the effects of breakage, leaking slides, dislodging of coverslips, chemical incompatibilities, or drying up. Much has been irretrievably lost already, through neglect and lack of support of historical references that cannot be valued in economic terms. To comply with ICZN (International Commission on Zoological Nomenclature) recommendations 72F.2 and 72F.3, however, institutions holding name-bearing types are challenged to (1) take all necessary steps for their safe preservation and to (2) make them accessible for study, respectively (ICZN, 1999: 79). At the ANSP all type specimens and high priority voucher specimens that were found in deteriorating condition have already been remounted (pure glycerin), and restoration of more slide material is planned. Although not a solution to the problem of deterioration, a complementary approach to preserve specimen information is to take digital images. This has the added benefits of being quick and inexpensive, and digital information can be made instantly accessible to the scientific community if published online. Moreover, by eliminating the need to physically handle a specimen, electronic products such as searchable virtual collections and illustrated online catalogs can catalyze systematic and biogeographic studies, and have an important potential for public outreach.

A searchable taxonomic database of the Academy's collection has recently been compiled and high resolution species plates were prepared of >700 species using an increased-depth-of-field imaging system. This illustrated catalog has been published on CD-ROM (Jersabek et al., 2003a) and it is also accessible on the Academy's website (Jersabek et al., 2003b). By making this major reference collection and type repository for the Rotifera electronically accessible to end-users, systematic research on it shall be promoted. This helps fulfill one of the primary missions to which the Academy of Natural Sciences is committed and shall also encourage researchers to contribute to this collection by depositing material, such as type and valuable voucher specimens.

References

Boyko, C. B., 1994. Catalog of recent type specimens in the Department of Invertebrates, American Museum of Natural History. I. Micro-Invertebrates (Phyla Sarcomastigophora, Gnathostomula, Gastrotricha, Rotifera, and Tardigrada). American Museum Novitates 3106: 1–44.

Harring, H. K. & F. J. Myers, 1922. The rotifer fauna of Wisconsin. Transactions of the Wisconsin Academy of Sciences 20: 553–662.

Harring, H. K. & F. J. Myers, 1924. The rotifer fauna of Wisconsin. II. A revision of the notommatid rotifers, exclusive of the Dicranophorinae. Transactions of the Wisconsin Academy of Sciences 21: 415–549.

Harring, H. K. & F. J. Myers, 1926. The rotifer fauna of Wisconsin. III. A revision of the genera *Lecane* and *Monostyla*. Transactions of the Wisconsin Academy of Sciences 22: 315–423.

Harring, H. K. & F. J. Myers, 1928. The rotifer fauna of Wisconsin. IV. The Dicranophorinae. Transactions of the Wisconsin Academy of Sciences 23: 667–808.

International Commission on Zoological Nomenclature (ICZN), 1999. International Code of Zoological Nomenclature (4th ed.). International Trust for Zoological Nomenclature, The Natural History Museum, London, 306 pp.

Jersabek C. D., Segers H. & B. J. Dingmann, 2003a. The Frank J. Myers Rotifera Collection. The whole collection in digital images. (CD-ROM). The Academy of Natural Sciences of Philadelphia, Special Publication 20.

Jersabek C. D., Segers H. & P. J. Morris, 2003b. An illustrated online catalog of the Rotifera in the Academy of Natural Sciences of Philadelphia (version 1.0: 2003-April-8). [WWW. database] URL http://data.acnatsci.org/biodiversity_databases/rotifer.php.

Myers, F. J., 1930. The rotifer fauna of Wisconsin. V. The genera *Euchlanis* and *Monommata*. Transactions of the Wisconsin Academy of Sciences 25: 353–413.

Myers, F. J., 1931. The distribution of Rotifera on Mount Desert Island. American Museum Novitates 494: 1–12.

Myers, F. J., 1933a. The distribution of Rotifera on Mount Desert Island. III. New Notommatidae of the genera *Pleurotrocha*, *Lindia*, *Eothinia*, *Proalinopsis* and *Encentrum*. American Museum Novitates 660: 1–18.

Myers, F. J., 1933b. The distribution of Rotifera on Mount Desert Island. Part II. New Notommatidae of the genera *Notommata* and *Proales*. American Museum Novitates 659: 1–26.

Myers, F. J., 1934a. The distribution of Rotifera on Mount Desert Island. Part IV. New Notommatidae of the genus *Cephalodella*. American Museum Novitates 699: 1–14.

Myers, F. J., 1934b. The distribution of Rotifera on Mount Desert Island. Part V. A new species of Synchaetidae, and new species of Asplanchnidae, Trichocercidae, and Brachionidae. American Museum Novitates 700: 1–16.

Myers, F. J., 1934c. The distribution of Rotifera on Mount Desert Island. Part VI. New Brachionidae of the genus *Lepadella*. American Museum Novitates 760: 1–10.

Myers, F. J., 1934d. The distribution of Rotifera on Mount Desert Island. Part VII. New Testudinellidae of the genus *Testudinella* and a new record of Brachionidae of the genus *Trichotria*. American Museum Novitates 761: 1–8.

Myers, F. J., 1936a. Psammolittoral rotifers of Lenape and Union Lakes, New Jersey. American Museum Novitates 830: 1–22.

Myers, F. J., 1936b. Three new brackish water and one new marine species of Rotatoria. Transactions of the American Microscopical Society 55: 428–432.

Myers, F. J., 1937. Rotifera of the Adirondack region of New York. American Museum Novitates 903: 1–17.

Myers, F. J., 1938. New species of Rotifera from the collection of the American Museum of Natural History. American Museum Novitates 1011: 1–17.

Myers, F. J., 1940. New species of Rotatoria from the Pocono Plateau, with note on distribution. Notulae Naturae of the Academy of Natural Sciences of Philadelphia 51: 1–12.

Myers, F. J., 1942. The rotatorian fauna of the Pocono Plateau and environs. Proceedings of the Academy of Natural Sciences of Philadelphia 94: 251–285.

Segers, H., 1997. Some Rotifera from the collection of the Academy of Natural Sciences of Philadelphia, including new species and new records. Proceedings of the Academy of Natural Sciences of Philadelphia 148: 147–156.

Segers, H., 2002. The nomenclature of the Rotifera: annotated checklist of valid family- and genus-group names. Journal of Natural History 36: 631–640.

Hydrobiologia (2005) 546:141–145
A. Herzig, R.D. Gulati, C.D. Jersabek & L. May (eds.) Rotifera X: Rotifer Research: Trends, New Tools and Recent Advances
DOI 10.1007/s10750-005-4111-8

Tale of a sleeping beauty: a new and easily cultured model organism for experimental studies on bdelloid rotifers

Hendrik Segers[1,*] & Russell J. Shiel[2]
[1]*Royal Belgian Institute for Natural Sciences, Freshwater Laboratory, Vautierstaat 29, B-1000, Brussels, Belgium*
[2]*Department of Environmental Biology, University of Adelaide, 5005, Adelaide, S.A, Australia*
*(*Author for correspondence: E-mail: Hendrik.Segers@naturalsciences.be)*

Key words: Rotifera, Bdelloidea, new species, Australia, anhydrobiosis

Abstract

We present the description of a new species of bdelloid rotifer, *Adineta ricciae* n. sp., which emerged from dry mud of Ryan's billabong, Victoria, Australia. Its conspicuous frontal eyes easily diagnose the species; it differs from *A. oculata* (Milne) by the position of the eyes and its general habitus. The animal came to our attention because it is exceptionally easy to culture, so that the species already is being used in diverse experimental studies utilising bdelloid rotifers as model organisms.

Introduction

Bdelloid rotifers have several features that make them outstandingly interesting model organisms. These include their ability to sustain prolonged dormancy periods as individual organisms (anhydrobiosis: see Ricci, 2001a), and their exclusive parthenogenetic reproduction (Mark Welch & Meselson, 2000), two key features explaining the ecological and evolutionary success of the group (Ricci, 1987). The study of bdelloid biology, however, is hampered by our poor understanding of the systematics of the group (Ricci, 2001b), and by the labour-intensive techniques required to culture the animals.

During a study on the hatchability of endemic Australian rotifers from resting eggs, we noted, in 1998, a rather peculiar bdelloid emerging from rehydrated mud samples. One specimen of the species (independent evolutionary lineage, in these asexual organisms) was placed in an embryo dish in some water extracted from the experimental vials, and stored in a wet chamber under room conditions. As bdelloid rotifers were not the main focus of the study, the embryo dish was left while attention was paid to the monogonont rotifers hatching from the mud. Knowing that many bdelloid rotifers are rather difficult or even impossible to culture (e.g., see Ricci, 1984), it was a surprise to find that a large population of the animal had developed after a few weeks. As it appeared impossible to match the specimens with any described species and as abundant material was available, a number of specimens were sent to Professor C. Ricci's laboratory in Milan, Italy. She confirmed our initial suspicion that the species was new to science, and even found it back personally and recorded it as *Adineta* cf. *oculata*, also from Ryan's III billabong (Ricci et a1., 2003).

Because the new species turned out to be extraordinarily easy to culture, it was used as model organism in several studies by C. Ricci and students, and by D.B. Mark Welch, to whom specimens were sent subsequently. Several studies involving the new species were presented at the Xth rotifer symposium, including one in which the 'sleeping beauty' strategy in responding to anhydrobiosis of the new species is demonstrated (Ricci & Covino, 2005, present volume). In view of these studies, we decided not to postpone the description of the species any longer. Therefore, we herewith present an account of the species, which we dedicate, with pleasure, to our friend and colleague Professor Claudia Ricci in recognition of her contributions to the knowledge of bdelloid biology.

Adineta ricciae new species

(Figures 1 a–e and 2 a–d)

Holotype and paratype: in the collection of the royal Belgian Institute of Natural Sciences, Brussels, Belgium (IG 30060 RIR 147–148). Specimens isolated from rehydrated dry mud of Ryan's billabong, collected by R.S. and L.W. Tan on June 26, 1998. Material consists of animals anaesthetised using Marcaine®, in permanent slides (Note: the colour of the eyes has faded in the type specimens).

Type locality: Ryan's III billabong, Bonegilla, Victoria, Australia. GPS 36° 06′ 32.1″ S/146° 58′ 37.5″ E.

Material: Abundant specimens of the species were present in the original culture. Populations of *A. ricciae* n. sp. are at present held at the Department of Biology, State University of Milan, Milano, Italy, and at the Josephine Bay Paul Center for Comparative Molecular Biology and Evolution, Marine Biological Laboratory, Woods Hole, USA. Material can be obtained on request from C. Ricci (Milano) or D.B. Mark Welch (Woods Hole).

Differential diagnosis: *A. ricciae* n. sp. is one of only two species in the genus *Adineta* Hudson & Gosse, 1886: that have distinct eyes. It can be distinguished from *A. oculata* (Milne, 1886) by the position of these eyes (frontal in *A. ricciae* n. sp., on the rostrum in *A. oculata*: Milne, 1886), and shape of foot (relatively longer in *A. ricciae* n. sp.) and spurs (generally triangular with set-off tips in *A. ricciae* n. sp., crescent-shaped in *A. oculata*). Also, *A. ricciae* n. sp. is much smaller than *A. oculata* (about 200 μm in *A. ricciae* n. sp., 500 μm in *A. oculata*. Sládeček (1969) records *A. oculata* of 350 μm long, and provides a report on trophi morphology, however, the taxonomic identity of his specimens is uncertain.

Morphologically, the new species appears close to *A. vaga minor* (Bryce, 1893), but this taxon lacks eyes, although it should be noted that colorless globules at more or less the same position of the eyes in *A. ricciae* n. sp. are sometimes represented in illustrations of *A. grandis* (Murray, 1910) and *A. vaga* (Davis, 1873) (see Fig. 200a and c in Donner, 1965). In contrast to *A. vaga minor*, the pseudosegmentation of the body and foot is relatively obvious in *A. ricciae* n. sp. Whereas the trophi of *A. oculata* is unknown, that of *A. vaga* (see Melone et al., 1998) appears to have a few more minor teeth than *A. ricciae* n. sp.

Description: Body elongate, pseudosegmented. Head short, about 10% of total body length; neck slightly longer, about 20% of body length; trunk and foot each about 35% of body length. Head slightly longer than wide, strongly flattened dorso-ventrally. Rotatory organ a ciliated ventral field, terminated by a pair of rakes (Melone & Ricci, 1995) consisting of four anterior-pointed teeth. Vestibulum and mouth aperture immediately posterior to the rakes. Two pigmented frontal eyes. Rostrum short, broad, with short lateral lamellas. Dorsal antenna two-pseudosegmented. Neck slightly narrower than the head anteriorly, widening to distally. Oesophagus relatively long, straight. Mastax small. Trunk filled with stomach and short intestine, paired vitellaria with 8 nuclei. Foot with cloaca, numerous foot glands and pseudosegments, these difficult to discern. Spurs short, triangular, with off-set tips; three toes.

Trophi (Figures 1d–e) ramate, with two pairs of major unci teeth. Nine pairs of minor unci teeth in the proximal, 12/13 pairs in the distal group (see Melone et al., 1998).

Additional information on the morphology of *A. ricciae* n. sp. is provided by Santo (in press), who studied the musculature of the species using confocal microscopy. Mark Welch & Meselson (2003) report on its genome size, Mark Welch (2005, present volume) uses a copy of its 82 kD heat-shock protein gene in a phylogenetic study of Rotifera, and Mark Welch & Mark Welch (2005, present volume) report on the use of DNA of *A. ricciae* n. sp. in microarrays.

Measurements: Total length 220 μm, head width 31 μm, length 43 μm, greatest trunk width 54 μm, spurs 10 μm (of creeping specimen). Trophi: length 13 μm, width (closed) 15 μm. Ramus: length 11 μm, major unci teeth length 4.8–5.4 μm.

Distribution and ecology: The new species is only known from its type locality. However, it may have been overlooked in the past: all *Adineta*

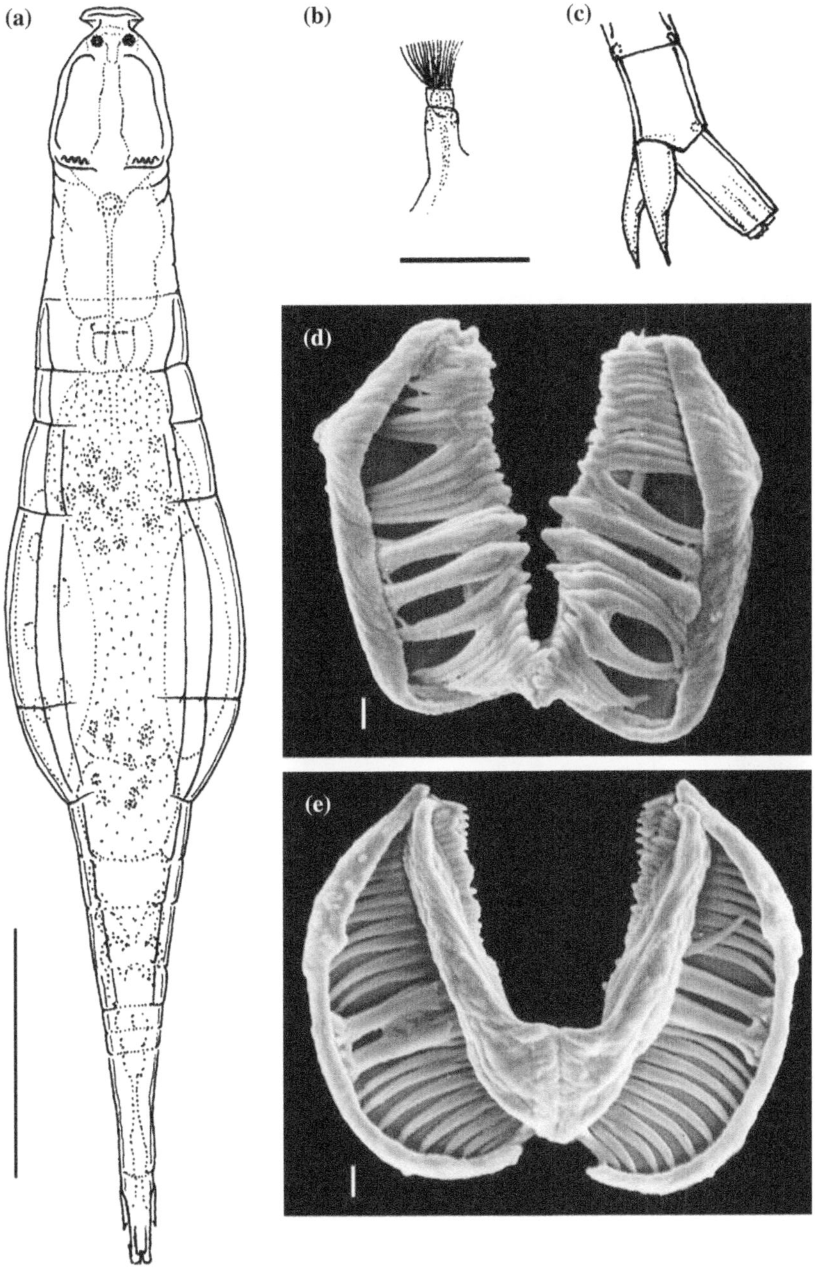

Figure 1. (a–e) *Adineta ricciae* n. sp. a: habitus, ventral view; (b) dorsal antenna, lateral; (c) foot and spurs, lateral; (d–e) trophi, SEM photographs; (d) frontal; (e) caudal. Scale bars: (a) 50 μm, (b–c) 10 μm, (d–e) 1 μm.

specimens with eyes have to date been identified as *A. oculata*, and there are strong indications that several different species are lumped under this name (e.g., *A. oculata* sensu Sládeček, 1969). The ease with which the species can be cultured indicates that it is probably eurytopic. *A. ricciae* n. sp. is recorded from Ryan's III billabong, a weedy temporary pond in north-eastern Victoria, Australia. This billabong is notable in that it has the greatest rotifer biodiversity yet known from a single site for the continent (>200 spp., cf. Brock et al., 2003; Shiel & Green, unpublished data). About >106 spp. of rotifers were recorded in a single net tow during a wetted phase (Shiel et al., 1998). New

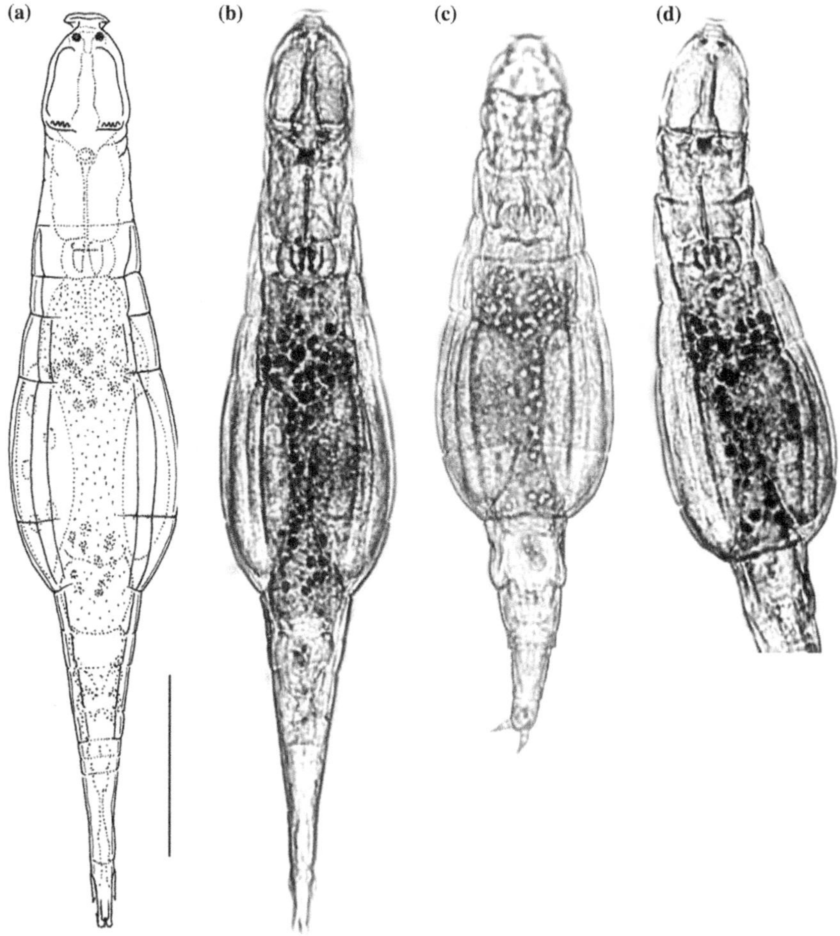

Figure 2. (a–d) *Adineta ricciae* n. sp. (a) habitus drawing; (b–d) light microscopy photographs (courtesy of G. Melone, Milano, Italy).

records for Australia, including new species, continue to be recorded there, even after >20 years relatively intensive sampling (Langley et al., 2001). The long evolutionary history of the River Murray floodplain is a likely major driver of high microfaunal diversity in both ephemeral and permanent waters in the region – the basin was in its present location >100 MY BP in Gondwana, in its present channel for >2 MY. Some R. Murray billabongs have been dated at ca. 8–10 000 years in their present locations (R. Ogden, pers. comm.).

Acknowledgements

We thank David Mark Welch for providing information on his use of *A. ricciae* n. sp., and for encouraging us to describe the animal. Lor Wai Tan assisted in the collection of sediment samples from the Ryan's Billabong series. David Mark Welch and Christian Jersabek are thanked for reviewing the manuscript.

References

Brock, M.A., D. L. Nielsen, R. J. Shiel, J. D. Green & J. M. Langley, 2003. Drought and aquatic community resilience: the role of eggs and seeds in sediments of temporary wetlands. Freshwater Biology 48: 1207–1218.

Donner J., 1965. Ordnung Bdelloidea (Rotatoria, Rädertiere). Bestimmungsbücher zur Bodenfauna Europas 6, Akademie-Verlag, Berlin, 297 pp.

Langley, J. M., R. J. Shiel, D. L. Nielsen & J. D. Green, 2001. Hatching from the sediment egg-bank or aerially-dispersed? – The use of mesocosms in assessing rotifer biodiversity. Hydrobiologia 446/447: 203–211.

Mark Welch, D. B., 2005. Bayesian and maximum likelihood analyses of rotifer- acanthocephalan relationships. Hydrobiologia 546: 47–54.

Mark Welch D. B. & J. L. Mark Welch, 2005. The potential of genomic approaches to rotifer ecology. Hydrobiologia 546: 101–108.

Mark Welch, D. B. & M. Meselson, 2000. Evidence for the evolution of bdelloid rotifers without sexual reproduction or genetic exchange. Science 288: 2111–2115.

Mark Welch, D. B. & M. Meselson, 2003. Genome size and GC content in rotifers of the anciently asexual Class Bdelloidea. Biological Journal of the Linnean Society 79: 85–91.

Melone, G. & C. Ricci, 1995. Rotatory apparatus in bdelloids. Hydrobiologia 313/314: 91–98.

Melone, G., C. Ricci & H. Segers, 1998. The trophi of Bdelloidea (Rotifera): a comparative study across the class. Canadian Journal of Zoology 76: 1755–1765.

Milne, W. M. A., 1886. On the defectiveness of the eye-spot as a means of generic distinction in the Philodinaea, with a description of two other Rotifera. Proceedings of the Philosophical Society of Glasgow 17: 134–145.

Ricci, C., 1984. Culturing of some bdelloid rotifers. Hydrobiologia 112: 45–51.

Ricci, C., 1987. Ecology of bdelloids: how to be successful. Hydrobiologia 147: 117–127.

Ricci, C., 2001a. Dormancy patterns in rotifers. Hydrobiologia 446/447: 1–11.

Ricci, C., 2001b. A reconsideration of the taxonomic status of *Macrotrachela quadricornifera* (Rotifera, Bdelloidea). Journal of Zoology, London 255: 273–277.

Ricci C. & C. Covino, 2005. Anhydrobiosis of *Adineta ricciae*: costs and benefits. Hydrobiologia 546: 307–314.

Ricci C., R. J. Shiel, D. Fontaneto, G. Melone, 2003. Bdelloid rotifers recorded from Australia, with description of *Philodinavus aussiensis* n. sp. Zoologischer Anzeiger 242: 241–248.

Santo, N., in press. Muscles and locomotion of three species of Rotifera Bdelloidea: a morpho-functional study. Zoomorphology.

Shiel, R. J., J. D. Green & D. L. Nielsen, 1998. Floodplain biodiversity: why are there so many species? Hydrobiologia 387/388: 39–46.

Sládeček, V., 1969. A note on the rotifer *Adineta oculata* (Milne). Věstnik Československé společsnosti zoologické (Acta Societatis Zoologicae Bohemoslovacae) 33: 369–371.

Hydrobiologia (2005) 546:147–157
A. Herzig, R.D. Gulati, C.D. Jersabek & L. May (eds.) Rotifera X: Rotifer Research: Trends, New Tools and Recent Advances
DOI 10.1007/s10750-005-4112-7

Life on the edge: rotifers from springs and ephemeral waters in the Chihuahuan Desert, Big Bend National Park (Texas, USA)

Robert L. Wallace[1], Elizabeth J. Walsh[2,*], M.L. Arroyo[2] & Peter L. Starkweather[3]
[1]*Department of Biology, Ripon College, Ripon, WI, 54971, USA*
[2]*Department of Biological Sciences, University of Texas – El Paso, El Paso, TX, 79968, USA*
[3]*Department of Biological Sciences, University of Nevada – Las VegasLas Vegas, NV, 89154, USA*
*(*Author for correspondence: E-mail: ewalsh@utep.edu)*

Key words: biogeography, huecos, seeps, tanks, tinajas

Abstract

Here we describe an on-going study of the rotifers inhabiting a sampling of springs (seeps), streams, ponds, tanks (diked ephemeral streams), and huecos and tinajas (small and large rock pools) of Big Bend National Park, a 3.23×10^5 ha region of the northern Chihuahuan Desert located in southwestern Texas (USA). We collected samples from planktonic, littoral, and benthic habitats comprising 92 sites representing 23 different aquatic systems. We documented 19 rotifer families (17 monongonont; 2 bdelloid) comprising 32 genera and 94 taxa. Of these, 70 were identified to species; 24 taxa (14 monogononts and 10 bdelloids) remain unidentified; several may be new to science. Redundancy Analysis revealed significant associations between environmental parameters and species distributions among water sources. Highest species richness was found in more permanent habitats such as ponds and springs while species associated with rock pools were associated with high conductivity and temperature.

Introduction

Extending from the north central plateau of Mexico into Arizona, New Mexico, and Texas, the Chihuahuan Desert has been recognized as one of the world's most unique and biologically diverse arid regions (Olsen & Dinerstein, 1998). Because of its unique features the World Wildlife Fund® has designated this desert as one of The Global 200: significant ecoregions identified for intensive conservation efforts (Anon., 2000). Like other desert ecosystems, the Chihuahuan is at significant risk from a variety of anthropogenic activities, including human habitation, industrial and recreational development, introduction of exotic species, overgrazing by non-native stock animals, atmospheric deposition of pollutants, and aquifer depletion (Shepard, 1993; Olsen & Dinerstein, 1998; Abell et al., 2000; Anon., 2000). Of these dangers, aquifer depletion is perhaps the most serious because of its extensive impact on the aquatic biota. Dinerstein et al. (2000) note that the Chihuahuan Desert likely has a high degree of local endemism of the freshwater biota; however, relatively few studies have been done on its aquatic systems and this supposition must be regarded as tentative. Thus, while these unique habitats with their associated biota potentially represent an important ecological sentinel for overall environmental health, there is only a sparse literature on the aquatic fauna of their springs and ephemeral waters (e.g., MacKay et al., 1990). Moreover, what work has been done has concentrated on the larger invertebrates, while neglecting microscopic forms, especially microfauna. These groups are important to desert aquatic communities because they comprise the food of invertebrate predators as well as small fishes. One taxon that has been nearly ignored in desert waters, but that is likely to demonstrate high species richness, is Phylum Rotifera (Williams, 2001). Here we describe an ongoing study of the rotifers that inhabit the rare and beautiful springs (seeps), streams, tanks (diked ephemeral streams), and huecos and tinajas (small and large rock pools) of Big Bend National Park (BBNP) (Texas, USA).

Study sites

Designated as a national park in 1944 and a Biosphere Reserve in 1979, BBNP comprises a large contiguous area of some 3.23×10^5 ha in southwest Texas, USA (Fig. 1). Environments of this region include broad, open flatlands, dominated by thorny desert plants including numerous cacti, eroded slopes of scrubby grasslands, arroyos and canyons with temporary streams, and rugged mountainous terrain. Park elevations range from 550 m along the Rio Grande to 2400 m in the Chisos Mountains. With an average rainfall of <25 cm/year (occurring mostly between May and October), the park is very dry; yet >300 springs, streams, and other water bodies, have been documented (e.g., Lind & Bane, 1980; National Geographic Society®, 1990; D.E. Bowles, pers. comm.).

Methods

Four extensive collecting trips into BBNP were undertaken over an 18-month period (Fig. 1; Appendix 1). We employed a variety of sampling techniques including using nets (64 μm), aspirating samplers for flocculent bottoms, and simple sediment corers, as well as taking grab samples (i.e., aquatic plants for sessile forms) (Nogrady et al.,

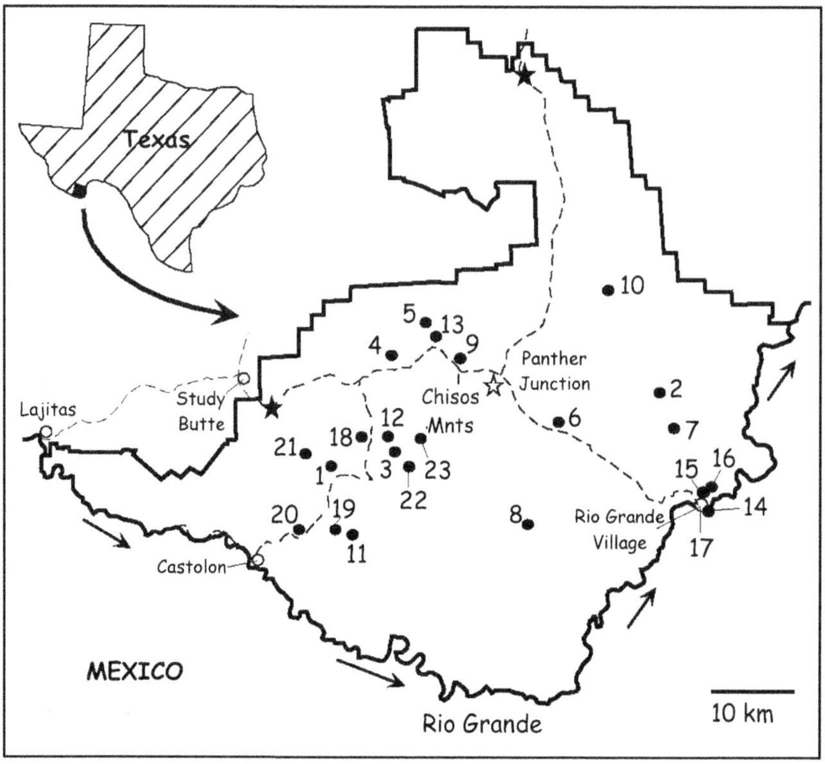

Figure 1. Map of the study area (Big Bend National Park, TX, USA; ca. 29° 15′ 00″ N; 103° 15′ 00″ W). Dashed lines are paved roads; unimproved roads, often passable only by four-wheel drive vehicles, are not illustrated. Arrows indicate direction of the Rio Grande flow. Open circles are local towns; open star is the park headquarters at Panther Junction; closed stars are park entrance stations; closed circles with numbers indicate the approximate location of our sampling sites, each with one or more distinct locations, either as an aquatic complex or multiple distinct sampling sites. Consult Appendix 1 for a key to these sites.

1993; Wallace & Snell, 2001; Wallace & Ricci, 2002). No samples were taken from hyporheic habitats. The equipment was cleaned using a distilled water rinse and, whenever possible, dried between uses in different systems. Although we usually took multiple samples at each site, we attempted to minimize environmental damage of the smaller systems by keeping the total amount of each sample to about 50 ml of source water. All samples were stored in Whirlpack® bags. The remoteness of many sites from serviceable roads required extensive hikes (up to 14 km round trip) along trails and, on occasion, across open desert. After the samples were collected we returned them to our support vehicle as soon as practical (usually <2 h) and placed them in a cooler at 5–10 °C, but avoided having the samples in direct contact with ice. Samples from Langford Hot Spring were stored at slightly warmer temperatures (ca. 10 °C).

Standard physical and chemical data were recorded at nearly every site sampled: i.e., water temperature, pH, DO, conductivity, total dissolved solids, oxidation reduction potential (ORP), habitat size (length, width, depth), vegetation, and general site and weather conditions. We recorded GPS coordinates using a Brunton Multi-Navigator® and took digital photographs to document each site.

Upon our return to the laboratory at the University of Texas – El Paso (UTEP) (<4 days) we stored all samples at ca. 4 °C until they were examined by at least two of us before being discarded. Samples from the hot spring were examined first. Pasteur micropipettes were flushed several times in boiling distilled water before being used to process samples. Only live animals were counted.

Keys to the Rotifera are nearly always regional in scope, thus we were cautious with their use. Species that did not match published descriptions were identified only to the level of genus or were assigned the designation of 'cf.' to indicate that while we could assign these specimens to a valid species name, there were minor differences from the published descriptions. The keys used in this study were as follows: Bdelloidea – Donner (1965), Koste & Shiel (1986), Ricci & Melone (2000); Monogononta – Edmondson (1949, 1959), Berzins (1951), Ruttner-Kolisko (1974), Koste (1978), Stemberger (1979), Nogrady et al. (1993, 1995), Segers (1995), De Smet (1996), De Smet & Pourriot (1997). Voucher specimens in the form of formalin and alcoholic specimens are deposited at the Laboratory for Environmental Biology (UTEP) and accessioned into the existing BBNP collection (cat (33728–33840).

Statistical analyses of patterns among habitat characteristics and species distributions were done using Redundancy Analysis in CANOCO (ter Braak & Smilauer, 1998–2002). For the analyses, water sources were scored for seven indices: (1) type (isolated cold springs, tinajas, huecos, stream, river/canal, hot spring, ponds, tanks, Cattail Spring flowage, or well); (2) complex (a interconnected group of water bodies or not); (3) sequence in the complex; (4) predominant vegetation (algae, allochthonous material, limited macrophytes, extensive macrophytes); (5) flow (stagnant or not); (6) season (summer, fall, winter); (7) exposure (full sun, partial sun, full shade).

Results and discussion

Overall species richness

A total of 92 sites comprising 23 aquatic systems (sites obviously sharing the same surface water source) were sampled in this study. This effort yielded 94 rotifer taxa, of which 70 were identified to species (Appendix 2). Although we did not ignore bdelloids in our samples, and we attempted to collect from habitats where these animals are usually well represented, we did not find many in our samples; only 2 of 12 bdelloids were identified to species. Of the 29 monogonont families of Rotifera recognized by Segers (2002), 17 were present in our samples.

It is doubtful that we have come close to exhausting the rotifer fauna of BBNP as we have only been able to survey fewer than 10% of the known springs, streams, and tinajas in the park. Moreover, a plot of the cumulative number of different taxa that we have recovered during the four collection trips does not appear to approach a plateau (Fig. 2). Nevertheless, the number of taxa that we have ascribed to the waters of BBNP fall well within the range (ca. 11–125) that have been reported in other studies of desert waters: e.g., Algeria: waters (De Ridder, 1991); ephemeral dune

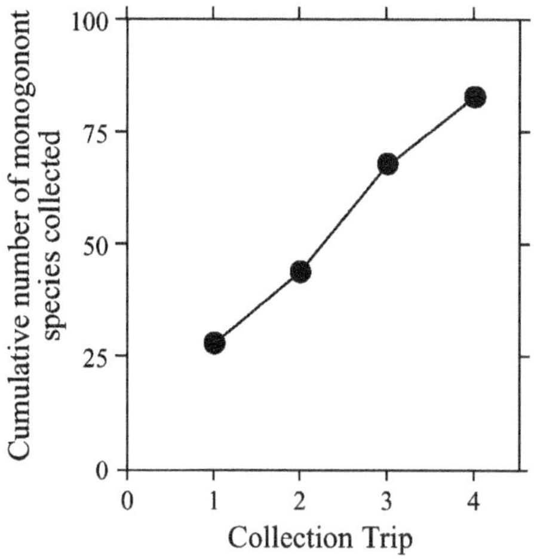

Figure 2. Cumulative number of monogonont taxa found in BBNP waters per sampling trip. Collection trips were as follows: (1) 11–19 July 2001; (2) 16–19 October 2001; (3) 11–15 July 2002; (4) 4–7 January 2003.

ponds, Doñana National Park, Spain (Mazuelos et al., 1993); Galapagos Islands (De Smet, 1989); Kalahari (Brain et al., 1995).

Comparison to other desert habitats

There has been little work done on the rotifers of arid regions, and so we are limited in our ability to make comparisons to similar habitats (e.g., Dumont & Coussement, 1976; Coussement & Dumont, 1980). To assess whether the rotifer fauna of BBNP is distinctive, we evaluated our findings against those reported by other researchers by comparing the percentage of species in the families present in our samples to that of other studies (Fig. 3). (In this figure we grouped all four bdelloid families into a single category.) Although we recognize that the results of these other studies may reflect differences in diversity of habitats sampled, relative sampling effort, sampling seasonality, and/or interannual variability – thus limiting this comparison in several respects – this approach leads to some interesting observations.

The total percentage of species present in each family varies considerably among the studies considered, with the compilation from Algerian desert waters being among those with greatest species richness (De Ridder, 1991), and those from the Kalahari most depauperate (Brain et al., 1995). These differences are, of course not surprising given the scope of these two studies. The Algerian data represent a large dataset from a wide variety of habitats encompassing sites ranging over an area >1/2 of the entire country. Nevertheless, there are only 21 families represented in Algerian desert waters. Thus, with a total of 19 families, the BBNP waters are nearly as diverse at the level of family as those of Algerian waters. On the other hand, the Doñana and Galápagos ecoregions, comprising 12 families each, are slightly less rich. With 21 families (Walsh, unpubl. data), the waters of Hueco Tanks State Historic Site (HTSHS), encompassing an area of only 348 ha, are as diverse, at the level of family, as those of Algeria and BBNP (Fig. 3).

The distribution patterns of species within rotiferan families in each of these desert regions are strikingly different. For example, in comparison to the other studies, our BBNP samples had no Asplanchnidae or Conochilidae and only a limited number of Brachionidae, Epiphanidae, Mytilinidae, and Synchaetidae, but they did possess more Collothecidae and Ituridae. Moreover, while four of the five studies summarized in Figure 3 reported members of the Hexarthridae, we can report only one species (*Hexarthra cf. brandorffi*) and that was found in a preliminary study, not from any of the samples reported here.

On the other hand, our BBNP samples had an interesting similarity to what Coussement & Dumont (1980) reported for the Moroccan Altas Mountains (MAM). In both of theses studies a substantial number of the species that were identified were of the genus *Lecane* (Lecanidae): ≈21% in the BBNP samples and ≈25% in the MAM samples. The only other genus in our samples that was a widely represented was *Cephalodella*: ≈16% of the identified species. We speculate that the dominance of *Cephalodella* and *Lecane* in BBNP waters may be due to the ability of these taxa for passive dispersal; indeed, 11 of the 15 species of *Lecane* identified from our samples are described by Segers (1996) as cosmopolitan or tropicopolitans (see also Langley et al., 2001).

When compared to an intensive 8-year sampling effort at another Chihuahuan desert site (HTSHS), Walsh (unpubl. data) found 74 species; of these, 38% overlapped with those found at BBNP. Rotifers found only at HTSHS (Asplanchnidae, Conochilidae, Testudinellidae)

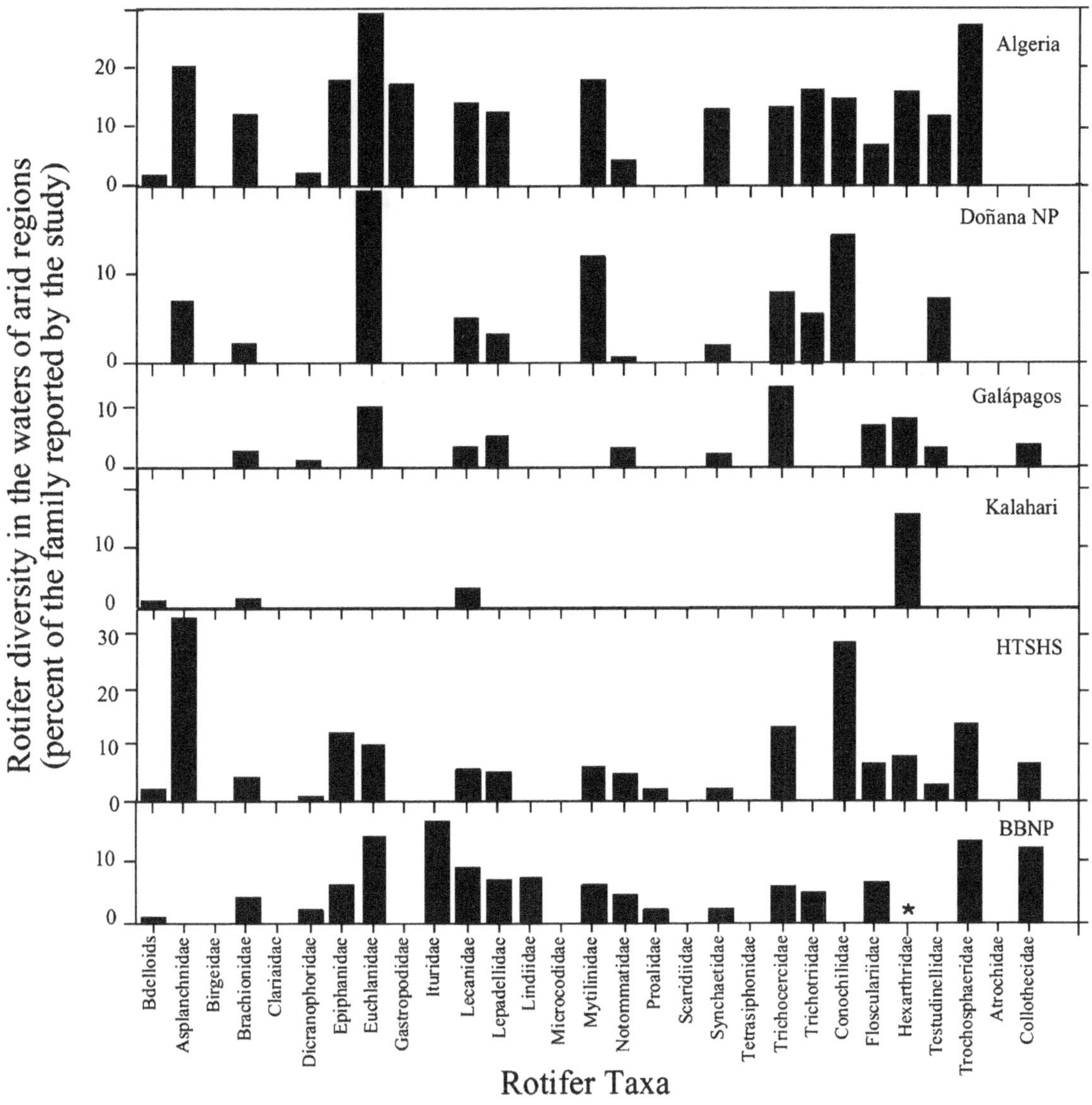

Figure 3. Rotifer diversity in the waters of six different arid regions. These studies are as follows: Algeria (De Ridder, 1991); ephemeral dune ponds, Doñana National Park, ES (Mazuelos et al., 1993); Galápagos (De Smet, 1989); Kalahari, Gemsbok National Park, RSA (Brain et al., 1995); HTSHS, US (Walsh, unpub. obs.); BBNP, US (this study). *, *Hexarthra brandorffi* was found in a preliminary study, but not in the samples reported herein. ‡, Here we follow Segers (2002) in subsuming Filiniidae within the Trochosphaeridae. Note that in this figure we grouped all four bdelloid families into a single category.

were primarily present in large natural and artificial playas, a habitat type not present in BBNP. However, of the species found in the rock pools alone, the overlap is still approximately one third.

We speculate that the differences found in these diverse habitats represent not only differences in the extent of the sampling efforts of each study, but more importantly they may represent fundamental differences in the habitats of the ecoregions examined. They also may reflect differences in the dispersal of the taxa, indicating that the rotifers of desert waters are probably not cosmopolitan.

Habitat type and species richness

A problem unique to the study of desert waters is the fact that some habitats are extremely ephemeral; these systems may contain standing water on one sampling date and then be dry the next. Thus, we believed it to be sensible not to include dry samples in our calculations of mean species richness (see below). Another problem with these systems is that spates sometimes restructure basin shape from one rainfall to the next. Thus, to estimate habitat size we used the average size of the site sampled for all wet periods.

Species richness varied considerably among habitat types (Fig. 4). Streams with short runs (<50 m), intermittent tinajas, small huecos, and wells had the fewest number of species, usually averaging ≤1.5 species per habitat. Isolated springs, Rio Grande (river/canal), streams with long continuous runs (>50 m), tinajas with longer lasting pools (Window Trail and Ernst Tinaja series), and tanks had higher species richness, averaging between 3.0 and 3.5 species per habitat. The single hot spring had 7 species. Species richness of the pools of Cattail Spring flowage was high, with a mean of 10.3 species, but these pools also had some of the greatest variability (range 2–25). Not surprisingly, the habitats with the greatest species richness were the two permanent ponds at Rio Grande Village (mean 22.5 species).

Analysis of environmental factors

Redundancy Analysis is a statistical technique that can be applied to make predictions of how entire assemblages of species respond to multiple environmental factors (Legendre & Anderson, 1999). We applied this technique to investigate relationships of physical characteristics of water sources and rotifer species found. In the analysis of the entire dataset, a significant relationship between environment parameters and species composition was found (RDA, F ratio = 1.76, p = 0.001). A further analysis focusing on the relationship between species composition and habitat type showed that ponds, Cattail Spring flowage, and tinajas accounted for most of the variation in species presence (Fig. 5a). This analysis indicated that many species were associated only with a particular habitat type. Most species of rotifers are associated with ponds, with a secondary trend toward high species richness in Cattail Spring flowage. This may reflect a combination of factors associated with permanence of the water source such as the development of complexity within the physical environment (for instance, permanent vegetation that can provide refugia from predators) and/or the presence of a temporal succession of species within a habitat. In contrast, there are

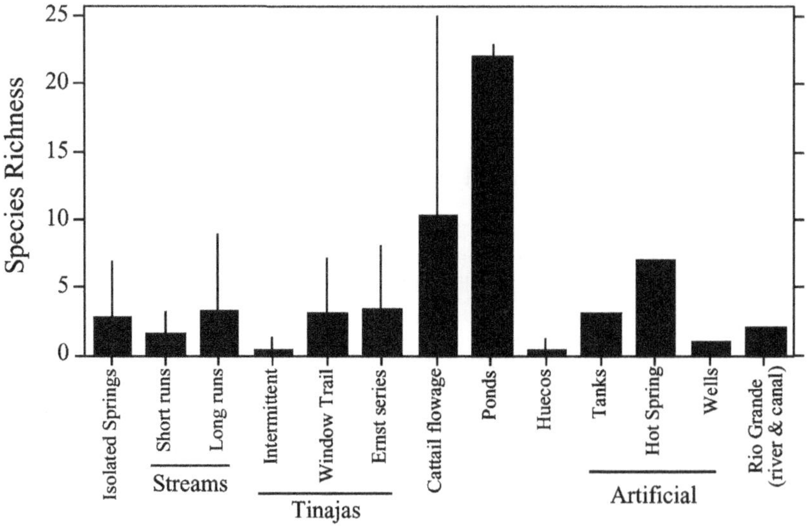

Figure 4. Mean species richness by habitat type. Means recorded here are for all samples taken for each habitat type. Thin vertical lines report maxima. Number of sites sampled within each habitat are: isolated springs (seeps), 7; short streams (<50 m), 2; long streams (> 50 m), 2; intermittent tinajas, 3; Window trail, 1; Ernst series, 1; Cattail flowage, 1; ponds, 2; huecos; 3; tanks, 2; hot spring, 1; wells, 2; Rio Grande (river/canal), 1. (NB: Some sites had multiple sampling stations.)

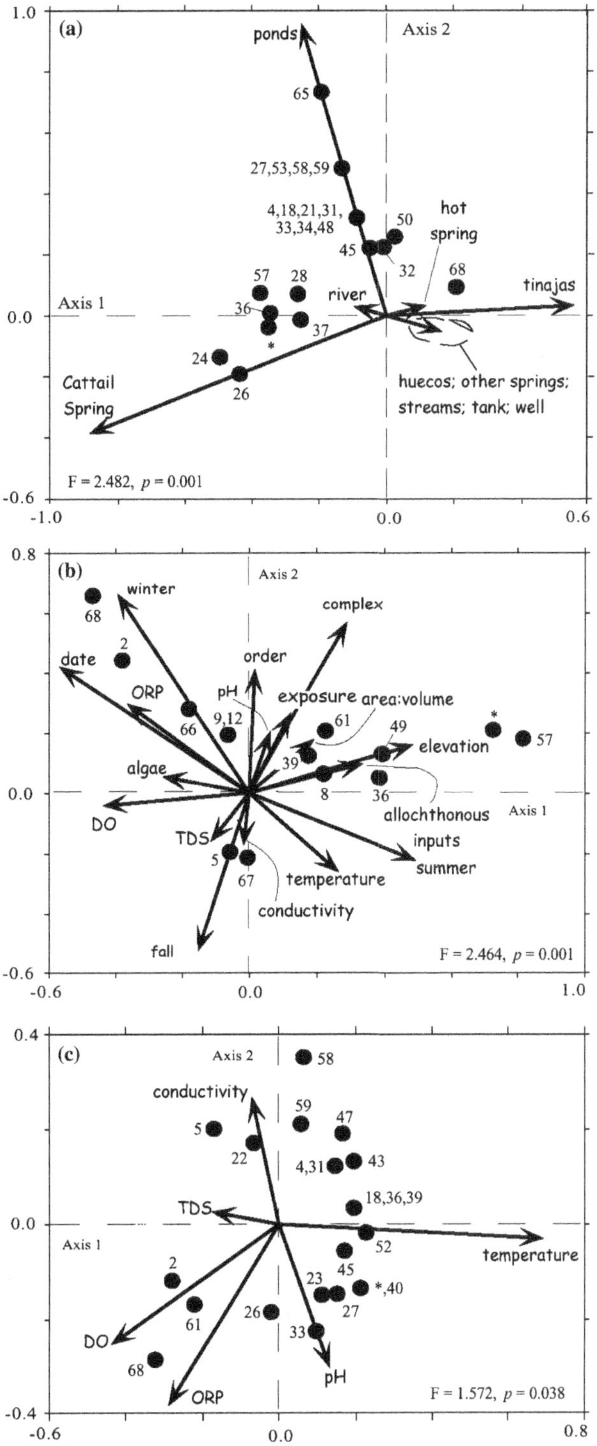

Figure 5. Redundancy Analyses of environmental parameters and species composition found in the water sources of BBNP. Numbers refer to the species listed in Appendix 2. These ordination diagrams present only those species that passed the inclusion rule of a 3–100% fit range. (a) Habitat type. (b) Tinajas only. (c) Water chemistry.

relatively few species associated with huecos, tinajas, and ephemeral streams which so typify North American deserts.

In addition to these general patterns, we noted that *Platyias quadricornis* (Ehrenberg, 1832), *Epiphanes senta* (O. F. Müller, 1773), *Itura viridis* (Stenroos, 1898), *Squatinella mutica* (Ehrenberg, 1832), *Notommata* sp., and *Macrochaetus sericus* (Thorpe, 1893) were found only in ponds while flosculariids (3 species of *Ptygura*), *Wierzejskiella vagneri* Koniar, 1955, and *Dipleuchlanis elegans* (Wierzejski, 1893) were found only in Cattail Spring flowage. *Lindia* cf. *pallida* Harring & Myers, 1922 was found only in the hot spring. Interestingly *Brachionus*, a member of a diverse and cosmopolitan family, is poorly represented in BBNP waters. *Brachionus* species were found only in ponds, tanks, and the canal/river.

In analysis of tinajas alone, most species are associated with high conductivities and TDS, high surface area:volume ratios, higher elevations, system exposures, and systems domination by allochthonous organic matter inputs (Fig. 5b). Most tinaja-specialist species also plot with winter, the season in which temporary ponds are more common and with ORP, this combination reflecting the pattern of rainfall and runoff in BBNP.

Examining the distribution of rotifer species based on water chemistry reveals two clusters of species (Fig. 5c). The first of these is associated with high conductivity and water temperature, two variables that strongly covary in desert systems. This major cluster of species also is arrayed along an axis representing habitat-specific pH, the significance of which is unclear but may reflect the heterogeneity of geological substrata in BBNP. A few species are strongly associated with high dissolved oxygen and ORP, these may be systems with particularly high autochthonous (and oxygenic) producers.

Conclusions

Our study demonstrates that aquatic systems in BBNP support diverse rotifer communities, but that these communities vary widely among habitat type. This variation appears to be due to the extensive influence exerted by the local edaphic conditions. Communities with the greatest species richness are those in more permanent habitats (e.g., Cattail Spring, permanent ponds). While we have shown that rotifers in desert systems are likely to have species richness comparable to that reported from other systems, our investigations have just begun to delve into these unusual habitats. Thus, we believe that a more comprehensive study of desert aquatic systems is likely to increase the number of rotifers known to science, provide a better understanding of their biogeography, and elucidate the fundamental structure of these communities.

Despite their unique biodiversity and diverse abiotic conditions, desert waters have been neglected by ecologists (Abell et al., 2000; Williams, 1985). Such inattentiveness is reasonable given the difficulty in locating and working in these systems. Nevertheless, such efforts offer a unique and valuable insight to the structure of these aquatic systems. It will only be with continued study of arid places that we will be able to understand the processes that lead to the development of their communities. Without an understanding of these processes we will be unable to assess the magnitude of ecological change taking place in ecoregions like BBNP.

Acknowledgements

We thank Carol Purchase, Raymond Skiles and the staff of BBNP for their cooperation and help in this undertaking. This research was carried out under BBNP permit BIBE-2001–SCI-0058. Field assistance was kindly provided by Judith Rios, Darrelle Leeming, Monica Sigla Arana, Gerardo Castellanos, Steve Aley, Brigitte and Jamie Newlin. We thank John Walton for statistical advice and two anonymous reviewers who improved the manuscript. Funding was provided by College of Science (UTEP), Faculty Development Funds (Ripon College), Department of Biological Sciences (UNLV) and NIH 5G12RR00B124.

References

Abell, R. A., D. M. Olson, E. Dinerstein, P. T. Hurley, J. T. Diggs, W. Eichbaum, S. Walter, W. Wettengel, T. Allnutt, C. J. Loucks & P. Heado, 2000. Freshwater Ecoregions of North America. A Conservation Assessment. Island Press, Washington D.C., 319 pp.

Anon, 2000. The Chihuahuan Desert: An Endangered Space. World Wildlife Fund, NW, Washington. DC, 1–14.

Berzins, B., 1951. On the Collothecacean Rotatoria, with special reference to the species found in the Aneboda district, Sweden. Arkiv för Zoologi 1: 565–592.

Brain, C. K., I. Flourie & R. J. Shiel, 1995. Rotifers of the Kalahari Gemsbok National Park, South Africa. Hydrobiologia 313/314: 319–324.

Coussement, M. & H. J. Dumont, 1980. Some peculiar elements in the rotifer fauna of the Atlantic Sahara and of the Atlas Mountains. Hydrobiologia 73: 249–254.

De Ridder, M., 1991. Rotifers from Algeria. Journal of African Zoology 105: 473–483.

De Smet, W. H., 1989. Rotifera uit de Galapagoseilanden. Natuurwetenschappelijk Tijdschrift 69: 110–131.

De Smet, W. H., 1996. Rotifera. Vol. 4: The Proalidae (Monogononta). Guides to the Identification of the Microinvertebrates of the Continental Waters of the World Vol. 9 SPB Acad. Publishing, The Hague, The Netherlands, 102 pp.

De Smet, W. H, 1997. Rotifera. Vol. 5: The Dicranophoridae (Monogononta) and: The Ituridae (Monogononta). Guides to the Identification of the Microinvertebrates of the Continental Waters of the World. Volume 12. SPB Acad. Publishing, The Hague, The Netherlands, 344 pp.

Dinerstein, E., D. Olson, J. Atchley, C. Loucks, S. Contreras-Balderas, R. Abell, E. Iñigo, E. Enkerlin, C. Williams & G. Castilleja, 2000. Ecoregion-based Conservation in the Chihuahuan Desert: A Biological Assessment. World Wildlife Fund, 1–128.

Donner, J., 1965. Ordnung Bdelloidea. Bestimmungsbücher zur Bodenfauna Europas. Akademie-Verlag, Berlin, 297 pp.

Dumont, H. J. & M. Coussement, 1976. Rotifers from Rio de Oro (North-Western Sahara). Hydrobiologia 51: 109–112.

Edmondson, W. T., 1949. A formula key to the rotatorian genus *Ptygura*. Transactions of the American Microscopical Society 68: 127–135.

Edmondson, W. T., 1959. Rotifera. In Edmondson, W. T. (ed.) Freshwater Biology. (2nd edn). John Wiley & Sons, Inc., New York (N.Y.), 420–494.

Koste W., 1978. Rotatoria. Die Rädertiere Mitteleuropas. 2 volumes. Gebrüder Borntraeger, Berlin, Stuttgart, Germany, Textband 673 pp., Tafelband 234 Tafeln.

Koste, W. & R. J. Shiel, 1986. Rotifera from Australian inland waters.I.Bdelloidea (Rotifera: Digononta). Australian Journal of Marine and Freshwater Research 37: 765–792.

Langley, J. M., R. J. Shiel, D. L. Nielsen & J. D. Green, 2001. Hatching from the sediment egg-bank, or aerial dispersing? The use of mesocosms in assessing rotifer biodiversity. Hydrobiologia 446/447: 203–211.

Legendre, P. & M. J. Anderson, 1999. Distance-based redundancy analysis: testing multispecies responses in multifactorial ecological experiments. Ecological Monographs 69: 1–24.

Lind O. T. &, C. A. Bane, 1980. Aquatic ecosystems of Big Bend National Park: Biological and chemical indicators of water quality. Proceedings of the Second Conference on Scientific Resources in National Parks. Vol. 2: Aquatic Biology, San Francisco, CA, 7–29.

MacKay, W. P., S. J. Loring, T. M. Frost & W.G. Whitford, 1990. Population dynamics of a playa community in the Chihuahuan Desert. Southwestern Naturalist 35: 393–402.

Mazuelos, N., J. Toja & C. Guisande, 1993. Rotifers in ephemeral ponds of Doñana National Park. Hydrobiologia 255/256: 429–434.

National Geographic Society, 1990. Big Bend National Park: trails illustrated map. National Geographic Maps, Evergreen, CO. ISBN 0–925873–77–2.

Nogrady, T, R. Pourriot & H Segers, 1995. Rotifera. Vol. 3: The Notommatidae and: The Scaridiidae. Guides to the Identification of the Microinvertebrates of the Continental Waters of the World Volume 6SPB Acad. Publishing, The Hague, The Netherlands, 226 pp.

Nogrady, T., R. L. Wallace & T. W. Snell, 1993. Rotifera. Vol. 1: Biology, Ecology and Systematics. Guides to the Identification of the Microinvertebrates of the Continental Waters of the World Volume 4SPB Acad. Publishing, The Hague, The Netherlands, 142 pp.

Olsen, D. M. & E. Dinerstein, 1998. The Global 200: a representation approach to conserving the Earth's most biologically valuable ecoregions. Biology 12: 502–515.

Ricci, C. & G. Melone, 2000. Key to the identification of the genera of bdelloid rotifers. Hydrobiologia 418: 73–80.

Ruttner-Kolisko, A., 1974. Planktonic Rotifers: Biology and Taxonomy. Die Binnengewässer (Supplement) 26: 1–146.

Segers, H., 1995. Rotifera. Vol. 2: The Lecanidae (Monogononta). Guides to the Identification of the Microinvertebrates of the Continental Waters of the World. SPB Acad. Publishing, The Hague, The Netherlands, 226 pp.

Segers, H., 1996. The biogeography of littoral *Lecane* Rotifera. Hydrobiologia 323: 169–197.

Segers, H., 2002. The nomenclature of the Rotifera: annotated checklist of valid family- and genus-group names. Journal of Natural History 36: 631–640.

Shepard, W. D., 1993. Desert springs – both rare and endangered. Aquatic Conservation: Marine and Freshwater Ecosystems 3: 351–359.

Stemberger, R. S., 1979. A Guide to Rotifers of the Laurentian Great Lakes. U.S. Environmental Protection Agency, Cincinnati, Ohio, PB80–101280.

ter Braak, C. J. F. & P. Smilauer, 1998–2002. CANOCO Reference manual and CanoDraw for Windows User's guide: Software for Canonical Community Ordination (version 4.5). Microcomputer Power, Ithaca, NY, USA, 500 pp.

Wallace, R. L. & C. Ricci, 2002. Rotifera. In Rundle, S. D., A. L. Robertson, & J. M. Schmid-Araya (eds) Freshwater Meiofauna: Biology and Ecology. Backhuys Publishers, Leiden, 15–44.

Wallace R. L. & T. W. Snell, 2001. Rotifera. In Thorp, J. & A. Covich (eds), Ecology and Classification of North American Freshwater Invertebrates, 2nd edn. Academic Press, 195–254.

Williams, W. D., 1985. Biotic adaptations to temporary lentic waters, with special reference to those in semi-arid and arid regions. Hydrobiologia 125: 85–110.

Williams, W. D., 2001. Biodiversity in temporary wetlands of dryland regions. Internationale Vereinigung für Theoretische und Angewandte Limnologie, Verhandlungen 27: 141–144.

Appendix 1

List of regions sampled in BBNP as depicted in Fig. 1. GPS coordinates were determined using a Brunton Multi-Navigator®. Sites within canyons were sometimes difficulty to obtain; all GPS coordinates were confirmed using USGS maps. **1.** Burro Spring (spring/seep and stream complex) [29° 14.250′ N; 103° 25.550′ W]; **2.** Carlota Tinaja (tinaja complex) [29° 16.745′ N; 103° 02.125′ W]; **3.** Cattail Spring flowage (spring/seep complex) [29° 16.745′ N; 103° 19.750′ W]; **4.** Croton Spring (tinajas + heuco) [ca. 29° 20.800′ N; 103° 20.750′ W]; **5.** Dripping Spring (spring/seep complex) [29° 24.291′ N; 103° 18.460′ W]; **6.** Dugout Wells (well) [29° 16.300′ N; 103° 08.190′ W]; **7.** Ernst Tinaja (tinaja complex) [29° 15.200–15.800′ N; 103° 00.935–01.600′ W]; **8.** Glenn Spring (spring/stream complex) [29° 09.400–10.525′ N; 103° 09.400–10.500′ W]; **9.** Government Spring (spring/seep) [29° 20.450′ N; 103° 15.325′ W]; **10.** McKinney Spring (spring/stream complex) [29° 24.442–24.612′ N; 103° 05.229–05.321′ W]; **11.** Mule Ears Spring (spring/seep complex) [29° 09.650′ N; 103° 24.450′ W]; **12.** Oak Creek (stream) [29° 16.962′ N; 103° 20.500′ W]; **13.** Paint Gap Cattle Tank (tank) [29° 23.276′ N; 103° 18.149′ W]; **14.** Rio Grande Village (Langford) Hot Spring (spring) [29° 10.771′ N; 102° 59.732′ W]; **15.** Rio Grande Village Pond (Lower) (pond) [29° 10.717′ N; 102° 57.226′ W]; 16. Rio Grande Village Pond (Upper) (pond) [29° 10.711′ N; 102° 57.197′ W]; **17.** Rio Grande (river & canal from river). **18.** Sam Nail Ranch (well) [29° 16.747′ N; 103° 22.200′ W]; **19.** Trap Spring (ephemeral stream) [29° 09.820′ N; 103° 25.164′ W]; **20.** Tuff Canyon (tinaja complex) [29° 09.050′ N; 103° 29.160′ W]; **21.** Tule Spring (complex of springs which feed a tank) [29° 14.525′ N; 103° 26.550′ W]; **22.** Ward Spring (spring/seep) [ca. 29° 14.500′ N; 103° 20.550′ W]; **23.** Window Trail (tinaja complex) [29° 16.800′ N; 103° 19.785–19.835′ W].

Appendix 2

List of species identified in this study. Taxa with the notation of 'cf.' indicate that while we could assign these specimens to a valid species name, there were minor differences from the published description. Thus, we are not absolutely sure of their identification and they may be new to science. Specimens from four other genera (*Encentrum, Notholca, Notommata, Synchaeta*) could not be identified to species. On a previous sampling trip one of us (EJW) collected *Hexarthra cf. brandorffi* from a tinaja at sampling site 4.

1 *Adineta vaga*
2 *Anuraeopsis fissa*
3 *Aspelta imbuta*
4 *Brachionus bidentatus*
5 *Brachionus dimidiatus*
6 *Brachionus urceolaris*
7 *Cephalodella catellina*
8 *Cephalodella compacta*
9 *Cephalodella doryphora*
10 *Cephalodella forficula*
11 *Cephalodella gibba*
12 *Cephalodella gracilis*
13 *Cephalodella* cf. *mira*
14 *Cephalodella sterea*
15 *Cephalodella tenuiseta*
16 *Cephalodella vacuna*
17 *Cephalodella vitella*
18 *Collotheca coronetta*
19 *Collotheca gracilipes*
20 *Collotheca ornata*
21 *Collotheca* cf. *paradoxa*
22 *Colurella colurus compressa*
23 *Colurella obtusa*
24 *Colurella uncinata*
25 *Dicranophorus haueri*
26 *Dipleuchlanis elegans*
27 *Epiphanes senta*
28 *Euchlanis dilatata*
29 *Euchlanis lyra*
30 *Euchlanis triquetra*
31 *Filinia longiseta*
32 *Filinia* cf. *novaezealandiae*
33 *Itura viridis*
34 *Lecane* cf. *abanica*
35 *Lecane bifurca*
36 *Lecane bulla*
37 *Lecane closterocerca*
38 *Lecane furcata*
39 *Lecane hamata*
40 *Lecane inermis*
41 *Lecane lateralis*

42 *Lecane luna*
43 *Lecane papuana*
44 *Lecane perpusilla*
45 *Lecane pyriformis*
46 *Lecane rudescui*
47 *Lecane tenuiseta*
48 *Lecane thalera*
49 *Lepadella ovalis*
50 *Lepadella patella*
51 *Lepadella pumilo*
52 *Lindia anebodica*
53 *Macrochaetus sericus*
54 *Monommata arndti*
55 *Monommata enedra*
56 *Mytilina mucronata*
57 *Philodina megalotrocha*
58 *Plationus patulus*
59 *Platyias quadricornis*
60 *Polyarthra dolichoptera*
61 *Proales daphnicola*
62 *Ptygura brevis*
63 *Ptygura crystallina*
64 *Ptygura longicornis*
65 *Squatinella mutica*
66 *Trichocerca collaris*
67 *Trichocerca marina*
68 *Trichocerca similis*
69 *Trichocerca tenuidens*
70 *Wierzejskiella vagneri*
* other bdelloids

Part IV
Morphology and Ultrastructure

Hydrobiologia (2005) 546:161–167
A. Herzig, R.D. Gulati, C.D. Jersabek & L. May (eds.) Rotifera X: Rotifer Research: Trends, New Tools and Recent Advances
DOI 10.1007/s10750-005-4113-6

Euryhaline *Brachionus* strains (Rotifera) from tropical habitats: morphology and allozyme patterns

Tomonari Kotani[1,*], Atsushi Hagiwara[2], Terry W. Snell[3] & Manuel Serra[4]
[1]*Nagasaki Industrial Promotion Foundation, Ikeda 2-1303-8, 856-0026 Oomura, Japan*
[2]*Graduate School of Science & Technology, Nagasaki University, Bunkyo 1-14, 852-8521 Nagasaki, Japan*
[3]*School of Biology, Georgia Institute of Technology, Atlanta, Georgia 30332-0230, USA*
[4]*Institut Cavanilles de Biodiversitat I Biologia Evolutiva, Universitat de Valencia, A.O. 22085, 46071 València, Spain*
*(*Author for correspondence: Tel.: +81-845-24-2933; Fax: +81-845-24-3449;*
E-mail: tkotani@ma.fuma.fukuyama-u.ac.jp)

Key words: Rotifera, S-morphotype *Brachionus*, allozyme pattern, isozyme analysis, genetic distance, dendrogram, morphology

Abstract

The euryhaline rotifer *Brachionus* is a complex of sibling species. Although many investigations have been carried out in the past, the relationships among the Spanish species, the tropical SS strains and the clusters previously described, remained unknown. In this study, allozyme data for five populations from the tropics and two from Spanish lagoons – one of them *B. ibericus* and the other *B. rotundiformis* – were combined with data from the previous studies. Cluster analysis based on genetic distance allowed the 74 strains to be divided into two major groups. One group was associated with *B. plicatilis*-like strains, and the other group was associated with *B. rotundiformis* and *B. ibericus*. This latter group was divided into two clades. One of these clustered most with the S-morphotype strains and the *B. ibericus* species. The other clustered most closely with the tropical (SS) strains and the *B. rotundiformis* Spanish species. These results show a correspondence between the species description based on Spanish strains and the allozyme groups identified in a larger collection of strains.

Introduction

The concept of morphospecies dominated animal taxonomy during the 19th and early 20th centuries. However, as some species were morphologically difficult to distinguish, Mayr (1942) introduced the idea of sibling species. Sibling species are common in aquatic invertebrates, where species recognition is often chemical (Knowlton, 1993; Serra et al., 1998). The failure of investigators to recognize sibling species has led to confusion in the interpretation of ecological processes (Paterson, 1991; Knowlton, 1993; Knowlton & Jackson, 1994).

Since euryhaline rotifers of the genus *Brachionus* are an essential live food in finfish larviculture, the correct determination of their taxonomy is important. Otherwise, their sibling species may confuse the determination of ecological tolerance limits and reproductive patterns. Initially, they were all classified into a single species, *B. plicatilis*. However, it was later discovered that *B. plicatilis* comprised two morphologically and ecologically distinct groups, one of which was called 'L-type' (large) and the other 'S- (small) type' (Oogami, 1976; Segers, 1995; Campillo et al., 2005). Oogami (1976) distinguished them by the shape of their lorica. Fu et al. (1991a, b) analyzed 67 *Brachionus* strains from a wide geographical area and found that these strains were morphologically and genetically divisible into two groups. In addition,

Rumengan et al. (1991) reported that the number of chromosomes differed between the S- and L-types. Moreover, it was shown that S- and L-types were reproductively isolated from each other (Fu et al., 1993; Gómez & Serra, 1995; Rico-Martínez & Snell, 1995; Gomez, 2005). As a result, Segers (1995) re-classified the S- and L-types as different species, and named the S-type *B. rotundiformis* (Tschugunoff, 1921) and the L-type *B. plicatilis* (O.F. Müller, 1786).

Some studies reported that some strains could not be easily classified as either *B. plicatilis* or *B. rotundiformis* (Carmona et al., 1995; Gómez et al., 1995; Gómez & Serra, 1995; Hagiwara et al., 1995). Rotifer strains in the lower size range of *B. rotundiformis* were discovered in tropical regions and were designated SS-type by aquaculturists because of their small size (Hagiwara et al., 1995). Hagiwara et al. (1995) suggested that SS-type rotifers belonged to *B. rotundiformis* because they were morphologically, ecologically and genetically similar and did not have any pre-mating reproductive isolation.

Three morphologically and genetically distinct types of euryhaline rotifers of the species *Brachionus* were also reported from a pond in Torreblanca Marsh in Castellon, Spain (Carmona et al., 1995; Gómez et al., 1995; Gómez & Serra, 1995). The two smaller strains (SM and SS) fitted within the morphospecies *B. rotundiformis*, but, in further studies, they differed morphologically and genetically, and were reproductively isolated from each other (Carmona et al., 1995; Gómez et al., 1995; Gómez & Serra, 1995; Ortells et al., 2000). It was proposed that these two strains were different species (Gómez et al., 1995; Serra et al., 1998). Ciros-Pérez et al. (2001) created a new species, *B. ibericus*, from the SM strain and retained the name *B. rotundiformis* for the SS strains.

The Spanish and newly discovered tropical strains were investigated for morphological and genetic characteristics (Hagiwara et al., 1995), and their genetic similarity was explored using allozyme markers (Ortells et al., 2000). However, the relationships among the Spanish- and tropical-strains and the 67 strains of Fu et al. (1991a, b) remained unclear. This study examines the relatedness of the small tropical rotifers and the Spanish strains by comparing their allozyme patterns at six loci.

Materials and methods

Isozyme analysis was performed on three strains of S morphotype *Brachionus*. These were collected from brackish ponds in Torreblanca Marsh, Spain (*B. rotundiformis* SS2, Gómez et al., 1995; Gómez & Serra, 1995), on the island of Langkawi, Malaysia, and on the island of Bali, Indonesia. The results were analyzed by cluster analysis, together with the allozyme patterns of 67 strains of Fu et al. (1991b), strains from Fiji and Thailand (Hagiwara et al., 1995; Kotani et al., 1997), a strain from Okinawa, Japan (Hagiwara et al., 1995), and *B. ibericus* strain SM1 (Kotani et al., 1997).

Allozyme analysis

Methods for analyzing *Brachionus* allozymes were described by Fu et al. (1991b). Rotifers were cultured under controlled conditions at 25 °C, in 22‰ diluted seawater in 5-l plastic beakers and fed with *Nannochloropsis oculata*. Each culture was started with an initial density of 1 ind ml^{-1}. Rotifer populations were harvested when they reached high densities (~200–300 ind ml^{-1}). To avoid contamination of rotifer enzymes with those of the food organisms, the rotifers were starved for 1 day before harvesting. The animals were filtered with a 43-μm mesh plankton net, and washed several times with clean diluted seawater. Then the rotifer samples were frozen immediately, and stored at −80 °C until they were used in electrophoresis.

The rotifer samples were thawed to provide a crude extract of enzymes just prior to electrophoresis. Approximately 5-μl of rotifer extract was loaded into each lane of a 12% horizontal starch gel using a filter paper (4 × 4 mm). Electrophoresis was carried out for 5 h at a constant current of 3.3 mA cm^{-2} of cross-section in a refrigerator at 5 °C. Six enzymes including LDH, MDH, 6PGD, SOD, PGM and GPI were characterized using the staining procedure described by Fu et al. (1991b). The buffer system for electrophoresis was 0.04 M citric acid, pH 6.9 (adjusted with *N*-(3-aminopropyl)morpholine and NaOH) for analyzing LDH, MDH, 6PGD, SOD and PGM, and 0.04 M citric acid, pH 8 (adjusted with Tris-(hydroxymethyl) methylamine) for GPI. As in the study of Fu et al. (1991b), genetic distance was calculated for six loci based on Rogers (1972). A dendrogram was constructed using the UPGMA procedure. The cluster analysis was conducted using MEGA version 2.1 (Kumar et al., 2001).

Morphological analysis

We compared the morphology of the groups of strains using data from Fu et al. (1991a) and Hagiwara et al. (1995). Seven characters were determined to obtain lorica length (A), lorica shape (C/A and B/C), and shape of anterior spines (E/D and G/F) (Fig. 1). For each strain, the values of seven characters were determined for 20 individuals. A Mann–Whitney's *U* test was performed to compare each index among groups and a scatter graph of A against B was plotted for each strain. Since there were no morphological data for the Indonesia and Malaysian strains, these were not included in the analysis. The two Spanish strains of Ciros-Pérez et al. (2001) were also excluded from the analysis because, although there are some biometrical data in the paper, no raw data are given.

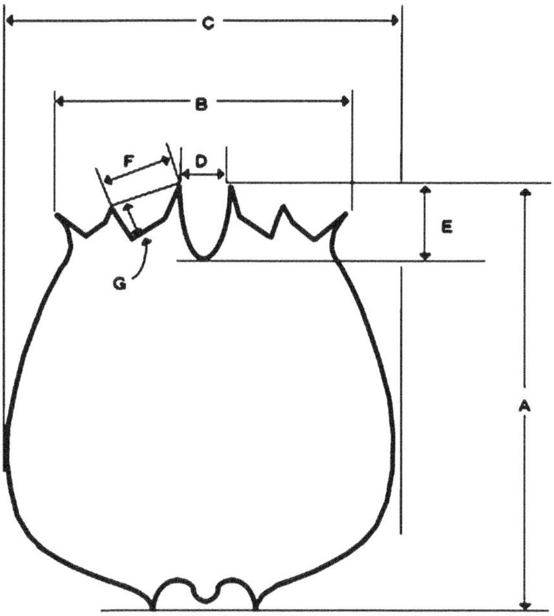

Figure 1. Seven characteristics (A–G) of the lorica of the euryhaline Brachionus, as measured by Fu et al. (1991a).

Results

Genotypes for each locus of the strains not described in Fu et al. (1991b) are indicated in Table 1. The dendrogram produced from the data of Fu et al. (1991b) and the data in Table 1 is indicated in Fig. 2. This dendrogram divides the strains into two groups. Group A is consistent with the L morphotype of *B. plicatilis*, and group B is consistent with the S morphotype of the *Brachionus* strains. The genetic distance between group A and group B was 0.349. Group B was subdivided further into two groups (C and D). Group C included the Japanese and *B. ibericus* SM1 strains, whereas group D included the Fiji, Thai, Malaysian, Indonesian and *B. rotundiformis* SS2 strains. The genetic distance between group C and D was 0.330.

The result of the morphological comparison between group C and group D is indicated in Table 2. There were significant differences in each index. This suggests that the lorica of group D is smaller (A) and rounder (C/A) than that of group C, the top opening of the lorica of group D is much narrower than that of group C (B/D), and that the anterior spine of group D is sharper than that of group C (E/D and G/F). Also, in the scatter graph, group D was more likely to be found in the region of smaller size (Fig. 3).

Table 1. Genetic constitution of seven strains detected from electrophoretical analysis of the following six enzymes: lactate dehydrogenase (LDH, EC 1.1.1.27), malate dehydrogenase (MDH, EC 1.1.1.37), 6-phosphogluconate dehydrogenase (6PGD, EC 1.1.1.44), superoxide dismutase (SOD, EC 1.15.1.1), phosophoglucomutase (PGM, EC 2.7.5.1) and glucose phosphate isomerase (GPI, EC 2.6.1.1). Symbols of alleles were based on Fu et al. (1991b). Alleles estimated at each locus were named alphabetically according to their mobility. The alphabetical order followed Fu et al. (1991b)

Strain	Locus and genotype					
	LDH	*MDH-1*	*6PGD*	*SOD*	*PGM*	*GPI*
SS2	HH	BB	BB	CC	DD	HH
Indonesia	HH	BB	FF	CC	DD	HH
Malaysia	HH	BB	FF	CC	DD	HH
Fiji[a, b]	HH	BB	DD	CC	GG	FF
Thailand[a, b]	HH	BB	DD	CC	GG	FF
Japan[a]	FF	AA	IL	CC	FF	BF
SM1[b]	CC	AA	BB	BB	AA	CC

[a] Hagiwara et al. (1995), [b] Kotani et al. (1997).

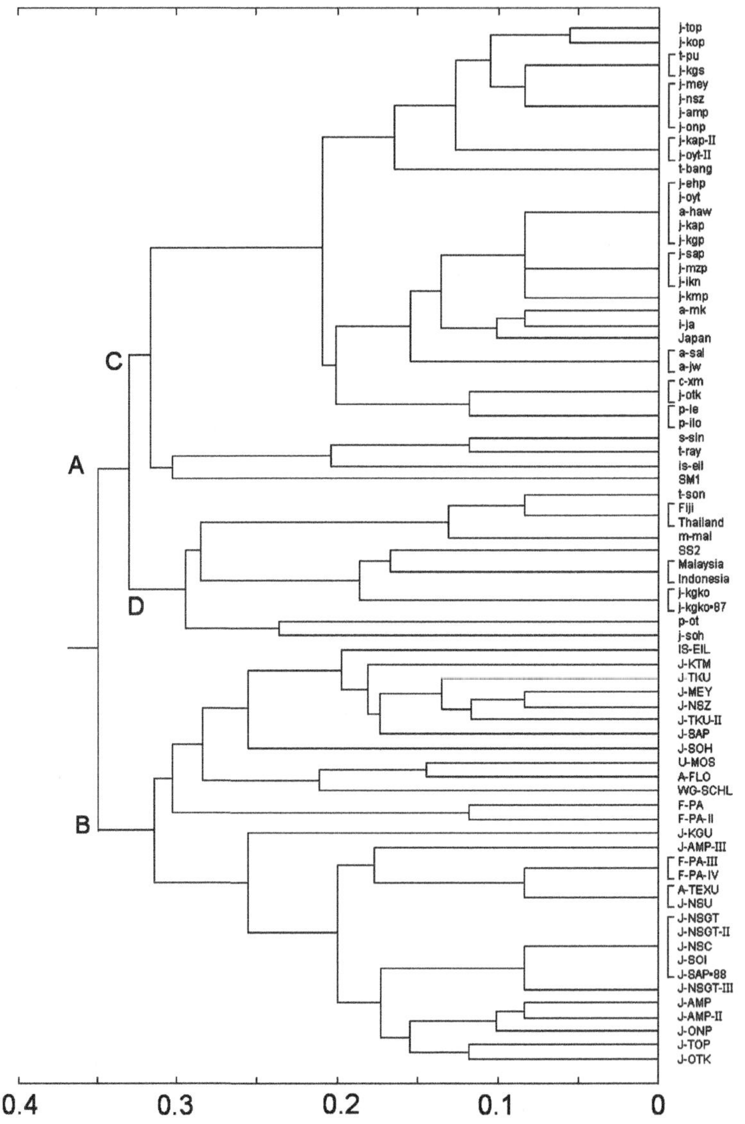

Figure 2. Dendrogram for 74 strains according to a genetic distance index. The dendrogram is the result of the UPGMA analysis of Rogers' genetic distance (1972). Lower case letters indicate the S morphotype; capital letters indicate the L morphotype.

Table 2. Mean and SD values of five variables for group C and D

Variable	Group C ($n = 32 \times 20$)		Group D ($n = 8 \times 20$)		Mann–Whitney's U test		<0.01
	Mean	SD	Mean	SD	U value	p value	
	A (μm)	214.9	19.1	189.3	13.2	14075.0	
C/A	0.805	0.035	0.841	0.035	23863.0	<0.01	
B/C	0.627	0.038	0.578	0.028	14468.5	<0.01	
E/D	1.063	0.226	1.000	0.213	44584.0	<0.05	
G/F	0.552	0.120	0.675	0.096	18288.5	<0.01	

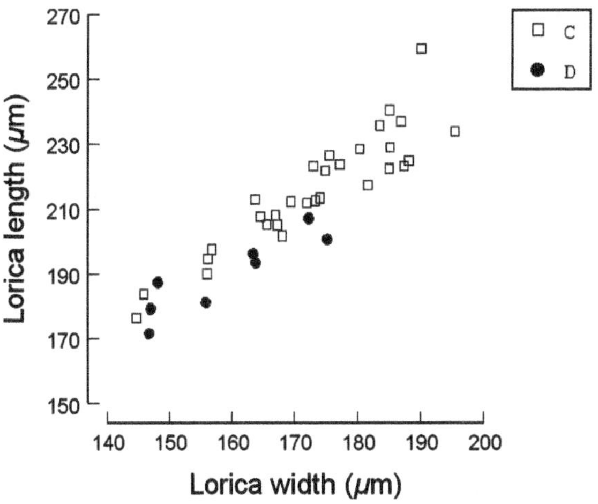

Figure 3. Scatter diagram of lorica length and width in relation to the groups identified by the cluster analysis. The open squares indicate the strains in group C and the solid circles indicate the strains in group D.

Discussion

Our study incorporated new strains into the previous set studied genetically and morphologically by Fu et al. (1991a, b), where euryhaline *Brachionus* was classified into two clades. In the genetic classification of Fu et al. (1991b), S morphotype strains formed two groups, one of them including six strains from tropical regions (t-son, Thailand; m-mal, Malaysia; j-kgko, j-kgko·87, Koshiki islands, Japan; p-ot, Philippines; and j-soh, Hamanako-lake, Japan). In our analysis, these tropical strains clustered into group D (Fig. 2), together with the four small rotifer strains discovered in the tropical regions (Fiji, Thailand, Indonesia and Malaysia) and the Spanish strain *B. rotundiformis* SS2. The latter five strains were newly added from our analysis. The other newly added strains (Japanese strain and *B. ibericus* SM1 strain) also clustered into group C (Fig. 2). The morphological classification of Fu et al. (1991a) divided S morphotype strains into two subgroups, but the morphological division was not consistent with the genetic grouping (See Fu et al., 1991a, b).

Since the results of Fu et al. (1991a, b), a number of analyses using a variety of approaches have been performed on the same and related strains of *Brachionus* for understanding their phylogeny. These studies provide support for both the similarity among strains belonging to the morphotype S (i.e. our groups C and D) and for the differentiation of clades within the morphotype S. An example is Rumengan's (1990) karyotype analysis on S morphotype strains that did not find differentiation in chromosome number among j-kgko, a-haw (Hawaii, USA) and j-onp (Okinawa, Japan), despite the former belonging to group D and the two latter to group C (Fig. 2). Some mating and copulation have been observed between strains belonging or related to groups C and D. Males of Koshiki (group D, same as j-kgko) strain copulated with the females of the Hawaii strain (group C, Rico-Martímez & Snell, 1995). The *B. ibericus* SM1 (group C) copulated with Fiji, Koshiki and *B. rotundiformis* SS strains (group D) (Gómez & Snell, 1996; Kotani et al., 1997, 2001). Moreover, Kotani et al. (1997, 2001) investigated the differences in the mate recognition pheromone (MRP) among various strains using the antibody of MRP of L and S morphotype *Brachionus*. They did not find differences among S morphotype *Brachionus*, although they studied strains related to both our groups C and D (tropical strains).

Evidence for the differences between groups C and D comes mainly from studies on the Spanish strains. Spanish sympatric *B. rotundiformis* (group D) and *B. ibericus* (group C) were reported to be reproductively isolated in the field, ecologically specialized, biometrically different, possessing strong homotypic mating preferences, and showing no hybridization in the laboratory (Gómez et al., 1995; Gómez & Serra, 1995; Serra et al., 1998, Ciros-Pérez et al., 2001). Molecular phylogenies of COI and ITS genes have shown deep genetic divergence between Spanish strains of *B. rotundiformis* and *B. ibericus* (Gómez et al., 2000). In addition, working with non-Spanish strains, Boehm et al. (2000) found genetic differences between S morphotype *Brachionus* strains and the tropical strains from the analysis of microsatellite DNA sequences. Genetic cohesion of group D is supported by (1) the genetic similarity between tropical small strains (Thailand and Fiji) and the Koshiki strain (Boehm et al., 2000), and (2) the lack of pre- or post-zygotic reproductive isolation (strains j-kgko·89, j-kgko·90, j-soh and m-mal; Fu et al., 1993).

After Segers (1995) re-classified euryhaline *Brachionus*, classification of the tropical strains was attempted by Hagiwara et al. (1995). They concluded that the tropical strains belonged to S morphotype *Brachionus* because their morphological, ecological, and genetic characters were more similar to S morphotype *Brachionus* than L morphotype *Brachionus*. They were also reproductively isolated from L morphotype *Brachionus* strains, but pre-zygotic isolation with other S morphotype *Brachionus* was incomplete. In this study, we were able to discriminate group C from D in the S morphotypes using cluster analysis of allozyme variation (Fig. 2). When combined with previous findings, our results show that S-morphotype strains tend to group together, and separately from L-morphotype strains. However, there is strong differentiation among S-morphotype strains, with the small strains, usually the tropical ones, being genetically differentiated into group D (*B. rotundiformis*). Despite their size differences, we found an extensive overlap in the lorica size of both groups C and D (Figure 3). Consequently, it is difficult to classify groups C and D with only biometrical methods. Our results emphasize the importance of using morphological, behavioral, and molecular approaches for resolving boundaries species (Knowlton, 2000).

Acknowledgement

T. Kotani is grateful to Japan Society for the Promotion of Science for research fellowship for young scientists. This study was partly supported by Grants-in-Aid from the Ministry of Education, Science and Culture, Japan, and from the Prefectural Collaboration of Regional Entities for the Advancement of Technological Excellence, JST. Authors would like to thank Aileen Choo, The Oceanic Institute, for comments.

References

Boehm, E. W. A., O. Gibson & E. Lubzens, 2000. Characterization of satellite DNA sequences from the commercially important marine rotifers *Brachionus rotundiformis* and *Brachionus plicatilis*. Marine Biotechnology 2: 38–48.

Campillo, S., E. M. García-Roger, D. Martinez-Torres & M. Serra, 2005. Morphological stasis of two species belonging to the L-morphotype in the *Brachionus plicatilis* species complex. Hydrobiologia 546: 181–187.

Carmona, M. J., A. Gómez & M. Serra, 1995. Mictic patterns of the rotifer *Brachionus plicatilis* Müller in small ponds. Hydrobiologia 313/314: 365–371.

Ciros-Pérez, J., A. Gómez & M. Serra, 2001. On the taxonomy of three sympatric sibling species of the *Brachionus plicatilis* (Rotifera) complex from Spain, with the description of *B. ibericus* n. sp. Journal of Plankton Research 23: 1311–1328.

Fu, Y., K. Hirayama & Y. Natsukari, 1991a. Morphological differences between two types of the rotifer *Brachionus plicatilis* O. F. Müller. Journal of Experimental Marine Biology and Ecology 151: 29–41.

Fu, Y., K. Hirayama & Y. Natsukari, 1991b. Genetic divergence between S and L type strains of the rotifer *Brachionus plicatilis* O. F. Müller. Journal of Experimental Marine Biology and Ecology 151: 43–56.

Fu, Y., A. Hagiwara & K. Hirayama, 1993. Crossing between seven strains of the rotifer *Brachionus plicatilis*. Nippon Suisan Gakkaishi 59: 2009–2016.

Gómez, A. & M. Serra, 1995. Behavioral reproductive isolation among sympatric strains of *Brachionus plicatilis* Müller 1786: insights into the status of this taxonomical species. Hydrobiologia 313/314: 111–119.

Gómez, A., M. Temprano & M. Serra, 1995. Ecological genetics of a cyclical parthenogen in temporary habitats. Journal of Evolutionary Biology 8: 601–622.

Gómez, A. & T. W. Snell, 1996. Sibling species and cryptic speciation in the *Brachionus plicatilis* species complex (Rotifera). Journal of Evolutionary Biology 9: 953–964.

Gómez, A., G. R. Carvalho & D. H. Lunt, 2000. Phylogeography and regional endemism of a passively dispersing zooplankter: mitochondrial DNA variation in rotifer resting egg banks. Proceedings of the Royal Society of London, Series B 267: 2189–2197.

Gómez, A., M. Serra, G. R. Carvalho & D. H. Lunt, 2002. Speciation in ancient cryptic complexes: evidence from the molecular phylogeny of *Brachionus plicatilis* (Rotifera). Evolution 56: 1431–1444.

Gómez, A., 2005. Molecular ecology of rotifers: from population differentiation to speciation. Hydrobiologia 546: 83–99.

Hagiwara, A., T. Kotani, T. W. Snell, M. Assava-Aree & K. Hirayama, 1995. Morphology, genetics, and mating behavior of small tropical marine *Brachionus* strains (Rotifera). Journal of Experimental Marine Biology and Ecology 194: 25–37.

Knowlton, N., 1993. Sibling species in the sea. Annual Review of Ecology and Systematics 24: 189–216.

Knowlton, N. & J. B. C. Jackson, 1994. New taxonomy and niche partitioning on coral reefs: jack of all trades or master of some?. Trends in Ecology & Evolution 9: 7–9.

Knowlton, N., 2000. Molecular genetic analysis of species boundaries in the sea. Hydrobiologia 420: 73–90.

Kotani, T., A. Hagiwara & T. W. Snell, 1997. Genetic variation among marine *Brachionus* strains and function of mate recognition pheromone (MRP). Hydrobiologia 358: 105–112.

Kotani, T., M. Ozaki, K. Matsuoka, T. W. Snell & A. Hagiwara, 2001. Reproductive isolation among geographically and temporally isolated marine *Brachionus* populations. Hydrobiologia 446/447: 283–290.

Kumar, S., K. Tamura, I. B. Jakobsen & M. Nei, 2001. MEGA2. Molecular evolutionary genetics analysis software. Bioinformatics, 17: 1244–1245.

Mayr, E., 1942. Systematics and the Origin of Species. Columbia University Press, New York, 334 pp.

Oogami, H., 1976. On the morphology of *Brachionus plicatilis*. Newsletter from Izu Branch, Shizuoka Prefectural Fisheries Research Center 184: 2–5.

Ortells, R., T. W. Snell, A. Gómez & M. Serra, 2000. Patterns of genetic differentiation in resting egg banks of a rotifer species complex in Spain. Archiv für Hydrobiologie 149: 529–551.

Paterson, H. E. H, 1991. The recognition of cryptic species among economically important insects. In Zalucki, P. (ed.), Heliothis: Research Methods and Prospects. Springer Nerlag, New York: 1–10.

Rico-Martínez, R. & T. W. Snell, 1995. Mating behavior and mate recognition pheromone blocking of male receptors in *Brachionus plicatilis* Müller (Rotifera). Hydrobiologia 313/314: 105–110.

Rogers, J. S, 1972. Studies in Genetics VII. University of Texas, University of Texas Publication, No. 7213, Austin, Texas, 145.

Rumengan, I. F. M., 1990. Studies on growth characteristics and karyotypes of S and L type rotifers, *Brachionus plicatilis*. Doctoral Thesis, Nagasaki University, 147 pp.

Rumengan, I. F. M., H. Kayano & K. Hirayama, 1991. Karyotypes of S and L type rotifers Brachionus plicatilis O. F. Müller. Journal of Experimental Marine Biology and Ecology 154: 171–176.

Segers, H., 1995. Nomenclature consequences of some recent studies on *Brachionus plicatilis* (Rotifera, Brachionidae). Hydrobiologia 313/314: 121–122.

Serra, M., A. Gómez & M. J. Carmona, 1998. Ecological genetics of *Brachionus* sympatric sibling species. Hydrobiologia 387/388: 373–384.

Hydrobiologia (2005) 546:169–179
A. Herzig, R.D. Gulati, C.D. Jersabek & L. May (eds.) Rotifera X: Rotifer Research: Trends, New Tools and Recent Advances
DOI 10.1007/s10750-005-4114-5

Morphological and morphometrical variations of selected rotifer species in response to predation: a seasonal study of selected brachionid species from Lake Xochimilco (Mexico)

Gabriela Garza-Mouriño[1,*], Marcelo Silva-Briano[2], S. Nandini[3], S.S.S. Sarma[4] & Maria Elena Castellanos-Páez[1]

[1]*Laboratory of Rotiferology and Molecular Biology of Plankton, Division of Biological Sciences and the Health, Autonomous Metropolitan University, Campus Xochimilco, Calzada del Hueso No. 1100, Villa Quietud, C.P. 04960, Mexico City, Mexico*

[2]*Centro de Ciencias Básicas, Departamento de Biologia, Universidad Autónoma de Aguascalientes, Av. Universidad No. 940, Fracc. Primo Verdad, C.P.20100, Aguascalientes, Ags, México*

[3]*UIICSE, Division of Research and Postgraduate Studies, National Autonomous University of Mexico, Campus Iztacala, Av. de Los Barrios No. 1, Los Reyes, Tlalnepantla, State of Mexico, Mexico*

[4]*National Autonomous University of Mexico, Campus Iztacala, Av. de Los Barrios No. 1, Los Reyes, AP 314, CP 54090, Tlalnepantla, State of Mexico, Mexico; E-mail: sarma@servidor.unam.mx*

*(*Author for correspondence: E-mail: spaez@correo.unam.mx)*

Key words: morphometric variation, rotifera, predation, SEM, xochimilco, mexico

Abstract

We observed different morphotypes of some species in the family Brachionidae from the seasonal plankton samples of Lake Xochimilco collected during 2002–2003. We measured the body length, width, and spine lengths (posterior and anterior spines) *of Brachionus havanaensis, Keratella americana, K. cochlearis* and *K. tropica* during the periods when the predator *Asplanchna brightwellii* was present in great abundance but also when it was nearly absent. In general, spines of most of the selected rotifer species were longer if *A. brightwellii* was abundant. Relatively, small spines were observed if the predator was rare. The body lengths of brachionid rotifers widely varied in samples with more and less abundant *A. brightwellii.* Morphometric data are interpreted in terms of morphological adaptations of *Brachionus* and *Keratella* in response to *Asplanchna* predation.

Introduction

Zooplankton are known to develop morphological defenses in response to predation pressure. The efficacy of spines, setae and movement patterns in avoiding predation has been well documented in laboratory experiments on predator feeding behaviour, functional response and numerical response (Sarma, 1993; Nandini & Sarma, 1999). Gilbert (1999) has reviewed the role of asplanchnin produced by *Asplanchna*, which induces elongation of lorica spines in some members of Brachionidae. In nature too, it has been observed that many brachionids develop long spines if *Asplanchna* densities are high (Green & Lan, 1974). Studies on the feeding preferences of *Asplanchna* have shown that this predator indeed avoids prey types with defenses such as spines (*Brachionus calyciflorus*), setae (*Filinia longiseta*) and a darting movement (e.g., *Hexarthra and Polyarthra*) (Iyer & Rao, 1996). The ingestion times on such prey types are also higher than on prey, which are preferred in the diet (Sarma, 1993). Nevertheless, it has also been observed that prey species with no apparent defenses such as *Brachionus budapestinensis* and *Lecane bulla* are avoided (Iyer & Rao, 1996).

Analyses of samples from Lake Xochimilco during several months revealed wide variations in the morphology of some rotifer species. Morphometric studies permit the evaluation of the range of taxonomical and morphological variations exhibited by rotifers in natural water bodies in response to biotic as well as abiotic factors (Green, 1981; Hillbricht-Ilkowska, 1983). In this regard, scanning electron microscopic information is highly useful, not only as a tool to confirm the identity of species, but also to highlight characteristics which could explain their resistance or susceptibility to particular conditions.

As a part of our ongoing research on the rotifers of Lake Xochimilco, we observed that, during certain months of the year, in response to the most numerically abundant predator *Asplanchna*, both the body and spine length of some brachionids appeared longer than in the other months. The aim of the present work was therefore to examine the range of morphological variations exhibited by some of these rotifer species from Lake Xaltocan in Xochimilco (Mexico) during both the periods of high abundance and near absence of *Asplanchna*. Detailed light microscopic and scanning electron microscopic observations were made for the common rotifer genera such as *Brachionus*, *Keratella*, and *Asplanchna*.

Materials and methods

Mexico City was built over an ancient lake system, a part of which is Lake Xochimilco. This water body is a complex system of canals and lakes supporting fisheries and local horticulture. These waterbodies are shallow, with a maximum depth of less than 2 m. In the summer, there is a high growth of cyanobacteria, particularly *Microcystis* and *Oscillatoria*. Several chlorophytes such as *Scenedesmus*, *Pediastrum*, *Eudorina*, and diatoms dominate the phytoplankton community during the rest of the year. In spite of the eutrophic nature (nitrogen and phosphorus levels can be as high as 21.4 mg l^{-1} and 5.8 mg l^{-1}, respectively) (S. Nandini, unpublished data), this waterbody supports a high diversity of rotifers (Flores-Burgos et al., 2003; Garza-Mouriño & Castellanos-Páez, 2003). Lake Xaltocan, a part of Lake Xochimilco is also a complex system of lakes and canals and is eutrophic since it receives effluents from domestic and horticultural wastes; nevertheless, it has a high rotifer species diversity. We observed at least 10 species of Brachionidae co-existing at any time in several sites of the lake. This, in spite of high densities of the predatory rotifer *Asplanchna brightwellii*, which at times reached more than 500 ind. l^{-1}.

Samples were collected from the surface in Lake Xaltocan from March 2002 to February 2003 by filtering 80l of water through a mesh of 40 μm and fixed immediately in 4% formalin. The rotifers were identified using standard literature (Koste, 1978; Koste & Shile, 1987). The mean density of the dominant species was derived from analyses of three aliquots of 1 ml each using a Sedgewick Rafter cell. The density of *Asplanchna* was lowest in June 2002 (4 ind. l^{-1}) and increased to about 50 and 135 ind. l^{-1} in July and August, respectively. The temperature was fairly constant from June to August, ranging between 19 and 21 °C. Hence we studied morphometric variations in some brachionids, often found in the stomach contents of *Asplanchna*.

The total length and width of parthenogenetic females of *Brachionus havanaensis, Keratella eachlearis, K. americana* and *K. tropica* were measured. Posterior spines were measured *for B. havanaensis. K. americana* and *K. tropica* (right posterior spine), but not for *K. cochlearis* var. *tecta* since they were absent. Anterolateral spine length and the distance between these spines were also measured for all species. Morphometric analysis was done using a BX50 Olympus, with Nomarsky illumination, using a spectral range of 420–480 nm with a planachromatic objective of 40× and a 22× ocular (maximum magnification, 880×). The photographs were taken using high resolution digital images (1300 × 1030 pixels) and were calibrated from a pixel of 6.7 μm^2, using an Olympus Mega Fire SP- Model S99810. The digital microanalysis was done using an Image-Pro Version 4.2 image analyzer.

For SEM photos, the following process was used: the specimens were dehydrated stepwise using from 70 to 100% alcohol, and for 24 h in absolute alcohol. The remaining water content from the specimens was removed using the

Plate 1. SEM photographs of *Keratella cochlearis* (a–c), *K. tropica* (d and e), *Brachionus havanaensis* (f and g) and *B. budapestinensis* (h). a: lorica, dorsal view, b: anterior spines, c: lateral antenna, d: dorsal view, e: dorsal lorica with folds, f: lorica, ventral view, g: anterior spines, h: ventral lorica with minute spinules.

TOUSIMIS critical dry point machine with CO_2 liquid. Later, the specimens were placed on aluminium tape with glue in upright position, and attached to a 1 cm high stub. The rotifers were gold-coated in DESK II machine, and then photographed using SEM JEOL LV 5900. Using the monitor EISO, it was possible to focus part of the specimens, which were of taxonomic importance.

Results

The ultrastructure of the specimens used in this work reveals several interesting aspects (Plate 1): lateral antennae of *Keratella cochlearis* and the ornamentation on the dorsal lorica of *K. tropica* were clearly visible. *Brachionus budapestinensis* has minute projections spread throughout the lorica, but more densely in the margins.

Morphometric data on the four most common species of Brachionidae (*B. havanaensis, K. americana, K. coahlearis* and *K. tropica)* are presented in Figures 1–4. One-way analysis of variance (ANOVA) revealed (Table 1) that lengths of lorica of *B. havanaensis*, both posterior spine length and distance between anterolateral spines were significantly influenced by the presence of the predatory *Asplanchna*. In *K. americana*, none of the variables tested were significantly

Table 1. Statistical evaluation using ANOVA on the selected morphometric variables of *K. cochlearis*, *K. americana*, *K tropica* and *B. havanaensis* collected at high and low abundances of the predator *A. brightwellii*

Source	DF	SS	MS	F
Keratella cochlearis				
Anterolateral spine length				
Predator's presence	2	50.69	25.34	4.79*
Error	67	353.96	5.283	
Distance between anterolateral pair				
Predator's presence	2	226.89	113.44	6.55**
Error	34	588.27	17.30	
Lorica width				
Predator's presence	2	34.51	17.25	0.90ns
Error	34	653.51	19.22	
Lorica total length				
Predator's presence	2	163.64	81.82	2.76ns
Error	34	1008.37	29.66	
Lorica length				
Predator's presence	2	66.278	33.14	3.17ns
Error	33	345.08	10.46	
K. americana				
Anterolateral spines				
Predator's presence	2	38.31	19.15	2.8ns
Error	50	332.82	6.65	
Distance between anterolateral pair				
Predator's presence	2	105.25	52.62	0.70ns
Error	25	1861.75	74.47	
Lorica width				
Predator's presence	2	0.012	0.006	0.0001ns
Error	25	1444.45	57.77	
Total lorica length				
Predator's presence	2	244.66	122.33	0.201ns
Error	25	15219.25	608.77	
Lorica length				
Predator's presence	2	391.41	195.70	1.10ns
Error	25	4407.91	176.31	
Posterior spine length				
Predator's presence	2	82.42	41.21	2.43ns
Error	17	288.15	16.95	
K. tropica				
Anterolateral spines				
Predator's presence	2	215.72	107.86	9.20***
Error	61	714.95	11.72	
Distance between anterolateral pair				
Predator's presence	2	932.77	466.38	14.34***
Error	31	1007.77	32.50	
Lorica width				
Predator's presence	2	20.52	10.26	0.26ns
Error	31	1194.23	38.52	

Continued on p. 173

Table 1. (Continued)

Source	DF	SS	MS	F
Total lorica length				
Predator's presence	2	7747.57	3873.78	16.026***
Error	31	7492.55	241.69	
Lorica length				
Predator's presence	2	1581.10	790.55	11.17***
Error	31	2193.50	70.75	
Posterior spine length				
Predator's presence	2	2870.76	1435.38	14.38***
Error	30	2993.64	99,78	
B. havanaensis				
Anterolateral spines				
Predator's presence	2	1189.80	594.90	19.31***
Error	59	1817.51	30.80	
Distance between anterolateral pair				
Predator's presence	2	799.17	399.58	5.75**
Error	28	1943.12	69.39	
Lorica width				
Predator's presence	2	1010.10	505.05	5.48**
Error	28	2576.23	92.01	
Total lorica length				
Predator's presence	2	19379.98	9689.99	12.14***
Error	28	22346.51	798.09	
Lorica length				
Predator's presence	2	3584.63	1792.31	11.06***
Error	28	4536.19	162.01	
Posterior spine length (smaller one)				
Predator's presence	2	3251.44	1625.2	17.34***
Error	28	2624.22	93.72	
Posterior spine length (larger one)				
Predator's presence	2	3138.36	1569.18	5.10*
Error	28	8612.66	307.59	

DF = degrees of freedom, SS = sum of squares, MS = mean square, F = *F*-ratio. Levels of significance: *** = $p < 0.001$; ** = $p < 0.01$; * = $p < 0.05$; ns = non-significant ($p > 0.05$).

($p > 0.05$) influenced by the predator. For *K. cochlearis* the anterolateral spine length and the distance between the anterolateral pair was significantly influenced by the presence of the predator. In *K. tropica* except for the lorica width, all other variables tested were significantly influenced by the presence of the predator.

Discussion

Field and laboratory data have established the existence of induced defenses in some genera of rotifer prey, e.g. *Brachionus* and *Keratella*, particularly in the presence of the predatory rotifer *Asplanchna* the anterior and posterior spines, lorica lengths and widths usually increase considerably (Green, 1981; Stemberger & Gilbert, 1984). The increase in lorica spines of both *Brachionus* and *Keratella* may either minimize or totally prevent predation by *Asplanchna* due to: (a) decreased capture success, (b) increased handling time, (c) damage to the predator's body, (d) reduced time available to capture alternate prey by interfering with the predators movements, or (e)

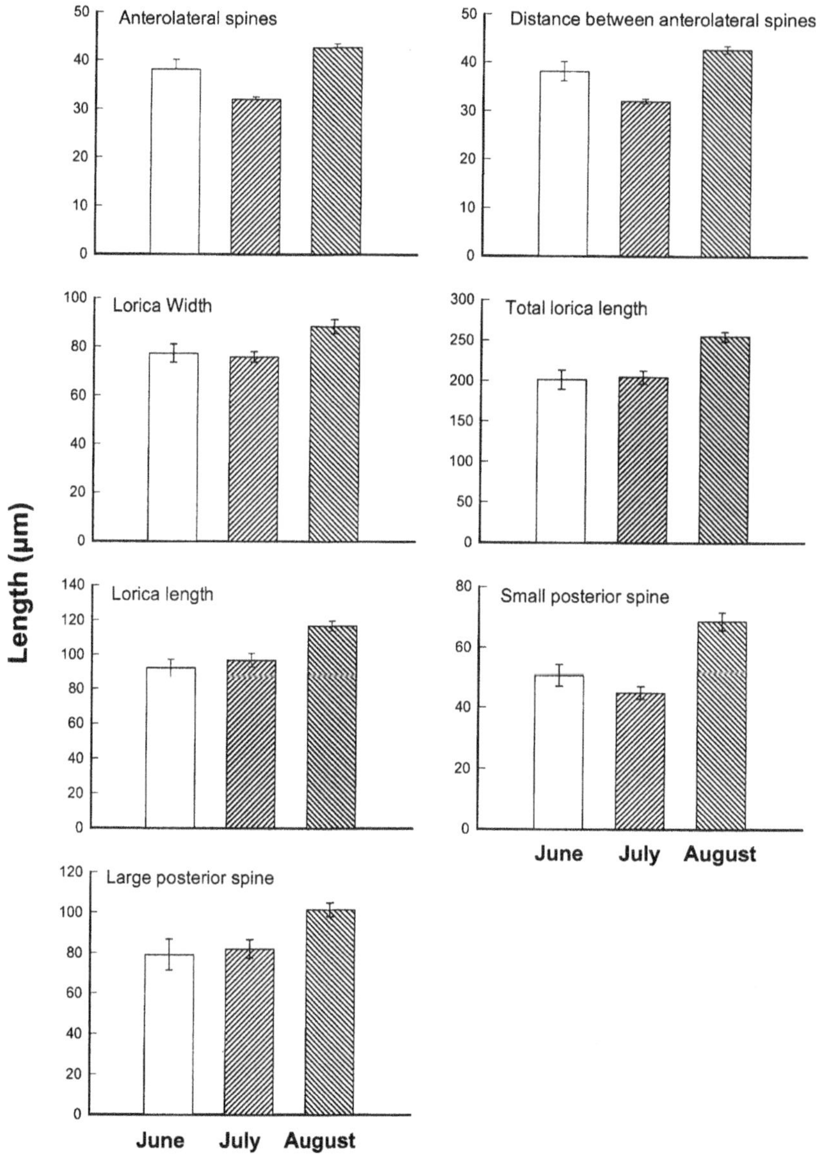

Figure 1. Morphometric data on the lorica of *B. havanaensis* measured in the presence of the predator *Asplanchna brightwellii*, which increased from June to August. For each morphometric variable, data represent mean ± standard error.

induction of resting egg production by the predator, thereby reducing its impact (Salt, 1987; Sarma, 1993; Iyer & Rao, 1995; Gilbert, 1999). Although for all these aspects, the importance of changes in the morphology of the prey in response to the presence of a predator is well recognized, morphometric data are rarely presented.

Sarma (1993) has shown that the handling time of *B. budapestinensis* by *Asplanchna brightwellii* is much longer than of *Brachionus calyciflorus*, in spite of the much smaller body size of the former (120 μm). Prey ingestion time is in general lower for the smaller than for the larger prey. It ranges between 4 ± 1 s for *A. fissa* (body length, 80 μm) to 23 ± 9 s for *B. calyciflorus* (180 μm) (Sarma et al., 1998). Avoidance of prey by *Asplanchna* has also been reported in literature, e.g., Iyer & Rao (1996) have observed that *B. budapestinensis was* avoided by *Asplanchna intermedia*. In our analysis of the stomach contents *of A. briglitwellii* collected from the field, *B. budapestinensis* was absent. Our SEM of *B. budapestinensis* (Plate 1) provides the

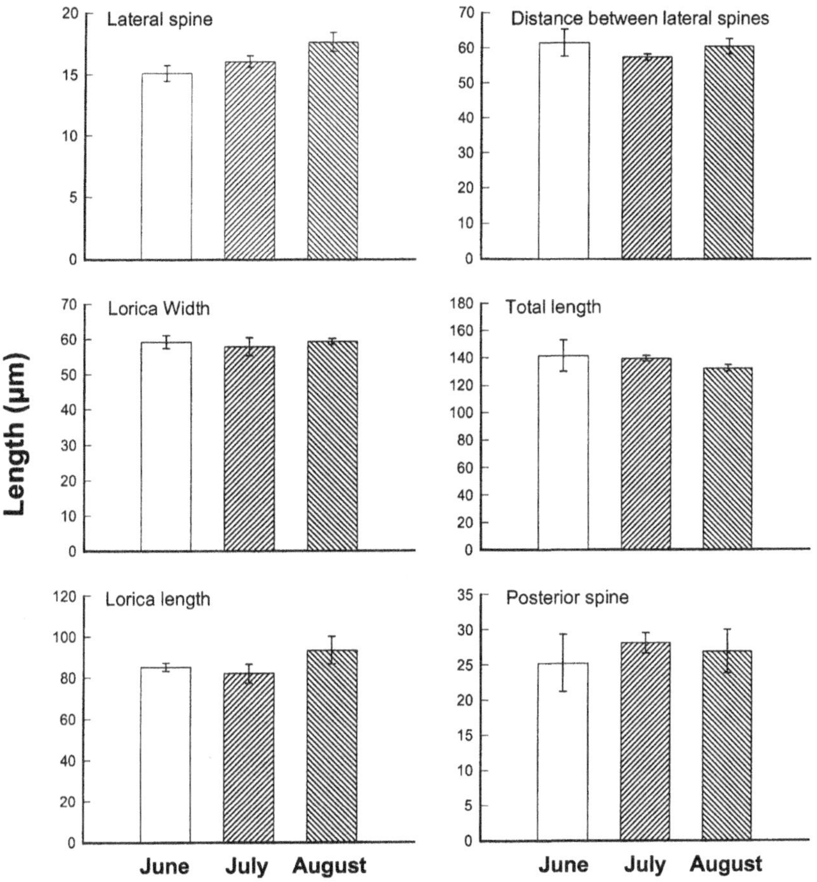

Figure 2. Morphometric data on the lorica of *K. americana* measured in the presence of the predator *Asplanchna brightwellii* which increased from June to August. For each morphometric variable, data represent mean ± standard error.

possible clue, i.e. the presence of numerous tiny spinules on the lorica. This SEM analysis is important for taxonomy but also to gain an insight into ecological relationships among microscopic organisms.

Rotifer prey defenses are largely related to the type of predator, with which they co-occur in nature. When fish are the dominant predators on rotifers, the strategy of the prey are to reduce the body size considerably, making predation energetically non-profitable (Zaret, 1980). This has been observed in Lake Parakrama Samudra (Duncan, 1984), a tropical waterbody, where fish is the dominant predator on zooplankton community throughout the year. *Brachionus* species such as *B. angularis* reduced the body size by about 15% in order to escape predation from fishes. On the other hand, if invertebrate predation is the main force structuring the zooplankton community in a water body, the rotifers tend to increase body size, as has been observed in many laboratory experiments on *Asplanchna* and *Brachionus* (Halbach & Halbach-Keup, 1974; Pourriot, 1974). However, both lorica length (or width) and spine lengths have certain limitations to grow in length. For example, Sarma (1985) has shown that by manipulation of food concentration it is possible to enhance the body size by about 15%. Therefore, if body size and spine lengths have reached their limits of elongation, other factors related to the morphology of lorica that could protect rotifers against invertebrate predation may become operational. These include widening both anterior and posterior spines. Both posterior and posterolateral spines in some species of *Brachionus* (e.g., *B. calyciflorus*) are known to extend laterally so that *Asplanchna* cannot easily

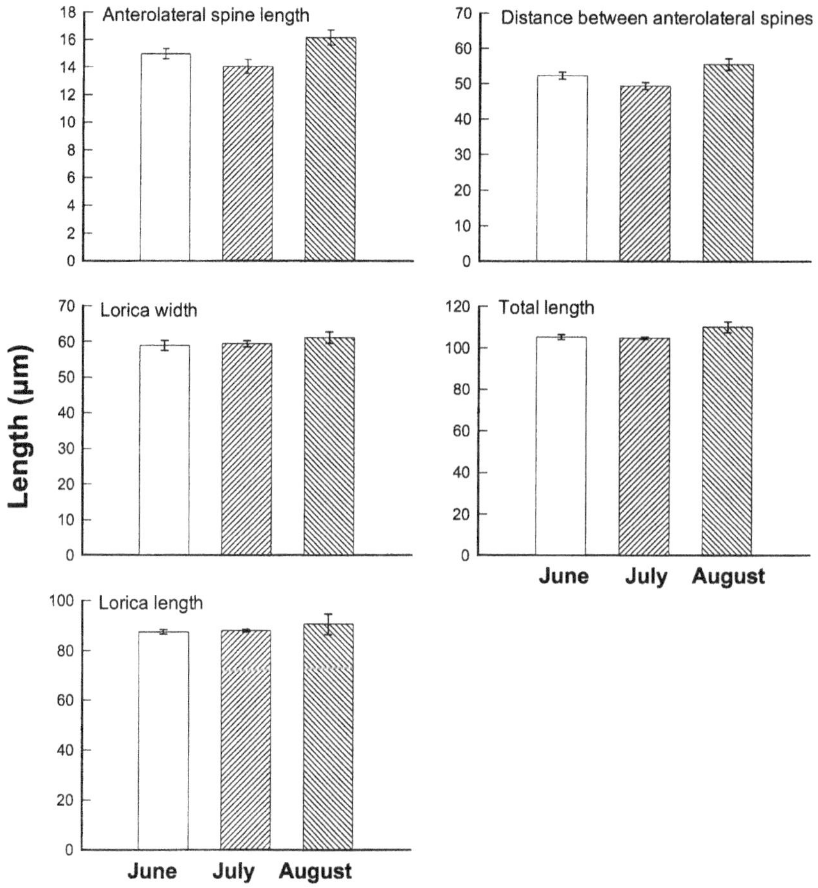

Figure 3. Morphometric data on the lorica of *K. cochlearis* measured in the presence of the predator *Asplanchna brightwellii*, which increased from June to August. For each morphometric variable, data represent mean ± standard error.

engulf them (Halbach & Halbach-Keup, 1974; Koste, 1978). Little information is, however, available with respect to anterior spines. It is evident from our study that *B. havanaensis* and *Keratella* spp., indeed, were with extended anterior spines in specimens collected from sampling sites with *Asplanchna*. Various invertebrate predators including *Asplanchna* are known to induce spine elongation of *Keratella* under experimental conditions (Stemberger & Gilbert, 1984). Our field data support this: we obtained individuals of *K. tropica* with long posterior spines when they co-occurred with *A. brightwellii*. *K. americana* did not show much variation in the presence of low and high abundances of *A. brightwellii*. Some members of Brachionidae possess prominent posterior and posterolateral spines even when *Asplanchna* was completely absent. For example Sarma & Nandini (2002) have cultured *B. macracanthus* for 2 years under laboratory conditions in the absence of *Asplanchna* and observed that long posterior and posterolateral spines are still prominent. The absence of significant morphological variations of *K. americana* in the presence of high and low abundances of *A. brightwellii* is similar to situation reported for *B. macracanthus* under laboratory conditions.

There are also other modes of avoidance by which brachionid rotifers escape predation from *Asplanchna* or other invertebrate predators. These include (a) production of resting eggs, which are either undigestible or have various morphological structures such as spines that deter consumption (Dumont et al., 2002), (b) switch to an epizoic mode as in *Brachionus rubens* by attaching to other zooplankton such as *Daphnia*, brachionids leave the water column so that planktonic species of Asplanchnidae find it difficult to encounter the

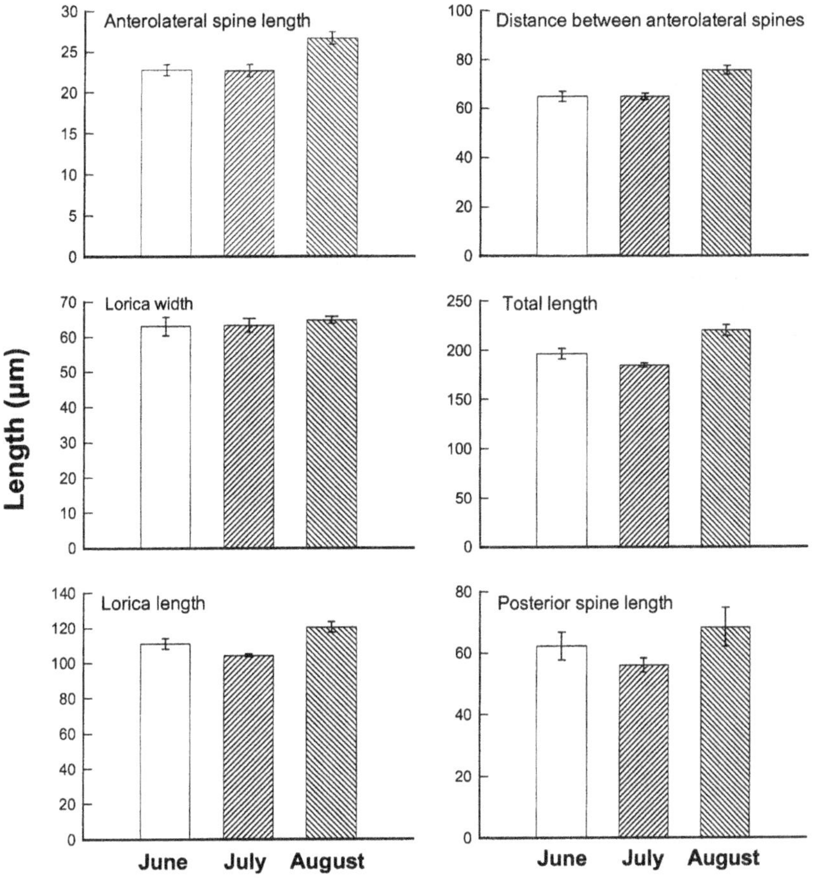

Figure 4. Morphometric data on the lorica of K. *tropica* measured in the presence of the predator *Asplanchna brightwellii*, which increased from June to August. For each morphometric variable, data represent mean ± standard error.

prey, or to swallow along with the host (Iyer & Rao, 1995) (c) large size due to colony formation or presence of toxic substances (e.g. *Sinantherina:* Felix et al., 1995), and (d) changes in morphology of lorica of the prey by formation of projections (e.g., *B. budapestinensis:* present study) or longitudinal folds (*Epiphanes macroura:* Sarma, 1993). It is, therefore, an integration of morphology (especially based on SEM observations) and morphometry of the prey that could help understand the role of *Asplanchna* predation on Brachionidae.

A preliminary analysis of the stomach contents of this predator did indicate that the long posterior spines of *Keratella tropica* and *K. americana* indeed protected them from predation, since we found several individuals of *K. cochlearis* but not so many *K. tropica or K. americana* in the stomachs of *Asplanchna* during these months (S. Nandini: unpublished data). It has been shown that *A. brightwellii* prefers short-spined (<15 µm) more frequently to long-spined (>20 µm) individuals of *K. cochlearis* (Conde-Porcuna et al., 1993). Iyer & Rao (1996) have also reported that *Asplanchna intermedia* shows a preference for *K. tropica* despite the latter's large posterior spines.

Although *K. tropica* measures up to 200 µm including the spines we often found it in the stomach contents. On the other hand, we rarely found *B. havanaensis* in the gut of *A. brightwellii*. This is probably due to the larger distance between the anterolateral spines in *B. havanaensis* compared with *K. tropica*. We, indeed, observed that the distance between the two anterolateral spines is significantly larger ($p < 0.001$, F-test) in *B. havanaensis* (74 ± 2 µm) than in *K. tropica* (69 ± 1 µm). Thus, despite its smaller body size than other brachionids, *B. havanaensis* is a well-protected species. *Asplanchna sieboldii*, which is one of the largest members of this family also has both low feeding rates and low preference for *B. havanaensis* compared with *B. calyciflorus*, *B. rubens* or *B. patulus* (Nandini et al., 2003). Nevertheless, in the absence of other prey types, several species of *Asplanchna* can feed on *B. havanaensis* (Nandini et al., 2003; Sarma et al., 2003).

In conclusion, our study shows that field collected samples of loricate rotifers exhibit strong morphological variations, mainly due to the presence *of Asplanchna*. Other invertebrate predators such as cyclopoid copepods may not have a significant impact since they were present in very low numbers (1–3 adult females per litre). The impact of *Asplanchna* on the other rotifer species too would be of considerable interest for understanding the range of morphological variations exhibited by prey in nature. The lack of preference or low capture rate of *B. budapestinensis* in our collections by *A. brighiwellii* is possibly due to the presence of minute spinules, which became evident only through SEM observations. Our study further reveals the need to consider the ultrastructural aspects of rotifer species in Lake Xochimilco, which could help to better understand prey adaptations against *Asplanchna* predation.

Acknowledgements

G.G.-M thanks Doctoral programme of UAM-X and CONACyT for scholarship (No. 142412). M.E.C.-P. and G. G-M. acknowledge project from P/FOMES 99-35-03 for financial support, and Olympus Mexico and Olympus Latin America for their economical and technical help. M. S.-B. thanks PROMEP UAAGS-PTC-24 and P/FOMES2000-01-13 for financial support. S.N. and S.S.S.S. thank financial support from UNAM (DGAPA) during their sabbatical year.

References

Conde-Porcuna, J. M., R. Morales-Baquero & L. Cruz-Pizarro, 1993. Effectiveness of the caudal spine as a defense mechanism in *Keratella cochlearis*. Hydrobiologia 255/266: 283–287.

Dumont, H. J., S. Nandini & S. S. S. Sarma, 2002. Cyst ornamentation in aquatic invertebrates: a defence against egg-predation. Hydrobiologia 486: 161–167.

Duncan, A., 1984. Assessment of factors influencing the composition, body size and turnover rate of zooplankton in Parakrama Salmdra, an irrigation reservoir in Sri Lanka. Hydrobiologia 113: 201–215.

Felix, A., M. E. Stevens & R. L. Wallace, 1995. Unpalatability of a colonial rotifer, *Sinantherina socialis*, to small zooplanktivorous fish. Invertebrate Biology 114: 139–144.

Flores-Burgos, J., S. S. S. Sarma & S. Nandini, 2003. Estudio preliminar sobre la fauna de rotiferos de Xochimilco (México). In Stephan-Otto, E. (ed.) El agua en la cuenca de México: Sus problemas históricos y perspectivas de solueión. Asociación Internacional de Investigadores de Xochimilco, A. C., Parque Ecológico de Xochimilco. Universidad Autónoma Metropolitana, Mexico City, Mexico, 163–171.

Garza-Mouriño, G. & M. E. Castellanos-Páez, 2003. Diversidad de rotiferos en los canales de la región noroeste de la zona chinampera de Xochimilco, Distrito Federal. In Stephan-Otto, E. (ed.) El agua en la cuenca de México: Sus problemas históricos y perspectivas de solución. Asociación Internacional de Investigadores de Xochimilco. A. C., Parque Ecológico de Xochimilco. Vol. 1 Universidad Autónoma Metropolitana, Mexico City, Mexico, 180–188.

Gilbert, J. J., 1999. Kairomone induced morphological defenses in rotifers. In Tollrian, R. & C. D. Harvell (eds.) The Ecology and Evolution of Inducible Defenses. Princeton University Press, NJ, 127–141.

Green, J. & O. B. Lan, 1974. *Asplanchna* and the species *Brachionus calyciflorus* in two Javanese sewage ponds. Freshwater Biology 4: 223–226.

Green, J., 1981. Altitude and seasonal polymorphism of *Keratella cochlearis* (Rotifera) in lakes of the Auvergne, Central France. Biological Journal of the Linnean Society, London 16: 55–61.

Halbach, U. & G. Halbach-Keup, 1974. Quantitative Beziehungen zwischen Phytoplankton and der Populationsdynamik des Rotators *Brachionus calyciflorus* Pallas. Befunde aus Laboratoriumsexperimenten und Freilanduntersuchungen. Archiv für Hydrobiologie 73: 273–309.

Hillbricht-llkowska, A., 1983. Morphological variation of, *Keratella cochlearis* (Gosse) in Lake Biwa, Japan. Hydrobiologia 104: 297–305.

Iyer, N. & T. R. Rao, 1995. The epizoic mode of life in *Brachionus rubens* Ehrenberg as a deterrent against predation by *Asplanchna intermedia* Hudson. Hydrobiologia 313/314: 377–380.

Iyer, N. & T. R. Rao, 1996. Responses of the predatory rotifer *Asplanchna intermedia* to prey species differing in vulnerability: laboratory and field studies. Freshwater Biology 36: 521–533.

Koste, W., 1978. Rotatoria. Die Rädertiere Mitteleuropas. Ein Bestimmungswerk begründet von Max Voigt. Bornträger, Stuttgart. I. Textband 673 pp., II. Tafelband, 234 Tafeln.

Koste, W. & R. J. Shiel, 1987. Rotifera from Australian inland waters. 2. Epiphanidae and Brachionidae (Rotifera: Monogononta). Invertebrate Taxonomy 7: 949–1021.

Nandini, S. & S. S. S. Sarma, 1999. Effect of starvation time on the prey capture behaviour, functional response and population growth of *Asplanchna sieboldi* (Rotifera). Freshwater Biology 42: 121–130.

Nandini, S., R. Pérez-Chávez & S. S. S. Sarma, 2003. The effect of prey morphology on the feeding behaviour and population growth of the predatory rotifer *Asplanchna sieboldi*: a case study using five species of Brachionus (Rotifera). Freshwater Biology 48: 2131–2140.

Pourriot, R., 1974. Relations predateur-proie chez les Rotifères: influence du predateur (*Asplanchna brightwellii*) sur la morphologie de la proie (*Brachionus bidentata*). Annales d'Hydrobiologie 5: 43–55.

Salt, G. W., 1987. The components of feeding behavior in rotifers. Hydrobiologia 147: 271–281.

Sarma, S. S. S., 1985. Effect of food density on the growth of the rotifer *Brachionus patulus* Müller. Bulletin of Botanical Society, Sagar 32: 54–59.

Sarma, S. S. S., 1993. Feeding responses of *Asplanchna brightwellii* (Rotifera): laboratory and field studies. Hydrobiologia 255/256: 275–282.

Sarma, S. S. S., H. J. Dumont & S. Nandini, 1998. Feeding preference and population growth of *Asplanchna brightwellii* (Rotifera) offered two non-evasive prey rotifers. Hydrobiologia 361: 77–87.

Sarma, S. S. S. & S. Nandini, 2002. Comparative life table demography and population growth of *Brachionus macracanthus* Daday, 1905 and *Platyias quadricornis* Ehrenberg, 1832 (Rotifera, Brachionidae) in relation to algal (*Chlorella vulgaris*) food density. Acta Hydrochimica et Hydrobiologica 30: 128–140.

Sarma, S. S. S., E. L. Pavón-Meza & S. Nandini, 2003. Comparative population growth and life table demography of the rotifer *Asplanchna girodi* at different prey (*Brachionus calyciflorus* and *Brachionus havanaensis*) (Rotifera) densities. Hydrobiologia 491: 309–320.

Stemberger, R. S. & J. J. Gilbert, 1984. Spine development in the rotifer *Keratella cochlearis:* induction by cyclopoid copepods and *Asplanchna*. Freshwater Biology 14: 639–647.

Zaret, T. M., 1980. Predation and Freshwater Communities. Yale University Press, New Haven, Connecticut.

Hydrobiologia (2005) 546:181–187
A. Herzig, R.D. Gulati, C.D. Jersabek & L. May (eds.) Rotifera X: Rotifer Research: Trends, New Tools and Recent Advances
DOI 10.1007/s10750-005-4120-7

Morphological stasis of two species belonging to the L-morphotype in the *Brachionus plicatilis* species complex

Sergi Campillo*, Eduardo M. García-Roger, David Martínez-Torres & Manuel Serra
Institut Cavanilles de Biodiversitat i Biologia Evolutiva, Universitat de València, València, Spain
*(*Author for correspondence: E-mail: sergi.campillo@uv.es)*

Key words: Rotifera, cytochrome oxidase I, sibling species, morphometry, taxonomy, discriminant analysis

Abstract

Detection and characterization of sibling species complexes in zooplankton are critical to understanding their ecological responses and patterns of evolution. The taxon *Brachionus plicatilis* is a complex of at least 14 species with three major, deeply diverged clades, which are morphologically distinct. We studied morphometric differences between two species – *B. plicatilis* sensu stricto and *B.* 'Manjavacas'– which belong to the L-(large) morphotype and often co-occur in ponds or lakes. *B. plicatilis* s.s. was on average 6% longer than *B.* 'Manjavacas'. They differed significantly in the measurements related to lorica spines. A significant discriminant function relating spine measurements was found, however, individuals from each species showed extensive overlap. Our morphometric data provide additional evidence for the species status of *B. plicatilis* s.s. and *B.* 'Manjavacas'. Since these are ancient species, our results support that a morphological stasis occurs in these taxa. We identified COI restriction sites for *Pvu*II and *Kpn*I which are diagnostic for *B.* 'Manjavacas' and *B. plicatilis* s.s., respectively. We conclude that morphometry is not useful in classifying the two species. At present, this can only be done reliably using molecular methods.

Introduction

Application of molecular markers to the study of aquatic invertebrates has shown that sibling species complexes are more common than previously thought (Knowlton, 1993; Hebert, 1998; Ortells et al., 2000; Gómez et al., 2002; Derry et al., 2003; Suatoni, 2003). In rotifers, genetic divergence resulting in speciation can occur independent of morphological divergence as mate recognition is not visual, and a remarkable morphological stasis can occur (Gómez et al., 2002). Detection and characterization of these species complexes is an important challenge in evolutionary ecology. Failure to do so means that species richness in a habitat may be underestimated and a species' ecological tolerance and habitat range could be overestimated.

The *Brachionus plicatilis* taxon is a complex of at least 14 sibling species (Suatoni, 2003). Most of them cluster into three major phylogenetic clades, which can be categorized as large, medium and small. Evidence for a large number of species comes mainly from molecular (allozyme and DNA) studies (Ortells et al., 2000; Gómez et al., 2002; Suatoni, 2003) and from mating behaviour analysis (Ortells et al., 2000; Suatoni, 2003). Detailed morphological studies generally confirm these species boundaries. An example is Ciros-Pérez et al. (2001) who used morphometry to describe the three formally named species in the complex. However, morphological differences between species in the same body size clade have not been studied yet, so it is not known whether subtle morphological differences can also be used to discriminate these species.

B. plicatilis sensu stricto and *B.* 'Manjavacas' both belong to the large (L-) morphotype (Gómez et al., 2002). Their species status is supported by the large genetic distance between them and by reproductive isolation in mating tests between sympatric population (Gómez & Snell, 1996; Ortells et al., 2000). The former has been reported in Europe, America, Asia and Oceania and the latter in Europe, America and Africa (Ortells et al., 2000; Gómez et al., 2002; Suatoni, 2003). They have been found in sympatry in 8 ponds of the Iberian Peninsula, while *B. plicatilis* s.s. alone has been reported in 14 ponds and *B.* 'Manjavacas' in 3 ponds (Ortells et al., 2000; Gómez et al., 2002).

We studied morphometry of these two species in order to test if differentiation in genetic markers is associated with morphological divergence, and to know whether we could identify characters that enable us to discriminate these species morphologically.

Methods

Clone collection

For species identification, a collection of 15 L-morphotype clones belonging to the *Brachionus plicatilis* complex was made from isolates collected in the Iberian Peninsula. These included five clones from the collection of the Institute Cavanilles of Biodiversity and Evolutionary Biology (University of Valencia, Spain): GALL1 from Laguna de Gallocanta, L1, L2 and L3 from Torreblanca Marsh (Poza Sur) and L5MAN from Laguna de Manjavacas. The other 10 clones originated from resting eggs that were hatched for this study, and were isolated from sediment samples collected in the following ponds: Salada de Chiprana, Laguna del Camino de Villafranca, Laguna de Pétrola and Laguna de Tírez. We followed the procedure in Gómez & Carvalho (2000) for the isolation of eggs, which were transferred individually to wells (Nunc™ polystyrene 96-well plates) containing 200 μl of 6 g l^{-1} artificial sea water (Instant Ocean™, Aquarium Systems). Hatching conditions were: temperature, 23 °C and light, approximately 150–170 μE m^{-2} s^{-1}. Once hatchlings reproduced parthenogenetically, the morphotypes were identified visually and only L-morphotype clones (Fu et al., 1991a) were selected. These were: ACHI1, ACHI2, ACHI3 and ACHI4 from Salada de Chiprana, ACVF1 and ACVF2 from Laguna del Camino de Villafranca, APET1 from Laguna de Pétrola and ATIR1, ATIR2 and ATIR3 from Tírez (for pond location see Ortells et al., 2000; Gómez et al., 2002). Clones were kept at 19 °C, 12 g l^{-1} of artificial sea water and fed the alga *Tetraselmis suecica*.

Clone identification and selection

The clones were tentatively identified as *B. plicatilis* s.s. or *B.* 'Manjavacas' by allozyme analysis using the loci *pgi*, *pgm*, *mdh-1* and *mdh-2* according to Ortells et al. (2000). For definitive identification, a restriction analysis on the mitochondrial gene COI was carried out on a subset of clones. Analysis of 112 COI sequences of *B. plicatilis* s.s. and *B.* 'Manjavacas' available at GenBank, after the extensive sequencing work by Dr. África Gómez and coworkers, revealed the presence of a *Kpn*I restriction site exclusive to *B. plicatilis* s.s. sequences and a *Pvu*II site only present in *B.* 'Manjavacas'. These two restriction enzymes were, therefore, chosen for species discrimination. For restriction analysis, rotifer DNA extraction was made using Chelex (6% Instagene™ Matrix, BioRad Laboratories; see Gómez et al., 1998). COI was amplified through Polymerase Chain Reaction using the following primers: COIdgF (5′-ggWATYTgAgCWggKCTYATTgg-3′) and COIdgR (5′-ggRTTACCTCCRCCKgCYggRTC-3′). These primers were designed based on the sequences available at GenBank for *B. plicatilis* s.s. and *B.* 'Manjavacas'. PCR was performed in vials containing 3 μl of template DNA, 0.2 mM of each nucleotide, 0.6 μM of each primer and 1.5 U of Taq-polymerase (Amersham Pharmacia Biotech). A Mastercycler™ (Eppendorf) was used for PCR using the following cycling profile: 3 min at 94 °C; 40 cycles of 30 s at 94 °C, 1 min at 48 °C and 1 min at 72 °C; a final step of 7 min at 72 °C was included after cycling. Restriction analysis for *Kpn*I and *Pvu*II was then performed using 15–20 ng of amplified DNA and 5–10 units of the corresponding enzyme in a final volume of 20 μl, and incubated at 37 °C for at least 1 h. Products were separated by standard agarose gel electrophoresis using 1.4% w/v agarose in 0.5 × TBE buffer. Gels were stained with ethidium bromide (0.1 μg ml^{-1}) and exposed to a UV transilluminator for direct DNA fragment visualization.

Morphometric analysis

Eight selected clones based on allozyme and DNA analysis were studied morphometrically. Two replicates per clone were established and grown independently in 250 ml of 12 g l^{-1} artificial sea water at 23 °C and constant illumination (ca. 100 μE m^{-2} s^{-1}) and fed 10^6 cells ml^{-1} of *Tetraselmis suecica*. Twice a week, a fraction of the rotifer culture was replaced with fresh medium (dilution rate: 0.77 $week^{-1}$). *Tetraselmis suecica* provided as food was grown in 22 g l^{-1} artificial sea water at 19 °C and constant illumination (ca. 100 μE m^{-2} s^{-1}) in a semicontinuous culture system (dilution rate: 0.65 day^{-1}). Before using them for rotifer feeding, the algal cells were pelleted by centrifugation (5 min at 3000 rpm), the supernatant was removed and the algal pellet was resuspended in 12 g l^{-1} artificial sea water.

From each replicate rotifer culture, 120 egg-bearing females were transferred to 1.5-ml wells (10 females per well) with 12 g l^{-1} artificial sea water and standard food conditions, and kept at 23 °C. After 5 h, at least 20 newborns per replicate were individually transferred to wells containing fresh medium and cultures as before. After 48 h, individuals were fixed with 4% formaldehyde and 10 females (age: 48–53 h) per replicate were randomly selected for morphometric measurements.

Nine morphometric characters were measured (Fig. 1). Morphometric characters (a–c) and (h) were measured at 400× magnification and (d–g) and (i) at 1000× magnification using a Nikon YS2 microscope. Morphometric characters (a–g) were selected based on Fu et al. (1991a) and (h–i) on Ciros-Pérez et al. (2001). For the statistical analysis of the differences ANOVA, MANOVA and a discriminant analysis were performed using SPSS (release 11.5. SPSS Inc., Chicago, IL).

Results

Species identification based on molecular markers

A restriction analysis of the COI gene was performed for definitive identification of 8 clones isolated from different sites. These clones had been tentatively identified by applying allozyme analysis to 15 L-morphotype clones (data not shown). Results of the digestion with *Kpn*I and *Pvu*II are shown in Figure 2. This restriction analysis allowed us to identify clones ATIR1, ACVF2 and L5MAN as *B.* 'Manjavacas' and GALL1, APET1, ACHI1, L3 and ACHI2 as *B. plicatilis* s.s. These clones were used for morphometric study.

Morphometric analysis

The average morphometric values for 5 *B. plicatilis* s.s. and 3 *B.* 'Manjavacas' clones are shown in

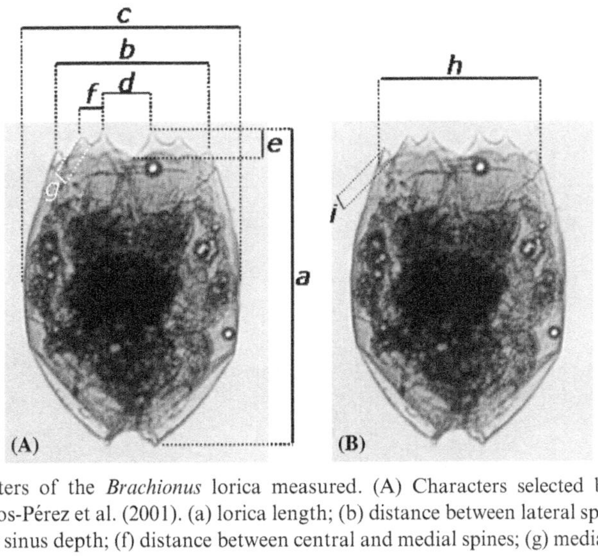

Figure 1. Morphometric characters of the *Brachionus* lorica measured. (A) Characters selected based on Fu et al. (1991a). (B) Characters selected based on Ciros-Pérez et al. (2001). (a) lorica length; (b) distance between lateral spines; (c) lorica width; (d) distance between central spines; (e) dorsal sinus depth; (f) distance between central and medial spines; (g) medial spine length; (h) head aperture; (i) lateral spine length.

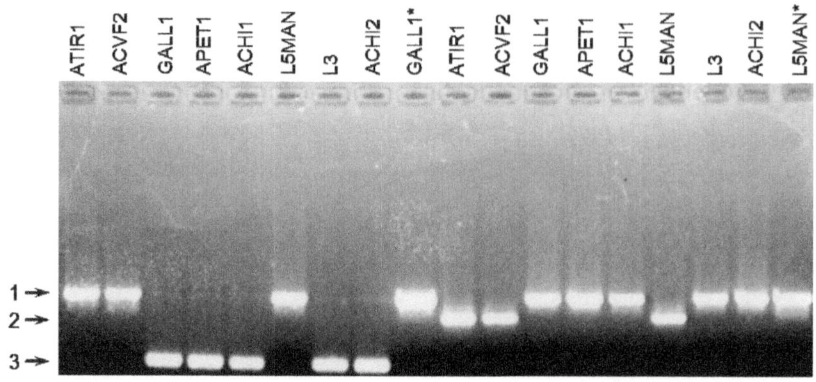

Figure 2. Restriction analysis of COI for eight clones belonging to *B. plicatilis* s.s. or *B.* 'Manjavacas'. A band located at (1) means that the enzyme had not a target; a band at (2) was found when *Pvu*II had a target (*B.* 'Manjavacas'); a band at (3) was found when *Kpn*I had a target (*B. plicatilis* s.s.). *Non-digested DNA.

Table 1. *B. plicatilis* s.s. is larger than *B.* 'Manjavacas' for all the measured lorica traits. The former species is about 6% larger based on major body measurements. However, only three of the nine measured characters (e), (f) and (i) showed significant statistical differences in characters relating to spine morphology, especially the lateral spine length [(i); Fig. 3 and Table 1].

A step-wise discriminant analysis with an associated one-way MANOVA ($F3,139 = 49.67$; $p < 0.01$) selected morphometric characters (e), (f) and (i) as being informative for species morphology discriminations. Thus, each one of these three measurements significantly improves the morphometric discrimination. The step-wise discriminant analysis combined the three significant measurements in a single discriminant axis, which is a new combined morphometric variable allowing for the best discrimination. When the distribution of individuals for this combined variable is plotted (Fig. 4), it is possible to observe a clear differential distribution of the species *B. plicatilis* s.s. and *B.* 'Manjavacas'. However, they extensively overlap in their distributions.

Discussion

The first evidence suggesting that *B. plicatilis* is actually a complex of sibling species comes from morphological data (Fu et al., 1991a) that distinguishes two morphotypes (L and S). According to

Table 1. Means and standard errors (S.E.) (in μm) of the morphometric characters for *B. plicatilis* s.s. and *B.* 'Manjavacas'

Morphometric character	*B. plicatilis* s.s.			*B.* 'Manjavacas'			ANOVA
	Mean	S.E.	*n*	Mean	S.E.	*n*	*p*
(a) Lorica length	385.7	2.8	100	360.7	3.6	60	0.118
(b) Distance between lateral spines	136.0	1.3	99	126.5	1.7	60	0.219
(c) Lorica width	298.2	1.9	100	280.8	2.8	59	0.080
(d) Distance between central spines	30.7	1.0	97	26.5	1.1	55	0.255
(e) Dorsal sinus depth	38.5	0.5	97	31.8	0.4	58	0.011*
(f) Distance between central and medial spines	35.3	0.4	100	31.1	0.5	60	0.021*
(g) Medial spine length	17.6	0.3	97	15.9	0.4	60	0.051
(h) Head aperture	161.3	1.2	100	152.8	1.1	60	0.171
(i) Lateral spine length	18.9	0.4	97	15.2	0.4	59	0.001**

p-values for between-species differences were computed from a four-level (species, clones, replicate, individual or error) nested ANOVA on log-transformed morphometric character (letters in parenthesis refer to Fig. 1). *n*: sample size. *$p < 0.05$ without Dunn–Sidak correction for multiple comparisons. **$p < 0.05$ with Dunn–Sidak correction.

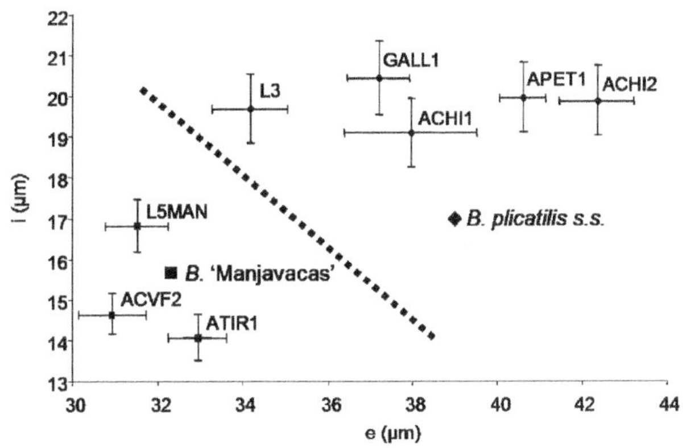

Figure 3. Mean and standard error of the morphometric characters (i) (lateral spine length) and (e) (dorsal sinus depth) for the eight clones. Characters were selected based on ANOVA results (Table 1).

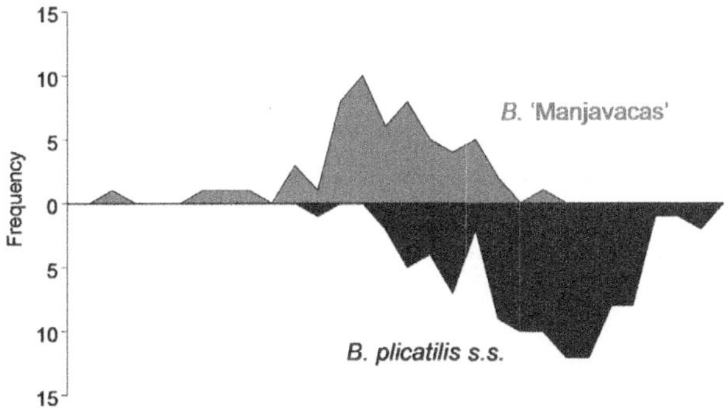

Figure 4. Distribution of individuals in the discriminant axis selected by a step-wise discriminant analysis on the morphometric characters. Characters (i), (e) and (f) were selected by the step-wise procedure.

phylogenetic analysis on DNA markers, the *B. plicatilis* complex is now recognized as a taxon composed of at least 14 species with deep genetic divergence between them (Gómez et al., 2002; Suatoni, 2003). The species are distinguished as three morphotypes: large (L), medium (SM), and small (SS) (Gomez & Serra, 1995; Ciros-Pérez et al., 2001; Suatoni, 2003), and each of these morphotypes corresponds to a major phylogenetic clade in the complex.

B. plicatilis s.s. and *B.* 'Manjavacas' both belong to L-morphotype. Maximum likelihood genetic distances between these two species, estimated on combined COI and ITS1 sequences available at GenBank using the GTR + I + G model, (Gómez et al., 2002), range from approximately 0.3 (strains 3MAN-L5 and AUSTRALIA1) to 0.4 (strains RUSSIA and 6TUR7). Mating behaviour suggests pre-zygotic reproductive isolation between the two species (Gómez & Snell, 1996; Ortells et al., 2000), and no evidence for hybridization in field studies exists, despite the fact that the species co-occur in some ponds in the Iberian Peninsula (Ortells et al., 2000). Our results show that *B. plicatilis* s.s. and *B.* 'Manjavacas' have different morphology, and that the former is on average larger. Traits related to spine morphology significantly differ. However, we do not rule out that there are minor size differences in other body characters that we did not study.

Our data provide additional support for considering *B. plicatilis* s.s. and *B.* 'Manjavacas' as different species, as well as show that if sibling species are first detected by either the molecular, morphological or mating behavioural methods, the divergence among species can be confirmed using the other approaches (Fu et al., 1991a, b; Gómez & Snell, 1996; Ortells et al., 2000; Gómez et al., 2002; Suatoni, 2003).

Phylogenetic analysis has recognized three major species clades in the *Brachionus plicatilis* species complex (Gómez et al., 2002). Each one of these clades corresponds to a clearly-defined different morphology related mainly to body size but also to spine shape (Fu et al., 1991a; Ciros-Pérez et al., 2001; Suatoni, 2003). We found that morphometric differences between these two relatively and closely related species in the complex, *B. plicatilis* s.s. and *B.* 'Manjavacas', although significant, are much smaller than those among the L-, SM- and SS-morphotypes. Therefore, phylogenetic divergence as inferred from molecular markers is correlated to morphological divergence in rotifer sibling species complexes when morphometry is studied in detail. However, the morphometric divergence found by us, though significant, is very small, which contrasts with the reported high genetic divergence based on molecular markers (0.3–0.4; Gómez et al., 2002). This genetic divergence suggests that the species are ancient (Gómez et al., 2002). Therefore, morphological stasis seems to be a characteristic in the evolution of these species. The causes of this stasis should form subject of future research.

The strains studied here were isolated in the Iberian Peninsula. So, they likely represent a small fraction of the diversity within *B. plicatilis* s.s. and *B.* 'Manjavacas'. On the other hand, as revealed by a discriminant analysis, these species show an extensive overlap in the measurements of lorica size and shape. Some individuals are less associated with the distribution of their own species than to the other species, and could be, therefore, easily misidentified based in their morphometric characters. This was observed despite our experimental design, which tended to minimize age and environmental effects. Therefore, we conclude that body measurements are unlikely to be reliable criteria for *B. plicatilis* species identification. As both species can co-occur, correct taxonomic classification is not possible, unless molecular analysis is performed.

Our study shows that deep genetic divergence is possible with high morphological stasis. Morphology is thought to be related to factors like predator vulnerability and feeding preferences. If morphology is similar, it is likely that these factors are also similar for both species. The striking morphological similarity between *B. plicatilis* s.s. and *B.* 'Manjavacas' prompts the question about how so similar species can coexist if differential predation or diet are unlikely to be mediating factors. We also do not know if these species respond differently to abiotic factors like temperature, salinity, pH and ionic composition. Moreover, it is not known whether co-occurrence of these two species in nature is transient and due to the length of the putative exclusion dynamics. Ecological studies and phylogeographic analysis on *B.* 'Manjavacas', similar to those performed on *B. plicatilis* s.s. (Gómez et al., 2000; Ortells et al., 2000; Suatoni, 2003) are needed to address these questions.

Acknowledgements

We thank Jorge Ciros and Raquel Ortells for their technical advice. Sara Lapesa, María José Carmona and Javier Armengol helped us in sediment sampling. África Gómez focused our attention on COI restriction analysis to differentiate *B. plicatilis* s.s. and *B.* 'Manjavacas'. We also thank María José Carmona, Terry W. Snell and Raquel Ortells for their valuable comments on the manuscript, as well as África Gómez and an anonymous reviewer. The study was supported by grant BOS2000-1451 from the Spanish Ministry of Science and Technology and SC was supported by grant *V Segles* of the University of Valencia.

References

Ciros-Pérez, J., A. Gómez & M. Serra, 2001. On the taxonomy of three sympatric sibling species of the *Brachionus plicatilis* (Rotifera) complex from Spain, with the description of *B. ibericus* n.sp. Journal of Plankton Research 23: 1311–1328.

Derry, A., P. D. N. Hebert & E. E. Prepas, 2003. Evolution in saline and subsaline lakes: a molecular phylogenetic approach. Limnology and Oceanography 48: 675–685.

Fu, Y., K. Hirayama & Y. Natsukari, 1991a. Morphological differences between two types of the rotifer *Brachionus plicatilis* O.F. Müller. Journal of Experimental Marine Biology and Ecology 151: 29–41.

Fu, Y., K. Hirayama & Y. Natsukari, 1991b. Genetic divergences between S and L type strains of the rotifer *Brachionus plicatilis* O.F. Müller. Journal of Experimental Marine Biology and Ecology 151: 43–56.

Gómez, A. & G. R. Carvalho, 2000. Sex, parthenogenesis and genetic structure of rotifers: microsatellite analysis of contemporary and resting egg bank populations. Molecular Ecology 9: 203–214.

Gómez, A., G. R. Carvalho & D. H. Lunt, 2000. Phylogeography and regional endemism of a passively dispersing zooplankter: mtDNA variation in rotifer resting egg banks. Proceedings of the Royal Society of London 267: 2189–2197.

Gómez, A. & C. Clabby & G. R. Carvalho, 1998. Isolation and characterization of microsatellite loci in a cyclically parthenogenetic rotifer, *Brachionus plicatilis*. Molecular Ecology 7: 1619–1621.

Gómez, A. & T. W. Snell, 1996. Sibling species and sibling speciation in the *Brachionus plicatilis* species complex (Rotifera). Journal of Evolutionary Biology 9: 953–964.

Gómez, A. & M. Serra, 1995. Behavioral reproductive isolation among sympatric strains of *Brachionus plicatilis* Müller 1786: insights into the status of this taxonomic species. Hydrobiologia 313/314: 111–119.

Gómez, A., M. Serra, G. R. Carvalho & D. Lunt, 2002. Speciation in ancient sibling species complexes: evidence from the molecular phylogeny of *Brachionus plicatilis* (Rotifera). Evolution 56: 1431–1444.

Hebert, P. D. N., 1998. Variable environments and evolutionary diversification in inland waters. In Carvalho, G. R. (ed.) Advances in Molecular Ecology. IOS Press, Amsterdam: 175–195.

Knowlton, N., 1993. Sibling species in the sea. Annual Review of Ecology and Systematics 24: 189–216.

Ortells, R., T. W. Snell, A. Gómez & M Serra, 2000. Patterns of genetic differentiation in resting egg banks of a rotifer species complex in Spain. Archiv für Hydrobiologie 149: 529–551.

Suatoni, L., 2003. Patterns of speciation in the rotifer species complex, *Brachionus plicatilis*. Yale University, New Haven 115 pp. (PhD disertation).

Hydrobiologia (2005) 546:189–196 © Springer 2005
A. Herzig, R.D. Gulati, C.D. Jersabek & L. May (eds.) Rotifera X: Rotifer Research: Trends, New Tools and Recent Advances
DOI 10.1007/s10750-005-4121-6

Morphological variation of *Keratella cochlearis* (Gosse) in a backwater of the River Thames

Jim Green
17 King Edwards Grove, TW11 9LY, Teddington, Middx
E-mail: jimgreen17keg@btinternet.com

Key words: Rotifera, variation, Lauterborn cycles, temperature

Abstract

The morphological variation of *Keratella cochlearis* in a Thames backwater has been studied over 4 years. There was a general inverse relationship between lorica length and temperature, but the annual cycle of change depended upon the rate of change of the temperature, and there was considerable variation between years. There was a similar inverse relationship between posterior spine length and temperature, and a shift in both relationships depending on whether the temperature was increasing or decreasing. As the water temperature increased from winter to summer the lorica and posterior spine were longer than at the same temperature as the water cooled from summer to winter. This shift can be modified by anomalous temperatures, such as a late spring or a cool summer. The form lacking a posterior spine usually, but not consistently, had a longer lorica than the typical spined forms. It usually disappeared from the samples at the end of November and did not reappear until March, although with a mild autumn and winter it persisted until January before disappearing. Forms without posterior spines did not all have the same origins.

Introduction

Any consideration of morphological variation in *Keratella cochlearis* must begin with the work of Lauterborn (1900, 1903). He established the basic aspects of variation in this species by arranging the forms in three main series: *tecta, hispida* and *irregularis*. The name he gave to the first series was unfortunate because it is also the name of the form without a posterior spine that occurs in warm, eutrophic conditions. This series, which I shall call the *typica* series, is characterised by large, long spined forms (*macracantha)* in winter, followed by progressively smaller forms with shorter posterior spines (*typica, micracantha, tuberculata)* until the unspined *tecta* form is reached in summer. Lauterborn found all possible intermediates in this series so that the names are useful descriptors but lack any precise taxonomic significance. The *hispida* series shows a thinning and shortening of the posterior spine, and characteristic development of pustules and spinules over the surface of the lorica. The *irregularis* series starts with long spined forms, with a kink in the median line of the dorsal sculpture, and proceeds through forms showing the progressive development of a median pentagon in the dorsal sculpture (*connectens, angulifera, irregularis*). A form lacking a posterior spine (*ecaudata*) also occurs. Lauterborn had thus shown two separate routes to unspined forms, so the use of the name *tecta* without careful examination of the dorsal sculpture is not valid.

Many aspects of Lauterborn's findings have been confirmed and extended by more recent work (Pejler, 1957, 1962, 1980; Hillbricht-Ilkowska, 1972; Ruttner-Kolisko, 1974; Koste, 1978; Hofmann, 1980, 1983; Eloranta, 1982). These authors also made emendments to Lauterborn's interpretations. For instance Hofmann (1980) found that in many Holstein lakes the typical *cochlearis, hispida* and *tecta* forms were clearly separated. He found no intermediates between typical *cochlearis* and *tecta*, and the lorica length of the latter was consistently greater, so could not be the end of a reduction series. The *tecta* forms were also abundant in autumn, when the spine length of *cochlearis* was increasing. This implies a third type without a posterior spine, not derived from the *typica* series. Some authors have also raised *irregularis* and *hispida* to separate species (e.g. Ahlstrom, 1943; Eloranta, 1982).

My observations on *K. cochlearis* in Broom Water began in August 1997, when Athene Jones drew my attention to an outburst of the freshwater medusa *Craspedacusta sowerbyi* (Lankester) (Green, 1998), but regular sampling did not begin until the end of 1998. The aim was to see how far the population in this part of the Thames conformed to the findings of Lauterborn and the other authors mentioned above.

The habitat

Broom Water is an artificial cut extending about 250 m from the main channel of the Thames above the weir at Teddington (Fig. 1). It is about 8 m wide and serves as a mooring area for numerous small boats of householders with gardens extending down to the water. The water is fresh, with conductivities ranging between 400 and 500 μS cm^{-1}. The water is open to fish movement from the main Thames, and serves as a spawning area for large carp (*Cyprinus carpio* Linn.) and as a nursery for young roach (*Rutilus rutilus* (Linn.)).

Methods

Samples were taken from the bank using a 55 μm meshed net. They were preserved in 4% formaldehyde, and subsamples were examined on a slide under a long coverslip. The slide was moved by a mechanical stage, so the sequence in which rotifers were encountered was essentially random. Measurements were made at a magnification of 400× using a calibrated eyepiece micrometer, using the limits shown in Green (1998). The normal sample size was 30, but this was sometimes increased to 50, and more rarely to 100. For most samples the standard errors of the means were <1 μm.

Results

Seasonal variation in lorica and spine length

Figure 2 shows variation in lorica length and spine length in specimens bearing any form of posterior spine over a period of 4 years. There was a regular decline in both dimensions during the summer. The most notable feature was the anomaly during the winter of 2001/2 when the lengths of the lorica and spine did not increase to the extent that they had in the previous three years.

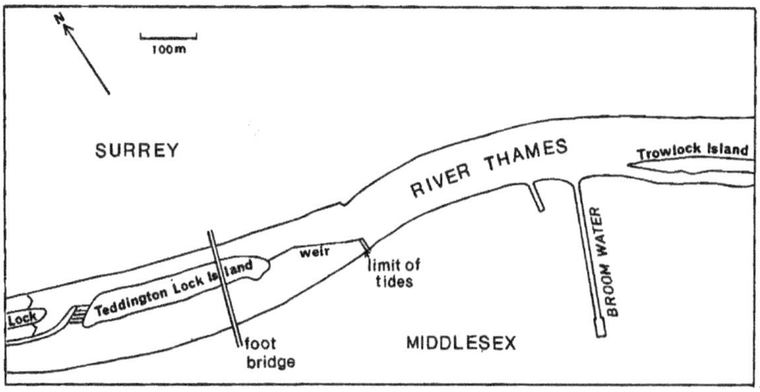

Figure 1. Sketchmap to show the location of Broom Water.

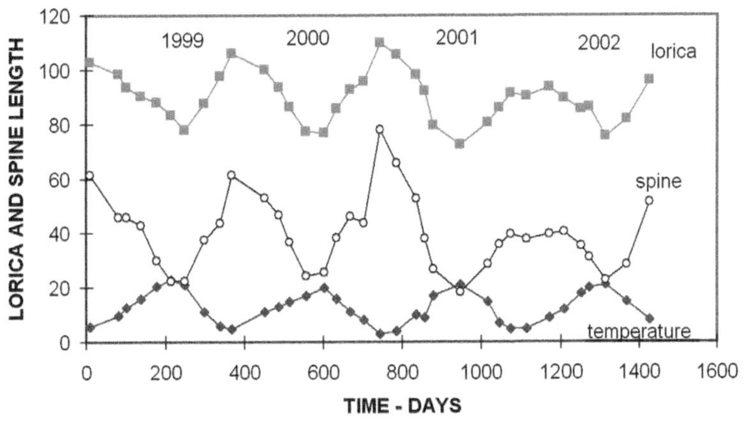

Figure 2. Temporal variation in temperature, lorica length and spine length (μm) in *Keratella cochlearis* in Broom Water over 4 years. The standard errors for each point are small enough to be contained within the symbol.

Lorica length and temperature

Figure 3 shows the general relationship for all 4 years combined. Both the lorica and the posterior spine decreased with increasing temperature. Figure 4 shows the sequential changes in each of the 4 years. In 1999 there was a smooth progression, with the lorica length decreasing steadily as the temperature increased. From September to December the lorica length increased steadily, but along a line below that of the changes in springtime. The other years do not show such a smooth relationship, partly because the temperature changed at different rates. For instance the September 1999 sample was taken when the water temperature was 21 °C, while in September 2000 the water temperature was 16 °C. These differences in the temperatures in different years altered the slope of the relationship between temperature and lorica length. Figure 5 shows the regression lines calculated for the 4 years. In 2000 and 2001 the slopes were significantly steeper than in 1999 and 2002. The regression coefficients for 1999 and 2000 differed at the level $p < 0.005$, >0.001, but there was no difference between 1999 and 2002.

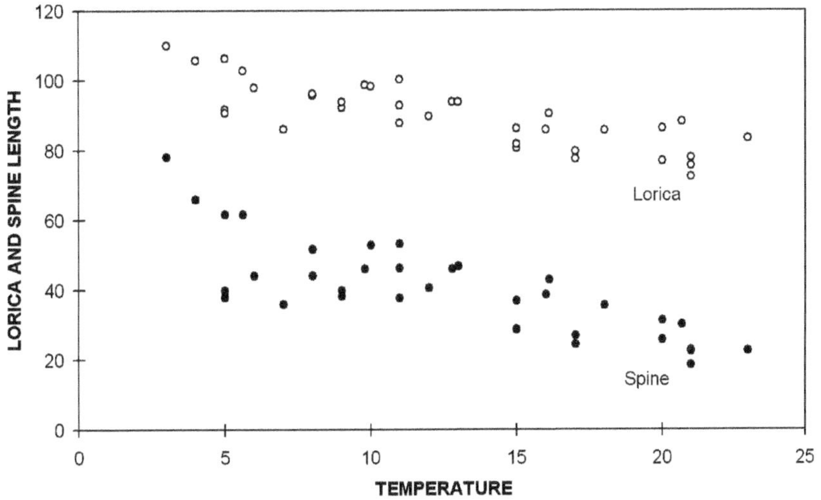

Figure 3. The general relationship between temperature and the lengths (μm) of the loricas and posterior spines of *Keratella cochlearis* in Broom Water.

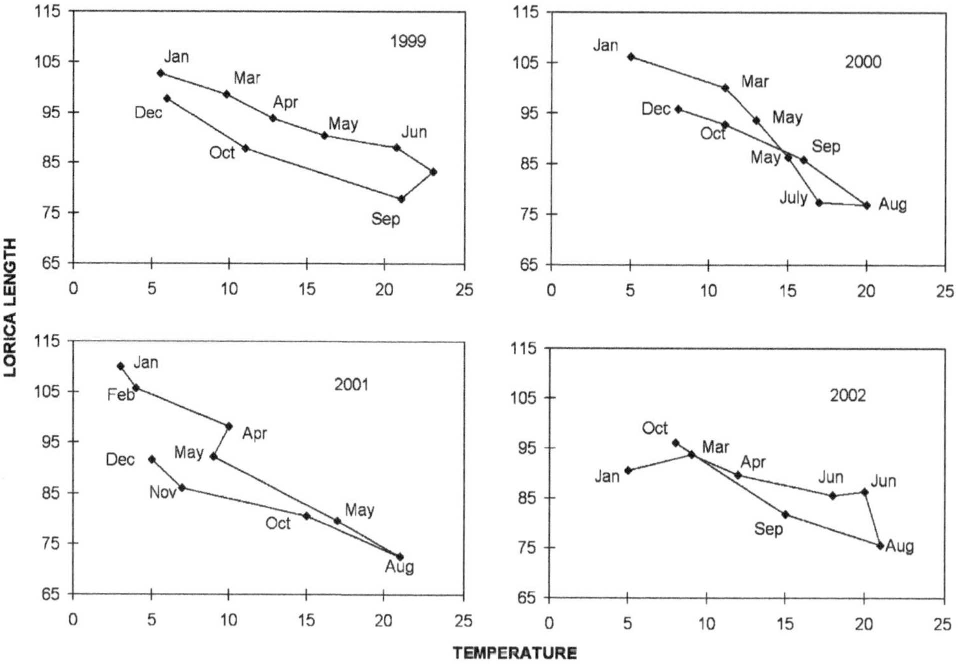

Figure 4. Sequential variation in the relationship between temperature and lorica length (μm) of *Keratella cochlearis* in Broom Water in 4 years.

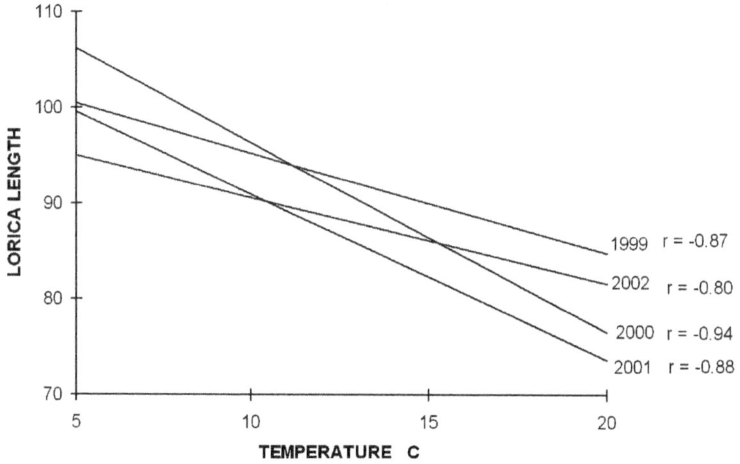

Figure 5. Regression lines for the 4 years shown in Fig. 4.

Lorica and posterior spine length

Figure 6 shows the relationship between lorica and posterior spine length for all 4 years combined. The correlation is high ($r = +0.956$).

Lorica lengths of spined and unspined forms

Figure 7 shows seasonal variation in the loricas over 4 years. Specimens without spines were absent in January–February and November–December 1999, January–February 2000, January–March 2001, and February–March 2002, but enough were found to measure in January 2002. The main point made by Figure 7 is that the unspined forms in Broom Water generally have longer loricas than the spined forms. This was particularly notable in December 2001 and January 2002.

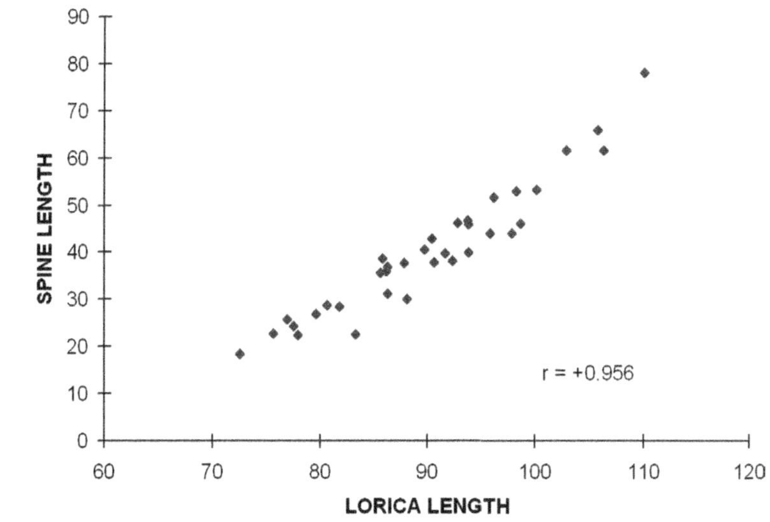

Figure 6. Keratella cochlearis in Broom Water, the relationship between lorica length and spine length (μm).

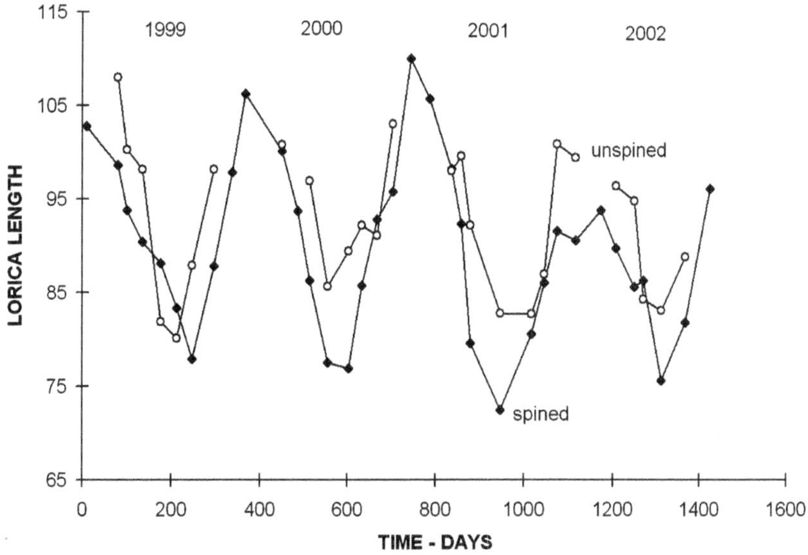

Figure 7. Temporal variation in lorica length (μm) of spined and unspined *Keratella cochlearis* in Broom Water. The standard errors of the means do not exceed one symbol width above and below each point.

Bimodality of unspined forms

Figure 8 shows the results from a sample taken on 29 September 2002. In this sample it was possible to distinguish forms with a somewhat more acuminate posterior end, like Lauterborn's figure 9. These had significantly shorter loricas than those with more rounded posterior ends.

Occurrence of irregularis *and* hispida *forms*

These forms were not usually common in Broom Water, but on 4 August 01 both were present: *irregularis* formed about 12% of the spined forms, and *hispida* about 0.5%. The *irregularis* were significantly larger (lorica length 83.3 $\pm$ 0.9 μm) than the typical form (lorica length 72.5 $\pm$ 0.3 μm).

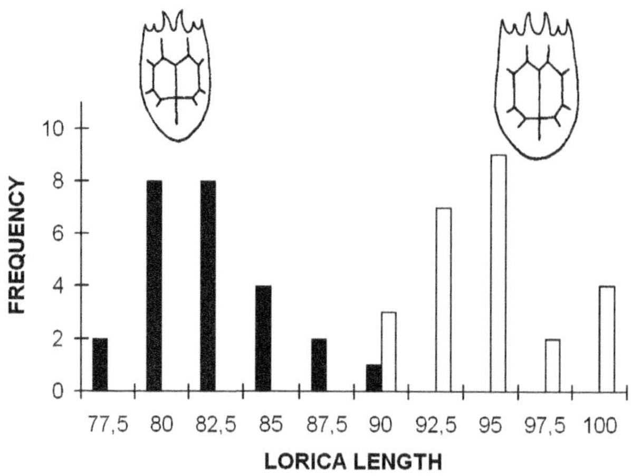

Figure 8. Bimodalism in lorica lengths (μm) of unspined *Keratella cochlearis* in Broom Water, 29 September 02.

Only one *hispida* was available for measurement. This had a lorica length of 85 μm, which was larger than any of the *typica* measured on the same day. In the same sample a single specimen without a posterior spine, but with an *irregularis* dorsal sculpture had a lorica length of 88 μm, which was larger than any of the *irregularis* with spines.

Discussion

The seasonal variations in lorica and posterior spine lengths of *K. cochlearis* in Broom Water are in general agreement with those observed by Lauterborn, and many other investigators, but Figure 2 shows an anomaly in the winter of 2001/2, when the spine lengths were only two thirds of those found in the previous 3 years. The water temperature had fallen to the same level as previous winters, so other factors must have been operating. The relationship between spine length and lorica length can be changed by variation in the food supply. Hillbricht-Ilkowska (1983) found an increase in lorica length from 89 to 94 μm between October and December, while the posterior spine decreased from 64 to 35 μm. This happened in the southern basin of Lake Biwa, Japan, during a period of nannoplankton abundance, and was accompanied by an increase in the fecundity of the rotifer.

The relationship between spine length and temperature can vary significantly from year to year. Predation pressure is a major factor contributing to this variation (Conde-Porcuna et al., 1993). At a given temperature when predation is high the spines are longer. Long spines give protection against predation by *Asplanchna*, but very short spines appear to offer less protection than being unspined. Conde-Porcuna et al. (1993) found positive selection by *Asplanchna* for individuals with spines less than 15 μm in length. This could be part of an explanation for the lack of intermediates between *micraspina* and *tecta* in some studies.

In Broom Water there was a strong correlation between lorica and spine length. Regression lines for each year show that spine length would reach zero at a lorica length between 60 and 65 μm. This was well below the actual lengths of the unspined forms in this locality. A precisely similar relationship was found by Bielanska-Grajner (1995) in Rybnik Reservoir, Poland. Figure 7 shows that the loricas of unspined forms in Broom Water were generally longer than those of the typical spined forms at any given time. But there were occasions when the unspined forms were as small, or even smaller than the typical forms. These could be produced from small (*micracantha* and *tuberculata*) parents at high temperatures. This was particularly well shown in June 1999, when the spineless loricas had a mean length of 82 (SE ± 0.7) μm while the typical forms were 88 (SE ± 1.04) μm. Very short spined forms (*micracantha* and *tuberculata*) were present at the time, and the water temperature was 20.7 °C. Evidence that unspined forms may have more than one origin is provided by Figure 8. The morphological differences between the two forms are subtle, but the lorica lengths fell into two clear groups. Overall the samples from Broom Water indicate three possible origins for forms without posterior spines:

(1) As the end of a reduction series – the true *tecta*, occurring from June to September.
(2) An unspined form with a longer lorica than co-occurring typical forms, and absent from samples only in January and February. Most authors would call this *tecta* but it is not the end of a reduction series, and showed seasonal changes paralleling those of the typical form (Fig. 7). The name *aspina* is proposed for these, as a descriptor without taxonomic significance, to separate them from the true *tecta*.
(3) An unspined form with the same dorsal sculpture, but having a longer lorica than co-occurring *irregularis*, and occurring in summer (*ecaudata*).

The origins of the true *tecta* are supported by the experimental work of Stemberger & Gilbert (1984) showing that unspined forms could be induced to develop posterior spines by a water soluble factor released by cyclopoid copepods. From the true *tecta* they induced *tuberculata, micracantha* and *typica*, showing that they can all be derived from a single genotype. Their figure of *tecta* corresponds to the acuminate form from Broom Water. The recent work of Derry et al. (2003) appears at first sight to be incompatible with the results of Stemberger & Gilbert. They found 'a deep genetic divergence' between spined and unspined forms of *K. cochlearis*. Unfortunately they did not figure the forms they analysed, so it is not known what the dorsal sculpture or the shape of the posterior part of the lorica were like. If the form they analysed was the second of those described above (i.e. *aspina*) it is perfectly possible for it to show genetic divergence from the spined form. A similar result could possibly be obtained with the *ecaudata* form of *irregularis*. On morphological grounds one would expect these to show some genetic divergence from *typica* and the true *tecta*.

Acknowledgements

This work would not have been possible without the hospitality and access to Broom Water generously provided by Ashley and Athene Jones. Comments by Christian Jersabek and an anonymous referee have helped to improve the final version of the paper.

References

Ahlstrom, E. H., 1943. A revision of the rotatorian genus *Keratella* with descriptions of three new species and five new varieties. Bulletin of the American Museum of Natural History 80: 411–457.

Bielanska-Grajner, I., 1995. Influence of temperature on morphological variation in populations of *Keratella cochlearis* (Gosse) in Rybnik Reservoir. Hydrobiologia 313/314: 139–146.

Carlin, B., 1943. Die Planktonrotatorien des Motalaström. Meddelanden från Lunds Universitets Limnologiska Institution 5: 1–255.

Conde-Porcuna, J. M., R. Morales-Baquero & L. Cruz-Pizarro, 1993. Effectiveness of the caudal spine as a defense mechanism in *Keratella cochlearis*. Hydrobiologia 255/256: 283–287.

Derry, A. M., P. D. N. Hebert & E. E. Prepas, 2003. Evolution of rotifers in saline and subsaline lakes: a molecular phylogenetic approach. Limnology and Oceanography 48: 675–685.

Eloranta, P., 1982. Notes on the morphological variation of the rotifer species *Keratella cochlearis* (Gosse) s.l. in one eutrophic pond. Journal of Plankton Research 4: 299–312.

Green, J., 1998. Plankton associated with medusae of the freshwater jellyfish *Craspedacusta sowerbyi* (Lankester) in a Thames backwater. Freshwater Forum 11: 69–76.

Green, J, 1999. Strategic variation of egg size in *Keratella cochlearis*. Hydrobiologia 387/388: 301–310.

Hauer, J., 1952. Pelagische Rotatorien aus dem Windgfällweiher, Schluchsee und Titisee im südlichen Schwarzwald. Archiv für Hydrobiologie, Supplement 20: 212–237.

Hillbricht-Ilkowska, A., 1972. Morphological variation of *Keratella cochlearis* (Gosse) (Rotatoria) in several Masurian Lakes of different trophic level. Polskie Archiwum Hydrobiologii. 19: 253–264.

Hillbricht-Ilkowska, A., 1983. Morphological variation of *Keratella cochlearis* (Gosse) in Lake Biwa, Japan. Hydrobiologia 104: 297–305.

Hofmann, W., 1980. On morphological variation in *Keratella cochlearis* populations from Holstein lakes (Northern Germany). Hydrobiologia 73: 255–258.

Hofmann, W., 1983. On temporal variation in the rotifer *Keratella cochlearis* (Gosse): the question of "Lauterborn-cycles". Hydrobiologia 101: 247–254.

Koste W., 1978. Rotatoria. Die Rädertiere Mitteleuropas. Ein Bestimmungswerk begründet von Max Voigt. Überordnung

Monogononta. Gebrüder Borntraeger, Berlin, Stuttgart, I. Textband 673 pp., II. Tafelband, 234 Tafeln.
Lauterborn, R., 1900. Der Formenkreis von *Anuraea cochlearis*. Ein Beitrag zur Variabilität bei Rotatorien I. Morphologische Gliederung des Formenkreises. Verhandlungen des Naturhistorisch-Medizinischen Vereins zu Heidelberg 6: 412–448.
Lauterborn, R., 1903. Der Formenkreis von *Anuraea cochlearis*. Ein Beitrag zur Variabilität bei Rotatorien. II. Die cyclische oder temporale Variation von *Anuraea cochlearis*. Verhandlungen des Naturhistorisch-Medizinischen Vereins zu Heidelberg. 7: 529–621.
Pejler, B., 1957. On variation and evolution in planktonic Rotatoria. Zoologiska Bidrag från Uppsala. 32: 1–66.
Pejler, B., 1962. On the variation of the rotifer *Keratella cochlearis* (Gosse). Zoologiska Bidrag från Uppsala. 35: 1–17.
Pejler, B., 1980. Variation in the genus *Keratella*. Hydrobiologia 73: 207–213.
Ruttner-Kolisko, A., 1974. Plankton Rotifers, Biology and Taxonomy. Binnengewässer 26 (Suppl 1): 1–146.
Stemberger, R. S. & J.J. Gilbert, 1984. Spine development in the rotifer *Keratella cochlearis*: induction by cyclopoid copepods and *Asplanchna*. Freshwater Biology 14: 639–647.

Hydrobiologia (2005) 546:197–202
A. Herzig, R.D. Gulati, C.D. Jersabek & L. May (eds.) Rotifera X: Rotifer Research: Trends, New Tools and Recent Advances
DOI 10.1007/s10750-005-4197-z

Trophi structure in bdelloid rotifers

Giulio Melone* & Diego Fontaneto
Department of Biology, University of Milan, I-20133, Milan, Italy
*(*Author for correspondence: E-mail: giulio.melone@unimi.it)*

Key words: Rotifera, Bdelloidea, jaws, SEM

Abstract

Bdelloids show a rather uniform morphology of jaws (trophi), named ramate. The most recognizable feature is the presence of a series of teeth forming unci plates. The unci are not uniform in size; each plate has 1–10 major median teeth. Using SEM pictures of trophi and data from the literature, we analyzed the number of major unci teeth in relation to trophi size, total number of teeth, and environmental features. Variability in the number of major unci teeth in bdelloids is not related to trophi size or to total number of unci teeth, while total number of unci teeth and trophi size seem to be related to each other: larger trophi in general have more teeth than smaller trophi. Few major teeth are more common in species living in water bodies where they possibly eat unicellular algae, while more major teeth are more common in species living outside water bodies, among mosses and lichens, where they possibly eat bacteria.

Introduction

Among other features, rotifers are characterized by their masticatory apparatus, the mastax. Its hard, sclerotized and articulated pieces are termed trophi. The general morphology of rotifer trophi was described by Gosse (1856), who distinguished a single median element, the fulcrum, with which articulate two symmetrical structures, the rami. Each ramus is associated with an uncus, which is connected to a manubrium. Totally, seven elements are present in rotifer trophi: one fulcrum and two paired rami, unci and manubria. Different models of trophi were described, but while trophi of monogononts are very variable in shape and size and are a widely used taxonomic feature (Koste, 1978; Markevich, 1989; Melone et al., 1998a; Sørensen, 2002), those of bdelloids show a rather uniform 'ramate' morphology (Melone et al., 1998b). Moreover, while for some bdelloid genera, such as *Pleuretra*, no taxonomic use is possible (Fontaneto & Melone, 2003), in other genera, such as *Rotaria*, trophi seem to have species-specific shape and size (Fontaneto et al., 2004).

The most recognizable feature in ramate trophi is the presence of a series of teeth forming unci plates. These teeth are not of uniform size; 1–10 major median teeth are present (Donner, 1965) (Fig. 1).

Because of their small size, trophi elements are difficult to observe by light microscopy only and an SEM approach is a more useful analytical tool for ultrastructure (e.g., Markevich, 1985, 1989; Markevich & Kutikova, 1989). The extensive SEM analyses of Melone et al. (1998b) confirmed the general homogeneity of the trophi structure of Bdelloidea, although minor differences among the three orders (Philodinavida, Adinetida, Philodinida) were evident. Some differences in trophi among species with regard to number of major unci teeth were known, and for some species this number can be used as a reliable taxonomic feature (Donner, 1965).

The aim of this study is to investigate the possible relationship of the number of major unci teeth with trophi size and total number of unci

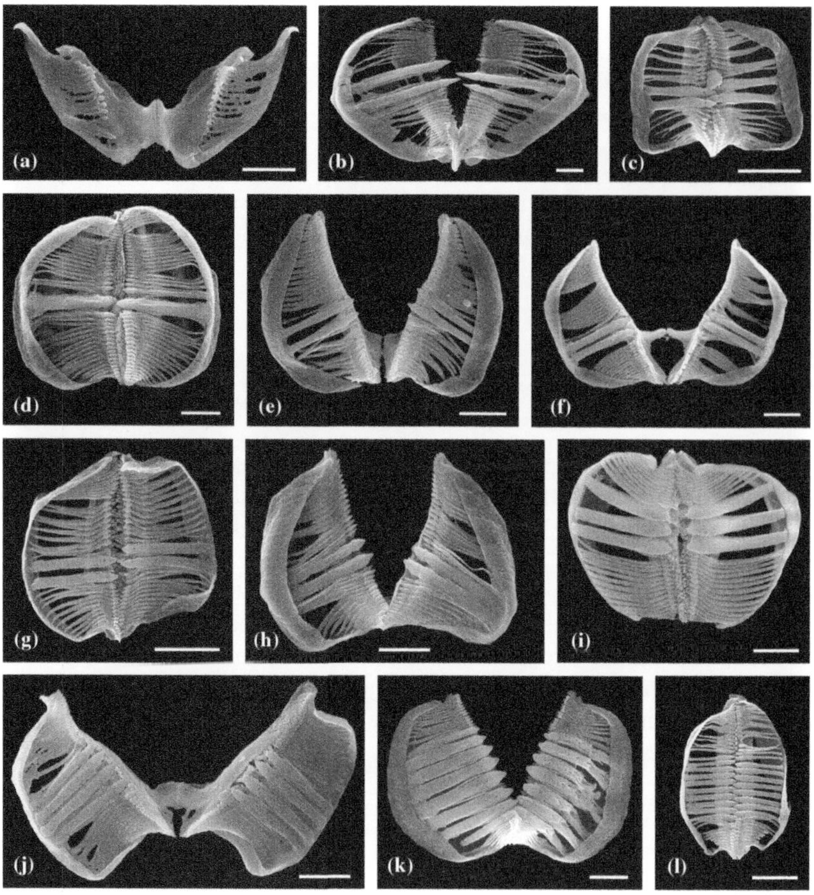

Figure 1. SEM pictures of trophi of bdelloid rotifers in cephalic view: (a and b) Philodinavida; (c) Adinetida; (d–l) Philodinida; (a) *Henoceros falcatus*; (b) *Abrochtha carnivora*; (c) *Adineta steineri*; (d) *Anomopus telphusae*; (e) *Macrotrachela crucicornis*; (f) *Philodina citrina*; (g) *Pleuretra reticulata*; (h) *Zelinkiella synaptae*; (i) *Dissotrocha aculeata*; (j) *Didymodactylos carnosus*; (k) *Mniobia magna*; (l) *Otostephanos donneri*. Scale bar = 5 μm.

teeth, and also with environmental requirements of the species, with special attention to different types of food. Moreover, we investigated the intraspecific variability in the way in which left and right uncus fit when the trophi close, with the major tooth closer to the rami articulation being on the left or on the right half depending on the specimen (Fontaneto et al., 2004).

Materials and methods

We directly investigated SEM pictures of the trophi of 285 bdelloid rotifer specimens, belonging to 27 species in 16 genera, representing all four families. We note that not all pictures gave us all the required information.

General data on features of the trophi of bdelloids is available for 381 species, the 374 recognized valid species cited by Segers (2002), and others not yet or only recently described (Ricci et al., 2003). Data on the number of major teeth came from Donner (1965) and from personal observations by light microscopy. Ecological requirements of bdelloids are reported upon by Donner (1965) and Bērziņš & Pejler (1987, 1989a, b), other than our unpublished data.

We considered as major teeth only the large major teeth distinguished by Melone et al. (1998b).

SEM pictures of isolated trophi were obtained following Segers' (1993) method of preparation on a circular cover slip by sequentially dissolving tissues in 5% NaOCl solution, and rinsing with distilled water.

Results

Depending on the species, the number of major unci teeth varies from 1 to 10; 'two teeth' is the most common configuration (38.2%), followed by 'three teeth' (22.1%). Other configurations are not so common (Figure 2) and they are not present in most genera (Table 1). Most species have a stable number of major teeth (73%), while other species are variable (27%), and different conspecific specimens may have a different number of teeth. We dealt with this intraspecific variability counting all the known configurations, and not the average number or the most common one, so some species may count for more than 1 in Table 1, as *Didymodactylos carnosus*, the only species of its genus, with three different configurations. This overcounting of some species leads to the fact that the total percentage obtainable from Figure 2 is more than 100.

Genera with the widest range in number of teeth are those with the greatest number of species

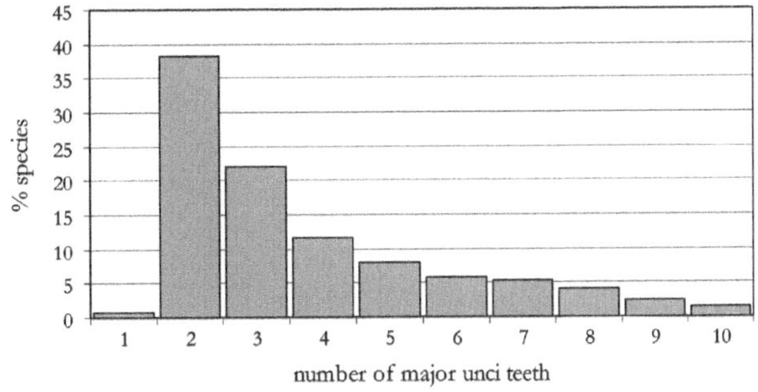

Figure 2. Percentage of bdelloid species (n = 381) with different number of major unci teeth.

Table 1. Percentage of stenoecious aquatic (grey), stenoecious terrestrial (white) or euryecious species (black), and number of species with different number of major unci teeth in all bdelloid genera

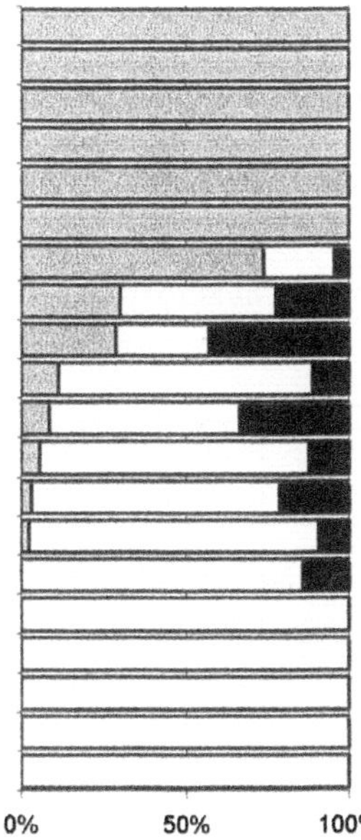

	N°of species	1	2	3	4	5	6	7	8	9	10
		number of teeth									
Anomopus	2	1	1								
Embata	5		5								
Henoceros	2	2	1								
Philodinavus	2			2	2						
Pseudoembata	1		1								
Zelinkiella	1		1								
Rotaria	24		23	1							
Philodina	44		32	15	3	1					
Dissotrocha	9			9							
Otostephanos	11			1			2	3	4	2	2
Adineta	14		14								
Macrotrachela	77	1	61	16	11	5			1		
Habrotrocha	109		26	42	34	26	19	16	11	8	5
Mniobia	47		17	22	8	6	4	5	3	2	1
Pleuretra	14		13	1							
Abrochtha	2		1	2							
Bradyscela	2		2								
Ceratotrocha	4		2	2							
Didymodactylos	1			1	1	1					
Scepanotrocha	10		1	2	2	3	6	4	2	1	
total N° of species	381	4	201	116	61	42	31	28	21	13	8

(*Habrotrocha*, *Macrotrachela* and *Mniobia*). Besides, some genera with few species have great diversification (*Otostephanos* and *Scepanotrocha*), while other ones are more conservative (*Adineta* and *Rotaria*). Usually the distribution of number of teeth inside each genus is continuous, except for two cases: *Macrotrachela faveolata*, with eight major teeth, while congeneric species have from one to five teeth, and *Otostephanos kostei*, with three teeth, while congeneric species have more than six.

A higher number of major unci teeth does not imply a greater size of the trophi (Fig. 3), on the contrary, both variables show a significant negative correlation ($r = -0.1827$, d.f. = 283, $p < 0.002$). Comparing the number of major unci teeth with rami lengths (from 11.1 to 34.1 μm) (Fig. 3), one single major tooth is present in large trophi longer than 20 μm, while more than four major teeth are present in trophi of intermediate size, between 14.8 and 23.7 μm. Ten major teeth are present in small trophi, with rami between 17.1 and 17.7 μm. Two major teeth are present across a great range of size, and both the smallest (*Adineta steineri*) and the largest trophi (*Abrochtha carnivora* and *Rotaria tardigrada*) have this configuration. Moreover, trophi with rami longer than 27 μm have only two major teeth.

The number of major unci teeth is not related to the total number of unci teeth (Fig. 4); no significant correlation was found ($r = -0.1260$, d.f. = 136, n.s.). One single major tooth is present in trophi with a high number of teeth (from 48 to 51), while 10 major teeth are present in trophi with few teeth (37 and 38). Extreme values of total number of teeth, 29 and 52, are present in trophi with 'two major teeth' configuration.

A significant positive relationship ($r = 0.3265$, d.f. = 134, $p < 0.001$) is shown between trophi size and total number of teeth (Fig. 5).

According to the environmental requirements, bdelloid species can be divided into three groups: (1) stenoecious aquatic, (2) stenoecious terrestrial, and (3) euryoecious aquatic–terrestrial. Stenoecious aquatic species ($n = 56$) have a significantly lower number of major teeth ($z = 3.91$, $p < 0.01$) than the stenoecious terrestrial species ($n = 239$). A significant difference ($z = 4.48$, $p < 0.01$) is

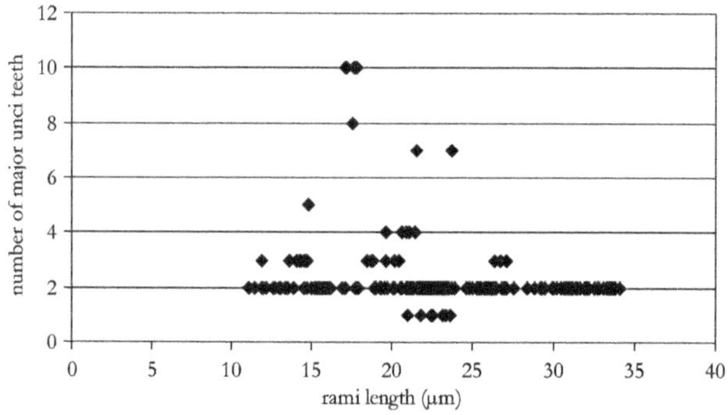

Figure 3. Distribution of number of major unci teeth in relation to rami length. Data from SEM pictures of 285 trophi.

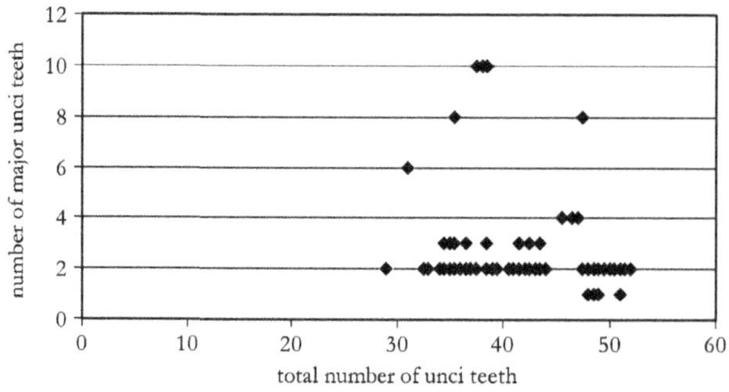

Figure 4. Distribution of number of major unci teeth in relation to total number of unci teeth. Data from SEM pictures of 138 trophi.

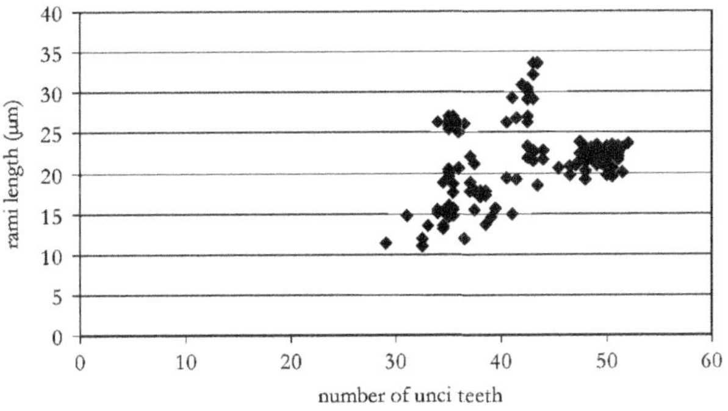

Figure 5. Distribution of total number of unci teeth in relation to rami length. Data from SEM pictures of 136 trophi.

shown also between aquatic species ($n = 56$) and euryoecious species ($n = 53$). More than 60% of stenoecious aquatic species have two major unci teeth (Table 1).

The same analyses at family level in the order Philodinida show similar trends for Philodinidae ($n = 215$), with aquatic species ($n = 45$) displaying a significantly lower number of major teeth than those exclusively terrestrial ($n = 142$) ($z = 5.40$, $p < 0.01$), and than those living in both environments ($n = 28$) ($z = 3.03, p < 0.01$). Only four species of the family Habrotrochidae ($n = 113$) live exclusively in water, so only comparison between exclusively terrestrial ($n = 88$) and aquatic–terrestrial ones ($n = 21$) is possible; in this case no significant difference is shown ($z = 0.62$, n.s.). Adinetida and Philodinavida are really conservative in number of major unci teeth and no trend can be demonstrated.

Because of the way the unci fit when trophi close, bdelloids can be divided into left- and right-handed, that is with the major tooth nearer to the articulation in the left (e.g., Fig. 1b, d, h and j) or in the right uncus (e.g., Fig. 1c, e, f, g, i, k and l). An analysis of one single population for different species with two major teeth produced no preference for left- or right-handedness (Table 2). Moreover, from an analysis of specimens belonging to seven different clonal lines of *M. quadricornifera*, no hereditary pattern was shown. All the daughters of the seven mothers were about 50% left- and 50% right-handed.

Discussion

A significant negative relationship seems to exist between number of major unci teeth and trophi size: larger trophi seem to have less major teeth than smaller trophi. The two analyzed variables, total number of unci teeth and trophi size, seem to be correlated: greater trophi have in general more teeth than the smallest trophi.

The habitat in which different bdelloid species live seems to influence the major teeth pattern, possibly related to different trophic availability in water and in aerophytic mosses and lichens. Few major teeth are more common in water dwelling

Table 2. Number of left- and right-handed specimens in different bdelloid species and statistical significance of the differences

Species	Left-handed	Right-handed	χ^2_c	Right/Left
Macrotrachela quadricornifera	67	78	0.69 n.s.	1.16
Rotaria macrura	9	14	0.69 n.s.	1.56
Rotaria magnacalcarata	11	20	2.06 n.s.	1.82
Rotaria neptunia	15	26	2.44 n.s.	1.73
Rotaria neptunoida	15	14	0.00 n.s.	0.93
Rotaria socialis	5	11	1.56 n.s.	2.20
Rotaria sordida	9	6	0.27 n.s.	0.67
Rotaria tardigrada	27	32	0.27 n.s.	1.19

species, where they possibly eat unicellular algae, while more major teeth are more common in species living outside water bodies, where maybe they eat bacteria. Unfortunately we never succeeded in maintaining living populations of any bdelloid species living in aerophytic mosses and lichens, so we do not know their real trophic requirements. The only notable feeding habit is that of *Abrochtha carnivora*, which possesses one of the largest trophi, possibly in relation to its predaceous habit, unique in bdelloid rotifers (Ricci et al., 2001).

Acknowledgements

We wish to thank Christian D. Jersabek, Russell J. Shiel, and two anonymous referees for their useful suggestions and for improving the English text.

References

Bērzinš, B. & B. Pejler, 1987. Rotifer occurrence in relation to pH. Hydrobiologia 147: 107–116.

Bērzinš, B. & B. Pejler, 1989a. Rotifer occurrence and trophic degree. Hydrobiologia 182: 171–180.

Bērzinš, B. & B. Pejler, 1989b. Rotifer occurrence in relation to oxygen content. Hydrobiologia 183: 165–172.

Donner, J., 1965. Ordnung Bdelloidea (Rotifera, Rädertiere). Akademie Verlag, Berlin 297.

Fontaneto, D. & G. Melone, 2003. Redescription of *Pleuretra hystrix*, an endemic alpine bdelloid rotifer. Hydrobiologia 497: 153–160.

Fontaneto, D., G. Melone, & A. Cardini, 2004. Shape diversity in the trophi of different species of *Rotaria* (Rotifera, Bdelloidea): a geometric morphometric study. Italian Journal of Zoology, 71: 63–72.

Gosse, P. H., 1856. On the structure, functions and homologies of the manducatory organs in the class Rotifera. Philosophical Transactions of the Royal Society of London 146: 419–452.

Koste, W., 1978. Rotatoria. Die Rädertiere Mitteleuropas. Borntraeger, Berlin, 673 pp.

Markevich, G. I., 1985. Ultrathin morphology of mastax of rotifers. I. Bdelloida [in Russian]. Bulletin of the Institute for the Biology of Inland Waters 68: 31–35.

Markevich, G. I., 1989. Morphology and the principle organization of sclerite system of the mastax in rotifers [in Russian]. Proceedings of the Institute for the Biology of Inland Waters 56: 27–82.

Markevich, G. I. & L. A. Kutikova, 1989. Mastax morphology under SEM and its usefulness in reconstructing rotifer phylogeny and systematics. Hydrobiologia 186/187: 285–289.

Melone, G., C. Ricci, H. Segers & R. L. Wallace, 1998a. Phylogenetic relationships of phylum Rotifera with emphasis on the families of Bdelloidea. Hydrobiologia 388: 101–107.

Melone, G., C. Ricci & H. Segers, 1998b. The trophi of Bdelloidea (Rotifera): a comparative study across the class. Canadian Journal of Zoology 76: 1755–1765.

Ricci, C., G. Melone & E. J. Walsh, 2001. A carnivorous bdelloid rotifer, *Abrochtha carnivora* n.sp. Invertebrate Biology 120: 136–141.

Ricci, C., R. J. Shiel, D. Fontaneto & G. Melone, 2003. Bdelloid rotifers recorded from Australia with description of *Philodinavus aussiensis* n.sp. Zoologischer Anzeiger 242: 241–248.

Segers, H., 1993. Rotifera of some lakes in the floodplain of the River Niger (Imo State, Nigeria). I. New species and other taxonomic considerations. Hydrobiologia 250: 39–61.

Segers, H., 2002. The nomenclature of the Rotifera: annotated checklist of valid family and genus–group names. Journal of Natural History 36: 631–640.

Sørensen, M. V., 2002. On the evolution and morphology of the rotiferan trophi, with a cladistic analysis of Rotifera. Journal of Zoological Systematics and Evolutionary Research 40: 129–154.

Hydrobiologia (2005) 546:203–211
A. Herzig, R.D. Gulati, C.D. Jersabek & L. May (eds.) Rotifera X: Rotifer Research: Trends, New Tools and Recent Advances
DOI 10.1007/s10750-005-4198-y

Study of the trophi of *Testudinella* Bory de St. Vincent and *Pompholyx* Gosse (Rotifera: Testudinellidae) by scanning electron microscopy

Willem H. De Smet
Laboratory of Polar Ecology, Limnology & Palaeobiology, Department of Biology, University of Antwerp, R.U.C.A. campus, Groenenborgerlaan 171, B-2020, Antwerpen, Belgium
E-mail: wides@ruca.ua.ac.be

Key words: Rotifera, Testudinellidae, *Testudinella*, *Pompholyx*, trophi structure, taxonomy

Abstract

The fine morphology of the trophi of *Pompholyx sulcata* and nine species of *Testudinella* (Rotifera, Monogononta, Flosculariacea, Testudinellidae) was studied by scanning electron microscopy. The number of unci teeth and arched rami scleropili, and the shape of the major unci teeth and fulcrum are considered to be reliable additional characters for identification.

Introduction

In Testudinellidae (Rotifera, Monogononta, Flosculariacea) three genera are recognized, *viz. Anchitestudinella*, *Testudinella*, and *Pompholyx*. To date, the taxonomy of the species exclusively relies (e.g., Kutikova, 1970; Koste, 1978) on the shape of the lorica, the shape and position of the foot opening, and the position of the lateral antennae. As has been shown for other Flosculariacea, e.g., *Filinia* (Sanoamuang, 1993, 2002), *Floscularia* (Segers, 1997), and Conochilidae (Segers & Wallace, 2001), trophi are fairly species-specific, and their study by scanning electron microscopy (SEM) is a helpful or even the most reliable tool for identification. Apart from the little documented information by Markevich (1989) on the trophi of *Pompholyx complanata* and *Testudinella patina*, a frontal view of the trophi of *Testudinella elliptica* shown in De Smet (1998), and a caudal view attributed to *T. truncata* figured in Sørensen (2002), extensive studies dealing with the detailed morphology of the trophi of Testudinellidae using SEM are lacking. I, therefore initiated a comparative study on the trophi morphology of the commonly found species belonging to the genera *Testudinella* and *Pompholyx*.

Materials and methods

Pompholyx sulcata and nine species of *Testudinella* were investigated (as shown in Table 1). Trophi were isolated using NaOCl, followed by repeated washing in distilled water (De Smet, 1998). The dried trophi were sputter coated with gold and examined using a Philips SEM 515 microscope, operated at 20 kV.

The terminology of the different sclerite elements and structures introduced by Markevich (1989), and Markevich & Kutikova (1989), is only followed partially.

Results

General description

The trophi of Testudinellidae belong to the malleoramate type, characteristic of the order Flosculariacea (see Remane, 1929; de Beauchamp, 1965; Nogrady et al., 1993). They are almost symmetrical and composed of paired manubria, rami and unci, and an unpaired fulcrum. The manubria are elongate crescentic elements without

Table 1. List of taxa examined

Taxon	Origin
Pompholyx sulcata (Hudson, 1885)	Belgium: Merksem, Fort; Willebroek, Lacourtvijver
Testudinella caeca (Parsons, 1892)	Belgium: Ekeren, Oude Landen
T. clypeata (Müller, 1786)	The Netherlands: Veerse Meer
T. elliptica (Ehrenberg, 1834)	Belgium: Ekeren, Oude Landen; Willebroek, Lacourtvijver; Opgrimbie, Zijpbeek
T. incisa (Ternetz, 1892)	Belgium: Begijnendijk, De Putten; Genk, Het Wik; Gent, Vinderhoutse bos
T. mucronata (Gosse, 1886)	Belgium: Landen, Beemden; Beringen-Koersel, Hemelbrug
T. parva (Ternetz, 1892)	Belgium: Kampenhout, Torfbroek; Begijnendijk, De Putten
T. patina (Hermann, 1783)	Alaska, Nome; Belgium: Brasschaat, Zeurt; St. Jan-Eremo, Roeselarekreek; Canada: Victoria Isl., Cambridge Bay; Congo: Bas-Congo, Muema & Kivu, Rubare; Galápagos: Santa Cruz; Greenland: Kangerlussuaq; Kenya: Lake Victoria; Morocco: Zeida-Midelt
T. truncata (Gosse, 1886)	Belgium: Ekeren, Muisbroek; Opgrimbie, Zijpbeek
T. sp. n	France: Mediterranean, Bay of Hyères

cauda, bordering the lateral edges of the unci. The unci are plate-shaped, composed of a great many differentiated and webbed teeth. The rami are elongate-triangular, and almost completely overlain by the unci; their inner edge and caudal part is provided with numerous rod-shaped sclerite bodies. The fulcrum is short and plate-shaped.

Manubria

The crescent manubria are composed of three superimposed chambers, the dorsal, median and ventral chamber (e.g., Fig. 2: dc, mc, vc), separated from each other by transverse walls, and opening caudally. The dorsal and ventral extremities of the manubria are more or less strongly tapering to inwardly directed projections. These projections are connected by tiny ligaments to the tip and base of the rami respectively (e.g., Fig. 7: l).

Unci

The unci plates consist of transversely placed, well-differentiated and more or less strongly webbed teeth, firmly connected to the inner edge of the frontal side of the manubria. The teeth of both unci plates are interlocking in the closed trophi. The size of the teeth shifts gradually from the large proximal teeth towards the smaller distal teeth. All species studied to date display three pairs of distinctly larger proximal teeth. Each uncus tooth shows a more or less strongly clubbed head, and a long shaft. The head of the major teeth is continuous with the shaft. The head of the smaller teeth is continuous with the shaft also (e.g., Fig. 3), or more or less strongly kinked towards the rami (e.g., Figs. 1 and 5). The shaft appears almost triangular in cross section, and shows a longitudinal ridge frontally, the sutura externa; a longitudinal seam is often distinct caudally (Figs. 16, 18). Structural differences between the proximal and distal teeth are usually small within the different species, but pronounced differences in morphology of the three major teeth are obvious between taxa. For example, the heads of the proximal teeth can be slender and weakly offset, and provided with two small lateral knobs at their base (e.g., *P. sulcata*, *T. elliptica*: Figs. 23 and 25), or stout and distinctly offset without small knobs (e.g., *T. clypeata*, *T.* sp. n.: Figs. 26 and 19). The shafts of the proximal teeth are more or less loosely connected (e.g., *P. sulcata*, *T. elliptica*: Figs. 23 and 25), or a large section of the shafts distally from the weakly offset heads, appears strongly fused into a plate (e.g., *T. caeca*, *T. incisa:* Figs. 9 and 24). A lateral semi-circular expansion at the free margin of the most proximal teeth is present in *T. caeca*, *T. incisa* and *T. parva* (Figs. 24, 9 and 13).

The subunci (Fig. 20: su) consist of a narrow rim, composed of small and fused scleropili, firmly attached to the uncinal plate at the height of the base of the heads of the teeth.

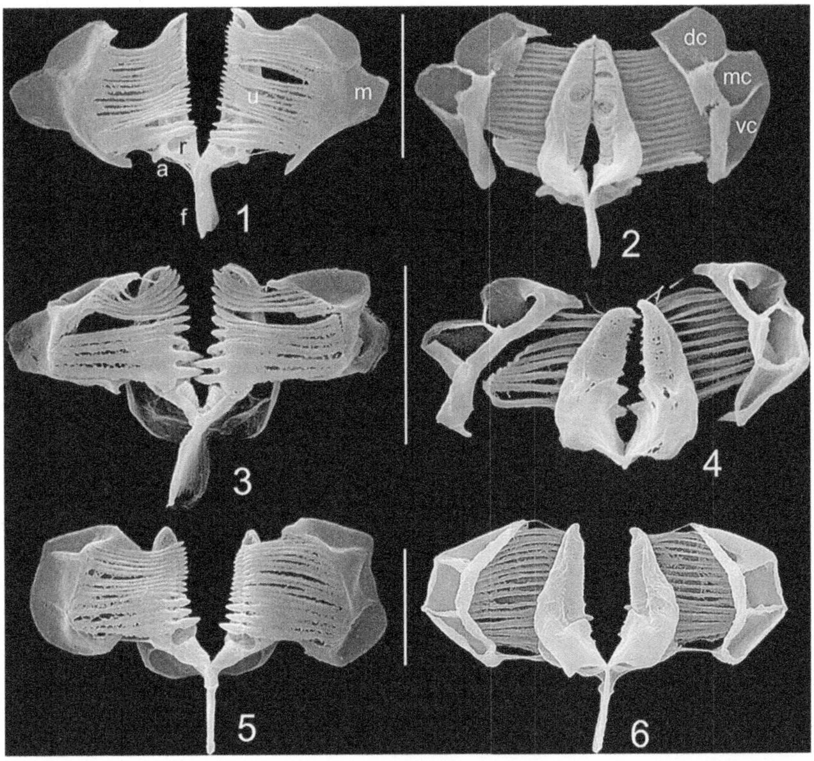

Figure 1–6. Pompholyx sulcata and *Testudinella* spp., SEM of trophi. 1, 2. *Pompholyx sulcata*. 3, 4. *T. caeca*. 5, 6. *T. clypeata*. 1, 3, 5. Frontal view; 2, 4, 6. caudal view. Scale bars: 10 μm. a, alula; dc, dorsal chamber; f, fulcrum; m, manubrium; mc, median chamber; r, ramus; u, uncus; vc, ventral chamber.

The number of unci teeth is rather constant in the different species (Table 2). It varies (right/left) from 9–11/9–11 (*T. caeca*) to 16–17/16–17 (*T. elliptica*) in *Testudinella*, and 17–20/18–21 in *Pompholyx sulcata*. Both symmetrical and asymmetrical unci configurations occur; asymmetrical ones predominate. Intra- and interpopulation variation in tooth number of each of the unci plates is fairly low, and not more than four teeth. The frequency distribution of unci teeth in a population of *P. sulcata* (N = 20; Fig. 27) shows a tendency for skewness: positive skew for the right uncus plate (mode 18 teeth), and negative skew for the left one (mode 20 teeth). Intrapopulation variation in *T. clypeata* (N = 29; Fig. 28) shows no pronounced maximum of frequency distribution in teeth number of the right uncus plate: almost all observations are more or less equally distributed over 14, 15 and 16 teeth. On the contrary, the left uncus plate shows a normal distribution of frequency of teeth number, with a mode of 15 teeth. The frequency histogram of *T. patina* (N = 33; Fig. 28) is based on specimens from different localities. No difference was found concerning the teeth number and the major biogeographical zones. A similar pattern as for *T. clypeata* was observed: the bulk of the frequencies for the right uncus plate is formed by more than one tooth, *viz.* 14 and 15 teeth, whereas in the left plate there is one considerable peak at 15 teeth. A relatively high frequency, both right and left, was also noted for tooth number 12, that could not be related to the morphology of the lorica nor to the origin of the specimens.

Rami

The rami are more or less elongate-triangular in frontal/caudal view. In lateral view the proximal enlarged part forms an obtuse angle with the distal part, and points ventro-caudally. The lateral proximal tips of the frontal walls are drawn out in

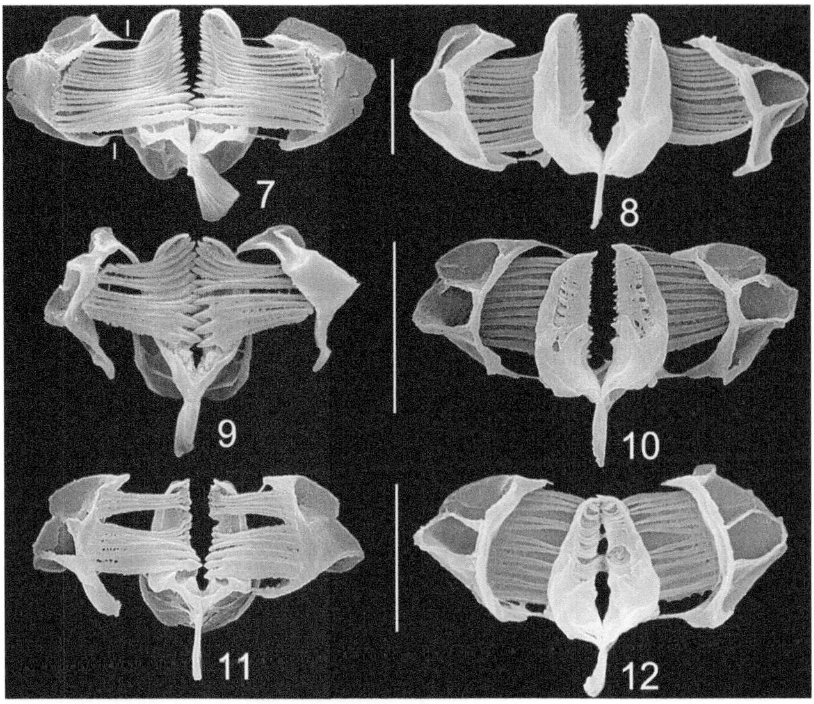

Figure 7–12. Testudinella spp., SEM of trophi. 7, 8. *T. elliptica*. 9, 10. *T. incisa*. 11, 12. *T. mucronata*. 7, 9, 11. Frontal view; 8, 10, 12. caudal view. Scale bars: 10 μm. l, ligament.

symmetrical alulae of variable length (e.g., Fig. 1: a); alulae can be absent also. The proximal part is largely hollow and shows a large lateral opening, continuing caudally. In ventral view, a shallow and oblique transverse septum (Figs. 30–32: s) can be seen, inserted on the inner side of the frontal wall and running dorsally, from the alula or caudo-lateral edge towards the trophi axis. The sections ventrally and dorsally from the septum probably represent the subbasal and basal rami chambers respectively (Figs. 30–32: sc, bc). The wall ventrally to the septum bears short sclerite bodies, which are probably part of the ramus apophysis.

In caudal view the inner margins of the proximal part of the rami bear asymmetrical and interlocking median apophyses, clearly composed of fused sclerite bodies (e.g., Fig. 35: ma). These apophyses are developed to a different degree in the different taxa studied. Openings in the proximal parts, probably median basifenestrae, are only found occasionally (e.g., *T. parva*, *T. truncata*: Figs. 14 and 18). The inner margins of the distal rami sections bear numerous transversely placed elongate and arched sclerite bodies, the arched rami scleropili (e.g., Fig. 33: rs), that can be more or less strongly webbed (e.g., *T. caeca*, *T. parva*: Figs. 4 and 14). The number of arched sclerite bodies varies according to the taxon (as shown in Table 2). In *Testudinella* the lowest number was noted in *T. caeca* (approximately 11/10), and the highest in *T. elliptica* (~46–66/50); in *P. sulcata* the numbers were 35–40/32–35.

In frontal view the rami show a proximal part bearing basal apophyses (e.g., Fig. 23: ba), and a gutter-shaped distal part terminating in more or less strongly hook-shaped rostella (e.g., Fig. 36: ro). The proximal margins may bear a more or less pronounced rim (e.g., Figs. 3, 9 and 13). The basal apophyses are more or less strongly developed, and consist of a varying number of more or less strongly fused rod-shaped sclerite bodies. The outer edge of the gutter-shaped section bears a fringe of fused sclerite bodies, the crista, at its inner side (e.g., Fig. 36: c). Near the inner margin of the gutter-shaped part several rows of shorter, more or less straight rami scleropili (e.g., Fig. 36: rs) are fairly loosely attached on the frontal wall of the gutter. They are probably grouped forming

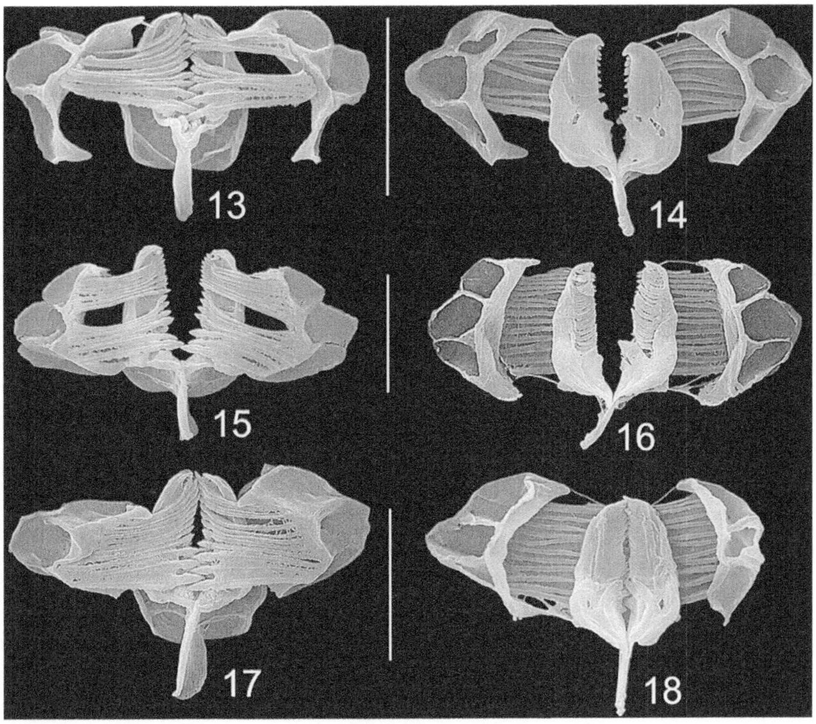

Figure 13–18. Testudinella spp., SEM of trophi. 13, 14. *T. parva*. 15, 16. *T. patina*. 17, 18. *T. truncata*. 13, 15, 17. Frontal view ; 14, 16, 18. caudal view. Scale bars: 10 μm.

some molar surface (e.g., *T. elliptica*, Fig. 38). Their distal end may be rounded (e.g., *T. elliptica*, Fig. 38), or acute (e.g., *P. sulcata*, Fig. 36; *T. truncata*, Fig. 40). The rami scleropili show weakly developed seams.

Fulcrum

The fulcrum is placed in the extension of the oblique proximal rami part, pointing ventro-caudally (Figs. 29 and 30). It is short, more or less trapezoid in lateral view, and uniformly thin with the exception of the frontal margin that can be thickened more or less (e.g., *T. incisa*, *T. truncata*). The fulcrum is apparently composed of a double layer of long and appressed sclerite bodies. Frontally a great number of these sclerite bodies are not involved in the formation of the junction with the rami. They can be appressed (e.g., *P. sulcata*, *T. elliptica*, Figs. 23, 25), or bordering a distinct opening proximally (*T. clypeata*, *T.* sp. n., Figs. 19, 22 and 26). A basal plate or hook is absent.

Discussion

Amongst the distinct challenges in rotifer research, reviewed by Wallace (2002), are taxonomic training and study, and phylogeny. The importance of detailed descriptions of the generally species-specific trophi based on SEM in taxonomic (e.g. Sanoamuang, 1993, 2002; Segers, 1995; De Smet, 1996, 1997; Segers & Wallace, 2001) and phylogenetic studies (e.g. Markevich, 1989; Segers & Wallace, 2001; Sørensen, 2002) has been established irrefutably. However, data from many taxa are completely lacking or based on light microscopy only. The results presented above for *Pompholyx sulcata* and nine species of *Testudinella* are the first to describe SEM trophi morphology of Testudinellidae in an extensive and detailed way. It is obvious that the overall similarity of the structure of the trophi in the Testudinellidae studied is very great. No fundamental differences are apparent between the genera *Pompholyx* and *Testudinella*, which in combination with the other

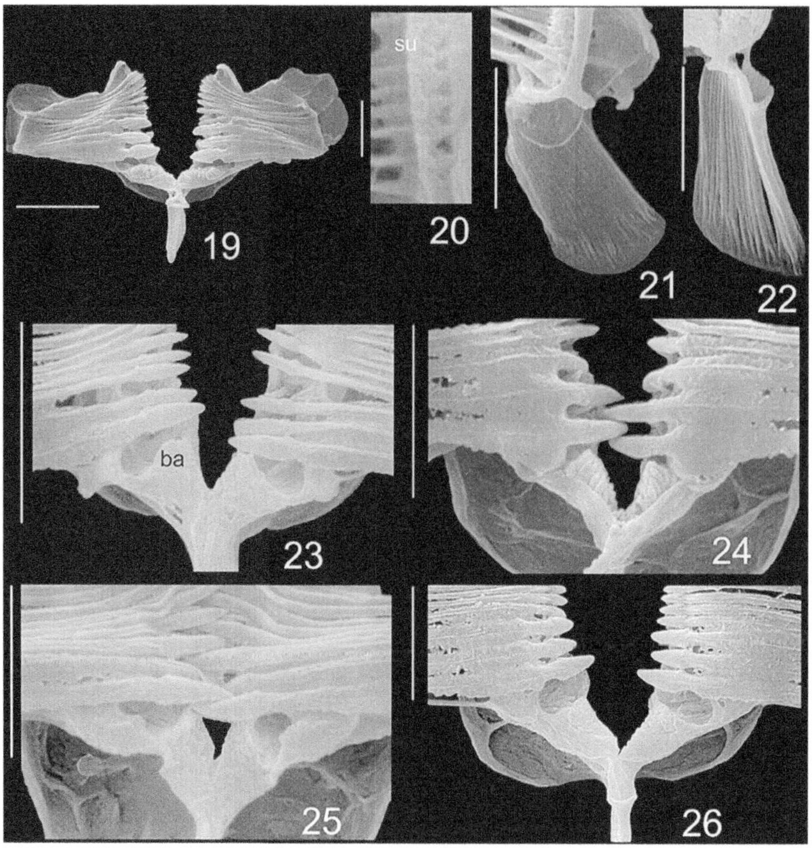

Figure 19–26. Pompholyx sulcata and *Testudinella* spp., SEM of trophi. 19. *Testudinella* sp. n., frontal view. 20. *P. sulcata*, detail subuncus. 21. *P. sulcata*, fulcrum, left lateral view. 22. *T. clypeata*, fulcrum, right lateral view. 23–24. Detail unci and rami, frontal view: 23. *P. sulcata*, 24. *T. caeca*, 25. *T. elliptica*, 26. *T. clypeata*. Scale bars: 19: 10 μm; 20: 1 μm; 21– 26: 5 μm. ba, basal apophysis; su, subuncus.

Table 2. Number of unci teeth and arched rami scleropili (right/left)

Taxon	Unci teeth	Arched rami scleropili
Pompholyx sulcata	17–20/18–21 (20)*	~35–40/32–35 (5)*
Testudinella caeca	9–11/9–11 (7)	~11/10 (2)
T. clypeata	14–16/14–16 (25)	~25–26/28–30 (5)
T. elliptica	16–17/16–17 (6)	~46–66/50 (3)
T. incisa	11/9–10 (8)	~18–25/20–25 (3)
T. mucronata	11–12/11–12 (8)	~15/20 (5)
T. parva	10/10 (8)	~19/18 (3)
T. patina	12–16/12–16 (26)	~30–31/31–*32* (4)
T. truncata	12/12 (5)	~25/22 (4)
T. sp. n.	10/10 (3)	–

*Number of specimens bracketed.

external morphological features, confirm them as monophyletic taxa.

To date, taxonomy of the genus *Testudinella* is exclusively based on features of the lorica (e.g. outline and cross-section, position of lateral antennae and foot opening), that are known to vary independently of each other, resulting in a large number of combinations (Ruttner-Kolisko, 1974; Koste, 1978). This led to the description of approximately 80 species and subspecies, of which several were synonymized or considered formae (e.g. Koste, 1978). In a recent publication, Segers (2002a) recognizes a tentative number of 40 valid species. The results of the SEM prove that trophi morphology in *Testudinella* is highly species-specific, and will be a most valuable feature for identification and revision of the family. Among

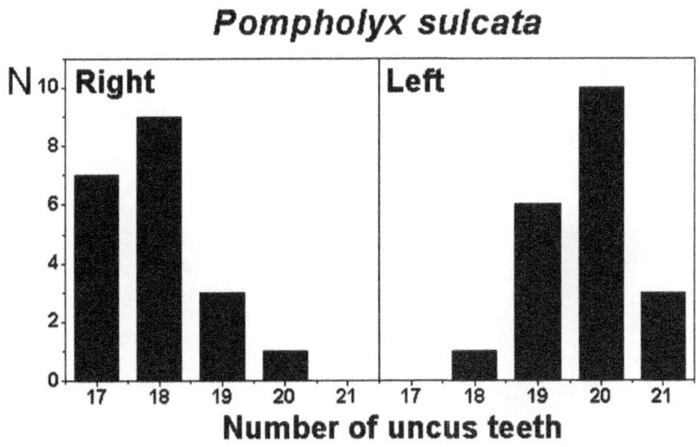

Figure 27. Frequency distributions of right and left uncus teeth in a population of *Pompholyx sulcata* (Fort Merksem, Belgium, 3 July 1985).

the characters considered to be most reliable and manageable for identification are: number of unci teeth and arched rami scleropili, shape of the major unci teeth, presence/absence of lateral expansion on the first major teeth, presence/absence of a proximal opening in the fulcrum, and presence/absence of alulae.

A detailed comparison of the trophi of Testudinellidae with these of the other families of Flosculariacea is difficult, since published records are still few (e.g., Sanoamuang, 1993, 2002: *Filinia*; Segers, 1997: *Floscularia*; Melone et al., 1998: *Sinantherina*; Segers & Wallace, 2001: Conochilidae, *Lacinularia*, *Ptygura*; Sørensen, 2002: *Floscularia*;

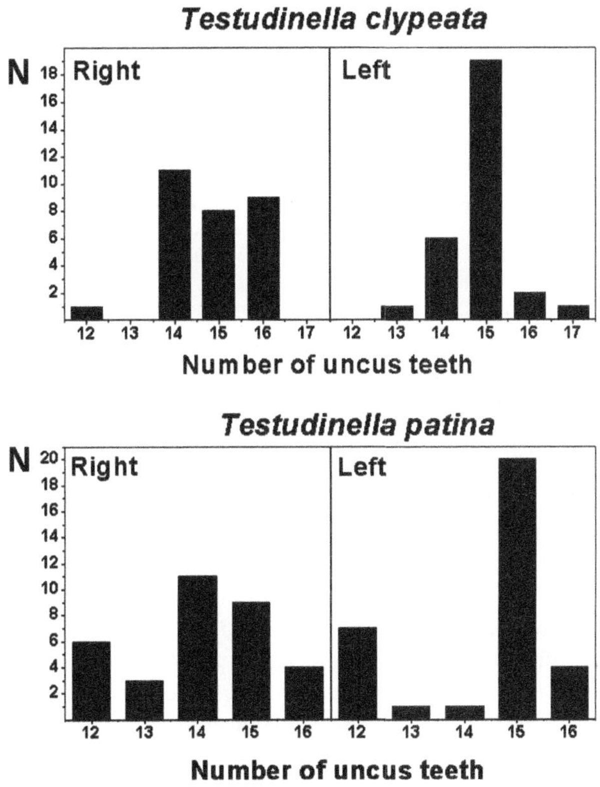

Figure 28. Frequency distributions of right and left uncus teeth in a population of *Testudinella clypeata* (Veerse Meer, The Netherlands, 8 September 2002), and in *T. patina* originating from different geographical localities.

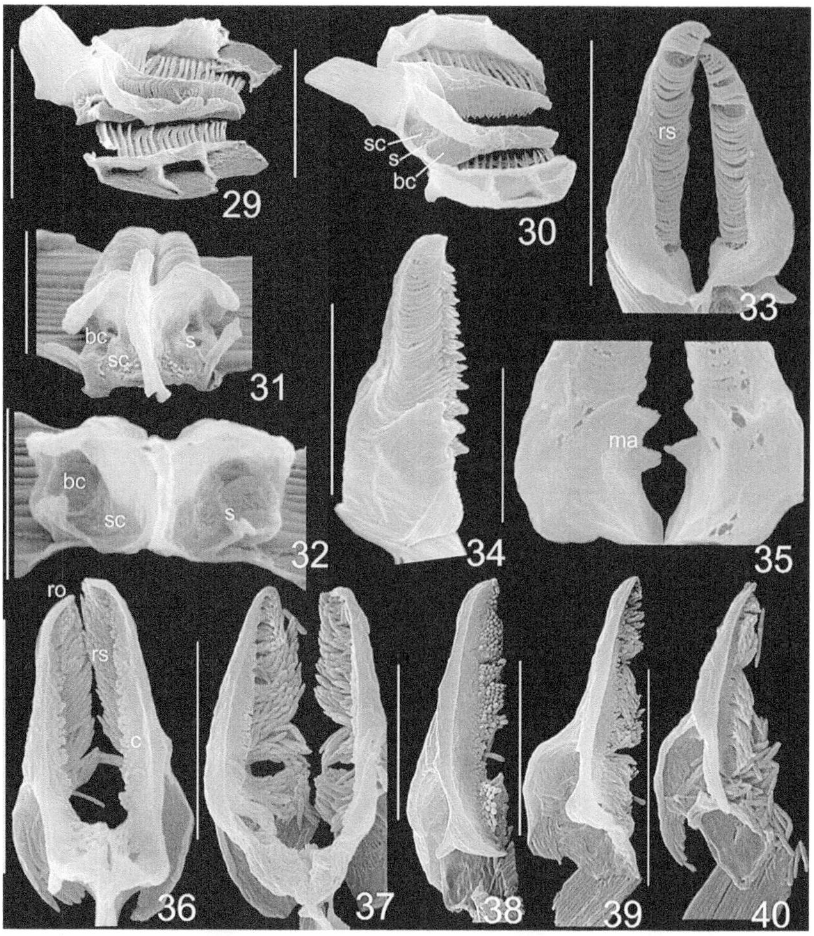

Figure 29–40. Pompholyx sulcata and *Testudinella* spp., SEM of trophi. 29. *P. sulcata*, caudo-lateral view. 30. *T. clypeata*, caudo-lateral view. 31. *P. sulcata*, rami caudo-ventral view. 32. *T. incisa*, rami caudo-ventral view. 33. *P. sulcata*, rami caudal view. 34. *T. patina*, ramus caudal view. 35. *T. caeca*, median apophyses, caudal view. 36–40. Rami, frontal view: 36. *P. sulcata*, 37. *T. clypeata*, 38. *T. elliptica*, 39. *T. patina*, 40. *T. truncata*. Scale bars: 29–31, 33, 34, 36–40: 10 μm; 32, 35: 5 μm. c, crista; bc, basal chamber; ma, median apophyses; ro, rostellum; rs, rami scleropili; s, septum; sc, subbasal chamber.

Segers, 2002b: *Trochosphaera*, *Horaella*). At this stage generalizations should be interpreted with caution, e.g., proximal expansions on the first major teeth reported (Segers & Wallace, 2001) to be present only in *Horaella brehmi* and *Filinia brachiata* (Trochosphaeridae), have now been demonstrated in *Testudinella* likewise. Within Flosculariacea trophi are fairly homogeneous. Supposed taxonomically significant differences in trophi structure between Testudinellidae and the other families concern the pseudoalulae, the shape of the fulcrum, and asymmetry of the unci. To date pseudoalulae have been reported in Trochosphaeridae only. Fulcra of Conochilidae, Hexarthridae and Flosculariidae usually bear a basal plate or hook, which is absent in Testudinellidae, and a pronounced asymmetry of the unci plates is characteristic for Conochilidae.

The fulcrum is an unpaired structure present in Monogononta and Seisonidea, but absent in Bdelloidea (de Beauchamp, 1965; Nogrady et al., 1993). Controversy exists on the origin of the fulcrum. According to Markevich (1985) and Markevich & Kutikova (1989) it is derived from scleropili on the inner margin of the rami. In an alternative hypothesis Segers & Melone (1998) suggest that the fulcrum is the median unpaired part of the original set of elements in rotifer trophi, which implies that the fulcrum has been lost in Bdelloidea. In Testudinellidae the fulcrum seems to arise by fusion of two sheet-like sets of sclerite bodies, which I suggest are derived from paired lateral elements (Segers & Melone, 1998), homologous to the elements responsible for the formation of the distal group of minor unci teeth in Bdelloidea.

Acknowledgements

I am indebted to the Laboratory of Cell Biology and Histology for access to the scanning electron microscope.

References

de Beauchamp, P., 1965. Classe des Rotifères. In P. -P. Grassé, Traité de Zoologie. T IV(3): 1225–1379. Masson et Cie. Ed., Paris.

De Smet, W. H., 1996. Rotifera 4. The Proalidae (Monogononta). In Nogrady, T. & H. J. F. Dumont (eds) Guides to the Identification of the Microinvertebrates of the Continental Waters of the World 9. SPB Academic Publishing bv, Amsterdam: 102 pp.

De Smet, W. H., 1997. Dicranophoridae (Monogononta) in Rotifera 5. The Dicranophoridae and the Ituridae (Monogononta). In Nogrady, T. & H. J. F. Dumont (eds) Guides to the Identification of the Microinvertebrates of the Continental Waters of the World 12. SPB Academic Publishing bv, Amsterdam: 1–325.

De Smet, W. H., 1998. Preparation of rotifer trophi for light and scanning electron microscopy. Hydrobiologia 387/388: 117–121.

Koste, W., 1978. Rotatoria. Die Rädertiere Mitteleuropas. Ein Bestimmungswerk, begründet von Max Voigt. Überordnung Monogononta. 2nd edn. I. Textband, 673 pp, II. Tafelband, 234 Taf., Gebrüder Borntraeger: Berlin, Stuttgart.

Kutikova, L. A., 1970. Kolovratki fauny SSSR (Rotatoria). Opredeliteli Faune SSSR 1049, 744 pp, Akademiya Nauk SSSR, Leningrad. (in Russian).

Markevich, G. I., 1985. Main trends of idioadaptive evolution of rotifers. Jaws. In Kutikova L. A. (ed), Proceedings of the Second All-Union Symposium on Rotifers, Leningrad, October 18–20, 1983. Akademiya Nauk SSSR, Zoologicheskaya Institut, 17–37.

Markevich, G. I., 1989. Morphology and principal organization of the sclerite system of the rotifer mastax. In Biologiya, Sistematika I Funksionalnya Morfologiya Presnovodnyk Zhivotnyk. Trudy Instituta Biologii Vnutrennikh Vod. Akademiya nauk S.S.S.R. 56: 27–82. (in Russian).

Markevich, G. I. & L. A. Kutikova, 1989. Mastax morphology under SEM and its usefulness in reconstructing rotifer phylogeny & systematics. Hydrobiologia 186/187: 285–289.

Melone, G., C. Ricci & H. Segers, 1998. The trophi of Bdelloidea (Rotifera): a comparative study across the class. Canadian Journal of Zoology 76: 1755–1765.

Nogrady, T., R. L. Wallace & T. W. Snell, 1993. Rotifera 1. In Nogrady, T. & H. J. F. Dumont (eds), Biology, Ecology and Systematics. Guides to the Identification of the Microinvertebrates of the Continental waters of the World. SPB Academic Publishing, The Hague, 142 pp.

Remane, A., 1929–1933. Rotatoria. In Bronn's Klassen und Ordnungen des Tier-Reichs, Bd. 4, Abt. II/1. Akademische Verlagsgesellschaft m.b.H., Berlin, 577 pp.

Ruttner-Kolisko, A., 1974. Plankton rotifers. Biology and Taxonomy. Die Binnengewässer 26: 99–234.

Sanoamuang, L., 1993. Comparative studies on scanning electron microscopy of trophi of the genus *Filinia* Bory De St. Vincent (Rotifera). Hydrobiologia 264: 115–128.

Sanoamuang, L., 2002. Genus *Filinia* Bory de St. Vincent, 1824. In Nogrady, T., H. Segers & H. J. F. Dumont (eds), Rotifera 6. Asplanchnidae, Gastropodidae, Lindiidae, Microcodidae, Synchaetidae, Trochosphaeridae and *Filinia*. Guides to the Identification of the Microinvertebrates of the Continental Waters of the World, 224–257.

Segers, H., 1995. A reappraisal of the Scaridiidae (Rotifera, Monogononta). Zoologica Scripta 24: 91–100.

Segers, H., 1997. Contribution to a revision of *Floscularia* Cuvier, 1798 (Rotifera: Monogononta): notes on some Neotropical taxa. Hydrobiologia 354: 165–175.

Segers, H., 2002a. The nomenclature of the Rotifera: annotated checklist of valid family- and genus-group names. Journal of Natural History 36: 631–640.

Segers, H., 2002b. Family Trochosphaeridae Harring, 1913. In Nogrady, T., H. Segers & H. J. F. Dumont (eds), Rotifera 6. Guides to the Identification of the Microinvertebrates of the Continental waters of the World, 214–223.

Segers, H. & G. Melone, 1998. A comparative study of trophi morphology in Seisonidea (Rotifera). Journal of Zoology, London 244: 201–207.

Segers, H. H. & R. L. Wallace, 2001. Phylogeny and classification of the Conochilidae (Rotifera, Monogononta, Flosculariacea). Zoologica Scripta 30: 37–48.

Sørensen, M. V., 2002. On the evolution and morphology of the rotiferan trophi, with a cladistic analysis of Rotifera. Journal of Zoological Systematics and Evolutionary Research 40: 129–154.

Wallace, R. L., 2002. Rotifers: exquisite metazoans. Integrative and Comparative Biology 42: 660–667.

Hydrobiologia (2005) 546:213–221
A. Herzig, R.D. Gulati, C.D. Jersabek & L. May (eds.) Rotifera X: Rotifer Research: Trends, New Tools and Recent Advances
DOI 10.1007/s10750-005-4199-x

Do rotifer jaws grow after hatching?

Diego Fontaneto & Giulio Melone*
Department of Biology, State University of Milan, via Celoria 26, I-20133 Milan, Italy
*(*Author for correspondence: E-mail: giulio.melone@unimi.it)*

Key words: Rotifera, Bdelloidea, Monogononta, trophi, post-embryonic development, geometric morphometrics, size

Abstract

The hard articulated jaws of some pseudocoelomate metazoans were recently used in reconstructing their phylogenetic relationships, but we still do not know if these structures could change in size and shape during the life of individuals, and experimental data are lacking on their post-embryonic development. Rotifers are one of the groups in which hard articulated jaws, called trophi, are well known, and are widely used taxonomically. Here we report on SEM study of trophi of rotifers of different ages, to determine if the trophi structures change in shape and/or in size during post-embryonic development. We used linear measurements and geometric morphometrics analyses from scanning electron microscopic pictures of trophi of *Cupelopagis vorax*, *Dicranophorus forcipatus*, *Macrotrachela quadricornifera*, *Notommata glyphura*, *Rotaria macrura*, *R. neptunoida*, and *R. tardigrada*. Results for these species show that trophi do not change after hatching, either in size or in shape. In contrast, data on *Asplanchna priodonta reveal* trophi growth after hatching.

Introduction

Hard articulated jaws are present in various groups of pseudocoelomate metazoans as, for instance, Gnathostomulida, Micrognathozoa and Rotifera, which recently have been grouped (together with Acanthocephala) in the monophylum Gnathifera (Ahlrichs, 1997). Using scanning electron microscopy, analyses of fine morphology of jaws were carried out for gnathostomulids (Sørensen, 2000, 2002b; Sørensen & Sterrer, 2002), for micrognathozoans (Kristensen & Funch, 2000; De Smet, 2002; Sørensen, 2003) and among rotifers, mostly monogononts (Nogrady et a1., 1995; Segers, 1995; De Smet & Pourriot, 1997, Segers, 1997; Melone, 2001; Segers & Wallace, 2001) but also seisonids (Segers & Melone, 1998) and bdelloids (Melone et al., 1998a). Shape and organization of rotifer jaws (trophi) was considered an important trait in taxonomy and phylogeny (Markevich, 1989; Segers & Melone, 1998; Melone et al., 1998b; Sørensen, 2002a).

Nine types of trophi were attributed to rotifers (Gosse, 1856; Nogrady et al., 1993). The ramate trophi of bdelloids hardly ever have traits that can be considered species-specific (Melone et al., 1998a; Ricci et al., 2001; Fontaneto & Melone, 2003; Fontaneto et al., 2004). Trophi usually consist of 7 different pieces: a single fulcrum, connected to 2 rami, on which lie 2 unci, articulated to 2 manubria (Nogrady et al., 1993). The ramate trophi typical of bdelloids (Fig. 1a–h) show a rather uniform morphology, no fulcrum, rami articulated together and connected with the apical parts of unci teeth, and with curved manubria at the base of the unci teeth (Melone et al., 1998a).

Jaws of rotifers are made of chitin and scleroproteins (Klusemann et al., 1990), the trophi development during embryogenesis is not well known and the only studies dealing with this point are dated (Zelinka, 1891; Tannreuther, 1920). Data regarding post-embryonic growth are available for some gnathostomulids only, in which jaws seem not to grow after hatching (Sørensen &

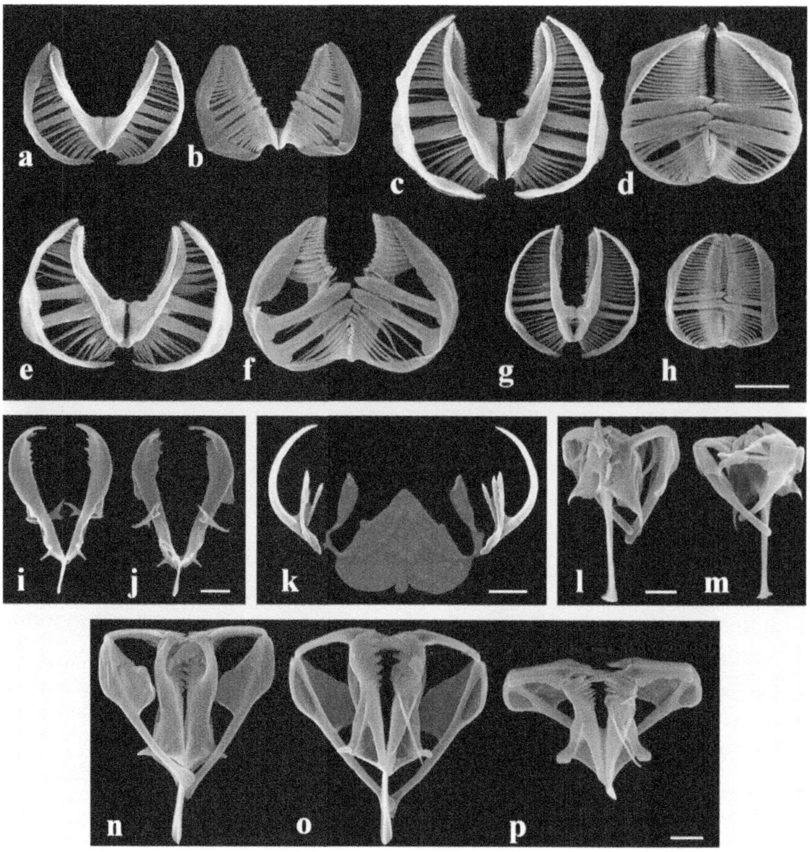

Figure 1. SEM pictures of trophi: (a) *Macrotrachela quadricornifera*, caudal view; (b) idem, cephalic view; (c) *Rotaria tardigrada*, caudal view; (d) idem, cephalic view; (e) *Rotaria macrura*, caudal view; (f) idem, cephalic view; (g) *Rotaria neptunoida*, caudal view; (h) idem, cephalic view; (i) *Asplanchna priodonta*, ventral view; (j) idem, dorsal view; (k) *Cupelopagis vorax*; (l) *Notommata glyphura*, ventral view; (m) idem, dorsal view; (n) *Dicranophorus forcipatus*, dorsal view; (o) idem, ventral view; (p) idem, frontal view, Scale bar = 10 μm.

Sterrer, 2002). Because of the taxonomic and phylogenetic relevance of shape and size of trophi in rotifers, assessing if and how their forms change during growth is of paramount importance.

The present study was aimed at detecting possible post-embryonic changes in size and shape of trophi, both of bdelloids and of monogononts. We focused mainly on bdelloids, characterized by obligatory parthenogenesis and absence of meiosis (Mark Welch & Meselson, 2000). Taxonomy and systematics of the taxon are not well resolved (Donner, 1965; Ricci & Melone, 2000). Thus, morphological analyses of trophi could help in finer discrimination among bdelloids, as has been recently applied to *Rotaria* (Fontaneto et al., 2004).

Materials and methods

Analysed species

We used four bdelloid species of the family Philodinidae: *Macrotrachela quadricornifera* (Milne, 1886), *Rotaria macrura* (Schrank, 1803), *Rotaria neptunoida* Harring, 1913, and *Rotaria tardigrada* (Ehrenberg, 1832), and four monogonont species: *Asplanchna priodonta* (Gosse, 1850), *Cupelopagis vorax* (Leidy, 1857), *Dicranophorus forcipatus* (O.F. Müller, 1786), and *Notommata glyphura* Wulfert, 1938.

All bdelloids have ramate trophi. *Macrotrachela quadricornifera* is a benthic filter-feeder which reproduces via ovipary; its life-cycle lasts about

30 days and somatic growth occurs mainly during the first 10 days of life (Ricci & Fascio, 1995). We isolated eggs from the population cultivated in the laboratory and selected animals at 1st, 5th and 10th day of age to extract trophi.

Species of the genus *Rotaria* are also benthic filter-feeders, but they are viviparous and the developing embryos can be easily seen inside the maternal body (Donner, 1965). We isolated individuals with mature embryos inside and could get trophi of a number of mother–daughter pairs for all three species. We used specimens directly collected in the field in different freshwater environments in the Piedmont region of northern Italy.

Asplanchna priodonta (Asplanchnidae) is a planktonic predator with incudate trophi, showing a wide trophic spectrum, feeding on rotifers, protists, and various unicellular algae (Josè de Paggi, 2002). We collected this species in Endine Lake (Bergamo). This species is viviparous and the developing embryos can be seen inside the maternal body; we isolated individuals with mature embryos and extracted the trophi of mother–daughter pairs.

Cupelopagis vorax (Atrochidae) is a sessile, viviparous predator with uncinate trophi. This species came from Robert L. Wallace's laboratory cultures at Ripon, U.S.A. We compared trophi from mature embryos and large adult specimens, all from fixed material.

Dicranophorus forcipatus (Dicranophoridae) is a benthic predator with forcipate trophi (De Smet & Pourriot, 1997). We collected this species in Piedmont. We compared trophi from eggs with mature embryos, newly hatched juveniles and adult specimens, obtained from laboratory cultures, feeding them various small monogononts and bdelloids.

Notommata glyphura (Notommatidae) is a benthic predator with virgate trophi (Nogrady et al., 1995). We collected it in Piedmont. We compared trophi from mature embryos and from adult specimens, obtained from laboratory cultures. This species can be cultivated provided it is fed live food such as small monogononts and bdelloids.

Trophi images

Trophi were isolated following Segers' (1993) and De Smet's (1998) method with preparation on a circular cover slip by sequentially dissolving tissues in a 5% NaOCl solution, and rinsing with distilled water. Dried trophi were then coated with gold and observed with a LEO 1430 scanning electron microscope.

Rotifer trophi prepared for SEM observation, depending on the preparation, can be seen in caudal or cephalic view, in dorsal or ventral or lateral view (Fig. 1). This situation can affect the trophi measurements. We decided to measure rami length, as reference of the size for comparisons, in ramate, incudate, uncinate, and forcipate trophi. Virgate trophi have a more complex three-dimensional shape, and only fulcrum length was available for comparisons.

In bdelloids, measures of rami length from SEM images of trophi in caudal or cephalic view were not directly comparable because they lay on different points; so caudal and cephalic lengths were analysed separately. For the mother–daughter pairs of the genus *Rotaria*, we could not distinguish which SEM picture was of the mother's trophi, and which belonged to the daughter, except for those in which embryos were not so mature, and in that case trophi of the daughters were not completely formed and so they could not be used in this study.

Images were processed with Adobe Photoshop 5.0.

Statistical analyses

We used various measurements for analysis of size and we carried out statistical comparisons using *t*-test for paired data to analyse left/right symmetries in trophi of bdelloids. After checking for symmetries, we used the mean value of each pair of rami in bdelloids and different single measurements in monogononts, to compare lengths at different ages. We used Mann–Whitney *U*-test, *t*-test, or ANOVA for comparison of the size data (Sokal & Rohlf, 1995). We gave mean values ± standard deviations for all species.

Geometric morphometrics

Geometric morphometric techniques are generally used to asses sexual dimorphism, growth and allometry in skeletal elements of vertebrates (e.g., Rohlf, 1998, Cardini & Tongiorgi, 2003) or in hard chitinous parts of arthropods (e.g. Adams & Funk, 1997; Klingenberg et al., 1998). These techniques were recently applied also in the analysis of SEM pictures of hard elements of microinvertebrates (Fontaneto et al., 2004).

In the present study, trophi of *M. quadricornifera* of different ages were compared in a geometric morphometric analysis (Rohlf & Marcus, 1993), which employs the Cartesian coordinates of a set of topographically corresponding landmarks to capture the morphological information of the specimens under study. Six landmarks were digitized on SEM pictures of the cephalic view of trophi (Fig. 2). As trophi used were symmetrical structures, landmarks were digitized on one half only to avoid redundant information. However, because of the way in which unci are fitted when trophi are closed, the disposition of the major teeth is asymmetrical: the major tooth can be closer to the proximal articulation of the rami in either the left or the right half depending on the specimen. We chose pictures with the first proximal major tooth in the left half; pictures of trophi with that tooth in the right half were electronically minor-reflected. This reflection allowed us to obtain comparable specimens always having major teeth in the same half as shown in Figure 2.

Landmark (L) description (uncus in the right half, after reflection of the picture, when necessary): L1, point where the first proximal minor tooth is connected with the ramus; L2, tip of first major tooth; L3, tip of second major tooth; L4, point where the most distal minor tooth is connected with the ramus; L5, distal point of the base of the second major tooth; L6, proximal point of the base of the first major tooth (Fig. 2).

Generalized Procrustes analysis (GPA) removes differences due to the specimen position during data collection, and separates size and shape components of form by scaling the landmark configurations to the same size, centring them at their origin and rotating them to minimize the distances among corresponding landmarks. After the GPA, each landmark configuration corresponds to a point in a curved shape space, which is then projected in a tangent Euclidean space to perform statistical analyses (Rohlf, 1998). The landmark coordinates are transformed into a set of shape variables (linear combinations of the original coordinates) describing those morphological features that do not change with scale, translation and orientation (Bookstein, 2000). Centroid size (CS) measures the overall size of the landmark configuration (before GPA), and it is computed as the square root of the sum of squared distances from the landmarks to their centroid.

Analysis of variance (ANOVA) and canonical variate analysis (CVA) were employed to test the significance of size and shape differences among the trophi of the different age classes.

Geometric morphometric analyses were performed with the computer programs of the TPS series (Rohlf, 2002a). Statistical analyses were done with NTSYS (Rohlf, 2002b) computer programs and with routines available in the geometric morphometric programs.

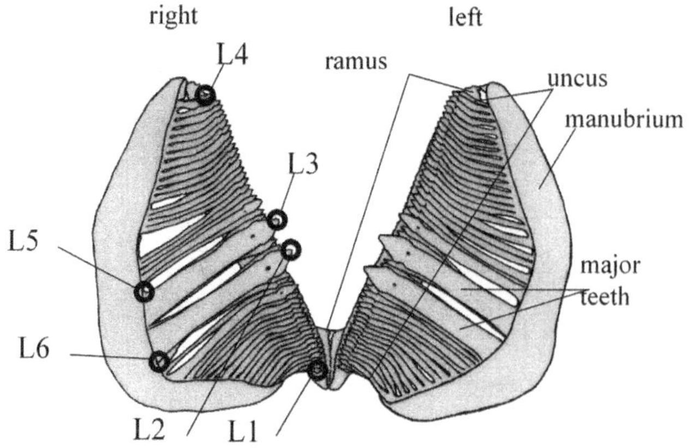

Figure 2. Landmark configuration for geometric morphometric analyses of trophi of *Macrotrachela quadricornifera* in cephalic view.

Results

Bdelloids

We obtained 23 pictures of trophi from *R. macrura*, 10 from *R. neptunoida*, 36 from *R. tardigrada*, and 55 from *M. quadricornifora* (Tables 1, 2).

Rotaria macrura: (Fig. 1e, f) rami were symmetrical (t = 1.08, d.f. = 21, n.s.) with maximum left–right discrepancy of 0.35 μm (1.4%). Rami lengths in caudal and cephalic view were significantly different (t = 3.38, d.f. = 20, $p < 0.01$). We obtained four mother–daughter pairs images in cephalic view, one in caudal view, five of mixed view. One group of one mother and two daughters had trophi in cephalic view. Mean difference in length of rami in each pair (cephalic view) was 0.52 ± 0.15 μm. The greatest observed discrepancy was 0.70 μm (2.6%), but in this case longer rami belonged to the trophi of the daughter. Caudally viewed pairs had a mean difference of 0.42 μm (1.6%), while the mixed pairs could not be used.

Rotaria neptunoida: (Fig. 1g, h) the 4 analysed mother–daughter pairs were all seen caudally; one pair was unfit for use because one specimen was in caudal view and the other in cephalic view. In the four caudally viewed pairs, rami were symmetrical (t = 2.24, d.f. = 7, n.s.) with maximum left–right discrepancy of 0.14 μm (0.6%). Length of rami seemed not to change: differences in each pair were from 0.02 μm (0.1%) to 0.07 μm (0.3%).

Table 1. Mean values of length of rami in the genus *Rotaria*

	Caudal view (μM)	Cephalic view (μm)
Rotaria macrura	26.50 ± 0.21	26.04 ± 0.58
Rotaria neptunoida	22.38 ± 0.28	
Rotaria tardigrada	33.06 ± 0.57	32.02 ± 1.62

Table 2. Mean values of length of rami of trophi of *Macrotrachela quadricornifera* at different ages

	Cephalic view			Caudal view		
Day of age	1st	5th	10th	1st	5th	10th
Number of trophy	8	6	7	10	14	10
Mean length of rami (μm)	21.49	22.49	22.51	22.75	22.77	22.80
Standard deviation (μm)	0.86	0.66	0.45	0.91	0.55	0.63
CV (s.d./mean)	1.34	1.09	0.47	1.33	0.56	1.04

Rotaria tardigrada: (Fig. 1c, d) rami were symmetrical (t = 0.49, d.f. = 29, n.s.) with maximum left–right discrepancy of 0.21 μm (0.6%). Rami lengths in caudal and cephalic view were significantly different (t = 2.77, d.f. = 32, $p < 0.01$). We obtained 9 mother–daughter pairs in caudal view, 1 in cephalic view and 8 of mixed view. The difference in length of the rami in caudal view averaged 0.90 ± 0.39 μm. The greatest observed discrepancy was 1.54 μm (4.5%). Difference in the pair in cephalic view was 1.99 μm (6.2%).

Macrotrachela quadricornifera: (Fig. 1a, b) lengths of left and right rami were similar in each specimen and without statistical difference in each age class, both in caudal (t_1 = 0.22, d.f. = 9, n.s.; t_5 = 0.22, d.f. = 13, n.s.; t_{10} = 0.31, d.f. = 9, n.s.) and cephalic view (t_1 = 1.53, d.f. = 7, n.s.; t_5 = 0.07, d.f. = 5, n.s.; t_{10} = 0.98, d.f. = 6, n.s.). Nevertheless maximum discrepancy between the lengths of left and right rami, expressed as percentage of the major ramus, was 0.9% in caudal view (mean 0.4%) and 0.9% in cephalic view (mean 0.5%). When mean length of left and right ramus of each specimen was used (Table 2), size apparently did not change between trophi of individuals of the 1st and the 5th day of age, neither in caudal (U_5 = 69, n_1 = 10, n_5 = 14, n.s.), nor in cephalic view (U_1 = 10, n_1 = 8, n_5 = 6, n.s.), as between animals of the 5th and the 10th day, both in caudal (U_{10} = 69, n_5 = 14, n_{10} = 10, n.s.) and cephalic view (U_{10} = 19, n_5 = 6, n_{10} = 7, n.s.). Nine pictures in cephalic view of trophi of individuals of 1st day of age, six of individuals of 5th day and eight of 10th day were employed in geometric morphometric analyses. Either ANOVA for the size ($F_{2,20}$ = 0.090, p = 0.914) or the CVA of the shape variables were not significant (wilks' Λ = 0.44633, $F_{16,26}$ = 0.80735, p = 0.6665). The greatest variability in the CS was shown by specimens of the 1st day of age (Fig. 3).

Monogononts

We obtained 24 pictures of trophi from *A. priodonta*, 24 from *C. vorax*, 15 from *D. forcipatus*, and 11 from *N. glyphura*.

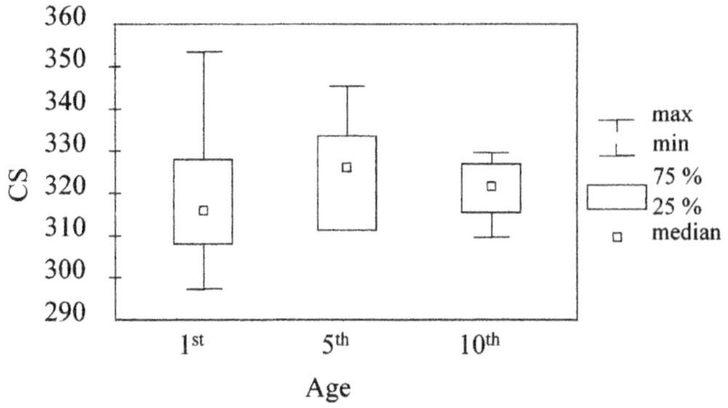

Figure 3. Box plot for the trophi centroid size (CS) in specimens of *Macrotrachela quadricornifera* of the 1st, 5th and 10th day of age.

Asplanchna priodonta: (Fig. 1i, j) significant differences were found in lengths of left (t = 11.0, d.f. = 11, $p < 0.001$) and right rami (11.8, d.f. = 11, $p < 0.001$) in newborns and in their mothers. Merged left–right lengths of newborns were 42.53 ± 1.47 μm, while those of their mothers were 49.44 ± 1.99 μm.

Cupelopagis vorax: (Fig. 1k) no differences (t = 0.08, d.f. = 22, n.s.) were present in compared rami lengths of embryos (18.61 ± 0.14 μm) and adults (18.57 ± 0.20 μm).

Dicranophorus forcipatus: (Fig. 1n–p) no differences were present in compared lengths of fulcrum, rami or manubria of embryos and adults (Table 3).

Notommata glyphura: (Fig. 1l, m) no differences (t = 1.47, d.f. = 9, n.s.) were present in compared fulcrum lengths of newborns (37.48 ± 0.47, n = 6) and adults (37.90 ± 0.36, n = 5).

Discussion

Bdelloids

From our statistical analyses we conclude that rami are symmetrical structures in bdelloids. Possibly real, although very modest, left/right differences maybe masked/emphasized by artefacts or instrumental errors. Besides this symmetry in size, some asymmetry in structure and disposition of teeth is present (Melone & Fontaneto, this volume).

Application of geometric morphometric analyses showed that neither length of rami nor trophi size apparently changed after hatching. The only feature that may change from very young animals to adults is the behaviour of the trophi during desiccation when preparing for SEM observation. Trophi of young animals can be easily bent, indeed. In fact, in *M. quadricornifera* values of the 1st day had a greater coefficient of variation than those of the 5th and 10th day. Moreover, when trophi were seen in caudal view they lay on the unci teeth, which formed a plane structure, while when they were seen in cephalic view they lay on the rami, which could be bent. This fact explains the low mean value of rami length in cephalic view of animals of the 1st day, in which rami seemed to be shorter than those of specimens of the 5th and 10th day. Also for *R. macrura* and *R. tardigrada*

Table 3. Mean values of length of trophi pieces of *Dicranophorus forcipatus* at different life-stages, with statistical significance of observed differences

	Dicranophorus forcipatus					
	n	Fulcrum	Ramus dx	Ramus sn	Manubrium dx	Manubrium sn
Embryo	5	24.57 ± 0.24	37.69 ± 0.93	37.77 ± 0.87	53.03 ± 2.51	52.08 ± 2.07
Newborn	5	24.06 ± 0.31	37.80 ± 0.61	37.71 ± 0.47	53.99 ± 1.56	52.86 ± 1.74
Adult	5	24.18 ± 0.46	37.74 ± 0.63	37.73 ± 0.53	53.81 ± 1.83	53.51 ± 0.70
$F_{2,12}$		2.82	0.03	0.01	0.32	0.98
p		n.s.	n.s.	n.s.	n.s.	n.s.

measurements in cephalic view showed a larger standard deviation. So we think that measurements obtained in caudal view are more accurate than those obtained in cephalic view, and we will, thus, discuss only the former.

The difference in rami length in caudal view between the 1st and 10th day in *M. quadricornifera* was 0.05 μm (0.22%), while body volume in the same laboratory population between the same ages grew up to over 5 times the starting volume (Ricci & Fascio, 1995).

Differences in rami length in mother–daughter pairs in caudal view in *R. neptunoida* were not significant (less than 0.3%), while they were larger for *R. macrura* and *R. tardigrada.* This outcome can be explained by the differently shaped major teeth of their unci: the first species had smaller major teeth than the other two. In the latter two species also trophi lying on unci could be bent because of the obstacle formed by great major teeth. The observation that trophi of daughters were sometimes found to be larger than those of their mothers supports this explanation likewise. We think that data from *R. neptunoida* better reflect the real size of their trophi, while those of the other *Rotaria* species are more susceptible to preparation artefacts.

Monogononts

Data from benthic predators revealed no change in size of trophi during post-embryonic development, as was previously known for the sessile filter-feeder *Floscularia ringens* (Fontaneto et al., 2003).

Only in *Asplanchna* we found that trophi of newborn were significantly smaller than those of their mothers. A great variability in trophi size in this genus is known (Josè de Paggi, 2002), mainly in *Asplanchna sieboldi*, in which trophi size was shown to be a hereditary feature (Badino & Robotti, 1975). Our results showed that trophi could grow. Only further more detailed analyses will show if this feature could be related to their food kind, to the induction of growth of spines in their brachionid preys (Gilbert, 2001), and to the presence of giant specimens with giant trophi (Koste & Shiel, 1980; Josè de Paggi, 2002).

Conclusion

Trophi of all filter-feeder rotifers, independently from their structure, apparently do not change in size after hatching; maybe dimensional invariance of their masticatory apparatus can be related to dimensional invariance of their food particles, which in filter-feeders rotifers is known to be in the range from 3 to 17 μm (Nogrady, 1982). Invariance of food prey size can be the reason of invariance of trophi of predators too. Maybe size variance in the omnivourous *Asplanchna* can be related to its great euryphagy.

Acknowledgements

We would like to thank Maddalena Bertapelle who helped us in the preparation of trophi of *M. quadricornifera* for SEM observation, Robert L. Wallace for giving us specimens of *C. vorax*, Letizia Garibaldi for her plankton samples with *A. priodonta*, Andrea Cardini for his invaluable help in geometric morphometrics analyses, Christian D. Jersabek, Claudia Ricci, Russell J. Shiel, Willem H. De Smet, and two anonymous referees for their suggestions and for correcting the English text.

References

Adams, D. C. & D. J. Funk, 1997. Morphometric inferences on sibling species and sexual dimorphism in *Neochlamisus bebbianae* leaf beetles: multivariate applications of the thin-plate spline. Systematic Biology 46: 180–194.

Ahlrichs, W. H., 1997. Epidermal ultrastructure of *Seison nebaliae* and *Seison annulatus*, and a comparison of epidermal structures within Gnathifera. Zoomorphology 117: 41–48.

Badino, G. & C. Robotti, 1975. Selection in parthenogenetic lines of *Asplanchna sieboldi* (Leydig) 1854 (Rotatoria). Experientia 31: 298–299.

Bookstein, F. L., 2000. Morphometrics. In {\sl Encyclopedia of Life Sciences,} MacMillan, {\tt www.els.net}.

Cardini, A. & P. Tongiorgi, 2003. Yellow-bellied marmots (*Marmota flaviventris*) 'in the shape space' (Rodentia, Sciuridae): sexual dimorphism, growth and allometry of the mandible. Zoomorphology 122: 11–23.

De Smet, W. H. & R. Pourriot, 1997. Rotifera. Vol. 5: The Dicranophoridae and the Ituridae. Guides to the identification of the microinvertebrates of the continental waters of the world, Vol. 12. SPB Academic Publishing, Amsterdam, 344 pp.

De Smet, W. H., 1998. Preparation of rotifer trophi for light and scanning electron microscopy. Hydrobiologia 388: 117–121.

De Smet, W. H., 2002. A new record of *Limnognathia maerski* Kristensen & Funch, 2000 (Micrognathozoa) from the subantarctic Crozet Islands, with redescriptions of the trophi. Journal of Zoology, London 258: 381–393.

Donner, J., 1965. Ordnung Bdelloidea (Rotifera, Rädertiere). Akademie Verlag, Berlin, 297.

Fontaneto, D. & G. Melone, 2003. Redescription of *Pleuretra hystrix*, an endemic alpine bdelloid rotifer. Hydrobiologia 497: 153–160.

Fontaneto, D., G. Melone & R. L. Wallace, 2003. Morphology of *Floscularia ringens* (Rotifera Monogononta) from egg to adult. Invertebrate Biology 122: 231–240.

Fontaneto, D., G. Melone & A. Cardini, 2004. Shape diversity in the trophi of different species of Rotaria (Rotifera, Bdelloidea): a geometric morphometric study. Italian Journal of Zoology, 71: 63–72.

Gilbert, J. J., 2001. Spine development in *Brachionus quadridentatus* from an Australian billabong: genetic variation and induction by *Asplanchna*. Hydrobiologia 446: 19–28.

Gosse, P. H., 1856. On the structure, functions and homologies of the manducatory organs in the class Rotifera. Philosophical Transactions of the Royal Society of London 146: 419–452.

Josè de Paggi, S., 2002. Family Asplanchnidae. In Nogrady, T. & H. Segers (eds), Rotifera. Vol. 6: Aspelianchnidae, Gastropodidae, Lindiidae, Microcodidae, Synchaetidae, Trochosphaeridae and Filinia. Guides to the identification of the microinvertebrates of the continental waters of the world, Vol. 18. Backhuys Publishers, Leiden, 1–24.

Klingenberg, C. P., G. S. McIntyre & S. D. Zaklan, 1998. Left-right asymmetry of fly wings and the evolution of body axes. Proceedings of the Royal Society of London Series B-Biological Sciences 265: 1255–1259.

Klusemann, J., W. Kleinow & W. Peters, 1990. The hard parts (trophi) of the rotifer mastax do contain chitin: evidence from studies on *Brachionus plicatilis*. Histochemistry 94: 277–283.

Koste, W. & R. J. Shiel, 1980. Preliminary remarks on the rotifer fauna of Australia (Notogea). Hydrobiologia 73: 221–227.

Kristensen, R. M. & P. Funch, 2000. Micrognathozoa: a new class with complicated jaws like those of Rotifera and Gnathostomulida. Journal of Morphology 246: 1–49.

Mark Welch, D. & M. Meselson, 2000. Evidence for the evolution of bdelloid rotifers without sexual reproduction or genetic exchange. Science 288: 1211–1215.

Markevich, G. I., 1989. Morphology and the principle organisation of sclerite system of the mastax in rotifers. [in Russian.] Proceedings of the Institute Biology of Inland Waters 56: 27–82.

Melone, G., 2001. *Rhinoglena frontalis* (Rotifera, Monogononta): a scanning electron microscopic study. Hydrobiologia 446: 291–296.

Melone G. & D. Fontaneto, 2004. Trophi structure in bdelloid rotifers. Hydrobiologia 546: 197–202.

Melone, G., C. Ricci & H. Segers, 1998a. The trophi of Bdelloidea (Rotifera): a comparative study across the class. Canadian Journal of Zoology 76: 1755–1765.

Melone G., C. Ricci, H. Segers & R. L. Wallace, 1998b. Phylogenetic relationships of phylum Rotifera with emphasis on the families of Bdelloidea. Hydrobiologia 388: 101–107.

Nogrady, T., 1982. Rotifera. In Parker, S.P. (ed.), Synopsis and Classification of Living Organisms. McGraw-Hill, New York, 865–872.

Nogrady, T., R. Pourriot & H. Segers, 1995. Rotifera, Vol 3. The Notommatidae and the Scaridiidae. Guides to the identification of the microinvertebrates of the continental waters of the world. Vol 8. SPB Academic Publishing, The Hague, 248.

Nogrady, T., R. L. Wallace & T. W. Snell, 1993. Rotifera. Vol I. Biology, Ecology and Systematics. Guides to the identification of the microinvertebrates of the continental waters of the world 4. SPB Academic Publishers, The Hague, 142.

Ricci, C. & U. Fascio, 1995. Life-history consequences of resource-allocation of 2 bdelloid rotifer species. Hydrobiologia 299: 231–239.

Ricci, C. & G. Melone, 2000. Key to the identification of the genera of bdelloid rotifers. Hydrobiologia 418: 73–80.

Ricci, C., G. Melone & E. J. Walsh, 2001. A carnivorous bdelloid rotifer, *Abrochtha carnivora* n.sp. Invertebrate Biology 120: 136–141.

Rohlf, F. J., 1998. On applications of geometric morphometrics to studies of ontogeny and phylogeny. Systematic Biology 47: 147–158.

Rohlf, F. J., 2002a. Tps Series. Department of Ecology and Evolution, State University of New York, Stony Brook, New York. http://life.bio.sunysb.edu/morph/.

Rohlf, F. J., 2002b. NTSYS-pc, version 2.10z. Exeter Software, Setauket, New York.

Rohlf, F. J. & L. F. Marcus, 1993. A revolution in morphometrics. Trends in Ecology and Evolution 8: 129–132.

Segers, H., 1993. Rotifera of some lakes in the floodplain of the River Niger (Imo State, Nigeria). I New species and other taxonomic considerations. Hydrobiologia 250: 39–61.

Segers, H., 1995. Rotifera. Vol. 2: The Lecanidae (Monogononta). Guides to the identification of the microinvertebrates of the continental waters of the world. 6. SPB Academic Publishing, The Hague, 226.

Segers, H., 1997. Contribution to a revision of *Floscularia* Cuvier, 1798 (Rotifera: Monogononta): notes on some Neotropical taxa. Hydrobiologia 354: 165–175.

Segers, H. & G. Melone, 1998. A comparative study of trophi morphology in Seisonidea (Rotifera). Journal of Zoology 244: 201–207.

Segers, H. & R. L. Wallace, 2001. Phylogeny and classification of the Conochilidae (Rotifera, Monogononta, Flosculariacea). Zoologica Scripta 30: 37–48.

Sokal R. R. & F. J.Rohlf, 1995. Biometry, the Principles and Practice of Statistics in Biological Research, 3rd edn. Freeman and co, New York, 887 pp.

Sørensen, M. V., 2000. An SEM study of the jaws of *Haplognathia rosea* and *Rastrognathia macrostoma*

(Gnathostomulida), with a preliminary comparison with the rotiferan trophi. Acta Zoologica, Stockholm 81: 9–16.

Sørensen, M. V., 2002a. On the evolution and morphology of the rotiferan trophi, with a cladistic analysis of Rotifera. Journal of Zoological Systematics and Evolutionary Research 40: 129–154.

Sørensen, M. V., 2002b. Phylogeny and jaw evolution in Gnathostomulida, with cladistic analysis of the genera. Zoologica Scripta 31: 461–480.

Sørensen, M. V., 2003. Further structures in the jaw apparatus of *Limnognathia maerski* (Micrognathozoa), with notes on the phylogeny of the Gnathifera. Journal of Morphology 255: 131–145.

Sørensen, M. V. & W. Sterrer, 2002. New characters in the gnathostofmulid mouth parts as revealed by scanning electron microscopy. Journal of Morphology 253: 310–334.

Tannreuther, G. W., 1920. The development of *Asplanchna ebbesbornii* (Rotifer). Journal of Morphology 33: 389–437.

Zelinka, C., 1891. Studien über Räderthiere III. Zur Entwicklungsgeschichte der Räderthiere nebst Bemerkungen über ihre Anatomie and Biologie. Zeitschrift für wissenschaftliche Zoologie 53: 1–159.

Hydrobiologia (2005) 546:223–229
A. Herzig, R.D. Gulati, C.D. Jersabek & L. May (eds.) Rotifera X: Rotifer Research: Trends, New Tools and Recent Advances
DOI 10.1007/s10750-005-4200-8

External morphology and muscle arrangement of *Brachionus urceolaris*, *Floscularia ringens*, *Hexarthra mira* and *Notommata glyphura* (Rotifera, Monogononta)

Nadia Santo[1,*], Diego Fontaneto[2], Umberto Fascio[1], Giulio Melone[2] & Manuela Caprioli[2]
[1]*C.I.M.A., Centro Interdipartimentale di Microscopia Avanzata, Università degli Studi di Milano, via Celoria 26, 20133 Milan, Italy*
[2] *Department of Biology, Università degli Studi di Milano, via Celoria 26, 20133 Milan, Italy*
*(*Author for correspondence: E-mail: nadia.santo@unimi.it)*

Key words: Ploima, Flosculariacea, actin filaments, phalloidin, CLSM, SEM

Abstract

We studied four monogonont rotifers (*Brachionus urceolaris, Floscularia ringens, Hexarthra mira, Notommata glyphura*) using two different techniques of microscopy: (1) the presence of filamentous actin was examined using phalloidin-fluorescent labelled specimens and a confocal laser scanning microscope (CLSM); (2) external morphology was investigated using a scanning electron microscope (SEM). *B. urceolaris, F. ringens*, and *N. glyphura* showed similar patterns of muscle distribution: a set of longitudinal muscles acting as head and foot retractors, and a set of circular muscles. However, the size and distribution of circular muscles differed among these species. *H. mira* differed from the other species in that it lacked circular muscles but possessed strong muscles that extended into each arm. The study showed that using both CLSM and SEM provides better resolution of the anatomy and external morphology of rotifers than using one of these techniques alone. This can facilitate better understanding of the complicated anatomy of these animals.

Introduction

Although our knowledge of rotifer morphology has increased significantly since the first International Rotifer Symposium (e.g., Clément 1977), modern techniques can add greatly to our understanding of their anatomy. This is especially true for internal structures such as the musculature. These structures have a complex, three-dimentional arrangement that can only be appreciated with accurate and detailed representations. Using classical histological methods, the description of muscle arrangement in rotifers received much attention at the beginning of 20th century (Martini, 1912; Remane, 1929–1933; Peters, 1931; Stoßberg, 1932; Kükenthal & Krumbach, 1933; Dehl, 1934). After some decades, electron microscopy showed the ultrastructural organisation of muscle fibers (Amsellem & Clément, 1977, 1988; Clément & Amsellem, 1989). Recently, thanks to F-actin staining and fluorescence microscopy, the pattern of musculature in whole mounts has been revealed (Hochberg & Litvaitis, 2000; Kotikova et al., 2001; Sørensen et al., 2003).

In this paper, we describe the external morphology using a scanning electron microscope (SEM) and the muscle arrangement using a confocal laser scanning microscope (CLSM) of four species of Rotifera Monogononta.

Materials and methods

We studied *Brachionus urceolaris* O.F. Müller, 1773, *Floscularia ringens* (Linnaeus, 1758),

Hexarthra mira (Hudson, 1871), and *Notommata glyphura* Wulfert, 1938. All specimens were collected from a pond near Novara (Piedmont, northern Italy) and prepared for observations as described below.

For SEM investigation, animals were anaesthetised with marcain, fixed in 2% OsO_4 solution and picric acid–formaldehyde at 240 mosM, dehydrated in graded ethanol solutions, and critical-point-dried with CO_2 (Melone & Ricci, 1995; Melone, 1998). Specimens were then mounted on stubs, coated with gold, and observed under a LEO 1430 scanning electron microscope.

To study body-wall musculature, all specimens were anaesthetised with marcain at room temperature and were fixed with 4% paraformaldehyde in 0.11 M phosphate-buffered saline (PBS; pH 7.4) for 1 h. After several rinses with PBS, the body wall was permeabilised in PBS containing 0.25% Triton X-100 and 0.1% Tween for 20 min. Musculature was revealed by staining the animals with phalloidin-TRITC labelled (Sigma) in PBS (0.5 $\mu g\ ml^{-1}$) overnight at 4 °C. Specimens were mounted in DABCO (Aldrich) and Mowiol 4–88 (Calbiochem) on microscope slides and examined on the confocal microscope Leica TCSNT equipped with laser Argon-Krypton 75 mW multilane. Series of optical sections, attained by scanning whole specimens, were projected into a 3D image to analyse spatial disposition of muscles. For each species, a total of about 30 specimens was processed for SEM and CLSM observations.

Results

Brachionus urceolaris

The cingulum of *Brachionus urceolaris* is arranged into three lobes, with the unpaired one on the ventral side; the pseudotrochus consists of three tufts of cilia (cirri). The foot is long, transversely wrinkled, and retractile. It terminates in a couple of acute, mobile toes, perforated at the tips by the ducts of the pedal glands (Fig. 1a, e).

The circular musculature consists of four pairs of dorso-ventral muscles lying in the lateral side of

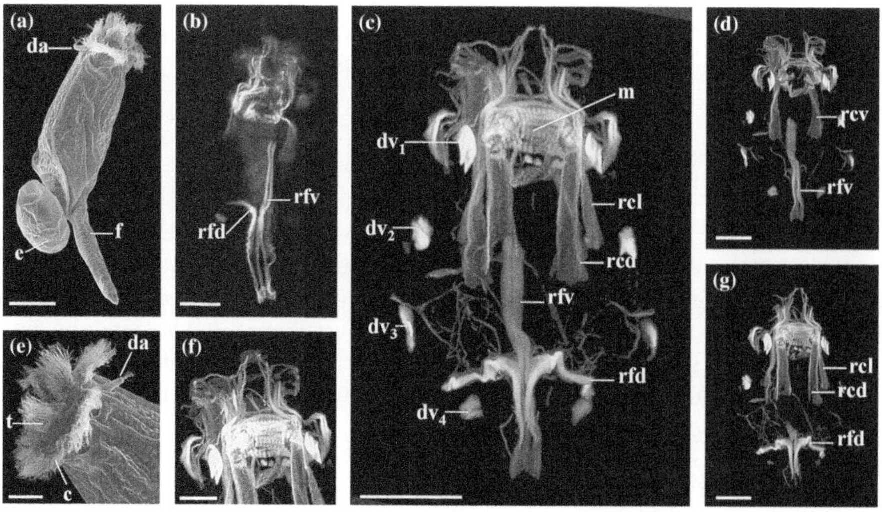

Figure 1. Brachionus urceolaris. (a) SEM micrograph of a specimen. Lateral view, *da* dorsal antenna, *e* egg, *f* foot. Scale = 50 μm. (b) Lateral optical section of whole mount stained by phalloidin-TRITC, *rfd* dorsal retractor of foot, *rfv* ventral retractor of foot. Scale = 40 μm.(c) Max-projection of whole mount stained by phalloidin-TRITC, dv_{1-4} dorso-ventral muscles, *m* mastax, *rcd* dorsal retractor of corona, *rcl* lateral retractor of corona, *rfd* dorsal retractor of foot, *rfl* lateral retractor of foot. Scale = 50 μm. (d) Ventral longitudinal optical section of image A, *rcv* ventral retractor of corona, *rfv* ventral retractor of foot, Scale = 50 μm. (e) SEM micrograph of the head, *c* cingulum, *da* dorsal antenna, *t* pseudotrochus. Scale = 20 μm. (f) Max-projection of the head stained by phalloidin-TRITC. Scale = 20 μm. (g) Dorsal longitudinal optical section of image A, *rcd* dorsal retractor of corona, *rcl* lateral retractor of corona, *rfd* dorsal retractor of foot. Scale = 50 μm.

the body extending posteriorly to the corona up to the junction between trunk and foot (Fig. 1c). Longitudinal muscles are arranged into three pairs of retractor muscles inserted at the base of the corona and reaching halfway along the trunk. Ventral, dorsal and lateral pairs of retractor muscles can be distinguished (Fig. 1c, d, g). The foot has two pairs of retractor muscles. The ventral ribbon inserts at the middle of the trunk on the ventral side and reaches the toes (Fig. 1b, d); the dorsal pair originates anteriorly to the last circular muscle and extends to the foot (Fig. 1b, g).

Phalloidin staining showed thin filaments of F-actin at the base of the cilia and in the posterior half of the trunk (Fig. 1c, f). In addition, the most well developed muscles of the mastax were highlighted (Fig. 1c, f).

Floscularia ringens

Adults of the sessile species *Floscularia ringens* produce their own tubes with gelatinous material and numerous small pellets that are arranged like bricks into a small tower (Wright, 1950; Tiefenbacher, 1972). The corona bears four wide lobes; the trunk is short, with the cloaca opening in the upper dorsal side; the foot is very long (Fig. 2a, c, d). The adult body morphology observed by SEM is described in Fontaneto et al. (2003).

Behind the corona, in the collar region, a set of circular muscles of varying sizes are located. Short and thin muscles arranged in arcs were evident on the ventral and dorsal sides, while complete circular muscles were detected on the anterior and posterior boundary of the collar region (Fig. 2i). It appears that, when the corona is retracted, the anterior circular muscle contracts too, acting like a sphincter (Fig. 2f). Posteriorly to this group, a set of circular muscles is arranged into several thin dorsal semicircular arcs that extend along the trunk, posteriorly to the cloaca all the way to the beginning of the foot (Fig. 2g). Longitudinal muscles are in three pairs, extending from the basis of the corona to the foot (Fig. 2b, g); these are U-shaped in transverse section (Fig. 2e). Along the length of the foot these muscles intertwine with each other (Fig. 2b).

Hexarthra mira

The trochus of *Hexarthra mira* consists of two symmetrical ciliated C-shaped bands (Fig. 3e); the cingulum is a ciliated band surrounding the mouth, consisting of cilia smaller than those of the trochus (Fig. 3g). The body is conical in shape and possesses six appendages, two median (respectively dorsal and ventral), two antero-lateral, and two postero-lateral (Fig. 3a, c). The terminal part of the body has two tubular processes, which are ciliated at the apex (Fig. 3b).

In the head region, a complex network of muscles was observed, corresponding to each area surrounded by the trochus bands (Fig. 3f). Some of these muscles branch at the anterior end behind the corona and then continue to the arms. Muscles, detected in each arm, differ in morphology and number (Fig. 3d). Two of these muscles run in parallel into the ventral arm, extending to the proximal end of the bristles (Fig. 3j); a thicker muscle was evident in the lateral arms (Fig. 3d). A pair of muscles runs along the length of the body and bifurcates at the distal end, where it is intercalated by numerous thinner fibres (Fig. 3i). Dorsally to the mastax musculature a series of differently sized muscle bands were evident (Fig. 3h).

Notommata glyphura

The corona consists of a ciliated ventral field and of characteristic retractile lateral auricles with long cilia in the upper lateral part of the head. The body is saccate and the foot is short (Fig. 4a, b, i).

Along the medial line of the body, four pairs of longitudinal ribbon muscles parallel each other to the foot (Fig. 4c, f). More externally and ventrally, a pair of thin muscles originates at the cephalic region and runs laterally to the mastax reaching down the trunk, where it branches in the posterior half of the body (Fig. 4c, d). On the lateral side, a third group of longitudinal muscles is found; they consist of a pair of muscles that branch into three thin bundles directed to the junction between trunk and foot (Fig. 4c, e). A circular musculature was observed in the posterior half of the trunk; it consists of three thin complete muscles that have an insertion point on the lateral side (Fig. 4c). In addition to the anterior part of the ventral muscles,

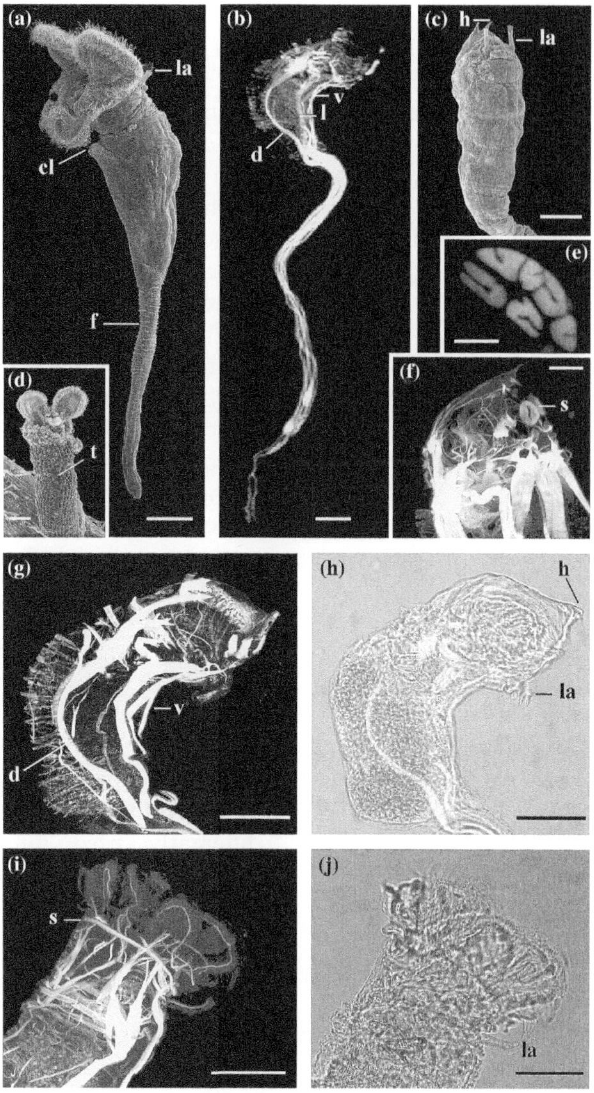

Figure 2. Floscularia ringens. (a) SEM micrograph of a specimen. Lateral view, *cl* cloaca, *f* foot, *la* lateral antenna. Scale = 50 μm. (b) Max-projection of a lateral view of whole mount stained by phalloidin-TRITC, *d* dorsal muscle pair, *l* lateral muscle pair, *v* ventral muscle pair. Scale = 50 μm. (c) SEM micrograph of a contracted specimen. Ventral view, *h* hooks, *la* lateral antennae. Scale = 50 μm. (d) SEM micrograph of a specimen inside its tube (t). Scale = 50 μm. (e) Transverse section of muscles into the foot. Scale = 5 μm. (f) Max-projection of a lateral view of a retracted head stained by phalloidin-TRITC, *s* sphincter. Scale = 25 μm. (g) Max-projection of a lateral view of a specimen with retracted head stained by phalloidin-TRITC, *d* dorsal muscle pair, *v* ventral muscle pair. Scale = 50 μm. (h) Light micrograph of image (g) *h* hooks, *la* lateral antennae. Scale = 50 μm. (i), Max-projection of a lateral view of a specimen with extended corona stained by phalloidin-TRITC, *s* sphincter. Scale = 25 μm. (j) Light micrograph of image (i) *la* lateral antenna. Scale = 25 μm.

a thin ring-like muscle is present in the head on the ventral side and external to the ciliated field (Fig. 4c). A network of thin bundles also was seen; these include those running to the auricles (Fig. 4g). The well-developed musculature of the mastax was revealed (Fig. 4c).

Discussion

The musculature of four monogonont species was studied using fluorescently labelled whole mounts to demonstrate filamentous actin. This technique resolves body-wall and visceral muscles, and also

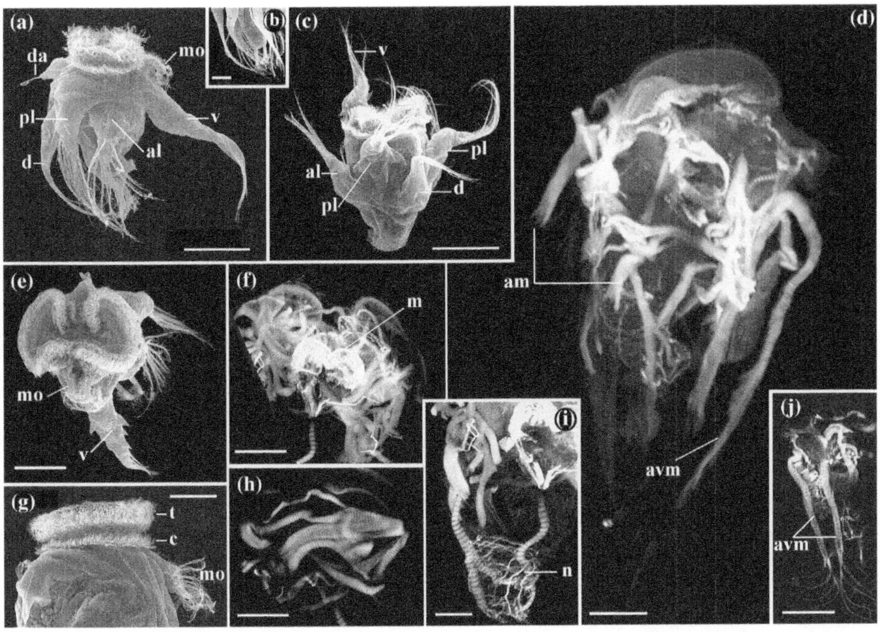

Figure 3. Hexarthra mira. (a, c) SEM micrographs of whole mount in a lateral view and a dorsal view respectively, *al* antero-lateral arm, *d* dorsal arm, *da* dorsal antenna, *mo* mouth, *pl* postero-lateral arm, *v* ventral arm. Scale = 50 μm. (b), SEM micrograph of the body terminal part with ciliated processes. Scale = 25 μm. (d), Max-projection of whole mount stained by phalloidin-TRITC, *am* muscle of lateral arm, *avm* muscle of ventral arm. Scale = 25 μm. (e) SEM micrograph of the anterior view of a specimen, *mo* mouth, *v* ventral arm. Scale = 50 μm. (f) Max-projection of the head stained by phalloidin-TRITC, *m* mastax. Scale = 25 μm. (g) SEM micrograph of the lateral view of the head, c cingulum, *mo* mouth, *t* trochus. Scale = 25 μm. (h) Muscles posterior to the mastax. Scale = 25 μm. (i) Muscles running the body length and bifurcating at the distal end, where is a network of thinner fibres (*n*). Scale = 25 μm. (j) Muscles into the ventral arm (*avm*). Scale = 50 μm.

thin filaments, which are often arranged in a network lying in different parts of the body.

Excluding *H. mira*, the presence of outer circular and inner longitudinal fibres can be considered to represent the basic pattern of muscle arrangement in the Rotifera. Longitudinal muscles originate at the head, run through the body cavity and terminate at different levels of the body length. They act as retractors of the rotatory apparatus. This is the case in *Brachionus*. According to previous data on this genus (Stoßberg, 1932; Kotikova et al., 2001), paired dorsal, lateral and ventral retractors were shown to have insertion points at the basis of the corona and on the integument on the dorsal, lateral and ventral side, approximately at the mid point of the body. Longitudinal muscles acting as foot retractors can originate at the head region or directly in the posterior half of the trunk. In the former case, they have an insertion point at the midpoint of the body as observed in *N. glyphura* and *F. ringens*. This may permit the muscles to retract the extremities independently. In *B. urceolaris* the foot retractors are restricted to the foot region and consist of three differently sized paired muscles, as already observed for other Brachionidae (Stoßberg, 1932; Kotikova et al., 2001).

Among the species studied, circular muscles exhibit different size and distribution. In *F. ringens* circular musculature consists of several muscles extending from the basis of the corona to the junction between trunk and foot. They are arranged in arcs in both the trunk and collar regions, where complete circular muscles were also seen. One of these has been found to contract after corona retraction, acting as a sphincter. Circular muscles are less developed in loricate species, like *Brachionus*, where they have been seen limited to dorso-ventral bands extending along the lateral sides posteriorly to the corona. They serve to bring the plates of the lorica closer together (Hyman, 1951). *Notommata glyphura* also has only few thin

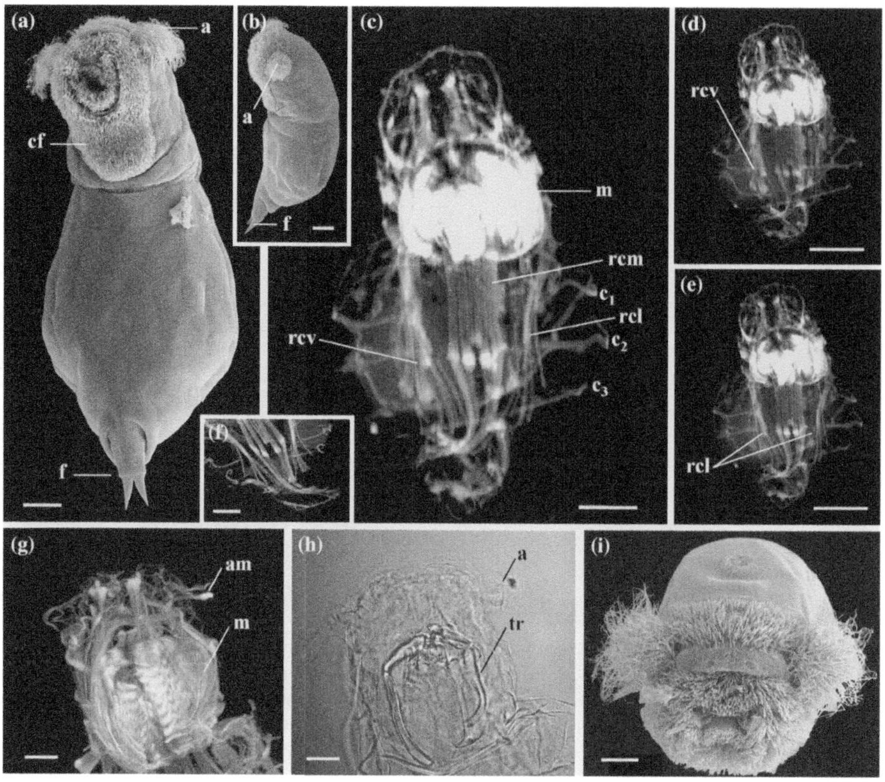

Figure 4. Notommata glyphura. (a, b) SEM micrographs of a specimen in a ventral and lateral view respectively, *a* auricles, *cf* ciliated field, *f* foot. Scale = 20 μm. (c) Max-projection of whole mount stained by phalloidin-TRITC, c_{1-3} circular muscles, m mastax, *rcl* lateral retractor of corona, *rcm* median retractor of corona, *rcv* ventral retractor of corona. Scale = 50 μm. (d) Ventral longitudinal optical section of image C, *rcv* ventral retractor of corona. Scale = 80 μm. (e) Dorsal longitudinal optical section of image C, *rcl* lateral retractor of corona. Scale = 80 μm. (f) Muscles into the foot. Scale = 40 μm. (g) Max-projection of the head of a specimen stained by phalloidin-TRITC, *am* muscles into auricles, *m* mastax. Scale = 20 μm. (h) Light micrograph of image G, *a* auricle, *tr* trophi. Scale = 20 μm. (I) SEM micrograph of a anterior view of the head. Scale = 10 μm.

circular bundles encircling the posterior half of the trunk. No circular muscles were detected in the foot of any of the species we examined. The circular muscles function in body extension, contracting against the hydrostatic pressure of the body cavity, causing body elongation. In creeping bdelloids, circular muscles have been seen to be well-developed, extending from the head to the foot and encircling the body (Zelinka, 1886; Brakenhoff, 1937; Hochberg & Litvaitis, 2000). *Brachionus urceolaris* and *N. glyphura* move mainly by ciliary beating of the corona; in these species circular muscles are very poorly developed. On the other hand, the sessile rotifer *F. ringens* has a well-developed circular musculature either behind the corona or along the trunk. This is probably related to the extension of the large corona by increasing the hydrostatic pressure in the body cavity.

In *H. mira*, the analysis of all optical sections revealed a complex arrangement of muscles that differs from the basic muscle pattern of the other species. This could be due to the presence of the arms and the jumping movements performed by these. It has been suggested that arm muscles have been derived from the circular muscles (Hyman, 1951), but we have no evidence to support this hypothesis.

In addition to the bundles of the muscular system, several small fibers containing F-actin were shown to be present in the head and foot region. Phalloidin staining showed the presence of F-actin at the basis of cilia of the corona. An ultrustructural study on the rotatory apparatus of *B. plicatilis* showed the presence of microfilaments associated with the microvilli (Dallai & Lupetti, 1994). We speculate that these head fibers may be responsible for some aspects of coronal coordination.

Acknowledgement

Thanks are due to Russ Shiel, for reading an earlier draft of the paper and correcting the English, and to Robert L. Wallace, for his comments on the manuscript. Financial support was provided by the State University of Milan, grant "Giovani Ricercatori", to N.S.

References

Amsellem, J. & P. Clément, 1977. Correlations between ultrastructural features and contraction rates in rotiferan muscle I. Preliminary observations on longitudinal retractor muscles in *Trichocerca rattus*. Cell and Tissue Research 181: 81–90.

Amsellem, J. & P. Clément, 1988. The muscles of a monogonont rotifer, *Trichocerca rattus* II. The central retractor muscles. Tissue and Cell 20: 89–108.

Brakenhoff, H., 1937. Zur Morphologie der Bdelloidea. Zoologische Jahrbücher Abteilung für Anatomie 63: 9–182.

Clément, P., 1977. Ultrastructural research on rotifers. Archiv für Hydrobiologie Beiheft Ergebnisse der Limnologie 8: 270–297.

Clément, P. & J. Amsellem, 1989. The skeletal muscles of rotifers and their innervation. Hydrobiologia 186/187: 255–278.

Dallai, R. & P. Lupetti, 1994. Ciliary and microvillar specializations in the corona of *Brachionus plicatilis* (Rotifera, Monogononta). Journal of Submicroscopic Cytology and Pathology 26: 497–506.

Dehl, E., 1934. Morphologie von *Lindia tecusa*. Zeitschrift für Wissenschaftliche Zoologie 145: 169–219.

Fontaneto, D., G. Melone & R. L. Wallace, 2003. Morphology of *Floscularia ringens* (Rotifera Monogononta) from egg to adult. Invertebrate Biology 122: 231–240.

Hochberg, R. & M. K. Litvaitis, 2000. Functional morphology of the muscles in *Philodina* sp.1 (Rotifera: Bdelloidea). Hydrobiologia 432: 57–64.

Hyman L. H., 1951. The Invertebrates Acanthocephala, Aschelminthes and Entoprocta. Volume III. McGraw-Hill, Inc. New York, 572 pp.

Kotikova, E. A., O. I. Raikova, L. P. Flyatchinskaya, M. Reuter & M. K. S. Gustafsson, 2001. Rotifer muscles as revealed by phalloidin-TRITC staining and confocal scanning laser microscopy. Acta Zoologica 82: 1–9.

Kükenthal W., T. Krumbach, 1933. Handbuch der Zoologie. Eine Naturgeschichte der Stämme des Tierreiches. II B and. Berlin und Leipzig, Walter de Gruyeter and Co.

Martini, E., 1912. Studien über die Konstanz histologischer Elemente III *Hydatina senta*. Zeitschrift für Wissenschaftliche Zoologie 102: 425–645.

Melone, G., 1998. The rotifer corona by SEM. Hydrobiologia 387/388: 131–134.

Melone, G. & C. Ricci, 1995. Rotatory apparatus in bdelloids. Hydrobiologia 313/314: 91–98.

Peters, F., 1931. Untersuchungen über Anatomie und Zellkonstanz von Synchaeta (*S. grimpei* Remane, *S. baltica* Ehr., *S. tavina* Hood and *S. triophthalma* Laut.) Ein Beitrag zur Frage der Artunterschiede bei konstantzelligen Tieren. Zeitschrift für Wissenschaftliche Zoologie 139: 1–119.

Remane A., 1929–1933. Rotatoria. In Bronn HG (ed) Klassen und Ordnung des Tierreichs, vol 4, sect II, book 1, part 3, Akademische Verlagsgesellschaft, Leipzig, 577 pp.

Sørensen, M. V., P. Funch, M. Hooge & S. Tyler, 2003. Musculature of *Notholca acuminata* (Rotifera: Ploima: Brachionidae) revealed by confocal scanning laser microscopy. Invertebrate Biology 122: 223–230.

Stoβberg, K., 1932. Zur Morphologie der Rädertiergattungen Euchlanis, Brachionus und Rhinoglena. Zeitschrift für Wissenschaftliche Zoologie 142: 313–424.

Tiefenbacher, L., 1972. Beiträge zur Biologie und Ökologie sessiler Rotatorien unter besonderer Berücksichtigung des Gehäusebaues und der Regenerationsfähigkeit. Archiv für Hydrobiologie 71: 31–78.

Wright, H. G. S., 1950. A contribution to the study of *Floscularia ringens*. Journal of the Queckett Microscopical Club Series 4, 3: 103–116.

Zelinka, C., 1886. Studien über Räderthiere I Über die Symbiose und Anatomie von Rotatorien aus dem Genus *Callidina*. Zeitschrift für Wissenschaftliche Zoologie 44: 396–507.

Hydrobiologia (2005) 546:231–238
A. Herzig, R.D. Gulati, C.D. Jersabek & L. May (eds.) Rotifera X: Rotifer Research: Trends, New Tools and Recent Advances
DOI 10.1007/s10750-005-4202-6

The musculature of *Testudinella patina* (Rotifera: Flosculariacea), revealed with CLSM

Martin Vinther Sørensen
Invertebrate Department, Zoological Museum, University of Copenhagen, Universitetsparken 15, DK-2100, Copenhagen Ø, Denmark
Tel: +45-35321116; Fax: +45-35321010; E-mail: mvsorensen@zmuc.ku.dk

Key words: confocal microscopy, morphology, muscles, Rotifera, phalloidin

Abstract

The musculature of *Testudinella patina* was visualized using phalloidin-linked fluorescent dye by confocal laser scanning microscopy. The conspicuous broad retractors appear to be made up of five separate fibers, of which three anchor in the neck region whereas two extend into the corona. Besides the broad retractors, a total of five paired longitudinal retractors are present and all of them extend into the corona. Incomplete circular muscles are found in groups in the neck region and in the medial and posterior parts of the trunk. The foot musculature comprises eight thin ventral foot muscles and six thicker dorsal foot muscles that all extend from the foot basis to the distal part of the foot. At the basis of the foot, each of the dorsal foot muscles anchors on a smaller, S-shaped subterminal foot muscle. The foot musculature furthermore comprises one pair of paraterminal foot muscles that each anchors basally on a subterminal foot muscle, extends into the most proximal part of the foot and attaches on one of the dorsal foot muscles. The visceral musculature is composed of extremely delicate fibers and is restricted to an area around and posterior to the foot opening. The presence of incomplete circular muscles supports that these muscles are a basal trait for Rotifera, whereas the morphology of the broad retractors and foot muscles is much more specialized and may be autapomorphic for *Testudinella* or alternatively for this genus and its closest relatives. The present results stress that revealing muscles by staining may produce new information from even well-investigated species, and that this information may contribute to a better understanding of functional as well as phylogenetic aspects of rotifer biology.

Introduction

From time to time, technological advances provide new tools and methods for the investigation of the fine morphology in microinvertebrates. One of the most recent techniques that have been applied to the study of rotifer morphology is immunostaining combined with confocal laser scanning microscopical (CLSM) investigations. Fluorescently tagged phalloidin binds to filamentous actin and is therefore useful for investigating musculature. When an animal is stained with such a fluorescent probe and subsequently is observed with an epiflourescence microscope or CLSM, the musculature will appear as bright bands and even the finest fibers that often have been overlooked in conventional light microscopy will be visible.

Until now most CLSM-based studies on the rotifer musculature did focus on bdelloid rotifers (Hochberg & Litvaitis, 2000; Santo, 2001) or rotifers belonging to the monogonont families Euchlanidae and Brachionidae (Kotikova et al., 2001; Sørensen et al., 2003), and even though these studies have produced much new information, data from further taxa are required before the information can be used in a broader systematic or phylogenetic context. This paper gives the first CLSM-based description of the musculature in a species from the order Flosculariacea. Whole mounts of *Testudinella patina* stained with phalloidin linked with fluorescent dye were investigated with CLSM and their musculature is compared with that of other rotifers and with earlier investigations of *T. patina* (see Seehaus, 1930).

Materials and methods

The investigated specimens were collected in a small lake, Lake Mossø, in Rold Forrest, Denmark (56° 49′ 26.2″ N, 009° 53′ 8.8″ E). The sample was taken with a conical plankton net with a diameter of 30 cm and a mesh size of 40 μm that was dragged through the bottom vegetation. The concentrate was sorted out under a dissecting microscope, and specimens for CLSM were placed in a drop of freshwater and relaxed by adding a few cocaine crystals to the drop. The anaesthetized specimens were fixed for 1 h in phosphate-buffered 4% formaldehyde, rinsed in PBS, made permeable by exposure to 0.1% Triton X-100 in PBS for 1 h, stained 40 min in Alexa-488-phalloidin, rinsed in PBS, and then mounted in Fluoromount-G on a cover slip. Images under blue-excitation wavelength were obtained with an Olympus BX50WI light microscope equipped with Ultra View LCI confocal imaging system.

Results

The obtained results are compared carefully with the descriptions made by Seehaus (1930). Wherever possible the terminology introduced by Seehaus (1930) is applied, whereas other and more precise names are suggested in cases where the new information has made Seehaus' nomenclature ambiguous.

The broad retractors and corona retractors

The most conspicuous muscles in *T. patina* are the broad retractors (lat. musculus retractor latus) that run from the head or neck region and attach posteriorly on the ventral lorica plate (Figs. 1–3). In the light microscope, the broad retractors resemble one or two paired muscles, but in CLSM each muscle appears to be made up of five broad, cross-striated muscle fibers (Figs. 2 and 4a). Three of the fibers attach anteriorly in the neck, whereas two extend into the corona and attach inside the trochal discs (Fig. 3). The broad retractors are the only conspicuously cross-striated muscles in *T. patina* (Fig. 1).

The remaining corona musculature comprises five much thinner muscles: the dorsomedial, the ventromedial, the dorsolateral, the ventrolateral and the lateral corona retractors (Figs. 1–3). The dorsomedial corona retractors are clearly longer than the ventrolateral ones, and attach posteriorly, near the foot opening. Anteriorly, they extend into the trochal discs, and bend outwards, following the dorsal rim of the trochus (Figs. 1 and 3). The ventromedial corona retractors also extend into the corona where they ramify twice (Figs. 2a, 3 and 4c). The most basal branch appears to run laterally into the cingulum while the two more apical branches attach on the trochus rim (Fig. 4c). In the pharyngeal area, very close to the ventromedial corona retractors, a pair of short mastax retractors (Fig. 3) extend posteriorly (other muscles in the mastax are arranged as described by Seehaus, 1930).

A pair of dorsolateral and ventrolateral corona retractors are present, approximately in the same position as the broad retractors (Figs. 1–3). These fibers are slightly thicker than the medial ones, but not as strong as the fibers in the broad retractors. The dorsolateral corona retractors are a pair of unbranched fibers that anchor in a position posterior to the mastax and run anteriorly into the corona where they attach at the dorsal rim of the trochus (Fig. 3). The ventrolateral retractors are located ventral to the broad retractors but otherwise have the same position as the dorsolateral ones. Anteriorly they are bifurcate with a fiber attaching ventrally and laterally on the rim of the trochus (Fig. 4c).

Only one pair of lateral corona retractors is present. The muscles are relatively short, same length as the ventromedial retractors, and anchor posteriorly on the ventral lorica plate (Figs. 1–3). Anteriorly, they extend into the corona where they attach laterally.

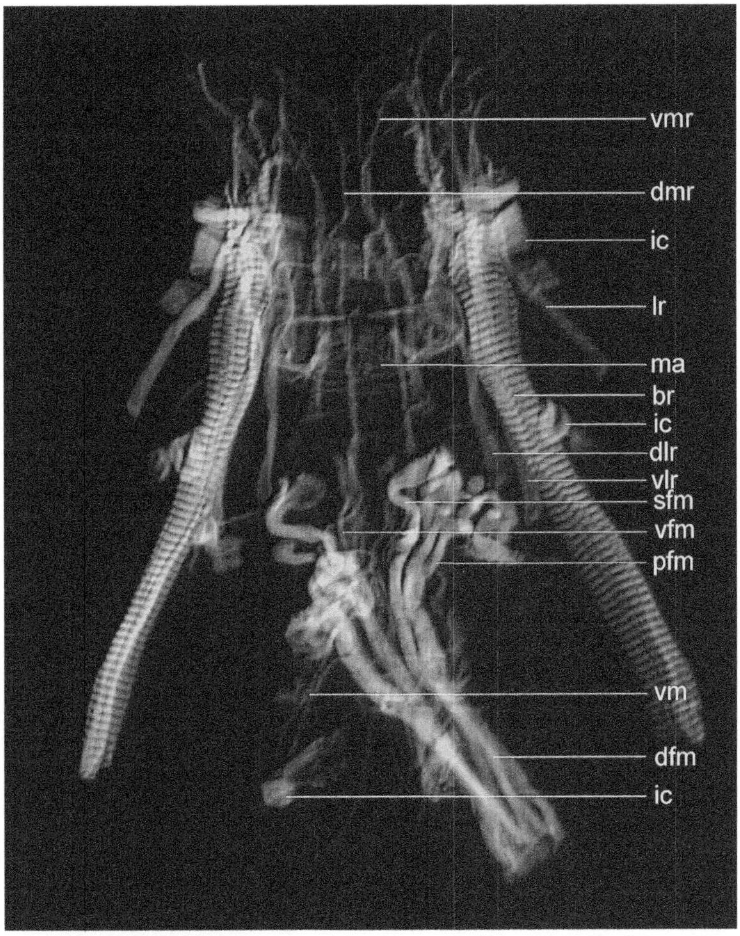

Figure 1. Fluorescing musculature in *Testudinella patina* in merged stack with 39 serial sections taken from ventral to dorsal side of specimen. br – broad retractor, dfm – dorsal foot muscle, dlr – dorsolateral corona retractor, dmr – dorsomedial corona retractor, ic – incomplete circular muscle, lr – lateral corona retractor, ma – mastax, pfm – paraterminal foot muscle, sfm – subterminal foot muscle, vfm – ventral foot muscle, vlr – ventrolateral corona retractor, vm – visceral musculature, vmr – ventromedial corona retractor.

Incomplete circular muscles

T. patina contains three groups of more or less modified incomplete circular muscles (Figs. 1–3). The anterior group is positioned in the neck area and consists of three muscles, a large median one, and two smaller anterior and posterior ones. All three muscles are bent as a U, and wrap around the broad retractors, the dorsolateral and ventrolateral corona retractors and the lateral corona retractors. The median group has two muscles and is located more posteriorly, but still in the anterior part of the animal. These muscles are J-shaped, anchored to the dorsal lorica plate and are wrapped around the broad retractors and the dorsolateral and ventrolateral corona retractors, but not around the lateral retractors. The posterior group also contains two pairs of muscles, and is located behind the foot opening, close to the stomach muscles (Fig. 4d). Like the muscles in the median group, they are partly modified into dorsoventral muscles with one end anchored to the dorsal lorica plate and the other inward bent.

Foot musculature

The foot contains six dorsal foot muscles and eight thinner ventral foot muscles (Figs. 1–3). At the foot opening, each of the dorsal foot muscles joins

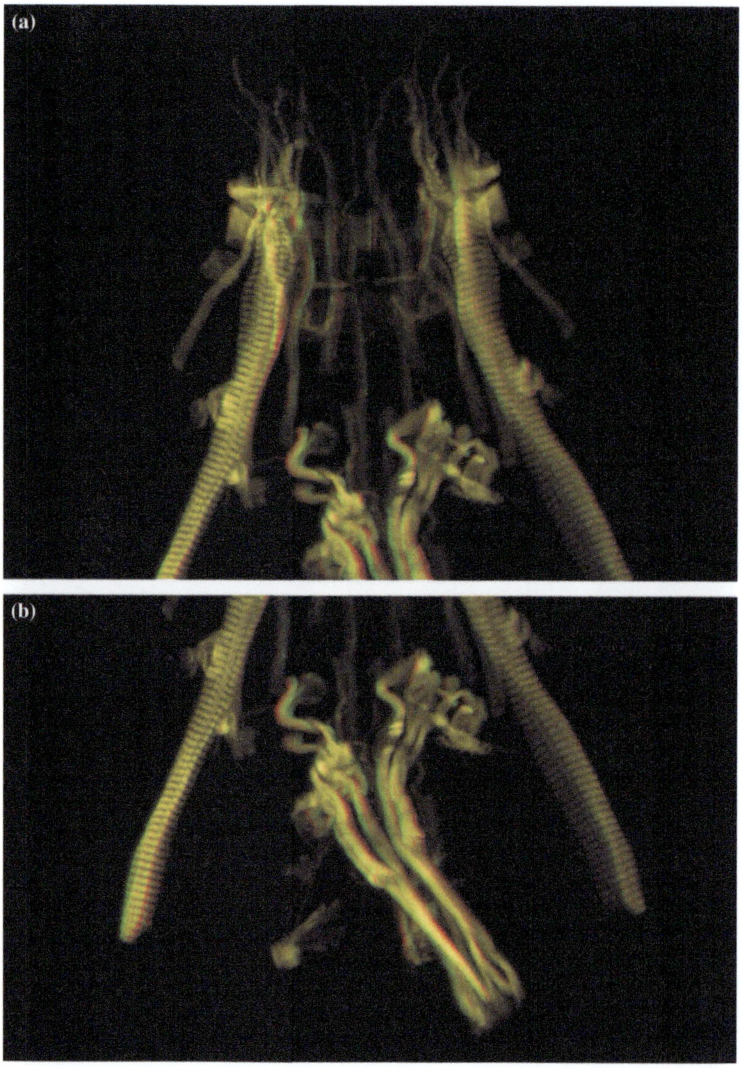

Figure 2. Stereo-pairs of whole mount of *Testudinella patina* showing fluorescing musculature (should be viewed with red/green 3D glasses). (a) Anterior end. (b) Posterior end.

a short S-shaped muscle, named the subterminal foot muscles (Fig. 4b). Likewise inside the foot, the muscles are densely packed in this area, but at least for some of the fibers there was clearly a myojunction between the foot muscles and the smaller subterminal muscles (Fig. 4b). Besides the dorsal foot muscles and their associated subterminal muscles, a pair of paraterminal foot muscles is present (Fig. 3). Like the dorsal foot muscles they have an associated subterminal muscle, but they only extend into the most proximal part of the foot where each of them anchors on a dorsal foot muscle (Fig. 3).

The ventral foot muscles are considerably thinner than the dorsal ones. They extend to the tip of the foot and basally they anchor directly on the lorica and not, such as the dorsal foot muscles, via inserted subterminal muscles.

Visceral musculature

Visceral musculature is only present in the posterior part of the animal (Fig. 3). It comprises a network of extremely fine, and very lightly stained, fibers (Fig. 4d). It is probably associated with the stomach and the germovitellarium.

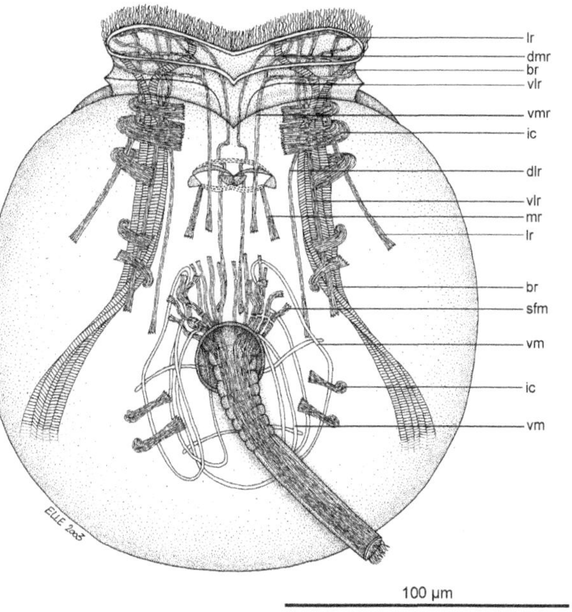

Figure 3. Schematic line drawing of the musculature in *Testudinella patina*, ventral view. br – broad retractor, dlr – dorsolateral corona retractor, dmr – dorsomedial corona retractor, ic – incomplete circular muscle, lr – lateral corona retractor, mr – mastax retractors, sfm – subterminal foot muscle, vlr – ventrolateral corona retractor, vm – visceral muscles, vmr – ventromedial corona retractor.

Discussion

Comparison with earlier descriptions

T. patina was described by Seehaus (1930), who made complete section-series of the species and used these to reconstruct most of its internal morphology. However, although the work of Seehaus provided much detailed information, the introduction of CLSM-based investigations has revealed much additional data.

The broad retractors are the most conspicuous muscles in *T. patina*. According to Seehaus (1930), each muscle is formed by a single fiber that is folded and held together with sarcoplasmatic connections, which make it appear as two separate fibers. However, the CLSM pictures show five distinct fibers in each retractor (Fig. 4a), and two of these anchor in the corona whereas three attach in the neck area (Fig. 3). Based on the phalloidin stainings, it cannot be certified whether the retractor is a single-folded muscle or five separate but closely associated fibers. It is, however, certain that the broad retractors are more complex than previously described, and that such muscles have not been reported earlier from other rotifer genera.

The phalloidin staining also provided new information about the nature of the smaller longitudinal retractors. According to Seehaus (1930), *T. patina* possesses paired dorsal, ventral, central and lateral retractors, but the CLSM investigations could furthermore detect the presence of paired dorsolateral and ventrolateral corona retractors, located very close to the broad retractors. The dorsal and ventral retractors, reported by Seehaus (1930), certainly correspond to the dorsomedial and ventromedial corona retractors found in this study. I chose to change the names of these muscles to stress their location near the midline of the animal and to avoid confusion with the above-mentioned dorso- and ventrolateral corona retractors. The central retractors reported by Seehaus (1930) were not found in the present study. They could, of course, have been overlooked, but since the other muscles in the anterior region are detected very easily, I find it more likely that it was a misinterpretation made by Seehaus.

The incomplete circular muscles are generally arranged as described by Seehaus (1930). The sixth pair reported by Seehaus, the pair located very close to the foot opening (see Figure 52 in Seehaus, 1930), was not found, but it could be hidden

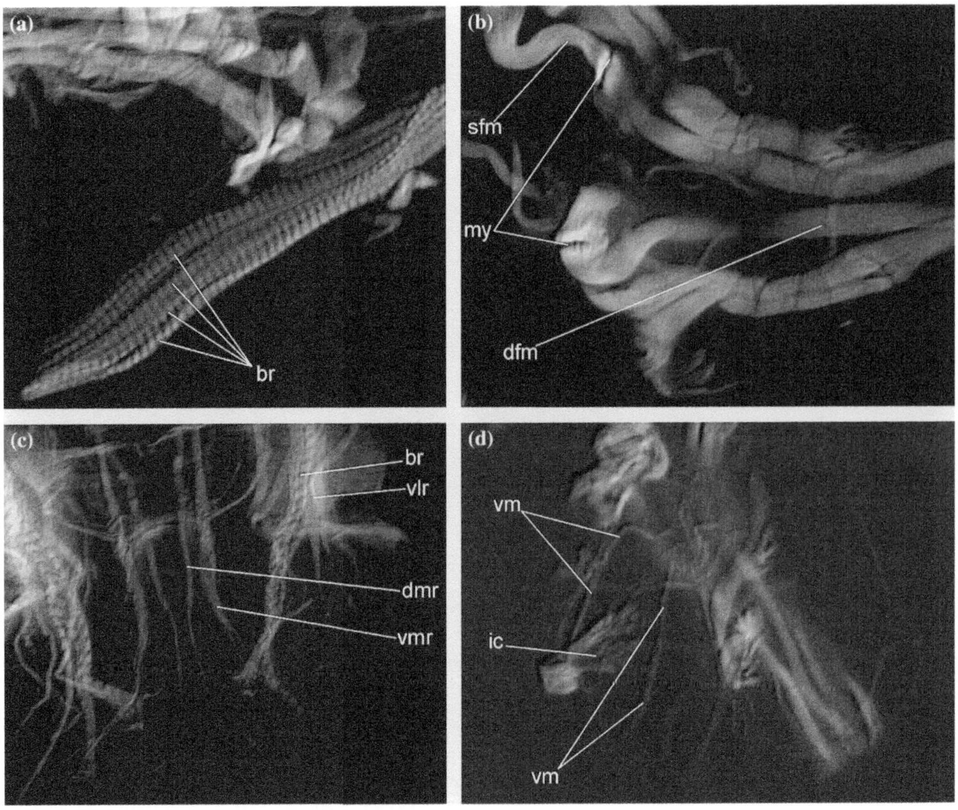

Figure 4. CLSM images showing details in the musculature of *Testudinella patina*. (a) Broad retractor. (b) Basis of foot. (c) Corona. (d) Stomach. br – broad retractors, dfm – dorsal foot muscle, dmr – dorsomedial corona retractor, ic – incomplete circular muscle, my – myojunction, sfm – subterminal foot muscle, vlr – ventrolateral corona retractor, vm – visceral musculature, vmr – ventromedial corona retractor.

between the four pairs of subterminal foot muscles. Another possibility is, however, that Seehaus (1930) confused the sixth pair with one of the subterminal foot muscles. Interestingly, Seehaus does not mention these special, inserted muscles at all, even though they are rather distinct.

Also the visceral musculature, located in the posterior part of the animal, has been omitted from Seehaus' description. The visceral muscle system comprises several extremely delicate fibers that were difficult to detect even at high magnifications. They are extremely flexible, and were never orientated in the same way in any of the investigated specimens.

In conclusion, this comparison illustrates that the use of CLSM studies on specimens stained with phalloidin may reveal much new data, even from previously well investigated species, and that this technique certainly will provide much new information that may be used in a systematic as well as in a more functionally orientated context.

Comparison with other taxa

Detailed descriptions of the musculature in the free-swimming Gnesiotrocha are scarce. Remane (1929–1933) and Wiszniewski [1933, ref. De Beauchamp (1965)] address some comments on the musculature in *Hexarthra*, but both studies only describe the most conspicuous fibers. The genera in Ploima are much better described in older (see Remane, 1929–1933) as well as more modern studies (Clément & Amsellem, 1989; Kotikova et al., 2001; Sørensen et al., 2003). Comparison with these studies suggests that some muscles may be considered rather common in rotifers while others are unique for *Testudinella* or its closest relatives. Most studies on the rotifer musculature reports the presence of incomplete circular muscles, in some studies referred to as dorso–ventral retractors. Such muscles are present in most monogonont species where they often appear in groups (Kotikova et al., 2001; Sørensen et al., 2003; this study), but they are also found as more serially distributed fibers in bdelloid rotifers (Hochberg & Litvaitis, 2000; Santo, 2001) and even in Seisonidea (De Beauchamp, 1965; Ahlrichs, 1995). Hence, the presence of incomplete circular muscles can be considered a basal trait in Rotifera.

It is more difficult to compare the longitudinal corona retractors with muscles found in other taxa. All coronal muscles in *T. patina* are direct extensions of the body musculature and this seems to deviate from the pattern found in other monogononts. In *Notholca acuminata* the coronal musculature primarily comprises delicate fibers in the head and none of the muscles in the trunk extend beyond the neck (Sørensen et al., 2003). Kotikova et al. (2001) describe the coronal musculature in *Brachionus quadridentatus* in less detail, but it generally appears to follow the same pattern as in *N. acuminata*. However, some of the muscles in *Epiphanes senta* actually seem to follow the same pattern as those in *T. patina* (see Martini, 1912), and this could also be the case in the more closely related genus *Conochiloides* (see Hyman, 1951, Figure 39). However, more information is needed to establish the systematic aspects of the coronal musculature.

The present study reveals two new muscular features that, to my knowledge, have not been described previously. These are the morphology of the broad retractors and the arrangement of the foot musculature. The presence of broad and strong retractors are known from several monogonont taxa such as Epiphanidae, Euchlanidae, Brachionidae and Synchaetidae (Martini, 1912; Peters, 1931; Clément & Amsellem, 1989; Kotikova et al., 2001; Sørensen et al., 2003), but the special arrangement found in *T. patina* has not been reported earlier. Whereas the main retractors often are one or two pairs of single, strong muscle fibers, the broad retractors in *T. patina* are formed by five closely associated but still separated fibers or alternatively one-folded fiber. It is not certain whether this trait is autapomorphic for *Testudinella* or whether it is found in other taxa, but it certainly may be phylogenetically significant.

Also the morphology of the foot musculature appears to be rather special. In most monogonont species, the movement of the foot is controlled by pairs of foot retractors that attach directly to the body wall (see for example Remane, 1929–1933; Kotikova et al., 2001), whereas each of the retractors in *T. patina* attaches via the inserted subterminal foot muscles. Like the broad retractors, our knowledge is still too insufficient to explain whether this trait is special feature for *Testudinella* or if it is present in other, possibly closely related, taxa as well, but it certainly stresses the importance of further investigations of the musculature in a much broader variety of rotifers.

Conclusions

The use of CLSM investigations of specimens of *T. patina* stained with phalloidin-linked fluorescent dye has revealed new data about the arrangement of the musculature in this species. Comparison with previous descriptions show that this technique can provide new information that formerly has been overlooked or misinterpreted. Our knowledge on the variability of the rotifer musculature is still too insufficient to form the basis for any systematic or phylogenetic conclusions, but the technique has proven to be capable of producing valuable information, which stresses the importance of further studies on other species to accumulate more information about the morphological variation in the rotifer musculature.

Acknowledgements

This work could not have been done without support from the Marine Biological Laboratory in Helsingør that provided all the microscopical facilities. I am indebted to Tom Fenchel for allowing me to use the CLSM and to Michael Kühl for spending much of his time introducing me to the microscope. I also thank Stine Elle for producing the line art illustration and Mary E. Petersen for correcting the language in the manuscript. This work was supported by the Carlsberg Foundation (Grants No. ANS-0178/20 and ANS-0724/20) and the Danish Natural Science Research Council (Grant No. 51-00-0278). The CLSM was financed by the Danish Natural Science Research Council (Grants No. 9700549 and 21-01-0211).

References

Ahlrichs, W. H., 1995. Ultrastruktur und Phylogenie von Seison nebaliae (Grube 1859) und Seison annulatus (Claus 1876). Cuvillier Verlag, Göttingen 310pp.

Clément, P. & J. Amsellem, 1989. The skeletal muscles of rotifers and their innervation. Hydrobiologia 186/187: 255–278.

De Beauchamp, P. M., 1965. Classe des rotifères. In Grassé, P.-P. (ed.), Traité de Zoologie. Anatomie, Systematique, Biologie. Masson et Cie, Paris, 1225–1379.

Hochberg, R. & M. K. Litvaitis, 2000. Functional morphology of the muscles in *Philodina* sp. (Rotifera: Bdelloidea). Hydrobiologia 432: 57–64.

Hyman, L. H., 1951. The Invertebrates: Acanthocephala, Aschelminthes, and Entoprocta. The Pseudocoelomate BilateriaVol. 3. McGraw-Hill Inc., New York 572 pp.

Kotikova, E. A., O. I. Raikova, L. P. Flyatchinskaya, M. Reuter & M. K. S. Gustafsson, 2001. Rotifer muscles as revealed by phalloidin-TRITC staining and confocal scanning laser microscopy. Acta Zoologica 82: 1–9.

Martini, E., 1912. Studien über die Konstanz histologischer Elemente III. *Hydatina senta*. Zeitschrift für wissenschaftliche Zoologie 102: 425–645.

Peters, F., 1931. Untersuchungen über Anatomie und Zellkonstanz von *Synchaeta* (*S. grimpei* Remane, *S. baltica* Ehrenb., *S. tavina* Hood und *S. triophthalma* Lauterborn). Zeitschrift für wissenschaftliche Zoologie 139: 1–119.

Remane A., 1929–1933. Rotatoria. In Bronn, H. G. (ed.), Klassen und Ordnungen des Tier-reichs, Vol. 4. Vermes. Akademische Verlagsgesellschaft mbH, Leipzig: 1–576.

Santo, N., 2001. Anidrobiosi in rotiferi bdelloidei: approccio morfo-funzionale all'indagine degli adattamenti caratterizzanti la fase preparatoria. Ph.D. thesis, University of Milan, 114 pp.

Seehaus, W., 1930. Zur Morphologie der Rädertiergattung *Testudinella* Bory de St. Vincent (= *Pterodina* Ehrenberg). Zeitschrift für wissenschaftliche Zoologie 137: 175–272.

Sørensen, M. V., P. Funch, M. Hooge & S. Tyler, 2003. Musculature of *Notholca acuminata* (Rotifera: Ploima: Brachionidae) revealed by confocal scanning laser microscopy. Invertebrate Biology 122: 223–230.

Hydrobiologia (2005) 546:239–248
A. Herzig, R.D. Gulati, C.D. Jersabek & L. May (eds.) Rotifera X: Rotifer Research: Trends, New Tools and Recent Advances
DOI 10.1007/s10750-005-4203-5

Rotifer nervous system visualized by FMRFamide and 5-HT immunocytochemistry and confocal laser scanning microscopy

Elena A. Kotikova[1], Olga I. Raikova[1], Maria Reuter[2,*] & Margaretha K. S. Gustafsson[2]
[1]*Zoological Institute, Russian Academy of Sciences, 199034, St. Petersburg, Russia*
[2]*Department of Biology, Abo Akademi University, FIN-20520, Åbo, Finland*
*(*Author for correspondence: Tel.: 358-2-215-4603; Fax: 358-2-215-4748; E-mail: mreuter@ra.abo.fi)*

Key words: Rotifera, nervous system, serotonin, 5-HT, FMRFamide, immunocytochemistry

Abstract

We present the first results of immunocytochemical (ICC) observations on serotonin (5-HT) and FMRFamide (Phe–Met–Arg–Phe–NH_2) immunoreactivity patterns in the rotifer nervous system investigated using a confocal laser scanning microscope (CLSM). Three species of rotifers are studied: *Platyias patulus* (*Plationus patulus*, Segers et al., 1993; Hydrobiologia 268: 1–8), *Euchlanis dilatata*, and *Asplanchna herrickii.* Independently from their systematic position, these species possess similar nerve structures. However, some differences were observed in the innervation of the corona, mastax, foot, and mostly in the pattern of the cerebral neurons. The general numbers of 5-HT-immunoreactive (IR) and FMRFamide-IR neurons are low (10–34), but constant for each species. The sizes of the neurons vary from 2, 5 to 10 μm. From 4 to 14 cerebral neurons lie at different levels and are arranged into an X- or a ring- or a curved arch shape. One or two pairs of neurons are localized along longitudinal nerve cords. Double staining of 5-HT and FMRFamide-IR elements shows no co-localization.

Introduction

High degree of tissue specialization and eutely is characteristic for rotifers. The total number of cells in the rotifer body can reach 1000 (Martini, 1912). Classic morphological studies have described the nervous system pattern of several species of rotifers and revealed different types of nerve cells in the brain (Nachtwey, 1925; Remane, 1929–1933). Electron microscopic observations have demonstrated that only few brain neurons are unipolar (Clément, 1980). The brain itself demonstrates considerable diversity of synaptic vesicles (Villeneuve & Clément, 1971; Clément, 1977; Wurdak et al., 1983). This is an indirect evidence of a variety of neuronal signal substances in the nervous system of these animals. Further histochemical and biochemical studies supported the existence of such a variety. The cholinergic (ChE) neuronal system was first described in 12 species of rotifers (Nogrady & Alai, 1983). Two types of ChE neurons have been distinguished in the brain of Bdelloidea (Raineri, 1984). Some years later the presence of catecholamines (CA) was noted in the nervous system of rotifers (Keshmirian & Nogrady, 1987, 1988). The mapping of the CA distribution in the nervous system of rotifers (Kotikova, 1994, 1995, 1997) should be regarded as a next study aspect.

At present immunocytochemical methods (ICC) are widely used to study different neuronal signal substances and to localize the corresponding neurons in the nervous system of invertebrates (Reuter & Halton, 2001; Reuter et al., 2001a, b). However, no such studies have been performed on rotifers. The aim of this paper is to investigate patterns of immunoreactivity in the nervous system of three species of rotifers. We have studied the localization of two major neuronal signal substances: FMRFamide and 5-HT. FMRFamide (Phe–Met–Arg–Phe–NH_2) is a tetrapeptide neurotransmitter, a member of the family of RFamide peptides, all sharing the same C terminal RFamide sequence. Originally identified in mollusc ganglia (Price & Greenberg, 1977), the FaRP family of peptides has been found in every major metazoan phylum, from coelenterates to chordates (e.g. Schneider & Taghert, 1988; Aarnisalo & Panula, 1995; Grimmelikhuijzen & Westfall, 1995). FMRFamide was demonstrated to be cardioexcitatory in molluscs and mammals and strongly myoactive in helminths (Geary et al., 1999).

5-HT (5-hydroxytryptamine), commonly known as serotonin, belongs to the biogenic amines, signalling molecules with established neurotransmitter and hormonal roles. Serotonin is an evolutionarily conserved neurotransmitter, found in both invertebrates and vertebrates, and involved in locomotor and behavioural roles. In mammals it occurs in both the central and peripheral nervous systems and is implicated in a variety of physiological tasks, including learning and memory in the central nervous system and emesis and peristaltic reflex in the enteric nervous system. 5-HT appears to be the dominant biogenic amine in all of the flatworm taxa examined, and it serves a variety of functions, most notably that of exitatory neurotransmission, inducing motility in muscles (Reuter & Halton, 2001).

The present study is the first attempt to study rotifer nervous system using ICC methods and confocal laser scanning microscopy (CLSM).

Materials and methods

Three species of Rotifera Monogononta have been investigated: *Platyias patulus* (Muller, 1786) (*Plationus patulus*, Segers et al., 1993), *Euchlanis dilatata* (Ehrenberg, 1832) and *Asplanchna herrickii;* (Guerne, 1888). *P. patulus* and *E.dilatata* are loricate rotifers feeding by filtration, while *A. herrickii* is an illoricate predatory rotifer. The material was collected in shallow pools at Borok (Yaroslavl District of Russia). The collected rotifers were fixed in Stefanini's fixative (2% paraformaldehyde and 15% picric acid in 0.1 M Na–phosphate buffer) at pH 7.6. They were stored for several days in fixative and rinsed for 24–48 h in 0.1 M Na–phosphate buffer (pH 7.6) containing 20% sucrose, then processed in 1.5 ml Eppendorf microfuge tubes. Prior to staining, the animals were immersed in phosphate-buffer saline containing 0.2% Triton X-100 (PBS-T). Non-specific antigens were blocked with 2% Bovine Serum Albumin (BSA) in PBS-T. Double-staining was carried out according to the indirect immunofluorescence method of Coons et al. (1955). Incubations were performed with a mixture of goat anti-5-HT (INCSTAR) antiserum and rabbit anti-FMRFamide (INCSTAR) antisera diluted 1:40 in PBS-T. The rotifers were incubated in 1.5 ml microfuge tubes on a shaker for 4–7 days at 10 °C. After incubation with the primary antibodies, the animals were rinsed 3 × 5 min in PBS-T and incubated for 1–2 h at room temperature with FITC-labelled swine-anti-goat (TACO) and swine anti-rabbit TRITC (DAKO) secondary antibodies (dilution 1:30), rinsed 3 × 5 min in PBS, mounted in 50% glycerol-PBS and stored in the dark at −20 °C. The controls for specificity included: (1) omitting the primary antibody, and (2) using non-immune serum.

The animals were examined with a confocal laser scanning microscope (CLSM) LEICA TCS 4D. Some specimens were observed from the dorsal side, and others from the ventral side. Usually 16–25 optical sections 0.8–2.5 μm thick were obtained while scanning through the specimen. The projection option was used to make reconstructions from all the optical sections in a series. The files obtained were processed with Adobe Photoshop 5.5 software.

Results

Platyias patulus

Pattern of FMRFamide immunoreactivity (Figs. 1a; 2a–c and e; Table 1)

The central part of the brain (the neuropile) comprises several FMRFamide-immunoreactive (IR) brain commissures about 1.6 μm thick (Figs. 1a; 2a and c). Two pairs of big (up to 10 μm) unipolar strongly IR neurons (na) lie in front of

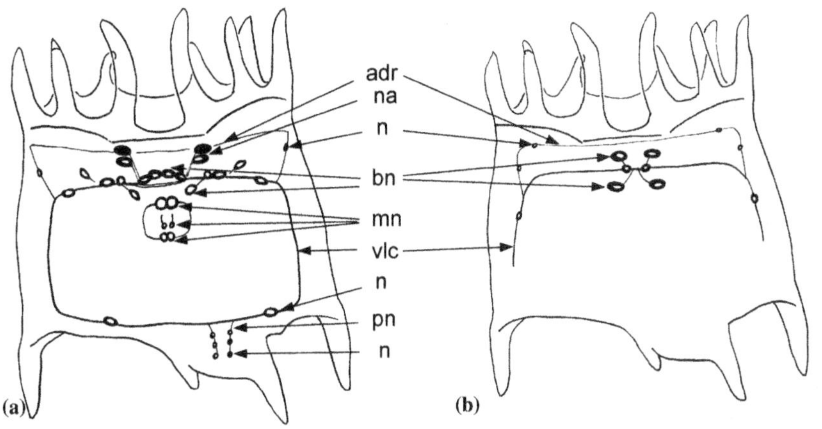

Figure 1. Platyias patulus, patterns of FMRFamide (a) and 5-HT immunoreactivity (b). Dorsal view. adr – anterior dorsal semi-ring; bn – brain neuron; mn – mastax neuron; n – neuron; na – neuron of antenna; pn – pedal nerve; vlc – ventro-lateral nerve cord.

the neuropile close to the proximal end of the anterior dorsal antenna (Fig. 2a–c; Table 1). On the dorsal side of the neuropile, six pairs of brain neurons (bn) are localized in two rows (Fig. 2b and c; Table 1). Two pairs of adjacent neurons (8 μm) of the anterior row lie dorsally to the middle part of the neuropile. Three pairs of smaller (5 μm) neurons flank the neuropile on lateral sides. Paired unipolar neurons of the posterior row send their processes to the smaller lateral brain neurons (Fig. 2c). Paired longitudinal nerve cords (vlc) extend from the brain towards the lateral sides of the body, turn in caudal direction, pass along the ventral side and join at the posterior end of the body (Fig. 2a and b). Two pairs of neurons (n), 6.6 μm in size, are associated with the proximal and distal parts of the ventro-lateral cords (Figs. 1a; 2e; Table 1). The ventro-lateral cords form anterior branches going to the lateral denticles of the lorica on the dorsal side. Joining each other, the anterior branches of the ventro-lateral cords form the anterior dorsal semi-ring (adr) associated with a pair of neurons 3.3 μm in size. These neurons lie in the area of the corona and possibly innervate the coronal tufts of cirri (Table 1). The mastax is surrounded by a ring of nerve fibres associated with four neurons (mn). Paired anterior mastax neurons reach 8 μm, while the posterior ones are smaller (6.5 μm). In the centre of the mastax lie two unipolar neurons (3.3 μm), with short processes oriented towards the brain (Fig. 2e; Table 1). A pair of foot nerves (pn) associated with three pairs of small (3.3 μm) neurons start from the longitudinal nerve cords in the posterior region of the trunk.

Pattern of 5-HT immunoreactivity (Figs. 1b; 2d and f; Table 1)

One thick 5-HT-IR transversal nerve commissure is revealed in the brain neuropile. Six strongly IR

Table 1. Number of neurons in the nervous system of the rotifers studied

	Platyias patulus		*Euchlanis dilatata*		*Asplanchna herricki*	
	FMRF amide	5-HT	FMRF amide	5-HT	FMRF amide	5-HT
Neuropile	12	6	4	4	14	4
Coronal tufts of cirri	2	4	2			
Dorsal neurons of antenna	4		2			
Nerve cords	4	2	4	4	2	4
Mastax	6		10	2	2	4
Foot	6		8			
Total	34	12	30	10	18	12

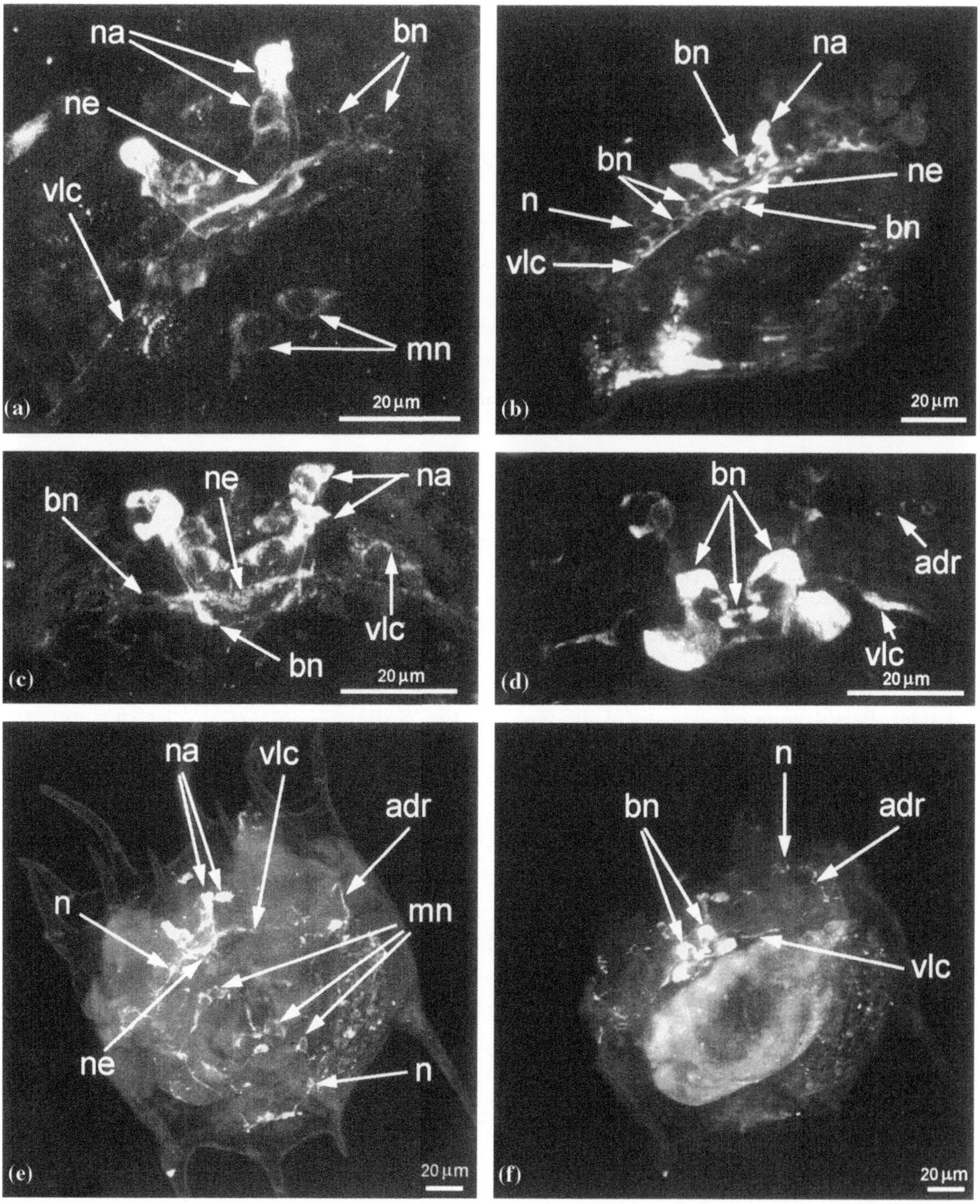

Figure 2. Platyias patulus, patterns of FMRFamide (a–c and e) and of 5-HT (d and f) immunoreactivity. Projections of CSLM optical sections. (a) FMRFamide-IR pattern in the brain with brain neurons (bn), neuropile (ne), ventro-lateral nerve cords (vlc) originating from it, two pairs of strongly IR anterior dorsal neurons of the antenna (na) and paired anterior mastax neurons (mn). (b) General FMRFamide-IR pattern. (c) FMRFamide-IR pattern in the brain of a double-stained specimen. (d) Corresponding 5-HT-IR pattern in the brain of the same specimen as (c). Note the absence of cross-reactivity of FMRFamide and 5-HT. (e) General FMRFamide-IR pattern in a double-stained specimen. (f) The corresponding general 5-HT-IR pattern in the same specimen as (e). Note the ventro-lateral nerve cords and anterior dorsal semi-ring (adr) with neurons (n).

brain neurons (bn) lie in three rows: anterior, middle and posterior, forming an X-shaped pattern (Figs. 1b; 2f; Table 1). In the middle of the brain, a pair of small multipolar neurons (5 μm in size) lies on the brain commissure (Fig. 2d). Two pairs of big (10 μm in size) unipolar neurons lie anterior and posterior to the commissure, being connected with the latter by their processes (Fig. 2d and f). The 5-HT-IR fibres of ventro-lateral cords (vlc) run along the FMRFamide-IR ones described above, but the 5-HT-IR fibres could be followed only till the mid-body. The only pair of neurons (5 μm) associated with the 5-HT-IR cords lies at a considerable distance from the brain (Fig. 1b). The anterior dorsal branches of the nerve cords form a semi-ring (adr), in the region of the corona. At least two pairs of small neurons (3 μm) are associated with the rings, possibly innervating the coronal tufts of cirri (Fig. 2f; Table 1).

Euchlanis dilatata

Pattern of FMRFamide immunoreactivity (Figs. 3a; 4a, c and d; Table 1)

Several IR nerve fibres (0.8–1 μm thick) associated with two pairs of neurons (bn) are present in the arch-shaped brain (Fig. 2a; Table 1), A pair of strongly IR neurons (6.6 μm in size) and a pair of weaker stained but larger (8 μm) neurons lie in the middle of the brain. Paired nerve fibres start from the lateral sides of the neuropile and run anteriorly towards the lateral sides of the corona. Each fibre ends with a strongly IR multipolar neuron (nct) probably innervating the coronal tufts of cirri (Figs. 3a; 4a; Table 1). The neurons (8.3 μm in size) are spindle-shaped and have two short lateral dendrites. Paired unipolar neurons (na), 3.3 μm in size, with laterally oriented processes lie in front of the strongly IR brain neurons close to the base of the dorsal antenna (Figs. 3a; 4a; Table 1). From the larger brain neurons paired ventro-lateral cords (vlc) run caudally and reach the posterior end of the body (Figs. 3a; 4d). Two pairs of neurons (10 μm in size) are associated with the cords (Fig. 4a; Table 1). Few plexus fibres occur between the cords. The mastax is surrounded by a net of nerve fibres associated with about a dozen of small (5 μm) neurons (mn) (Figs. 3a; 4c). Paired pedal nerves (pn) run from the longitudinal cords to the foot. The pedal nerves are associated with four pairs of neurons: the pair of larger (5 μm) neurons lies at the base of the foot (Fig. 4d; Table 1), while another three pairs of smaller (2 μm) neurons are evenly distributed along the foot.

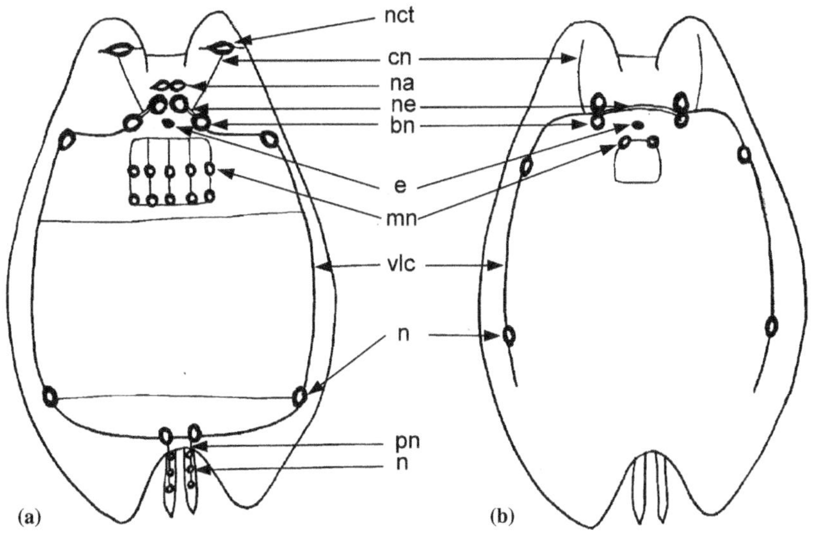

Figure 3. Euchlanis dilatata, patterns of FMRFamide (a) and 5-HT immunoreactivity (b) Dorsal view. bn – brain neuron; cn – coronal nerve ; e – eye; mn – mastax neuron; n – neuron; na – neuron of antenna; nct – neuron of the coronal tuft; ne – neuropile; pn – pedal nerve; vlc – ventro-lateral nerve cord.

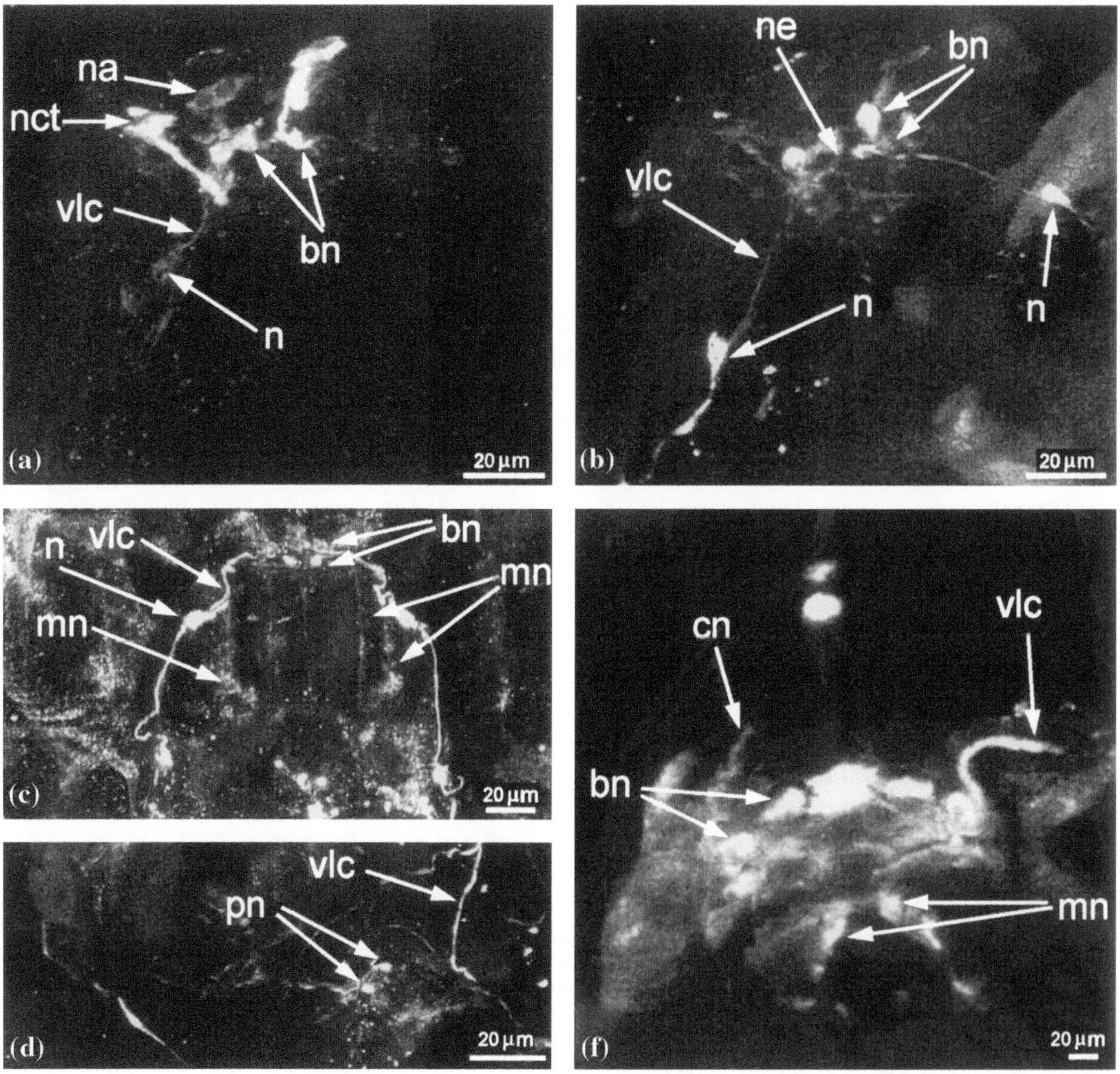

Figure 4. Euchlanis dilatata, patterns of FMRFamide (a, c and d) and of 5-HT (b and e) immunoreactivity. Projections of CSLM optical sections. (a) FMRFamide-IR pattern. Note the brain neurons (bn), paired neurons of the anterior dorsal antenna (na), strongly IR neurons of the coronal tufts (nct); ventro-lateral nerve cords (vlc) associated with neurons (n). (b) Corresponding 5-HT-IR pattern in the same double-stained specimen. Note the neuropile (ne) with brain neurons (bn) and ventro-lateral nerve cords (vlc) with proximal neurons, (c) FMRFamide-IR pattern in the mastax region, ventral view. Note brain neurons (bn), ventro-lateral cords (vlc) with proximal and distal neurons (n) and mastax neurons (mn). (d) FMRFamide-IR pattern in the posterior part of the body, dorsal view, showing the pedal nerves (pn). (e) General 5-HT-IR pattern in the anterior part of the body, dorsal view, showing brain neurons (bn), coronal nerve (on), ventro-lateral cords (vlc) and mastax neurons (mn).

Pattern of 5-HT immunoreactivity (Figs. 3b; 4b and e; Table 1)

Two pairs of IR neurons (bn), anterior and posterior, lie laterally from the neuropile (ne) of the brain. Anterior neurons are 8.3 μm in size, but the posterior ones are smaller (5 μm) (Fig. 4e; Table 1). Ventro-lateral cords (vlc) can be followed till 2/3 of the body length and are associated with two pairs of neurons 6.6 μm in size (Fig. 4b; Table 1). Short nerves (cn) starting from the lateral sides of the neuropile innervate the coronal tufts of cirri. A pair of neurons (mn),

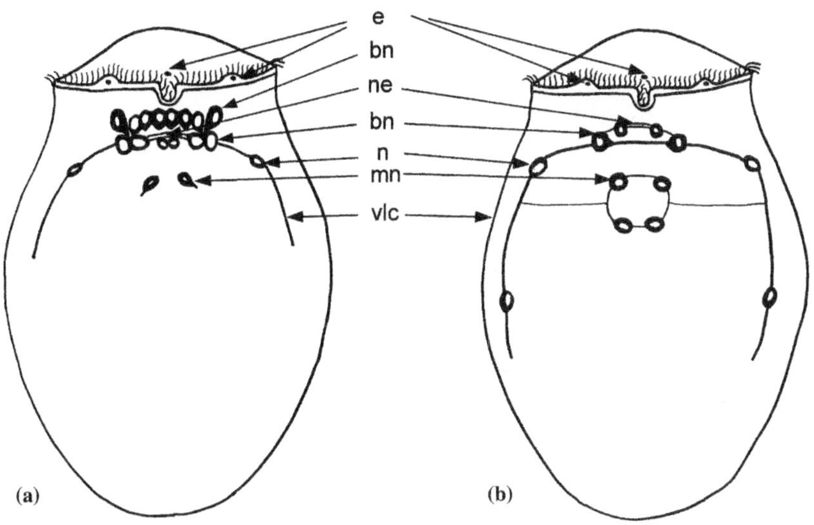

Figure 5. Asplanchna herrickii, patterns of FMRFamide (a) and 5-HT immunoreactivity (b). Dorsal view, bn – brain neuron; e – eye; mn – mastax neuron; n – neuron; ne – neuropile; vlc – ventro-lateral nerve cord.

5 μm in size, lies near the proximal end of the mastax (Fig. 4e; Table 1). Their processes form a ring around the mastax.

Asplanchna herrickii

Pattern of FMRFamide immunoreactivity (Figs. 5a; 6a and b; Table 1)

FMRFamide-IR elements in the brain of *A. herrickii* form a ring-shaped pattern composed of 14 neurons (bn) (Fig. 5a; Table 1). Six of the neurons (size 5.5 μm) are hexagonal in shape (Fig. 6b). They form the ventral side of the brain ring and lie closer to the anterior end of the animal than the other eight neurons forming the dorsal side of the ring. Paired large (8 μm), strongly IR drop-shaped neurons lie in dorso-lateral regions of the brain (Fig. 6a and b). The large neurons are flanked by two pairs of smaller (4.5 μm) spherical neurons. A pair of small (2 μm) weakly stained neurons lies immediately behind the median part of the neuropile (ne) (Fig. 5a). Paired ventro-lateral cords (vlc) run caudally and can be followed till one-third of the body. The proximal regions of the cords are associated with paired neurons (2.6 μm) (Fig. 6a). A pair of unipolar neurons (mn), 7 μm in size, with short processes is present in the mastax region.

Pattern of 5-HT immunoreactivity (Figs. 5b; 6c and d; Table 1)

The arch-shaped neuropile (ne) of the brain comprises few 5-HT-IR fibres (Fig. 5b). Of the two pairs of immunoreactive neurons (bn) present in the brain (Fig. 6c and d; Table 1), the median one is smaller (3.5 μm) and shows stronger immunoreactivity than the outer pair of larger (5 μm) neurons (Fig. 6d). Paired ventro-lateral longitudinal cords (vlc) that can be followed till mid-body are associated with two pairs of neurons: proximal (6.6 μm) and distal (8 μm). Two pairs of neurons (mn), 7 μm in size, are present in the mastax region (Fig. 5b; Table 1).

Discussion

The first attempt to use ICC methods to study rotifer nervous system has revealed 5-HT-IR and FMRFamide-IR elements in all parts of nervous system of the three species investigated. The total numbers of 5-HT-IR and FMRFamide-IR neurons are low (10–34), but seem to be always constant for each species. There are evidently too few data available for discussing about a possible correlationship between the number of neurons and the systematic position or the way of life of the rotifers studied.

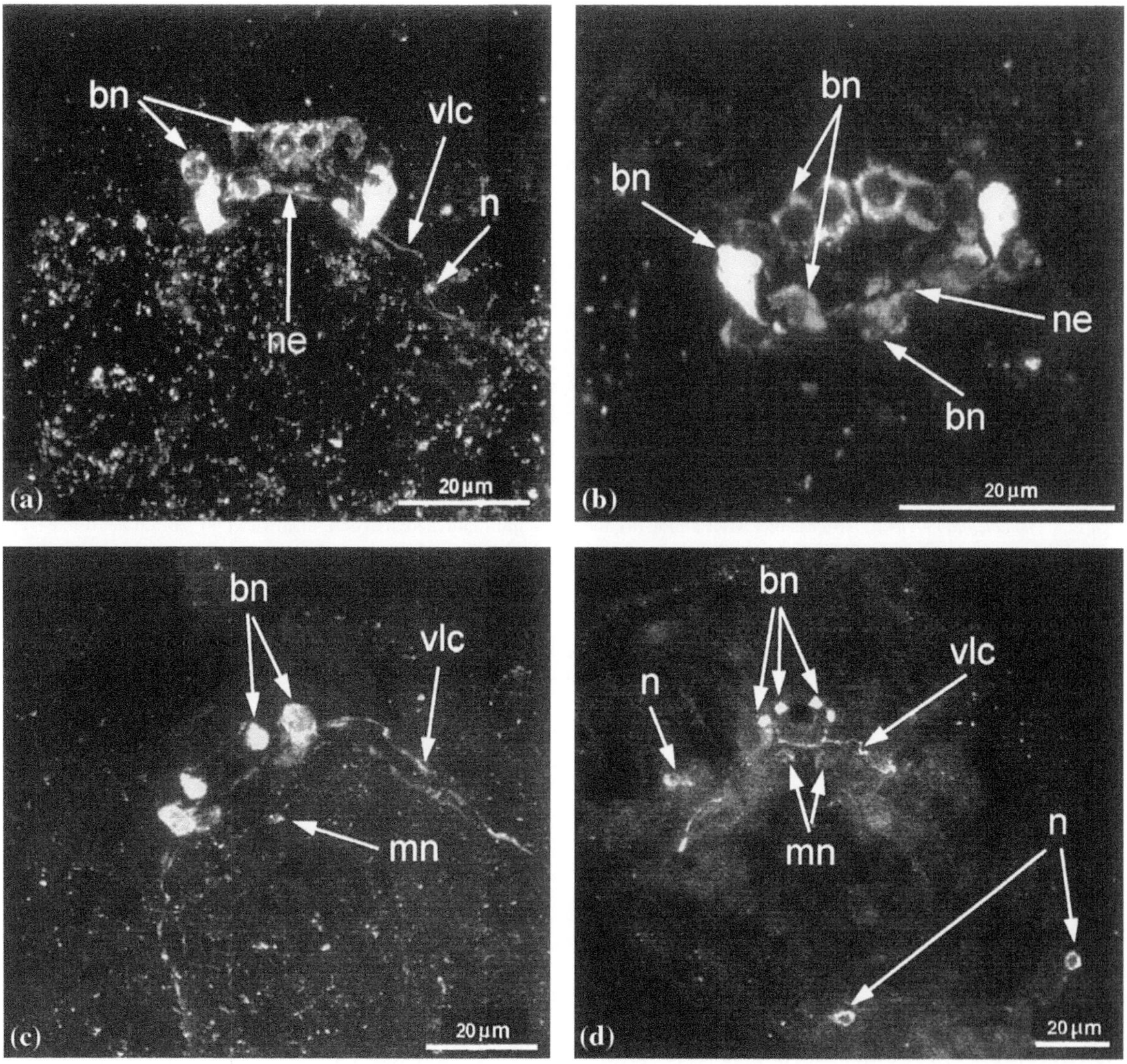

Figure 6. Asplanchna herrickii, patterns of FMRFamide (a and b) and of 5-HT (b and e) immunoreactivity. Projections of CSLM optical sections. (a and b) General FMRFamide-IR pattern in the brain of two different specimens. Note that the brain neurons (bn) are arranged in a ring, neuropile (ne), ventro-lateral cords (vlc) with neurons (n). (c and d) General 5-HT-IR pattern in two different specimens with brain neurons (bn), ventro-lateral cords (vlc) with proximal and distal neurons (n) and mastax neurons (mn).

In all parts of the nervous system described above, the presence of CAE was recorded formerly (Kotikova, 1995, 1997, 1998). In rotifers, brain neurons constitute 20% of the total number of cells in the body (Martini, 1912). The brain neurons of *Asplanchna brightwellii* number 200 cells, demonstrating a perfect bilateral symmetry (Ware & Lopresti, 1975). The total number of CAE, 5-HT-IR and FMRFamide-IR brain neurons reaches 8–11.5% of the total number of brain cells.

The present study has revealed the precise pattern of IR neurons in the brain of the three species studied. The brain neurons are either evenly distributed (*P. patulus*, *A. herrickii* and *E. dilatata*, FMRFamide-IR) or concentrated in the lateral regions of the brain (*E. dilatata*, 5-HT-IR brain neurons). Usually all the brain neurons occur on the dorsal side of the body only, forming X-shaped or arch-shaped patterns. The exception is the ring-shaped pattern of FMRFamide-IR elements in the brain *of A. herrickii*, where 6 of the

14 brain neurons lie at the ventral side. A ring-shaped brain pattern of CA-ergic neurons occurs also in *Dicranophorus forcipatus* and *Brachionus quadridentatus* (Kotikova, 1998). These rotifers are systematically quite distant from each other and from *A. herrickii*. They differ in their feeding behaviour and way of locomotion. Therefore, it is likely that the ring-shaped brain pattern has evolved in parallel in these three species.

The presence of paired strongly stained FMRFamide-IR neurons, lying anteriorly to the neuropile at the lateral sides of the brain, in all the three species studied is an interesting observation. In *Platyias patulus*, these neurons seem to innervate the antenna (na in Fig. 1), in *Euchlanis dilatata*, they possibly innervate the coronal tufts of cirri (nct in Fig. 3), and in *Asplanchna herrickii*, similar but smaller neurons constitute a part of the brain ring itself (bn in Fig. 5). The strong staining by anti-FMRFamide antibody and the similar form of the cells suggest a possible homology. However, a large number of species needs to be investigated for a definite conclusion.

One or two pairs of neurons occur along longitudinal ventro-lateral cords originating from the brain. The mastax region is characterized by a considerable diversity in both the number of neurons and their distribution pattern. In the corona, isolated neurons and nerves as well as a dorsal anterior semicircle with 1 or 2 pairs of neurons have been found. These structures most probably innervate the coronal tufts of cirri. In the regions of the dorsal antenna and in the foot only FMRFamide-IR neurons have been observed.

Double immunostaining with both 5-HT and FMRFamide shows no co-localization of these neuronal signal substances in any of the cells or fibres. In the single pair of longitudinal cords, elements with different immunoreactivity run parallel to each other. FMRFamide-IR neurons and fibres mostly lie beneath the 5-HT-IR ones. The absence of co-localization of 5-HT and FMRFamide was also observed in several groups of early bilaterians: acoels (Reuter et al., 2001a, b; Raikova et al., 2004), nemertodermatids (Raikova et al., 2000a), enigmatic xenoturbellids (Raikova et al., 2000b), as well as in some Rhabditophoran Platyhelminthes (Kotikova et al., 2002).

The simple structure of the nervous system of rotifers (Remane, 1929–1933) and the constancy of its cellular composition (Martini, 1912; Ware & Lopresti, 1975) should allow more detailed studies on other rotifer species in order to localize other neuronal signal substances. A detailed knowledge of pattern of localization of neurons of different immunoreactivity will provide a new insight into rotifer neuroanatomy.

Acknowledgements

Financial support was received from the Russian Basic Research Foundation grant No.: 02-04-48583. We thank JSC "Kinex Sankt-Petersburg" for financing the participation of the authors in the Rotifer Symposium X at Illmitz, Austria. We thank the Institute for Biology of Inland Water RAS (Borok) for the opportunity afforded to collect and study the material, the Research Institute of the Åbo Akademi University Foundation, Svenska Kulturfonden, Magnus Ehrnrooths foundation for grants to M. Gustafsson. The authors are also deeply grateful to Professor L.A. Kutikova for advice and helpful comments on this manuscript.

References

Aarnisalo, A. A. & P. Panula, 1995. Neuropeptide FF containing efferent projections from the medial hypothalamus of rat: a *Phaseolus vulgaris* leucoagglutinin study. Neuroscience 65: 175–192.

Coons, A. H., E. H. Leduc & J. M. Conolly, 1955. Studies on antibody production. 1. A method of the histochemical demonstration of specific antibody and its application to a study of the hyperimmune rabbit. Journal of Experimental Medicine 102: 49–60.

Clément, P., 1977. Ultrastructural research on rotifers. Archiv für Hydrobiologie 8: 270–297.

Clément, P., 1980. Phylogenetic relationship of rotifers, as derived from photoreceptor morphology and other ultrastructural analyses. Hydrobiologia 73: 93–117.

Geary, T. G., N. J. Marks, A. G. Maule, J. W. Bowman, S. J. Alexander-Bowman, T. A. Day, M. J. Larsen, T. M. Kubiak, J. P. Davis & D. P. Thompson, 1999. Pharmacology of FMRFamide-related peptides in helminths. Annals of the New York Academy of Sciences 897: 212–227.

Grimmelikhuijzen, C. J. & J. A. Westfall, 1995. The nervous systems of cnidarians. Experientia Supplement 72: 7–24.

Keshmirian, J. & T. Nogrady, 1987. Histofluorescent labelling of catecholaminergic structures in rotifers (Aschelminthes) in whole animals. Histochemistry 87: 351–357.

Keshmirian, J. & T. Nogrady, 1988. Histofluorescent labelling of catecholaminergic structures in rotifers (Aschelminthes). II Males of *Brachionus plicatilis* and structures from sectioned females. Histochemistry 89: 189–192.

Kotikova, E. A., 1994. Distribution of catecholamines in the nervous system of Bdelloida (Rotifera). In Sakharov, D. A. (ed.), Simple Nervous Systems. ISIN, Pushcino: 21–22.

Kotikova, E. A., 1995. Localization and neuroanatomy of catecholaminergic neurons in some rotifer species. Hydrobiologia 313/314: 123–127.

Kotikova, E. A., 1997. Localization of catecholamines in the nervous system of the order Transversiramida. Doklady Rossiiskoi Akademii Nauk (in Russian) 353: 841–843.

Kotikova, E. A., 1998. Catecholaminergic neurons in the brain of rotifers. Hydrobiologia 387/388: 135–440.

Kotikova, E. A., O. I. Raikova, M. Reuter & M. K. S. Gustafsson, 2002. The nervous and muscular systems in the free-living flatworm *Castrella truncata* (*Rhabdocoela*): an immunocytochemical and phalloidin fluorescence study. Tissue and Cell 34: 365–374.

Martini, E., 1912. Studien über die Konstanz histologischer Elemente. III. *Hydatina senta*. Zeitschrift für wissenschaftliche Zoologie 102: 425–645.

Nachtwey, R., 1925. Untersuchungen über die Keimbahn Organogenese und Anatomie von *Asplanchna priodonta* Gosse. Zeitschrift für wissenschaftliche Zoologie 126: 239–492.

Nogrady, T. & M. Alai, 1983. Cholinergic neurotransmission in rotifers. Hydrobiologia 104: 149–153.

Price, D. A. & M. A. J. Greenberg, 1977. Structure of a molluscan cardioexcitatory neuropeptide. Science 197: 670–671.

Raikova, O. I., M. Reuter, U. Jondelius & M. K. S. Gustafsson, 2000a. The brain of Nemertodermatida as revealed by anti-5-HT and anti-FMRFamide immunostainings. Tissue and Cell 32: 358–365.

Raikova, O. I., M. Reuter, U. Jondelius & M. K. S. Gustafsson, 2000b. An immunocytochemical and ultrastructural study of the nervous and muscular systems *of Xenoturbella westbladi* (Bilateria inc.sed). Zoomorphology 120: 107–118.

Raikova, O. I., M. Reuter, M. K. S. Gustafsson, A. G. Maule, D. W. Halton & U. Jondelius, 2004. Evolution of the nervous system in *Paraphanostoma* (Acoela). Zoologica Scripta 33: 71–88.

Raineri, M., 1984. Histochemical investigations of Rotifera Bdelloida. I. Localization of cholinesterase activity. Histochemical Journal 16: 601–616.

Remane, A., 1929–1933. Rotatorien. In Bronn, H. G. (ed.), Klassen and Ordnungen der Tierreichs, IV. Winter, Leipzig: 1–576.

Reuter, M. & D. W. Halton, 2001. Comparative neurobiology of Platyhelminthes. In Littlewood, D. T. J. & R. A. Bray (eds), Interrelationships of the Platyhelminthes. The Systematics Association Special, Vol. 60. Taylor and Francis, London & New York: 239–249.

Reuter, M., O. I. Raikova & M. K. S. Gustafsson, 2001a. Patterns in the nervous and muscle systems in the lower flatworms. Belgian Journal of Zoology 131: 47–53.

Reuter, M., O. I. Raikova, U. Jondelius, M. K. S. Gustafsson, D. W. Halton, A. G. Maule & C. Shaw, 2001b. Organisation of the nervous system in the Acoela: an immunocytochemical study. Tissue & Cell 33: 119–128.

Schneider, L. E. & P. H. Taghert, 1988. Isolation and characterisation of a *Drosophila* gene that encodes multiple neuropeptides related to Phe–Met–Arg–Phe–NH_2 (FMRFamide). Proceedings of the National Academy of Sciences USA 85: 1993–1997.

Segers, H., G. Murugan & H. J. Dumont, 1993. On the taxonomy of the Brachionidae: description of *Plationus* n. gen. (Rotifera, Monogononta). Hydrobiologia 268: 1–8.

Villeneuve, J. & P. Clément, 1971. Le neuropile du cerveau de Rotifère: observations ultrastructurales préliminaires. Journal de Microscopie Français 11: 108.

Ware, R. W. & V. Lopresti, 1975. Three-dimensional reconstruction from serial sections. International Review of Cytology 40: 325–440.

Wurdak, E. S., P. Clément & J. Amsellem, 1983. Sensory receptors involved in the feeding behaviour of the rotifer *Asplanchna brightwelli*. Hydrobiologia 104: 203–212.

Hydrobiologia (2005) 546:249–253
A. Herzig, R.D. Gulati, C.D. Jersabek & L. May (eds.) Rotifera X: Rotifer Research: Trends, New Tools and Recent Advances
DOI 10.1007/s10750-005-4204-4

Identification of acetylcholinesterase receptors in Rotifera

Arikitza Pineda-Rosas, G.E. Santos-Medrano, M.F. Zavala-Reynoso & R. Rico-Martínez*
Departamento de Química, Universidad Autónoma de Aguascalientes, Centro Básico, Avenida Universidad 940, 20100, Aguascalientes, Ags, C.P., México
*(*Author for correspondence: E-mail: rrico@correo.uaa.mx)*

Key words: α- bungarotoxin, β- bungarotoxin, acetylcholinesterase, cholinergic system, muscarinic receptors, nicotinic receptors, nervous system

Abstract

We have identified acetylcholinesterase (AChE) receptors in six freshwater rotifers. Using β-bungarotoxin labelled with fluoresceinisothiocyanate (FITC), muscarinic and nicotinic receptors were found in *Brachionus quadridentatus* (females and males), *Lecane luna, Lecane quadridentata, Plationus patulus*, and *Rotaria neptunia.* Using α-bungarotoxin-FITC, nicotinic receptors were identified in *B. quadridentatus, Lecane bulla, L. luna, L. quadridentata, P. patulus* and *R. neptunia*. Concentrations as low as 1.5 nM of β-bungarotoxin, and 5 nM of α-bungarotoxin identified receptors in the digestive tract. Higher concentrations of both toxins identified additional receptors associated with the lorica. A preliminary analysis of fluorescence intensity in *L. quadridentata* showed that response to α-bungarotoxin increases with age from newborn to 48-h old, but not in older individuals, thus suggesting an increase in binding sites, and possibly in number of nicotinic receptors, during the first 48-h of life. Our study extends the number of rotifer species in which AChE receptors have been reported.

Introduction

Although much is known about rotifer anatomy and behavior, relatively little is understood about their neurophysiology and what we do know comes from the work of only a few investigators (Nogrady et al., 1993). For example, Kotikova (1995, 1998) and Keshmirian & Nogrady (1987, 1988) studied catecholaminergic systems, while Nogrady & Alai (1983) investigated acetylcholinesterase (AChE). Raineri (1984) reported the presence of acetyl- and butyryl-cholinesterase activities in secretory and gonad cells of four species of bdelloids. Nogrady & Keshmirian (1986a) also investigated the effect of acetylcholine on egg retention in *Philodina acuticornis* and in a companion paper (Nogrady & Keshmirian 1986b) reported that acetylcholine had a synergistic effect increasing local anesthesia in *Brachionus calyciflorus.* Here we report our efforts to map AChE receptors in six monogonont rotifers: *Brachionus quadridentatus* (females and males), *Lecane bulla, Lecane luna, Lecane quadridentata, Plationus patulus*, and *Rotaria neptunia.* To do this we used two powerful toxins (α- and β-bungarotoxin), which are antagonists of AChE activity. Thus, our study expands the number of rotifers for which AChE receptors are known and it provides comparative information regarding the differential response to rotifers to α- and β-bungarotoxins.

Materials and methods

Collection and culture of freshwater rotifers

Rotifers were collected from several ponds around the city of Aguascalientes (Mexico). The samples were collected with a Wisconsin-type planktonic net (53 μm mesh), transported back to the laboratory as soon as possible, and identified using the keys of Koste (1978), Segers (1995) and Stemberger (1979). Rotifers were cultured in EPA medium (US Environmental Protection Agency, 1985) prepared with deionized water (16–18 megaΩ) obtained through a Water Pro PS Purification System (Labconco Co., USA). Animals were fed with 1×10^5 cells ml^{-1} of the microalgae *Nannochloropsis oculata* (UTEX LB 2164), acclimated to freshwater medium and grown in Bold's Basal Medium (Nichols, 1973). Rotifer cultures were kept in a bioclimatic chamber (REVCO Co, USA) at a temperature of 25 ± 2 °C, under fluorescent lamps at 600–1100 lux.

Choice of bungarotoxins

The bungarotoxins used in this study are found in venom of the snake *Bungarus multicinctus*. α-bungarotoxin (hereafter α-B) is a 74-amino acid polypeptide that blocks neuromuscular junctions by binding to each of the α-subunits of the postsynaptic nicotinic AChE receptors (Chiapinelli et al., 1981). Although it can bind to some of muscarinic AChE receptors, binding of α-B is mostly restricted to nicotinic receptors and has been used widely to map nicotinic receptors in several organisms (Stiles, 1993; Barrantes et al., 1995; Horch & Sargent, 1995). In contrast, β-bungarotoxin (hereafter β-B) is a two-chain phospholipase A2 neurotoxin that blocks voltage-gated K^+ channels of presynaptic membranes of motor nerve terminals. The A subunit has 120 amino acids and the B subunit has 60 amino acids giving a combined molecular weight of 21,800 Da (Rowan, 2001). Blockage of muscarinic receptors by β-B has been reported by Ochillo et al. (1981).

Assay for AChE receptors

We transferred up to 30 neonates (≤24-h old) of each species into a 3 ml well of a 24-well plate and exposed them to either α-B or β-B labeled with Fluoresceinisothiocyanate (FITC) (Sigma Co., USA) according to the following program. Species exposed to α-B were *Brachionus quadridentatus* (females and males), *Lecane bulla*, *Lecane quadridentata* (individuals of 6, 24, 48 and 72-h old), *Lecane luna*, *Plationus patulus*, and *Rotaria neptunia*. Species exposed to β-B were *Brachionus quadridentatus* (females and males), *Lecane luna*, *Lecane quadridentata*, *Plationus patulus*, and *Rotaria neptunia*. The concentrations used for α-B ranged from 1.5 to 150 nM; those for β-B ranged from 5 to 250 nM. Typical exposure times ranged from 5–180 min.

Results

Rotifers of all six species examined showed a positive reaction to at least one of the AChE binding toxins within 30 min. We examined these reactions more closely by exposing neonates (≤3-h old) of *L. quadridentata*, *L. luna*, and *L. bulla* for 30 min to various concentrations of the labels (Table 1). Results of experiments that examined the differential response of neonates of *L. quadridentata* to

Table 1. Responses of AChE receptors in rotifers: concentrations of α- and β-B producing observable reactions[a]

Lecane quadridentata				*Lecane luna*		*Lecane bulla*	
α-B		β-B		α-B		α-B	
[nM]	Reaction	[nM]	Reaction	[nM]	Reaction	[nM]	Reaction
0	A	0	A	0	A	0	A
5	D	1.5	A	5	D	5	A
10	B	3	A	10	B	10	A
50	B	15	A	50	B	50	B
100	B	30	D	100	B	100	B
250	B	75	B	250	B	250	B
–	–	150	B	–	–	–	–

[a]A = absent; D = digestive track; B = all over the body. In each experiment, 20–24-h old neonates were used.

Table 2. Results of experiments to determine the differential response to different concentrations of α-B among 24-h neonates of *Lecane quadridentata*

Concentration of α-B (nM)	Mean	SD	N
Control	0	0	20
10	3.66	2.56	8
50	6.23	5.51	15
100	6.02	3.91	15

Table 3. Results of experiments to determine the differential response between adults (48–72 h) and neonates (0–24 h) of *Lecane quadridentata* at 100 nM α-B

Age groups	Mean	SD	N
Control	0	0	20
Neonates (6–20 h-old)	6.02	5.245	15
Adults (48–72 h-old)	12.34	3.908	16

the concentration of α-B are reported in Table 2, while the response of neonates vs. adults are reported in Table 3.

Reaction to α-bungarotoxin

There was a differential response to exposure to α-B depending on the species (Table 1). *Lecane luna* and *L. quadridentata* were the most sensitive species, reacting after 30-min exposure at concentrations as low as 5 nM. For example, in *L. quadridentata* and *L. luna* exposure to 5 nM of α-B for 30 min result in an observable reaction in the digestive tract, while at this concentration no reaction was observed in *L. bulla*. However, when exposure to a concentration 100 nM for 1-h, fluorescence was observed all over the body of *L. quadridentata*. This pattern is documented in Figure 1. *Bracbionuis quadridentatus* (females and males), *P. patulus* and *R. neptunia* all required a concentration of 50 nM or greater initiate an observable reaction of fluorescence. In all species and all individuals, AChE receptors were found all over the body with particularly high densities

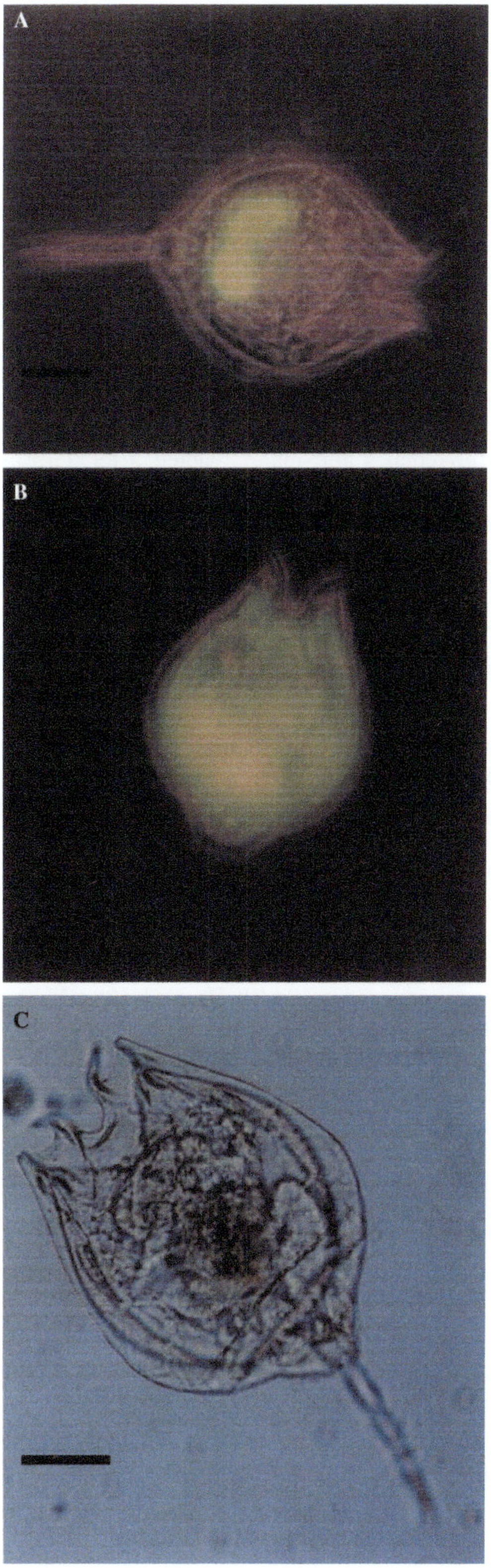

Figure 1. Photomicrographs illustrating the reactions of *Lecane quadridentata* exposed to α-bungarotoxin, a) 5 nM for 30 min; b) 100nM for 1-h exposure; c) Control; (all scale bars = 50 μm).

found in the gonad and intestinal tract region. In fact, in *L. luna* and *L. quadridentata* exposed to 5 nM, only the gonads and intestinal tract region reacted with the α-B (compare Fig. 1a and c).

Reaction to β-bungarotoxin

Differential responses in relation to exposure time and concentrations were also recorded for exposure to β-B. *L. quadridentata* showed an observable reaction after 30-min exposure at 30 nM (Table 1). All of the other species (including *L. luna)* required a concentration of 100 nM to react: *P. patulus* reacted after 15-min; *B. quadridentatus* females and males reacted after 25-min; *L. luna* and *R. neptunia* reacted after 30-min. When *L. quadridentata* were exposed to concentrations of 5, 10 and 50 nM at 25 min exposure time, reactivity was observed in the region of the digestive tract and the gonads. When exposed to 100 nM or higher there was reactivity all over the body. The pattern of AChE receptors in the males of *B. quadridentata* was quite interesting: fluorescence was concentrated around the corona and penis, and along the body.

Kodak 2-D Image Analysis Software was used to measure the mean fluorescence intensity inside the body of *L. quadridentata*. In this analysis the mean fluorescence intensity is an indirect measure of the number of AChE receptors. Little differences were seen when 24-h neonate *L. quadridentata* were exposed to 0, 10, 50, and 100 nM α-B (Table 2). However, additional results showed that 48-h and 72-h old *L. quadridentata* adults have a larger amount of fluorescence, and therefore of AChE receptors (mean = 12.34), than neonates of 6-h and 20-h old (mean = 6.02; Table 3; df = 6; $p < 0.05$). No significant differences were found between 24-h and 72-h adults.

Discussion

Nogrady & Alai (1983) reported the presence of AChE receptors in eleven species of freshwater rotifers: *Asplanchna brightwelli*, *Asplanchna priodonta*, *Conochilus natans*, *Euchlanis dilatata*, *Euchlanis incisa*, *Keratella earlinae*, *Lepadella* sp., *Polyarthra euryptera*, and *Synchaeta oblonga*. Here, we provide information on six additional species (*Brachionus quadridentatus*, *Lecane bulla*, *L. quadridentata*, *L. luna*, *Plationus patulus*, and *Rotaria neptunia*).

Concentrations as low as 1.5 nM of β-B and 5 nM of α-B identified the receptors located mainly in the area of the digestive tract. Higher concentrations of both toxins identified further receptors along the lorica. An analysis of fluorescent intensity in *Lecane quadridentata* showed that the response to α-B increased with age from newborn up to 48-h old. This suggested an increase in the number of binding sites and possibly in the number of nicotinic receptors over this period. The responses between α- and β-B among the species studied in this work did not allow us to separate nicotinic from muscarinid AChE receptors.

Acknowledgments

The authors acknowledge with great appreciation the help of Linda May and three anonymous reviews.

References

Barrantes, G. E., A. T. Rogers, J. Lindstrom & S. Wonnacott, 1995. Alpha-Bungarotoxin binding sites in rat hippocampal and cortical cultures: initial characterization, co-localization with alpha 7 subunits and up-regulation by chronic nicotine treatment. Brain Research 672: 228–236.

Chiapinelli, V. A., J. B. Cohen & R. E. Zigmond, 1981. The effect of alpha-neurotoxins from the venoms of various snakes on transmission in autonomic ganglia. Brain Research 211: 107–126.

Horch, H. L. & P. B. Sargent, 1995. Perisynaptic surface distribution of multiple classes of nicotinic acetylcholine receptors on neurons in the chicken ciliary ganglion. Journal of Neuroscience 15: 7778–7795.

Nogrady, T., 1987. Histofluorescent labelling of catecholaminergic structures in rotifer (Aschelminthes) in whole animals. Histochemistry 87: 351–357.

Nogrady, T., 1988. Histofluorescent labelling of catecholaminergic structures in rotifer (Aschelminthes). II, Males of *Brachionus plicatilis* and structures from sectioned females. Histochemistry 89: 189–192.

Koste, W., 1978. Rotatoria, Die Rädertiere Mittleleuropas. Überordnung Monogononta. Begründet von Max Voigt. Gebruder Borntraeger, Berlin, Stuttgart, Textband 673 pp, Tafelband, 234 Tafeln, 673 pp.

Kotikova, E. A., 1995. Localization and neuroanatomy of catecholaminergic neurons in some rotifer species. Hydrobiologia 313/314: 123–127.

Kotikova, E. A., 1998. Catecholaminergic neurons in the brain of rotifers. Hydrobiologia 387/388: 135–140.

Nichols, H. W., 1973. Growth Media-Freshwater. In J. R. Stein (ed.), Handbook of phycological methods. Cambridge University Press: 7–24.

Nogrady, T. & M. Alai, 1983. Cholinergic neurotransmission in rotifers. Hydrobiologia 104: 149–153.

Nogrady, T. & J. Keshmirian, 1986a. Rotifer neuropharmacology I. Cholinergic drug effects on oviposition of *Philodina acuticornis* (Rotifera, Aschelminthes). Comparative Biochemistry & Physiology C 83: 335–338.

Nogrady, T. & J. Keshmirian, 1986b. Rotifer neuropharmacology. II Synergistic effects on local anesthetic activity in *Brachionus calyciflorus* (Rotifera, Aschelminthes). Comparative Biochemistry & Physiology C 83: 339–344.

Nogrady, T., R. L. Wallace & T. W. Snell, 1993. Rotifera 1. Biology, Ecology and Systematics. Guides to the Identification of the Microinvertebrates of the Continental Waters of the WorldVolume 4, SPB Academic Publishing, New York, 142 pp.

Ochillo, R. F., C. S. Tsai & M. H. Tsai, 1981. Mechanism of action of muscarine on the longitudinal muscle of the guinea-pig isolated ileum. British Journal of Pharmacology 72: 225–232.

Raineri, M., 1984. Histochemical investigations of Rotifera Bdelloidea. I. Localization of cholinesterase activity. Histochemical Journal 16: 601–616.

Rowan, E. G., 2001. What does beta-bungarotoxin do at the neuromuscular junction? Toxicon 39: 107–118.

Segers, H., 1995. Rotifera 2. The Lecanidae (Monogononta). Guides to the Identification of the Microinvertebrates of the Continental Waters of the World. Volume 6 SPB Academic Publishing, The Hague, The Netherlands, 226 pp.

Stemberger, R. S., 1979. A Guide to Rotifers of the Laurentian Great Lakes. Environmental Protection Agency, EPA-600/4-79-021, Cincinnati, Ohio, 186 pp.

Stiles, B. G., 1993. Acetylcholine receptor binding characteristics of snake and cone snail venom postsynaptic neurotoxins: further studies with a non-radioactive assay. Toxicon, 825–834.

Part V
Mating, Resting Eggs, Diapause, Anhydrobiosis, Embryonic Development

Hydrobiologia (2005) 546:257–265
A. Herzig, R.D. Gulati, C.D. Jersabek & L. May (eds.) Rotifera X: Rotifer Research: Trends, New Tools and Recent Advances
DOI 10.1007/s10750-005-4205-3

Brachionus calyciflorus is a species complex: Mating behavior and genetic differentiation among four geographically isolated strains

John J. Gilbert[1] & Elizabeth J. Walsh[2,*]
[1]*Department of Biological Sciences, Dartmouth College, Hanover, New Hampshire, 03755, U.S.A.*
[2]*Department of Biological Sciences, University of Texas at El Paso, 500 West University Avenue, El Paso, Texas, 79968, U.S.A.*
(**Author for correspondence: E-mail: ewalsh@utep.edu*)

Key words: cryptic species, genetic distance, mating behavior, reproductive isolation, rotifers

Abstract

Four geographic strains of *B. calyciflorus* are investigated regarding their genetic similarity and ability to cross-mate. DNA sequence analysis of the mitochondrial *cox*1 gene (694 bp) and the nuclear ribosomal ITS region (735 bp) showed that the Florida and Georgia strains were very similar to each other (0.3% sequence divergence for the 1429 bp) and different from the Texas and Australia strains (~7% and 9% sequence divergence for the 1429 bp, respectively). Consistent with this genetic relatedness, cross-copulation occurred only between the Florida and Georgia strains. Thus, *B. calyciflorus* is a complex of cryptic species. While the Florida, Texas and Australia strains were reproductively isolated from one another, most combinations of cross-strain mating tests showed intense and prolonged male circling behavior following male–female encounters. This suggests that precopulatory male circling and copulation are two separate behaviors that may be controlled by different female chemicals and male coronal receptors. In some cross-strain mating tests, females regularly retracted their corona when circled by a male, indicating that they can recognize 'foreign' males and actively interfere with copulation.

Introduction

Recent studies have shown that zooplankton species previously considered to be cosmopolitan can be complexes of sibling species. This has been clearly demonstrated for the copepod *Eurytemora affinis* (Lee, 2000; Lee & Frost, 2002) and for the rotifer *Brachionus plicatilis* (Ciros-Peréz et al., 2001; Gómez et al., 2002). For the *B. plicatilis* complex within the Iberian Peninsula, three species have been described (Ciros-Peréz et al., 2001) and three more have been identified (Gómez et al., 2002). There is no evidence for any hybridization or introgression among these species, despite frequent sympatry (Gómez et al., 2002).

Brachionus calyciflorus exhibits considerable geographic variation in morphology (Kutikova & Fernando, 1995) and may be a complex of sibling species. The present study begins to address this possibility by examining the affinities of strains from Florida, Georgia, Texas and Australia using both mating tests and DNA sequence analysis. Results linking reproductive isolation and genetic distance can then be compared to those obtained for *B. plicatilis* (Gómez et al., 2002) and *E. affinis* (Lee, 2000).

Materials and methods

Cultures

Four strains of *B. calyciflorus*, from Florida, Georgia, Texas and Australia, were used in this study. The Florida strain was derived from a pond on the University of Florida campus (Gainesville). Resting eggs were purchased from Florida Aqua Farms Inc. (Dade City). The Georgia strain originated from Piedmont Park Pond (Atlanta). Resting eggs were kindly provided by Terry W. Snell. The Texas strain was collected from a temporary pond at the Hueco Tanks State Historic Site (El Paso Co.) by one of us (E.J.W.) and maintained in culture. The Australian strain was obtained from resting eggs in sediment collected by one of us (J.J.G.) in May 1997 from above the water level of a River Murray billabong-Ryan's 2 (Wodonga, Victoria). Clones of the Florida, Georgia and Australian strains were derived from single resting eggs; a clone of the Texas strain was derived from a single amictic female. The clones tested in this study were: FL 23, 40 and 51 from the Florida strain; GA 3 and 5 from the Georgia strain; TX 1 from the Texas strain, and AUS 5 and 20 from the Australian strain. All clones were cultured on *Cryptomonas erosa* var. *reflexa* in MBL medium at 20 °C in a photoperiod (L:D 16:8), as described elsewhere (Gilbert, 2002, 2004). Mictic females were induced in all strains by crowding (Gilbert, 2002, 2003, 2004).

Live females of the North American strains were extremely similar to each other, but morphologically distinct from those of the Australian strain. Australian females were more bladder-like than those of the North American strains, especially as juveniles (Fig. 1). They were also slightly larger; the mean lorica lengths of live, adult amictic females of one Australian and one Florida clone were 220 μm and 168 μm, respectively (Gilbert, 2003). The Australian strain appeared to be at the low end of the size range recorded for the *B. calyciflorus* from that continent (200–400 μm; Koste & Shiel, 1987).

Mating tests

Young males were obtained by isolating single, ovigerous mictic females in depressions of 96-well tissue culture plates with 200 μl of a suspension of *C. erosa*. Young females were obtained by isolating a group of ovigerous amictic females in 15 ml of a suspension of *C. erosa* and then removing their neonate daughters. All males were less than 22 h (usually 8 h) old and probably had not copulated prior to the tests as they were exposed only to their mothers. All females were less than 8 h (usually 4 h) old and had not been exposed to males prior to the tests. Young individuals of both sexes were used to maximize the potential for mating responses (Gilbert 1963; Snell & Childress, 1987; Snell & Hoff, 1987; Gómez & Serra, 1996).

Mating tests were conducted at room temperature (19–23 °C) using a Wild M 5 stereomicroscope. The tests determined the probability that males would 1) initiate circling behavior with encountered females and 2) copulate with females after circling them. For each replicate, a single female was introduced into a depression of a 96 well plate containing 3–8 males produced there by a single mictic mother. Just before introduction of the newborn female, the mictic mother of the males and about 100 μl of medium were removed, leaving the group of males in about 100 μl of medium. Each newborn female and group of males was used only once. Observation of a group of males ceased either after the first copulation event, or after a set number of encounters (usually 4, but sometimes 10). The number of replicates for a test varied from 3 to 19.

An encounter was said to occur when a male physically contacted the body of a female with the center of his corona. Male precopulatory circling behavior was initiated after an encounter if the male turned around and around the female's body while arching his body and touching her with both his corona and penis. Copulation was said to occur following circling behavior if the male attached his penis to the female (in this study always to her corona) and then released coronal contact with her body while remaining attached to her for some time by his penis before swimming away. In addition, notes usually were made on the approximate duration of circling behavior before copulation occurred or the male swam away, and on the tendency of the female to retract her corona while being circled by the male.

Eight mating-test experiments were conducted at different times. As males of the Florida strain mated as intensively with females of the Georgia strain as with those of their own strain, only the Florida strain was tested with the Texas and Australian strains. In each mating test, males from a clone of one strain were paired with females from a clone of a different strain. More than one clone of the Florida and Australian

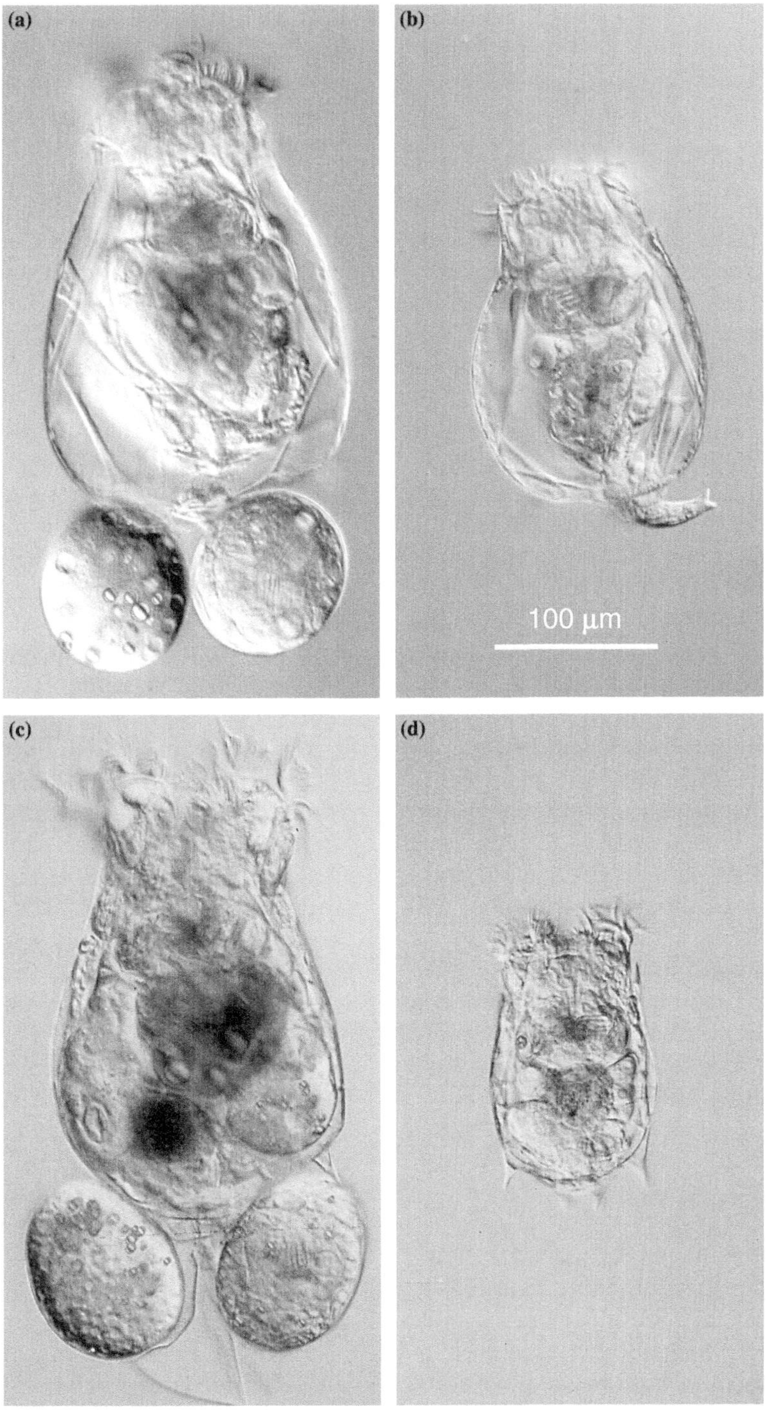

Figure 1. Photographs of live *Brachionus calyciflorus* (Nomarski interference contrast optics and flash illumination) showing difference in shape between females from Australia clone 20 (a, adult; b, juvenile) and Florida clone 40 (c, adult; d, juvenile).

strains were used because cultures of some clones were discontinued. No tests paired different clones within strains. In most experiments, mating behaviors of males with females of a different strain were compared to those with females of the same strain. For these pair-wise comparisons, probabilities of male circling behavior and copulation were statistically evaluated using Mann–Whitney tests corrected for ties (Zar, 1999).

DNA sequence analysis

DNA was isolated and amplified from axenized fresh or ethanol preserved animals by homogenizing 100 to 150 individuals in 100–150 μl 5% Chelex-100 (BioRad) followed by boiling for 7 min. Primers used to amplify the mitochondrial *cox*1 gene (LCO1490: 5′-GGTCAACAAATCATAAAGATATTGG-3′, HCO2198: 5′-TAAACT TCAGGGTGACCAAAAAATCA-3′; Folmer et al., 1994) and the nuclear ribosomal ITS region (ITS4: 5′-TCCTCCGCTTATTGATATGC-3′, ITS5; 5′-GGAAGTAAAAGTCGTAACAAGG-3′; White et al., 1990) each amplify approximately 800 bp. The ITS primers amplify ITS1, 5.8S, and ITS2 regions of the nuclear ribosomal gene complex. All amplifications included a negative control to detect amplification of contaminating DNA. Amplification products were examined by electrophoresis to verify their size and cleaned with GeneClean kits (Bio101) before sequencing. Amplified products were sequenced directly using SequiTherm kits (EpiCentre Tech) and run on a LI-COR 4200 Series automated sequencer. All gene regions or genes were sequenced at least twice in both directions. Sequences were aligned using Clustal W (Thompson et al., 1994) and then adjusted by eye. Uncorrected Genetic distance ("p") calculations were done using PAUP*4.0b10 (Swofford, 2002) using UPGMA.

Results

Results of the eight mating experiments are shown in Table 1 and summarized in Table 2. In crosses between the Florida and Georgia strains, male circling behavior was initiated after each encounter, and the males always copulated with the females (Table 1), usually after no more than several seconds. In most between-strain crosses with the Florida, Texas and Australian strains, male circling behavior was observed but never led to copulation (Tables 1 and 2). Despite this complete absence of copulation, male circling behavior in some of the crosses was very strong and often persisted for more than a minute before the male released contact with the female. Such strong circling behavior typically was observed when Florida males encountered Texas or Australian females, and when Texas males encountered Australian females. In all within-strain crosses using the Florida, Texas and Australian strains, the probability of copulation was very high (0.65–1.0; Table 1), and copulation usually occurred shortly after initiation of male circling behavior.

Relatively low probabilities of male circling behavior occurred only when Australian males encountered Florida or Texas females (Table 1). The probability of AUS 5 males circling FL 23 females was 0.49 (vs. 1.0 for AUS 5 females), and the probability of AUS 20 males circling FL 40 females was 0 (vs. 1.0 for AUS females). Similarly, the probability of AUS 20 males circling TX 1 females was only 0.25 (vs. 0.86 for AUS 20 females). In all of these pair-wise comparisons, the difference in probabilities was statistically significant.

Females from the Texas and Australian strains showed a pronounced tendency to retract their corona when circled by Florida males. The same was true when Australian females were circled by Texas males. For example, in Experiment 7 with Texas females and Florida males (Table 1), the Texas females retracted their corona throughout prolonged circling behavior in 14 of the 16 cases, and intermittently during this behavior in the other 2 cases. In contrast, Florida females rarely retracted their corona when circled by Texas or Australian males. Similarly, female coronal retraction was rare when Texas or Australian females were circled by males of their own strain.

A total of 1429 bp were sequenced, aligned and analyzed from each of the strains (Table 3). The *cox*1 region showed slightly higher levels of sequence divergence (mean 11.5%) as compared to the ITS region (mean 5.2%). The Georgia and Florida strains were identical in their *cox*1 sequences and showed only 5 differences in the ITS region. Compared to these two strains, the Texas strain showed 4% and 10% sequence divergence in the ITS and *cox*1 regions, respectively. The Australian strain showed the greatest sequence divergence, ranging from approximately 6% (ITS) to 13% (*cox*1).

Table 1. Results of eight mating-test experiments using strains of *Brachionus calyciflorus* from Florida (FL), Georgia (GA), Texas (TX) and Australia (AUS)

Experimental and statistical test		Strain and clone		Mating test		
		Male	Female	Number of replicates	p (cir)	p (cop)
1		GA 5	FL 40	5	1.00	1.00
		FL 40	GA 3	5	1.00	1.00
2	a	AUS 5	FL 23	8	0.49^{**}	0^{**}
	a		AUS 5	8	1.00	0.94
	b	FL 23	AUS 5	8	0.91^{*}	0^{**}
	b		FL 23	8	1.00	0.85
3	c	AUS 20	FL 40	4	0^{**}	
	c		AUS 20	4	1.00	1.00
		FL 40	AUS 20	4	1.00	0
4			Fl 40	19	0.83	0.65
5	d	TX 1	FL 40	4	0.88^{NS}	0^{*}
	d		TX 1	3	1.00	0.83
6	e	TX 1	FL 51	4	0.75^{NS}	0^{**}
	e		TX 1	4	1.00	0.81
		FL 40	TX 1	4	1.00	0
	f	FL 51	TX 1	4	0.94^{NS}	0**
	f		Fl 51	4	1.00	0.88
			FL 40	4	1.00	1.00
7		TX 1	FL 40	4	0.75	0
		FL 40	TX 1	7	1.00	0
8	g	TX 1	AUS 20	7	0.89^{NS}	0^{**}
	g		TX 1	5	0.90	0.67
	h	AUS 20	TX 1	7	0.25^{**}	0^{**}
	h		AUS 20	7	0.86	1.00

Probabilities (p) are of male circling behavior (cir) after an encounter with a female, and copulation (cop) following circling behavior. Pair-wise comparisons (Mann–Whitney tests) of probabilities within experiments are indicated by letters (a–h); superscripts indicate statistical significance: NS ($p > 0.05$), *($p \leq 0.05$), **($p \leq 0.01$).

Table 2. Summary results for probabilities (p) of male circling behavior (cir) after encounter with a female, and copulation (cop) following circling behavior, in mating-test experiments using strains of *Brachionus calyciflorus* from Florida (FL), Texas (TX) and Australia (AUS)

Male strain	Female strain					
	FL		TX		MAUS	
	p (cir)	p (cop)	p (cir)	p (cop)	p (cir)	p (cop)
FL	0.96 (0.04)	0.85 (0.07)	0.98 (0.02)	0	0.96 (0.05)	0
TX	0.79 (0.04)	0	0.97 (0.03)	0.77 (0.05)	0.89	0
AUS	0.25 (0.25)	0	0.25	0	0.95 (0.05)	0.98 (0.02)

Values are means (±1 SE) when more than one experiment was conducted (see Table 1).

Discussion

Affinities among the four strains of B. calyciflorus

The results of the mating tests showed that the Florida and Georgia strains of *B. calyciflorus* readily cross-copulate (Table 1). The ability of such crosses to result in the production of viable fertilized eggs and hybrids is probable but has not been determined. The DNA analysis demonstrated the similarity of these two strains (Table 3).

In contrast, the Florida, Texas and Australian strains did not cross-copulate and thus were completely reproductively isolated. The distinctiveness of these strains is consistent with their considerable geographic isolation, with the unique morphology of the Australian strain (Figure 1), and with the inability of a high density of females of the Australian strain to induce mictic-female production in the Florida and Georgia strains (Gilbert, 2003). The DNA analysis confirms the genetic distinctiveness of the Florida, Texas and Australian strains. There were from 95 to 132 nucleotide differences between each of these strains in the *cox*1 gene and ITS region (Table 3). Consistent with its geographic separation and distinct morphology, the Australian strain was more genetically different from the Florida and Georgia strains than was the Texas strain.

Using molecular characters, a variety of morphologically described cosmopolitan species have been found to be complexes of cryptic species. These species, although morphologically similar, typically have 5% or greater sequence divergence – the level typically found to occur among related species across a wide variety of taxa (Avise, 1994; Moriyama & Powell, 1996; Avise & Walker, 1999). Within the Family Brachionidae, DNA sequence divergences between species range from 8 to 24% and from 22 to 29% for different genera (Gómez et al., 2002; Derry et al., 2003, Walsh & De La Riva, unpublished). Among the closely related *B. plicatilis*, *B. ibericus*, and *B. rotundiformis*, lineages show >12% sequence divergence (Gómez et al., 2002). Many of the strains in the *B. plicatilis* species complex have recently been designated as separate species (Segers, 1995; Ciros-Pérez et al., 2001). This complex includes some co-occurring strains that are demarcated by morphological differences, sequence divergence and behavioral reproductive isolation.

Table 3. Genetic distance among *B. calyciflorus* strains

Strain	Georgia	Florida	Texas	Australia
A. ITS – 735 bp				
Georgia	–	5	32	46
Florida	0.00685	–	28	42
Texas	0.04391	0.03844	–	42
Australia	0.06371	0.05824	0.05795	–
B. *Cox*1 – 694 bp				
Georgia		0	66	82
Florida	0.00000	–	67	87
Texas	0.10258	0.09654	–	90
Australia	0.12727	0.12536	0.12968	–
C. Combined – 1429 bp				
Georgia	–	5	98	128
Florida	0.00685	–	95	129
Texas	0.04391	0.03844	–	132
Australia	0.06371	0.05824	0.05795	–

ITS (A) and *cox*1 (B) sequence distances are calculated as uncorrected "*p*" below the diagonal, and as absolute number of differences above the diagonal. Combined distances are shown in (C). Genbank accession numbers are: Georgia DQ071668 (ITS), DQ071672 (*cox*1); Florida DQ071669 (ITS), DQ071673 (*cox*1); Texas DQ071670 (ITS), DQ071674 (*cox*1); and Australia DQ071671 (ITS), DQ071675 (*cox*1).

Sequence divergence among the strains of *B. calyciflorus* examined in the present study ranged from 0–13% depending on the region sequenced. Among the three strains reproductively isolated from one another (Florida, Texas, Australia), sequence divergences were about 6% for the ITS region and 9–13% for the *cox*1 region. Similarly, among strains of the *B. plicatilis* species complex, Gómez et al. (2002) reported sequence divergences ranging from 0 to 20%. *Cox*1 sequence divergence of the four *B. calyciflorus* strains was nearly identical to within-clade divergences in this region for the *B. plicatilis* complex (0–12%). Levels of variation in our ITS sequences (0–6.8%) are slightly higher than those reported by Gómez et al. (2002) for *B. plicatilis* complex strains (0–2%), most likely due to our inclusion of the more variable ITS2 region. In addition, using the molecular clock calculations in Gómez et al. (2002), the Australian strain of *B. calyciflorus* has been diverging from the other lineages for at least 3 million years. This is ample time for the evolution of reproductive isolating mechanisms, and is consistent with the reproductive isolation observed.

Mating behavior in Brachionus

When Florida males encountered Texas or Australian females, or when Texas males encountered Australian females, there was a very high probability of intense and prolonged circling behavior which never progressed to copulation (Tables 2 and 3). Absence of copulation in these tests cannot be attributed to the condition of the males or females, because the probability of copulation was very high in control crosses with males and females of the same strain (Tables 1 and 2). An implication of these results is that mating bioassays for determining affinities or genetic relatedness should not rely on male precopulatory circling behavior. Male circling may be a meaningful criterion when the probabilities of this behavior in between-group crosses are lower than those in within-group crosses, but it is clear that high probabilities in between-group crosses do not necessarily mean close affinities.

Gómez & Serra (1995) were the first to warn about using only male circling behavior for affinity testing. They noted that copulations between clonal groups of the *B. plicatilis* complex generally occurred when the probability of initiating circling behavior was high, but that this pattern did not always hold. They observed copulations in some crosses when circling behavior was unlikely, and few copulations in other crosses when the probability of circling was high. Gómez & Serra (1996) found a similarly weak coupling between the probability of male circling and copulation when they examined the effect of female age on male mating behavior. Males initiated circling behavior with old females, although to a lesser extent than with young females, but copulated only with young females.

Prolonged male circling behavior without copulation in some between-strain crosses in *B. calyciflorus* has important implications regarding the control of mating behavior. Clearly, in some cases a strong signal for the initiation of male circling behavior never triggers copulation. Therefore, these two behaviors appear to be separate and may involve the recognition of different female signals by different male receptors. One female signal could initiate male circling behavior after an encounter with a female. A second female signal could induce circling males to copulate.

The signal that induces males to circle females after they contact them is a chemical in the female that is recognized by receptors on the male corona (Gilbert, 1963). This chemical, called a mate-recognition pheromone (MRP), has been isolated in *B. plicatilis* and characterized as a 29 kD glycoprotein concentrated in the female corona (Snell et al., 1988, 1995; Snell, 1998). Recognition of the MRP by a male should depend on the chemical structure of the MRP and the male contact receptors. The signal initiating copulation in males that are circling females could be a different female chemical, or pheromone, detected by male coronal receptors. Thus, in the between-strain crosses where males initiate circling behavior but never copulate, male receptors for copulation may not recognize the copulation pheromone.

Observations of mating behavior in the present study indicate that females circled by males can distinguish males of different strains and actively respond to strains other than their own by retracting their corona. Female coronal retraction was particularly pronounced when Texas or Australian females were circled by Florida males, and when Australian females were circled by Texas males. Florida, Texas and Australian females rarely retracted their corona when circled by males of their own strain. Also, Florida females rarely retracted their corona when circled by Texas males. The mechanism by which these females can distinguish males of different strains is not known. Females may recognize some male substance via receptors. Alternatively, or in addition, they may retract their corona in response to prolonged male circling. Female coronal retraction during male circling behavior was first noted in *B. plicatilis* by Gómez & Serra (1995).

Absence of copulation between Florida males and Texas or Australian females, and between Texas males and Australian females, cannot be attributed to female coronal retraction. Reciprocal crosses showed that copulation never occurred even when females had a fully extended corona, as in tests with Texas or Australian males with Florida females. Also, Florida males failed to copulate with Australian females when these females were narcotized with carbonated water and unable to retract their corona (Gilbert, unpublished). Finally, female coronal retraction does not necessarily prevent copulation. In some within-strain crosses with the Florida strain, copulation frequently occurred despite female coronal retraction (Gilbert, unpublished). However, retraction of the female corona in many of these cases did interfere with the ability of a male to attach his penis to the female corona and appeared to prevent copulation.

Conclusions

The DNA sequence analysis and mating tests in the present study show that *B. calyciflorus* is a species complex in which some geographically and genetically distinct strains are reproductively isolated from one another. This species complex is similar to one described in more detail for *B. plicatilis* (Gómez et al., 2002). It also is similar to the species complex in the copepod *Eurytemora affinis* (Lee, 2000). However, it is interesting that the mechanism of reproductive isolation is behavioral in *Brachionus* and postzygotic in the copepod. Copepod populations having *cox*1 sequence divergences ranging from 0.15 to 17.1% cross-mated but showed hybrid breakdown in the *F1* or *F2* generations (Lee, 2000).

Observations of mating behavior in the present study indicate that the signals and responses leading to copulation in rotifers may be more complex than previously suspected. Male circling behavior and copulation appear to be two separate behaviors and may involve the recognition of separate female chemicals by different male receptors. Furthermore, females may play an active role in mating. They can recognize males of different strains, and can interfere with copulation attempts by retracting their corona.

Acknowledgements

We thank Raquel Garcia for laboratory assistance, Ryan A. Thum and two anonymous referees for improving the manuscript, and the staff at Hueco Tanks State Historic Site (collecting permits (66–99; 07–02). J.J.G. thanks Mr and Mrs Ernest Magnien for supporting the mating-test research. The molecular portion of this research was supported by NSF HRD (9628568) and NIH 5G012RR008124.

References

Avise, J. C., 1994. Molecular Markers, Natural History and Evolution. Chapman & Hall Inc, New York NY.

Avise, J. C. & D. Walker, 1999. Species realities and numbers in sexual vertebrates: perspectives from an asexually transmitted genome. Proceedings of the National Academy of Sciences USA 96: 992–995.

Ciros-Pérez, J., A. Gómez & M. Serra, 2001. On the taxonomy of three sympatric sibling species of the *Brachionus plicatilis* (Rotifera) complex from Spain, with the description of *B. ibericus* n. sp. Journal of Plankton Research 23: 1311–1328.

Derry, A. M., P. D. N. Hebert & E. E. Prepas, 2003. Evolution of rotifers in saline and subsaline lakes: a molecular phylogenetics approach. Limnology and Oceanography 48: 675–685.

Folmer, O., M. Black, W. Hoeh, R. Lutz & R. Vrijenhoek, 1994. DNA primers for amplification of mitochondrial cytochrome c oxidase subunit 1 from diverse metazoan invertebrates. Molecular Marine Biology and Biotechnology 3: 294–299.

Gilbert, J. J., 1963. Contact chemoreception, mating behaviour, and sexual isolation in the rotifer genus *Brachionus*. Journal of Experimental Biology 40: 625–641.

Gilbert, J. J., 2002. Endogenous regulation of environmentally induced sexuality in a rotifer: a multigenerational parental effect induced by fertilisation. Freshwater Biology 47: 1633–1641.

Gilbert, J. J., 2003. Specificity of crowding response that induces sexuality in the rotifer *Brachionus*. Limnology and Oceanography 48: 1297–1303.

Gilbert, J. J., 2004. Population density, sexual reproduction and diapause in monogonont rotifers: new data for *Brachionus* and a review. Journal of Limnology 63 (Suppl. 1): 32–36.

Gómez, A. & M. Serra, 1995. Behavioral reproductive isolation among sympatric strains of *Brachionus plicatilis* Müller 1786: insights into the status of this taxonomic species. Hydrobiologia 313/314: 111–119.

Gómez, A. & M. Serra, 1996. Mate choice in male *Brachionus plicatilis* rotifers. Functional Ecology 10: 681–687.

Gómez, A., M. Serra, G. R. Carvalho & D. H Lunt, 2002. Speciation in ancient cryptic species complexes: evidence from the molecular phylogeny of *Brachionus plicatilis* (Rotifera). Evolution 56: 1431–1444.

Koste, W. & R. J. Shiel, 1987. Rotifera from Australian inland waters II. Epiphanidae and Brachionidae (Rotifera: Monogononta). Invertebrate Taxonomy 1: 949–1021.

Kutikova, L. A. & C. H. Fernando, 1995. *Brachionus calyciflorus* Pallas (Rotatoria) in inland waters of tropical latitudes. Internationale Revue der gesamten Hydrobiologie 80: 429–441.

Lee, C. E., 2000. Global phylogeography of a cryptic copepod species complex and reproductive isolation between genetically proximate "populations". Evolution 54: 2014–2027.

Lee, C. E. & B. W. Frost, 2002. Morphological stasis in the *Eurytemora affinis* species complex (Copepoda: Temoridae). Hydrobiologia 480: 111–128.

Moriyama, E. N. & J. R. Powell, 1996. Intraspecific nuclear DNA variation in *Drosophilia*. Molecular Biology and Evolution 13: 261–277.

Segers, H., 1995. Nomenclatural consequences of some recent studies of *Brachionus plicatilis* (Rotifera, Brachionidae). Hydrobiologia 313/314: 121–122.

Snell, T. W., 1998. Chemical ecology of rotifers. Hydrobiologia 387/388: 267–276.

Snell, T. W. & M. Childress, 1987. Aging and loss of fertility in male and female *Brachionus plicatilis* (Rotifera). International Journal of Invertebrate Reproduction and Development 12: 103–110.

Snell, T. W., M. J. Childress & B. C. Winkler, 1988. Characteristics of the mate recognition factor in the rotifer *Brachionus plicatilis*. Comparative Biochemistry and Physiology 89A: 481–485.

Snell, T. W. & F. H. Hoff, 1987. Fertilization and male fertility in the rotifer *Brachionus plicatilis*. Hydrobiologia 147: 329–334.

Snell, T. W., R. Rico-Martinez, L. N. Kelly & T. E. Battle, 1995. Identification of a sex pheromone from a rotifer. Marine Biology 123: 347–353.

Swofford, D. L., 2002. PAUP*. Phylogenetic analysis using parsimony (* and other methods), vers. 4.0. Sinauer Associates, Sunderland, MA.

Thompson, J. D., D. G. Higgings & T. J. Gibson., 1994. Clustal W: improving the sensitivity of progressive multiple sequence alignment through sequencing weighting, position specific gap penalties and weight matrix choice. Nucleic Acids Research 22: 4676–4680.

White T. J., J. Bruns, S. Lee & J. Taylor, 1990. Amplification and direct sequencing of fungal ribosomal RNA genes for phylogenetics. In Innis, M., D. Gelfand, J. Sninsky & T. White (eds), PCR Protocols: A Guide to Methods and Applications.

Zar J. H., 1999. Biostatistical Analysis. Prentice Hall.

Hydrobiologia (2005) 546:267–274
A. Herzig, R.D. Gulati, C.D. Jersabek & L. May (eds.) Rotifera X: Rotifer Research: Trends, New Tools and Recent Advances
DOI 10.1007/s10750-005-4207-1

Removal of surface glycoproteins and transfer among *Brachionus* species

Terry W. Snell[1,*] & Claus-Peter Stelzer[2]
[1]*School of Biology, Georgia Institute of Technology, Atlanta, Georgia, 30332–0230, USA*
[2]*Department of Evolutionary Biology Institute of Animal Evolution and Ecology, Westfälische Wilhelms Universität Münster, Hüfferstr. 1, 48148, Münster, Germany*
*(*Author for correspondence: E-mail: terry.snell@biology.gatech.edu)*

Key words: rotifers, surface proteins, *Brachionus plicatilis*, EDTA, EGTA, mating signals, mate recognition

Abstract

Glycoproteins on the body surface of females of the rotifer *Brachionus plicatilis* are a key signal in their mate recognition system. When *B. plicatilis* Russian strain females were exposed to 50 mM EDTA or EGTA, several surface glycoproteins were removed. Females exposed to EDTA died, but remained intact and were used in mating bioassays with conspecifics males. Live control females elicited a male mating response in 21% of encounters, freeze-killed control females elicited responses in 23%, but EDTA extracted females elicited a mating response in only 5% of encounters. At least some of the EDTA-extractable proteins on the surface of females appear to be critical to male mate recognition. EDTA treated females could be exposed to proteins extracted from other females and some proteins re-attached to their body surface, restoring their attractiveness to males. SDS-PAGE of these proteins revealed 15–17 prominent bands, most ranging in molecular mass from 66 to 12 kD. The EDTA-extractable proteins were separated using ion exchange chromatography and each fraction was tested for its ability to restore female attractiveness. When proteins in fraction 22 were bound to females, they restored 80% of the females' ability to elicit male mating responses. Exposing EDTA treated females to bovine serum albumin or casein had no effect on their attractiveness to males. EDTA treated females from different *Brachionus* clades and species were exposed to proteins from fraction 22. Female attractiveness could be restored in most clades of *B. plicatilis*, but no transfer of mating attractiveness was observed to *B. rotundiformis* or *B. ibericus* females. Conspecific males treated with EDTA and exposed to proteins in fraction 22 could not be feminized and made attractive to other males. A sexual dimorphism in surface proteins therefore exists between *B. plicatilis* females and males. Successful transfer of glycoproteins critical in mate recognition is dependent on signal glycoprotein structure and the structure of the other proteins present on the surface of females.

Introduction

The taxon *Brachionus plicatilis* is actually a complex of several cryptic species (Gómez & Snell, 1996; Serra et al., 1997; Serra et al., 1998). Prior to 1995 only one of these was named (Segers, 1995), but now three are recognized by systematists (Ciros-Perez et al., 2001). Phylogenetic analysis of COI and ITS gene sequences suggests that there could be at least 11 more as yet unnamed species (Gómez et al., 2002; Derry et al., 2003; Suatoni, 2003). Many of these species are sympatric (Gómez & Serra, 1995; Gómez & Carvalho, 2000; Ortells et al., 2000), yet there is no evidence of hybridization or introgression between them (Ortells et al., 2000). These observations suggest that reproductive barriers among *B. plicatilis* species are well developed and effective. It also raises questions about the nature of the reproductive barriers, how they arose, and how they are maintained.

A key element in the sexual reproductive system of *B. plicatilis* is mate recognition by males of conspecific females. This is accomplished by contact chemoreception of a glycoprotein signal on the body surface of females (Snell & Hawkinson, 1983; Snell, 1989; Snell et al., 1995). There has been a long term effort to isolate and characterize the glycoprotein signal and its receptor, and to clone the underlying genes (Snell, 1998). In this paper, we describe how surface glycoproteins can be stripped from females by treatment with the chelator EDTA and re-attached to similarly treated females of the same or different species. This removal and re-attachment bioassay was developed for testing ion exchange chromatography and HPLC fractions for male mating activity. However, the technique also has allowed us to probe the phylogenetic distances over which transfer of these signal glycoproteins can occur and to develop hypotheses about the limitations on this transfer. We further have tested whether females of one species could be made attractive to males of a different species and whether males could be feminized by changing their surface glycoprotein profile. Such experiments are analogous to the strikingly successful epicuticular hydrocarbon transfer experiments in *Drosophila* (Coyne & Charlesworth, 1997; Blows & Allan, 1998; Etges & Ahrens, 2001). This approach has demonstrated that epicuticular hydrocarbons are essential elements of the *Drosophila* mate recognition system, that they are used as contact sex pheromones with information encoded in hydrocarbon composition and structure, and that there is epicuticular sexual dimorphism between males and females. We report here results of the application of a similar technique in rotifers.

Methods

The rotifers used in these experiments are part of the *B. plicatilis* species complex described by Gómez et al. (2002). The strains designated RUS, GP, AUS, CH, and L1 were originally collected from the Azov Sea (Russia), Gaynor Pond (Colorado, USA) Obere Halbjochlacke (Austria), Tianjin (China), and Torreblanca (Spain), respectively, and maintained in the lab for many years as resting eggs. All are currently classified as members of the *B. plicatilis* morphospecies, but some clades are likely to be independent species. Strains LFL and IR2 were originally collected from Little Fish Lake, Nevada, and Indian Rocks Beach (Florida, USA, GPS 27.77° N, 82.68° W) [TS1] and are currently classified in the *B. ibericus* morphospecies. The ITS1 sequence of IR2 showed 100% similarity with the Californial populations of the ‚Almenara' clade (Gómez et al., 2002, Genebank accession AF387222). The HAW strain is currently classified as *B. rotundiformis* morphospecies and was obtained from the Oceanic Institute in Hawaii, but its original collection site is unknown.

Rotifers were hatched from resting eggs and cultured in 15 ppt artifical seawater (Instant Ocean) at 25 °C on a diet of *Tetraselmis suecica* in 5 l bags that were lightly aerated. Constant fluorescent illumination of approximately 2000 lux was provided. Males and females in log-phase populations were filtered from about 200 ml of culture using a 68 μm screen and re-suspended in clean seawater. Experimental animals were isolated under a stereomicroscope at 10× magnification using a narrow bore micropipet and separated into Petri dishes according to sex in 5 ml seawater. Only vigorous, fast swimming males (ages unknown) were isolated and mated with young (<24 h old), non-ovigerous females.

The positive control mating bioassay was performed by placing 7–10 males and 4–6 live females into about 50 μl of seawater on the inverted top of a 96-well plate, which provides a flat, clear viewing surface. Mating behavior was videotaped for 5 min under a stereomicroscope at 10× magnification using a CCD camera. The number of male–female encounters and the number of matings initiated (circlings) by males were recorded in three replicate trials for each treatment. A second positive control was conducted using females that were killed by freezing at −80 °C for 1 h. Male matings with control females were compared to matings with females exposed to a variety of treatments. Surface proteins were removed from females by exposing them to 100 mM EDTA prepared in 2 ppt seawater for 15 min. Approximately 50 females were pipetted into a minimum volume in a nine spot glass depression plate. We used glass so that the plates could be baked overnight at 100 °C between experiments. Addition of 1 ml of 100 mM EDTA to the rotifers in the well dilutes it to about 50 mM EDTA which is enough to cause most females to stop swimming and fall to the bottom after about 15 min. As much of the solution as possible was then removed, being careful not to remove rotifers, and replaced with fresh EDTA solution for another 15 min of incubation. Females were then washed by serial transfer through three rinses of 2 ml of clean seawater, being careful to transfer minimum volumes to each well. At this point, females were immobile and 4–6 were transferred for a final wash to a well containing 2 ml clean seawater. EDTA treated females have a strong tendency to pick up proteins, so glass micropipets must be baked overnight at 100 °C between experiments and changed between each treatment. The 4–6 females then were transferred in about 20 μl to a spot on the 96-well lid to begin the bioassay. Care was taken so that females were arrayed towards the middle of the spot and not trapped in the surface tension or along the edges. About 7–10 young, fast males were transferred in minimum volume to the spot and excess seawater removed so that the spot was flat. If EDTA had not been completely removed in the washing steps, male swimming markedly slowed, rendering the replicate unusable. Mating behavior was videotaped for 5 min and scored as described above.

Females treated with EDTA were exposed to ion exchange fractions (see below) containing proteins to test their ability to elicit male mating. In this experiment, 6–8 EDTA treated females were transferred in minimum volume to a well in a 96-well plate. About 1–2 μl of the test fraction was added, followed by 20 μl of seawater to mix thoroughly, and incubated for 5 min. Females then were transferred to a well containing 2 ml of seawater for washing. Finally, they were transferred in minimum volume to the lid of a 96-well plate, males added, and the bioassay was conducted as described above. Treatments with EGTA, bovine serum albumen, and casein followed similar protocols. The active ion exchange fraction number 22 was boiled for 10 min to test its thermal stability.

Approximately 20–30 g wet-weight RUS clade biomass was filtered from mass cultures and re-suspended in 5 l of clean seawater for 3–4 h with aeration. Seawater was replaced with clean seawater every hour so that the rotifer guts were cleared. Proteins for ion exchange chromatography were extracted from rotifer biomass using 2× volume of 100 mM EDTA in 2 ppt seawater containing a cocktail of protease inhibitors (Roche Complete Mini protease inhibitor cocktail, 1 tablet/7 ml). The biomass was shaken on a rotary shaker for 1 h to solubilize surface proteins. Rotifers were separated from soluble proteins by decanting off the liquid, then centrifuging at 20,000 × g for 30 min at 4 °C. Supernatant was collected and EDTA was removed by ultrafiltration using a 10,000 Dal molecular weight cut-off filter that retained the proteins of interest. Proteins were re-suspended from the membrane in a 20 mM Tris–HCl buffer, pH 8.0, containing 50 mM NaCl. This solution was applied to Q Sepharose (Amersham Pharmacia) high performance ion exchange resin packed in a 1 cm diameter glass column to a height of about 3 cm. Proteins were eluted with a linear gradient from 50 to 1000 mM NaCl over 40 min in 1 ml fractions collected each minute. Samples were stored at –80 °C until tested for mating activity.

EDTA-extractable proteins were separated and visualized by SDS-polyacrylamide gel electrophoresis performed according to the protocol described by Snell et al. (1995). EDTA or NaCl was removed from electrophoresis samples by centrifugation with 10 000 Da MWCO filters. Proteins were re-suspended in DI water, then electrophoresis sample solution was added in a ratio of 1 to 3 parts sample volume. Proteins were separated on 12% acrylamide gels and visualized with Sypro Orange protein gel stain (Molecular Probes) according to the manufacturer's protocol.

Results

When *B. plicatilis* Russian females were treated with either EDTA or EGTA they elicited about 4-fold fewer mating responses from conspecific males (Fig. 1). Males initiated mating (circled) freeze-killed females with the same propensity as live females. When EDTA treated females were exposed to proteins in ion exchange fraction 22 (see below), their ability to elicit male mating

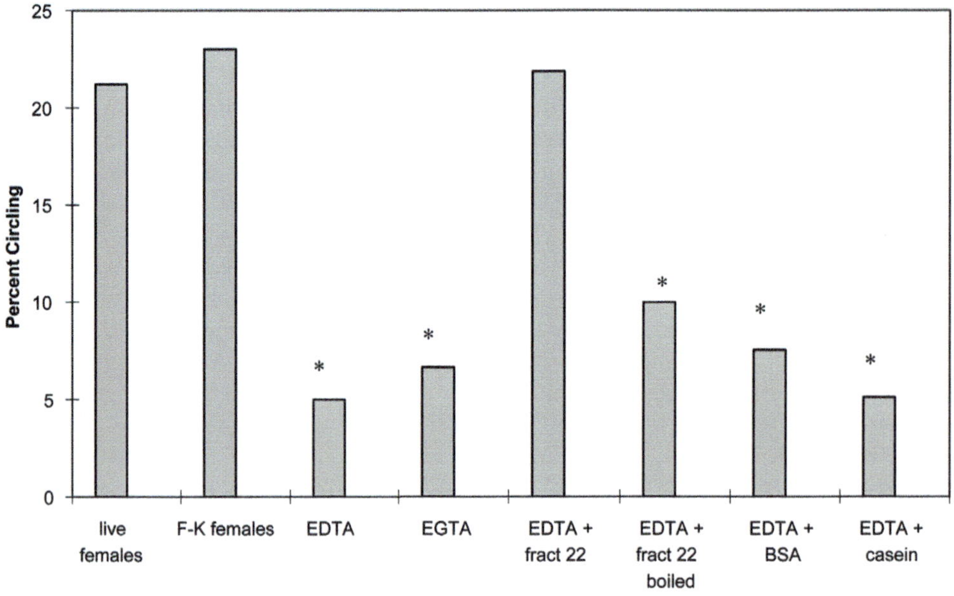

Figure 1. Effects of EDTA/EGTA extraction of female surface proteins on male mate recognition. F-K are freeze-killed females. BSA is bovine serum albumen. *indicates a significant difference from live female control, Fisher's exact test, $p < 0.05$. Percent circling is the proportion of male–female encounters that resulted in males initiating mating behavior.

responses was restored to that of live females. However, if fraction 22 was boiled for 10 min, it lost its activity. If EDTA treated females were exposed to the proteins BSA or casein, there was no restoration of their ability to elicit male mating responses.

The mating bioassay demonstrated that treatment of female rotifers with EDTA extracted surface proteins that are involved in mate recognition. This enabled us to attempt to re-attach these proteins to other EDTA treated females from different geographic populations and species (Fig. 2). Russian females served as the positive control against which all other strains were compared. Treatment of Russian females with EDTA significantly reduced by several fold their ability to elicit male mating responses, but female attractiveness was restored by exposure to ion exchange fraction 22 (Tables 1 and 2). Russian males attempted to mate with live GP females with the

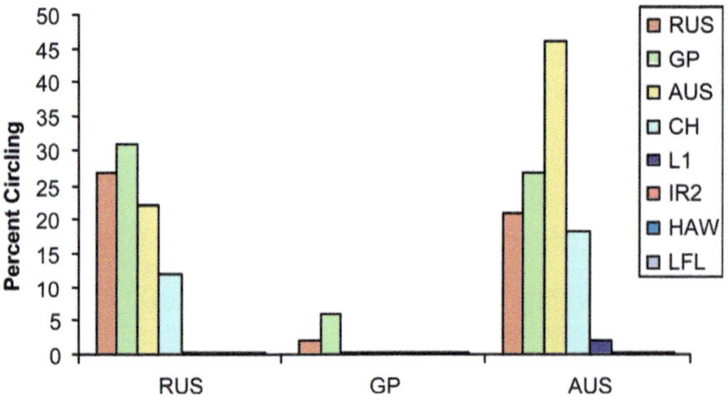

Figure 2. Comparison of EDTA extraction and reattachment of ion exchange fraction 22 to females of different clades and species. Live females were untreated, EDTA females were exposed to EDTA for 30 min, and EDTA + 22 females were exposed to EDTA then fraction 22. Percent circling is the proportion of male–female encounters that resulted in males initiating mating behavior. Results of statistical tests are presented in Table 1.

Table 1. Fisher's exact test comparing RUS male mating response (circling) to various homogamic and heterogamic females

Control female	Vs. Comparison female	Fisher's exact test p
RUS LIVE	RUS EDTA	<0.001
RUS live	RUS EDTA + F22	0.26
RUS live	GP live	0.35
GP live	GP EDTA	0.001
GP live	GP EDTA + F22	0.68
RUS live	AUS live	0.71
AUS live	AUS EDTA	0.001
AUS live	AUS EDTA + F22	0.061
RUS live	CH live	0.05
CH live	CH EDTA	0.008
CH live	CH EDTA + F22	0.45
RUS live	L1 live	<0.001
L1 live	L1 EDTA	>0.999
L1 live	L1 EDTA + F22	0.524
RUS live	LFL live	<0.001
LFL live	LFL EDTA	>0.999
LFL live	LFL EDTA + F22	>0.999

p is the probability of obtaining the result by chance.

same frequency as their own RUS females. Likewise, EDTA treatment significantly reduced GP female attractiveness, but it could be restored by exposure to fraction 22 (Table 1). A similar pattern was observed for AUS females, but exposure to fraction 22 seemed to render these females even more attractive than live AUS females, a result that is near significant by Fisher's exact test at $p = 0.061$ (Table 1).

RUS males initiated mating with live CH females at only one half the frequency as live RUS females. This response was eliminated by EDTA treatment, but restored by exposure to fraction 22. RUS males did not attempt to mate with live females of the L1, IR2, HAW, and LFL populations. More importantly, these females could not be made attractive to RUS males by exposure to fraction 22 (Table 2). We attempted to feminize RUS males by treatment with EDTA followed by expose to fraction 22. We hypothesized that conspecific males would detect them as 'females' and attempt to mate. All males thus treated failed to elicit any male mating responses.

The proteins extracted from the surface of RUS females by EDTA treatment were visualized on an SDS-PAGE gel stained with Sypro (Fig. 3). A few high molecular weight proteins (>66 kD) are present, but most of the approximately 17 prominent bands fall within the 66–12 kD range. These proteins were separated by ion exchange chromatography and the 40 one ml fractions were tested using the standard mating bioassay. Significant activity was found only in fractions 22 and 23 (Fig. 4), with Fisher's exact test $p < 0.05$. Visualization of these proteins on a SDS-PAGE gel stained with Sypro stain revealed about 10 prominent bands (Fig. 5). A 24 kD band was conspicuous in fractions 21, 22, and 23 and much reduced in other fractions.

Table 2. Mating bioassay of RUS males with females of various clades and species

Female	Live		EDTA		EDTA + F22		*Brachionus* morphospecies	COI Clade
	E	C	E	C	E	C		
RUS	230	59	303	7	206	43	*Plicatilis*	Manjavacas[a]
GP	67	21	48	3	45	12	*Plicatilis*	?
AUS	45	10	41	0	25	12	*Plicatilis*	Austria[a]
CH	43	5	66	0	85	15	*Plicatilis*	Austria[a]
L1	52	0	29	0	84	2	*Plicatilis*	*Plicatilis*[a]
LFL	76	0	65	0	34	0	*Ibericus*	Almenara[a]
IR2	30	0	34	0	49	0	*Ibericus*	Almenara[b]
HAW	34	0	53	0	55		*Rotundiformis*	*Rotundiformis*[b]

E – male–female encounters, C – circlings, F22 – ion exchange fraction 22, COI – cytochrome C oxidase subunit I gene.
[a]Gómez et al. (2002), [b]Snell & Stelzer, unpublished.

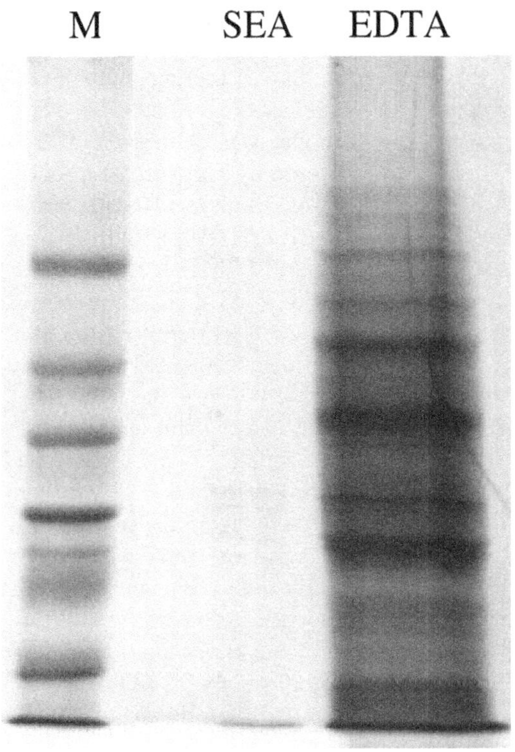

Figure 3. SDS-PAGE of EDTA-extractable surface proteins from *B. plicatilis*. M is Mark VII molecular weight markers 66–14 kD (Sigma Chemical Company). SEA is extraction with seawater, and EDTA is extraction with seawater containing 100 mM EDTA.

Discussion

The abundance of glycoproteins on the surface of rotifers was demonstrated by the binding of several fluorescently labeled lectins (Snell & Nacionales, 1990; Snell et al., 1993). A variety of lectins was tested, but only those with glucose/mannose affinity like Con A, *Lens culinaris, Vicia fava*, and *Pisum sativum* bound to females. Localization was primarily in the corona region where they bound to ciliary membranes. When these lectins were bound, females elicited significantly fewer mating responses from males. This lectin blocking of mate recognition demonstrated the functional significance of these surface glycoproteins as signals used by males to recognize mating partners. Cleavage of surface proteins by proteinase K also rendered females significantly less attractive to males (Snell et al., 1988), as did cleavage of N-linked oligosaccharrides by the glycohydrolase N-glycanase (Snell et al., 1995). These observations clearly implicated surface glycoproteins as having a key role in rotifer mate recognition.

Treatment of *B. plicatilis* females with the detergent CHAPS and EDTA removed surface proteins and eliminated the male mating response (Snell & Nacionales, 1990). The EDTA effect was especially interesting because it often did not kill the females, yet removed surface proteins critical for mate recognition. How EDTA removes proteins from membranes is not well understood, yet it is routinely used in extraction procedures for isolating active membrane proteins (Nomura & Suzuki, 1995; Ziola et al., 2000). As a strong chelator of Ca^{++} and Mg^{++} ions, EDTA disrupts electrostatic binding between peripheral membrane proteins and proteins more firmly anchored in the membrane. This selectively releases and solubilizes the peripheral membrane proteins,

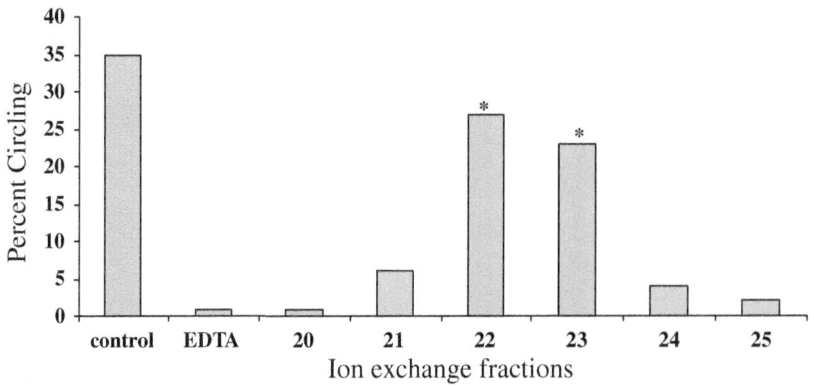

Figure 4. Ion exchange chromatography of EDTA-extractable rotifer proteins. Control is live females, numbers 20–25 refer to the ion exchange fractions. * indicates a significant difference from EDTA treated females, Fisher's exact test, $p < 0.05$. Percent circling is the proportion of male–female encounters that resulted in males initiating mating behavior.

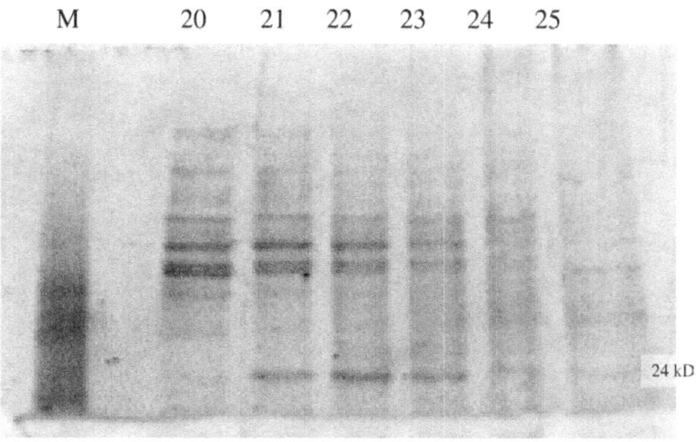

Figure 5. SDS-PAGE of ion exchange fractions of EDTA-extractable surface proteins from *B. plicatilis*. M is molecular weight markers 66–14 kD. Numbers 20–25 refer to the ion exchange fractions.

leaving the remaining membrane intact. The activity of this class of proteins in mating bioassays has demonstrated their role in rotifer mate recognition. EGTA is a more selective chelator than EDTA, chelating only Ca^{++}. EGTA's effectiveness in removing peripheral membrane proteins critical in mate recognition demonstrates that Ca^{++} and not Mg^{++} mediates the binding of these proteins to rotifer body surfaces. It further emphasizes that these proteins are only loosely associated with the body surface and are not covalently linked or transmembrane proteins.

Surface glycoproteins were removed from *B. plicatilis* RUS females by our EDTA treatment and successfully transferred to other *B. plicatilis* strains, but not all. RUS females are in the Manjavacas clade (Table 2) and their mate recognition proteins could be transferred to AUS and CH strains which are in the Austrian clade. The Manjavacas and Austrian clades are closely related phylogenetically, separated by only about 18 ITS nucleotide substitutions and about 150 (21%) COI nucleotide substitutions (Gómez et al., 2002). However, RUS mate recognition proteins could not be transferred to the L1 strain which is in the *B. plicatilis sensu strictu* clade. The *B. plicatilis* clade is separated from the Manjavacas clade by about 32 ITS nucleotide substitutions and about 235 (33%) COI nucleotide substitutions. Furthermore, no successful transfers of mate recognition proteins were made from the Manjavacas clade to any *B. ibericus* and *B. rotundiformis* species. The conclusion is that successful transfer of mate recognition proteins among rotifer species is possible only when they are quite closely related.

What mechanism limits the transfer of mate recognition proteins to only very closely related species? The model that we envision is one where a primary signal glycoprotein on females interacts species-specifically with a male receptor. This triggers the male mating response when the stimulus is sufficiently intense. This signal glycoprotein is only loosely bound to the body surface of females, but it is oriented so that its oligosaccharrides are accessible to males. Surrounding surface proteins on females provide anchor sites and hold this primary signal glycoprotein in the right orientation to be detected by male receptors. These surrounding proteins are important because their structure determines the affinity of the body surface for the signal glycoprotein and its orientation to the external environment. As species diverge, small changes in the structure of their surface proteins can reduce affinity for the signal glycoprotein or alter its orientation. Likewise, small changes in the structure of the signal glycoprotein itself could modify its ability to interact appropriately with the other body surface proteins on females.

Transfer of fraction 22 proteins to EDTA treated RUS females restored their attractiveness to conspecific males. However, we were unable to transfer these proteins to EDTA treated males and render them attractive to other conspecific males. *B. plicatilis* males almost never attempt to mate with other males and this aversion could not be overcome by transferring female proteins to males. This suggests a sexual dimorphism in the body surface proteins of *B. plicatilis* males and females.

Snell et al. (1995) described a rotifer glycoprotein called gp29 that seemed to be a prime candidate for the primary signal glycoprotein. It was purified from *B. plicatilis* using lectin affinity chromatography and its activity was probed using a polyclonal antibody raised against this protein. Unfortunately, not enough protein was isolated to obtain any amino acid sequence and the polyclonal antibody has been expended. Consequently, there is no way to relate gp29 to the EDTA extracted proteins described in this paper that have clear activity in mate recognition. The proteins described here should be regarded as an independent approach to mating protein isolation that hopefully will lead to the characterization and sequencing of proteins and genes responsible for mate recognition in rotifers.

Acknowledgements

We thank Manuel Serra and John Gilbert for comments that improved this paper. We acknowledge financial support to C.P.S. by Deutsche Forschungsgemeinschaft (Grant STE 1021/1–1).

References

Blows, M. W. & R. A. Allan, 1998. Levels of mate recognition within and between two *Drosophila* species and their hybrids. American Naturalist 152: 826–837.

Ciros-Perez,, A. Gómez & M. Serra, 2001. On the taxonomy of three sympatric species of the *Brachionus plicatilis* (Rotifera) complex from Spain, with the description of B. ibericus n. sp. Journal of Plankton Research 23: 1311–1328.

Coyne, J. A. & B. Charlesworth, 1997. Genetics of a pheromonal difference affecting sexual isolation between *Drosophila mauritiana and D. sechellia Genetics*. 145: 1015–1030.

Derry, A., P. D. N. Hebert & E. E. Prepas, 2003. Evolution of rotifers in saline and subsaline lakes:a molecular genetic approach. Limnology and Oceanography 48: 675–685.

Etges, W. J. & M. A. Ahrens, 2001. Premating isolation is determined by larval-rearing substrates in cactophilic *Drosophila mojavensis*. V. Deep geographic variation in epicuticular hydrocarbons among isolated populations. American Naturalist 158: 585–598.

Gómez, A. & M. Serra, 1995. Behavioral reproductive isolation among sympatric strains of *Brachionus plicatilis* (Müller, 1786) insights into the status of this taxonomic species. Hydrobiologia 313/314: 111–119.

Gómez, A. & T.W. Snell, 1996. Sibling species and cryptic speciation in the *Brachionus plicatilis* species complex (Rotifera). Journal of Evolutionary Biology 9: 953–964.

Gómez, A. & G. R. Carvalho, 2000. Sex, parthenogenesis and the genetic structure of rotifers: microsatellite analysis of contemporary and resting egg bank populations. Molecular Ecology 9: 203–214.

Gómez, A., M. Serra,, G. R. Carvalho & D. H. Lunt, 2002. Speciation in ancient cryptic species complexes: evidence from the molecular phylogeny of *Brachionus plicatilis* (Rotifera). Evolution 56: 1431–1444.

Nomura, K. & N. Suzuki, 1995. Sea-urchin ovoperoxidase-solubilization and isolation from the fertilization envelope,some structural and functional properties, and degradation by hatching enzyme. Archives of Biochemistry and Biophysics 319: 525–534.

Ortells, R., T. W. Snell, A. Gómez & M. Serra., 2000. Patterns of genetic differentiation in resting egg banks of a rotifer species complex in Spain. Archiv für Hydrobiologie 149: 529–551.

Segers, H., 1995. Nomenclatural consequences of some recent studies on *Brachionus plicatilis* (Rotifera, Brachionidae). Hydrobiologia 313/314: 121–122.

Serra, M., A. Galiana & A. Gómez, 1997. Speciation in monogonont rotifers. Hydrobiologia 358: 63–70.

Serra, M., A. Gómez & M. J. Carmona, 1998. Ecological genetics of *Brachionus* sibling species. Hydrobiologia 387/388: 373–384.

Snell, T. W. & C. A. Hawkinson, 1983. Behavioral reproductive isolation among populations of the rotifer *Brachionus plicatilis*. Evolution 37: 1294–1305.

Snell, T. W., M. J. Childress & B. C. Winkler, 1988. Characteristics of the mate recognition factor in the rotifer *Brachionus plicatilis*. Comparative Biochemistry and Physiology 89A: 481–485.

Snell, T. W., 1989. Systematics, reproductive isolation and species boundaries in monogonont rotifers. Hydrobiologia 186/187: 299–310.

Snell, T. W. & M. A. Nacionales, 1990. Sex pheromone communication in *Brachionus plicatilis* (Rotifera). Comparative Biochemistry and Physiology 97A: 211–216.

Snell T. W., Morris P. D. & 1993. Sexual communication in copepods and rotifers. Hydrobiologia 255/256: 109–116.

Snell, T. W., P. D. Morris & G. A. Cecchine, 1993. Localization of the mate recogntion pheromone in Brachionus plicatilis (O.F. Müller) (Rotifera) by fluorescent labeling with lectins. Journal of Experimental Marine Biology and Ecology 165: 225–235.

Snell, T. W., R. Rico-Martinez, L. N. Kelly & T. E. Battle, 1995. Identification of a sex pheromone from a rotifer. Marine Biology 123: 347–353.

Snell, T. W., 1998. Chemical ecology of rotifers. Hydrobiologia 387/388: 267–276.

Suatoni, L. A., 2003. Phylogenetic and biogeographic patterns in the rotifer species group *Brachionus plicatilis*. Ph.D. Dissertation, Yale University, New Haven, Connecticut, USA.

Ziola, B., L. Gee, N. N. Berg & S. Y. Lee, 2000. Serogroups of the beer spoilage bacterium Megasphaera cerevisiae correlate with the molecular weight of the major EDTA-extractable surface protein. Canadian Journal of Microbiology 46: 95–100.

Hydrobiologia (2005) 546:275–279

A. Herzig, R.D. Gulati, C.D. Jersabek & L. May (eds.) Rotifera X: Rotifer Research: Trends, New Tools and Recent Advances
DOI 10.1007/s10750-005-4208-0

Maternal effect by stem females in *Brachionus plicatilis*: effect of starvation on mixis induction in offspring

Atsushi Hagiwara[1,*], Yoji Kadota[1] & Akinori Hino[2]
[1]*Graduate School of Science & Technology, Nagasaki University, Bunkyo 1-14, 852-8521, Nagasaki, Japan*
[2]*Graduate School of Agricultural and Life Sciences, The University of Tokyo, Bunkyo, 113-8657, Tokyo, Japan*
*(*Author for correspondence: E-mail: hagiwara@net.nagasaki-u.ac.jp)*

Key words: rotifera, *Brachionus plicatilis*, stem female, maternal effect, starvation, mixis

Abstract

We examined whether starvation during the initial period of life in stem females affected reproductive characteristics of the offspring. Starvation treatment had different effects on rotifers hatched from resting eggs and those hatched from amictic eggs. When stem females experienced starvation after hatching, this induced a higher percentage of mixis in their offspring. When the same starvation treatment was applied to rotifers hatched from amictic eggs, there was no effect on the induction of mixis. It is probable that stem females hatched from resting eggs have specific features that are vulnerable to unfavorable environmental conditions, and that these features can be inherited by their offspring through the maternal cytoplasm.

Introduction

Rotifers hatched from resting eggs are diploid amictic females. These are known as stem females. Stem females are difficult to distinguish from asexually produced amictic females by their morphology alone. The only observable difference is the presence of some yolk derived from the cytoplasm of the resting egg for a few hours after hatching.

Several biological features of stem females have been reported. For example, external conditions such as incubation temperature and salinity during resting egg formation affect the frequency of mixis of a rotifer clone initiated by stem females (Hino & Hirano, 1985, 1988). Variations in food availability during population growth and resting egg formation affect the hatchability of resting eggs and the population growth rate of the derived clones (Hagiwara & Hino, 1990). External conditions during initial developmental stages to complete resting egg maturation affect the hatching patterns of the eggs (either sporadic or simultaneous), frequency of mixis induction and the population growth rate of the derived clones (Hagiwara & Hino, 1989).

It is possible that stem females are especially vulnerable to environmental effects, and that these features may be inherited by their offspring through the maternal cytoplasm. In support of this hypothesis, Hino & Hirano (1977) found that a change of food decreased mixis induction in *Brachionus plicatilis* cultures, except when the food was changed to that ingested by stem females. In order to investigate this hypothesis further, we conducted research to determine whether the starvation of stem females immediately after hatching affects the reproductive characteristics of their offspring.

Materials and methods

We used *Brachionus plicatilis* Russian strain, which was obtained from Terry W. Snell, Georgia Institute of Technology, USA. This strain was selected for the experiments because, among all of the rotifer strains maintained in our laboratory, it shows the highest percentage of mixis induction.

In order to obtain resting eggs, the Russian strain was clonally cultured on *Nannochloropsis oculata* at 25 °C in 18 ppt diluted seawater. *N. oculata* was grown in modified Erd-Schreiber medium (Hagiwara et al., 1994). Resting eggs were kept in total darkness at 25 °C for more than 2 weeks prior to use in the experiments.

We obtained stem females by incubating resting eggs at 25 °C under continuous light. One set of six stem females were starved for 12 h after hatching, and then fed *N. oculata* at 7×10^6 cells ml^{-1}. A second set (control), was not starved but continuously fed *N. oculata* from hatching. Amictic females were obtained by isolating egg bearing females from a 500 ml rotifer batch culture, and shaking them to remove the amictic eggs. Two sets of six of these neonates were subjected to the same two treatments as the two sets of stem females.

The 24 rotifers were individually cultured in glass Durham tubes (0.2 ml in water volume) for 10 successive generations. The salinity of the culture medium was maintained at 18 ppt, with the opening of the tubes was sealed with Parafilm (American Can Company) to prevent any evaporation losses that might cause a salinity change in the medium. The cultures were kept at 25 °C in total darkness.

For each generation, the first offspring hatched within 48 h after birth were used for the next generation, according to the method reported by Hino & Hirano (1977, 1985). When the first daughter turned out to be a mictic female, the first amictic females in subsequent daughters were used for the next generation. Every 24 h, neonates were isolated and reared in a Petri dish until they laid eggs and their female types were identified based on egg morphology (Hagiwara et al., 1988). Female types were identified in at least the first ten neonates. Maternal rotifers were transferred to a new tube containing fresh food. After isolating the 10th offspring, we also isolated additional neonates if they were born.

The ratio of mictic females to total females (mixis, %) was calculated from data on rotifer female types. A χ^2 contingency test was performed to see whether the starvation treatments had any different effect on neonates hatched from resting eggs in comparison with those hatched from amictic eggs.

Results

Percentage of mixis in females derived from stem females was compared between treatments (starved and non-starved) (Fig. 1). Mixis induction was found to be higher from the third generation onwards in rotifers derived from starved stem females. In contrast, in rotifers initiated from non-starved stem females, mixis induction started to decrease from the 8th generation onwards. In addition, rotifers derived from starved stem

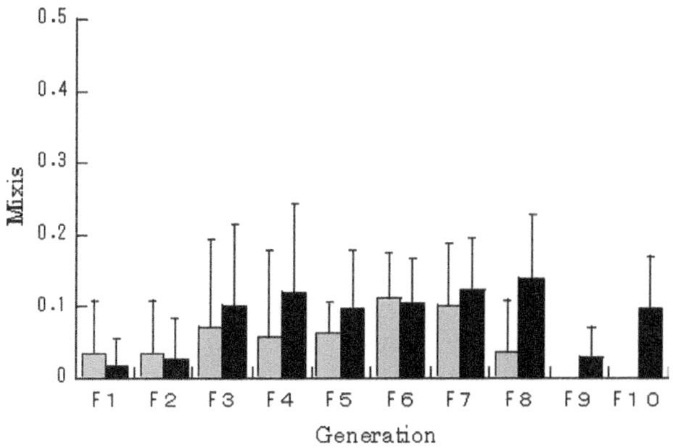

Figure 1. Changes in rate of mixis for 10 generations (F_1 to F_{10}) of *Brachionus plicatilis* derived from stem female hatched from resting eggs and reared under different feeding regimes: □ – Stem female starved for 0 h from hatching. ■ – Stem female starved for 12 h from hatching. Each column indicates an average of six replicates, and vertical bars indicate standard deviations.

Table 1. Percentage of mictic females among individuals from 10 generation of *B. plicatilis* derived from stem females, in relation to feeding regime

Feeding regime	Total number of offspring	Number of mictic females	Mixis (%)	χ^2 value	*p*
Starved for 0 h	533	26	4.9		
Starved for 12 h	639	56	8.8	5.882	0.0153

females induced a significantly higher percentage of mixis, than those derived from non-starved stem females (Table 1).

Figure 2 shows the same comparison in females derived from starved and non-starved females hatched from amictic eggs. In both treatments, mixis induction decreased in the last five generations compared to the first five generations. Starvation treatments with females hatched from amictic eggs did not affect mixis induction (Table 2).

Discussion

Different phytoplankton species have different nutrient composition. When given as food to rotifers, these differences affect lifespan and fecundity resulting to different parthenogenetic population growth (Hirayama et al., 1979; Korstad et al., 1989), sexual reproduction (Snell & Hoff, 1985; Snell, 1986; Hamada et al., 1993) and resting egg hatching (Hagiwara & Hino, 1990). However, the way in which the food environment of stem females affects the reproductive characteristics of derived clones is not yet understood. The present study suggests that newly recruited rotifers from resting eggs receive specific signals from the environment that determines the mixis responsiveness of successive generations.

Starvation treatment had different effects on rotifers hatched from resting eggs and those hatched from amictic eggs. When stem females experience starvation, their offspring experienced a higher percentage of mixis. However, when the same starvation treatment was applied to rotifers hatched from amictic eggs, this did not result in elevated mixis in later generations.

Previous research reported that mixis in rotifer populations is induced under moderate environmental conditions (Lubzens et al., 1980; Snell, 1986; Hagiwara et al., 1988), through the accumulation of chemical substances in water that are derived from the conditioning of rotifers (Hino & Hirano, 1976; Carmona et al., 1993; Stelzer & Snell, 2003). Hagiwara et al. (1994) reported that a water-soluble extract from rotifers or bacteria in water induces mixis.

In the heterogony of rotifers, sexual reproduction to produce fertilized resting eggs is generally recognized as an ecologically important process that enable these animals to avoid unfavorable

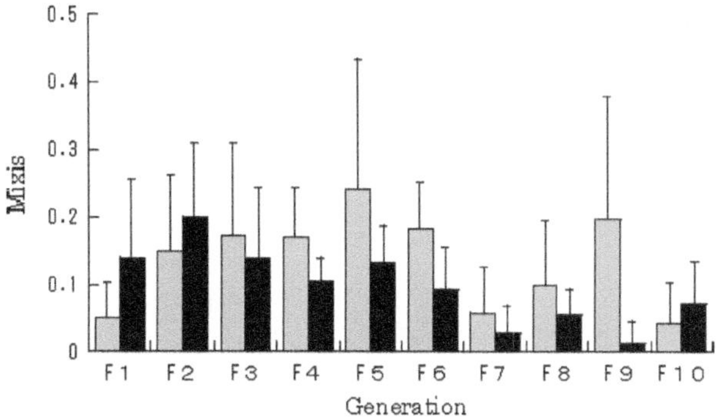

Figure 2. Changes in rate of mixis for 10 generations (F_1 to F_{10}) of *Brachionus plicatilis* derived from rotifers hatched from amictic eggs and reared under different feeding regimes: □ – Rotifer starved for 0 h from hatching from amictic egg. ■ – Rotifer starved for 12 h from hatching from amictic eggs. Each column indicates an average of six replicates, and vertical bars indicate standard deviations.

Table 2. Percentage of mictic females among individuals from 10 generations of *B. plicatilis* derived from females hatched from parthenogenetic eggs, in relation to feeding regime

Feeding regime	Total number of offspring	Number of mictic females	Mixis (%)	χ^2 value	*p*
Starved for 0 h	675	92	13.6		
Starved for 12 h	686	68	9.9	3.578	0.585

environments. In some invertebrate species, unfavorable environmental conditions are the signal to switch to a different reproductive mode and produce resting eggs. These include cladoceran, where sexual reproduction is triggered by adverse environmental signals, such as shorter photoperiod, starvation and a chemically mediated crowding stimulus (Kleiven et al., 1992) and timed to occur when the environment is about to change. However, most of the work to date, suggests that sexual reproduction in rotifers is induced when the environment is moderate. During periods of starvation, mixis is suppressed (Snell & Boyer, 1988). One fertilized mictic female can produce up to 6–8 resting eggs under moderate environmental conditions (Hagiwara & Hino, 1990). Sexual reproduction from mixis induction to resting egg formation in *B. plicatilis* takes about 5 days at 25 °C (Hagiwara et al., 1988). However, when environmental conditions turn unfavorable, sexual reproduction and the production of resting eggs tends to be delayed. In addition, resting eggs formed in toxic environments, have low hatchability (Marcial et al., 2005).

There is no evidence in rotifers that sexual reproduction occurs in response to stress, or that rotifer sexual reproduction is induced by adverse conditions, as is found in cladocerans, although asexual diapause eggs in *Synchaeta pectinata* are produced in response to food limitation (Gilbert & Schreiber 1995, 1998). Our results suggest that starvation at an early age affects stem females resulting in higher mixis in later generations, and such maternal effects continue for at least 10 generations. Even when exposed to starvation, mictic female production by amictic females is unchanged when compared to a moderate environment.

The ability to induce mixis is primarily determined genetically (Hino & Hirano, 1976). This is illustrated by the fact that the Russian strain used in this study consistently shows the highest level of mixis among the 70 strains of *B. plicatilis* complex maintained in our laboratory.

In rotifer cultures initiated from stem females, an accumulation of parthenogenetic generations after hatching from resting egg is necessary for mixis induction (Hino & Hirano, 1977; Gilbert, 2002). The current experiments confirm and extend these results. In nature, the appearance of a rotifer population is often limited to a certain period of the year, and such a response by stem females will, effectively, enable the rotifer population to avoid an unfavorable fluctuating environment.

Acknowledgements

This study was partly supported by Grants-in-Aid from the Ministry of Education, Science and Culture, Japan, and from the prefectural collaboration of regional entities for the advancement of technological excellence, JST. Authors would like to express thanks to Linda May for improving the manuscript.

References

Carmona, M. J., M. Serra & M. R. Miracle, 1993. Relationships between mixis in *Brachionus plicatilis* and preconditioning of the culture medium by crowding. Hydrobiologia 255/256: 145–152.

Gilbert, J. J. & D. K. Schreiber, 1995. Induction of diapausing amictic eggs in *Synchaeta pectinata*. Hydrobiologia 313/314: 345–350.

Gilbert, J. J. & D. K. Schreiber, 1998. Asexual diapause induced by food limitation in the rotifer *Synchaeta pectinata*. Ecology 79: 1371–1381.

Gilbert, J. J., 2002. Endogenous regulation of environmentally induced sexuality in a rotifer: a multigenerational parental effect induced by fertilization. Freshwater Biology 47: 1633–1641.

Hagiwara, A., A. Hino & R. Hirano, 1988. Effects of temperature and chlorinity on resting egg formation in the rotifer *Brachionus plicatilis*. Nippon Suisan Gakkaishi 54: 569–575.

Hagiwara, A. & A. Hino, 1989. Effect of incubation and preservation on resting egg hatching and mixis in the derived clones of the rotifer *Brachionus plicatilis*. Hydrobiologia 186/187: 415–421.

Hagiwara, A. & A. Hino, 1990. Feeding history and hatching of resting eggs in the marine rotifer *Brachionus plicatilis*. Nippon Suisan Gakkaishi 56: 1901–1907.

Hagiwara, A., K. Hamada, S. Hori & K. Hirayama, 1994. Increased sexual reproduction in *Brachionus plicatilis* (Rotifera) with the addition of bacterial and rotifer extracts. Journal of Experimental Marine Biology and Ecology 181: 1–8.

Hamada, K., A. Hagiwara & K. Hirayama, 1993. Use of preserved diet for rotifer *Brachionus plicatilis* resting egg formation. Nippon Suisan Gakkaishi 59: 85–91.

Hino, A. & R. Hirano, 1976. Ecological studies of the mechanism of bisexual reproduction in the rotifer *Brachionus plicatilis*-I. General aspects of bisexual reproduction. Nippon Suisan Gakkaishi 42: 1093–1099.

Hino, A. & R. Hirano, 1977. Ecological studies of the mechanism of bisexual reproduction in the rotifer *Brachionus plicatilis*-II. Effects of cumulative parthenogenetic generation on the frequency of bisexual reproduction. Nippon Suisan Gakkaishi 43: 1147–1155.

Hino, A. & R. Hirano, 1985. Relationship between the temperature given at the time of fertilized egg formation and bisexual reproduction pattern in the deriving strain of the rotifer *Brachionus plicatilis*. Nippon Suisan Gakkaishi 51: 511–514.

Hino, A. & R. Hirano, 1988. Relationship between water chlorinity and bisexual reproduction rate in the rotifer *Brachionus plicatilis*. Nippon Suisan Gakkaishi 54: 1329–1332.

Hirayama, K., K. Takagi & H. Kimura, 1979. Nutritional effect of eight species of marine phytoplankton on population growth of the rotifer, *Brachionus plicatilis*. Nippon Suisan Gakkaishi 45: 11–16.

Kleiven, O. T., P. Larsson & A. Hobaek, 1992. Sexual reproduction in *Daphnia magna* requires three stimuli. Oikos 65: 197–206.

Korstad, J., Y. Olsen & O. Vadstein, 1989. Life history characteristics of *Brachionus plicatilis* (Rotifera) fed different algae. Hydrobiologia 186: 46–50.

Lubzens, E., R. Fishler & V. Berdugo-White, 1980. Induction of sexual reproduction and resting egg production in *Brachionus plicatilis* reared in sea water. Hydrobiologia 73: 55–58.

Marcial, H. S., A. Hagiwara & T. W. Snell, 2005. Effect of some pesticides on reproduction of euryhaline rotifer *Brachionus plicatilis* Müller. Hydrobiologia 546: 569–575.

Snell, T. W., 1986. Effect of temperature, salinity and food level on sexual and asexual reproduction in *Brachionus plicatilis* (Rotifera). Marine Biology 92: 157–162.

Snell, T. W. & F. H. Hoff, 1985. The effect of environmental factors on resting egg production in the rotifer *Brachionus plicatilis*. Journal of World Mariculture Society 16: 484–497.

Snell, T. W. & E. M. Boyer, 1988. Thresholds for mictic female production in the rotifer *Brachionus plicatilis*. Journal of Experimental Marine Biology and Ecology, 124: 73–85.

Stelzer, C. P. & T. W. Snell, 2003. Induction of sexual reproduction in *Brachionus plicatilis* (Monogononta, Rotifera) by a density-dependent chemical cue. Limnology and Oceanography 48: 939–943.

Hydrobiologia (2005) 546:281–289
A. Herzig, R.D. Gulati, C.D. Jersabek & L. May (eds.) Rotifera X: Rotifer Research: Trends, New Tools and Recent Advances
DOI 10.1007/s10750-005-4209-z

Restoration of tropical peat swamp rotifer communities after perturbation: an experimental study of recovery of rotifers from the resting egg bank

Supenya Chittapun[1,*], Pornsilp Pholpunthin[2] & Hendrik Segers[3]
[1]*Department of Biotechnology, Faculty of Science and Technology, Thammasat University, Rangsit Center, Pathumthani Province, 12121, Thailand*
[2]*Department of Biology, Faculty of Science, Prince of Songkhla University, 90112, Hat Yai, Songkla, Thailand*
[3]*Royal Belgian Institute of Natural Sciences, Freshwater Biology, Vautierstraat 29, B-1000, Brussels, Belgium*
*(*Author for correspondence: E-mail: Chittapun@yahoo.com)*

Key words: resting eggs, rotifer, tropical, peat swamp, disturbance

Abstract

In order to assess the recovery potential of tropical freshwater communities after disturbance, we performed an experimental study on the effects of exposure conditions and durations of storage on hatching of rotifer resting eggs in sediment. Well-mixed surface sediment samples from Mai Khao peat swamp on Phuket Island, Thailand, were stored under three conditions (cold –4 °C & dark: CD; ambient –32–42 °C & dark: AD; and ambient & daylight conditions: AL), for different periods of time (1, 2, 4, 6, 12, 18 and 24 months).The number of species hatching from the sediment was significantly affected by treatment for both short- (1–6 months) and long-term (6–20 months) exposure. Significant effects of short- and long-term exposure within treatments were also present. Both factors interacted significantly. Regarding numbers of specimens hatching, no short-term effects of differences in treatment condition were found, but increasing the duration did have an effect. Significant effects of treatment occurred after 6 months, in addition to prolonged effects of duration. Again, both factors interacted significantly. These experiments indicate that exposure time has a strong impact on the viability of resting eggs, whereas, an effect of exposure condition appears only after 6 months. So, recovery of rotifer communities from resident sediment egg banks in disturbed peat swamps can only be effectively attained when restoration occurs within a relatively short period after perturbation.

Introduction

Presently, the most severe threat to the world's wetlands is posed by land uses that destroy or severely damage habitats (Finlayson & Moser, 1991). Human-induced pressures affect ecosystem functioning as well as biodiversity at all levels, from ecosystems to organisms, Whereas populations of some organisms are irreversibly affected, others may be able to recover from the effects of disturbances. This resilience results at least partly from their potential to survive periods of adverse conditions through resistant, dormant stages.

Monogonont Rotifera, being of prime ecological importance in freshwater ecosystems, has resting eggs or cysts as dormant stages (Gilbert, 1974). These are diapausing embryos produced by fertilized mictic females. Sexual reproduction is induced by a variety of cues including the occurrence of environmental changes associated with habitat deterioration. Hatching of these resting eggs generally occurs in coincidence of favorable conditions in the habitat, and results in the re-establishment of populations (Pourriot & Snell, 1983; Ricci, 2001). Resting eggs thus represent a biodiversity bank, as they can assure genetic continuity through periods of hazardous environmental conditions and offer a recolonization resource when favorable conditions return (Pourriot & Snell, 1983; Ricci, 2001).

So far, the majority of studies on rotifer resting eggs consist of investigations on resting egg production and hatching, often in relation to the use of rotifers as food source in aquaculture (Lubzens et al., 1980, 1993; Pourriot et al., 1980; Minkoff et al., 1983; Serrano et al., 1989). There are few studies on resting eggs in natural rotifer populations. Ito (1958) and Nipkow (1961) were amongst the first to study incubation of rotifer resting eggs from sediments (May, 1987). Pourriot et al. (1984) and Gilbert & Wurdak (1978) compared the morphology of resting eggs of different taxa. May (1987) performed a quantitative study of rotifers hatching from sediments from Loch Leven, Scotland, and recorded species-specific effects of temperature on the emergence of rotifers, and showed that all pelagic rotifer species found in the lake could be hatched from the sediment egg bank. Langley et al. (2001) investigated the relative importance of recruitment from the resting egg bank vs. passive dispersal in the recolonization of temporary ponds, and found that the former is by far the most important source. These studies clearly show the potential importance of resting egg banks in the restoration of rotifer communities after disturbance. However, there are several hiatuses remaining (see Ricci, 2001). For instance, no information is available on rotifer resting egg banks in tropical habitats, and little is known on any but pelagic rotifer taxa.

As for the diversity of tropical habitats, one of the most intriguing habitats is that of peat swamp forest. Previous studies on the diversity of monogonont Rotifera in peat swamps suggest that this ecosystem type has a diverse rotifer fauna, as a result of its long history and unique ecological characteristics (Chittapun et al., 1999, 2002, 2003; Chittapun & Pholpunthin, 2001; Segers & Chittapun, 2001). Unfortunately, these habitats are seriously threatened by human activities such as agriculture (e.g., transformation to arable land, eutrophication) and aquaculture (e.g., salinization resulting from discharge of saltwater from shrimp farms). These activities constitute serious threats to the general biodiversity, and diversity of Rotifera Monogononta in particular, of these ecosystems in Thailand. In order to assess if, and to what extent rotifer communities can recover after restoration of these peat swamps, we studied the recruitment of rotifers from the sediment resting egg bank stored under different condition and duration of exposition.

Materials and methods

Study area

Mai-Khao is one of the six remaining peat swamps located along Mai-Khao coast on Phuket Island, Southern Thailand (Fig. 1). Historically, the different peat swamps in the area were connected, but they are now isolated and diversified ecologically due to different human activities in each fragment. Mai-Khao peat swamp has recently become brackish as it received discharged saltwater from nearby aquaculture farming. As a result, the once thriving macrophyte vegetation has disappeared, and the accumulated layer of peat is decomposing. Because of the decline of macrophytes, the sediment in its shallow areas is now exposed to direct sunlight during the dry season.

Sediment collection, treatment and incubation

Sediment including resting eggs was collected randomly from a dry area of Mai-Khao peat swamp on 27 February 2000, yielding a total of approximately 5 kg of material. To avoid excessive differences among resting egg ages, only the top 1 cm of sediment was scraped off from the soil. The sediment was allowed to dry further under a paper cover for a month. Then, it was homogenized by removing large pieces of plant material, grinding and passing it through a 0.5 cm mesh sieve. The sediment was then divided in three equal parts, which were subjected to different treatments:

- Cold-Dark (CD): Sediment stored in an opaque box, and kept in a refrigerator (24 °C). This condition was assumed to reflect the optimal condition to retain viability in the resting eggs;
- Ambient-Dark (AD): Sediment stored in an opaque box, under ambient temperature. Condition reflects that of resting eggs deep in the sediment;

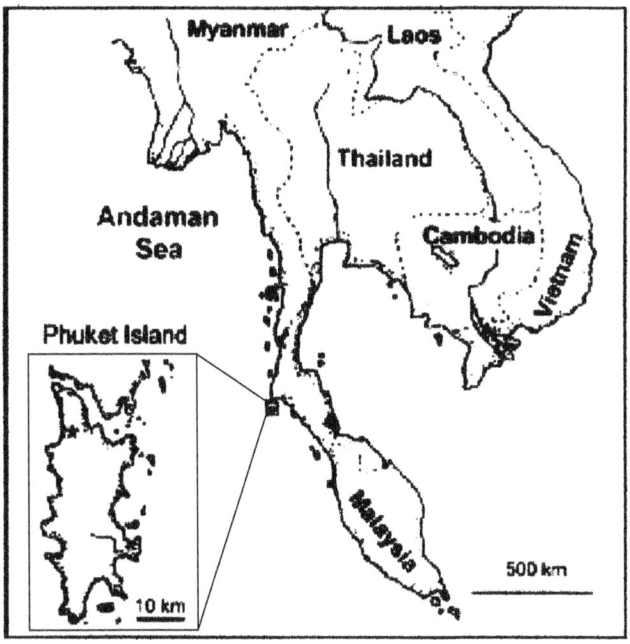

Figure 1. Location of Mai-Khao peat swamp (*) on Phuket Island, Southern Thailand.

- Ambient-Light (AL): Sediment stored in a translucent box, under ambient temperature. Condition reflects that of completely exposed resting eggs.

The sediment boxes in the 'ambient' treatments were placed together in a water bath, in order to keep the temperature in the boxes similar, and placed under a clear roof in the culture laboratory, Temperature varied in this treatment, ranging from 27 to 42 °C, but remained similar in both boxes. The duration of the treatments varied from 0 months (initial experiment), to 1, 2, 4, 6, 12, 18 and 24 months (Fig. 2).

Hatching of rotifer resting eggs was tested by placing exactly 20 g of sediment into 250 ml beakers, and adding 150 ml of distilled water. Each test was replicated four times. The beakers were placed in an incubator (Jermaks) at 28 °C, with a 12 h light–12 h dark light regime. Every four days during 3 months, the water in these beakers was poured out into a different vial, and topped back to the same level in the original

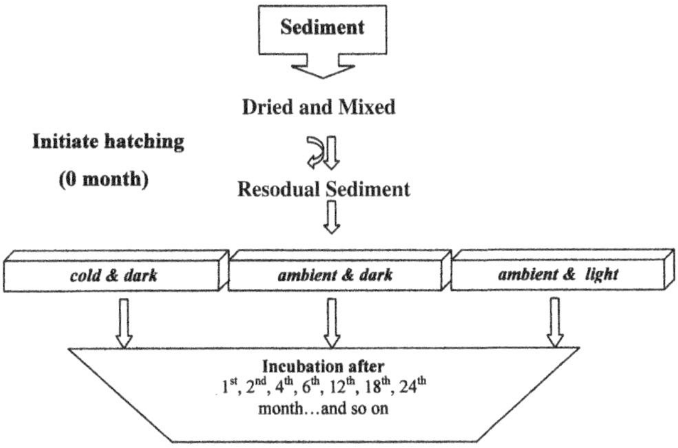

Figure 2. Flow-chart of the experiment.

beaker. To collect and count the rotifers hatched during each 4-days interval, formalin was added to the collected water to a final concentration of about 5%. Rotifers were then sorted and counted under an Olympus VM dissecting microscope, and identified using an Olympus CH-2 compound microscope. As few individuals, and no additional species were found to hatch after 3 months in an initial test experiment, the observations were stopped after this period.

Data analysis

Data on emerging of rotifers after different exposure conditions and durations were analyzed by applying Repeated Measures Analysis (SPSS statistical package for Window, Release 11.0.1). Analyses were performed on two different aspects of data: number of species emerging, and number of individual specimens hatching. To examine the combined effect of exposure condition and duration on the diversity of the rotifers hatching, each group of data was separated into two time periods. First, we tested the data for a short-term effect, by analyzing results from 2 months intervals over a period of 6 months (0, 2, 4 and 6 months); second, we tested for long-term effects by analyzing results from six months intervals over a total period of 24 months (0, 6, 12, 18 and 24 months).

Results

We analyzed the effect of treatment and exposure time on rotifer hatching by considering two aspects of diversity, viz. number of species and number of individuals.

Number of species

The number of species hatching from the sediment was affected significantly by exposure conditions, both in the short- and the long term (F = 4.97 and 10.37, $p < 0.05$, df = 2). Significant effects of short- and long-term exposure within treatments were also present (F = 20.94 and 66.25, $p < 0.01$, df = 3 and 4, respectively). Both factors interacted significantly (short-term: $F = 4.60$, $p < 0.01$, df = 6; long-term: $F = 2.68$, $p < 0.01$, df = 8) (Table 1).

The results demonstrate that rotifer species diversity was affected by exposure conditions. The highest number of species hatched from sediment kept under cold and dark conditions, fewer hatched from sediments kept in ambient temperature and in the dark, and the lowest number was recorded from sediment kept in ambient and light conditions (Table 2). This effect is significant even after short-term storage, but is especially obvious when comparing long-term effects (Fig. 3). After 24 months of storage, hatching of rotifers could only be observed from sediments stored under cold and dark conditions.

The number of individuals hatching

The number of rotifers hatching initially from the sediment amounts to 470–956 per gram. No short-term effects of differences in treatment conditions on the numbers of rotifers hatching were found ($F = 0.68$, $p > 0.05$, df = 2), although an

Table 1. Repeated measurement analysis of the short- and long-term effect of exposure condition and duration on the number of species hatching

Source	Type III Sum of squares	df	Mean square	F	Sig.
Short-term effect					
Duration	26.896	3	8.965	20.935	0.000
Exposure condition	2.260	2	1.130	4.969	0.035
Duration*Exposure condition	11.792	6	1.965	4.598	0.002
Long-term effect					
Duration	206.100	4	71.304	66.246	0.000
Exposure condition	3.227	2	1.613	10.371	0.005
Duration*Exposure condition	16.700	8	1.965	2.684	0.020

Table 2. Rotifer species hatching from the sediment exposed to different conditions

Species	Start	CD	AD	AL
Brachionus rotundiformis		+		
B. urceolaris	+	+	+	+
Cephalodella gibba	+	+		
C. innesi		+		
Encentrum pornsilpi		+	+	
Floscularia conifera	+	+	+	
Hexartha mira				+
Lecane bifurca		+	+	+
L. bulla	+	+	+	+
L. inermis		+	+	+
L. ludwigii		+	+	
L. obtusa	+	+	+	+
L. tenuiseta	+	+	+	+
L. unguitata	+			
Lindia torulosa		+		
Trichocerca pusilla		+		
T. tenuior		+		
	7	15	9	7

increase in duration did have an effect (F = 6.55, p < 0.01, df = 3). Significant effects of treatment occurred after 6 months (F = 14.83, p < 0.01, df = 2), in addition to prolonged effects of duration (F = 42.00, p < 0.01, df = 4). Again, both factors interacted significantly (short-term: F = 0.54, p < 0.01, df = 6; long-term: F = 9.05, p < 0.01, df = 8) (Table 3).

The results point out that time also has a significant effect on rotifer diversity in terms of number of specimens hatching. An additional effect of exposure condition only becomes significant after 6 months. As before, cold and dark conditions appears to affect hatching the least (Fig. 4).

Discussion

Species composition

Throughout the 2 years of the experiment, 17 rotifer species emerged from the sediment (Table 1). This equals to only 23.5% of the total rotifer record from the swamp. One species, *Lindia torulosa*, emerged from the sediment but was never found in regular plankton samples collected in the swamp. This discrepancy is not unexpected. Evidently, it reflects the difference in sampling intensity between the zooplankton survey (ca. 10 vertical hauls in different parts of the swamp monthly, over a period of 16 months) and the collection of sediment for the experiment (point sample). Moreover, it is unlikely that the single sediment sample adequately reflects the habitat heterogeneity of a shallow peat swamp in the composition of its resting egg bank. It should also be noted that the majority of species recorded in the zooplankton samples are littoral or benthic animals, and it is known that at least some of these attach their resting eggs to a substratum, or are otherwise selective in this respect. Hence, some rotifers inhabiting Mai-Khao peat swamp may not have been present as resting eggs in the sediment collected for the experiment. Additionally, as we collected exposed sediment, it cannot be excluded that particularly vulnerable taxa may already have been eliminated from the active resting egg bank. Finally, the incubation procedure applied in the experiment may not have generated the necessary cue for hatching of some taxa.

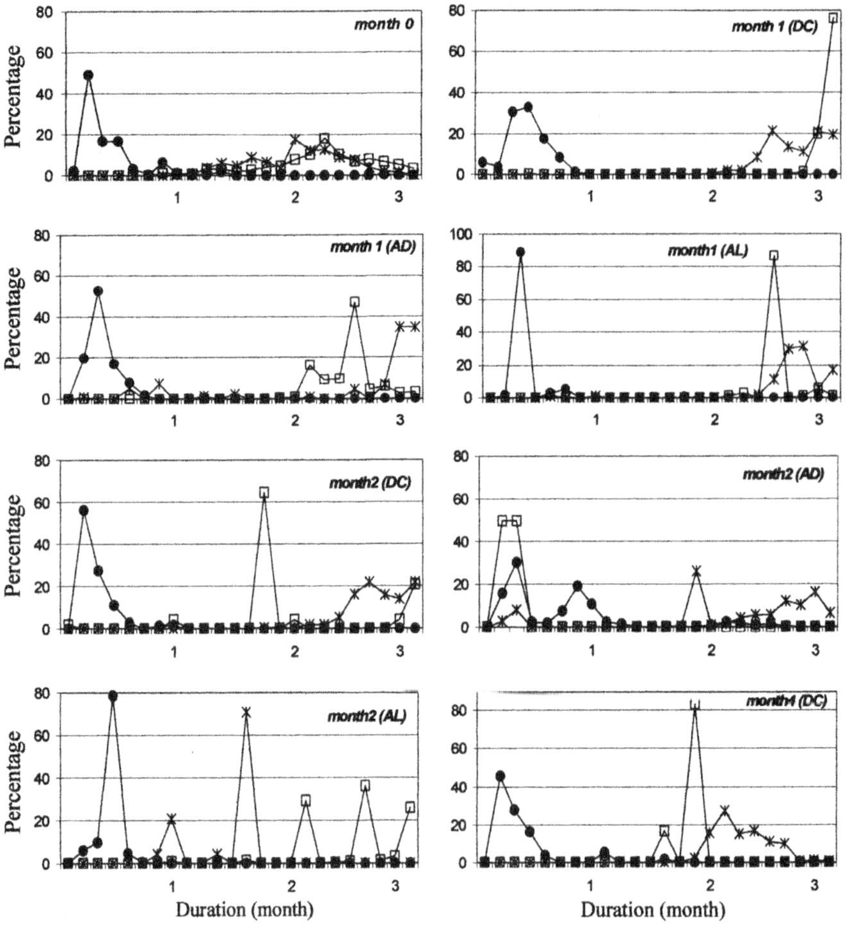

Figure 3. The number of species hatching from different exposure conditions and durations (mean ± 1SE.) (□ = cold and dark, □ = ambient and dark and ▲ = ambient and light).

Table 3. Repeated measurement analysis of the short- and long-term effect of exposure conditions and duration on the number of rotifers hatching

Source	Type III sum of squares	df	Mean square	*F*	Sig.
Short-term effect					
Duration	48.474	3	16.158	6.546	0.002
Exposure condition	0.267	2	0.134	0.681	0.530
Duration*Exposure condition	7.971	6	1.329	0.538	0.774
Long-term effect					
Duration	515.447	4	128.862	42.004	0.000
Exposure condition	25.949	2	12.974	14.826	0.001
Duration*Exposure condition	148.523	8	18.565	9.052	0.000

A striking observation is that the first species to emerge from the sediment invariably turned out to be *B. urceolaris* (Fig. 5). More than 50% of *B. urceolaris* individuals hatched within two weeks of incubation. Both observations support the hypothesis that *B. urceolaris* is a pioneer species, and suggest that the species responds relatively quickly to environmental cues.

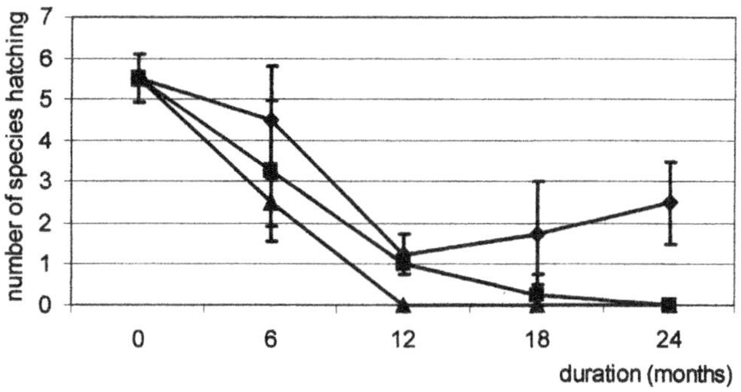

Figure 4. The number of specimens hatching of all species and of the three commonest species from different exposure conditions and durations (mean ± 1SE.): 5a = all species, 5b = *B. urceolaris*, 5c = *L. bulla*, 5d = *L. obtusa* (□ = cold and dark, □ = ambient and dark and ▲ = ambient and light).

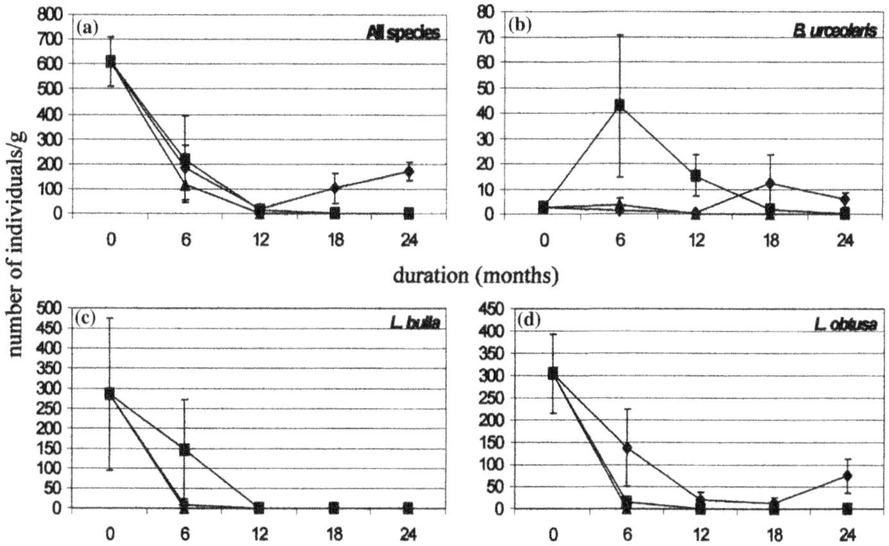

Figure 5. Percentage of hatching of the three commonest species for 3 months (● = *Brachionus urceolaris*, □ = *Lecane bulla*, * = *L. obtusa*).

Effects of treatment and exposure time

Exposure time plays an important role in the recovery of rotifer diversity from the sediment egg bank. The longer the sediment egg bank is stored, the lower the number of species and individuals of rotifer that emerge. Our observations clearly demonstrate that resting eggs have a limited, and probably species-specific viability. The results we obtained for various rotifer species are in contrast with the report on *Brachionus plicatilis*-group, in which 100% of resting eggs desiccated for up to 6 months can be made to hatch (Lubzens et al., 1980). This variability in resting egg duration is further illustrated by Kotani et al., (2001), who report hatching of resting eggs of *B. rotundiformis* of over 60 years old. In addition, we here present the first quantitative data indicating that the time lapse between dehydration and effective hatching also varies between species.

Exposure conditions have obvious effects after 6 months of storage. There is a significant difference in the number of species and individuals hatching after exposure to cold and dark conditions, in comparison to resting eggs exposed to ambient temperatures and light conditions. That cool and dark conditions extend diapause, and increase the viability of stored rotifer resting eggs, has been reported by many researchers (Pourriot et al., 1980; Minkoff et al., 1983; Pourriot & Snell, 1983; Hagiwara & Hino, 1989). The lower temperature and absence of light may prevent degradation of compounds, and/or inhibit bacterial development damaging the resting eggs.

Conclusions

Our results demonstrate a strong effect of duration on diversity both in terms of species richness and in number of specimens hatching. Exposure conditions start having significant effects after periods as short as 6 months. This contrasts with general views that rotifer resting eggs are effective for long-term survival of rotifers (e.g., Nogrady et al., 1993). It should be borne in mind that most studies on rotifer diapause are conducted on material stored under optimal conditions (cold and dark), which may not realistically reflect natural conditions, especially when dealing with tropical organisms. This may result in over-estimating the significance of resting egg banks as source for re-establishing populations in nature. The results presented here show that rotifer resting eggs have only a limited viability, and may not be effective in serving as source for recovery of rotifer diversity, even for short-term disturbances.

So, recovery of rotifer communities from sediment egg banks in disturbed peat swamps can only be effectively attained when restoration occurs within a relatively short period after perturbation.

Acknowledgements

This work was supported by Royal Golden Jubilee Ph.D. Program No. 4.B.PS/42 and partially supported by TRF/BIOTEC Special Program for Biodiversity Research and Training grant BRT 541051. We thank two anonymous referees for their valuable suggestions.

References

Chittapun, S., P. Pholpunthin & H. Segers, 1999. Rotifera from peat-swamps in Phuket Province, Thailand, with the description of a new *Colurella* Bory de St. Vincent. International Review of Hydrobiology. 84: 587–593.

Chittapun, S. & P. Pholpunthin, 2001. The rotifer fauna of peat-swamps in southern Thailand. Hydrobiologia 446/447: 255–259.

Chittapun, S., P. Pholpunthin & H. Segers, 2002. Rotifer diversity in a peat-swamp in southern Thailand (Narathiwas Province) with the description of a new species of *Keratella* Bory de St. Vincent. Annales de Limnologie. 38: 185–190.

Chittapun, S., P. Pholpunthin, & H. Segers, 2003. Contribution to the knowledge of Thai microfauna diversity: notes on rare peat swamp Rotifera, with the description of a new *Lecane* Nitzsch, 1872. Hydrobiologia. 501: 7–12.

Finlayson M., & M. Moser, 1991. Wetland. International Waterfowl and Wetlands Research Bureaw (IWRB). Hong Kong, 224 pp.

Gilbert, J. J., 1974. Dormancy in rotifers. Transactions of the American Microscopical Society 93: 490–513.

Gilbert, J. J. & E. S. Wurdak, 1978. Species-specific morphology of resting eggs in the rotifer *Asplanchna*. Transactions of the American Microscopical Society 97: 330–339.

Hagiwara, A. & A. Hino, 1989. Effect of incubation and preservation of resting egg hatching and mixis in the derived clones of the rotifer *Brachionus plicatilis*. Hydrobiologia 186/187: 415–421.

Ito, T., 1958. Studies on the 'Mizukawari' in eel culture ponds. X. The density of dormant eggs of rotifer on bottom deposits in eel culture ponds. Report of the Faculty of Fisheries, Prefectoral University Mie. 3: 170–177.

Kotani, T., M. Ozaki, K. Matsuoka, T. W. Snell & A. Hagiwara, 2001. Reproductive isolation among geographically and temporally isolated marine *Brachionus* strains. Hydrobiologia 446/447: 283–290.

Langley, J. M., R. J. Shiel, D. L. Nielsen & J. D. Green, 2001. Hatching from the sediment egg-bank, or aerial dispersing? – the use of mesocosms in assessing rotifer biodiversity. Hydrobiologia 446/447: 203–211.

Lubzens, E., R. Fishler & V. Berdugo-White, 1980. Induction of sexual reproduction and resting egg production in *Brachionus plicatilis* reared in sea water. Hydrobiologia 73: 55–58.

Lubzens, E., Y. Wax, G. Minkoff & F. Adler, 1993. A model evaluating the contribution of environmental factors to the production of resting eggs in the rotifer *Brachionus plicatilis*. Hydrobiologia 255/256: 127–138.

May, L., 1987. Effect of incubation temperature on the hatching of rotifer resting eggs collected from sediments. Hydrobiologia 147: 335–338.

Minkoff, G., E. Lubzens & D. Kahan, 1983. Environment factors affecting hatching of rotifer *(Brachionus plicatilis)* resting eggs. Hydrobiologia 104: 61–69.

Nipkow, F., 1961. Die Rädertiere im Plankton des Zürichsees und ihre Entwicklungsphasen. Schweizerische Zeitschrift für Hydrologie 23: 398–461.

Nogrady, T., R. L. Wallace & T. W. Snell, 1993. Rotifera vol. 1: Biology, Ecology and Systematics. Guides to the Identification of the Microinvertebrates of the Continental Waters of the World 4 (H. J. Dumont ed.),. SPB Academic Publishing bv, The Hague, 142 pp.

Pourriot, R. & T. W. Snell, 1983. Resting eggs in rotifers. Hydrobiologia 104: 213–224.

Pourriot, R., C. Rougier & D. Benest, 1980. Hatching of *Brachionus rubens* O. F. Müller resting eggs (Rotifers). Hydrobiologia 73: 51–54.

Pourriot, R., D. Benest, P. Clément & C. Rougier, 1984. Morphologie comparée d'oeufs de durée de brachionides. Bulletin de la Société Zoologique de France 109: 231–138.

Ricci, C., 2001. Dormancy patterns in rotifers. Hydrobiologia 446/447: 1–11.

Serrano, L., M. Serra & M. R. Miracle, 1989. Size variation in *Brachionus plicatilis* resting eggs. Hydrobiologia 186/187: 381–386.

Segers, H. & S. Chittapun, 2001. The interstitial Rotifera of a tropical freshwater peat swamp on Phuket Island, Thailand. Belgian Journal of Zoology 131: 25–31.

Hydrobiologia (2005) 546:291–306 © Springer 2005
A. Herzig, R.D. Gulati, C.D. Jersabek & L. May (eds.) Rotifera X: Rotifer Research: Trends, New Tools and Recent Advances
DOI 10.1007/s10750-005-4235-x

Diapause in monogonont rotifers

Thomas Schröder
Department of Biological Sciences, Dartmouth College, Hanover, NH, 03755, USA
Present Address: Department of Biological Sciences, University of Texas at El Paso, El Paso, TX, 79968, USA
E-mail: th.schroeder@gmx.net

Key words: diapause, diapausing eggs, resting eggs, sexual reproduction, cyclic parthenogenesis

Abstract

This review focuses on more recent findings on the diapause in Monogonont rotifers, since the major reviews by Pourriot & Snell (1983, Hydrobiologia 104: 213–224) and Gilbert (1992, Rotifera. In Adiyodi, K. G. & R. G. Adiyodi (eds), Reproductive Biology of Invertebrates, Vol. 5 – Sexual Differentiation and Behaviour. IBH Publishing Co., Oxford: 115–136; Vol. 6A – Asexual Propagation and Reproductive Strategies. IBH Publishing Co., Oxford: 231–263.). It covers diapausing egg formation, diapausing egg survival, and diapausing egg hatching as well as possible strategies involved in these processes. Data from laboratory and field studies exist, but little information is available on diapausing egg hatching in the field. Resting or diapausing egg production can be mictic and in some cases amictic. Mictic diapausing egg production depends on the balance between cues promoting and inhibiting mictic female production. Such cues can be either environmental or endogenous. Our knowledge on factors inducing mixis is still limited to a few species, but effects of crowding on mixis induction may be more widespread. Recent results also show that male mating behavior may play an important role in the successful production of diapausing eggs. Hatching may be induced by factors related to temperature and light conditions; also desiccation may have a role. However, desiccation of temporary habitats may also be damaging to diapausing stages. Although few data are available, those existing point to distinct differences between populations and the importance of clonal variation within populations regarding the initiation as well as the termination of diapause.

Introduction

Diapause has been most intensively studied in insects, where it is defined as a dynamic state of low metabolic activity associated with reduced morphogenesis and increased resistance to environmental extremes. It occurs during a genetically defined stage of the life cycle, usually in response to a number of environmental stimuli that precede unfavorable conditions. Once diapause has begun, metabolic activity is suppressed, even if conditions favorable for development prevail (Tauber et al., 1986). This form of dormancy is in contrast to quiescence, which is a state of suppressed metabolism directly provoked by conditions unfavorable for growth and reproduction, and which can occur during any stage of the life cycle.

Diapause and quiescence are widespread forms of dormancy among animal phyla (Cáceres, 1997). In the phylum Rotifera diapause is limited to the Monogononta, whereas quiescence is found in the Bdelloidea (Ricci, 2001). As in crustaceans, dormant stages of rotifers can serve two purposes, dispersal in space and time (Hairston, 1998): dormancy allows survival during periods of harsh environmental conditions which do not permit the existence of active stages (dispersal in time) and dormant stages are the main propagule stages by which new habitats can be colonized (dispersal in space).

The life cycle of the Monogononta is characterized by cyclic parthenogenesis. Amictic females reproduce parthenogenetically, producing female offspring (amictic reproduction). Diapause is initiated by environmental cues when part of a population starts to reproduce sexually (mictic reproduction). Mictic females are produced, whose oocytes undergo meiosis. Unfertilized mictic females parthenogenetically produce haploid male offspring. If a mictic female is fertilized by a male, it produces diploid diapausing eggs, also called resting eggs. In some species amictic (parthenogenetic) production of diapausing eggs has been reported.

Research on diapause in monogonont rotifers has covered formation, hatching, and survival of resting eggs as well as possible strategies involved in these processes. Factors controlling initiation of mixis in monogonont rotifers have been an important area of diapause research. The second focus of research has been on factors controlling resting egg hatching and affecting resting egg survival. Both processes affect the function of the resting egg bank, which ensures the survival of the population during periods of harsh environmental conditions. This review covers new results on diapause in monogonont rotifers that have been published since the last major reviews by Pourriot & Snell (1983) and by Gilbert (1992, 1993). Resting egg hatching and unique features of the stem females hatching from resting eggs are covered in detail in another review (Gilbert & Schröder, 2004).

Mictic reproduction

Diapause in monogonont rotifers depends on the production of resting eggs by fertilized mictic females. The factors initiating mixis in rotifer populations, but also the frequencies of fertilizable mictic females and males, as well as the ratio of unfertilized male-producing females to resting egg producing females may be important and influence the production of resting eggs. Also mating behavior must be considered in this context because it may critically affect successful resting egg production.

Mixis initiation

Factors that induce the production of mictic females are still poorly known for most species. They include photoperiod in species of *Notommata* and *Trichocerca*, the uptake of α-tocopherol enriched food in many species of the genus *Asplanchna*, and factors related to density in several other species. The work on photoperiod and uptake of α-tocopherol as mixis-inducing factors has been reviewed in detail by Pourriot & Snell (1983) and Gilbert (1992, 1993) and will not be covered here.

Increasing population density induces mixis in species of several genera. Crowding has long been known to affect mictic female production in the *Brachionus plicatilis* species complex and in *B. calyciflorus* (Pourriot & Snell, 1983; Gilbert, 1992, 1993); however, recent studies have demonstrated that increasing population density also induces mictic female production in *Epiphanes brachionus* (Pourriot & Rougier, 1999) and in *Brachionus angularis*, *Epiphanes senta*, and *Rhinoglena frontalis* (Schröder & Gilbert, 2004). Carmona et al. (1993) provided some evidence that mixis in *B. plicatilis* is induced by a chemical cue which is present in the water of dense populations. Stelzer & Snell (2003) definitely showed that this is the case. Their results strongly support the hypothesis that the animals produce a chemical substance which accumulates in the water at high population densities and then induces mixis. This hypothesis would also explain the results of Hagiwara et al. (1994) who found that the addition of water soluble extracts of conspecific rotifers from mass cultures increases mixis in *B. plicatilis*.

Although it is still unknown whether a chemical is also involved in the density-dependent mixis induction of the other species mentioned above, such a possibility is likely. Physical contacts between females at high densities can be excluded as an obligatory mixis stimulus because mixis could be induced in females kept individually in a small volume of medium. In *E. brachionus*, Pourriot & Rougier (1999) found no significant differences in the proportion of mictic offspring of females which were kept individually and females which were grouped with other females at the equivalent density. The density-dependent stimulus appears to be very specific at least in some species, but it may be less specific in others. Gilbert (2003b) found that mixis is not induced in single females of a *B. calyciflorus* strain from Florida if they were crowded with females from a reproductively isolated Australian *B. calyciflorus* strain. However, the mixis inducing signal may be quite unspecific in *B. plicatilis*. Carmona et al. (1993) reported that mixis could be induced by a medium that was preconditioned by an *Artemia* culture.

The induction of mixis at high densities of conspecific females has been considered to be adaptive in several aspects. First, it has been argued that high densities increase the probability of male–female encounters, which seem to be random and depend on males recognizing females only upon contact (Snell, 1998). Second, if high densities are reached by amictic reproduction, and subsequently the reproductive mode switches to mictic reproduction, at a given fertilization rate more resting eggs can be produced simply because the number of sexual females will be higher (Snell & Boyer, 1988; Serra & King, 1999). Gilbert (1993) postulates that factors inducing sexual reproduction reflect favorable conditions which allow rapid population growth, high population densities, and the production of large numbers of energy-rich resting eggs. On the other hand, it has been argued that high population densities may lead to over-exploitation of food resources and therefore cause deteriorating conditions in the near future (Serra & Carmona, 1993; Aparici et al., 1996; Ciros-Pérez et al., 2002). In this case, increasing densities would lead to the production of resting eggs in anticipation of the deterioration of the habitat. However, threshold densities inducing mixis are usually well below observed maximum population densities (Gilbert, 2002; Gilbert & Schröder, unpublished data), indicating that resting eggs should be produced when food resources are still sufficient.

While unfavorable environmental conditions such as food limitation may directly induce sexual reproduction in cladocerans (Kleiven et al., 1992; Alekseev & Lampert, 2001; LaMontagne & McCauley, 2001), there has been little experimental evidence so far that this also may be the general case in monogonont rotifers. Mictic reproduction seems to be more inhibited by unfavorable conditions than amictic reproduction (Snell & Boyer, 1988; Snell & Carmona, 1995), and conditions enhancing mictic female production also increase fecundity of fertilized mictic females (Hagiwara et al., 1988). Also, crowding, which is known to induce mixis in several species, was not found to be associated with reduced fecundity (Schröder & Gilbert, 2004). All these findings indicate that mictic reproduction and resting egg production are more likely to take place under favorable conditions, well before the habitat deteriorates and conditions become unfavorable for population growth. However, the possibility that food deprivation associated with imminent habitat deterioration may directly influence mictic reproduction has been raised (Carmona et al., 1993; Aparici et al., 1996; Ciros-Pérez et al., 2002), and it has been shown for *B. plicatilis* that starvation of the stem females promotes mixis induction in later generations (Hagiwara, this volume Part V). This question certainly requires further research, especially since factors inducing mixis are still unknown for most monogonont species.

Exogenous and endogenous factors modifying the mixis response

The response to the mixis cue can be modified by endogenous as well as exogenous factors. First it can be influenced by exogenous environmental factors which inhibit or increase the effect of the actual mixis stimulus in some way. Snell & Boyer (1988) found that mictic female production in *B. plicatilis* is more sensitive to food limitation and to increased ammonia concentration than amictic female production: females individually cultured in small volumes comparable to high population density cease to produce mictic offspring at low food concentrations and high ammonia concentrations, while they are still producing amictic offspring under these conditions. Snell & Hoff (1985, 1987) found that the type of diet also had a large effect on mictic female production in *B. plicatilis* as well as on the fertility of the male offspring produced by unfertilized mictic females. Mictic female production was highest when food was a combination of *Chlorella vulgaris* and Baker's yeast, while on a diet of the cyanobacterium *Schizotrix calcicola* mictic females produced male offspring with significantly reduced fertility.

It has been demonstrated in several species that mixis induction is affected by temperature. In experimental cultures the proportion of mictic offspring of *E. brachionus* females increases to 50% of the population size with increasing population density at 14 °C, whereas at 10 °C it remains at a constant low rate of 5–8% even with increasing density (Pourriot & Rougier, 1999). As a result of the prevailing amictic reproduction at the lower temperature, the population growth rate was higher at 10 °C than at 14 °C. In experiments with females of this species individually isolated in small volumes, the proportion of mictic offspring was higher at 20 °C than at 14 °C. In contrast, Hagiwara et al. (1988) found that the proportion of mictic females of *B. plicatilis* in experimental culture is higher at a temperature of 15 °C than at 20, 25 and 30 °C. Fecundity of fertilized mictic females (i.e. the number of resting eggs produced) increased with decreasing temperatures. However, mictic female production was significantly increased when females were moved from 10 °C to 25 °C (Kogane et al., 1997).

Also varying salinities have a similar effect as varying temperature in *B. plicatilis*: the proportion of mictic females as well as the fecundity of fertilized mictic females was highest at a salinity of 4‰ and decreased with increasing salinity (Hagiwara et al., 1988). Similar results were obtained by Lubzens et al. (1985) and by Pozuelo & Lubián (1993). *Brachionus rotundiformis* displays the reverse trend (Hagiwara et al., 1989): the proportion of mictic females and the fecundity of fertilized mictic females increases with increasing salinity. It should be pointed out, however, that it is difficult to separate effects of salinity and food type and concentration on the one side and effects of crowding on the other. Suboptimal salinity and low food concentration may decrease the crowding response simply because they decrease population growth and metabolic rate relative to controls. Thus they may operate indirectly only by affecting population density.

Hagiwara et al. (1994) demonstrated that the presence of certain bacteria led to increased mictic reproduction in *B. plicatilis*, suggesting that the synthesis of vitamins by the bacteria could enhance mictic female production.

Endogenous factors – Carmona et al. (1994) noted that the proportion of mictic offspring in *B. plicatilis* is also dependent on the age of the mother. The proportion of mictic offspring in young females – during the first 2 days of the reproductive period – was higher than the proportion of mictic offspring in older females.

Endogenous factors may also affect the extent to which amictic females respond to a given mixis cue. It has first been demonstrated for *B. calyciflorus* that the first generations that follow the ex-resting-egg female (stem female) respond not at all, or to a much less extent, to a high population density than do later generations (Gilbert, 2002, 2003a). The offspring of these early generation females are mostly amictic, even when they are crowded. Females of the 12th and later generations produce 50% or more mictic offspring when they are produced in a comparable high density environment. This delay in the mictic response is also found in *B. angularis*, *E. senta*, and *R. frontalis* (Schröder & Gilbert, 2004), all of which respond to population density as a mixis cue.

The mechanisms of these endogenous effects on mictic reproduction are still unknown. However, it has been recently demonstrated that common invertebrate hormones are also present in rotifers (Gallardo et al., 2000a). Serotonin (5-HT) influences mictic female production in *B. plicatilis* (Gallardo et al., 2000b), suggesting that hormonal activity in combination with external mixis stimuli affects mictic female production.

It is important to note that a large amount of variation within many species has been found in the response to mixis cues. Usually mixis stimuli only induce females to produce a certain proportion of mictic offspring and only in a few cases is this close to 100% (Pourriot, 1965; Buchner, 1992; Aparici et al., 1996). The proportion of mictic female offspring can vary remarkably. Variation can be found at three different levels of organization. First there is intraclonal variation: it can often be observed in experiments with a single clone that a few females do not produce any mictic offspring under conditions that otherwise induce mixis. Then there is variation among clones within a population. In *B. calyciflorus*, *B. angularis*, *E. senta*, and *R. frontalis*, the propensity to produce mictic offspring varies significantly among clones of the same strain (Gilbert, 2002; Schröder & Gilbert, in press). Similar results were also found in comparisons of *B. plicatilis* clones from the same strain (Hino & Hirano, 1977). Aparici et al. (2001) detected large variation in the time of mixis initiation after hatching from the resting egg as well in the densities at which mictic females appeared in a number of clones from a *B. plicatilis* strain collected in the Torreblanca Marsh (Spain). However, only a small part of that variation was found to be heritable and therefore clone specific; most of the detected variation was assumed to be phenotypic and of environmental origin.

Not surprisingly, there is variation among strains of different populations. Carmona et al. (1994) found that mictic rates in 13 clones of *B. plicatilis* collected at different sites varied significantly when they were exposed to the same high density stimulus. In the study of mixis delay in *B. calyciflorus*, *B. angularis*, *E. senta*, and *R. frontalis* (Schröder & Gilbert, 2004), significant variation was found among strains: a delay of mixis in the early generation was found in one strain of *B. calyciflorus* from a temporary pond in Florida, but it was absent in another strain from a permanent pond in Georgia. The same was observed for a floodplain strain of *R. frontalis* and a strain from a permanent pond, but strains of *E. senta* from the same habitats as the *Rhinoglena* strains both displayed a delay of mixis in the early generations (Schröder & Gilbert, 2004).

Cyclic parthenogenesis in the life cycle of monogonont rotifers includes important trade-offs between parthenogenetic and sexual reproduction. Population growth and increases in frequency of individual clones relative to other clones within the population depend on parthenogenetic reproduction of amictic females. Therefore, allocating resources to the production of mictic offspring decreases a clone's intrinsic growth rate and thus its frequency within the population (Snell, 1987; Serra & Carmona, 1993; Serra & King, 1999). Results from long-term batch and chemostat cultures show that the propensity for mictic reproduction in a population declines and finally disappears in continuous cultures so that reproduction is completely amictic within 2 months to 3 years (Boraas, 1983; Bennett & Boraas, 1989; Buchner, 1992; Fussmann et al., 2003) and indicate that mixis is under strong selection. Mixis is favored if periods of adverse environmental conditions can only be survived by diapausing resting eggs. But, in the absence of such periods clones with high propensity for mixis will be selected against because they will display a reduced growth rate compared to clones with low propensity for mixis. Selection for exclusive amictic reproduction may also be acting in natural populations. Reproduction was found to be purely amictic in populations of *Keratella cochlearis*, which were perennial in large lakes, whereas in populations inhabiting small ponds sexual periods did occur (Wesenberg-Lund, 1930).

Ciros-Pérez et al. (2002) have also demonstrated trade-offs between mictic and amictic reproduction. They have shown that mictic reproduction associated with reduced population growth affects competitive ability when closely related species compete for food. They found in resource competition experiments with pairs of experimental populations of *B. plicatilis*, *B. rotundiformis* and *B. ibericus* that one species was excluded by the other when its mixis investment was high where that of the other species was low. If mixis investment was intermediate in both species, they were able to coexist. However, mixis investment was always maximal when a species was the inferior competitor and its exclusion was predicted by Tilman's resource competition theory (Tilman, 1982). This led the authors to raise the question whether increased mixis levels are to some extent an adaptive response of the competing species which is excluded. This would require a cue for sexual reproduction which is a predictor of the effects of interspecific competition. The authors conclude that food deprivation could be a possible cue, arguing that in exploitative interspecific competition the species with the lower competitive capability will be food deprived. This argument is consistent with results on *B. plicatilis* (Hagiwara et al., this volume Part V), showing that starvation of stem females positively influences mixis in later generations.

Mictic patterns and timing of mixis

Recent efforts have been made to theoretically predict the optimal time of mixis induction and optimal mictic ratios as well as sex allocation and threshold age of fertilization, after which mictic females can no longer be fertilized and produce only haploid male offspring. The timing of mictic reproduction as well as mictic ratios are critical for the optimal production of resting eggs. Serra & Carmona (1993), Aparici et al. (1996), King & Serra (1998) and Serra & King (1999) theoretically explored whether the optimal mixis strategy for monogonont rotifers is a polyphasic pattern of resting egg production or a 'bang-bang' strategy, and whether it depended on the predictability of the environment. Following Ricci's definition (2001), in a polyphasic pattern resting egg production would start immediately at the beginning of population development with a low proportion of mictic females and continue until the habitat deteriorates. In a 'bang-bang' strategy, resting eggs are produced by a large proportion of the population only after a period of exclusive amictic population growth and just before the environment deteriorates. Continuous mixis starting at low densities decreases population growth rates, but guarantees a continuous production of resting eggs as soon as critical male–female encounter densities are reached. It is therefore not so much dependent on long-term habitat stability as a 'bang-bang' strategy with short periods of mictic reproduction at high population densities only, even though the latter strategy may lead to a higher yield of resting eggs. Aparici et al. (1996) and Spencer et al. (2001) come to the conclusion that polymorphisms in the timing of mixis induction and mixed strategies may be important in environments of high unpredictability.

Clonal diversity in the timing of mixis and the mictic ratio within a population may be maintained by the storage effect of the resting egg bank. Fluctuating selection may favor clones with late mixis initiation in some years and those with early mixis initiation in others, leading to varying success in sexual reproduction and varying recruitment of the different phenotypes to the resting egg bank in different years. The resting egg bank may then act as a buffer in years of poor recruitment of new resting eggs produced by the active stages of the population. In theory, the storage effect can explain the maintenance of genetic variation within a population of a cyclic parthenogen (Hairston et al., 1996) and thus the coexistence of competing clones in a temporally varying environment. A population of coexisting clones with different mixis patterns which may be favorable in some seasons and disadvantageous in others depending on fluctuating selection, may display a mixis pattern that is intermediate between a 'bang-bang' strategy and a recurrent mictic pattern.

Optimal sexual reproduction is dependent on the ratio between males and females. Sex allocation theory predicts that the optimal sex ratio is 1:1 (Charnov, 1982). However, in haplodiploid cyclic parthenogens such as monogonont rotifers the number of males is irrelevant; rather, the ratio of unfertilized and fertilized mictic females is important and should be even (Aparici et al., 1998, 2002; Calsina et al., 2000; Calsina & Ripoll, 2002). In order to maintain an even ratio between the two types of mictic females, the threshold age of fertilization of mictic females is critical. A young threshold age of fertilization will shift the ratio towards male producing females, an old threshold age will cause the opposite. The theoretical findings for sex allocation in rotifers are supported by field and laboratory data (Aparici et al., 2002), although evidence is scarce. Laboratory populations of *B. plicatilis* displayed ratios of fertilized to unfertilized mictic females close to 0.5 at the end of the exponential growth phase; however, it was only close to 0.5 in one of the two natural populations that were investigated. The authors attributed this deviation from the predicted value to environmental random variation. Aparici et al. (2002) argue that insights from the application of sex allocation theory to monogonont rotifers exclude one reason frequently invoked why the induction of mixis is related to high population densities in several species: the argument that higher densities will increase encounter probabilities between males and females and lead to higher insemination rates and subsequently to a higher number of resting egg producing females, is inconsistent with their results. According to the sex allocation theory and their empirical evidence, the number of resting egg producing females integrated over time should always be half the total number of sexual females.

Mictic patterns in natural populations

Only few studies addressed mictic patterns and mictic strategies in natural populations. Very high mictic ratios (close to 100%) seem to be rare in natural populations; they have been reported in *B. plicatilis* (Aparici et al., 1996), in *B. urceolaris* (Buchner, 1992), and in *B. calyciflorus* (Pourriot, 1965). More often, mictic ratios are around 20–30% (Carmona et al., 1995; Miracle &

Armengol-Díaz, 1995). Field data indicate that mixis already occurs at much lower density levels than the laboratory data would suggest as threshold levels (Carmona et al., 1995; Schröder, 2001).

Miracle & Armengol-Díaz (1995) observed different mixis patterns in two species which co-occur in the oxicline of a Spanish lake. Sexual reproduction in *Filinia hofmanni* was observed continuously over the whole period of population development with maxima at the time of highest densities. *Anuraeopsis fissa* produced resting eggs only at the very end of the population development.

Carmona et al. (1995) identified different mictic patterns for three species of the *Brachionus plicatilis* complex in small marsh ponds in Spain. *B. plicatilis* displays a more continuous mictic pattern with sexual reproduction throughout the time of its presence in the ponds. This may be a conservative bet-hedging strategy – suitable in an unpredictable environment – as this species appears in winter and spring, when sudden unpredictable flooding of the marsh with sharp decreases in salinity may occur. In contrast, *B. rotundiformis* and *B. ibericus* showed a punctuated mictic pattern. Especially in *B. rotundiformis*, higher mictic ratios were observed only for brief periods towards the end of population development. Both species are spring-summer species, whose existence in the ponds is limited by gradual desiccation in summer or decreasing temperatures in the fall. Desiccation and falling temperatures are more predictable than floods and the authors hypothesize that a punctuated mixis pattern would be the optimal strategy in these habitats.

Male mating behavior

Recent studies have shown that male mating behavior may also play an important role in maximizing resting egg production. Mictic females are fertilizable only for a short period after hatching (Buchner et al., 1967; Snell & Childress, 1987; Hagiwara et al., 1988); after that period, they are only able to produce males by parthenogenesis. Males who copulate with older females or with amictic females are wasting their sperm, which may be relevant at least in some species such as *B. plicatilis* where male sperm reserves are limited (Snell & Childress, 1987). Therefore, there should be a strong selection for male copulation only with fertilizable females. Gómez & Serra (1996) have shown that males of *B. plicatilis* initiate mating behavior significantly more often when they encounter young females rather than females aged 1 day or older and they rarely finish mating and copulate with females that are 1 day or older. Males also tend to prefer mictic rather than amictic females. Mating behavior is initiated significantly less often, when males encounter amictic females. However, they still start to exhibit mating behavior and copulate with a large proportion of amictic females they encounter.

Males of *Epiphanes senta* display a behavior that resembles mate-guarding known from many arthropods (Schröder, 2003). Males who encounter a female egg attend it until the female hatches and then copulate with the newborn. However, this behavior differs from mate-guarding in arthropods in that *Epiphanes* males do not show any antagonistic behavior against other males who approach the egg. Males do not attend male or resting eggs. Also, they are able to discriminate between female eggs which have just recently been deposited and mature eggs which will hatch within a short period of time. By preferentially attending eggs that are going to hatch soon, males minimize the amount of time they spend waiting on the eggs. Such a behavior could have two interpretations. First, it would ensure that males only mate with females in their susceptible period, since they only mate with newborn females. Second, this behavior could be a time investment strategy to maximize mating success, if the probability of encountering a fertilizable female during the time period necessary to wait on the egg is lower than 1. Interestingly, males are not able to discriminate between eggs of mictic and amictic females. Similarly, males of *Asplanchna brightwelli* and *Brachionus calyciflorus* seemed unable to discriminate between mictic and amictic females (Gilbert, 1963; Aloia & Moretti, 1973).

Amictic diapausing eggs

Amictic production of diapausing eggs has only been documented for very few species. Ruttner-Kolisko (1946) reported the production of diapausing eggs without fertilization by males in *Keratella hiemalis*. These amictic diapausing eggs are morphologically very similar to fertilized resting eggs, dark in color and with a multi-layered shell. Factors that induce the production of these amictic diapausing eggs as well as conditions of hatching, are unknown. Production of amictic diapausing eggs that are morphologically similar to the fertilized resting eggs has also been observed in *Notholca squamula* (Schröder, 1999). The amictic diapausing eggs of *N. squamula* hatch several weeks after being produced when kept in the same culture conditions as those in which the eggs were produced. The factors inducing the production of amictic diapausing eggs in *N. squamula* are also unknown.

The production of amictic diapausing eggs in *Synchaeta pectinata* is better understood. These diapausing eggs differ from the fertilized resting egg (Gilbert, 1995). Their shells consist only of a single layer. Females produce as many amictic diapausing eggs as amictic subitaneous eggs (Gilbert, 1995), indicating that the production of amictic diapausing eggs does not require more resources than the production of subitaneous eggs.

The amictic diapausing egg in *S. pectinata* is the only diapausing stage in monogonont rotifers known to be directly induced by deteriorating environmental conditions. Amictic females that are starved for a certain time period are induced to produce diapausing eggs (Gilbert, 1995; Gilbert & Schreiber, 1995). Short periods of food limitation already induce some females of a clone to produce diapausing eggs and longer periods of more severe food limitation lead to strong diapause response, where 75% of all eggs produced are diapausing (Gilbert & Schreiber, 1998). Food concentrations inducing the production of diapausing eggs are still above threshold concentrations for population growth, but reduced food concentrations may be predicting deteriorating resource conditions (Gilbert & Schreiber, 1998). Amictic females which are cultured in a food-limiting environment vary in their response of diapausing egg production. While some females of a clone are induced to produce some or only diapausing eggs, others continue to produce only subitaneous eggs under the same inducing conditions; this may be considered a bet-hedging strategy (Gilbert, 1998; Gilbert & Schreiber, 1998). It may allow positive population growth if food conditions remain above threshold concentration. At the same time the production of diapausing eggs provides a refuge if food concentrations fall below levels sustaining positive population growth.

There is also considerable genetic variation in the production of amictic diapausing eggs (Fradkin, 1997). *S. pectinata* produces two types of diapausing eggs that differ in their morphology and in the length of diapause. The production of different diapausing egg types is clone specific. Some clones only produce long-term diapausing eggs with a diapause of at least 3 months, while others produce short-term diapausing eggs which invariably hatch after a diapause of only 2 weeks. The diapausing egg types differ in their number of blastomeres. Both have thicker shells than the subitaneous eggs, but the shell of the long-term diapausing egg is much thinner compared to that of the short-term diapausing egg (Fradkin, 1997). Fradkin (1997) found that both types of diapausing eggs were produced in a population in Star Lake (Vermont). The population collapsed over a 3-month period in winter when food concentrations were very low. Over 80% of the population produced long-term diapausing eggs at the beginning and at the end of that period.

Clones also vary in their propensity to produce diapausing eggs, some producing 82–100% diapausing eggs when induced by starvation and others producing only 16–38% after a starvation period, the latter having significantly higher net reproductive rates as a consequence (Fradkin, 1997). This creates the potential of an ecological trade-off, where different clones may be favored depending on food level. Fradkin (1997) showed that clones with a low propensity are favored due to their higher growth rates, if food resources reach inducing levels but remain above threshold levels. If food levels drop from inducing concentrations to levels below threshold concentrations, clones with a strong propensity are favored, because they produce more diapausing eggs at inducing food levels, and these escape periods of very low food that inflict high mortality on non-diapausing stages.

Resting eggs

The fertilized resting eggs of monogonont rotifers do not seem to fall in two distinct categories of short-term diapause and long-term diapause as the amictic diapausing eggs of *Synchaeta pectinata*. However, the latency period, i.e. the minimum period of dormancy during which hatching is not possible, seems to be very variable both among and within species. Resting eggs may hatch after a latency period of only a few days as in *Brachionus quadridentatus* from an Australian billabong (Gilbert, 2001), but in other species the latency period may last as long as 5–6 months such as in some species of the genus *Polyarthra* (Nipkow, 1952). Pourriot et al. (1982) found differences in the latency period of two clones of *B. calyciflorus*: in one clone, the period was only a day, whereas the resting eggs of the other clone had a latency period of 1 week.

Resting egg hatching and hatching patterns

Factors influencing or inducing hatching of fertilized resting eggs have been extensively investigated experimentally in *Brachionus rubens*, *B. calyciflorus*, *B. angularis*, *B. budapestinensis*, and *B. plicatilis*. The major factors are temperature and light (Pourriot et al., 1980, 1981, 1982, 1983; Blanchot & Pourriot, 1982a, b) and salinity for the *B. plicatilis* species complex (Blanchot & Pourriot, 1982b; Minkoff et al., 1983; Hagiwara & Hino, 1989; Hagiwara et al., 1989). These studies have been reviewed in detail by Pourriot & Snell (1983) and Gilbert (1993).

Rotifers hatch within certain ranges of temperatures which are species specific and correspond to the thermal preference range of these species. For example, resting eggs of the thermophilic species *B. budapestinensis* do not hatch at temperatures of 5 °C and below, and resting eggs of a cold-water strain of *B. angularis* show decreased hatching rates at 18 °C and above (Pourriot et al., 1983). May and coworkers (May, 1987; May et al., 2001) showed that different temperatures induce different species to emerge from sediment samples of Loch Leven/Scotland. Temperature-dependent hatching rates also reflected seasonality and temperature preferences of the different species. Perennial species as *Keratella cochlearis*, *Synchaeta grandis* and *S. kitina* hatched over the whole range of temperatures tested, whereas hatching of the spring and summer species *Asplanchna priodonta*, *Trichocerca pusilla*, and *Pompholyx sulcata* was restricted to temperatures of 10 °C and above. The cold-stenotherm species *Notholca squamula* and *Polyarthra dolichoptera* whose populations almost always developed in the lake at temperatures not higher than 15 °C, only hatched in majority at the colder temperatures tested (5 and 10 °C).

In other species, however, there is no evidence for a connection between temperatures that promote hatching and temperatures suitable for population development. *Cephalodella hoodi*, inhabiting an acidic mining lake in East Germany, is most abundant in spring and summer, but it could only be induced to hatch from sediments at 5 °C, but not at 20 °C (Bell & Weithoff, 2003).

Recent investigations have shown that temperature conditions inducing resting egg hatching may involve the temperature changes which occur across seasons. *Rhinoglena frontalis* is a cold-stenotherm species present in winter and spring in small ponds and in temporary floodplain habitats when water temperatures range between 1–17 °C (Wesenberg-Lund, 1930; Schröder, 2001). In the floodplains of the Oder River, resting eggs are produced in April and May shortly before the waters dry up (Schröder, 2001). Resting eggs produced by a floodplain clone hatched at 6 °C, but only after they were exposed to an intermittent period of high temperature. Virtually no hatching occurred, if the resting eggs were kept constantly at 6 °C, or at 20 °C (Schröder, 1999). The requirement of a high-temperature signal could ensure that resting eggs produced in the spring would not hatch before the flooded areas dried out, but only when they are flooded again the following season.

Also, resting eggs of species inhabiting temporary habitats may be induced to hatch in the next season by the intermittent terrestrial period, independent of temperatures. Experiments with resting eggs produced by a clone of *B. calyciflorus* from the temporary Oder River floodplain showed that these eggs would hatch after they had been exposed to terrestrial conditions at a relative humidity ranging from 55 to 100% for 10–30 days. Only very few eggs hatched when the resting eggs were kept continuously in water (Schröder, 1999).

Light may be another factor which influences hatching of resting eggs. It has little or no effect on hatching in *B. angularis* and *B. budapestinensis*, but it is important in *B. plicatilis* (Blanchot & Pourriot, 1982b; Minkoff et al., 1983; Hagiwara & Hino, 1989) and *B. rubens* (Pourriot et al., 1980, 1981). Hatching increases under short wavelengths 400–480 nm in *B. rubens* (Blanchot & Pourriot, 1982a) and 347–400 nm in *B. plicatilis* (Hagiwara et al., 1995), but UV light had no effect in *B. calyciflorus* (Schröder, 1999).

Light also increased resting egg hatching in *Epiphanes senta* strains from the temporary floodplains of the Oder River/Germany and neighboring permanent ponds (Schröder, unpublished data), although the main factor that induced hatching was low temperature. *Epiphanes senta* occurs in floodplain waters in winter and spring but is absent in summer (Schröder, 1999). Resting eggs produced by several clones were induced to hatch after 3.5 months at 20 °C if exposed to 10 °C. However, resting eggs produced by some clones could also be induced to hatch at 20 °C, if they were exposed to light, although the hatching rates were much lower than at 10 °C and did not exceed 15%. Hatching rates at the low temperature were significantly different when the resting eggs produced by the different clones were compared. They fell into two groups: hatching rates of eggs produced by some clones reached 90–95%, whereas hatching rates of egg produced by the other clones ranged between 45 and 70%. Resting eggs produced by some clones hatched synchronously within a few days, while resting eggs produced by other clones hatched over a longer period of time. Parental clones of both groups could be found within the same strain, indicating the existence of a genetic polymorphism in the hatching pattern of the strain (Schröder, unpublished data).

Studies estimating the densities of resting eggs in the sediments by counting eggs recovered with sugar floatation methods or by measuring emergence of rotifers from flooded sediments in the laboratory, show that the egg bank in the sediments can contain very large numbers of resting eggs (Nipkow, 1961; Snell et al., 1983; May 1986, 1987; Mnatsakanova & Polishchuk, 1996; Schröder, 2001; Duggan et al., 2002). Snell et al. (1983) recorded as many as 194 *B. plicatilis* resting eggs cm^{-3} at the sediment surface in a brackish-water pond in Florida. However, when densities derived from resting egg counts are compared with numbers of emerging rotifers, densities exceed hatching rates by 1–2 orders of magnitude. Duggan et al. (2002) counted both numbers of resting eggs in the sediments and recorded hatching rates from sediment samples of two New Zealand lakes; they found a twofold to 10-fold difference, depending on the incubation temperature of flooded sediments. Such a difference may reflect the function of the egg bank: if only some of the resting eggs are induced to hatch at a time, those remaining dormant carry the population through years of poor recruitment to the egg bank (Hairston & de Stasio, 1988; de Stasio, 1989).

There may be several reasons why only a fraction of the eggs hatches at a given time. One of them certainly is that a large number of eggs may face conditions that inhibit hatching. Such conditions could be low oxygen concentrations, low temperatures, lack of light, or high concentrations of hydrogen sulphide and methane and may vary with sediment depth. Thus, hatching of buried eggs may require that the eggs are suspended in the water by strong currents or bioturbation (Gilbert & Schröder, 2004).

Another reason may be a bet-hedging strategy involved in the hatching of resting eggs. Experiments with resting eggs derived from single clones of *R. frontalis* and *B. calyciflorus* have shown that conditions inducing hatching in the two species do not induce all exposed resting eggs to hatch. However, if the still dormant eggs are exposed to the same conditions again later, another fraction of the dormant resting eggs could be induced to hatch (Schröder, 1999). Such a bet-hedging mechanism, which could be caused either by a genetic polymorphism or by a polyphenism, could contribute to the long-term survival of populations if conditions for population growth and production of new resting eggs are unpredictable. It should be expected that hatching rates would be lower in less predictable habitats. However, there is almost no evidence to support this prediction. Pourriot et al. (1982) compared hatching rates of two clones of *B. calyciflorus* and attributed large differences in hatching fractions to different strategies in a predictable and an unpredictable habitat. They did not, however, take into account the possibility of existing polymorphisms in hatching patterns within the populations, since they only examined one clone of each population. That such a polymorphism within a population may exist, has been shown in *E. senta* as described above (Schröder, unpublished data).

In most laboratory studies, synchronous hatching of a significant fraction of resting eggs was observed under certain environmental conditions, but in other cases sporadic hatching over extended periods was found (Pourriot & Snell, 1983). In shallow temporary waters, where recolonization after reflooding obviously has to take place via hatching from resting eggs, it seems likely that large numbers of resting eggs hatch within a short period of time, as soon as the previously dried sediments are submerged again. The flooding of the dried sediments of shallow temporary waters constitutes an environment with well oxygenated water and light which should be favorable for hatching (Gilbert, 2001), and a large diversity of rotifers has been found to hatch from dried sediments of floodplains and billabongs (Boulton & Lloyd, 1992; Nielsen et al., 2000; Langley et al., 2001; Schröder, 2001). The fact that Gómez & Carvalho (2000) found an extraordinarily high genetic diversity in planktonic populations of *B. plicatilis* in a shallow temporary pond (Poza Sur/Spain) also supports the hypothesis of synchronous hatching of many different clones, rather than the hatching of a few stem females which then found a large population consisting only of a small number of clones.

Still the question remains whether synchronous mass hatching over a short period of time also occurs in field populations of large permanent lakes, where a large fraction of the resting egg bank is contained in sediments at greater depth, or whether continuous hatching of only few resting eggs takes place over longer periods of time, especially since field studies on resting egg hatching are rare (Gilbert & Schröder, in press). Indirect evidence exists for massive resting egg hatching in the littoral induced by drought-associated desiccation, intense light and high oxygen concentrations at the sediment surface. Arnott et al. (2001) observed an increase in species richness associated with a drought event. This may be indicating that substantial numbers of resting eggs hatched from previously dried sediments after they were reflooded.

Resting egg survival

Little is known about the time periods for which resting eggs remain viable in sediment egg banks. In sediments of permanent habitats, they may still hatch after several decades (Marcus et al., 1994; Viitasalo & Katajisto, 1994). Fu (1991, in Kotani et al., 2001) was even able to hatch resting eggs of *B. rotundiformis* from sediments up to 100 years old. Marcus et al. (1994) were able to hatch resting eggs of *Brachionus* sp. from sediments that were more than 40 years old. However, hatching rates of 40-year old eggs represented only 5–10% of the hatching rates of resting eggs in the sediment egg bank which were less than 15 years old. Chittapun et al. (this volume Part V) demonstrated that resting egg viability in the sediments of a tropical swamp was already significantly reduced after resting eggs had been buried in the sediments for only 12–24 months. The factors that affect long-term survival of rotifer resting eggs in sediments of permanent waters are mostly unknown. In cultures of *B. plicatilis* Balompapueng et al. (1997a, b) found that resting eggs whose surface is colonized by bacteria during diapause show drastically decreased hatching rates. Hatching rates increase when bacteria are removed by chemical treatment. The authors conclude that bacteria affect the hatchability possibly by clogging pores of the egg shell and interfering with the gas exchange of the developing embryo after diapause has been terminated. It is possible that bacterial growth may also affect the viability of resting eggs in natural sediments. Nothing is known about the impact of predation on survival of resting eggs in sediment egg banks.

In sediments of temporary waters resting eggs may lose viability after much shorter periods than in permanent waters. Boulton & Lloyd (1992) found no hatching of rotifers from floodplain sediments of the River Murray (Australia) which were flooded only on an average of every 22 years, whereas rotifers hatched from sediments that were flooded annually or on an average of 7 or 11 years. The viability of resting eggs of some rotifer species occurring in temporary habitats may be significantly affected by severe drought stress. Experiments on *Rhinoglena frontalis*, *Brachionus urceolaris* and *B. calyciflorus*, exposing resting eggs to terrestrial conditions at controlled relative humidity for different times revealed that the eggs of *B. urceolaris* were much more vulnerable to drought stress than those of *B. calyciflorus* and *R. frontalis* (Schröder, 1999). Hatching rates of

B. urceolaris eggs decreased significantly even after short exposure to dry periods lasting 1–10 days at low relative humidities of 76 and 33%, compared to hatching after terrestrial conditions at a relative humidity of 100% and to hatching of resting eggs which were kept in water permanently. Resting eggs of *B. calyciflorus* and *R. frontalis* proved to be much more resistant to drought stress. Resting eggs of *R. frontalis* were not negatively affected by terrestrial periods of 20 days at relative humidities of 55.5 and 33%, but almost no hatching occurred if the treatment was prolonged for 60 days. Resting egg hatching in *B. calyciflorus* was not negatively affected by terrestrial periods of 30 days at 55.5% relative humidity, but hatching rates decreased significantly after a 30 day terrestrial period at 33% relative humidity.

The viability of resting eggs may also be affected by the conditions under which they were produced. Hagiwara & Hino (1990) have shown that resting egg hatching of *B. plicatilis* and *B. rotundiformis* is strongly affected by the diet of the resting egg producing mothers. Resting eggs produced in cultures that were fed with *Chlamydomonas* sp. would hatch at a lower percentage than resting eggs produced by females fed *Tetraselmis tetrahele*. Also, the hatching females differed significantly in their fecundity: those hatching from resting eggs produced on the *Chlamydomonas* diet were less fecund than those hatching from resting eggs produced on the *Tetraselmis* diet.

The long-term survival of rotifer populations in unpredictable environments relies on the existence of a resting egg bank, but the work reviewed above suggests that the functioning of the resting egg bank may be limited in many cases. Local extinction of subpopulations and recolonization events within the metapopulation may be important, but hardly any data exist for monogonont rotifers and more research is needed here.

Conclusions

It seems evident that monogonont rotifers exhibit an immense variety of diapausing patterns with a number of different ways to induce mixis and subsequent resting egg production, as well as varying hatching patterns. This holds true even if the fact is taken into account that most research has been done on a small group of species; it can be expected that diversity will increase if more species are considered. Interestingly, of the relatively few species that have been investigated, many respond to high densities as a cue for mictic female production and subsequent production of resting eggs. In these cases high density is not associated with reduced fecundity due to food limitation. This is in agreement with the argument made by Gilbert (1993) and others that resting egg production in monogonont rotifers is likely to take place under favorable conditions in advance of habitat deterioration. However, the results of Hagiwara et al. (this volume Part V) indicate that starvation as caused by habitat deterioration can also trigger resting egg production in other cases. It seems possible that both factors are synergistic, since crowding may lead to overexploitation of food resources.

Since the production of resting eggs involves a trade-off between mictic and amictic reproduction, populations of the same species are likely to display different mixis patterns depending on the specific conditions of the habitat. Selection for early sexual reproduction should be more intense in populations in short-lived temporary ponds than in seasonal populations of large lakes. The more unpredictable a habitat is, the more likely are bet-hedging strategies involved in mictic and amictic reproduction as well as genetic polymorphisms maintained by storage effects of the sediment egg bank, which may also lead to a wide variety of mixis patterns.

Resting egg hatching patterns may also involve bet-hedging strategies and may be influenced by genetic polymorphisms. Therefore, synchronous hatching within a short time period, continuous hatching over a longer period or a combination of both – some clones of a population producing resting eggs that hatch synchronously and other clones producing resting eggs that hatch erratically – may be found in different populations. The sediment egg bank obviously plays a crucial role for the persistence of populations in temporary habitats. However, the long-term viability of resting eggs may be more limited than previously thought and differ significantly among species. Some species seem to be more vulnerable than others to adverse conditions affecting the sediment egg bank such as drought or an anoxic environment. Thus, specific species assemblages in temporary habitats may be determined by environmental conditions during diapause as well as during population development of active stages. The relationships between environment and population dynamics of the sediment egg bank (i.e. recruitment, hatching rates and mortality rates) are poorly understood. Exploring these relationships would contribute to a better understanding of diapause patterns in monogonont rotifers.

Acknowledgments

I thank John Gilbert and an anonymous referee for valuable comments that helped to improve the manuscript. The work was funded by a postdoctoral fellowship from the Deutsche Forschungsgemeinschaft.

References

Alekseev, V. & W. Lampert, 2001. Maternal control of resting-egg production in *Daphnia*. Nature 414: 899–901.

Aloia, R. C. & R. L. Moretti, 1973. Mating behavior and ultrastructural aspects of copulation in the rotifer *Asplanchna brightwelli*. Transactions of the American Microscopical Society 92: 371–380.

Aparici, E., M. J. Carmona & M. Serra, 1996. Polymorphism in bisexual reproductive patterns of cyclical parthenogens – A simulation approach using a rotifer growth model. Ecological Modelling 88: 133–142.

Aparici, E., M. J. Carmona & M. Serra, 1998. Sex allocation in haplodiploid cyclical parthenogens with density-dependent proportion of males. American Naturalist 152: 652–657.

Aparici, E., M. J. Carmona & M. Serra, 2001. Variability for mixis initiation in *Brachionus plicatilis*. Hydrobiologia 446/447: 45–50.

Aparici, E., M. J. Carmona & M. Serra, 2002. Evidence for an even sex allocation in haplodiploid cyclical parthenogens. Journal of Evolutionary Biology 15: 65–73.

Arnott, S. E., N. Yan, W. Keller & K. Nicholls, 2001. The influence of drought-induced acidification on the recovery of plankton in Swan Lake (Canada). Ecological Applications 11: 747–763.

Balompapueng, M. D., A. Hagiwara, Y. Nozaki & K. Hirayama, 1997a. Preservation of resting eggs of the euryhaline rotifer *Brachionus plicatilis* O.F.Müller by canning. Hydrobiologia 358: 163–166.

Balompapueng, M. D., N. Munuswamy, A. Hagiwara & K. Hirayama, 1997b. Effect of disinfectants on the hatching of marine rotifer resting eggs *Brachionus plicatilis* Muller. Aquaculture Research 28: 559–565.

Bell, E. M. & G. Weithoff, 2003. Benthic recruitment of zooplankton in an acidic lake. Journal of Experimental Marine Biology and Ecology 285/286: 205–219.

Bennett, W. N. & M. E. Boraas, 1989. A demographic profile of the fastest growing metazoan – a strain of *Brachionus calyciflorus* (Rotifera). Oikos 55: 365–369.

Blanchot, J. & R. Pourriot, 1982a. Effets de l'intensité d'éclairement et de la longueur d'onde sur l'éclosion des aeufs de durée de *Brachionus rubens* (Rotifère). Comptes rendus hebdomadaires des séances de l'académie des sciences 295: 123–125.

Blanchot, J. & R. Pourriot, 1982b. Influence de trois facteurs de l'environnement, lumière, température et salinité, sur l'éclosion des aeufs de durée d'un clone de *Brachionus plicatilis* (O.F.Müller) Rotifère. Comptes rendus hebdomadaires des séances de l'académie des sciences 295: 243–246.

Boraas, M. E., 1983. Population dynamics of food-limited rotifers in 2-stage chemostat culture. Limnology and Oceanography 28: 546–563.

Boulton, A. J. & L. N. Lloyd, 1992. Flooding frequency and invertebrate emergence from dry floodplain sediments of the River Murray, Australia. Regulated Rivers, Research, Management 7: 137–151.

Buchner, H., 1992. Studies on the control of heterogonous reproduction in rotifers. IV. The reactivation of mictic potential in *Brachionus urceolaris*. Zoologische Jahrbücher, Abteilung für allgemeine Zoologie und Physiologie der Tiere 96: 97–165.

Buchner, H., C. Mutschler & H. Kiechle, 1967. Die Determination der Männchen- und Dauereiproduktion bei *Asplanchna sieboldi*. Biologisches Zentralblatt 86: 599–621.

Cáceres, C. E., 1997. Dormancy in invertebrates. Invertebrate Biology 116: 371–383.

Calsina, A. & J. Ripoll, 2002. Hopf bifurcation in a structured population model for the sexual phase of monogonont rotifers. Journal of Mathematical Biology 45: 22–36.

Calsina, A., J. M. Mazon & M. Serra, 2000. A mathematical model for the phase of sexual reproduction in monogonont rotifers. Journal of Mathematical Biology 40: 451–471.

Carmona, M. J., M. Serra & M. R. Miracle, 1993. Relationships between mixis in *Brachionus plicatilis* and preconditioning of culture-medium by crowding. Hydrobiologia 255/256: 145–152.

Carmona, M. J., M. Serra & M. R. Miracle, 1994. Effect of population density and genotype on life-history traits in the rotifer *Brachionus plicatilis* O.F. Müller. Journal of Experimental Marine Biology and Ecology 182: 223–235.

Carmona, M. J., A. Gómez & M. Serra, 1995. Mictic patterns of the rotifer *Brachionus plicatilis* Müller in small ponds. Hydrobiologia 313/314: 365–371.

Charnov, E. L., 1982. The theory of sex allocation. Princeton University Press, Princeton N.J.

Ciros-Pérez, J., M. J. Carmona & M. Serra, 2002. Resource competition and patterns of sexual reproduction in sympatric sibling rotifer species. Oecologia 131: 35–42.

Chittapun, S., P. Pholpunthin & H. Segers, 2005. Restoration of tropical peat swamp rotifer communities after perturbation: an experimental study of recovery of rotifers from the resting egg bank. Hydrobiologia 546: 281–289.

De Stasio, B. T., 1989. The seed bank of a freshwater crustacean: copepodology for the ecologist. Ecology 70: 1377–1389.

Duggan, I. C., J. D. Green & R. J. Shiel, 2002. Rotifer resting egg densities in lakes of different trophic state, and their assessment using emergence and egg counts. Archiv für Hydrobiologie 153: 409–420.

Fradkin, S. C., 1997. Asexual diapause in the rotifer *Synchaeta pectinata*: fitness costs and trade-offs associated with phenotypic variation in a natural population. Ph.D. Dissertation. Dartmouth College, 132 pp.

Fussmann, G. F., S. P. Ellner & N. G. Hairston Jr., 2003. Evolution as a critical component of plankton dynamics. Proceedings of the Royal Society of London, Series B 270: 1015–1022.

Gallardo, W. G., A. Hagiwara, K. Hara, K. Soyano & T. W. Snell, 2000a. GABA, 5-HT and amino acids in the rotifers *Brachionus plicatilis* and *Brachionus rotundiformis*. Comparative Biochemistry and Physiology Part A 127: 301–307.

Gallardo, W. G., A. Hagiwara & T. W. Snell, 2000b. Effect of juvenile hormone and serotonin (5-HT) on mixis induction of the rotifer *Brachionus plicatilis* Muller. Journal of Experimental Marine Biology and Ecology 252: 97–107.

Gilbert, J. J., 1963. Contact chemoreception, mating behavior, and sexual isolation in the rotifer genus *Brachionus*. Journal of Experimental Biology 40: 625–641.

Gilbert, J. J., 1992. Rotifera. In Adiyodi, K. G. & R. G. Adiyodi (eds.), Reproductive Biology of Invertebrates, Vol. 5 – Sexual Differentiation and Behaviour. Oxford & IBH Publishing Co, New Delhi: 115–136.

Gilbert, J. J., 1993. Rotifera. In Adiyodi, K. G. & R. G. Adiyodi (eds.), Reproductive Biology of Invertebrates, Vol. 6A – Asexual Propagation and Reproductive Strategies. Oxford & IBH Publishing Co, New Delhi: 231–263.

Gilbert, J. J., 1995. Structure, development and induction of a new diapause stage in rotifers. Freshwater Biology 34: 263–270.

Gilbert, J. J., 1998. Asexual diapause in the rotifer *Synchaeta:* diversified bet-hedging, energetic cost and age effects. Archiv für Hydrobiologie, Special Issues: Advances in Limnology 52: 97–107.

Gilbert, J. J., 2001. Spine development in *Brachionus quadridentatus* from an Australian billabong: genetic variation and induction by *Asplanchna*. Hydrobiologia 446/447: 19–28.

Gilbert, J. J., 2002. Endogenous regulation of environmentally induced sexuality in a rotifer; a multigenerational parental effect induced by fertilisation. Freshwater Biology 47: 1633–1641.

Gilbert, J. J., 2003a. Environmental and endogenous control of sexuality in a rotifer life cycle: developmental and population biology. Evolution & Development 5: 19–24.

Gilbert, J. J., 2003b. Specificity of crowding response that induces sexuality in the rotifer *Brachionus*. Limnology and Oceanography 48: 1297–1303.

Gilbert, J. J. & D. K. Schreiber, 1995. Induction of diapausing amictic eggs in *Synchaeta pectinata*. Hydrobiologia 313/314: 345–350.

Gilbert, J. J. & D. K. Schreiber, 1998. Asexual diapause induced by food limitation in the rotifer *Synchaeta pectinata*. Ecology 79: 1371–1381.

Gilbert, J. J. & T. Schröder, 2004. Rotifers from diapausing, fertilized eggs: unique features and emergence. Limnology and Oceanography 49: 1341–1354.

Gómez, A. & G. R. Carvalho, 2000. Sex, parthenogenesis and genetic structure of rotifers: microsatellite analysis of contemporary and resting egg bank populations. Molecular Ecology 9: 203–214.

Gómez, A. & M. Serra, 1996. Mate choice in male *Brachionus plicatilis* rotifers. Functional Ecology 10: 681–687.

Hagiwara, A. & A. Hino, 1989. Effect of incubation and preservation on resting egg hatching and mixis in the derived clones of the rotifer *Brachionus plicatilis*. Hydrobiologia 186/187: 415–421.

Hagiwara, A. & A. Hino, 1990. Feeding history and hatching of resting eggs in the marine rotifer *Brachionus plicatilis*. Nippon Suisan Gakkaishi 56: 1965–1971.

Hagiwara, A. & C. S. Lee, 1991. Resting egg formation of the L-type and S-type rotifer *Brachionus plicatilis* under different water temperature. Nippon Suisan Gakkaishi 57: 1645–1650.

Hagiwara, A., A. Hino & R. Hirano, 1988. Studies on the formation and hatching of fertilized eggs of the rotifer *Brachionus plicatilis*. 2. Effects of temperature and chlorinity on resting egg formation in the rotifer *Brachionus plicatilis*. Nippon Suisan Gakkaishi 54: 569–575.

Hagiwara, A., Y. Kadota & A. Hino, 2005. Maternal effect by stem females in *Brachionus plicatilis:* effect of starvation on mixis induction in offspring. Hydrobiologia 546: 275–279.

Hagiwara, A., C. S. Lee, G. Miyamoto & A. Hino, 1989. Resting egg formation and hatching of the S-type rotifer *Brachionus plicatilis* at varying salinities Marine Biology 103: 327–332.

Hagiwara, A., K. Hamada, S. Hori & K. Hirayama, 1994. Increased sexual reproduction in *Brachionus plicatilis* (Rotifera) with the addition of bacteria and rotifer extracts. Journal of Experimental Marine Biology and Ecology 181: 1–8.

Hagiwara, A., N. Hoshi, F. Kawahara, K. Tominaga & K. Hirayama, 1995. Resting eggs of the marine rotifer *Brachionus plicatilis* Müller: development, and effect of irradiation on hatching. Hydrobiologia 313/314: 223–229.

Hairston, N. G. & Jr., 1998. Time travelers: What's timely in diapause research?. Archiv für Hydrobiologie, Special Issues: Advances in Limnology 52: 1–15.

Hairston, N. G., Jr. & B. T. Stasio, 1988. Rate of evolution slowed by a dormant propagule pool. Nature 336: 239–242.

Hairston, N. G., Jr., S. Ellner & C. M. Kearns, 1996. Overlapping generations: the storage effect and the maintenance of biotic diversity. In Rhodes, O. E., R. K. Chesser, & M. H. Smith (eds.), Population Dynamics in Ecological Space and Time. University of Chicago Press, Chicago: 109–145.

Hino, A. & R. Hirano, 1977. Ecological studies on the mechanisms of bisexual reproduction in the rotifer *Brachionus plicatilis* – II. Effects of cumulative parthenogenetic generation on the frequency of bisexual reproduction. Bulletin of the Japanese Society of Scientific Fisheries 43: 1147–1155.

King, C. E. & M. Serra, 1998. Seasonal variation as a determinant of population structure in rotifers reproducing by cyclical parthenogenesis. Hydrobiologia 387/388: 361–372.

Kleiven, O. T., P. Larsson & A. Hobaek, 1992. Sexual reproduction in *Daphnia magna* requires three stimuli. Oikos 65: 197–206.

Kogane, T., A. Hagiwara & K. Imaizumi, 1997. Temperature conditions enhancing resting egg production of the euryhaline rotifer *Brachionus plicatilis* O.F.Müller (Kaimiura strain). Hydrobiologia 358: 167–171.

Kotani, T., M. Ozaki, K. Matsuoka, T. W. Snell & A. Hagiwara, 2001. Reproductive isolation among geographically and temporally isolated marine *Brachionus* strains. Hydrobiologia 446/447: 283–290.

LaMontagne, J. M. & E. McCauley, 2001. Maternal effects in *Daphnia*: what mothers are telling their offspring and do they listen? Ecology Letters 4: 65–71.

Langley, J. M., R. J. Shiel, D. L. Nielsen & J. D. Green, 2001. Hatching from the sediment egg-bank, or aerial dispersing? The use of mesocosms in assessing rotifer biodiversity. Hydrobiologia 446/447: 203–211.

Lubzens, E., G. Minkoff & S. Marom, 1985. Salinity dependence of sexual and asexual reproduction in the rotifer *Brachionus plicatilis*. Marine Biology 85: 123–126.

Marcus, N. H., R. Lutz, W. Burnett & P. Cable, 1994. Age, viability, and vertical distribution of zooplankton resting eggs from an anoxic basin – evidence of an egg bank. Limnology and Oceanography 39: 154–158.

May, L., 1986. Rotifer sampling – a complete species list from one visit. Hydrobiologia 134: 117–120.

May, L., 1987. Effect of incubation-temperature on the hatching of rotifer resting eggs collected from sediments. Hydrobiologia 147: 335–338.

May, L., A. E. Bailey-Watts & A. Kirika, 2001. The relationship between *Trichocerca pusilla* (Jennings), *Aulacoseira* spp. and water temperature in Loch Leven. Hydrobiologia 446/447: 29–34.

Minkoff, G., E. Lubzens & D. Kahan, 1983. Environmental factors affecting hatching of rotifer (*Brachionus plicatilis*) resting eggs. Hydrobiologia 104: 61–69.

Miracle, M. R. & X. Armengol-Díaz, 1995. Population dynamics of oxiclinal species in lake Arcas-2 (Spain). Hydrobiologia 313/314: 291–301.

Mnatsakanova, E. A. & L. V. Polishchuk, 1996. Role of parthenogenetic natality and emergence from diapausing eggs in the dynamics of some rotifer populations. Hydrobiologia 320: 169–178.

Nielsen, D. L., F. J. Smith, T. J. Hillman & R. J. Shiel, 2000. Impact of water regime and fish predation on zooplankton resting egg production and emergence. Journal of Plankton Research 22: 433–446.

Nipkow, F., 1961. Die Rädertiere im Plankton des Zürichsee und ihre Entwicklungsphasen. Schweizerische Zeitschrift für Hydrologie 23: 398–461.

Pourriot, R. & C. Rougier, 1999. Température, démographie et mixis chez un rotifère héléoplanctonique, *Epiphanes brachionus* (Ehrb.). Annales de Limnologie 35: 167–172.

Pourriot, R. & T. W. Snell, 1983. Resting eggs in rotifers. Hydrobiologia 104: 213–224.

Pourriot, R., C. Rougier & D. Benest, 1980. Hatching of *Brachionus rubens* O.F. Müller resting eggs (rotifers). Hydrobiologia 73: 51–54.

Pourriot, R., C. Rougier & D. Benest, 1981. Rôle de la lumière et de la température dans l'éclosion des oeufs de durée de *Brachionus rubens* Ehr. (Rotifère). Netherlands Journal of Zoology 31: 637–649.

Pourriot, R., D. Benest & C. Rougier, 1982. Processus d'éclosion des oeufs de durée de *Brachionus calyciflorus* Pallas (Rotifère). Comparaison de deux clones. Vie et Milieu 32: 83–87.

Pourriot, R., D. Benest & C. Rougier, 1983. Effet de la température sur l'éclosion d'aeufs de durée provenant de populations naturelles. Bulletin de la Societé Zoologique de France 108: 59–66.

Pourriot, R., C. Rougier & D. Benest, 1986. Food quality and mictic female control in the rotifer *Brachionus rubens*, Ehr. Bulletin de la Societé Zoologique de France 111: 105–111.

Pozuelo, M. & L. M. Lubian, 1993. Asexual and sexual reproduction in the rotifer *Brachionus plicatilis* cultured at different salinities. Hydrobiologia 255/256: 139–143.

Ricci, C., 2001. Dormancy patterns in rotifers. Hydrobiologia 446/447: 1–11.

Ruttner-Kolisko, A., 1946. Über das Auftreten unbefruchteter "Dauereier" bei *Anuraea aculeata* (*Keratella quadrata*). Österreichische Zoologische Zeitschrift 1: 425–468.

Schröder, T., 1999. Lebenszyklusstrategien planktischer Rotatorien (Monogononta, Rotifera) im Zusammenhang mit den saisonalen Überflutungen in der Flussaue des Unteren Odertals. Ph.D. Dissertation. Freie Universität Berlin, 183 pp.

Schröder, T., 2001. Colonising strategies and diapause of planktonic rotifers (Monogononta, Rotifera) during aquatic and terrestrial phases in a floodplain (Lower Oder Valley, Germany). International Review of Hydrobiology 86: 635–660.

Schröder, T., 2003. Precopulatory mate guarding and mating behaviour in the rotifer *Epiphanes senta* (Monogononta, Rotifera). Proceedings of the Royal Society of London, Series B. 270: 1965–1970.

Schröder, T. & J. J Gilbert, 2004. Transgenerational plasticity for sexual reproduction and diapause in the life cycle of monogonont rotifers: intraclonal, intraspecific and interspecific variation in the response to crowding. Functional Ecology 18: 458–466.

Serra, M. & M. J. Carmona, 1993. Mixis strategies and resting egg production of rotifers living in temporally-varying habitats. Hydrobiologia 255/256: 117–126.

Serra, M. & C. E. King, 1999. Optimal rates of bisexual reproduction in cyclical parthenogens with density-dependent growth. Journal of Evolutionary Biology 12: 263–271.

Snell, T. W., 1986. Effect of temperature, salinity and food level on sexual and asexual reproduction in *Brachionus plicatilis* (Rotifera). Marine Biology 92: 157–162.

Snell, T. W., 1987. Sex, population dynamics and resting egg production in rotifers. Hydrobiologia 144: 105–111.

Snell, T. W., 1998. Chemical ecology of rotifers. Hydrobiologia 388: 267–276.

Snell, T. W. & E. M. Boyer, 1988. Thresholds for mictic female production in the rotifer *Brachionus plicatilis* (Muller). Journal of Experimental Marine Biology and Ecology 124: 73–85.

Snell, T. W. & M. J. Carmona, 1995. Comparative toxicant sensitivity of sexual and asexual reproduction in the rotifer

Brachionus calyciflorus. Environmental Toxicology and Chemistry 14: 415–420.

Snell, T. W. & M. Childress, 1987. Aging and loss of fertility in male and female *Brachionus plicatilis* (Rotifera). International Journal of Invertebrate Reproduction and Development 12: 103–110.

Snell, T. W. & F. H. Hoff, 1985. The effect of environmental factors on resting egg production in the rotifer *Brachionus plicatilis*. Journal of the World Mariculture Society 16: 484–497.

Snell, T. W. & F. H. Hoff, 1987. Fertilization and male fertility in the rotifer *Brachionus plicatilis*. Hydrobiologia 147: 329–334.

Snell, T. W., B. E. Burke & S. D. Messur, 1983. Size and distribution of resting eggs in a natural population of the rotifer *Brachionus plicatilis*. Gulf Research Reports 7: 285–287.

Spencer, M., N. Colegrave & S. S. Schwartz, 2001. Hatching fraction and timing of resting stage production in seasonal environments: effects of density dependence and uncertain season length. Journal of Evolutionary Biology 14: 357–367.

Stelzer, C. P. & T. W. Snell, 2003. Induction of sexual reproduction in *Brachionus plicatilis* (Monogononta, Rotifera) by a density-dependent chemical cue. Limnology and Oceanography 48: 939–943.

Tauber, M. J., C. A. Tauber & S. Masaki, 1986. Seasonal adaptations of insects. Oxford University Press, New York.

Tilman, D., 1982. Resource Competition and Community Structure. Princeton University Press, Princeton N.J.

Viitasalo, M. & T. Katajisto, 1994. Mesozooplankton resting eggs in the Baltic Sea: identification and vertical distribution in laminated and mixed sediments. Marine Biology 120: 455–465.

Wesenberg-Lund, C., 1930. Contributions to the Biology of the Rotifera, Vol. II. The Periodicity and Sexual Periods. A.F. Hoest & soen, Copenhagen.

Hydrobiologia (2005) 546:307–314
A. Herzig, R.D. Gulati, C.D. Jersabek & L. May (eds.) Rotifera X: Rotifer Research: Trends, New Tools and Recent Advances
DOI 10.1007/s10750-005-4238-7

Anhydrobiosis of *Adineta ricciae:* costs and benefits

Claudia Ricci* & Cesare Covino
Department of Biology, State University of Milan, via Celoria 26, 20133, Milano, Italy
*(*Author for correspondence: E-mail: claudia.ricci@unimi.it)*

Key words: bdelloid rotifers, anhydrobiosis, life history, fecundity, longevity, maternal age effect

Abstract

To study the effect of anhydrobiosis on the rotifer life cycle, we dried a bdelloid species, *Adineta ricciae*, and determined the life-history traits of 1) the rehydrated animals and 2) the offspring produced after a period of dormancy. In the first experiment, a cohort was dried when 8-days-old, kept dry for 7 days and then rehydrated. Recovery was about 75%. The recovered rotifers had similar longevity and significantly higher fecundity than the hydrated controls. The time spent dry was completely disregarded by the anhydrobiotic rotifers, as predicted by the 'Sleeping Beauty' model. In the second experiment, we recorded the life-history traits of the orthoclones produced by recovered mothers aged 11 days, and 18 days. These orthoclones were coupled to controls that had been established from hydrated mothers aged 11 days and 18 days. The orthoclones produced after dormancy had significantly higher fecundity and longevity than the controls of same maternal age. Maternal age had a marked effect on the life-history traits of the orthoclones of both lines, causing the same loss of fitness in both.

Introduction

To survive in habitats that are exposed to occasional drought, some organisms have acquired the capacity to enter a reversible state of suspended life. This is called anhydrobiosis (Gilbert, 1974; Crowe et al., 1992; Clegg, 2001; Ricci, 2001). During anhydrobiosis, any activity is suspended and metabolism stops (Clegg, 2001); both of these features are restored at the end of the period of dormancy. The effect of desiccation on plants and animals has been explored in terms of their physiological and chemical adaptations to water loss (e.g., Clegg, 1997, 2001; Bartels & Salamini, 2001; Jönsson & Bertolani, 2001; Wright, 2001; Guidetti & Jönsson, 2002). Among the animals, one of the commonest anhydrobionts is the bdelloid rotifer, but very little is known about how they are adaptated to survive this process. This is because, although it is known that bdelloid rotifers consist of species that are capable of anhydrobiosis at any stage during their lifetime (Ricci, 1998), their morphological and biochemical adaptations have attracted little interest among investigators until relatively recently (Ramløv et al., in preparation; Ricci et al., 2003a; Tunnacliffe & Lapinski, 2003). Little has been discovered so far, except that, in contrast to most other anhydrobionts, bdelloids do not synthesize trehalose as osmoprotectant sugar (Caprioli et al., 2004; Tunnacliffe & Lapinski, 2003; see Tunnacliffe et al., 2005).

The effect of anhydrobiosis on bdelloids has been explored mostly in terms of recovery capacity. This varies with species (Ricci, 1998), age (Ricci et al., 1987; Örstan, 1995), dormancy duration (Ricci et al., 1987; Caprioli & Ricci, 2001) and desiccation procedure (Ricci et al., 2003a). In one bdelloid species, *Macrotrachela quadricornifera* Milne, anhydrobiosis has been found to represent a simple interlude in the animals life history after which it has the same fecundity as its hydrated control (Ricci et al., 1987). This type of response has been compared to the tale of 'Sleeping Beauty', in which everyone falls asleep until the 'princess' is awakened by the 'prince'. When awakened, nobody can recall the time spent sleeping. In contrast, another taxon, a nematode appears to age during dormancy, behaving like the picture of Dorian Gray in the novel by Oscar Wilde. In this story, the picture becomes older, while Mr. Gray keeps his young age (Ricci & Pagani, 1997). Recovered nematodes were also found to have lower fecundity than the hydrated controls.

Although recovered *M. quadricornifera* had the same fecundity as the hydrated rotifers, other bdelloid populations that have recovered from anhydrobiosis have been reported to show an increase in their birth rate. However, there are no data in the literature supporting this observation (Dobers, 1915; Hickernell, 1917). There is also evidence from another bdelloid rotifer (*Philodina roseola*) that fecundity and longevity may increase after a period of desiccation, in comparison with a parallel population maintained in the hydrated state (Ricci, 1987). So, the evidence that exists at the moment suggests that anhydrobiosis could keep bdelloid net fecundity rate at the same level as the species' potential, improve the net fecundity rate of the animal itself or its progeny, or, eventually, decrease it. The latter scenario has never been reported. Whether this is a general effect of anhydrobiosis on bdelloids, or what the likely mechanisms responsible for it might be, are unexplored issues at present.

Here, we investigate the effect of anhydrobiosis on life history traits by inducing dormancy in a bdelloid species, *Adineta ricciae*, and studying the life table of the recovered rotifers. The aim is to test 1) if the 'Sleeping Beauty' model describes the response of any more species of bdelloids, and 2) if anhydrobiosis can affect the fitness of the progeny of desiccated rotifers.

Material and methods

The species studied belongs to the family Adinetidae and was originally collected from the dry sediment of an Australian billabong (Ryan's 3) by Russ Shiel and one of the present authors (C.R.) (Ricci et al., 2003b). It is a new species and is first described in this volume as *Adineta ricciae* (Segers & Shiel, 2005).

Adineta ricciae is about 200–250 μm long and, like all species of this genus, moves by crawling. It feeds by scraping the substrate with the rakes (cuticular thickening lateral to mouth) (Melone & Ricci, 1995). The species has been studied for molecular analysis by Mark Welch & Meselson (2003), and has been cultivated for over 3 years with deionized water as a medium and a suspension of powdered fish pellets (Friskies) as food. The cultures are maintained at the Rotifer Laboratory, Department of Biology, University of Milan, Italy.

Life table experiments

All of the life table experiments were run at 22 °C. Every day, eggs produced were counted and removed, deaths were recorded, and culture medium and food were renewed; each rotifer received about 5 μg of food per day. A life table (Birch, 1948) was compiled for each experiment, and age-specific fecundity rate (m_x) (eggs produced by a female aged x), age-specific survival rate (l_x) (ratio between females aged x and females aged 0), mean lifetime fecundity and mean life span were calculated. Every experiment was replicated two or three times.

Desiccation procedure

Anhydrobiosis was induced in 8-day-old rotifers and lasted for 7 days. The recovery percentage was recorded about 24 h after water addition.

Rotifers to be dried were transferred, either singly (exp. 1) or in groups of 12 (exp. 2), to a filter paper substrate. After any excess water had been removed, this was placed in an embryo dish contained in a Petri dish. The dish was placed in a humido–thermostatic chamber, where relative humidity (RH) was set at 97% for 3 days, then decreased to 40% over a 4h period and then maintained at 40% until the day 7 (Protocol C of Ricci et al., 2003a).

Experiment 1: Effect of anhydrobiosis on the life history of Adineta ricciae

Following the life table experimental design, a cohort of *A. ricciae* aged 8-days was dried for

7 days and then rehydrated. The life-history traits of the recovered rotifers were then analysed. Each experiment was replicated three times, and the data from all of the replicates were pooled and analysed as a single cohort once statistical analysis had shown that there was no significant difference in response among the replicates. Each cohort was paired with a control cohort to compare with the survival and fecundity of the recovered rotifers. The effect of anhydrobiosis on these life table characteristics was assessed by matching the values of age-specific fecundity (m_x) of the recovered cohorts to their paired control, either including or excluding the days spent in the anhydrobiotic state.

Experiment 2: Effect of anhydrobiosis on the progeny of Adineta ricciae

Two parallel cohorts were established to serve as parental groups. Orthoclones (a cohort obtained from a group of isogenetic mothers of the same age) from both of these cohorts were set up, cultured and studied (Fig. 1). When the cohorts were 7-days-old, eggs produced by both parental groups were collected and hatched. The newborns formed the 7-day orthoclones, A_1 and A_2, which served as a reference for further comparisons. One parental cohort was dried at the age of 8 days, kept dry for 7 days and then rehydrated. The rotifers were

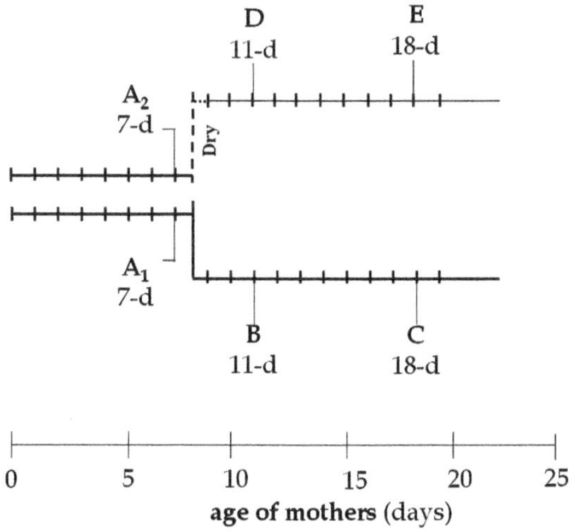

Figure 1. Design of Experiment 2 showing the orthoclones established from the parental cohort (A_2, D, E) undergoing anhydrobiosis at the age of 8-days. A_1, B and C orthoclones were set up from the hydrated parental cohort.

desiccated in groups of 12. Two orthoclones were established from rehydrated bdelloids at the ages of 11 days (D) and 18 days (E), respectively. Since the days spent in a dehydrated state were not included in the rotifers' ages, parallel orthoclones were also isolated from the hydrated parental cohort at the ages of 11 days (B) and 18 days (C). All orthoclones were studied following the life table experimental schedule and their fecundity and survival rates were calculated.

Orthoclones established as outlined above differed in age of the mother (7-days: A_1, A_2; 11-days: B, D; 18-days: C, E) and in the anhydrobiosis experienced by the maternal cohort (B, C vs. D, E). The experiment was replicated twice, and the results were pooled after a statistical check confirmed that there was no significant difference between the replicate orthoclones. To document the effect of maternal age on orthoclone life traits, we calculated the correlation coefficients of the regression lines obtained by plotting the mean fecundity of orthoclones B, C and D, E against maternal ages. The resulting slope coefficients were compared to assess the degree parallelism of the lines. Mean longevity and mean fecundity were compared by the t-test.

Results

When cultivated at 22 °C and fed 5 μg of fish pellets daily, individuals of *A. ricciae* lived for about 35–40 days and produced about 32 eggs whose embryos developed in about 40 h. The capacity of the species to recover after a 7-day period of desiccation ranged from 70% when single animals were desiccated to 85%, when groups of 12 rotifers were desiccated. In all experiments, the species resumed reproduction soon after the addition of the water medium.

Experiment 1: Effect of anhydrobiosis on the life history of A. ricciae (Table 1)

Life span did not differ between recovered rotifers and their controls ($p > 0.05$, at t-test) if the dormancy period was excluded from the age of the recovered group. However, the two cohorts differed significantly ($p \leq 0.001$) if the age of the rotifers did include the anhydrobiosis days. After

Table 1. Experiment 1: Fecundity (eggs*rotifer*lifetime) and longevity (days) ($\bar{X} \pm$ SE) of *Adineta ricciae* after recovery from 7-days of anhydrobiosis

Cohorts	#	Mean fecundity (±ES)	Mean longevity (±ES) [incl. 7d anhydrobiosis]	Mean longevity (±ES) [excl. 7d anhydrobiosis]
Recovered	34	37.6*** (±0.4)	48.2*** (±1.2)	$41.2^{n.s.}$ (±1.2)
Hydrated	24	31.9 (±0 .4)	37.7 (±1.2)	37.7 (±1.2)

The age for computation of longevity either includes or excludes the dry days.

recovery, *Adineta* increased its mean fecundity significantly in comparison with the hydrated control cohort (37.6 vs. 31.9; $p \leq 0.001$). Age-specific (excluding the 7-day anhydrobiosis) fecundity (m_x) and survival (l_x) rates of recovered and control bdelloids are presented in Figure 2. The correlation test applied to m_x values of recovered and control cohorts demonstrated a strong similarity between their fecundity patterns ($r = 0.82$, $p < 0.0001$) if the 'dry days' were disregarded. However, if the 'dry days' were added to the age of recovered rotifers, the correlation was poor ($r = 0.39$, $p > 0.1$). After recovery, the rotifers reproduced for longer and reached higher m_x values than their respective hydrated controls.

Experiment 2: Effect of anhydrobiosis on the progeny of A. ricciae (Table 2)

Orthoclones A_1 and A_2 (7-day orthoclones) had similar life-history traits, age-specific fecundity and survival rates, and remarkably similar patterns of m_x among orthoclones of same maternal age (Fig. 3 a, b, c). Using orthoclone A_2 as the reference for comparison of the orthoclones established from recovered mothers, cohort D (11-day orthoclone) had a similar mean fecundity ($p > 0.5$). The same could be found by comparing A_2 with E (18-day orthoclone), which had slightly lower fecundity ($0.05 < p < 0.01$). No difference was found among life spans of the orthoclones.

Cohort A_1 (7-day orthoclone) was the reference for orthoclones established from hydrated mothers. A_1 fecundity was slightly higher than that of cohort B ('hydrated' 11-day orthoclone) ($0.05 < p < 0.01$), and significantly higher than cohort C ('hydrated' 18-day orthoclone) ($p < 0.0001$). When mean fecundity values for the orthoclones were plotted against the ages of their mothers, mean fecundity was found to decrease markedly with increasing maternal age (Fig. 4). This decreasing trend was almost in parallel for the

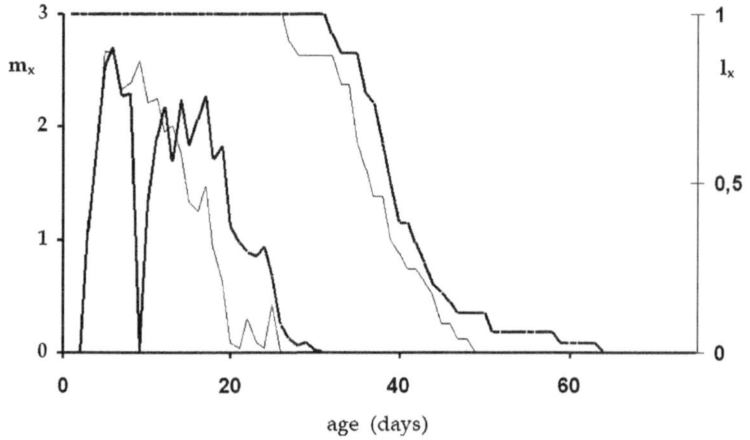

Figure 2. Experiment 1: Age-specific fecundity rates (m_x), on left axis, and age-specific survival rates (l_x), on right axis, of controls (thin line) and of the cohort desiccated at the age of 8-days (thick line). The days spent in anhydrobiosis were excluded from rotifer age.

Table 2. Experiment 2: Fecundity (eggs*rotifer*lifetime) and longevity (days) (±SE) of the orthoclones of *Adineta ricciae* established from mothers kept hydrated (A_1, B, C) and those desiccated at the age of 8-days (A_2, D, E)

Orthoclones	#	Mean fecundity (±SE)	Mean longevity (±SE)
A_1 (7-d)	30	27.5 (±0.7)	45.1 (±2.8)
B (11-d)	21	24.8 (±1.6)	42.7 (±3.0)
C (18-d)	45	20.1 (±0.5)	34.1 (±1.7)
A_2 (7-d)	24	28.2 (±0.5)	47.9 (±3.4)
D (11-d)	21	29.2 (±1.3)	43.8 (±2.3)
E (18-d)	24	25.4 (±0.6)	39.8 (±3.1)

two series of orthoclones (B–C and D–E) , and the two slope coefficients (−0.67, −0.55) differed by less than 1σ ($0.46\ \sigma$).

The mean fecundities of orthoclones of equal maternal age showed that the 'desiccated' orthoclones produced 5 more eggs than their respective 'hydrated' controls. In contrast, no difference was noted in their mean longevity, except in orthoclone C, which lived for a significantly shorter time than other cohorts.

Discussion

Recovery of *A. ricciae* after 7 days of desiccation indicates that the species is well adapted to anhydrobiosis. This can be related to its natural habitat, a water body prone to drying known as a billabong (Ricci et al., 2003b). The recovery rates obtained were very similar among replicates, but differed between experiments 1 and 2. The major difference between the experiments concerned the procedure followed for drying the rotifers, singly in experiment 1 and in groups of 12 in experiment 2. This difference probably accounts for the variation in recovery rates observed, because groups of rotifers desiccate more slowly than single rotifers (Ricci et al., 2003a).

The replicates of each experiment gave very similar results in terms of fecundity and lifespan and so the life table data were pooled into a single, bigger cohort. In close agreement with the results obtained in a previous study (Ricci et al., 1987), no difference was found between the life spans of desiccated and hydrated *A. ricciae*. Also, the pattern of age-specific fecundity of the two experimental groups was very similar if the dryness days were not included in the estimated age of the rotifer. A similar indication about rotifer age after anhydrobiosis was given indirectly by the comparison of life traits between the orthoclones. Maternal age is known to affect the life traits of rotifers (e.g., Lansing, 1947, 1954). The orthoclones compared for the pattern of age-specific fecundity rates were obtained from mothers that were of equal age only if the desiccation period was disregarded. Thus, cohort D was matched to cohort B, and both were established from 11-day-old mothers; but D would be compared to C (18-days orthoclone), if the mother's age were increased to take account of the desiccation period. The pattern of age-specific fecundity rate of D was very close to B. The same was true between C and E traits, but not between D and C. Thus the mothers, also, seemed not to age during anhydrobiosis, supporting the results obtained in Experiment 1.

Maternal age produced similar effects in all orthoclones, whether the mothers were desiccated or not. As the mother's age increased, the mean fecundity of orthoclones B and C decreased as much as those of D and E, and the differences between the two orthoclones for each group were similar. These results indicate that the time of anhydrobiosis had no effect on the age of *A. ricciae*, confirming the results obtained for *M. quadricornifera*.

Systematically, *A. ricciae* and *M. quadricornifera* are not closely related taxa (Melone et al., 1998), nevertheless both suspend life during desiccation and do not 'age' during anhydrobiosis. This meets the prediction of the 'Sleeping Beauty' model. The alternative would have been for the animal to age during anhydrobiosis, as has been found in a species of nematode (Ricci & Pagani, 1997). Although the 'Sleeping Beauty' model seems to apply to *A. ricciae* and *M. quadricornifera*, it is too early to say whether it also applies to other bdelloids.

One aim of this study was to test the effect of anhydrobiosis on the fecundity of the rotifer or its offspring, following suggestions present in the literature (Dobers, 1915; Hickernell, 1917; Ricci, 1987). The results obtained for *M. quadricornifera* did not show an increase in fecundity after anhydrobiosis (Ricci et al., 1987). In contrast, the recovered *A. ricciae* had a higher fecundity than

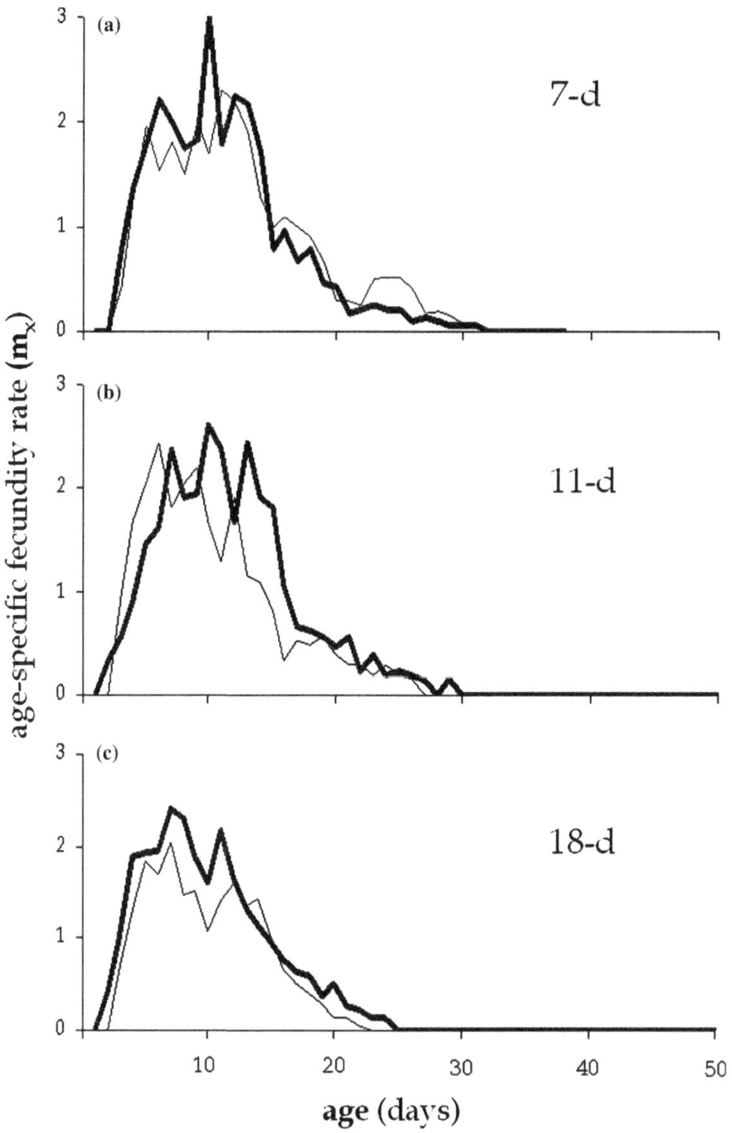

Figure 3. Experiment 2: Age-specific fecundity rates (m_x) of the orthoclones set up from mothers kept hydrated (controls, thin line) and desiccated at the age of 8-days (thick line). a) Orthoclones A_1 (control) and A_2 ('dry'), set up from mothers that were 7-days-old before anhydrobiosis b) Orthoclones B (control) and D ('dry'), set up from 11-day old mothers c) Orthoclones C (control) and E ('dry'), set up from 18-day-old mothers.

the hydrated controls, and their offspring showed increased mean fecundities too. The increase was about 5 eggs female^{-1} life time^{-1} in all of the experiments. It is possible that the constancy of exactly 5 eggs among the different independent experiments is due only to chance, however this is still a significant difference as a 5-egg increase represents as much as 15% of the mean fecundity of *A. ricciae*. At present we have no explanations to offer for the different effect of anhydrobiosis on the fecundity of *Adineta* and *M. quadricornifera*. In the latter species, offspring fecundity was not recorded so these results cannot be compared.

How can a bdelloid rotifer increase its fecundity? It must be recalled that rotifers are strictly eutelic and bdelloids were found to possess a fixed number of ovocytes at birth to be used during life (Pagani et al., 1993). Thus, we can hypothesis that

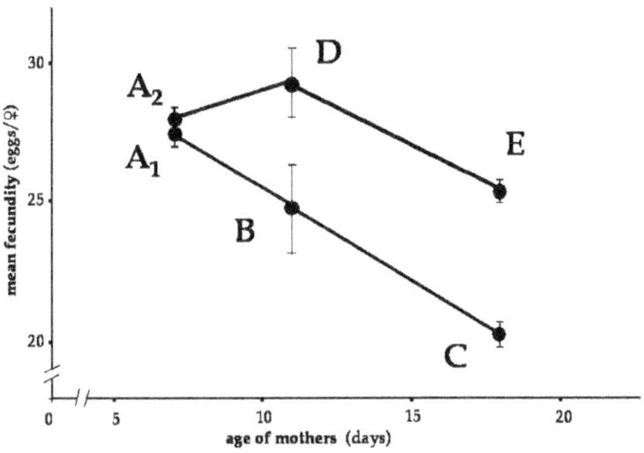

Figure 4. Experiment 2. Mean fecundities of orthoclones plotted against the age of the mothers at their birth. A_1, B, C were established from the hydrated control cohort. A_2, D, E were established from the cohort desiccated at the age of 8-days. The bars represent the standard errors of each mean fecundity value.

Adineta does not use all ovocytes present in its germarium under normal conditions, but that the remainder are made accessible only after anhydrobiosis. Alternatively, the rotifer may produce new ovocytes after anhydrobiosis, thus making the eutely of the germarium reversible. In a newborn *Adineta*, a preliminary count of the nuclei number in the germarium gave about 30 nuclei, lending some support to a cell reproduction hypothesis. This aspect deserves further study.

Other hypotheses can be proposed to explain how a rotifer could benefit from anhydrobiosis. For instance, we could imagine that rotifer viruses or parasites may be less desiccation-tolerant than the animals themselves, and that desiccation affects the rotifers and their 'guests' in different ways, perhaps relieving some of the pathogenic load. This idea could also be functional to the 'unique' continuous parthenogenesis of bdelloids, as one hypothesis for the role of sex is that it allows the organism to stay one step ahead of its parasites by modulating its immune system (Maynard Smith, 1978). Also, if anhydrobiosis is better tolerated by the rotifer than by its parasites, this might have allowed bdelloids to rule mixis out permanently. One apparent limit to this speculation is that the increased fecundity was ascertained in *A. ricciae*, but not in *M. quadricornifera* (Ricci et al., 1987).

Despite the differences between the two bdelloids investigated so far, it seems that both disregard the time spent in an anhydrobiotic state and that the only cost of anhydrobiosis is paid in terms of survival after dormancy. The event of anhydrobiosis, on the other hand, might 'switch' some physiological process, giving some 'advantage' to the rotifer after dormancy. It is not uncommon that cultures of animals capable of anhydrobiosis, i.e. bdelloids and tardigrades, cannot be maintained for several generations if they are not dried out occasionally (Kristensen, personal communication). Any positive effect of dormancy on animals that will colonize an empty environment would be promptly selected for, but the nature of that 'positive effect' is an open field of research.

Acknowledgements

Thanks are due to an anonymous referee for his stimulating comments that greatly improved the speculative parts of the discussion. Russ Shiel and one of the Editors, Linda May, revised an earlier version of this manuscript for language. Financial support came from MIUR and ASI grants to C.R.

References

Bartels, D. & F. Salamini, 2001. Desiccation tolerance in the resurrection plant *Craterostigma plantagineum*. A contribution to the study of drought tolerance at the molecular level. Plant Physiology 127: 1346–1353.

Birch, L.C., 1948. The intrinsic rate of natural increase of an insect population. Journal of Animal Ecology 17: 15–26.

Caprioli, M. & C. Ricci, 2001. Recipes for successful anhydrobiosis in bdelloid rotifers. Hydrobiologia 446/447: 13–17.

Caprioli, M., A. Krabbe Katholm, G. Melone, H. Ramløv, C. Ricci & N. Santo, 2004. Trehalose in desiccated rotifers: a comparison between a bdelloid and a monogonant species. Comparative Biochemistry and Physiology. Part A 139: 527–532.

Clegg, J. S., 1997. Embryos of *Artemia franciscana* survive four years of continuous anoxia: the case for complete metabolic rate depression. Journal of Experimental Biology 200: 467–475.

Clegg, J. S., 2001. Cryptobiosis – a peculiar state of biological organization. Comparative Biochemistry and Physiology 128: 613–624.

Crowe, J., F. Hoekstra & L. Crowe, 1992. Anhydrobiosis. Annual Review of Physiology 54: 579–599.

Dobers, H., 1915. Über die Biologie der Bdelloidea. Internationale Revue der gesamten Hydrobiologie und Hydrographie 7: 1–128.

Gilbert, J. J., 1974. Dormancy in rotifers. Transactions of the American Microscopical Society 93: 490–513.

Guidetti, R. & K. I. Jönsson, 2002. Long-term anhydrobiotic survival in semi-terrestrial micrometazoans. Journal of Zoology, London 257: 181–187.

Hickernell, L. M., 1917. A study of desiccation in the rotifer *Philodina roseola*, with special reference to cytological change accompanying desiccation. Biological Bulletin 32: 343–407.

Jönsson, K. I. & R. Bertolani, 2001. Facts and fiction about long-term survival in tardigrades. Journal of Zoology, London 255: 121–123.

Lansing, A., 1947. A transmissible, cumulative and reversible factor in aging. Journal of Gerontology 2: 228–239.

Lansing, A., 1954. A nongenic factor in the longevity of rotifers. Annals of the New York Academy of Sciences 57: 455–464.

Mark Welch, D. & M. Meselson, 2003. Oocyte nuclear DNA content and GC proportion in rotifers of the anciently asexual Class Bdelloidea. Biological Journal of the Linnean Society 79: 85–91.

Maynard Smith, J., 1978. The Evolution of Sex. Cambridge University Press, 222 pp.

Melone, G. & C. Ricci, 1995. Rotatory apparatus in Bdelloids. Hydrobiologia 313/314: 91–98.

Melone, G., C. Ricci, H. Segers & R. L. Wallace, 1998. Phylogenetic relationships of phylum Rotifera with emphasis on the families of Bdelloidea. Hydrobiologia 387/388: 101–107.

Örstan, A., 1995. Desiccation survival of the eggs of the rotifer *Adineta vaga* (Davis, 1873). Hydrobiologia 313/314: 373–375.

Pagani, M., C. Ricci & C. A. Redi, 1993. Oogenesis in Macrotrachela quadricornifera. I. Germarium nuclei, caryotype and DNA content. Hydrobiologia, 255/256: 225–230.

Ricci, C., 1987. Ecology of bdelloids: how to be successful. Hydrobiologia 147: 117–127.

Ricci, C., 1998. Anhydrobiotic capabilities of bdelloid rotifers. Hydrobiologia 337/338: 321–326.

Ricci, C., 2001. Dormancy patterns in rotifers. Hydrobiologia 446/447: 1–11.

Ricci, C. & M. Pagani, 1997. Desiccation of *Panagrolaimus rigidus* (Nematoda): survival, reproduction and the influence on the internal clock. Hydrobiologia 347: 1–13.

Ricci, C., L. Vaghi & M. Manzini, 1987. Desiccation of rotifers (*Macrotrachela quadricornifera*): survival and reproduction. Ecology 68: 1488–1494.

Ricci, C., G. Melone, N. Santo & M. Caprioli, 2003a. Morphological response of a Bdelloid rotifer to desiccation. Journal of Morphology 257: 246–253.

Ricci, C., R. J. Shiel, D. Fontaneto & G. Melone, 2003b. Bdelloid rotifers recorded from Australia with description of *Philodinavus aussiensis* n.sp. Zoologischer Anzeiger 242: 241–248.

Segers, H. & R. J. Shiel, 2005. Tale of a sleeping beauty: a new and easily cultured model organism for experimental studies on bdelloid rotifers. Hydrobiologia 546: 141–145.

Tunnacliffe, A. & J. Lapinski, 2003. Resurrecting van Leeuwenhoek's rotifers: a reappraisal of the role of disaccharides in anhydrobiosis. Philosophical Transaction of the Royal Society of London B 358: 1755–1771.

Tunnacliffe, A., J. Lapinski & B. McGee, 2005. A putative LEA protein, but no trehalose, is present in anhydrobiotic bdelloid rotifers. Hydrobiologia 546: 315–321.

Wright, J.C., 2001. Cryptobiosis 300 years on from van Leeuwenhoek: what have we learned about tardigrades? Zoologischer Anzeiger 240: 563–582.

Hydrobiologia (2005) 546:315–321 © Springer 2005
A. Herzig, R.D. Gulati, C.D. Jersabek & L. May (eds.) Rotifera X: Rotifer Research: Trends, New Tools and Recent Advances
DOI 10.1007/s10750-005-4239-6

A putative LEA protein, but no trehalose, is present in anhydrobiotic bdelloid rotifers

Alan Tunnacliffe*, Jens Lapinski & Brian McGee
Institute of Biotechnology, University of Cambridge, Tennis Court Road, Cambridge, CB2 1QT, UK
*(*Author for correspondence: E-mail: at10004@biotech.cam.ac.uk)*

Key words: late embryogenesis abundant protein, trehalose, trehalose-6-phosphate synthase, LEA protein, desiccation tolerance

Abstract

Some eukaryotes, including bdelloid rotifer species, are able to withstand desiccation by entering a state of suspended animation. In this ametabolic condition, known as anhydrobiosis, they can remain viable for extended periods, perhaps decades, but resume normal activities on rehydration. Anhydrobiosis is thought to require accumulation of the non-reducing disaccharides trehalose (in animals and fungi) or sucrose (in plant seeds and resurrection plants), which may protect proteins and membranes by acting as water replacement molecules and vitrifying agents. However, in clone cultures of bdelloid rotifers *Philodina roseola* and *Adineta vaga*, we were unable to detect trehalose or other disaccharides in either control or dehydrating animals, as determined by gas chromatography. Indeed, trehalose synthase genes (*tps*) were not detected in these rotifer genomes, suggesting that bdelloids might not have the capacity to produce trehalose under any circumstances. This is in sharp contrast to other anhydrobiotic animals such as nematodes and brine shrimp cysts, where trehalose is present during desiccation. Instead, we suggest that adaptations involving proteins might be more important than those involving small biochemicals in rotifer anhydrobiosis: on dehydration, *P. roseola* upregulates a hydrophilic protein related to the late embryogenesis abundant (LEA) proteins associated with desiccation tolerance in plants. Since LEA-like proteins have also been implicated in the desiccation tolerance of nematodes and micro-organisms, it seems that hydrophilic protein biosynthesis represents a common element of anhydrobiosis across several biological kingdoms.

Introduction

Although anhydrobiosis ('life without water') was first described by van Leeuwenhoek some three centuries ago in a species of bdelloid rotifer (van Leeuwenhoek, 1702), the biochemical and genetic mechanisms involved are still unclear. In the last 30 years, however, increasing emphasis has been placed on the role of non-reducing disaccharides. For example, the anhydrobiotic nematode *Aphelenchus avenae* produces trehalose as up to 10–15% of the animal's dry weight, before it becomes fully desiccation tolerant (Madin & Crowe, 1975; Perry, 1999). Similarly, in the budding yeast *Saccharomyces cerevisiae* accumulation of high concentrations of trehalose correlates with increased survival of desiccation (Gadd et al., 1987; Hottiger et al., 1987). Synthesis of high concentrations of sucrose coincides with maturation and the acquisition of desiccation tolerance in plant seeds (Oliver et al., 2000), and in the resurrection plant *Craterostigma plantagineum*, sucrose represents 90% of total sugar content in dried leaf tissues (Bianchi et al., 1991). These data suggest that disaccharides protect against stresses imposed by extreme water loss, a hypothesis which is supported by many *in vitro* drying experiments showing stabilisation of labile proteins and membrane systems by sugars (Crowe et al., 1998). Such has been the importance attached to disaccharides, that some workers have proposed that not only are they necessary for anhydrobiosis, but that their presence may also be sufficient (Crowe & Crowe, 1992).

One test of such hypotheses would be to use trehalose or sucrose to attempt to confer desiccation tolerance on sensitive cell types, such as mammalian cells in tissue culture, an approach we have termed anhydrobiotic engineering (García de Castro et al., 2000). However, recent attempts at anhydrobiotic engineering have met with limited success (Guo et al., 2000; García de Castro & Tunnacliffe, 2000; García de Castro et al., 2000; Chen et al., 2001). One exception might be the stabilisation of platelets with retention of function (Wolkers et al., 2001a), although these are non-proliferating cell fragments and might have less stringent requirements than replication-competent cells. Furthermore, in naturally anhydrobiotic organisms, the correlation of the acquisition of desiccation tolerance with disaccharide biosynthesis is weak, as we have argued (Tunnacliffe & Lapinski, 2003). For example, in anhydrobiotic nematodes, trehalose biosynthesis in the early stages of dehydration reaches maximal concentrations before full desiccation tolerance is attained, implying that additional physiological adaptations are required (Higa & Womersley, 1993; Womersley & Higa, 1998; Perry, 1999). In *Arabidopsis thaliana*, the seeds of abscisic acid (ABA)-insensitive mutants in the *abi3* locus lack desiccation tolerance, although the accumulation of sucrose is unhindered (Ooms et al., 1993). Such reports imply that disaccharides are not sufficient to achieve anhydrobiosis. In this review, we discuss evidence from our laboratory showing that, in two species of bdelloid rotifer, *Philodina roseola* and *Adineta vaga*, disaccharides are apparently also unnecessary for anhydrobiosis. Instead, a protein related to highly hydrophilic plant proteins associated with desiccation tolerance is induced by dehydration.

Anhydrobiosis without trehalose in bdelloid rotifers

When clonal cultures of bdelloid rotifers, *P. roseola* and *A. vaga*, were subjected to a slow drying regime similar to that described (Ricci, 1998), some 75% of the animals were recovered after rehydration, in line with Ricci's results for these species. In contrast, more rapid drying led to a markedly lower survival, suggesting that rotifers, like nematodes, need time to effect certain physiological changes to be able to undergo anhydrobiosis. In support of this was the fact that starved rotifers, compared with well-fed individuals, showed a reduced ability to enter anhydrobiosis, which indicated an energy expensive process (Lapinski & Tunnacliffe, 2003).

Thermogravimetric analysis of viable rotifers, dried according to a protocol giving high survival rates, showed that a residual moisture content of between 6 and 10% of dry weight is reproducibly achieved (Lapinski & Tunnacliffe, 2003). It was therefore anticipated that significant quantities of trehalose, which is thought to act as a bioprotectant in the dry state, would be present in the desiccated animals. To attempt to quantitate trehalose and any other simple sugars present, carbohydrate extracts of dried, viable rotifers were analysed by gas chromatography (GC), with samples from the anhydrobiotic nematode *A. avenae* serving as reference. GC data showed the expected high level of trehalose in dried *A. avenae*, but no such accumulation in either of the two rotifer species examined; indeed, no trehalose was observed at all (Fig. 1). Apart from glucose, which was also found in similar quantities in hydrated rotifers (not shown), no other simple sugar was present in significant quantities during dehydration: mono-, di- and small oligosaccharides would be apparent on the chromatograph. Sucrose, when added to the samples as a standard prior to extraction (not shown in the Fig.), and intracellular glucose, are easily detected in this system. Glucose constitutes approximately 1% dry weight in the nematode *A. avenae* (Browne, 2001), and similar quantities were present in the rotifers *P. roseola* and *A. vaga*, whereas trehalose comprises at least 10% of dry weight in desiccated nematodes. The GC procedure followed is therefore sufficiently sensitive to detect disaccharide levels at least in the concentration range expected in anhydrobiotic animals, even the somewhat modest quantities produced by tardigrades (2% dry weight; Westh & Ramløv, 1991).

In an attempt to further characterise the metabolite profile of dehydrating rotifers, proton nuclear magnetic resonance spectroscopy of

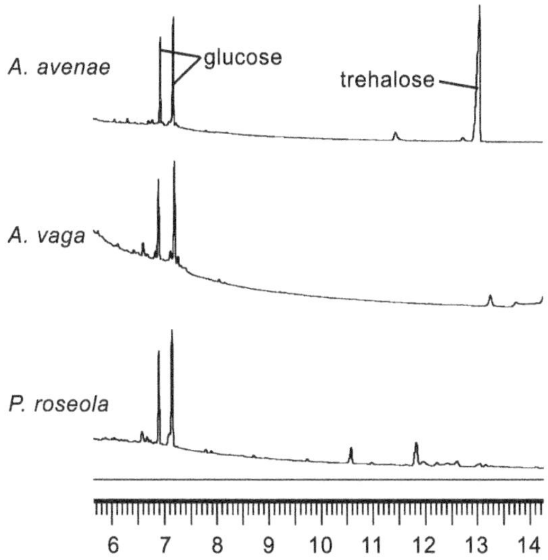

Figure 1. Gas chromatography analysis of derivatised carbohydrate extracts from dried nematode and rotifers. In the nematode *A. avenae*, exposure to 98% relative humidity and subsequent drying strongly induces trehalose biosynthesis. Bdelloid rotifers *A. vaga* and *P. roseola* show no trehalose or comparable carbohydrate accumulation on drying, although glucose was detected. The scale shows the time in minutes of emergence of the derivatised carbohydrates from the GC column. Reproduced from Lapinski & Tunnacliffe (2003), with permission.

control and dried *P. roseola* has been performed (in collaboration with Dr Kevin Brindle, Department of Biochemistry, University of Cambridge). While still preliminary, these results indicate that relatively subtle changes in small molecule populations occur during drying, consistent with a general decrease in metabolism, with no hint of a dramatic shift analogous to the trehalose accumulation seen in nematodes.

If rotifers are able to produce trehalose, they are likely to use the two-step pathway found in all other eukaryotes, whose first step is the transfer of glucose from UDPG to glucose-6-phosphate to yield trehalose-6-phosphate. This step is catalysed by trehalose-6-phosphate synthase (EC 2.4.1.15), an enzyme whose sequence is highly conserved from Gram-negative bacteria to yeast, nematodes and higher plants (Vogel et al., 1998; Blázquez et al., 1998). A polymerase chain reaction (PCR) method, which uses degenerate oligonucleotide primers based on conserved protein sequence, was devised to clone the respective gene (*tps*) from a wide range of species. Low stringency conditions were applied to allow for amplification of fragments with moderate sequence similarity to *tps* genes. Fragments of *tps* genes, confirmed by sequencing, were isolated using this technique from species including *S. cerevisiae, Escherichia coli, Drosophila melanogaster* and *A. avenae*, which are known to contain such sequences in their genomes. As expected, we were unable to amplify *tps* gene fragments from three negative control species (*Pseudomonas putida, Mus musculus* and *Homo sapiens*) whose genomes do not harbour *tps* genes. Crucially, *tps* gene fragments were also not obtained from the two bdelloid rotifer species, *P. roseola* and *A. vaga*, consistent with the absence of such genes from their genomes (Fig. 2). Two weakly amplified fragments from *P. roseola* were present in the experiment shown, but sequencing of more than 50 clones of these fragments confirmed that they did not correspond to *tps* genes (Lapinski & Tunnacliffe, 2003). The larger size of amplified fragments from eukaryotic species, and particularly *A. avenae*, is due to the presence of introns. It is conceivable that large introns are present in putative rotifer *tps* genes, and that

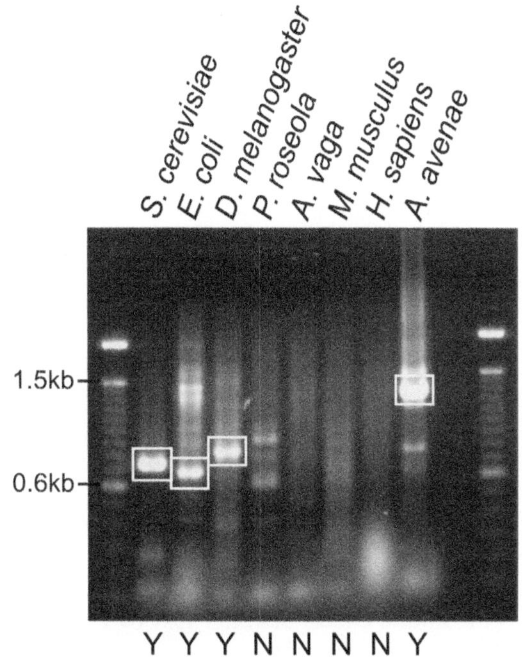

Figure 2. PCR amplification of trehalose-6-phosphate synthase (*tps*) genes of different species. Y denotes a successful *tps* amplification, N a negative result. Amplified fragments corresponding to *tps* sequences are enclosed in a box. Reproduced from Lapinski & Tunnacliffe (2003), with permission.

they are therefore not amplified efficiently by conventional PCR. However, a second series of experiments was performed using PCR primers which amplified much smaller gene fragments of ~250 bp in the eukaryotes containing *tps* genes, and which do not contain introns. These primers also failed to direct amplification of *tps* homologues from bdelloid rotifer DNA (not shown). Therefore, it seems that, not only do bdelloid rotifers not produce trehalose in preparation for desiccation, but they might not have the capacity to produce this disaccharide under any conditions since they lack the necessary genes.

A LEA protein homologue in *P. roseola*

The absence of trehalose or an analogue in bdelloid rotifers challenges our understanding of anhydrobiosis, since the various hypotheses which attempt to explain the phenomenon all require a major role for non-reducing disaccharides; there are no fully developed alternative hypotheses. Clearly, physiological changes associated with anhydrobiosis must take place in bdelloid rotifers and one possibility is that adaptations involving proteins are of more importance than those involving small biochemicals. In plants (Bartels & Salamini, 2001), and more recently in nematodes (Solomon et al., 2000; Browne et al., 2002), production of hydrophilic proteins known as late embryogenesis abundant (LEA) proteins has been associated with water stress. As their name suggests, LEA proteins are produced in abundance during seed development, comprising up to 4% of cellular protein (Roberts et al., 1993), with maximum expression post-abscission during desiccation (Hughes & Galau, 1989). At least five different groups of LEA proteins have been defined on the basis of expression pattern and sequence, and their expression is linked to the acquisition of desiccation tolerance in orthodox seeds, pollen and anhydrobiotic plants. They have been variously proposed to protect cellular structures from the effects of water loss by action as a hydration buffer, by sequestration of ions, by direct protection of other proteins or membranes, or by renaturation of unfolded proteins (Bray, 1993; Cuming, 1999). These proposals have been supported by relatively little evidence, but the discovery of LEA proteins in anhydrobiotic nematodes raises the possibility of similar mechanisms of desiccation tolerance in plants and animals.

Recent work with the Group 3 LEA protein, AavLEA1, from the anhydrobiotic nematode *A. avenae*, has uncovered a remarkable property, which might be characteristic of at least a subset of these proteins (Goyal et al., 2003). Under fully hydrated conditions, AavLEA1 is unstructured throughout its length, but, when dried, Fourier transform infra-red spectroscopy shows it to become folded, adopting an α-helical conformation. Moreover, these α-helices may then form superhelical structures, such as coiled coils. This is a most unusual observation, since drying of proteins usually leads to unfolding and aggregation (Dong et al., 1995), rather than increased structural organisation. A partially characterised Group 3 LEA-like protein from bullrush pollen may also behave similarly to AavLEA1 (Wolkers et al., 2001b). If LEA proteins are able to form α-helical coiled coils on drying, they may also form higher order supramolecular assemblies similarly to the way keratins, neurofilament proteins and lamins, for example, form intermediate filaments (IFs). These cytoskeletal components add mechanical strength to the eukaryotic cells in which they are found, i.e. vertebrates and some invertebrates such as nematodes, but not fruit flies; intriguingly, they are not found at all in plant species (Fuchs & Cleveland, 1998). A hypothesis we are currently testing is that Group 3 LEA proteins allow cells to resist the physical stresses imposed during desiccation in a dehydration-inducible fashion.

Given the occurrence of LEA-like proteins in non-plant species (Dure, 2001), it was of interest to test for their presence in bdelloid rotifers, particularly in the light of the lack of trehalose in these animals. Accordingly, a polyclonal antibody raised against recombinant AavLEA1 protein was used to probe protein extracts from *P. roseola* in Western blots, using methods described (Goyal et al., 2003). The degree of sequence conservation between LEA proteins of different species led us to expect that, if a similar protein were present in rotifers, then the polyclonal antiserum should detect it. Indeed, a clear signal corresponding to a protein of approximately 12 kDa is seen in total *P. roseola* protein extracts (Fig. 3). This protein is

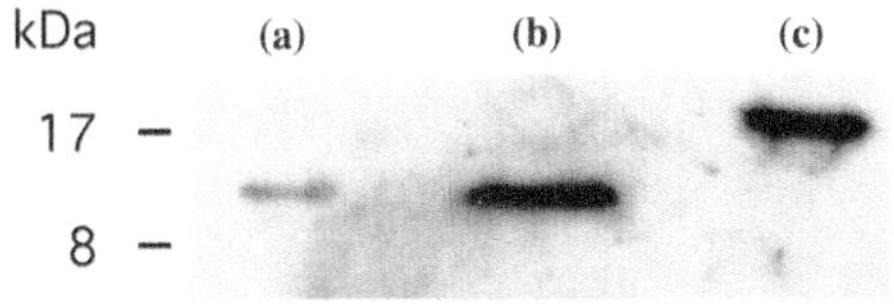

Figure 3. Western blot of protein extracts from *P. roseola*. (a) Control, fully hydrated rotifers; (b) Rotifers dried for 48 h; (c) Recombinant LEA protein (rAavLEA1) from nematode *A. avenae*. Positions and sizes of molecular weight standards are shown on the left of the figure. The filter was probed with a rabbit polyclonal antiserum raised against rAavLEA1. Equivalent amounts of protein (10 μg) were loaded in lanes a and b.

somewhat smaller than the corresponding nematode homologue at 16 kDa in its native form – the recombinant form of AavLEA1 shown in the Figure is slightly larger – but within the observed range of Group 3 LEA proteins; the bullrush pollen protein reported by Wolkers et al. (2001b) is 8 kDa, for example. The rotifer protein, which we tentatively name ProLEA1, is also apparently upregulated by dehydration: the drying protocol described previously leads to a marked increase in signal strength in the Western blot (Fig. 3).

Conclusions and perspectives

While confirmation of the presence of a dehydration-inducible LEA protein in a bdelloid rotifer will await sequencing of purified ProLEA1, it seems likely that the occurrence of this fascinating group of proteins may be widespread in organisms other than plants, and may have particular relevance for anhydrobiosis. For example, Battista et al. (2001) have shown that mutagenesis of either of two LEA-like protein genes in the highly desiccation tolerant micro-organism *Deinococcus radiodurans* leads to decreased survival of drying. It is probably the case, however, that multiple adaptations are required to achieve the anhydrobiotic state, and that expression of LEA proteins alone is insufficient.

As we have stated previously (Tunnacliffe & Lapinski, 2003), there is very little conclusive evidence from living systems which points to particular adaptations being essential for anhydrobiosis. Enormous attention has been directed at the non-reducing disaccharides, yet there are no *in vivo* data – for instance, from drying experiments after mutagenesis or functional inactivation of disaccharide synthase genes – which demonstrate their absolute requirement. Indeed, in some organisms, notably bdelloid rotifers, but also Gram-positive bacteria (Linders et al., 1997) and some plant seeds (Bewley & Black, 1994), disaccharides are absent. There has been an understandable assumption that, because trehalose, in particular, is an excellent biostabiliser *in vitro* (e.g. Colaço et al., 1992), it must also perform the same function *in vivo*. This need not be the case, however, since other reagents can give comparable *in vitro* stabilisation (Crowe et al., 2001). Trehalose may well be a general stress molecule, being induced by osmotic stress, heat stress, oxidative stress, stationary phase and cold stress in micro-organisms (Larsen et al., 1987; Gadd et al., 1987; Hottiger et al., 1987; Eleutherio et al., 1997; Benaroudj et al., 2001; Kandror et al., 2002). Its main role may be, therefore, to counteract these various stresses as a thermodynamic stabiliser of macromolecule structure *in vivo* (Singer & Lindquist, 1998). This function would undoubtedly also be important during the early stages of desiccation, but other so-called compatible solutes might also perform similarly. Clearly, however, because some anhydrobiotes do not use disaccharides does not mean that these sugars are not important in protecting other organisms from desiccation damage. There may well be multiple mechanisms, which enable desiccation tolerance, just as there are for tolerance of e.g. salt, or hyper-osmotic, stress in different organisms, but it is crucial to test putative mechanisms appropriately and objectively.

We feel that the discovery of the lack of trehalose in bdelloid rotifers exposes the poverty of our understanding of anhydrobiosis, and it now seems important to begin a systematic analysis of its biochemistry and genetics. The search for the 'anhydrobiotic gene set', and corresponding 'anhydrobiotic protein set', has been underway for many years in plants (Bartels & Salamini, 2001; Oliver et al., 2000). A similar strategy in bdelloid rotifers and other animals should elucidate mechanisms of anhydrobiosis in these organisms and lead to a fuller understanding of this remarkable phenomenon.

Acknowledgements

A.T. is the Anglian Water Fellow in Biotechnology, and J.L. is the Anglian Water Research

Student, of Pembroke College, Cambridge. This work was funded by The Royal Society research grant no. 22084, and by a BBSRC CASE studentship (B.M.) in collaboration with Integrin Advanced Biosystems Ltd.

References

Bartels, D. & F. Salamini, 2001. Desiccation tolerance in the resurrection plant *Craterostigma plantagineum*. A contribution to the study of drought tolerance at the molecular level. Plant Physiology 127: 1346–1353.

Battista, J. R., M.-J. Park & A. E. McLemore, 2001. Inactivation of two homologues of proteins presumed to be involved in the desiccation tolerance of plants sensitizes *Deinococcus radiodurans* R1 to desiccation. Cryobiology 43: 133–139.

Benaroudj, N., D. H. Lee & A. L. Goldberg, 2001. Trehalose accumulation during cellular stress protects cells and cellular proteins from damage by oxygen radicals. Journal of Biological Chemistry 276: 24261–24267.

Bewley, J. D. & M. Black, 1994. Seeds: Physiology of Development and Germination. Plenum Press, New York.

Bianchi, G., A. Gamba, C. Murelli, F. Salamini & D. Bartels, 1991. Novel carbohydrate metabolism in the resurrection plant *Craterostigma plantagineum*. Plant Journal 1: 355–359.

Blázquez, M. A., E. Santos, C.-L. Flores, J. M. Martínez-Zapater, J. Salinas & C. Gancedo, 1998. Isolation and molecular characterization of the *Arabidopsis TPS1* gene, encoding trehalose-6-phosphate synthase. Plant Journal 13: 685–689.

Bray, E. A., 1993. Molecular responses to water deficit. Plant Physiology 103: 1035–1040.

Browne, J. A., 2001. An investigation into the physiological and genetic changes associated with anhydrobiosis in the nematodes *Panagrolaimus* sp. and *Aphelenchus avenae*. PhD Thesis. National University of Ireland, Maynooth.

Browne, J., A. Tunnacliffe & A. Burnell, 2002. Plant desiccation gene found in a nematode. Nature 416: 38 .

Chen, T., J. P. Acker, A. Eroglu, S. Cheley, H. Bayley, A. Fowler & M. L. Toner, 2001. Beneficial effect of intracellular trehalose on the membrane integrity of dried mammalian cells. Cryobiology 43: 168–181.

Colaço, C., S. Sen, M. Thangavelu, S. Pinder & B. Roser, 1992. Extraordinary stability of enzymes dried in trehalose: simplified molecular biology. Bio/Technology 10: 1007–1011.

Crowe, J. H. & L. M. Crowe, 1992. Membrane integrity in anhydrobiotic organisms: toward aa mechanism for stabilizing dry cells. In: Somero, G. N. & C. B. C. L. Osmond Bolis (eds), Water and Life. Springer-Verlag, Berlin, Germany: 87–103.

Crowe, J. H., J. F. Carpenter & L. M. Crowe, 1998. The role of vitrification in anhydrobiosis. Annual Review of Physiology 60: 73–103.

Crowe, J. H., L. M. Crowe, A. E. Oliver, N. Tsvetkova, W. Wolkers & F. Tablin, 2001. The trehalose myth revisited: introduction to a symposium on stabilization of cells in the dry state. Cryobiology 43: 89–105.

Cuming, A. C., 1999. LEA proteins. In: Shewry, P. R. & R. Casey (eds), Seed Proteins. Kluwer Academic, Dordrecht, NL: 753–780.

Dong, A. C., S. J. Prestrelski, S. D. Allison & J. F. Carpenter, 1995. Infrared spectroscopic studies of lyophilization-induced and temperature-induced protein aggregation. Journal of Pharmaceutical Sciences 84: 415–424.

Dure, III & L. , 2001. Occurrence of a repeating 11-mer amino acid sequence motif in diverse organisms. Protein and Peptide Letters 8: 115–122.

Eleutherio, E. C. A., F. M. Maia, M. D. Pereira, R. Degre, D. Cameron & A. D. Panek, 1997. Induction of desiccation tolerance by osmotic treatment in *Saccharomyces uvarum* var. carlsbergensis. Canadian Journal of Microbiology 43: 495–498.

Fuchs, E. & D. W. Cleveland, 1998. A structural scaffolding of intermediate filaments in health and disease. Science 279: 514–519.

Gadd, G. M., K. Chalmers & R. H. Reed, 1987. The role of trehalose in dehydration resistance of *Saccharomyces cerevisiae*. FEMS Microbiology Letters 48: 249–254.

García de Castro, A. & A. Tunnacliffe, 2000. Intracellular trehalose improves osmotolerance but not desiccation tolerance in mammalian cells. FEBS Letters 487: 199–202.

García de Castro, A., J. Lapinski & A. Tunnacliffe, 2000. Anhydrobiotic engineering. Nature Biotechnology 18: 473.

Goyal, K., L. Tisi, A. Basran, J. Browne, A. Burnell, J. Zurdo & A. Tunnacliffe, 2003. Transition from natively unfolded to folded state induced by desiccation in an anhydrobiotic nematode protein. Journal of Biological Chemistry 278: 12977–12984.

Guo, N., I. Puhlev, D. R. Brown, J. Mansbridge & F. Levine, 2000. Trehalose expression confers desiccation tolerance on human cells. Nature Biotechnology 18: 168–171.

Higa, L. M. & C. Z. Womersley, 1993. New insights into the anhydrobiotic phenomenon – the effects of trehalose content and differential rates of evaporative water-loss on the survival of *Aphelenchus avenae*. Journal of Experimental Zoology 267: 120–129.

Hottiger, T., T. Boller & A. Wiemken, 1987. Rapid changes of heat and desiccation tolerance correlated with changes in trehalose content in *Saccharomyces cerevisiae* cells subjected to temperature shifts. FEBS Letters 220: 113–115.

Hughes, D. W & G. A. Galau, 1989. Temporally modular gene expression during cotyledon development. Genes and Development 3: 358–369.

Kandror, O., A. DeLeon & A. L. Goldberg, 2002. Trehalose synthesis is induced upon exposure of *Escherichia coli* to cold and is essential for viability at low temperatures. Proceedings of the National Academy of Sciences USA 99: 9727–9732.

Lapinski, J. & A. Tunnacliffe, 2003. Anhydrobiosis without trehalose in bdelloid rotifers. FEBS Letters 553: 387–390.

Larsen, P. I., L. K. Sydnes, B. Landfald & A. R. Strøm, 1987. Osmoregulation in *Escherichia coli* by accumulation of organic osmolytes: betaines, glutamic acid, and trehalose. Archives of Microbiology 147: 1–7.

Linders, L. J. M., W. F. Wolkers, F. A. Hoekstra & K. van't Riet, 1997. Effect of added carbohydrates on membrane

phase behaviour and survival of dried *Lactobacillus plantarum*. Cryobiology 35: 31–40.
Madin, K. A. C. & J. H. Crowe, 1975. Anhydrobiosis in nematodes: carbohydrate and lipid metabolism during dehydration. Journal of Experimental Zoology 193: 335–342.
Oliver, M. J., Z. Tuba & B. D. Mishler, 2000. The evolution of vegetative desiccation tolerance in land plants. Plant Ecology 15: 85–100.
Ooms, J. J. J., K. M. Léon-Kloosterziel, D. Bartels, M. Koornneef & C. M. Karssen, 1993. Acquisition of desiccation tolerance and longevity in seeds of *Arabidopsis thaliana*. Plant Physiology 102: 1185–1191.
Perry, R. N., 1999. Desiccation survival of parasitic nematodes. Parasitology 119: S19–S30.
Ricci, C., 1998. Anhydrobiotic capabilities of bdelloid rotifers. Hydrobiologia 387/388: 321–326.
Roberts, J. K., N. A. DeSimone, W. L. Lingle & L. III Dure, 1993. Cellular concentrations and uniformity of cell-type accumulation of two LEA proteins in cotton embryos. Plant Cell 5: 769–780.
Singer, M. A. & S. Lindquist, 1998. Multiple effects of trehalose on protein folding *in vivo* and *in vitro*. Molecular Cell 1: 639–648.
Solomon, A., R. Salomon, I. Paperna & I. Glazer, 2000. Desiccation stress of entomopathogenic nematodes induces the accumulation of a novel heat-stable protein. Parasitology 121: 409–416.
Tunnacliffe, A. & J. Lapinski, 2003. Resurrecting van Leeuwenhoek's rotifers: a reappraisal of the role of disaccharides in anhydrobiosis. Philosophical Transactions of the Royal Society, London B 358: 1755–1771.
Van Leeuwenhoek, A. Letter to Hendrik van Bleyswijk, dated 9th February 1702. In van Rijnberk, G. & L. C. Palm, 1999 (eds), The collected letters of Antoni van Leeuwenhoek. Vol. 14, 1701–1704. Swets and Zeitlinger, Amsterdam: 55–73.
Vogel, G., R. A. Aeschbacher, J. Müller, T. Boller & A. Wiemken, 1998. Trehalose-6-phosphate phosphatases from *Arabidopsis thaliana*: identification by functional complementation of the yeast *tps2* mutant. Plant Journal 13: 673–683.
Westh, P. & H. Ramløv, 1991. Trehalose accumulation in the tardigrade *Adorybiotus coronifer* during anhydrobiosis. Journal of Experimental Zoology 258: 303–311.
Wolkers, W. F., N. J. Walker, F. Tablin & J. H. Crowe, 2001a. Human platelets loaded with trehalose survive freeze-drying. Cryobiology 42: 79–87.
Wolkers, W. F., S. McReady, W. F. Brandt, G. G. Lindsey & F. A. Hoekstra, 2001b. Isolation and characterization of a D-7 LEA protein from pollen that stabilises glasses *in vitro*. Biochimica Biophysica Acta 1544: 196–206.
Womersley, C. Z. & L. M. Higa, 1998. Trehalose: its role in the anhydrobiotic survival of *Ditylenchus myceliophagus*. Nematologica 44: 269–291.

Hydrobiologia (2005) 546:323–331 © Springer 2005
A. Herzig, R.D. Gulati, C.D. Jersabek & L. May (eds.) Rotifera X: Rotifer Research: Trends, New Tools and Recent Advances
DOI 10.1007/s10750-005-4241-z

The development of a bdelloid egg: a contribution after 100 years

Chiara Boschetti[1,*], Claudia Ricci[1], Cristina Sotgia[1] & Umberto Fascio[2]
[1]*Department of Biology, Università degli Studi di Milano, via Celoria 26, 20133 Milan, Italy*
[2]*C.I.M.A. Centro Interdipartimentale di Microscopia Avanzata, Università degli Studi di Milano, via Celoria 26, 20133 Milan, Italy*
*(*Author for correspondence: E-mail: chiara.boschetti@unimi.it)*

Key words: Bdelloid rotifers, *Macrotrachela quadricornifera*, embryo development, cytoskeleton, mastax, confocal microscopy

Abstract

Rotifer development has received very little attention: studies date back to the 19th century and to the first half of 20th century, and very limited contributions have been added in recent times. All information we have on rotifer embryology is mostly based on *in vivo* observation of developing embryos by light microscopy, and only in a minor way by classical histology. The study of rotifer embryogenesis is approached here using *in vivo* observation and laser confocal microscopy. We revealed cytoskeletal components (filamentous actin and tubulin) and nuclear DNA of the embryos to draw the pattern of the early development of *Macrotrachela quadricornifera*. Our results were then compared to the literature data, to determine a development pattern that can be generalized to the whole rotifer group. On the whole, our results agree with the general description provided by previous authors, i.e. the holoblastic unequal segmentation, the transverse furrow of the first division, the typical 16-cell stage, and the early gastrulation by epiboly. A peculiar pattern could also be seen that was interpreted as the formation of the mastax; it seemed to start from a mould of actin, visible by confocal only. The present study provides a preliminary contribution to a too-long-neglected aspect of rotifer biology.

Introduction

Like several lower invertebrates, rotifers possess a determinative spiral cleavage, meaning that the fate of the cells is established very early in development. Moreover, they are eutelic (e.g. Clément & Wurdak, 1991): throughout all life rotifers possess a constant number of cells, or better, of nuclei, because several tissues are syncytial. Consequently, cell divisions occur during embryogenesis only, when the egg divides several times and gives rise to a fixed cell number that is species-specific. At the end of development, the newborn rotifer possesses the same number of cells as the adult, and the cells will only increase in size during life. In eutelic animals, tissues and organs of the animal are thus formed during embryogenesis and will not be modified during the animal's life; and any anomaly could have consequences for the integrity of the adult. Development is therefore a very delicate process, and its knowledge is very important.

Unfortunately, the study of rotifer development received very little attention: studies date back to the 19th century and to first half of 20th century (e.g. Tessin, 1886; Zelinka, 1892; Jennings, 1896; Tannreuther, 1920; Nachtwey, 1925; Remane, 1929–1933; de Beauchamp, 1956; Pray, 1965; Lechner, 1966). In recent times very limited contributions appeared (e.g. Plasota & Plasota, 1980; Castellano Paez et al., 1988). Rotifer embryology was approached through *in vivo* observation of living specimens by light microscopy or through classical histology. Only Lechner

(1966) followed a different approach, exposing embryos at various stages of development to UV microbeams. He intended to destroy selected blastomeres and to analyse later embryos, with a view to reconstructing the cell lineage. These two approaches, the descriptive and the experimental, produced all the information we have at present. However, the use of new techniques developed in the last decades provide the opportunity to explore in greater detail the developmental process.

Outline of developmental pattern

The steps through which the zygote becomes a complete pluricellular animal are typically 3: cleavage, gastrulation and organogenesis (e.g. Wolpert, 2002). During cleavage, the zygote gives rise to smaller and smaller cells, called blastomeres, that do not grow after each mitosis. The egg cytoplasm, which is not homogeneous because of morphogenetic determinants (molecules that will contribute to the different fate of the blastomeres), segregates differently in the blastomeres. After the undifferentiated mass of cells, called blastula, is formed, gastrulation starts and the three germ layers (ectoderm, endoderm and mesoderm) are defined. Organogenesis consists of the differentiation of each germ layer to form the definitive tissues and organs. In different taxa, different patterns of development are recognized. Cleavage of the rotifers can be ascribed to a modified spiral type (as summarised by Gilbert, 1989).

Review of the literature

Very few authors studied rotifer embryology; major contributions concerned monogonont species and are very old (see Gilbert, 1989). In addition, the different studies dealt with different species, making any comparison difficult (Table 1). The bdelloids were studied by Zelinka (1892) in detail and, subsequently, by Plasota & Plasota (1980), only.

Authors agree on the description of the rotifer egg as oligolecitic, i.e. with a moderate amount of yolk, that does not interfere with cleavage pattern (Remane, 1929–1933). The egg is laid unsegmented, and extrudes the polar body (two polar bodies only in Hsu, 1956a, b) before undergoing the first cleavage. Cleavage is holoblastic (total) and cell divisions are unequal: the blastomeres differ in size and in amount of cytoplasm. Development is determinative: each blastomere will form only a given part of the adult and cannot change its fate (Lechner, 1966).

The different authors used different nomenclature when referring to a same blastomere. When possible, the different nomenclature is here unified to present a common pattern that should serve as

Table 1. List of studies on rotifer embryo development

Author	Year	Studied species
Tessin	1886	*Brachionus urceolaris* O.F. Müller, 1773
		Eosphora najas (as *E. digitata*) Ehrenberg, 1830
		Rotaria rotatoria (as *Rotifer vulgaris*) (Pallas, 1766)
Zelinka	1892	*Mniobia russeola* (as *Callidina russeola*) (Zelinka, 1891)
Jennings	1896	*Asplanchna herrickii* de Guerne, 1888
Tannreuther	1920	*Asplanchna sieboldii* (as *A. ebbesbornii*) (Leydig, 1854)
Nachtwey	1925	*Asplanchna priodonta* Gosse, 1850
Remane	1929–1933	Review
Hyman	1951	Review
de Beauchamp	1956	*Ploesoma hudsoni* (Imhof, 1891)
Pray	1965	*Lecane cornuta* (as *Monostyla cornuta*) (O.F. Müller, 1786)
Lechner	1966	*Asplanchna girodi* de Guerne, 1888
Plasota & Plasota	1980	*Habrotrocha rosa* Donner, 1949
Castellano Paez et al.	1988	*Brachionus plicatilis* O.F. Müller, 1786
Gilbert	1989	Review

background for further comparison. We shall follow the nomenclature used by Jennings (1896), Nachtwey (1925) and Lechner (1966) (hereafter JNL), that is similar to that currently used in modern embryology. For precision, we shall report two more ,names' within brackets; these refer to the nomenclature of Zelinka (1892) and of Tannreuther (1920), respectively. Thus, for example, the first 2 blastomeres are respectively AB2 and CD2 according to JNL, but A and I by Zelinka (1892), AB and CD by Tannreuther (1920). These are here reported as AB2 (A, AB) and CD2 (I, CD), respectively (Fig. 1). Other authors do not name each blastomere after second cleavage.

The first two cleavage divisions are unequal and the produced cells differ in size. The first cleavage produces a small AB2 (A, AB) and a large CD2 (I, CD). de Beauchamp (1956) and Pray (1965) found that the yolk granules are mostly attributed to AB2 (A, AB), and a little amount is maintained by CD2 (I, CD). The cleavage furrow is commonly reported as transverse to the major axis of the egg. Only Zelinka (1892) found that the furrow is initially almost longitudinal, and becomes transverse later on.

Except Pray (1965), all authors report that the second cleavage division occurs in the two cells asynchronously, starting in CD2. This gives a larger D3 (I, D) and a smaller C3 (II, C). Cell AB2 then divides into A3 (a, A) and B3 (b, B), of similar sizes. Thus the cells A3, B3, and C3 have same size, while D3 is larger. At first, the four blastomeres are arranged asymmetrically, and become symmetrical very shortly (e.g. Zelinka, 1892). According to de Beauchamp (1956) and Pray

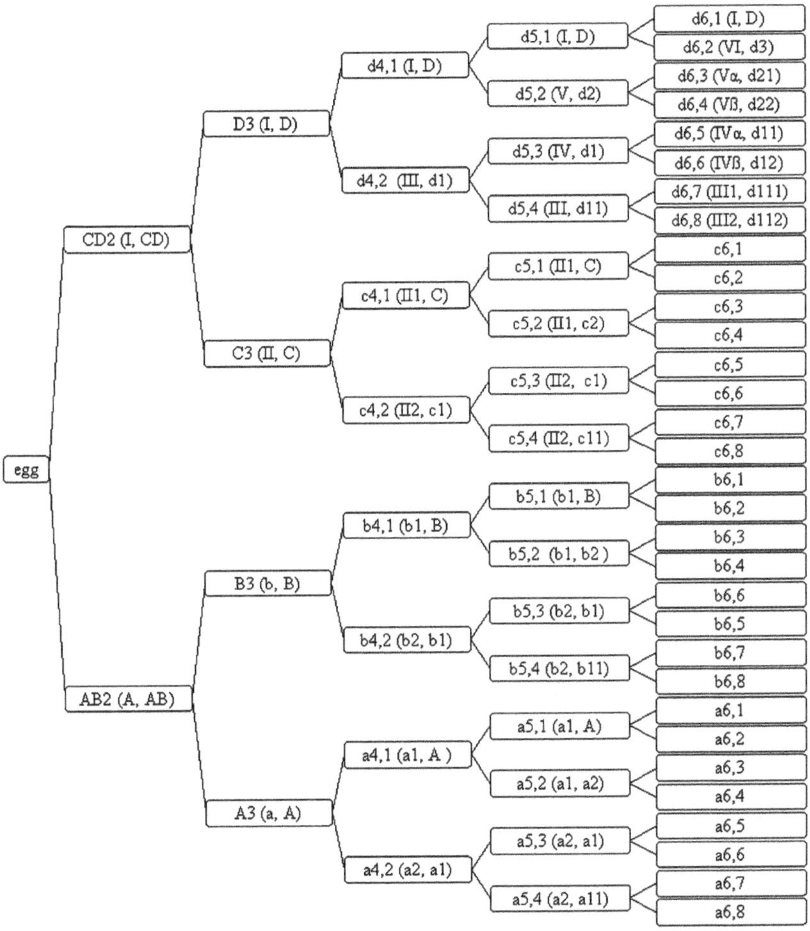

Figure 1. Scheme presenting the cleavage divisions of rotifer embryos. See text for names of blastomeres.

(1965), most yolk granules are segregated in micromeres A3, B3, and C3. The polar body is recognized at the contact between the four cells.

The third cleavage also is not synchronous in the 4 cells, and produces one larger cell (macromere), called d4,1 (I, D), and 7 smaller ones (micromeres). Immediately after division, the cells are not symmetrically arranged, but become symmetrical soon after (e.g. Zelinka, 1892).

Then the macromere d4,1 (I, D) divides into a large d5,1 (I, D) and a small d5,2 (V, d2). The other micromeres divide asynchronously and produce the 16-cell stage.

This stage is typical of all rotifers: four rows of four cells each are easily recognized (i.e. Tessin, 1886; Zelinka, 1892; Jennings, 1896; Tannreuther, 1920; de Beauchamp, 1956; Pray, 1965; Lechner, 1966). Each row is produced by one cell at the 4-cell stage; the row derived from D3 (I, D) contains larger cells. Very minor differences about this stage are present among the different studies. Some discrepancies concern the way this stage is attained, and the number of mitoses required (see Remane, 1929–1933). According to de Beauchamp (1956) and Pray (1965), the yolk granules are now concentrated in one cell for each row; this produces a ring of yolk-rich dark cells that surround the large blastomere d5,1 (I, D).

Gastrulation is assumed to start as early as the 16-cell stage is attained (Zelinka, 1892; Nachtwey, 1925; de Beauchamp, 1956; Pray, 1965; Lechner, 1966; Gilbert, 1989). Only Tannreuther (1920) recognized gastrulation at a later stage (32–64 cells). Gastrulation consists of movements of the cells that first encircle the large blastomere (d5,1) through epiboly and then invaginate.

According to Zelinka (1892) the blastomere d5,1 (I, D) divides into d6,1 (I, D) and to d6,2 (VI, d3). The blastomere d6,1 produces, through 3 subsequent divisions, eight cells. Of them, the two anterior ones, d9,1 (E1) and d9,3 (E2), are responsible for mid gut, the two central ones, d9,2 (ϵ1) and d9,4 (ϵ2), are the origin of the reproductive system (germarium and vitellarium), the four posterior cells, d9,5 (eo1), d9,6 (eo2), d9,7 (eu1) and d9,8 (eu2), will give the posterior gut. Gastrulation concerns also the three cells a5,1 (a1, A), b5,1 (b1, B) and c5,1 (II1, C), that invaginate at the blastopore giving rise to the stomodeum (mouth, pharynx and then mastax). Therefore Zelinka (1892) reported the common origin of the digestive and reproductive system from the cell d5,1 (I, D).

The other authors gave similar descriptions of the gastrulation process, but discrepancies concerned the fate of the cells. Some authors (Jennings, 1886; Tannreuther, 1920) stated the common origin of the digestive and reproductive system from the cell d5,1 or its derivatives, in accord with Zelinka (1892). Other authors (Nachtwey, 1925; de Beauchamp, 1956; Pray, 1965; Lechner, 1966) reported that the large cell d5,1, when invaginating, gives rise to the reproductive system only. The digestive system is formed by other small cells produced by cell d4,2 (III, d1) and by other quadrants (cells stemmed from blastomeres A3, B3 and C3). Lechner (1966) described gastrulation as a two-stage process. The first is the epibolic growth of the blastoderm cell (little cells coming from the A, B and C quadrants) around the d5,1 cell, that will be pushed into the embryo. The second stage is a further epibolic growth and the involution of the blastoderm to form the blastopore. During this process the cells that migrate inward form the endoderm and, later, the digestive system.

Also the position of the blastopore and the origin of the mouth are doubtful. The blastopore does not originate directly from the stomodaeum: it will be formed either near (de Beauchamp, 1956; Pray, 1965) or at the opposite side of the embryo (Tannreuther, 1920). In contrast Zelinka (1892), Nachtwey (1925) and Lechner (1966) reported that the blastopore gives rise to the stomodaeum and the mouth, as expected in protostomes.

Organogenesis has not been studied in great detail, and there are discrepancies on this stage also. The main difference concerns the different origin of digestive and reproductive systems. Gilbert (1989) reports that ectoderm gives rise to stomodaeum and pharynx, nervous system, but also excretory system and muscles. The mesoderm gives rise to the reproductive system, germarium and vitellarium only. Endoderm gives the digestive system.

Present contribution to rotifer embryology

Using a confocal laser microscope, we observed the morphology of early embryo stages of a bdelloid species. Here we present the pattern of the early development and compare our results with literature data with a view to finding a development pattern that can be generalized to the whole rotifer group.

Material and methods

The experimental model is a bdelloid rotifer, *Macrotrachela quadricornifera* Milne, 1886. For many years it has been reared under laboratory conditions, at 22 °C (details in Ricci et al., 1999).

Eggs at various stages of development were collected and observed on a Leica TCSNT confocal microscope equipped with Argon–Krypton laser (CLSM). To visualize cytoskeletal elements, actin and tubulin of embryos at different stages of development were stained using fluorescent markers.

The shell of *M. quadricornifera* egg is transparent, very resistant and impermeable to many molecules. It was therefore necessary to make the eggshell permeable to the large molecules used to stain the cytoskeleton. After several attempts with different chemicals, e.g. NaOCl, NaOH, HCl, we used a mixture of 1% thioglycolic acid and 0.05% pronase in TRIS buffer (Tris(hydroxymethil)aminomethane, 200 mM, pH8.5) for 90′. The permeabilized embryos were then fixed with 4% paraformaldehyde in PBS (phosphate-buffered saline, 110 mM, pH 7.4) for 1 h, permeabilised with 0.25% Triton X–100 and 0.1% Tween in PBS for 20′, and then processed for visualizing actin (microfilaments) and tubulin (microtubules). Actin was resolved by staining the embryos overnight at +4°C with phalloidin conjugated with rodamin (Sigma, 0.5 μg/ml in PBS). To resolve microtubules, the embryos were incubated with the primary antibody anti α-tubulin (Sigma, clone n. B–5–1–2, 1:500 in PBS) overnight at +4°C, treated again with 0.25% Triton X–100, and exposed to the secondary antibody (fluorescein conjugated antibody anti-mouse IgG, Molecular Probes, 1:50 in PBS) overnight at +4°C. After rinsing in PBS, embryos were stained with DAPI (Sigma, 0.5 μg/ml in PBS) for 20′ to visualize DNA, mounted on microscope slides with DABCO (Aldrich) and MOWIOL 4–88 (Calbiochem) and observed by CLSM.

Results

The egg of *M. quadricornifera* is laid at one-cell stage (Fig. 2a). At 22 °C, after about one hour, before undergoing first cleavage, it extrudes one polar body, recognizable as a nucleus close to the shell (Fig. 2b). The first cleavage occurs about 3 h later. Its furrow is transverse to the major axis of the egg. The cell divides asymmetrically into a small blastomere, AB2, and a large blastomere, CD2, and the polar body remains along the furrow (Fig. 2c). About 90′ later, the second cleavage produces four cells. The blastomere CD2 divides into a small C3 and a large D3. Subsequently, AB2 divides into two cells of equal size: A3 and B3. At the 4-cell stage (Fig. 2d), three small cells of the same size (A3, B3 and C3,) and one large cell, D3, are recognizable. About 75′ later, the third cleavage follows. Similar to previous cleavage, the divisions are not synchronous. Initially the large D3 divides into large d4,1 and small d4,2. Then C3 follows, and subsequently both B3 and A3 divide. The fourth cleavage produces the 16-cell stage (Fig. 2e, f). Blastomeres keep dividing, but the cells resulting from each blastomere cannot be distinguished. At this time, the cells start moving. The blastomeres derived from the D3 (called D quadrant) divide and migrate inward, and the small superficial cells wrap them (Fig. 2g). Judging from the cell displacement, the embryo is starting gastrulation. All cells continue divisions, and become smaller and smaller. Many small cells form a thin layer at one pole of the egg, and wrap the large internal cells (Fig. 2h).

Cells keep dividing, but it is very difficult to track each single stage. Tissues starts organizing but single structures are hardly distinguishable, except for the invagination of the cells that will form the pharyngeal pouch. In the middle of the embryo a double ring of cells is recognisable, because of the concentration of actin filaments. These are mostly visible between the two cell rings, along the cell membranes of the outer ring (Fig. 3a). Some hours later, in the same region, embryos often present the gross design of a mastax (Fig. 3b) where the teeth are clearly visible if stained with phalloidin (Fig. 3c). It must be stressed that by light microscopy the same embryos do not reveal any structure (Fig. 3d). In a later embryo (Fig. 3e), the mastax structure also is visible by light microscopy (Fig. 3f). In embryos at this stage also

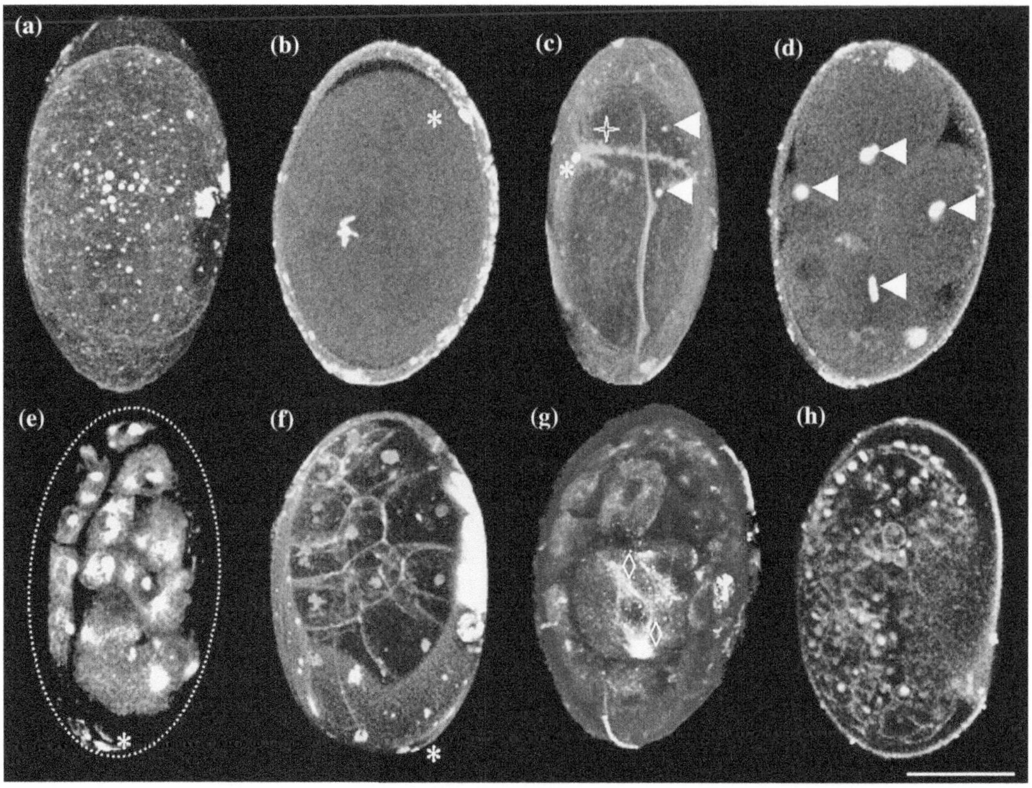

Figure 2. Cleavage and gastrulation process of *M. quadricornifera*'s embryo demonstrated by confocal microscopy. (a) Unsegmented egg. (b) Unsegmented egg and its polar body. (c) First cleavage division: nuclei and actin ring prior to cytodieresis. (d) 4-cell embryo. (e) Embryo between 3rd and 4th cleavage divisions: small blastomeres from A, B and C quadrants on left, large blastomeres derived from D quadrant on right. (f) The 16-cell stage with 4 rows of cells (3 visible). (g) Gastrulation: the mitotic spindle present in d5,1 blastomere. (h) Late gastrulation: small blastomeres externally and large blastomeres internally (one optical section). Scale bar: 40 μm. Arrowhead: nucleus. Fourpoint star: actin ring. Asterisk: polar body. Rhombus: mitotic spindle. a, b, c, d, f, h: rhodaminated phalloidin and DAPI. e: antibody anti-α-tubulin and DAPI. g: antibody anti-α-tubulin.

other structures are recognizable: circular muscles, similar to those present in the adult (Santo, personal communication), are already visible around the embryo body (Fig. 3g). Also the mastax musculature can be distinguished (Fig. 3h), and the trophi are clearly visible (Fig. 3i).

When the embryo is completely formed, it remains in the egg for about 24 h for this species. Structures are complete and fully functional, since it is common to see the trophi pieces moving inside the egg.

Discussion

The study of bdelloid developmental pattern seems very important and in need of much more attention, as is the study of embryology of the rotifers in general. The whole field has been silent for a very long time (the last detailed study was published by Lechner in 1966) and it is unbelievable that the new techniques developed in recent times, from electron microscopy to molecular studies, were never applied to rotifer embryology. Reasons can be found in the difficulty of the material, the resistance of the eggshell, whose chemical composition is almost unknown (Depoortere & Magis, 1967; Piavaux, 1970; Piavaux & Magis, 1970), the unfamiliarity of rotifer scientists with embryology, and of embryologists with rotifers. Nevertheless, the parthenogenetic reproduction of most rotifers, both monogononts and bdelloids, gives the opportunity to investigate the mechanisms responsible for zygote activation and embryo orientation, in the absence of fertilization.

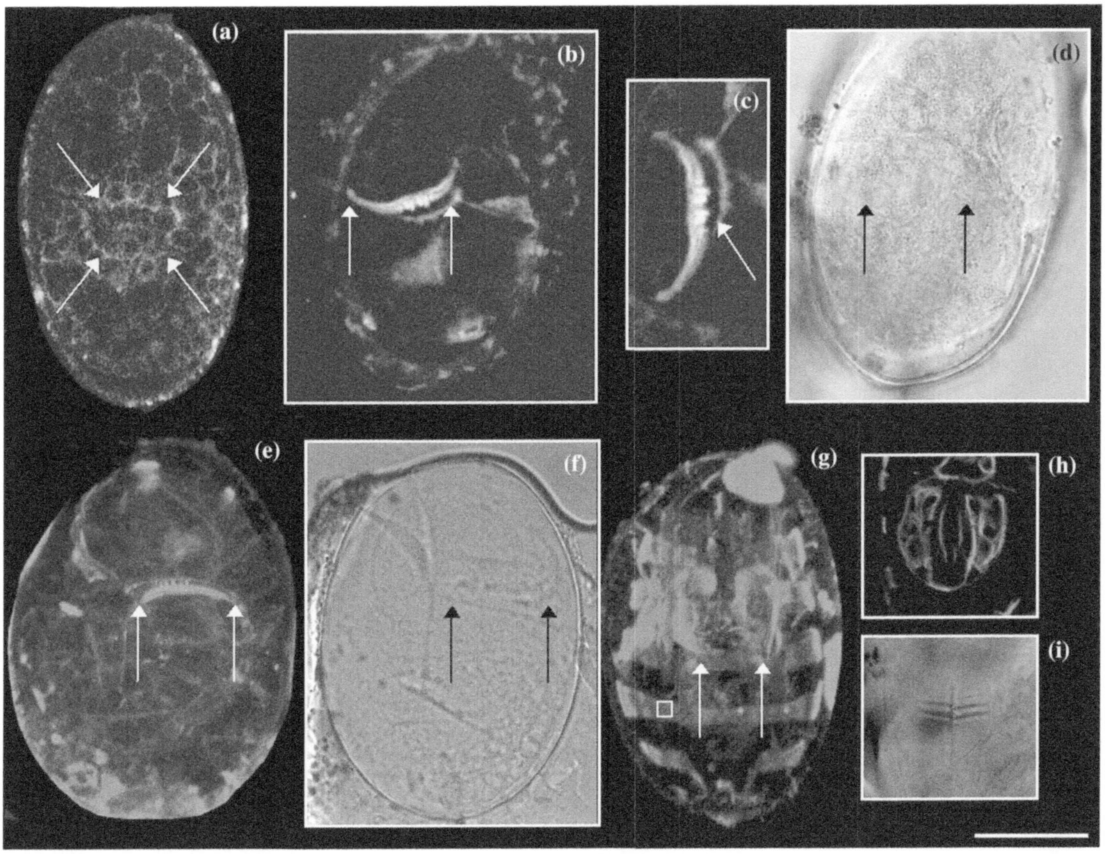

Figure 3. Early organogenesis of *M. quadricornifera*'s embryo demonstrated by confocal microscopy. (a) Actin filaments preliminary to mastax mould formation (see text for explanation). (b) Mould of mastax. (c) 2× magnification of b. (d) Same as b under light microscopy. (e) Mastax in a later embryo. (f) Same as e under light microscopy. (g) Muscles of a developed embryo. (h) Single section of mastax musculature. (i) Same as h under light microscopy. a, e, f, g, h, i: scale bar 40 μm. b, d: scale bar 35 μm. c: scale bar 20 μm. a, b, c, e, g, h rhodaminated phalloidin. d, f, i light microscopy. Arrow: mastax. Square: circular body muscle.

The detailed contributions by early authors on the developmental processes of rotifers give an excellent background to study rotifer development through a more experimental approach. However several difficulties have to be overcome. First, data coming from different studies have to be compared, and nomenclature used by each author have to be unified. Our interpretation of the names of blastomeres comes from a detailed analysis of the literature, which was not always successful. Second, old results need to be confirmed, and a general pattern must be found that can be extended to the rotifers as a group. Alternatively, differences in development between species should be checked and understood. At present, it seems impossible to state if the differences in the literature are between the species or between the authors or both, because the studies were run by different authors on different species (see Table 1).

In this paper we report the study of embryology of the bdelloids using a very informative instrument, the confocal microscope. It permits study of optical sections, avoids treatment artifacts, and allows reconstruction of the whole embryo into 3-dimensional images. We dealt with the early process of bdelloid development through a morphological approach only.

We documented that in *M. quadricornifera* the extrusion of one polar body occurs after laying, right before the first cleavage division. Hsu (1956a, b) reported two equational divisions prior to the formation of the egg, with the extrusion of two polar bodies by *Habrotrocha tridens* (Milne, 1886) and *Philodina roseola* Ehrenberg, 1832.

In contrast, all other authors, including Zelinka (1892), mentioned a single polar body. On the other hand, for monogononts authors report the extrusion of a single polar body during the formation of amictic eggs (Gilbert, 1989). The presence of one or two polar bodies in the apomictic parthenogenesis of bdelloids can be relevant for understanding the oogenesis process itself. Whether the bdelloids differ from the monogononts for the number of polar bodies is unclear. We cannot contribute on this point, except for the fact that before the first cleavage division only one polar body is visible in the laid egg. Whether the one we observed is the second one or the only one, is impossible to state at this stage, and deserves further investigation. Another possibility, that we consider quite remote, is that *M. quadricornifera* differs from *H. tridens* and *P. roseola* on this aspect.

Worthy of note is the result on the mould of the mastax. It must be recalled that the newborn rotifer already possesses its definitive mastax structure and that trophi pieces of bdelloids do not change their size during life (Fontaneto & Melone, this volume). The ,mould of trophi teeth' is visible by fluorescence observation but disappears by light microscopy. Phalloidin resolves filamentous actin, that is present in muscles, but also in cytoskeleton elements. We have no evidence to state that the mastax muscles, at formation, have such a shape as to make teeth visible, or that a sort of mould of actin for trophi pieces is prepared during development. In analogy, a mould of actin very similar in shape to the definitive hard structure, was found during the formation of a nematode cuticle (Costa et al., 1997). In any case, the trophi and the mastax in general are among the first structures that can be recognized during embryo development.

On the whole, our results conform to the general pattern as outlined by the previous authors (Tessin, 1886; Zelinka, 1892; Jennings, 1896; Tannreuther, 1920; Nachtwey, 1925; Remane, 1929–1933; de Beauchamp, 1956; Pray, 1965; Lechner, 1966), i.e. the holoblastic unequal cleavage, the transverse furrow of the first division, the typical 16-cell stage, and the early gastrulation by epiboly. However, there are still points that remain unclear and require further resolution; among them cell-lineage, apoptosis and gene expressions.

Acknowledgements

Nadia Santo and Manuela Caprioli helped at different stages of present research. Russ Shiel corrected the English style. Financial support came from ASI grant to C.R.

References

Castellano Paez, M. E., H. Kurokura & S. Kasahara, 1988. Embryonic development of amictic eggs of a rotifer *Brachionus plicatilis*. Journal of the Faculty of Applied Biological Science, Hiroshima University 27: 93–99.

Clément, P. & E. Wurdak, 1991. Rotifera. In Harrison, F. W. & Y. E. E. Ruppert (eds), Microscopic Anatomy of Invertebrates. Vol. 4, Aschelminthes, Wiley Liss, Inc., 219–297.

Costa, M., B. W. Draper & J. R. Priess, 1997. The role of actin filaments in patterning the *Caenorhabditis elegans* cuticle. Developmental Biology 184: 373–384.

de Beauchamp, P., 1956. Le développement de *Plœsoma hudsoni* (Imhof) et l'origine des feuillets chez les rotifères. Bulletin de la Société Zoologique de France 81: 374–383.

Depoortere, H. & N. Magis, 1967. Mise en evidence, localisation et dosage de la chitine dans la coque des oeufs de *Brachionus leydigii* Cohn et d'autres rotiferes. Annales de la Société Royale Zoologique de Belgique 97: 187–195.

Fontaneto, D. & G. Melone, 2005. Do rotifer jaws grow after hatching? Hydrobiologia 546: 213–221.

Gilbert, J. J., 1989. Rotifera. In Adiyodi, K. G. & R. G. Adiyodi (eds), Reproductive biology of invertebrates. Fertilization, Development, and Parental Care, Vol. IV, Part A. Oxford & IBH Publishing Co. Pvt. Ltd., 179–199.

Hsu, W. S., 1956a. Oogenesis in the bdelloidea rotifer *Philodina roseola* Ehr. La Cellule 57: 283–296.

Hsu, W. S., 1956b. Oogenesis in *Habrotrocha tridens* (Milne). Biological Bulletin 111: 364–374.

Hyman, L. H., 1951. The Rotifera. In The Invertebrates: Acanthocephala, Aschelminthes, and Entoprocta. The pseudocelomate Bilateria, Vol. 3., Mc-Graw Hill Book Company, New York, USA. 59–151.

Jennings H. S., 1896. The early development of *Asplanchna herrickii* de Guerne. Bulletin of the Museum of Comparative Zoology, Harvard College 30: 1–117.

Lechner, M., 1966. Untersuchungen zur Embryonalentwicklung des Rädertieres *Asplanchna girodi* de Guerne. Roux' Archiv für Entwicklungsmechanik 157: 117–173.

Nachtwey, R., 1925. Untersuchungen über die Keimbahn, Organogenese und Anatomie von *Asplanchna priodonta* Gosse. Zeitschrift für wissenschaftliche Zoologie 126: 239–492.

Piavaux, A. & N. Magis, 1970. Données complémentaires sur la localisation de la chitine dans les enveloppes des oeufs de rotifères. Annales de la Société Royale Zoologique de Belgique 100: 49–59.

Piavaux, A., 1970. Origine de l'enveloppe chitineuse des oeufs de deux rotifères du genre *Euchlanis* Ehrenberg. Annales de la Société Royale Zoologique de Belgique 100: 129–137.

Plasota, K. & M. Plasota, 1980. Some problems in the embryogenesis of *Habrotrocha rosa* Donner 1949. Hydrobiologia 70: 39–41.

Pray, F. S., 1965. Studies on the early development of the rotifer *Monostyla cornuta* Müller. Transactions of the American Microscopical Society 84: 210–216.

Remane, A., 1929–1933. Rotatoria. In H. G. Bronn (ed.), Klassen und Ordnungen des Tier-reichs, Vol. 4, Part 2, Sections 1–4. C.F. Winter, Leipzig: 1–576.

Ricci, C., N. Santo, E. Radaelli & A. M. Bolzern, 1999. Epigenetic inheritance systems in bdelloid rotifers. I. Maternal-age-related biochemical effects. Italian Journal of Zoology 66: 333–339.

Tannreuther, G. W., 1920. The development of *Asplanchna ebbesbornii* (Rotifer). Journal of Morphology 33: 389–437.

Tessin, G., 1886. Über Eibildung und Entwicklung der Rotatorien. Zeitschrift für wissenschaftliche Zoologie 44: 273–302.

Wolpert, L., 2002. Principles of development. Oxford University Press, New York, 542 pp.

Zelinka, C., 1892. Studien über Räderthiere. III. Zur Entwicklungsgeschichte der Räderthiere nebst Bemerkungen über ihre Anatomie und Biologie. Zeitschrift für wissenschaftliche Zoologie 53: 1–159.

Part VI
Population and Community Ecology

Hydrobiologia (2005) 546:335–346 © Springer 2005
A. Herzig, R.D. Gulati, C.D. Jersabek & L. May (eds.) Rotifera X: Rotifer Research: Trends, New Tools and Recent Advances
DOI 10.1007/s10750-005-4243-x

Evolution of rotifer life histories

Claus-Peter Stelzer
Department of Evolutionary Biology, Institute of Animal Evolution and Ecology, Westphalian Wilhelms-University Münster, Hüfferstrasse 1, D-48149 Münster, Germany
E-mail: cpstelzer@web.de

Key words: life history theory, egg size, body size, trade-off, constraint, adaptation

Abstract

When compared to most other multicellular animals, rotifers are all relatively small, short-lived and fast-reproducing organisms. However among and within different rotifer species there is a large variation in life history patterns. This review accounts for such variation in rotifers, with a strong focus on monogonont rotifers. As the life cycle of monogonont rotifers involves both asexual and sexual reproduction, life history patterns can be examined on the level of the genetic individual, which includes all asexual females, sexual females and males that originated from one resting egg. This concept has been applied successfully in many areas, for example in predicting optimal levels of mictic reproduction or sex allocation theory. The benefits and implications of the view of the genetic individual are discussed in detail. Rotifer life histories can also be viewed on the level of physiological individuals. A large part of this review deals with the life histories of individual amictic females and addresses life history traits like body size, egg size and resource allocation patterns. It asks which trade-offs exist among those traits, how these traits change under the influence of environmental factors like food availability or temperature, and whether these changes can be interpreted as adaptive.

Introduction

The aim of life history theory is to understand how the life history traits (see Table 1) of organisms have evolved. In other words, life history theory is an attempt to account for the tremendous variation of life histories within and among different groups of organisms (Stearns, 1992). For example, why do some organisms take years to mature and produce just one offspring every other year, while others produce a huge number of offspring after a relatively short juvenile phase? In a gross comparison with other animal groups, rotifers seem relatively uniform with regard to their life histories. They all are small, short lived and fast reproducing invertebrates (Allan, 1976). However, within the phylum Rotifera there is a large variation in life histories. One example is body size, which spans almost three orders of magnitude when different rotifer species are considered.

This review addresses life history variation in Rotifera, mostly in monogonont rotifers. The first part will introduce some basic concepts of life history theory, like constraints and trade-offs, and give examples for those in Rotifera. Another part will address the problem that genetic identity and physiological identity are not congruent entities in rotifers (as they are in most sexual diploid organisms). Thus in rotifers it is possible to distinguish between the life history of a physiological individual (individual amictic females, mictic females or males) and a genetic individual (all physiological individuals that derived from a single resting egg). The remaining parts of the review will be devoted to the life history of amictic females, as most of the previous empirical work has focussed

Table 1. The basic life-history traits

The basic life history traits
Egg size, size at birth
Growth pattern (e.g., determinate vs. indeterminate growth)
Age at maturity
Size at maturity
Number, size and sex ratio of offspring
Age- and size-specific reproductive investments
Age- and size-specific mortality schedules
Length of life

on them. Adaptation with respect to several life history traits, like body size or egg size, will be considered here. Finally some recent studies on life history adaptation to resource limitation will be summarized and discussed.

General considerations

Variation in life histories can be assessed on different levels. One example for life history variation at the level of individuals is phenotypic plasticity. Individual females of one clone may show variation in a life history trait depending on the environment they grew up in. Such phenotypic plasticity may be adaptive, as it can result in phenotypes fitter than alternative ones in the development-controlling environment. Life history variation can also be assessed at the population genetic (microevolutionary) level. This would involve a comparison of many different clones of the same rotifer species in a standardised environment. Differences among these clones have a genetic basis and would thus respond to selection. In a comparative analysis, life history traits are compared among the members of some higher level of phylogenetic organisation (e.g., monogononts vs. bdelloids). This type of analysis looks for general patterns that are characteristic for higher taxonomic ranks.

Constraints and trade-offs

A central tenet of life history theory is that individual traits cannot vary independently, but are subject to constraints and trade-offs. An example for a phylogenetic constraint that probably applies to all rotifers is eutely. The fixed cell number at birth sets upper limits to the evolution of the trait body size, since cells cannot grow indefinitely large. Additionally, ciliary propulsion, the basic mode of locomotion in rotifers, may impose a constraint on maximum body size, as it becomes energetically inefficient at large body sizes (Sleigh & Blake, 1977; Epp & Lewis, 1984; Stemberger & Gilbert, 1987). Physiological constraints can occur in individuals within a species. For example, a decrease in egg size at very low food concentrations may be inevitable if there is not enough material for a larger egg. One constraint that applies for all rotifers is that clutch size is fixed at one egg. Even bdelloid rotifers, which possess two ovaries, produce their eggs sequentially, alternating between the two ovaries (Ricci, 1995). Pseudoclutches as they can be observed in egg-bearing genera (e.g., *Brachionus*) are sequentially produced eggs of different ages, whereas a real clutch would be parallel produced offspring. Pseudoclutches can arise when the egg-laying interval is shorter than the period of embryonic development.

Trade-offs arise when an increase in one trait, e.g., current reproduction, causes a decrease in another trait, e.g., future survival. On the physiological level a trade-off will arise when a limited amount of resources has to be allocated among competing processes, such as somatic growth, reproduction or the accumulation of storage tissue. On the genetic level, trade-offs can be caused by antagonistic pleiotrophy, e.g., there may be genes that increase fecundity at an early adult age but are also detrimental to survival at an older age. Trade-offs may also arise as a consequence in particular ecological situations, e.g., if bearing eggs (reproduction) decreases survival due to a higher susceptibility to predators. Trade-offs, where an increase in current reproduction is paid by a decrease of future survival and/or fecundity are often called "costs of reproduction" (e.g., Snell & King, 1977; Sarma et al., 2002). Trade-offs can manifest themselves in negative phenotypic correlations between two life history traits. However, as correlation does not prove causation, a negative correlation itself does not prove the existence of a trade-off. To be sure that two functions are traded off against each other, direct experimental manipulations are necessary (Reznik, 1992). In this case, an artificially induced increase in one trait will result in a significant decrease of the other trait. This can happen either on the physiological level (e.g., manipulations of brood size or egg size) or on the genetic level (e.g., artificial selection experiments). None of these manipulations have been carried out on rotifers yet, although artificial selection experiments seem feasible in species that readily produce sexual stages (e.g., *Brachionus*).

The genetic individual and the physiological individual

In monogonont rotifers there are different perspectives from which life histories can be viewed. Traditionally, most of the work has been done on the life history of amictic females (this topic will be addressed later in this review). Other studies have compared the life histories of amictic females with those of mictic females (King, 1970) or males (Snell & Childress, 1987). All these studies focus on physiological individuals, i.e., units that are physically separated and have their own metabolism. A completely different way of looking at rotifer life histories is to focus on the genetic individual, which in case of a monogonont rotifer are all descendents of a single resting egg (i.e., all amictic females, mictic females and males). All these physiological individuals share the same genome (although males have just half of it), barring mutations that may have occurred in the germ line during asexual reproduction.

There are several analogies between the genetic individual of a monogonont rotifer and a typical diploid sexual organism. Amictic females represent the somatic tissue whereas mictic females and males represent the reproductive tissue. This analogy has been pointed out by Serra & King (1999). Whereas the life of a physiological individual ends with the death of the body, a genetic individual would die when the last amictic female of the clone has died. Genetic individuals therefore live considerably longer. In pond species of temperate regions this may be up to one season and in lake species which are not subject to draught or freezing, it may be even longer.

One may argue that over several successive asexual generations rotifers will accumulate mutations, which is at odds with the concept of genetic identity (Loxdale & Lushai, 2003). Strictly speaking, genetic identity is destroyed after only one mutation in the germline. However, from the standpoint of evolutionary adaptation within a rotifer clone these changes are completely irrelevant. Even with some mutations that may have occurred during asexual reproduction, the coefficient of relatedness among members of a clone is still essentially one. The genetic divergence due to mutations on these time scales is too minor to cause a conflict among the physical individuals of one clone.

The view of the genetic individual and the view of the physiological individual are not mutually excluding concepts. They just focus on different aspects and time scales. However they imply different measures of fitness. In the physiological individual (usually the amictic female) the population growth rate r is the relevant measure of fitness, whereas for the clone the relative genetic contribution (relative contribution of haploid genomes) to the resting egg bank is more appropriate.

Life history theory of the genetic individual

There are several studies on monogonont rotifers that incorporate both asexual and sexual stages of the life cycle. In these studies the concept of the genetic individual is often implicit. One example is a theoretical analysis of optimal mictic ratios in the rotifer *Brachionus plicatilis* by Serra & King (1999). If a rotifer clone induces mixis very late (at high population densities), it may enjoy a much higher population size when it finally starts to produce mictic females and may therefore increase its contribution to the resting egg bank. However, it may also risk its own extinction if the habitat deteriorates before resting eggs can be produced. This conflict bears striking similarity with the trade-off between early vs. late reproduction in other organisms. Another conflict that can arise is the question whether the descendants of a single resting egg, once they switched to sexual reproduction, should produce 100% mictic daughters or continue to produce also some amictic ones. In an environment where density-dependent growth can occur, the model of Serra & King (1999) predicts mixis to be induced as soon as the population comes close to its carrying capacity and intermediate mixis ratios thereafter. However, the exact conditions that rotifers will encounter in their natural habitats are temporally and spatially highly variable. Hence it is not surprising that there is considerable variation in the propensity of different *Brachionus* clones to produce mictic daughters (e.g., Gilbert, 2003).

Another phenomenon where the analogy between the genetic individual and a "real" individual comes in mind, are trans-generational effects in successive generations of amictically reproducing *Brachionus*. Gilbert (2002) found that females newly hatched from resting eggs showed extremely low propensity for mixis induction by the mixis factor, and that this effect gradually disappeared in successive generations. Only after 12 parthenogenetic generations, the different clones reached their maximum propensity for mixis induction. This pattern can be interpreted as a developmental programme of the genetic individual to delay its own reproduction.

Another example where it makes more sense to look at the genetic individual is sex allocation. In rotifers, a treatment of sex allocation is actually impossible from the perspective of the physiological individual, as one mictic female will either produce male or female offspring (Serra & Snell, 1998), but not both (except for the rare cases of amphotheric reproduction, see King & Snell, 1977). The life history trait that affects sex allocation in monogonont rotifers is the threshold age at which a mictic female can be fertilised (Aparici et al., 1998). If a female is fertilised before this age, she will produce (female) resting-eggs, otherwise she will produce male offspring. Aparici et al. (1998) developed a model to predict the evolutionary stable strategy for the threshold age of fertilisation in a population at equilibrium (zero net-growth, but constant supply of mictic females). They found that the only evolutionary stable strategy was a threshold age of fertilisation where the number of fertilised and unfertilised females was equal. This means that selection will favour clones which invest equally into male and female function. This theoretical prediction was validated in a subsequent experimental study (Aparici et al., 2002).

The genetic individual is a simultaneous hermaphrodite, as it possesses both male and female reproductive function (i.e., unfertilised and fertilised mictic females respectively, see Aparici et al., 1998). Unfertilised mictic females are analogous to the male reproductive organs. This is because meiosis, the process that initiates the production of male gametes happens already in the body of mictic females. Although the haploid egg cells first develop into males, there is no relevant additional event from the genetic view. Meiotic recombination and segregation happen when the mictic female produces her haploid egg cells. Each male develops by mitotic cell divisions only, thus its sperm are genetically identical. The analogue to a rotifer male in a diploid sexual organism would be an individual spermatozoon, because genetically male rotifers are the vehicle for one single spermatozoon that has been multiplied 20-fold (there are about 20 sperms per one male *Brachionus plicatilis*, Snell & Hoff, 1987).

From the view of the physiological individual the fertilisation mode in rotifers is internal fertilisation. On the other hand, the gametes of one genetic individual are widely dispersed in the environment and fertilisation among them is mostly governed by random encounters. In a rotifer clone, the female gametes are the haploid egg cells inside the body of a mictic female (the ones that will be fertilised). In *Brachionus*, a fertilised mictic female can typically grow just 2–3 of her fertilised oocytes into a resting egg (Snell & Childress, 1987). In some species within the *Brachionus plicatilis* species complex, the resting egg stays inside the body of the female (Gomez & Snell, 1996). In this case just one resting egg is produced per female, hence every mictic female body carries just one gamete. Thus, although the gametes of rotifers are not shed free into the water, genetically the fertilisation system in rotifers bears more resemblance to broadcast spawning than to internal fertilisation. At one time, fertilisation can occur massively parallel, among and also within many different genetic individuals.

Life history theory of amictic females

The following paragraphs deal with the life histories of amictic females only. Amictic females are the life history stage of rotifers that is exposed to natural selection most of time. Thus it is of great importance to study the life history adaptations that allow these units to efficiently replicate themselves.

Life table analysis

A life table summarises several life history traits, such as age-specific mortalities and fecundities, including the age at first reproduction. Life tables have been a popular tool since the early days of rotifer biology. They are conducted over the life span of a cohort of amictic females and the data are usually obtained in daily intervals. The data can be used to calculate important measures, such as the population growth rate r but also for more specific questions about survival patterns and trade-offs between fecundity and survival (Snell & King, 1977; Sarma et al., 2002).

The calculation of the population growth rate r can be done in different ways. The simplest way uses the formulae (Begon et al., 1991):

$$r \approx \frac{\ln R_0}{T} \quad \text{and} \quad T \approx \frac{\sum x \cdot l_x \cdot m_x}{\sum l_x \cdot m_x}$$

where R_0 is the average number of offspring per female per lifetime, x is the age class (e.g., 1 for the first 0–24 h), m_x are the age-specific fecundities (offspring per female per 24 h), and l_x is the age-specific survivorship (proportion of females still alive at age class x). As rotifers are continuously producing offspring, it is actually more accurate to use $x - 0.5$ instead of x (when calculating r based on daily observations). This assumes that all births happened exactly in between the observation intervals, whereas using x assumes that all births and deaths happened just before the census.

A more accurate method to calculate r is the Lotka–Euler equation:

$$1 \approx \sum e^{-rx} \cdot l_x \cdot m_x$$

This equation has to be solved iteratively, usually by the use of a computer. An example for an algorithm in the computer language PASCAL can be found in Stearns (1992) but other computer languages, or even calculations in Excel spreadsheets can be used for this purpose.

The population growth rate r can also be calculated from Leslie matrices (Caswell, 1989; Stelzer, 2002). Although computationally more elaborate than the above methods, Leslie matrices allow a more detailed analysis of the life table data. For example, the sensitivity of r to changes in vital rates of different ages can be calculated analytically (using specialised software packages like MATLAB®). This analysis allows identifying which of the many differences in age-specific survivorship and fecundity between two life tables are actually responsible for the observed difference in the population growth rate. Additionally, Leslie matrices can also be used to model the influence of age-structure fluctuations on population dynamics (Caswell, 1989).

For comparison of survivorship data (e.g., to compare starvation times or life spans), specialised statistical methods have been developed (e.g., Serra et al., 1994). Since survivorship data are not normally distributed, standard parametric tests are not appropriate (Fox, 2001). Alternative tests have been used in several studies with rotifers (Kirk, 1997a; Stelzer, 2001; Yoshinaga et al., 2003).

To test whether the population growth rates obtained from two life tables are significantly different from each other, confidence intervals for r have to be calculated (Conde-Porcuna, 1998; Stelzer, 2002). These can be estimated by re-sampling methods (the Bootstrap or the Jackknife method, Meyer et al., 1986) For these methods the use of a desktop computer is inevitable, as for each life table r has to be calculated multiple times from a subset of the data (Meyer et al., 1986; Caswell,1989).

Life tables will continue to be an important tool in the research on rotifers. Often they are the first type of analysis that is done with a newly cultured species. However, if the only relevant measure taken out of such a study is the population growth rate r, it is often easier to directly measure r from the growth of semi-continuous cultures (for methods, see Rothhaupt, 1990).

Body size

Body size is an adaptive trait, shaped by various selective pressures during its evolution (Roff, 1981). During juvenile growth many organisms face a trade-off between reproducing early at a smaller size and reproducing late at a larger size. Rotifers grow by increase in cell size only. The fact that there are no cell divisions after hatching certainly sets upper limits to body size. Despite this rather theoretical constraint there is a large variation in body size among different rotifer species

(Stemberger & Gilbert, 1985; Telesh et al., 1998), among species within genera (Rougier et al., 2000), among clones (Snell & Carrillo, 1984), and within a clone due to phenotypic plasticity (Stelzer, 2002).

Size differences among species of different genera can be immense. In terms of body volume, *Asplanchna priodonta* is 2000-fold bigger than *Keratella cochlearis tecta* (Telesh et al., 1998). However, large rotifers often contain more intercellular space, whereas the tissues of small rotifers are more "dense" in terms of the ratio between cellular and intercellular space. In the carbon content the difference between *A. priodonta* and *K.c. tecta* is only 87-fold (Telesh et al., 1998).

Since most rotifer species, especially those of different genera, have evolved under very different ecological conditions, it is hard to establish a general relationship between body size and its selective value. For example, whereas a large body size can be expected to protect against predation by invertebrate predators like *Asplanchna* or copepods (Lapesa et al., 2002) and against interference competition by *Daphnia* (Gilbert, 1989), there are also examples for alternative solutions (e.g., the escape response of *Polyarthra*) that retain a relatively small body size (Wickham & Gilbert, 1991).

One popular generalisation is that animals with a large body size have a higher fecundity than small animals. Thus, during juvenile growth there is a trade-off between reproducing early and staying small and reproducing later at a larger body size while having a higher fecundity (Roff, 1981). The mechanistic explanation for the higher fecundity in larger females is often that they can carry more offspring per brood (e.g., in *Daphnia*, Glazier, 1992; Lampert, 1993). In rotifers the situation is more complex. As eggs are produced individually and sequentially, the argument with the larger brood is not valid. One advantage of a large body size could be that it allows more eggs to be attached to the outside of the body (Walz et al., 1995). On the other hand, there are many rotifer species that do not carry their eggs but release them into the water. So the question remains, whether larger females in a particular rotifer species can produce offspring at a higher speed than smaller females. One way to test this hypothesis is to compare the fecundity of differentially sized females within one species (i.e., either clones of different size or individuals with phenotypically induced size-differences within one clone). To my knowledge, no such study has been performed yet. There are, however, studies that report positive correlations between body size and birth rate in comparisons among species of different genera (Lewis, 1979; Walz, 1995).

Reproduction and its costs

There are two characteristics of rotifer reproduction that are fundamentally different from many other organisms. First, the number of oocytes is fixed at birth. This feature can sometimes be observed in life table experiments due to the existence of a distinct post-reproductive phase, often after a characteristic number of eggs have been laid (e.g., ~20 in *Brachionus*, Halbach, 1970). Even when cultured under the most benign conditions, rotifers stop reproduction at some age, presumably because they have depleted their reservoir in oocytes. It is currently unknown whether the limited oocyte supply has any significance under natural conditions. Another question is, whether there is variability in total oocyte number (i.e., maximum possible number of offspring per female) among different species. It is possible that in some rotifer species, where the expected number of offspring per female is low under natural conditions, the number of oocytes is also reduced. A second characteristic of rotifer reproduction is that there is no parallel production of offspring (clutch size = 1). This complicates the analysis of trade-offs between offspring size and number. Theoretically a trade-off between offspring size and the speed of reproduction is possible, since offspring size may only be increased at the cost of longer egg laying intervals. However, confounding variables such as age-related changes in food intake or other age-specific effects may influence this trade-off or even influence egg size directly.

Several studies have addressed costs of reproduction in rotifers, both on the physiological and on the population genetic level. Demonstrations of reproductive costs often involve life table analyses. Current reproductive output is correlated with measures of later reproduction and/or survival. A negative correlation indicates reproductive costs. In their pioneering study Snell & King (1977) showed that in *Asplanchna brightwelli* both the residual reproductive value and survival after 24 h were negatively correlated with age-specific fecundity. The study involved 21 different clones and Snell & King (1977) extracted two basic life history patterns: some clones produced a lot of offspring in a short time and had a short lifespan, while others reproduced slower but lived longer. A similar study by Sarma et al. (2002) addressed the cost of reproduction in four rotifer species and found evidence in the form of negative correlations between fecundity and survival in the majority of cases. Stelzer (2001) measured the costs of reproduction arising from resource allocation in individual *Synchaeta pectinata*. *Synchaeta* can use the cytoplasm in their vitellarium for either the production of eggs and or to maintain their metabolism during starvation. In animals with a similar amount of cytoplasm in the vitellarium, those that reproduced during starvation died on average one day earlier than those that did not reproduce. This demonstrated the cost of reproduction on the level of individual physiology.

A cost of reproduction that specifically applies to egg-bearing species may be a decrease in swimming performance when the number of carried eggs is high (J. J. Gilbert, personal communication). Decreased swimming speed will result in lower energy intake rates because the volume searched for food decreases. Yufera et al. (this volume Part V) report that ovigerous females exhibit higher oxygen consumption than it is expected from the sum of the oxygen consumption of non-ovigerous females and that of isolated eggs. Additionally, the absolute swimming speed declined in all treatments (although not significantly) in ovigerous females. Both findings indicate that carrying a large number of eggs may be an additional cost that has to be considered in egg-bearing species. However, before final conclusions can be drawn, a wider range in the number of eggs carried needs to examined. In the study of Yufera et al. (this volume Part V) up to two eggs per female were considered, whereas under good conditions *Brachionus* females often carry 3–4 eggs.

Egg size

Egg size is under the simultaneous influence of many factors (Bernardo, 1996). For example it may co-vary with factors like maternal age or size. Also environmental factors, such as food availability or temperature can influence egg size. It is an important question whether this variability in egg size has any adaptive significance.

Egg size and offspring fitness

It is often assumed that offspring size and offspring fitness are positively correlated and that this results in a trade-off between producing few large or many small offspring (Smith & Fretwell, 1974; Glazier, 1992). This hypothesis has been addressed in rotifers with regard to egg size (e.g., Kirk, 1997b; Orsenigo et al., 1998). Egg size within a rotifer clone is known to vary with environmental factors like food concentration or temperature. One question would be whether this phenotypic plasticity is adaptive. For example, it may be beneficial to produce larger, starvation resistant offspring at low food concentrations (Glazier, 1992). This problem has actually two parts. First, are larger offspring produced at low food concentrations and, second, are these larger offspring fitter at low food concentrations/under starvation? These two hypotheses have been addressed by Kirk (1997a) in a study using *B. calyciflorus* and *S. pectinata*. The two species responded differently along the gradient of food concentrations. Egg size in *Brachionus* increased with food concentration (see also Sarma & Rao, 1987; Walz & Rothbucher, 1991), whereas it stayed relatively constant in *Synchaeta*. Constant egg sizes along a wide range of food concentrations in *Synchaeta* were also observed by Stelzer (2001). In *Synchaeta*, egg size seems to increase not as a function of food level, but as a function of the mother's age and possibly size. For example three day old females produce larger eggs than two day old females (Kirk, 1997b). Both in *Brachionus* and *Synchaeta*, a larger egg size seems to be advantageous, as offspring from large eggs had longer starvation times (Kirk, 1997b). Similar questions were addressed in a study on the bdelloid rotifer *Macrotrachela quadricornifera* (Santo et al., 2001). In this study differentially sized eggs were produced by culturing the mothers at high vs. low food concentrations. Large eggs were obtained at the higher food concentration. Offspring from large eggs developed faster, having a significantly shorter embryonic development and juvenile period. Recovery from desiccation, a fitness-related trait in bdelloids, was not affected by the difference in egg size (Santo et al., 2001). The shorter embryonic development of large eggs as observed by Santo et al. (2001) is not a general pattern though. Studies with other rotifers have shown that embryonic development time can increase or also be unaffected by egg size (summarised in Walz, 1995).

Temperature is another factor that can influence egg size in rotifers (Green, 1998; Stelzer, 2002). Large eggs are usually produced at low temperatures, whereas small eggs are produced at high temperatures. Stelzer (2002) showed that when *S. pectinata* were cultured at 4 °C they produced eggs that were 35% larger than those of animals cultured at 12 °C. This is in strong contrast to the apparent constancy of *Synchaeta*'s egg size across different food concentrations (see above). However, offspring from large eggs did not exhibit a greater fitness than those of small eggs (Stelzer, 2002). In fact it seemed that offspring from large and small eggs had a slightly higher growth rate at the temperature at which they were produced. Offspring from small eggs were growing significantly faster at high temperatures than offspring from large eggs (Stelzer, 2002). Offspring from large eggs did slightly better at low temperatures, but the difference was non-significant. Therefore the higher investment into individual offspring at low temperatures does not seem to be rewarded by higher offspring fitness.

In summary, although there is considerable phenotypic plasticity for egg size in rotifers, the question whether these patterns are adaptive remains unanswered. Currently there is no good support for the claim that offspring of a certain egg size do best in the environment in which they were induced. There may be several explanations for this. First, constraints may prevent rotifers from adjusting their egg size optimally (Kirk, 1997a). For example, it could be that *Synchaeta* already produces the minimum viable egg size (for a given temperature), so that it would be very disadvantageous to decrease egg size further at high food concentrations. In fact, offspring from small eggs (produced at extremely low food concentrations) often develop very slowly and have a reduced fecundity (C.P. Stelzer, personal observation). Another possibility that may explain the lack of fit between theory and data lies in the nature of the optimality model itself. Models on optimal offspring investment usually assume equilibrium, i.e., offspring will experience the same environment as their mother. This may not be true for rotifers. Especially resource conditions can be extremely variable, both in space and in time (Steward & Wetzel, 1986). The results of studies on offspring fitness also depend on the measure of fitness being employed. These measures have been quite diverse: neonate starvation time (Kirk, 1997a), population growth rate (Stelzer, 2002), duration of embryonic and juvenile development, and resistance to desiccation (Santo et al., 2001).

Large-scale patterns in relative egg size

Comparative analysis has revealed an interesting large scale pattern in rotifers. Although the volume of amictic eggs increases with body volume in a comparison across species, the ratio of egg volume to body volume (REV = relative egg volume) does not stay constant. Instead, small rotifer species consistently show a larger REV than large species (Walz et al., 1995). This indicates that small rotifers have a higher reproductive effort (Walz et al., 1995). The basic pattern was independently confirmed in other studies, e.g., in the taxa Collothecidae and Flosculariidae (Wallace et al., 1998), and in several marine *Synchaeta* species (Rougier et al., 2000). For bdelloid rotifers too few data are available at this time. In one study, involving a comparison between two differentially sized species, the smaller one had a larger REV (Ricci & Fascio, 1995), whereas in another study with two bdelloid species REV was equal (Ricci, 1995). The pattern of REV indicates that small rotifers show, on average, a higher investment in individual offspring than large rotifers. The ubiquity of this phenomenon suggests a common explanation. Current explanations for the higher REV in small rotifers relate the phenomenon to selection for low food demands in small rotifers (Walz et al., 1995). One advantage that small rotifers may gain through their large REV is that they have to grow less to reach adult size (Walz et al., 1995).

Life history adaptations to resource limitation

Resource limitation is one of the most important environmental factors experienced by rotifers (Gonzalez & Frost, 1992; Merriman & Kirk, 2000). Several studies have addressed life history adaptations to conditions of low resource availability on both theoretical and empirical grounds. Kirk (1997b) examined the life history responses to starvation in nine different rotifer species. Starvation times ranged from 0.4 to 5 days. Some species, like *Keratella* or *Brachionus*, immediately stopped allocating resources towards reproduction after food deprivation, whereas others, like *Synchaeta* or *Asplanchna*, continued to produce offspring, sometimes at an even higher rate than the fed controls, but with reduced survival. The differences in reproductive behaviour explained the variation in starvation times among species much better than other variables, e.g., body size. The main finding of this study was that species that reproduced after being deprived of food had considerably shorter starvation times than those which stopped reproduction. Kirk et al. (1999) extended these findings by showing that respiration rates of non-reproducing rotifers drop shortly after food deprivation whereas those of reproducing rotifers stay high for a longer time.

The two diametrically different responses to starvation were confirmed in studies with *Brachionus plicatilis* (Yoshinaga et al., 2003) and *Synchaeta pectinata* (Stelzer, 2001). When *B. plicatilis* was deprived of food at an age of 2 days they not only did stop to lay eggs, but also lived significantly longer than the fed controls (Yoshinaga et al., 2003). Periodical starvation, i.e., feeding *Brachionus* for just 1–3 h and starving them for the rest of each day resulted in lifespans 2–3 times longer than those of non-starved animals (Yoshinaga et al., 2000). According to the authors this strategy may pay off because it increases the probability that females survive adverse conditions to reproduce later in life when food concentrations recover. In contrast to *Brachionus*, *S. pectinata* will continue reproducing under starvation, provided that they have an ovary size large enough to produce at least one more egg (Stelzer, 2001). Although this behaviour decreases the survival chances of the mother, it also helps the clone to persist through periods of starvation. First, because the embryo stage itself is protected from starvation (it does not need to feed) and second, because juvenile rotifers can starve longer than adults (Kirk, 1997b). As *S. pectinata* do not carry their eggs, egg development time may be further prolonged if the eggs sink to the colder depths in a stratified lake.

The length of the expected starvation period, in the natural systems where the species have evolved, may explain the difference between the two different responses of *Brachionus* and *Synchaeta*. The "*Brachionus* response" (stop reproduction and lower the metabolism in order to preserve the mothers body) will only pay off if the starvation period is quite short. At 20 °C adult *B. calyciflorus* can starve for two days at maximum (Kirk, 1997b). If the starvation period is longer, the mother will die without leaving any further offspring. Longer starvation periods seem to be the conditions that favour the "*Synchaeta* response". Depending on the duration of embryonic development, Stelzer (2001) estimated that using this strategy asexually reproducing *Synchaeta* can span up to 1 week of adverse resource conditions. Even longer periods of starvation can be bridged if embryonic development is "intrinsically" prolonged. This is the case for some strains of *S. pectinata*, which have evolved amictic eggs with a diapausing stage that can be induced by low food concentrations (Gilbert & Schreiber, 1998).

In a modelling study, Shertzer & Ellner (2002) tried to predict optimal allocation among somatic growth, reproduction and storage under conditions of variable resource supply. Life history data from *B. calyciflorus* was used in the model and some of its boundary conditions reflect this, e.g., it was assumed that allocation to reproduction stopped when food intake dropped below the maintenance threshold. Shertzer & Ellner's model predicts that the optimal body size should be smaller if animals evolve under variable resource concentrations and that optimal body size decreases further as the length of "bad periods" increases. Other adaptations to variable resource conditions were increased allocation to storage, delayed maturity, and indeterminate growth. The model correctly predicted starvation times in animals that were starved at different age classes or acclimatised to different food levels prior to starvation. However, the predictions regarding the above mentioned life history characteristics still await empirical validation. This could be done by artificial selection experiments in which *B. calyciflorus* is cultured under constant vs. variable resource conditions.

A more general model for resource allocation under variable food conditions would need more relaxed boundary conditions than the one of Shertzer & Ellner (2002). First, the model should allow allocation to reproduction to continue, even when animals are starved (as this seems to be the case in some rotifer species). Second, it would be more realistic if stored energy could be used both for maintenance of the metabolism and for reproduction in animals that experience starvation. Third, longer periods of starvation should be examined. In Shertzer & Ellner (2002) they are only up to 2 days (this is approximately the starvation time of *Brachionus*). If a starvation-protected embryonic stage would be incorporated, the model might produce a similar dichotomy than we see in the starvation response of planktonic rotifers (Kirk, 1997b).

Conclusions and future perspectives

Life history theory in rotifers has made much progress in recent years. It has been increasingly common to consider the adaptive significance of life history patterns and to relate them to selective pressures that rotifers are exposed to in their natural habitats. In some cases it has been relatively easy to find plausible adaptive explanations (e.g., resource allocation, mixis strategies). In other cases, the adaptive significance is still unclear (e.g., egg size). Such traits may be determined mainly by constraints and some of these constraints may be very specific to rotifers. There is certainly more work needed on the influence of constraints on rotifer life histories.

Several lines of research seem particularly promising for future studies:

Due to their asexual mode of reproduction, rotifers are ideal study objects to look at the reaction norms of genotypes (clones). In most studies so far, phenotypic plasticity has been assessed only for one or few clones. This is probably insufficient, since a lot of genetic variation for the norm of reaction probably exists in nature. Therefore, several clones of a species should be tested when studying a particular question. Since some rotifers form cryptic species complexes (Gomez et al., 2002), an analysis of one or two phylogenetic informative genes may be necessary to confirm that the studied clones do indeed belong to the same species.

In the future it may be worthwhile to adopt techniques that are already standard in other organisms, for instance artificial selection experiments. *Brachionus* would be a suitable organism for this approach as they are easy to culture and readily reproduce sexually. Interesting selective regimes would be environments with constant vs. variable resources (to test the predictions of Shertzer & Ellner, 2002) or the selection for large vs. small body size/egg size.

Life history studies are usually performed with monogonont rotifers or bdelloid rotifers, but not with both. It would be interesting to have more data on both of these groups, especially those bdelloid and monogonont species that occur in the same habitat. Such studies could establish the basic differences in life history evolution of facultative vs. obligate asexual organisms. Rotifers are probably the only phylum where such a comparison is possible.

Acknowledgements

I would like to thank John Gilbert, Claudia Ricci, Hendrik Segers, Manuel Serra and Manuel Yufera for discussions on the material of this manuscript during and after the Rotifer Symposium. Two anonymous reviewers provided comments that improved the manuscript.

References

Allan, J. D., 1976. Life history patterns in zooplankton. American Naturalist 110: 165–180.

Aparici, E., M. J. Carmona & M. Serra, 1998. Sex allocation in haplodiploid cyclical arthenogens with density-dependent proportion of males. The American Naturalist 152: 652–657.

Aparici, E., M. J. Carmona & M. Serra, 2002. Evidence for an even sex allocation in haplodiploid cyclical parthenogens. Journal of Evolutionary Biology 15: 65–73.

Begon, M., J. L. Harper & C. R. Townsend, 1991. Ökologie. Birkhäuser, Basel.

Bernardo, J., 1996. The particular maternal effect of propagule size, especially egg size: atterns, models, quality of evidence and interpretations. American Zoologist 36: 216–236.

Caswell, H., 1989. Matrix Population Models. Sinauer, Sunderland.

Conde-Porcuna, J. M., 1998. Chemical interference by *Daphnia* on *Keratella*: a life table experiment. Journal of Plankton Research 20: 1637–1644.

Epp, R. W. & W. M. Lewis, 1984. Cost and speed of locomotion for rotifers. Oecologia 61: 289–292.

Fox, G. A., 2001. Failure time analysis: studying times-to-events and rates at which events occur. In Scheiner, S. M. & J. Gurevitch (eds), Design and Analysis of Ecological Experiments. Oxford University Press, Oxford, UK, 253–289.

Gilbert, J. J., 1989. The effect of *Daphnia* interference on a natural rotifer and ciliate community: Short-term bottle experiments. Limnology and Oceanography 34: 606–617.

Gilbert, J. J., 2002. Endogenous regulation of environmentally induced sexuality in a rotifer: a multigenerational parental effect induced by fertilization. Freshwater Biology 47: 1633–1641.

Gilbert, J. J., 2003. Environmental and endogeneous control of sexuality in a rotifer life cycle: developmental and population biology. Evolution and Development 5: 19–24.

Gilbert, J. J. & D. K. Schreiber, 1998. Asexual diapause induced by food limitation in the rotifer *Synchaeta pectinata*. Ecology 79: 1371–1381.

Glazier, D. S., 1992. Effects of food, genotype, and maternal size and age on offspring investment in *Daphnia magna*. Ecology 73: 910–926.

Gómez, A., M. Serra, G. R. Carvalho & D. H. Lunt, 2002. Speciation in ancient cryptic species complexes: evidence from the molecular phylogeny of *Brachionus plicatilis* (Rotifera). Evolution 56: 1431–1444.

Gómez, A. & T. W. Snell, 1996. Sibling species and cryptic speciation in the *Brachionus licatilis* species complex (Rotifera). Journal of Evolutionary Biology 9: 953–964.

Gonzalez, M. J. & T. M. Frost, 1992. Food limitation and seasonal population declines of rotifers. Oecologia 89: 560–566.

Green, J., 1998. Strategic variation of egg size in *Keratella cochlearis*. Hydrobiologia 387/388: 301–310.

Halbach, U., 1970. Einfluß der Temperatur auf die Populationsdynamik des planktischen Rädertieres *Brachionus calyciflorus*. Oecologia 4: 176–207.

King, C. E., 1970. Comparative survivorship and fecundity of mictic and amictic female rotifers. Physiological Zoology 43: 206–212.

King, C. E. & T. W. Snell, 1977. Genetic basis of amphotheric reproduction in rotifers. Heredity 39: 361–364.

Kirk, K. L., 1997a. Egg size, offspring quality and food level in planktonic rotifers. Freshwater Biology 37: 515–521.

Kirk, K. L., 1997b. Life-history responses to variable environments: starvation and reproduction in planktonic rotifers. Ecology 78: 434–441.

Kirk, K. L., J. Ellis & J. Taylor, 1999. Physiological responses to variable environments: storage and respiration in rotifers. Freshwater Biology 42: 637–644.

Lampert, W., 1993. Phenotypic plasticity of the size at first reproduction in *Daphnia*: the importance of maternal size. Ecology 74: 1455–1466.

Lapesa, S., T. W. Snell, D. M. Fields & M. Serra, 2002. Predatory interactions between a cyclopoid copepod and three sibling rotifer species. Freshwater Biology 47: 1685–1695.

Lewis, W. M., 1979. Zooplankton community analysis. Springer, New York.

Loxdale, H. D. & G. Lushai, 2003. Rapid changes in clonal lines: the death of a sacred cow. Biological Journal of the Linnean Society 79: 3–16.

Merriman, J. L. & K. L. Kirk, 2000. Temporal patterns of resource limitation in natural populations of rotifers. Ecology 81: 141–149.

Meyer, J. S., C. G. Ingersoll, L. L. McDonald & M. S. Boyce, 1986. Estimating uncertainty in population growth rates: Jackknife vs. Bootstrap techniques. Ecology 67: 1156–1166.

Orsenigo, S., C. Ricci & M. Caprioli, 1998. The paradox of bdelloid egg size. Hydrobiologia 387/388: 317–320.

Reznik, D., 1992. Measuring the costs of reproduction. Trends in Ecology and Evolution 7: 42–45.

Ricci, C., 1995. Growth pattern of four strains of a Bdelloid rotifer species: egg size and numbers. Hydrobiologia 313/314: 157–163.

Ricci, C. & U. Fascio, 1995. Life-history consequences of ressource allocation of two bdelloid rotifer species. Hydrobiologia 299: 231–239.

Roff, D. A., 1981. On being the right size. American Naturalist 118: 405–422.

Rothhaupt, K. O., 1990. Population growth rates of two closely related rotifer species: effects of food quality, particle size, and nutritional quality. Freshwater Biology 23: 561–570.

Rougier, C., R. Pourriot & T. Lam-Hoai, 2000. The genus *Synchaeta* (rotifers) in a north-western Mediterranean coastal lagoon (Etang de Thau, France): taxonomical and ecological remarks. Hydrobiologia 436: 105–117.

Santo, N., M. Caprioli, S. Orsenigo & C. Ricci, 2001. Egg size and offspring fitness in a bdelloid rotifer. Hydrobiologia 446/447: 71–74.

Sarma, S. S. S., S. Nandini & R. D. Gulati, 2002. Cost of reproduction in selected species of zooplankton (rotifers and cladocerans). Hydrobiologia 481: 89–99.

Sarma, S. S. S. & T. R. Rao, 1987. Effect of food level on body size and egg size in a growing population of the rotifer *Brachionus patulus* Muller. Archiv für Hydrobiologie 111: 245–253.

Serra, M., M. J. Carmona & M. R. Miracle, 1994. Survival analysis of three clones of *Brachionus plicatilis* (Rotifera). Hydrobiologia 277: 97–105.

Serra, M. & C. E. King, 1999. Optimal rates of bisexual reproduction in cyclical parthenogens with density-dependent growth. Journal of Evolutionary Biology 12: 263–271.

Serra, M. & T. W. Snell, 1998. Why are male rotifers dwarf ? Trends in Ecology and Evolution 13: 360–361.

Shertzer, K. W. & S. P. Ellner, 2002. State-dependent energy allocation in variable environments: Life history evolution of a rotifer. Ecology 83: 2181–2193.

Sleigh, M. A. & J. R. Blake, 1977. Methods of ciliary propulsion and their size limitations. In Pedley, J. R. (ed.), Scale effects in animal locomotion. Academic Press, New York, 243–256 pp.

Smith, C. C. & S. D. Fretwell, 1974. The optimal balance between size and number of offspring. American Naturalist 108: 499–506.

Snell, T. W. & K. Carrillo, 1984. Body size variation among strains of the rotifer *Brachionus licatilis*. Aquaculture 37: 359–367.

Snell, T. W. & M. Childress, 1987. Aging and loss of fertility in male and female *Brachionus licatilis* (Rotifera). International Journal of Invertebrate Reproduction and Development 12: 103–110.

Snell T. W. & F. H. Hoff, 1987. Fertilization and male fertility in the rotifer *Brachionus licatilis*. Hydrobiologia, 329–334.

Snell, T. W. & C. E. King, 1977. Lifespan and fecundity patterns in rotifers: the cost of reproduction. Evolution 31: 882–890.

Stearns, S. C., 1992. The evolution of life histories. Oxford University Press, Oxford.

Stelzer, C. P., 2001. Resource limitation and reproductive effort in a planktonic rotifer. Ecology 82: 2521–2533.

Stelzer, C. P., 2002. Phenotypic plasticity of body size at different temperatures in a lanktonic rotifer: mechanisms and adaptive significance. Functional Ecology 16: 835–841.

Stemberger, R. S. & J. J. Gilbert, 1985. Body size, food concentration, and population growth in planktonic rotifers. Ecology 66: 1151–1159.

Stemberger, R. S. & J. J. Gilbert, 1987. Rotifer threshold food concentrations and the size-efficiency hypothesis. Ecology 68: 181–187.

Steward, A. J. & R. G. Wetzel, 1986. Cryptophytes and other microflagellates as couplers in lanktonic community dynamics. Archiv für Hydrobiologie 106: 1–19.

Telesh, I. V., M. Rahkola & M. Viljanen, 1998. Carbon content of some freshwater rotifers. Hydrobiologia 387/388: 355–360.

Wallace, R. L., J. J. Cipro & R. W. Grubbs, 1998. Relative investment in offspring by sessile Rotifera. Hydrobiologia 387/388: 311–316.

Walz, N., 1995. Rotifer populations in plankton communities: Energetics and life history strategies. Experientia 51: 437–453.

Walz, N. & F. Rothbucher, 1991. Effect of food concentration on body size, egg size, and population dynamics of *Brachionus angularis* (Rotatoria). Internationale Vereinigung für Theoretische und Angewandte Limnologie, Verhandlungen 24: 2750–2753.

Walz, N., S. S. S. Sarma & U. Benker, 1995. Egg size in relation to body size in rotifers: an indication of reproductive strategy?. Hydrobiologia 313/314: 165–170.

Wickham, S. A. & J. J. Gilbert, 1991. Relative vulnerabilities of natural rotifer and ciliate communities to cladocerans: laboratory and field experiments. Freshwater Biology 26: 77–86.

Yoshinaga, T. & A. K. Hagiwara Tsukamoto, 2000. Effect of periodical starvation on the life history of *Brachionus plicatilis* O. F. Müller (Rotifera): a possible strategy for population stability. Journal of Experimental Marine Biology and Ecology 287: 253–260.

Yoshinaga, T., A. Hagiwara & K. Tsukamoto, 2003. Life history response and age-specific tolerance to starvation in *Brachionus plicatilis* O. F. Müller (Rotifera). Journal of Experimental Marine Biology and Ecology 287: 261–271.

Yufera, M., E. Pascual & J. M. Olivares, 2004. Factors affecting swimming speed in the rotifer *Brachionus plicatilis*. Hydrobiologia 546: 375–380.

Hydrobiologia (2005) 546:347–352

A. Herzig, R.D. Gulati, C.D. Jersabek & L. May (eds.) Rotifera X: Rotifer Research: Trends, New Tools and Recent Advances
DOI 10.1007/s10750-005-4244-9

Insulin-like growth factor signaling pathway involved in regulating longevity of rotifers

Tatsuki Yoshinaga[1,4,*], Gen Kaneko[2], Shigeharu Kinoshita[3], Satoshi Furukawa[2], Katsumi Tsukamoto[1] & Shugo Watabe[2]
[1]*Ocean Research Institute, The University of Tokyo, 164–8639, Tokyo, Japan*
[2]*Graduate School of Agricultural and Life Sciences, The University of Tokyo, 113-8657, Tokyo, Japan*
[3]*RIKEN Brain Science Institute, 351–0198, Saitama, Japan*
[4]*Present address: School of Fisheries Sciences, Kitasato University, 022-0101, Iwate, Japan*
*(*Author for correspondence: E-mail: yoshinaga@kitasato-u.ac.jp)*

Key words: *Brachionus plicatilis*, insulin-like growth factor, lifespan, LY294002, PI3-kinase, rotifer

Abstract

Rotifers have been used to study the mechanisms of ageing for more than a century, but the underlying molecular basis of ageing in rotifers is largely unknown. The insulin/insulin-like growth factor (IGF-1) signaling pathway has been found to regulate the lifespan of evolutionarily distinct eukaryotes from yeast to mammals. We therefore assume that the insulin/IGF-1 pathway is a candidate for regulating the rotifer's lifespan. Accordingly, we examined the action of an inhibitor to PI3-kinase involved in the pathway for the rotifer *Brachionus plicatilis* O. F. Müller. This kinase was first discovered as *age-1* to regulate the longevity of *Caenorhabditis elegans*. As expected, the inhibitor treatment resulted in the extension of lifespan by 30% compared to the reference group without the treatment, whereas reproductive characters were not apparently changed. These results were consistent with those observed in *C. elegans*, suggesting that the lifespan of *B. plicatilis* is likely to be regulated by the signaling pathway involving PI3-kinase.

Introduction

Ageing is an inevitable biological phenomenon for all organisms. Even if there were no extrinsic source of mortality such as predation and environmental extremes, cellular integrity would be gradually decreased by intrinsic sources of mortality, eventually resulting in death after a certain period (Stearns, 1992). The existence of a species-specific lifespan and tremendous variation of lifespan among species have attracted much attention to explore the phenomena from various kinds of view asking how ageing occurs and why it has evolved (Arking, 1998).

The genetic approaches on ageing in model organisms such as nematodes, fruitflies, and mice have identified insulin/insulin-like growth factor-1 (IGF-1) signaling pathway to have a significant role in the regulation of ageing (Guarente & Kenyon, 2000; Clancy et al., 2001; Holzenberger et al., 2003). As insulin/IGF-1 regulates the metabolic activity in mammals, it may have a similar function in lower invertebrate such as nematode *Caenorhabditis elegans*. During a larval stage, *C. elegans* has an alternative state called dauer in which the worm is developmentally arrested and can persist under stressful environments such as scarce food and crowding (Cassada & Russell, 1975). Because the length of dauer period does not affect subsequent lifespan in adulthood, this special stage has been considered as non-ageing state, and various mutants in dauer formation has been isolated (Riddle et al., 1981). Following identification of mutant genes has shown that these are homologues of genes, which function in mammalian insulin signaling pathway (Guarente & Kenyon, 2000). The ligand of this pathway is suggested as a hybrid molecule of insulin and IGF in *C. elegans* (Kawano et al., 2000). The loss-of-function mutation of a transmembrane receptor, DAF-2, of this ligand peptide extends *C. elegans* lifespan dramatically, indeed almost twice (Kenyon et al., 1993; Kimura et al., 1997). DAF-2 activates a downstream kinase, AGE-1, which encodes a phosphatidylino-sitol-3-OH kinase (PI3-kinase) (Morris et al., 1996). A reduced expression of AGE-1 also makes *C. elegans* to remain healthy for an extended period, resulting in longevity (Johnson, 1990). On the other hand, if DAF-16, which encodes a forkhead/winged helix family transcription factor, defects in the above-mentioned long-lived mutants, lifespan is not extended (Kenyon et al., 1993; Lin et al., 1997; Ogg et al., 1997). Thus, DAF-16 directly regulates the longevity.

DAF-16 is considered to induce an expression of various genes associating with longevity. Recently, Lee et al. (2003) carried out genome-wide screening to search for candidate target genes of DAF-16, and found that at least seven genes were regulated in *daf-2* pathway, mediating longevity, metabolism and development of *C. elegans*. Moreover, Murphy et al. (2003) investigated the genes that act downstream of DAF-16 using DNA microarray analysis, and found that some tens of genes may function to exert longevity in *C. elegans*. Among them one possible target of DAF-16 is *sod-3*, which encodes Mn-superoxide dismutase (SOD) (Honda & Honda, 1999). Mn-SOD mediates a conversion of reactive oxygen species (ROS) into harmless forms in mitochondria where a large amount of ROS is generated in the electron transport chain (Finkel & Holbrook, 2000). Accordingly it seems reasonable to suppose that in long-lived mutants the accumulation of damage by ROS is reduced by increased Mn-SOD, and this accomplishes, at least partly, cellular integrity for extended period and thus brings about the longevity. On the other hand, in the PI3-kinase deficient long-lived mutant (*age-1*) not only Mn-SOD but also cytosolic catalase have been found to be expressed at higher levels than in a wild type (Larsen, 1993). All long-lived mutants of *C. elegans* examined so far have been found to be tolerant against not only oxidative stress but also heat shock and UV radiation (Larsen, 1993; Lithgow et al., 1995; Murakami & Johnson, 1996), and thus Mn-SOD and other molecules function simultaneously for longevity.

A decade ago Enesco (1993) reviewed four major theories of ageing in rotifers, and among them the free radical theory of ageing, or oxidative stress theory, was concluded to be a promising one explaining well the ageing process of rotifers. The fact that the components of insulin/IGF-1 signaling pathway are well conserved among evolutionarily distinct eukaryotes such as yeast, nematode and even in mammals (Guarente & Kenyon, 2000), allows us to assume that this pathway could regulate the rotifer's lifespan. The direct approach to examine this hypothesis will be an observation of lifespan in genetically modified rotifers as has been applied in numerous organisms. However, it has been not feasible yet to manipulate a gene expression or to utilize mutants in rotifers.

We therefore employed in this study the use of a chemical named LY294002 which inhibits the action of PI3-kinase, and examined its effect on the lifespan as well as other life history parameters of the rotifer *Brachionus plicatilis* O. F. Müller to gain an insight into the molecular mechanism underlying the ageing of rotifer. This inhibitor has been commonly used to explore the function of PI3-kinase in a variety of studies. In this study, the inhibition of PI3-kinase was found to extend the lifespan of *B. plicatilis* by 30%, suggesting that the lifespan of rotifer is likely to be regulated by the PI3-kinase signaling pathway.

Materials and methods

The rotifer *B. plicatilis* Ishikawa strain was used in this study (Yoshinaga et al., 1999). The experimental animals are genetically identical because they originated from a single amictic female, and there is no mictic reproduction in our culture system. The rotifer culture was conducted in sterilized Brujewicz artificial seawater of salinity at 16 ppt (Yoshinaga et al., 1999). Dietary algae (Chlorella, Nikkai Center, Japan) was rinsed twice with fresh culture medium and fed at 430 μg dry weight/ml. Rotifers bearing three to four amictic eggs were inoculated at

10 individuals/ml and observed hourly to collect an experimental cohort.

Experimental design

Forty offspring (age <1 h) were randomly divided into five groups (n = 8) and assigned two control and three experimental groups. The treatment with LY294002 (Promega) was carried out at three concentrations of 1, 10 and 100 nM. The stock inhibitor solution was prepared in ethanol, and dissolved in the seawater medium. In a preliminary experiment, 1% of ethanol caused the abnormality in embryonic development (data not shown), so that the culture media of each treatment group was prepared to contain 0.002% of ethanol. One group was treated with ethanol only (solvent-control group).

The culture was carried out by an individual culture method. All individuals were transferred daily to newly prepared culture media. Observations were conducted daily, and terminated when the last animal died. Following life history parameters were calculated: ages at the first and last offspring production, reproductive period, post-reproductive period, and lifespan. The time of individual death was determined as the mean between those of observations before and after deaths.

Statistical analysis

To compare the lifespan among the three treatment groups and two controls, the survival analysis methods (Kaplan–Meier method) were carried out to compute product-limit estimates of survival functions. Differences between survival functions among the five groups were analyzed using log-rank test for homogeneity of survival functions.

One-way analysis of variance (ANOVA) and following *post-hoc* Fisher's PLSD test were carried out to compare the lifetime fecundity, ages at the first and last offspring production, reproductive period and post-reproductive periods.

All statistical analyses were carried out by StatView 5.0J (SAS Institute).

Results

Survival

The lifespan of the 1 nM group (mean product-limit estimate ± S.E.; 13.9 ± 1.3 days) was significantly longer by 30% than those of either the control and solvent-control groups with 9.6 ± 0.8 and 8.9 ± 0.9 days, respectively (Fig. 1b; log-rank test, $p < 0.01$ in each case). However, the lifespan of the 10- and 100-nM groups (9.9 ± 1.1 and 10.8 ± 1.0 days, respectively) were not significantly different from those of the control and solvent-control groups (Fig. 1c,d; $p > 0.2$ in each case). There was no difference in the lifespan between the control and solvent-control groups (Fig. 1a; $p = 0.8$).

The lifespan extension of the 1-nM group was caused by a prolonged post-reproductive period (Table 1). There was no difference in the pre-reproductive and reproductive periods among the five groups.

Fecundity

The pattern of age-specific fecundity did not differ among the two control and three treatment groups (Fig. 1e–h). Similarly, there was no difference in the lifetime fecundity among the five groups (Table 1).

Discussion

In this study, we examined whether the insulin/IGF-1 signaling pathway is involved in regulating lifespan of the rotifer *B. plicatilis*. We examined the effect of PI3-kinase inhibitor, LY294002, and observed that the treatment of the drug at 1 nM extended the lifespan of *B. plicatilis* by 30% (Fig. 1b). This is consistent with our assumption that the inactivation of PI3-kinase leads to the activation of longevity signaling in downstream. Accordingly it seems possible to conclude that the lifespan of *B. plicatilis* is regulated by PI3-kinase activity. The lifespan of *C. elegans* has been reported to be extended by LY294002 at concentrations of 0.8 and 16 μM in a liquid medium

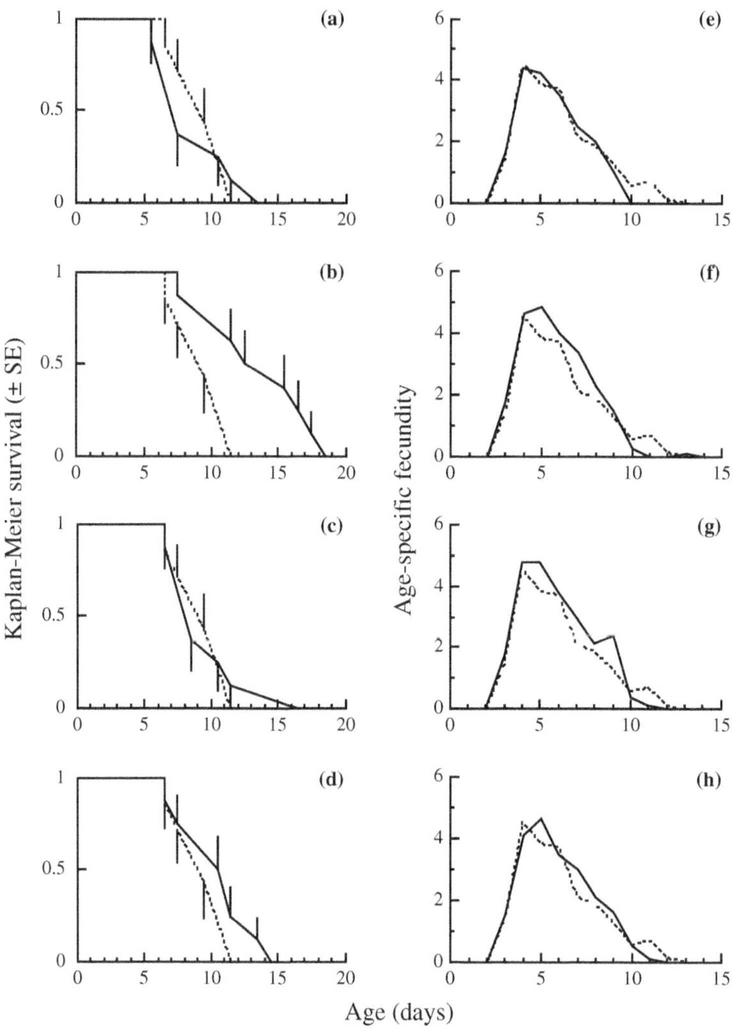

Figure 1. Kaplan–Meier survival curves (a–d) and age-specific fecundity (e–h) in the rotifer *Brachionus plicatilis*. Rotifers were treated with PI3-kinase inhibitor, LY294002, at concentrations of 0 (ethanol only; a and e), 1 nM (b and f), 10 nM (c and g) and 100 nM (d and h) as shown by solid lines. Broken lines indicate the control group without treatment. The 1 nM group survived significantly longer than the control group (log-rank test, $p < 0.01$), whereas the other three treatment groups did not.

(Babar et al., 1999), but in this study the effect did not appear at concentrations above 10 nM (Fig. 1b, c). Thus, the effective range of the drug dose for *C. elegans* did not overlap with that of *B. plicatilis*, possibly due to different uptake efficiencies and/or binding affinities of the drug between two species.

Although LY294002 does not inhibit the activity of PI4-k, PKC, PKA and MAPK which are all involved in important signaling pathways (Vlahos et al., 1994; Rameh & Cantley, 1999), it may influence the activity of PI3-k-like kinases. It is therefore impossible to conclude from our results that only PI3-kinase is responsible for the regulating in the lifespan of *B. plicatilis*. To cope with such uncertainty accompanied by the drug treatment, a gene-specific disruption approach will be necessary. Currently, no gene targeting method including the use of mutants is available in rotifers, therefore RNA-mediated interference (RNAi) may have a potential to inhibit the function of the gene. Even the availability of RNAi has been suggested to be dependent on the species or conditions (Braasch & Corey, 2002), it will be important to

Table 1. Effect of PI3-kinase inhibitor on the life history parameters of the rotifer *Brachionus plicatilis*

	Control	Solvent-control (0.002% ethanol)	LY294002 concentration (nM)		
			1	10	100
Lifetime fecundity (individuals)	20.29 (0.52)^{ns}	19.25 (0.35)^{ns}	22.75 (0.40)^{ns}	23.00 (0.25)^{ns}	21.13 (0.32)^{ns}
Age at first reproduction (days)	2.00 (0.00)^{ns}	2.00 (0.00)^{ns}	2.00 (0.00)^{ns}	2.00 (0.00)^{ns}	2.13 (0.13)^{ns}
Age at last reproduction (days)	7.86 (0.80)^{ns}	7.13 (0.35)^{ns}	7.88 (0.74)^{ns}	8.25 (0.41)^{ns}	8.00 (0.46)^{ns}
Reproductive period (days)	5.86 (0.80)^{ns}	5.13 (0.35)^{ns}	5.88 (0.74)^{ns}	6.25 (0.41)^{ns}	5.88 (0.40)^{ns}
Post reproductive period (days)	1.79 (0.42)^{a}	1.75 (0.67)^{a}	6.06 (1.44)^{b}	1.69 (1.01)^{a}	2.81 (0.81)^{a}

Values are means and standard errors in parentheses of eight replicates per group. Postscript letters indicate the result of statistical analyses (ns, no significant difference; a < b, $p < 0.05$, PLSD test).

challenge in rotifers to precisely identify the responsible genetic pathway regulating longevity.

The extension of lifespan by LY294002 resulted from an extended post-reproductive period (Fig. 1; Table 1). This is consistent with observation in PI3-kinase deficient mutant of *C. elegans* (Chen et al., 2001). Modifications of life history parameters during the post-reproductive period are likely to be less important under natural selection because these have substantially no contribution to the fitness (Stearns, 1992). However, the animals with extended post-reproductive period might have had higher stress tolerance during the reproductive period, and thus increased survival ability. In fact all long-lived mutants so far isolated in nematodes, fruitflies, and mice show high stress tolerance (Finkel & Holbrook, 2000). Therefore, one might imagine that long-lived trait will be favored by natural selection. However, it is not like that. In a mixed population of long-lived (*age-1*) and wild type *C. elegans*, both genotypes have similar fitness under food-rich environment (Walker et al., 2000). However, if this population exposed to a cyclic starvation, mimicking undernourished environments in nature, the *age-1* genotype disappeared after several cycles because this long-lived mutant failed to reproduce under starvation (Walker et al., 2000). It is noteworthy that insulin/IGF-1 signaling pathway is also regulated by signals from reproductive organs (Hsin & Kenyon, 1999). Even though it is not yet fully understood how this signaling pathway integrates reproduction and lifespan, it is crucial to consider life history evolution because the ability to respond to starvation is essential for all organisms. The cyclic starvation, usually termed as caloric or dietary restriction (CR/DR), causes the extension of lifespan in a variety of organisms from unicellular yeast to multicellular animals including rotifer, nematode, insects, and mammals (see reviews by Guarente & Kenyon, 2000, and by Kirk, 2001 in rotifers). *B. plicatilis* under cyclic starvation produces less than half the number of offspring than the well-fed animals, but it survives twice as long (Yoshinaga et al., 2000). Moreover, the offspring produced under cyclic starvation shows higher survival than those of well-fed mothers (Yoshinaga et al., 2001). Such alteration in reproductive strategy is essential to consider diverse ecological aspects such as population dynamics and life history evolution (Yoshinaga et al., 2003). Further analyses of PI3-kinase as well as other components in the insulin/IGF-1 pathway will provide molecular mechanisms involved not only in the regulation of lifespan but also in a variety of ecological phenomena.

Acknowledgements

We would like to thank Drs T. Nogrady and D. M. Welch for their critical reading of our manuscript. This study was partly supported by Grants-in-Aid for Creative Scientific Research No. 12NP0201 and Exploratory Research No. 14656080 from the Ministry of Education, Culture, Sports, Science and Technology of Japan. T. Y. and G. K. were supported by Research Fellowships of the Japan Society for the Promotion of Science for Young Scientists.

References

Arking, R., 1998. Biology of Aging. 2nd ed. Sinauer, MA.

Babar, P., C. Adamson, G. A. Walker, D. W. Walker & G. J. Lithgow, 1999. PI3-kinase inhibition induces dauer formation, thermotolerance and longevity in *C. elegans*. Neurobiology of Aging 20: 513–519.

Braasch, D. A. & D. R. Corey, 2002. Novel antisense and peptide nucleic acid strategies for controlling gene expression. Biochemistry 41: 4503–4510.

Cassada, R. C. & R. L. Russell, 1975. The dauer larva, a post-embryonic developmental variant of the nematode *Caenorhabditis elegans*. Developmental Biology 46: 326–342.

Chen, J., J. R. Carey & H. Ferris, 2001. Comparative demography of isogenic populations of *Caenorhabditis elegans*. Experimental Gerontology 36: 431–440.

Clancy, D. J., D. Gems, L. G. Harshman, S. Oldham, H. Stocker, E. Hafen, S. J. Leevers & L. Partridge, 2001. Extension of life-span by loss of CHICO, a *Drosophila* insulin receptor substrate protein. Science 292: 104–106.

Enesco, H. E., 1993. Rotifers in aging research: use of rotifers to test various theories of aging. Hydrobiologia 255–256: 59–70.

Finkel, T. & N. J. Holbrook, 2000. Oxidants, oxidative stress and the biology of ageing. Nature 408: 239–247.

Guarente, L. & C. Kenyon, 2000. Genetic pathways that regulate ageing in model organisms. Nature 408: 255–262.

Hsin, H. & C. Kenyon, 1999. Signals from the reproductive system regulate the lifespan of C. *elegans*. Nature 399: 362–366.

Holzenberger, M., J. Dupont, B. Ducos, P. Leneuve, A. Geloen, P. C. Even, P. Cerverak & Y. L. Bouc, 2003. IGF-1 receptor regulates lifespan and resistance to oxidative stress in mice. Nature 421: 182–187.

Honda, Y. & S. Honda, 1999. The *daf-2* gene network for longevity regulates oxidative stress resistance and Mn-superoxide dismutase gene expression in *Caenorhabditis elegans*. The FASEB Journal 13: 1385–1393.

Johnson, T. E., 1990. The increased life span of *age-1* mutants in *Caenorhabditis elegans* results from lowering the Gompertz rate of aging. Science 249: 908–912.

Kawano, T., Y. Ito, M. Ishiguro, K. Takuwa, T. Nakajima & Y. Kimura, 2000. Molecular cloning and characterization of a new insulin/IGF-like peptide of the nematode *Caenorhabditis elegans*. Biochemical and Biophysical Research Communications 273: 431–436.

Kenyon, C., J. Chang, E. Gensch, A. Rudner & R. Tabtiang, 1993. A C. *elegans* mutant that lives twice as long as wild-type. Nature 366: 461–464.

Kimura, K. D., H. A. Tissenbaum, Y. X. Liu & G. Ruvkun, 1997. *daf-2*, an insulin receptor-like gene that regulates longevity and diapause in *Caenorhabditis elegans*. Science 277: 942–946.

Kirk, K. L., 2001. Dietary restriction and aging: comparative tests of evolutionary hypotheses. Journal of Gerontology 56: 123–129.

Larsen, P. L., 1993. Aging and resistance to oxidative damage in *Caenorhabditis elegans*. Proceeding of the National Academy of Sciences of the United States of America 90: 8905–8909.

Lee, S. S., S. Kennedy, A. C. Tolonen & G. Ruvkun, 2003. DAF-16 target genes that control *C. elegans* life-span and metabolism. Science 300: 644–647.

Lin, K., J. B. Dorman, A. Rodan & C. Kenyon, 1997. *daf-16*: an HNF-3/forkhead family member that can function to double the life-span of *Caenorhabditis elegans*. Science 278: 1319–1322.

Lithgow, G. J., T. M. White, S. Melov & T. E. Johnson, 1995. Thermotolerance and extended life-span conferred by single-gene mutations and induced by thermal stress. Proceeding of the National Academy of Sciences of the United States of America 92: 7540–7544.

Morris, J. Z., H. A. Tissenbaum & G. Ruvkun, 1996. A phosphatidylinositol-3-OH kinase family member regulating longevity and diapause in *Caenorhabditis elegans*. Nature 382: 536–539.

Murakami, S. & T. E. Johnson, 1996. A genetic pathway conferring life extension and resistance to UV stress in *Caenorhabditis elegans*. Genetics 143: 1207–1218.

Murphy, C. T., S. A. McCarroll, C. I. Bargmann, A. Fraser, R. S. Kamath, J. Ahringer, H. Li & C. Kenyon, 2003. Genes that act downstream of DAF-16 to influence the lifespan of *Caenorhabditis elegans*. Nature 424: 277–283.

Ogg, S., S. Paradis, S Gottlieb, G. I. Patterson, L. Lee, H. A. Tissenbaum & G. Ruvkun, 1997. The fork head transcription factor DAF-16 transduces insulin-like metabolic and longevity signals in *C. elegans*. Nature 389: 994–999.

Rameh, L. E. & L. C. Cantley, 1999. The role of phosphoinositide 3-kinase lipid products in cell function. Journal of Biological Chemistry 274: 8347–8350.

Riddle, D. L., M. M. Swanson & P. S. Albert, 1981. Interacting genes in nematode dauer larva formation. Nature 290: 268–271.

Stearns, S. C., 1992. The Evolution of Life Histories. Oxford University Press: Oxford .

Vlahos, C. J., W. F. Matter, K. Y. Hui & R. F. Brown, 1994. A specific inhibitor of phosphatidylinositol 3-kinase, 2-(4-morpholinyl)-8-phenyl-4H-1-benzopyran-4-one (LY294002). Journal of Biological Chemistry 269: 5241–5248.

Walker, D. W., G. McColl, N. L. Jenkins, J. Harris & G. L. Lithgow, 2000. Evolution of lifespan in *C. elegans*. Nature 405: 296–297.

Yoshinaga, T., A. Hagiwara & K. Tsukamoto, 1999. Effect of conditioned media on the asexual reproduction of the monogonont rotifer *Brachionus plicatilis* O.F.Müller. Hydrobiologia 412: 103–110.

Yoshinaga, T., A. Hagiwara & K. Tsukamoto, 2000. Effect of periodical starvation on the life history of *Brachionus plicatilis* O.F.Müller (Rotifera): a possible strategy for population stability. Journal of Experimental Marine Biology and Ecology 253: 253–260.

Yoshinaga, T., A. Hagiwara & Tsukamoto, 2001. Effect of periodical starvation on the survival of offspring in the rotifer *Brachionus plicatilis*. Fisheries Science 67: 373–374.

Yoshinaga, T., G. Kaneko, S. Kinoshita, K. Tsukamoto & S. Watabe, 2003. Review: the molecular mechanisms of life history alterations in a rotifer: a novel approach in population dynamics. Comparative Biochemistry and Physiology, Part B, Physiology and Molecular Biology 136: 715–722.

Hydrobiologia (2005) 546:353–360
A. Herzig, R.D. Gulati, C.D. Jersabek & L. May (eds.) Rotifera X: Rotifer Research: Trends, New Tools and Recent Advances
DOI 10.1007/s10750-005-4245-8

Combined effects of algal (*Chlorella vulgaris*) food level and temperature on the demography of *Brachionus havanaensis* (Rotifera): a life table study

E. Lucía Pavón-Meza[1,*], S. S. S. Sarma[1] & S. Nandini[2]
[1]*Laboratory of Aquatic Zoology, Division of Research and Postgraduate Studies, National Autonomous University of Mexico Campus Iztacala, AP 314, CP 54090, Tlalnepantla, State of Mexico, Mexico; E-mail: sarma@servidor.unam.mx*
[2]*UIICSE, Division of Research and Postgraduate Studies, National Autonomous University of Mexico; Campus Iztacala, Av. de Los Barrios No. 1, AP 314 CP 54090, Tlalnepantla, State of Mexico, Mexico*
*(*Author for correspondence: E-mail: luciapav@hotmail.com)*

Key words: temperature, food concentration, life history, rotifer, *Brachionus*

Abstract

We evaluated the combined effects of food (0.5×10^6, 1.0×10^6 and 2.0×10^6 cells ml^{-1} of *Chlorella vulgaris*) and temperature (15, 20 and 25 °C) on life history variables of *B. havanaensis*. Regardless of *Chlorella* density there was a steep fall in the survivorship of *B. havanaensis* at 25 °C. Both food level and temperature affected the fecundity of *B. havanaensis*. At any given food level, rotifers cultured at 15 °C showed extended but low offspring production. At 25 °C, offspring production was elevated, the duration of egg laying reduced and the fecundity was higher during the latter part of the reproductive period. The effect of food level was generally additive, at any given temperature, and higher densities of *Chlorella* resulted in higher offspring production. Average lifespan, life expectancy at birth and generation time were 2–3 times longer at 15 °C than at 25 °C. At 20 °C, these remained at intermediate levels. The shortest generation time (about 4 days) was observed at 25 °C. Gross and net reproductive rates and the rate of population increase (r) increased with increasing temperature and generally, at any given temperature, higher algal food levels contributed to higher values in these variables. The r varied from 0.11 to 0.66. The survival patterns and lower rates of reproduction at 15 °C suggest that the winter temperatures (10–15 °C) prevailing in many waterbodies in Mexico City allow this species to sustain throughout the year under natural conditions.

Introduction

Brachionus havanaensis Rousselet is a widely distributed freshwater rotifer in the American continent including Mexico (Ahlstrom, 1940; De Ridder, 1981). Its ability to tolerate a wide range of temperature and food concentrations and its role as an important prey item for certain invertebrate predators, such as *Asplanchna*, make this rotifer an important zooplankton species in both tropical and high altitude freshwater water bodies of Mexico (Torres-Orozco & Zanatta, 1998). Information pertaining to the ecology of this species is mostly based on field collections. However, a few laboratory studies have focused on its role as prey for *Asplanchna* (Nandini et al., 2003; Sarma et al., 2003a). Like some other spined species of *Brachionus*, such as *B. calyciflorus* and *B. quadridentatus*, *B. havanaensis* too is capable of showing variation in posterior and anterior spine length in response to *Asplanchna* predation (Garza-Mouriño et al., 2005). However, survival of this species is related not only to its anti-predatory adaptations, but also to its high reproductive potential enabling it to compete with other members of Brachionidae (*Keratella cochlearis*, *K. tropica*, *B. budapestinensis* and *B. calyciflorus*) with which it often co-occurs.

Two important factors controlling life history traits of several rotifer genera including *Brachionus* are food and temperature (Edmondson, 1965). Food effects may be evaluated by considering nutritional quality or abundance (Enrique-García et al., 2003). Numerous studies on the effect of algal quality on herbivorous rotifers and other zooplankton species (Rothhaupt, 1990; Gulati & DeMott, 1999; DeMott et al., 2001; Ramos-Rodríguez & Conde-Porcuna, 2003) generally show that the algae grown in medium containing low amounts of certain elements (e.g. phosphorus), have low nutritional value for zooplankton. However, in many eutrophic tropical waterbodies it is doubtful if algae suffer from a lack of essential nutrients such as N, P and K (Ramírez-García et al., 2002). With regard to algal density, it is generally believed that an increase in the concentration of edible algal food enhances the offspring production. This is true for certain brachionid species, such as *B. calyciflorus* (Sarma et al., 1999). Sarma & Nandini (2001) have shown that increasing *Chlorella* density from 0.25×10^6 to 4.0×10^6 cells ml^{-1} caused decreased egg in the rotifer *B. variabilis*. Therefore, generalizations based on a few taxa may not be applicable to an entire family or even to a given genus. Thus, the evaluation of variable food levels for *B. havanaensis* is important to understand changes in life history parameters (such as lifespan, age-specific survivorship, fecundity and generation time).

For ectothermic organisms like rotifers, temperature affects metabolic processes (Halbach, 1973): higher temperature accelerates egg hatching, age at maturity and rate of egg production but shortens lifespan. Lower temperatures usually have the opposite effect on these variables (Sarma & Rao, 1991). In nature both food level and temperature act synergistically. While the relative allocation of energy intake to reproduction is a function of both the quantity of food available and food consumed, its magnitude varies in relation to temperature (Sarma & Rao, 1990).

The objective of the present study was to evaluate the combined effects of food level and temperature on survivorship and reproductive performance of *B. havanaensis*.

Materials and methods

A single parthenogenetic individual of *B. havanaensis* was isolated from the National Canal of Lake Xochimilco (Mexico City) and cultured on *Chlorella vulgaris* (Strain No. CL-V-3 Algal Department, CICESE, Ensenada, Baja California, Mexico) using moderately hardwater (EPA medium) prepared by dissolving 96 mg $NaHCO_3$, 60 mg $CaSO_4$, 60 mg $MgSO_4$ and 4 mg KCl in 1 l of distilled water (Anonymous, 1985). *Chlorella* was cultured in 2 l transparent bottles using Bold's basal medium (Borowitzka & Borowitzka, 1988). Log phase algae were harvested, concentrated by centrifugation at 3000 rpm and resuspended in distilled water. Algal density was measured using a haemocytometer. Rotifers were cultured in 20 l glass aquaria, fed every day using alga at a concentration of about 1×10^6 ml^{-1}. The medium in aquaria was changed every other day. Rotifer mass cultures were maintained at 22 ± 2 °C under continuous, diffused fluorescent illumination.

Prior to conducting experiments, small cultures (about 1 l) of *B. havanaensis* were maintained at least for a week at the three chosen temperatures (low, 15 °C; medium, 20 °C; and high, 25 °C). To obtain neonates of known age, we filtered rotifers during the exponential phase using 80 μm mesh and collected parthenogenetic eggs. Neonates and small individuals of *B. havanaensis*, if any, were removed immediately. Neonates hatching within 3 h following removal were used in the experiments. Standard cohort life table experiments were conducted at three temperatures and three *Chlorella* densities (low, 0.5×10^6; medium, 1.0×10^6; and high, 2.0×10^6 cells ml^{-1}). For each temperature and food level combination, we used six replicates. Thus, the experimental design consisted of 54 transparent jars of 50 ml capacity. We introduced 20 neonates into each jar for each temperature and *Chlorella* density combination. Following inoculation, we counted and removed neonates born during the successive observations at 12 h intervals. Simultaneously, dead adults were removed. The number of eggs carried by each female and loose eggs were also recorded. The surviving females were transferred every 24-h to fresh test jars containing *C. vulgaris* at the appropriate level and corresponding temperature. The experiments were discontinued when all *B. havanaensis* had died. The number of eggs and neonates of each 24 h interval were pooled and considered as the daily offspring production for calculating life history variables. We used standard formulae (Krebs, 1985) to derive the selected life history variables (survivorship and fecundity curves, average lifespan and net reproductive rate, generation time and rate of population increase). The survivorship and reproductive variables were analyzed statistically using two-way analysis of variance (ANOVA) following Sokal & Rohlf (2000).

Results

The survivorship curves (l_x) and age-specific offspring production (fecundity, m_x) of *B. havanaensis* grown under different algal food densities and temperatures are shown in Fig. 1. Regardless of *Chlorella* density there was a steep decline in survivorship of *B. havanaensis* at 25 °C. At 15 and 20 °C, although the survival of the test individuals was longer, the curves showed a nearly constant mortality with age. The effect of food level was not apparent, although an increase in *Chlorella* density did increase the longevity of rotifers at 20 °C.

Both food level and temperature affected the fecundity of *B. havanaensis*. At all three food levels, the rotifers cultured at 15 °C showed extended but low offspring production. With a temperature increase to 20 °C, the fecundity increased during the early reproductive period. With a further increase in temperature to 25 °C, offspring production was elevated during the latter part of the reproductive period and the duration of egg laying was reduced. The effect of food level was generally additive, in that at any given temperature, higher densities of *Chlorella* increased offspring production. Females carrying three eggs were observed only at 20 °C.

The age-specific life expectancy curves generally showed increased mortality with increasing age of rotifers at 15 and 25 °C but at the intermediate temperature (20 °C) there was a slight increase in the life expectancy after 10 days. There was no apparent effect of food level (Fig. 2). The average lifespan, life expectancy at birth and the generation times were all 2–3 times longer at 15 °C than at 25 °C. At 20 °C, these remained at intermediate levels. The shortest generation time, about 4 days, was observed at 25 °C. Gross and net reproductive rates increased with increasing temperature and also higher algal food levels generally

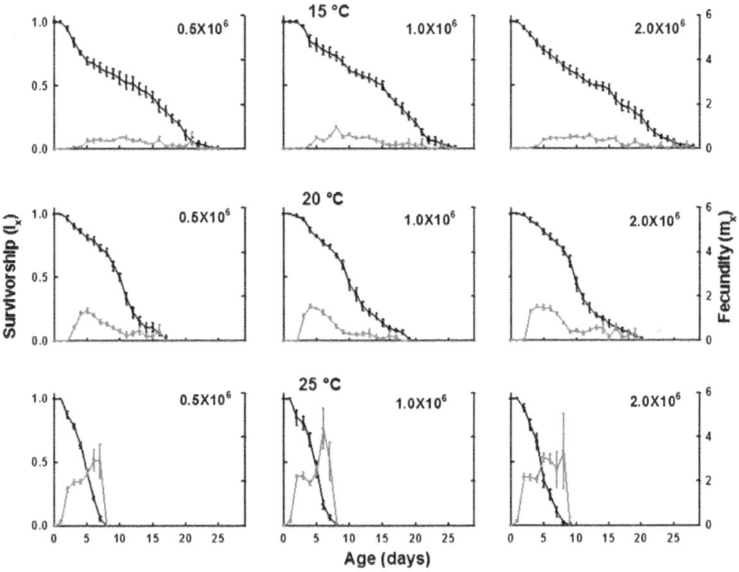

Figure 1. Age-specific survivorship (l_x) and fecundity (m_x) curves of *B. havanaensis* cultured using three different food densities (0.5×10^6, 1.0×10^6 and 2.0×10^6 cells ml^{-1} of *Chlorella*) and each of these at three temperatures (15, 20 and 25 °C). Shown are the mean ± SE of six replicates.

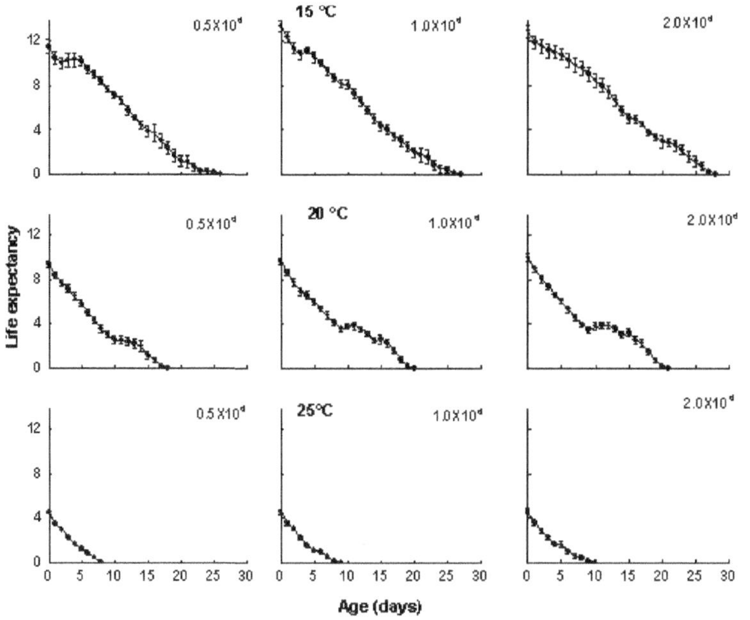

Figure 2. Age-specific life expectancy curves of *B. havanaensis* cultured using three different food densities (0.5×10^6, 1.0×10^6 and 2.0×10^6 cells ml^{-1} of *Chlorella*) and each of these at three temperatures (15, 20 and 25 °C). Shown are the mean ± SE of six replicates.

contributed to higher rates (Table 1). Regardless of food density and temperature, the rate of population increase (r) varied from 0.11 to 0.66. For all *Chlorella* densities, increase in temperature resulted in increased r. The effect of food level on r in general was that increased food level resulted in higher population growth rates (Fig. 3).

Both food density and temperature had statistically significant effect on average lifespan, gross and net reproductive rates, generation time and the rate of population increase ($p < 0.01$, Two way ANOVA, Table 2). However, the interaction between food level and temperature was not significant ($p > 0.05$) for the measured life history traits.

Discussion

Brachionus havanaensis has been recorded in high elevation ponds (2000 m above sea level) in and around Mexico City and in tropical regions in the southeastern states of Mexico (Torres-Orozco & Zanatta, 1998; Flores-Burgos et al., 2003). Temperature ranges from ~10 to 30 °C. Our study covers the wide temperature range as well as the range of food concentrations that occurs in the waterbody from which this rotifer was originally collected. In addition, these food concentrations have been used for life history assessment in many brachionid rotifers (e.g., *B. calyciflorus*: Sarma et al., 1999, *B. macracanthus*: Sarma & Nandini, 2002, *B. rubens*: Sarma et al., 2003b) thus facilitating comparison within this genus.

Among survivorship variables, median lifespan (or $0.5l_x$) and average lifespan ($= \Sigma l_x$) have received considerable attention because of their role in the evolution of lifespan. These two measures are not identical but are closely related (Sarma & Rao, 1991). King (1982) has hypothesized that for iteroparous organisms, including rotifers, the median lifespan is twice the generation time, regardless of culture conditions. When this was applied to some other rotifer species (Sarma & Rao, 1991) and crustaceans (Anaya-Soto et al., 2003), a significant correlation was, indeed, observed although the slope deviated from 2, as also found in the present study (figure not presented). Increases in temperature have been shown to reduce the lifespan in many rotifer species (Halbach, 1973; Sarma & Rao, 1991), a trend also observed in the present study. When temperature was increased

Table 1. Selected life history traits of *B. havanaensis* in relation to different temperatures and algal food (*Chlorella*) concentrations

Food density ($\times 10^6$ cells ml^{-1})	Average lifespan	Life expectancy at birth	Gross reproductive rate	Net reproductive rate	Generation time
15 °C					
0.5	12.03 ± 0.07	11.53 ± 0.61	5.61 ± 0.22	2.82 ± 4.38	9.71 ± 0.14
1.0	13.89 ± 0.47	13.39 ± 0.47	7.39 ± 0.49	4.38 ± 0.33	9.54 ±0.20
2.0	13.89 ± 0.47	12.94 ± 0.55	7.07 ± 0.41	3.59 ± 0.19	10.26 ± 0.27
20 °C					
0.5	9.89 ± 0.24	9.39 ± 0.24	9.16 ± 0.67	5.83 ± 0.42	6.13 ± 0.21
1.0	10.16 ± 0.30	9.66 ± 0.30	9.54 ± 0.65	6.54 ± 0.37	5.67 ± 0.12
2.0	10.49 ± 0.30	9.99 ± 0.30	12.18 ± 0.66	7.68 ± 0.28	5.99 ± 0.08
25 °C					
0.5	4.98 ± 0.08	4.48 ± 0.08	14.11 ± 0.65	6.25 ± 0.19	3.65 ± 0.09
1.0	5.02 ± 0.25	4.52 ± 0.25	16.55 ± 1.10	7.21 ± 0.71	3.53 ± 0.11
2.0	5.11 ± 0.25	4.61 ± 0.25	18.64 ± 2.77	7.61 ± 0.65	3.71 ± 0.06

Mean ± SE of six replicates.

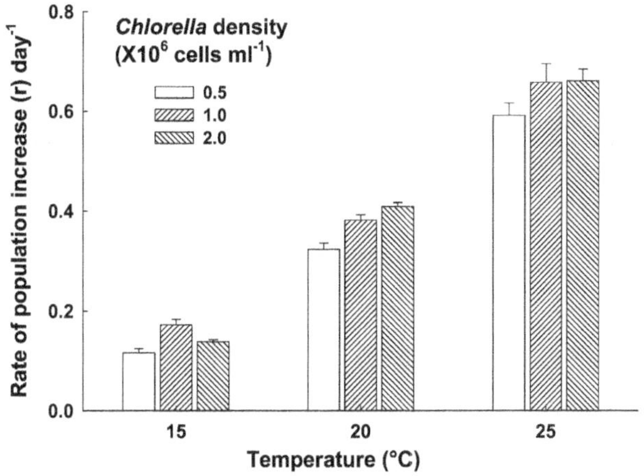

Figure 3. Rate of population increase (day^{-1}) of *B. havanaensis* cultured using three different food densities (0.5×10^6, 1.0×10^6 and 2.0×10^6 cells ml^{-1} of *Chlorella*) and each of these at three temperatures (15, 20 and 25 °C). Shown are the mean ± SE of 6 replicates.

from 15 to 25 °C, the average lifespan decreased by 2/3. Our observation that an increase in temperature by about 10 °C reduced the rotifer lifespan considerably is similar to reports on other rotifer species (Sarma & Rao, 1991; Sanoamuang, 1993). Regardless of food concentration and temperature, the average lifespan of brachionid rotifers appears to vary from 10 to 25 days at temperatures of 20–25 °C (Nogrady et al., 1993); our study indicates that *B. havanaensis* also falls within this range. Age-specific life expectancy generally decreases with increasing age of the individuals in a cohort population; in certain cases, however, elevated values at certain days during the lifespan have been documented (Sarma et al., 2003b). We also observed this for *B. havanaensis* at 20 °C, especially at higher food level.

Age-specific fecundity curves observed for *B. havanaensis* are similar to those for other brachionid rotifers, e.g. *B. calyciflorus* (Halbach, 1973) and *B. plicatilis* (King, 1982). The pattern of offspring production in rotifers appears to depend on the conditions in which they are reared (Nogrady et al., 1993). For example, at high temperatures, the duration of egg laying and the rate of egg production are rapid. On the other hand, at

Table 2. Statistical evaluation using two-way analysis of variance (ANOVA) on selected life history traits of *B. havanaensis* fed *Chlorella* at three densities under three temperatures (see Table 1)

Source	DF	SS	MS	*F*-ratio
Average lifespan				
Temperature (A)	2	601.148	300.57	383.67 ***
Food density (B)	2	5.749	2.87	3.67 **
Interaction of A × B	4	6.247	1.56	1.99 ns
Error	45	34.471	0.77	
Gross reproductive rate				
Temperature (A)	2	874.71	437.35	58.38 ***
Food density (B)	2	81.314	40.66	5.43 ***
Interaction of A × B	4	23.78	5.94	0.79 ns
Error	45	329.516	7.32	
Net reproductive rate				
Temperature (A)	2	128.304	64.15	63.4 ***
Food density (B)	2	17.991	9.00	8.89 ***
Interaction of A × B	4	5.729	1.43	1.42 ns
Error	45	44.52	0.99	
Generation time				
Temperature (A)	2	354.664	177.33	1135.73 ***
Food density (B)	2	1.387	0.69	4.44 **
Interaction of A × B	4	0.938	0.23	1.5 ns
Error	45	6.87	0.15	
Rate of population increase				
Temperature (A)	2	2.208	1.1	531.48 ***
Food density (B)	2	0.043	0.02	10.28 ***
Interaction of A × B	4	0.008	0.0	0.99 ns
Error	45	0.091	0.0	
Highest value of egg ratio (eggs/female)				
Temperature (A)	2	1.036	0.52	14.55 ***
Food density (B)	2	0.827	0.41	11.62 ***
Interaction of A × B	4	0.189	0.05	1.33 ns
Error	45	1.566	0.03	

DF = degrees of freedom, SS = sum of squares, MS = mean square, F = F-ratio..Levels of significance: ***$p < 0.001$; **$p < 0.01$; ns = non-significant ($p > 0.05$).

lower temperatures, both these are lower. This is evident in the fecundity of *B. havanaensis*, where egg production was observed for about 20 days at 15 °C and this was reduced to 15 and 10 days at 20 and 25 °C, respectively. The number of eggs daily produced by *Brachionus* varies from 2 to 6 eggs (Duncan, 1989). We observed about 1–5 eggs female^{-1} day^{-1}. Enhanced egg production with increasing food concentration, as observed in *B. havanaensis*, is similar to *B. calyciflorus* (Guisande & Mazuelos, 1991) but differs in *B. variabilis* where an inverse relation has been reported (Sarma & Nandini, 2001). Gross and net reproductive rates of *B. havanaensis* were lower than those reported for *B. calyciflorus* but similar to those reported for *B. macracanthus* (Sarma & Nandini, 2002). A 15 °C food level did not enhance gross or net reproductive rates even though the food was not limiting, which suggests that temperature had an overriding influence on the offspring production.

The rate of population increase (r) is considered to be influenced by factors like food level, temperature or toxicants (Forbes & Calow, 1999). While *Brachionus* has *r* values in the range of 0.1–2.0 day^{-1}, many species have $r < 0.5$ day^{-1} (Sarma et al., 2001). Our *r*-values varied by a factor of 6 (0.11–0.66 day^{-1}). Since both net reproductive rate and the *r* were positive, *B. havanaensis* appears to be capable of increasing its population density even at low temperature.

The lack of a significant influence of the interaction between food level and temperature on any of the tested life history traits of *B. havanaensis* is an interesting observation. In our life table study, neonates were continuously removed and hence the surviving adults did not experience resource limitations as observed in the case of population dynamics studies. Thus, even if both population growth and life table demography studies are conducted under similar conditions, some factors have a greater effect than the others e.g., rate of population increase, as documented for other brachionid rotifers (Sarma & Rao, 1990, 1991; Sarma et al., 2003b). The lack of a significant interaction of food level and temperature on the life history traits of *B. havanaensis* found by us may be attributed to this phenomenon. Lastly, the fact that *B. havanaensis* showed enhanced reproductive output at 20 and 25 °C but much lower output at 15 °C, suggests that this species is adapted to tropical conditions, though it is capable of maintaining itself at temperatures lower than 15 °C (Flores-Burgos et al., 2003).

The strain of *B. havanaensis* used in this study was apparently adapted to higher temperatures. The increase in temperature and the algal food level enhanced egg output which was eventually reflected in higher net reproductive and population growth rates. The survival patterns and lower rates of reproduction at 15 °C suggest that the winter temperatures (10–15 °C) prevailing in many waterbodies in Mexico City allow this species to flourish and maintain its population until spring and summer optima.

Acknowledgements

One of us (ELPM) thanks CONACyT (Mexico) for a scholarship and the Postgraduate Programme of Limnology and Marine Sciences (ICMyL) for travel grants for attending the conference.

References

Ahlstrom, E. H., 1940. A revision of the rotatorian genera *Brachionus* and *Platyias* with descriptions of the new species and two new varieties. Bulletin of the American Museum of Natural History 57: 143–184.

Anaya-Soto, A., S. S. S. Sarma & S. Nandini, 2003. Longevity of the freshwater anostracan Streptocephalus mackini (Crustacea: Anostraca) in relation to food (*Chlorella vulgaris*) concentration. Freshwater Biology 48: 432–439.

Anonymous, 1985. Methods of Measuring the acute Toxicity of Effluents to Freshwater and Marine Organisms. US Environment Protection Agency. EPA/600&4–85/013.

Borowitzka, M. A. & L. J. Borowitzka, 1988. Micro-algal Biotechnology. Cambridge University Press, London.

De Mott, W. R., R. D. Gulati & E. Van Donk, 2001. Effects of dietary phosphorus deficiency on the abundance, phosphorus balance, and growth of *Daphnia cucullata* in three hypereutrophic Dutch lakes. Limnology and Oceanography 46: 1871–1880.

De Ridder, M., 1981. Some considerations on the geographical distribution of rotifers. Hydrobiologia 85: 209–225.

Duncan, A., 1989. Food limitation and body size in the life cycle of planktonic rotifers and cladocerans. Hydrobiologia 186/187: 11–28.

Edmondson, W. T., 1965. Reproductive rate of planktonic rotifers as related to food and temperature. Ecological Monographs 35: 61–111.

Enrique-García, C., S. Nandini & S. S. S. Sarma, 2003. Food type effects on the population growth patterns of littoral rotifers and cladocerans. Acta Hydrochimica et Hydrobiologica 31: 120–133.

Forbes, V. E. & P. Calow, 1999. Is the per capita rate of increase a good measure of population-level effects in ecotoxicology? Environmental Toxicology and Chemistry 18: 1544–1556.

Garza-Mouriño, G., M. Silva-Briano, S. Nandini, S. S. S. Sarma & M. E. Castellanos-Páez, 2005. Morphological and morphometrical variations of selected rotifer species in responce to predation: a seasonal study of selected brachionid species from Lake Xochimilco (Mexico). Hydrobiologia 546 169–179.

Guisande, C. N. & Mazuelos, 1991. Reproductive pattern of *Brachionus calyciflorus* Pallas at different food concentrations. Journal of Plankton Research 13: 279–286.

Gulati, R. D. & W. R. DeMott, 1999. The role of food quality for zooplankton: remarks on the state-of-the-art, perspectives and priorities. Freshwater Biology 38: 753–768.

Halbach, U., 1973. Life table data and population dynamics of the rotifer *Brachionus calyciflorus* Pallas as influenced by periodically oscillating temperature. In Wieser, W (ed.), Effects of Temperature on Ectothermic Organisms. Springer Verlag, Berlin, 219–228 pp.

Flores-Burgos, F., S. S. S. Sarma & S. Nandini, 2003. Estudio preliminar sobre la fauna de rotíferos de Xochimilco (México). El agua de cuenca de México. Sus problemas históricos y perspectivas de solución. Proceedings of the International Conference on XochimilcoEcological Park of Xochimilco, U.A.M. Xochimilco, Mexico City, Mexico, Vol. 1, 163–171.

King, C. E., 1982. The evolution of lifespan. In Dingle Hegmann, H. J. P. (ed.) Evolution and Genetics of Life Histories. Springer Verlag, New York: 121–128 pp.

Krebs, C. J., 1985. Ecology (3rd edn). Harper & Row, New York, 800 pp.

Nandini, S., R. Pérez-Chávez & S. S. S. Sarma, 2003. The effect of prey morphology on the feeding behaviour and population growth of the predatory rotifer *Asplanchna sieboldi*: a case study using five species of *Brachionus* (Rotifera). Freshwater Biology 48: 2131–2140.

Nogrady, T., R. L. Wallace & T. W. Snell, 1993. Rotifera 1. Biology, Ecology and Systematics. Guides to the Identification of the Microinvertebrates of the Continental Waters of the World, Vol. 84. SBP Academic Publishers, The Hague 142 pp.

Ramírez-García, P., S. Nandini, S. S. S. Sarma, E. Robles-Valderrama, I. Cuesta & D. Hurtado-Maria, 2002. Seasonal variations of zooplankton abundance in the freshwater reservoir Valle de Bravo (Mexico). Hydrobiologia 467: 99–108.

Ramos-Rodríguez, E. & J. M. Conde-Porcuna, 2003. Nutrient limitation on a planktonic rotifer: life history consequences and starvation resistance. Limnology and Oceanography 48: 933–938.

Rothhaupt, K. O., 1990. Population growth rates of two closely related rotifer species: effects of food quantity, particle size, and nutritional quality. Freshwater Biology 23: 561–570.

Sanoamuang, L. O., 1993. The effect of temperature on morphology, life history and growth rate of *Filinia terminalis* (Plate) and *Filinia* cf. *pejleri* Hutchinson in culture. Freshwater Biology 30: 257–267.

Sarma, S. S. S. & T. R. Rao, 1990. The population dynamics of *Brachionus patulus* Müller in relation to food and temperature. Proceedings of the Indian Academy of Sciences (Animal Sciences) 99: 335–343.

Sarma, S. S. S. & T. R. Rao, 1991. The combined effects of food and temperature on the life history parameters of *Brachionus patulus* Muller (Rotifera). Internationale Revue der gesamten Hydrobiologie 76: 225–239.

Sarma, S. S. S. & S. Nandini, 2001. Life table demography and population growth of *Brachionus variabilis* Hampel, 1896 in relation to algal (*Chlorella vulgaris*) density. Hydrobiologia 446/447: 75–83.

Sarma, S. S. S. & S. Nandini, 2002. Comparative life table demography and population growth of *Brachionus macracanthus* Daday, 1905 and *Platyias quadricornis* Ehrenberg, 1832 (Rotifera, Brachionidae) in relation to algal (*Chlorella vulgaris*) food density. Acta Hydrochimica et Hydrobiologica 30: 128–140.

Sarma, S. S. S., M. A. Fernández-Araiza & S. Nandini, 1999. Competition between *Brachionus calyciflorus* Pallas and *Brachionus patulus* (Müller) (Rotifera) in relation to algal food concentration and initial population density. Aquatic Ecology 33: 339–345.

Sarma, S. S. S., P. S. Larios-Jurado & S. Nandini, 2001. Effect of three food types on the population growth of *Brachionus calyciflorus* and *Brachionus patulus* (Rotifera: Brachionidae). Revista de Biología Tropical 49: 75–82.

Sarma, S. S. S., E. L. Pavón-Meza & S. Nandini, 2003a. Comparative population growth and life table demography of the rotifer *Asplanchna girodi* at different prey (*Brachionus calyciflorus* and *Brachionus havanaensis*) (Rotifera) densities. Hydrobiologia 491: 309–320.

Sarma, S. S. S., H. E. Trujillo-Hernández & S. Nandini, 2003b. Population growth of herbivorous rotifers and their predator (*Asplanchna*) on urban wastewaters. Aquatic Ecology 37: 243–250.

Sokal, R. R. & F. J. Rohlf, 2000. Biometry. W.H. Freeman and Company, San Francisco.

Torres-Orozco, B. R. E. & S. A. Zanatta, 1998. Species composition, abundance and distribution of zooplankton in a tropical eutrophic lake: Lake Catemaco, Mexico. Revista de Biología Tropical 46: 285–296.

Hydrobiologia (2005) 546:361–373 © Springer 2005
A. Herzig, R.D. Gulati, C.D. Jersabek & L. May (eds.) Rotifera X: Rotifer Research: Trends, New Tools and Recent Advances
DOI 10.1007/s10750-005-4247-6

Factors affecting egg-ratio in planktonic rotifers

S.S.S. Sarma[1,*], R.D. Gulati[2] & S. Nandini[3]
[1]*Laboratory of Aquatic Zoology, Building UMF, Division of Research and Postgraduate Studies, National Autonomous University of Mexico, Campus Iztacala, AP 314, CP 54090, Los Reyes, Tlalnepantla, State of Mexico*
[2]*The Netherlands Institute of Ecology, Centre for Limnology, Rijksstraatweg 6, 3631 AC Nieuwersluis, The Netherlands*
[3]*UIICSE, Division of Research and Postgraduate Studies, National Autonomous University of Mexico, Campus Iztacala, AP 314, CP 54090, Los Reyes, Tlalnepantla, State of Mexico*
*(*Author for correspondence: E-mail: sarma@servidor.unam.mx)*

Key words: egg ratio, Rotifera, culture conditions, fecundity

Abstract

Edmondson's egg ratio (number of amictic eggs per female) is an important life history variable, which has been in wide use to understand and predict patterns of population growth in planktonic rotifers under field conditions. It is also useful as an indicator of the health of rotifers under culture conditions. Generally, an inverse relationship exists between the egg ratio and the density of females in a population. A number of abiotic and biotic factors influence the egg ratio. For example, temperature can cause marked changes in the egg ratio by influencing the frequency of egg production and the hatching times of parthenogenetic eggs. Also, preferential feeding on ovigerous females of rotifers by invertebrate predators such as *Asplanchna* will lower the egg ratios of the population. The easy detachment of eggs, as may be the case in some members of the Brachionidae especially during enhanced reproduction when food levels are high, may also cause an underestimation of the egg ratio. In this review, we discuss the egg ratio of selected rotifer species in relation to the role of diel changes in egg production, the frequency or the intensity of feeding, the problems of distinguishing between different egg types and the negative effect of stressors such as toxicants and diet quality.

Introduction

In freshwater lakes and ponds, rotifers, cladocerans and copepods are the most common groups of zooplankton (Thorp & Covich, 2001). On the basis of biomass, the metazoan zooplankton almost always dominates in freshwater bodies (Downing & Rigler, 1984). Since eggs are the potential offspring, their numbers could be used to understand trends in the population growth curves of zooplankton species. An inverse relationship exists between the depth of Secchi-disc transparency and rotifer densities in many waterbodies (Edmondson, 1957). The lower the Secchi transparency, the higher is the number of eggs per female. Thus, the egg numbers in zooplankton are largely related to the concentration, i.e. availability of food in the environment (Dolan & Gallegos, 1992; Urabe, 1992).

Edmondson (1960) was among the first to use the relationship between the number of parthenogenetic eggs and the number of females (called egg ratio, or ER) as an indicator for predicting changes in natural zooplankton populations (Bengtsson, 1993; Ohman et al., 1996; Razlutskij, 2000). Together with the developmental time of embryo, ER is used to calculate birth and death rates for deriving the population growth rates (Edmondson, 1965; Paloheimo, 1974). At present the ER method is considered to be a very important tool for estimating natality and mortality rates under field conditions, and it is probably among the most significant contributions of Edmondson to life history studies on zooplankton (Lehman, 2000).

Even though the ER method was primarily developed for use under field conditions, where its applicability is more reliable, laboratory studies have used this approach to compare the predicted and observed growth rates. This, coupled with mathematical modelling and computer simulations, have greatly improved the applicability of ER for estimating population growth rates (Keen & Nassar, 1981; Bennett & Boraas, 1989; Devetter & Sed'a, 2003). Population growth rate experiments in the laboratory are simplified and allow separate measurement of both the natality and mortality rates. There are two approaches for obtaining these rates (Krebs, 1985): the first involves the life-table method, where a cohort population is followed until all test individual have died and the number of eggs (offspring) produced have been counted and removed. The second approach involves a population-growth approach, in which the quantification of natality and mortality is more complicated because of the uncertainty of the age of reproducing females. Using a growth indicator, e.g. body-size, it is possible to distinguish between neonates and adults or even between ages of the adults after death (Threlkeld, 1983; Sarma, 1996). Both the life table and population-growth approaches can be used under field and laboratory conditions. However, under field conditions, only certain rotifer taxa (e.g. members of Collothecidae) are suitable for this purpose. This is due to the difficulties of quantifying egg production in certain taxa. For example, in non-planktonic taxa, such as *Lecane* or *Lepadella*, parthenogenetic eggs are deposited (e.g. on vegetation, benthos etc.) rather than carried by the female and are thus difficult to quantify.

Review of egg ratio studies from selected field and laboratory studies

Information on the use of and the problems associated with egg ratios for rotifers under field conditions are given in Edmondson (1965). Also, some other reviews (e.g. Dumont, 1977; Devetter & Sed'a, 2003) demonstrate that the usefulness of ERs based on field samples is limited by: (a) methodological difficulties with the data, (b) weak interpretation of the results due to an inadequate number of related studies and (c) the absence of affirmative laboratory tests. Nevertheless, some of the previous studies (e.g., Ohman et al., 1996; Razlutskij, 2000) formed a strong basis for constant improvement in the use of ER method for zooplankton.

Differences between field and laboratory studies

There is generally an inverse, linear or curvilinear relationship between population density and ER. At low population densities, when resources are not limiting, planktonic rotifer populations continue to produce between 3 and 5 eggs per female per day (Duncan, 1984; Sarma & Rao, 1991). However, as resources become limiting, population growth reaches a stationary phase and parthenogenetic egg production declines (Fig. 1), which finally results in an inverse relation between population density and ER (Ooms-Wilms et al., 1999). Even if food is added at regular intervals (e.g., 24 h), as is common in laboratory studies (Nandini et al., 2002), it is possible that the test animals face short periods of low food availability between food additions, particularly when population densities are high. Changes in other experimental conditions such as temperature and food concentrations may also affect the frequency of egg production and hatching. So, ER can vary periodically.

Field collections of rotifers sometimes yield very high or very low densities of rotifers if sampling is random and infrequent. For example, in Australian billabongs, *B. plicatilis* may occur in densities as high as 50,000 ind. l^{-1} (Shiel & Koste, 1986). If sampled inadequately, high or even low density estimates based on field samples may be mistaken for patchy distribution of zooplankton. Such a problem can be clarified using the ER method. If high population densities are correlated with a decrease in the ER, the role of patchy distribution could be ruled out (Fuller et al., 1977). In addition to the problems associated with sampling size and sampling frequency, rotifers in natural waterbodies may be infected with parasites. If such an infection is heavy, the egg output of the rotifers will be reduced, causing a decrease in the ER. Laboratory cultures are, however, usually free

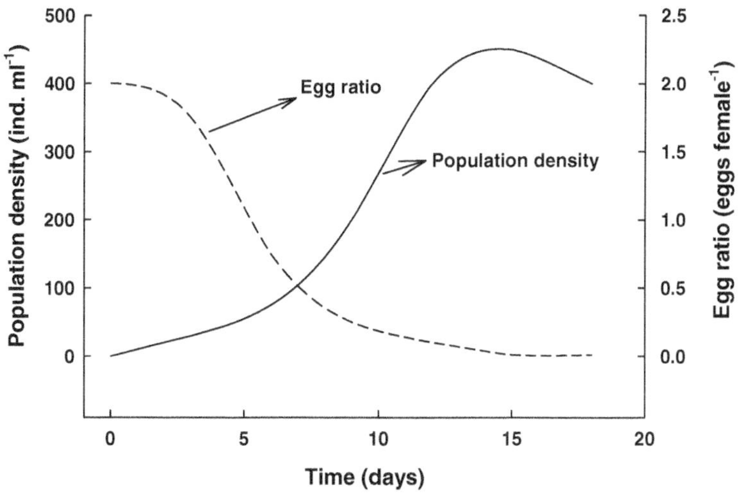

Figure 1. Schematic representation of changes in population density and egg ratio of brachionid rotifers in relationship to time in typical batch cultures.

from parasites and infection is not so severe as to adversely affect the egg output.

Comparison of fecundity and egg ratio

Egg ratio for a population is usually derived from the total number of parthenogenetic eggs and the total number of females in this population (Edmondson, 1960, 1965). So, this method includes all females irrespective of their age as the potential contributors of eggs. However, this assumption is not strictly true because not all females are capable of producing eggs (Sarma, 1996). Consequently, ERs based on both field or laboratory populations *per se* underestimate fecundity per female at the individual level. Some workers believe that ER derived from a field collection is closer to actual fecundity than that derived for a laboratory population because field populations tend to be younger at first reproduction and have a near-absence of post reproductive females for most genera of planktonic rotifers (Urabe, 1992). This is especially true at temperatures above 20 °C, where both the duration of egg hatching and time to maturity are short. At lower temperatures, more neonates (and eggs) than reproducing females may be found because of the delay in the onset of reproduction and egg hatching (see below). For temperate taxa, the influence of low temperature on egg output may differ from that observed for tropical species because cold stenothermal taxa have some adaptation to lower temperatures (Amren, 1964).

The problem of deriving true fecundity still remains unresolved, especially for field animals. Some approaches based on body size structure are probably reliable for loricate species. Sarma (1996) classified egg-bearing adults of five *Brachionus* species into different size frequencies using the length of the smallest egg-bearing female as the threshold size for egg bearing adults. He used this criterion to divide the entire population into adults and juveniles. The quantified egg numbers were then used to derive ER (Table 1). As expected, the ERs derived from the total rotifer population were lower than the actual fecundity values. However, this approach, too, has certain limitations (e.g., lorica sizes may vary depending on whether the neonates hatched from resting eggs or from parthenogenetic eggs – see Sarma, 1996). The data on ER derived using the conventional method and that from the modified method are strongly correlated (Fig. 2).

Life table compared to population growth studies

Egg per female ratios may also depend on the source of the data collected. ER data obtained from population growth give information on the eggs per female as well as the population density. During the exponential phase of growth rotifers have a higher egg production rate per female, so

Table 1. Egg ratios (eggs adult^{-1}) for some species of the Brachionidae. The percentage of adults was derived by size, taking into consideration the smallest egg-bearing individual from each species (Data after Sarma, 1996)

Species	Number of females	Number of adults	Total number of eggs	Modified egg ratio (eggs adults^{-1})
Brachionus calyciflorus	100	59	45	0.76
Brachionus diversicornis	100	69	78	1.13
Brachionus rubens	100	51	64	1.26
Kellicottia longispina	100	85	9	0.11
Keratella quadrata	100	90	25	0.28
Pompholyx sulcata	100	81	33	0.41

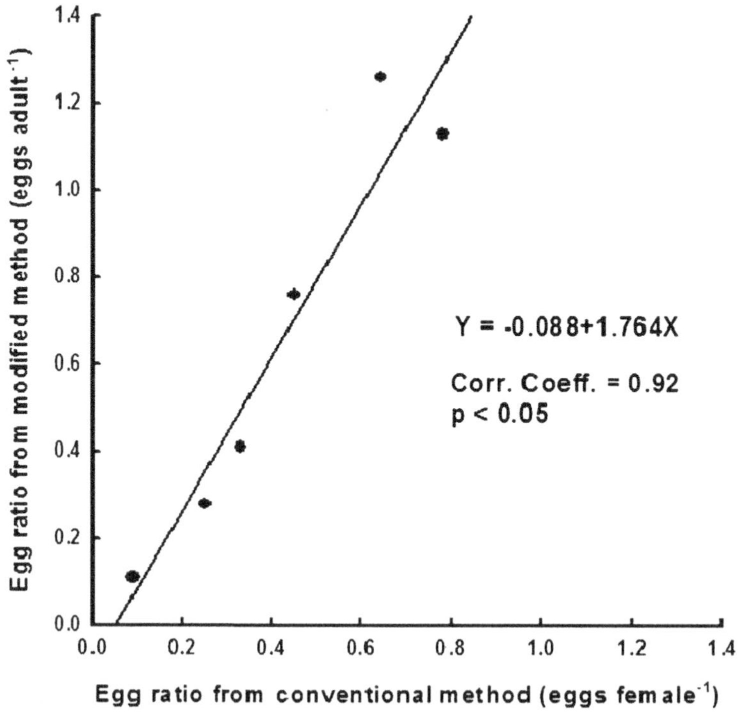

Figure 2. Relationship between egg-ratios derived using two different methods. In the conventional method, the egg ratio was calculated assuming that all females were contributors to egg production. In the second method, only adult females were considered (data source Table 1).

the ER also would be higher. In natural waters, ER usually ranges between almost 0 and 2, depending on the food abundance (Devetter & Sed'a, 2003). Under culture conditions, ER can range from nearly zero, as during the declining phase of the population, to as high as three, if food concentration, temperature and population density conditions are good (Sarma, 1987). Elevated ERs among rotifers are more frequently observed in life table studies because of the absence of intraspecific competition (Sarma & Rao, 1991). Thus, regardless of food concentration and temperature, ER from life tables generally exceeds 0.5. This is probably an underestimate at the individual level, because it also includes the surviving post-reproductive individuals.

While applying ER to rotifers, it is important to state explicitly if the study is aimed at the populations or individuals. For example, in the life table method involving individuals of the test species, the number of eggs produced reflects true fecundity at the individual level. In contrast, at the population level even after discarding juveniles, the derived ERs may not represent the true fecundity of an average individual, primarily because of the inclusion of non egg-bearing females. If the aim of a study is to calculate birth rates and death rates of a population, juveniles with zero fecundity must be included while deriving ER. For such a study, we must also recognize the factors that influence the egg count in rotifers. It is, therefore, important to consider the factors influencing egg production and retention by the female, as well as the female density. These can be easily understood from both life table and population methods.

Advantages and limitations of the ER method

The main advantage of the ER is that it enables us to predict the oscillations and growth rates of populations. Some workers (e.g., Polishchuk, 1978; Wolfinbarger, 1999) have shown that the measured growth rates for rotifer species and those predicted for them using the ER method are quite similar. This is probably because many planktonic species carry eggs either outside their body (such as many members of Brachionidae) or inside (e.g., *Asplanchna*). The egg output of some littoral taxa, e.g. tube-dwelling rotifers such as *Ptygura* and *Collotheca*, can also be measured since the number of offspring can be more or less correctly quantified (Edmondson, 1965). However, there are limitations in applying this method for many rotifer species. Some of these limitations are discussed here.

Differences in the reproductive strategies and mode of life

As a general rule, non-planktonic and littoral taxa do not carry eggs attached to the posterior end of the body, but planktonic taxa do (Gilbert, 1983). Members of the Euchlanidae, Lecanidae, Mytilinidae and Colurellidae, being predominantly littoral or benthic, deposit their eggs on substratum or attach them to vegetation (Koste, 1978). However, for some other families of rotifers, generalizations of egg-carrying trends (and hence ERs) are difficult to derive. An example of this is the genus *Epiphanes*. While *Epiphanes brachionus*, though planktonic, does not carry its eggs, both *Epiphanes clavulata* and *Epiphanes macrourus* do. *Epiphanes senta*, a non-planktonic species, deposits its eggs on macrophytes (Ruttner-Kolisko, 1974). A similar situation exists in the Brachionidae. Although planktonic, *Notholca* does not carry eggs but other planktonic genera such as *Brachionus*, *Keratella* and *Anuraeopsis* do. *Platyias*, being non-planktonic, does not carry its eggs (Pontin, 1978). Families such as Trichocercidae, though typically planktonic, which rarely carry their eggs, often attach them to other planktonic organisms. It is not known why certain species living in the water column carry their eggs, while others do not. Gilbert (1983) suggested some general advantages of egg carrying: (a) the eggs are more protected from predation, (b) the eggs are exposed to better conditions of dissolved oxygen through the constant swimming activities of the females, and (c) in deep lakes, the eggs are prevented from sinking into the anoxic bottom layers where, coupled with low temperatures, hatching success may be reduced.

Difficulties in identifying different egg types

Although there have been considerable advances in the use of the ER method for interpreting population growth rates, problems related to differentiating the egg types are still unsolved in many planktonic rotifer species (Koste, 1978). Since the hatching of parthenogenetic eggs causes rapid population increase (Nogrady et al., 1993), it is crucial to identify the eggs correctly for deriving ER. In general, most rotifers produce three types of eggs: parthenogenetic eggs, and eggs hatching into males and resting eggs. For most rotifer species, resting eggs are the largest in size, followed by the parthenogenetic eggs and male eggs (Gilbert, 1983). Unfortunately, for most rotifer species, information on the different egg types is either not reported or difficult to obtain. Often, even for the more common genera such as *Keratella* and *Kellicottia*, only parthenogenetic eggs are known (Ruttner-Kolisko, 1974). Another problem is the size of the rotifer itself. In general, smaller the rotifer species produce smaller eggs. This relationship is generally curvilinear (Walz et al., 1995). So, for many smaller species, the differences among egg types (parthenogenetic, male and resting eggs) are not easy to distinguish. For example,

Proalides (*Liliferotrocha*) has resting eggs no more than 40 μm in size, parthenogenetic eggs of 25 μm, and male eggs about 20 μm in size. Such small differences are not easy to observe, even by an experienced eye. Even in some larger species (e.g. *Epiphanes macrourus*), the size differences among the egg types are not easily discernible.

Neonates hatched from resting eggs differ from those of parthenogenetic eggs

Differences in food concentrations for field animals and those from laboratory cultures, as mentioned earlier, have implications for deriving a relationship between population density and ER. Although food concentrations under field conditions tend to remain almost stable for periods longer than the generation times of many planktonic rotifers (Lampert & Sommer, 1997), reproductive rates may vary dramatically within this time (Amren, 1964; Devetter & Sed'a, 2003). This is particularly true for rotifers in ephemeral waterbodies or in lakes that freeze in the winter months. Growth of natural rotifer population in temporary waterbodies typically starts with the hatching of resting eggs (Pourriot & Snell, 1983). The reproductive potential of these neonates differs from those produced through parthenogenesis. Generally neonates from resting eggs are larger and have higher reproductive growth rates than those from parthenogenesis (Sarma, 2000). So, first generation adults originating from resting eggs usually higher ERs, sometimes up to four parthenogenetic eggs per female in some genera such as *Polyarthra* (Amren, 1964), than those from parthenogenetic eggs. This may cause high variability in intraspecific growth rates, even under constant food concentration and temperature conditions. In addition, growth rates from laboratory and field populations may differ, especially during the first population cycle.

Role of mother's body volume and egg volume in the ER

Egg ratio does not account for the ratio between egg volume and that of the mother. For example, in relatively small-sized forms such as *Anuraeopsis*, the ratio of egg volume to mother's volume is much higher (0.18) than for larger taxa, e.g., *Epiphanes clavulata* where this ratio is 0.04% (Walz et al., 1995). If such discrepancies are offset by differences in egg-laying or hatching frequencies, ER can be used for different rotifer species regardless of the ratio of egg volume to mother's body volume. Data available in the literature on rates of egg laying (Pourriot, 1983) and egg hatching (Foran & King, 1981; Herzig, 1983) for different genera of rotifers are not sufficient to evaluate the significance of egg volumes in the use of ER method. However, this is an important aspect to consider for deriving ER for different rotifer species.

ER and population size

Egg ratio, being a relative index of egg-bearing females within a population, disregards population size and is, therefore, not a strict quantitative parameter for expressing the reproductive status or potential of a given population. There are also certain field observations where the density of rotifers is unrelated to the ER (Devetter & Sed'a, 2003). One can derive ER by quantifying the relative proportion of egg-bearing females within the total female population rather than derive it *per se* from a population that is collected quantitatively (e.g. see Sarma, 1996). The question of sample size and variability in the accuracy of ER, and its consequences on other variables such as death rates, have been addressed by DeMott (1980) and Razlutskij (2000). Even though a quantitative sample is not needed for deriving ER, one cannot ignore the fact that the relative proportion of females in the population must be derived from a larger, rather than a smaller, sample size. If this criterion is adequately met, ER will be a useful indicator for assessing the quantity of food available for rotifers under culture conditions (Snell et al., 1987).

Egg loss versus food concentration

Although egg ratios are quantifiable for most rotifer taxa under field conditions, the eggs are more easily detachable in some species (e.g., *Polyarthra*) than others (e.g., *E. macrourus*) and their numbers can be underestimated. There are few estimates of the loss of eggs in natural waters, but laboratory studies reveal that loose eggs may constitute up to 10% of the total production (Sarma, 1987). Under field conditions, this percentage could be much higher (Ooms-Wilms et al., 1999). Dumont et al. (1995), who cultured *Anuraeopsis fissa* using a wide range of algal (*Scenedesmus obliquus*) densities (0.5×10^6 to 8×10^6 cells ml^{-1}), found not only that ER decreased with increasing densities of *A. fissa*, but also frequency of egg detachment increased as food level increased.

Use of ER in aquaculture

The fact that ERs are directly related to the quantity of food available in the medium has found its application in aquaculture where it is used to assess rotifer health. Snell et al. (1987) were among the first to successfully use this approach. Korstad et al. (1995) observed that low egg ratios in *Brachionus plicatilis* (0–0.17 eggs $female^{-1}$) indicated reduced or even negative rates of population growth. This is important because, if growth rates decrease to below a certain threshold level, mortality exceeds reproduction and the population begins to crash. A declining ER clearly marks the start of such a situation (Threlkeld, 1983). In aquaculture, an ER of about 0.2 is considered to be a threshold level below which the population is likely to crash and above which there will be positive growth. It should be noted that the threshold ER value will depend not only on the rotifer species but also on food concentration, the consistency of food availability, and the density of animals in the culture. It is necessary to derive the threshold ER value for each rotifer species separately, since certain rotifer species such as *B. calyciflorus* (Sarma et al., 1999) always have higher ER than others (e.g., *A. fissa*: Dumont et al., 1995).

Post versus pre-reproductive females

The importance of population age structure in egg ratio studies has not received much attention. Age has a strong influence on ER. For example, under natural conditions, synchronous hatching of resting eggs can yield a large number of neonates that, upon maturing, may possess up to 4 eggs $female^{-1}$ (Amren, 1964). On the other hand, when food concentrations are very low, individuals may become too weak to reproduce, even on the return of favourable conditions. At this stage, ER becomes very low and may even reach zero. This is more easily quantified using a life-table approach: a cohort of one age group can become senile almost simultaneously (Castellanos-Páez et al., 1999).

Influence of abiotic and biotic factors on the egg ratio

Photoperiod, temperature and salinity influence egg production, hatching and maturity of a population of rotifers, thereby causing changes in the relationship between ER and the population density. Photoperiod largely influences the biological rhythm. In nature, certain rotifer species show diel changes in reproduction. For example, Bosch & Ringelberg (1985) found a diurnal pattern of ERs in two planktonic rotifer species (*Keratella cochlearis* and *Kellicottia longispina*) in Lake Maarsseveen (Netherlands). The highest ER values were recorded at night and the lowest in the afternoon. Whereas most laboratory populations show adaptation to light regimes, some species, such as *Hexarthra bulgarica*, have different patterns of offspring production during night and day. In the field, Cruz-Pizarro et al. (1998) noted that egg production increased during the night in the lakes of Sierra Nevada and that oviposition of amictic eggs by *H. bulgarica* was also higher in the pelagic zone than in the littoral area. There is some evidence that, if photoperiod is regulated, ERs may vary considerably (Ringelberg & Steenvoorden, 1986) regardless of the adaptation of cultures to light conditions. Ovigerous and non-ovigerous females respond differently to light resulting in different ER at the population level (Magnien & Gilbert, 1983).

More than 90% rotifer species live in freshwater, and salinities >2 g l^{-1} are associated with strong reductions in rotifer diversity and abundance in natural waters (Green, 1993). However, some rotifer species can gradually adapt to salinities >6 g l^{-1} (Herzig & Koste, 1989). Nevertheless, several rotifers, including *B. calyciflorus* and *B. patulus*, cannot produce eggs at such a high level of salinity (Peredo-Alvarez et al., 2003). Some rotifer species may, however, survive at higher salinities but without producing eggs, irrespective of their density. This is apparently because the costs associated with the maintenance of osmoregulation at higher salinities are so high that egg production is not feasible. Changes in other abiotic factors such as pH, dissolved oxygen levels and ammonia concentrations within the ranges generally prevalent in waterbodies do not, generally, exert a strong influence on the ER in zooplankton, including rotifers (Saunders et al., 1999). However, when close to their LC_{50} levels, these variables affect reproduction more drastically than the survival.

Temperature influences ER primarily through embryonic and post-embryonic developmental times (Herzig, 1983). At lower temperatures, there is a delay in the onset of reproduction, embryonic egg development and hatching time (Halbach & Halbach-Keup, 1974), all of which have an impact on ER; the longer the eggs take to hatch, the longer they are retained in the body. This changes the magnitude of the correlation between population density and ER (Fig. 3). At high temperatures, with egg hatching times less than 12 h, and a frequency of data collection of 24 h, the population increase will be rapid, despite the low number of eggs carried by a female (Sarma & Rao, 1990). Reliability of the ER will thus depend on how close to the hatching period the samples were taken. Magnien & Gilbert (1983) have reported that *Keratella crassa* shows synchronous hatching in summer, with up to 93% of eggs hatching within a 7 h period.

Among the biotic factors, the influence of predation, especially of egg-carrying females, on ERs is of considerable interest in rotifer studies. If predation on a rotifer population is random, the changes in the ERs will be small, but if the predator selectively preys on egg-bearing adults or on eggs, ER will be underestimated. Because rotifer eggs are attached loosely to the mother's body (Gilbert, 1983), the strong currents created by raptorial predators such as anostracans while capturing and eating their prey will result in detachment and loss of some eggs, which will then enter the water. Subsequently, these detached eggs may be counted and attributed to prey that have not been eaten, so that ER will be overestimated. Such discrepancies have come to light during observations on the feeding of the anostracan *Streptocephalus proboscideus* on *Anuraeopsis fissa* (Dierckens et al., 1995). Errors in the ER are more likely to occur in the preferred prey, i.e. *Brachionus* and *Keratella*. In addition, *Asplanchna girodi* feeding selectively on egg bearing females of *Keratella cochlearis* (Conde-Porcuna & Sarma, 1995) led to overestimates of ER compared with the ERs determined in the absence of predation.

Both food type and food concentration have a great impact on rotifer survival and reproduction. Increased egg output with increasing algal concentration is the general rule for most zooplankton species (Dumont et al., 1995; Nandini & Sarma, 2001). However, if food quality is poor (Halbach, 1972), or if the food alga develops defence mechanism such as toxic strains or spines, this positive relationship between food density and egg production will no longer hold. Cyanobacteria as food have been reported not only to inhibit reproduction but also to reduce the lifespan (Nandini, 2000). Feeding on certain food types, e.g. normal yeast for prolonged periods, may lead to the production of empty eggs as for *B. patulus* and *B. calyciflorus* (Sarma et al., 2001). Such eggs are not useful for deriving ERs. In contrast, adding selected bacteria to axenic diet, may cause enhanced ERs, as observed by Rombaut et al. (1999) for *Brachionus plicatilis*. Similarly, the addition of algal taxa *Isochrysis* or *Tetraselmis* to culture tanks of *B. plicatilis* may increase the ER compared with cultures maintained without these additional algae (Reitan et al., 1993). At low levels of the green alga *Monoraphidium minutum* and the filamentous cyanobacterium *Planktothrix agardhii*, both ERs and egg volume of *B. calyciflorus* were higher on the latter (Weithoff & Walz, 1995).

Periodicity of food supply is also reported to increase the frequency of ovigerous females in cultures. In chemostats and turbidostats, egg production is independent of time of day and, therefore, samples irrespective of their collection time could be nearly similar in terms of ER (Bennett & Boraas, 1993). On the other hand, in batch cultures where food supply is replenished periodically, e.g. every 12 h or 24 h, the error in ER estimated will increase as the time gap between food addition and sampling time increases, and *vice versa*. For *Brachionus*, the time lag between food assimilation and the release of eggs through the cloaca seems to vary from 8 to 18 h (Snell et al., 1987). Ideally, taking observations at 12 h

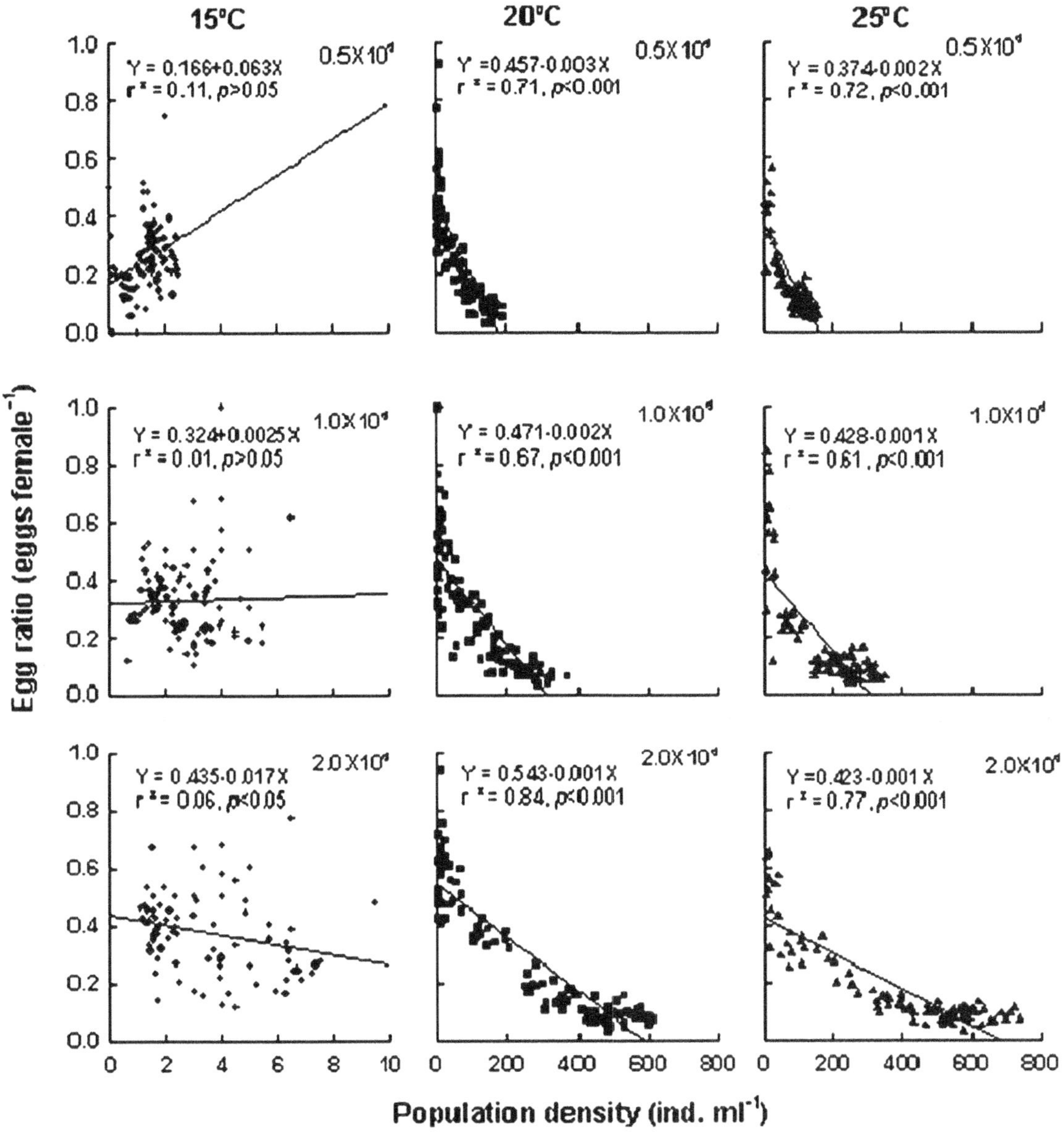

Figure 3. Combined effects of food level (0.5×10^6, 1.0×10^6 and 2.0×10^6 cells ml^{-1} of *Chlorella vulgaris*) and temperature (15, 20 and 25 °C) on the egg ratios of *Brachionus havanaensis*. Regressions of ER against population density under different test conditions are shown (Pavón-Meza et al., 2004).

and 18 h would give better estimates of ER than either of these periods alone; this is, however, laborious and increases the stress on the rotifer population being cultured.

Competition is another important biotic variable in nature that may affect the ER of zooplankton in ponds and lakes (reviewed in Gilbert, 1988). While exploitative competition between large cladocerans and small rotifers results in diminished rotifer abundance, mechanical interference directly affects the overall ER. In interference competition, cladocerans physically damage parthenogenetic eggs attached to rotifers, especially in relatively smaller-bodied rotifers. Eggs, being more vulnerable due to the absence of a strong lorica, are more likely to be damaged than the loricate adults. Thus, interference competition not only causes egg loss from the ovigerous females but also affects their low survival. Nandini et al. (2002) found that large cladocerans had no significant effect on the relationship between ER and population abundance of *B. patulus*. More work is needed to quantify the adverse effects of interference competition on life history parameters, especially the ER of rotifers.

Toxicant stress

Depending on the intensity of the toxicant stress, reproduction in rotifer populations may be inhibited, with or without affecting their survival (Snell & Janssen, 1995). If only reproduction is inhibited, the expected inverse relationship between rotifer density and ER would be lacking. Luna-Andrade et al. (2002) showed that when *B. plicatilis* was subjected to copper stress, regardless of the food (*Tetraselmis suecica*) level, ER was not related to the population density (Fig. 4). Thus, despite the abundant food levels, under stressful levels of copper the reproduction of the test population was more adversely affected than survival.

Experimental manipulation of waterbodies, e.g., the addition of toxic substances, can cause changes in the ER of zooplankton by differentially affecting egg development and fecundity. When egg mortality occurs, neonates become scarce in the population while the fecundity is unaffected, and a high ER is expected (Pfnur et al., 1991). Egg ratios of zooplankton species differed through competitive and predatory interactions in acidified lakes (Locke & Sprules, 1993). However, species resistant to acid stress remained unaffected by the pH changes (Havens & DeCosta, 1988).

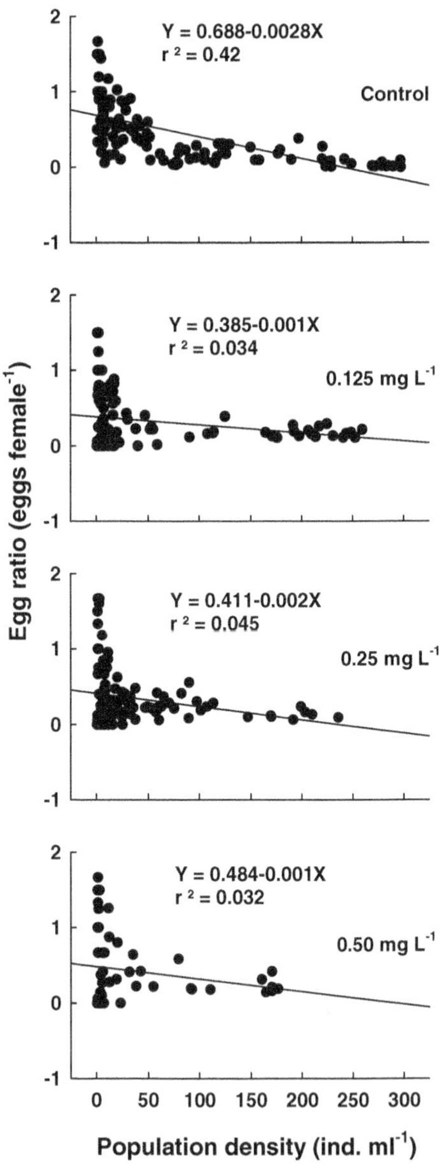

Figure 4. Effect of copper concentration on the relationship between egg ratio and population abundance in *B. plicatilis* (after Luna-Andrade et al., 2002).

Conclusions and general remarks

Egg ratio method is a useful tool for measuring population growth rates, predicting oscillations in a population, and can be used as an early warning system in aquaculture tanks and as indicator of stress in ecotoxicological studies on rotifers. Such a ratio also has implications for measuring swimming speeds of adults, biomass or dry weight estimations and for quantifying prey consumption by predators. Although an inverse relationship between egg ratio and population density is commonly observed in zooplankton, there have been few attempts to critically analyse the changes in these ratios. We demonstrate that several factors influence the ratio of parthenogenetic eggs to females in planktonic rotifers. In addition, frequency of sampling also influences the derived ER values: the higher the sampling frequency the greater the chances of repeatedly counting the same eggs more than once, at lower temperatures, i.e., if the sampling frequency is shorter than the duration of egg laying, we are liable to overestimate the egg ratio by counting the same egg in successive samples. In contrast, if the sampling interval is greater than the duration of egg laying, some eggs may not be counted at all leading to an underestimate of the ER. In addition, both abiotic and biotic factors modify ER through egg output, hatching or differential predation on ovigerous females. Abiotic and biotic factors modify ER under both laboratory and field conditions. Man-made influences such as toxicant stresses also change the egg ratios in planktonic species. Changes in ER are eventually reflected in population growth rates.

Acknowledgements

We thank two anonymous reviewers for improving our manuscript. Linda May made editorial and linguistic corrections. SSSS and SN are thankful to National System of Researchers (SNI –18723 & 20520). Travel grants to attend the conference were provided by FES-Iztacala.

References

Amren, H., 1964. Ecological studies of zooplankton populations in some ponds of Spitsbergen. Zoologiska Bidrag fran Uppsala 36: 161–191.

Bengtsson, J., 1993. Interspecific competition and determinants of extinction in experimental populations of three rockpool *Daphnia* species. Oikos 67: 451–464.

Bennett, W. N. & M. E. Boraas, 1989. An experimental test of the egg-ratio method with instantaneous birth rate as an independent variable. Limnology and Oceanography 34: 1120–1125.

Bennett, W. N. & M. E. Boraas, 1993. Rotifer culture in the turbidostat In Walz, N. (eds), Plankton Regulation Dynamics. Experiments and Models in Rotifer Continuous Cultures. Springer Verlag, Berlin. Ecological Studies 98: 30–38.

Bosch, F. V. D. & J. Ringelberg, 1985. Seasonal succession and population dynamics of *Keratella cochlearis* (Ehrb.) and *Kellicottia longispina* (Kellicott) in Lake Maarsseveen 1 (Netherlands). Archiv für Hydrobiologie 103: 273–290.

Castellanos-Páez, M., G. Garza-Mouriño & S. Marañón-Herrera, 1999. Aislamiento, Caracterización, Biología y Cultivo del rotífero *Brachionus plicatilis* (O.F. Müller U.A.M. Xochimilco, Mexico City, Mexico, 121 pp.

Conde-Porcuna, J. M. & S. S. S. Sarma, 1995. Prey selection by *Asplanchna girodi* (Rotifera): the importance of prey defence mechanisms. Freshwater Biology 33: 341–348.

Cruz-Pizarro, L., J. M. Conde-Porcuna & P. Carrillo, 1998. Diel variation in the egg ratio of *Hexarthra bulgarica* in the high mountain lake La Caldera (Spain). Hydrobiologia 387/388: 295–300.

DeMott, W. R., 1980. An analysis of the precision of birth and death rate estimates for egg-bearing zooplankters. In Kerfoot, W. C. (ed.), Evolution and Ecology of Zooplankton Communities. University Press of New England, Hanover, New Hampshire and London, England, 337–345.

Dierckens, K. R., S. S. S. Sarma, J. Mertens & H. J. Dumont, 1995. Feeding the fairy shrimp *Streptocephalus* (Anostraca–Crustacea) with the rotifer *Anuraeopsis*. Hydrobiologia 308: 29–33.

Devetter, M. & J. Sed'a, 2003. Rotifer fecundity in relation to components of microbial food web in a eutrophic reservoir. Hydrobiologia 504: 167–175.

Dolan, J. R. & C. C. Gallegos, 1992. Trophic role of planktonic rotifers in the Rhode River estuary, spring–summer 1991. Marine Ecology Progressive Series 85: 187–199.

Downing, J. A. & F. H. Rigler (eds), 1984. A Manual for the Methods of Assessment of Secondary Productivity in Fresh Waters (2nd edn). IBP Handbook 17. Blackwell Scientific Publishers, London, 501 pp.

Dumont, H. J., 1977. Biotic factors in the population dynamics of rotifers. Archiv für Hydrobiologie, Beihefte 8: 98–122.

Dumont, H. J., S. S. S. Sarma & A. J. Ali, 1995. Laboratory studies on the population dynamics of *Anuraeopsis fissa* (Rotifera) in relation to food density. Freshwater Biology 33: 39–46.

Duncan, A., 1984. Factors influencing the composition, body size and turnover rate of zooplankton in Parakarama Samudra, an irrigation reservoir in Sri Lanka. Hydrobiologia 113: 210–215.

Edmondson, W. T., 1957. Trophic relations of the zooplankton. Transactions of the American Microscopical Society 76: 225–246.

Edmondson, W. T., 1960. Reproductive rates of rotifers in natural populations. Memorie dell'Istituto Italiano di Idrobiologia 12: 21–77.

Edmondson, W. T, 1965. Reproductive rate of planktonic rotifers as related to food and temperature. Ecological Monographs 35: 61–111.

Foran, J. A. & R. H. King, 1981. Summer production estimates for the rotifer *Polyarthra vulgaris* in a northern Michigan bog lake. Journal of Freshwater Ecology 1: 3–11.

Fuller, D. R., R. S. Stemberger & J. E. Gannon, 1977. Limnetic rotifers as indicators of trophic change. Journal of the Elisha Mitchell Scientific Society 93: 104–113.

Gilbert, J. J., 1983. Rotifera, Oogensis, oviposition and oosorption. In Adiyodi, K. G. & R. G. Adiyodi (eds) Reproductive Biology of Invertebrates. vol. I. John Wiley & Sons, New York, 181–209.

Gilbert, J. J., 1988. Suppression of rotifer populations by *Daphnia*: a review of the evidence, the mechanisms, and the effects on zooplankton community structure. Limnology and Oceanography 33: 1286–1303.

Green, J., 1993. Zooplankton associations in East African lakes spanning a wide salinity range. Hydrobiologia 267: 249–256.

Halbach, U., 1972. Influence of food quality and quantity on the population dynamics of the planktonic rotifer *Brachionus calyciflorus* in laboratory experiments and in the field. Verhandlungen der Deutschen Zoologischen Gesellschaft 35: 83–88.

Halbach, U. & G. Halbach-Keup, 1974. Quantitative Beziehungen zwischen Phytoplankton und der Populationsdynamik des Rotators *Brachionus calyciflorus* Pallas. Befunde aus Laboratoriumsexperimenten und Freilanduntersuchungen. Archiv für Hydrobiologie 73: 273–309.

Havens, K. & J. DeCosta, 1988. An experimental analysis of the acid sensitivity of the common planktonic rotifer *Keratella cochlearis*. Internationale Revue der gesamten Hydrobiologie 73: 407–416.

Herzig, A., 1983. Comparative studies on the relationship between temperature and duration of embryonic development of rotifers. Hydrobiologia 104: 237–246.

Herzig, A. & W. Koste, 1989. Development of *Hexarthra* spp. in a shallow alkaline lake. Hydrobiologia 186/187: 129–136.

Keen, R. & R. Nassar, 1981. Confidence intervals for birth and death rates estimated with the egg-ratio technique for natural populations of zooplankton. Limnology and Oceanography 26: 131–142.

Korstad, J., A. Neyts, T. Danielsen, I. Overrein & Y. Olsen, 1995. Use of swimming speed and egg ratio as predictors of the status of rotifer cultures in aquaculture. Hydrobiologia 395–398: 313–314.

Koste, W., 1978. Rotatoria. Die Rädertiere Mitteleuropas. Ein Bestimmungswerk begründet von Max Voigt. Gebrüder Borntraeger, Berlin, Stuttgart. Textband, 673 pp., Tafelband 234 pp.

Krebs, C. J., 1985. Ecology: The Experimental Analysis of Distribution and Abundance (3rd edn). Harper and Row, New York, 800 pp.

Lampert, W. & U. Sommer, 1997. Limnoecology. The Ecology of Lakes and Streams. Oxford University Press, New York, 382 pp.

Lehman, J. T., 2000. In memoriam: W. Thomas Edmondson (1916–2000). Limnology and Oceanography 45: 1448 .

Locke, A. & W. G. Sprules, 1993. Effects of experimental acidification on zooplankton population and community dynamics. Canadian Journal of Fisheries and Aquatic Sciences 50: 1238–1247.

Luna-Andrade, A., R. Aguilar-Duran, S. Nandini & S. S. S. Sarma, 2002. Combined effects of copper and microalgal (*Tetraselmis suecica*) concentrations on the population growth of *Brachionus plicatilis* Müller (Rotifera). Water, Air and Soil Pollution 141: 143–153.

Magnien, R. E. & J. J. Gilbert, 1983. Diel cycles of reproduction and vertical migration in the rotifer *Keratella crassa* and their influence on the estimation of population dynamics. Limnology and Oceanography 28: 957–969.

Nandini, S., 2000. Responses of rotifers and cladocerans to *Microcystis aeruginosa* (Cyanophyceae): a demographic study. Aquatic Ecology 34: 227–242.

Nandini, S. & S. S. S. Sarma, 2001. Population growth of *Lepadella patella* (O.F. Müller, 1786) at different algal (*Chlorella vulgaris*) densities and in association with *Philodina roseola* Ehrenberg, 1832. Hydrobiologia 446/447: 63–69.

Nandini, S., S. S. S. Sarma & M. D. Hurtado-Bocanegra, 2002. Effect of four species of cladocerans (Crustacea) on the population growth of *Brachionus patulus* (Rotifera). Acta Hydrochimica et Hydrobiologica 30: 101–107.

Nogrady, T., R. L. Wallace & T. W. Snell, 1993. Rotifera 1 Biology, Ecology and Systematics. Guides to the Identification of the Microinvertebrates of the Continental Waters of the World vol. 4SBP. Academic Publishers, The Hague 142.

Ohman, M. D., D. L. Aksnes & J. A. Runge, 1996. The interrelationship of copepod fecundity and mortality. Limnology and Oceanography 41: 1470–1477.

Ooms-Wilms, A. L., G. Postema & R. D. Gulati, 1999. Population dynamics of planktonic rotifers in Lake Loosdrecht, the Netherlands, in relation to their potential food and predators. Freshwater Biology 42: 77–97.

Paloheimo, J., 1974. Calculation of instantaneous birth rate. Limnology and Oceanography 19: 692–694.

Pavón-Meza, E. L., S. S. S. Sarma & S. Nandini, 2004. Combined effects of food (*Chlorella vulgaris*) concentration and temperature on the population growth of (*Brachionus havanaensis*) (Rotifera: Brachionidae). Journal of Freshwater Ecology 19: 521–530.

Peredo-Alvarez, V. M., S. S. S. Sarma & S. Nandini, 2003. Combined effect of concentrations of algal food (*Chlorella vulgaris*) and salt (sodium chloride) on the population growth of *Brachionus calyciflorus* and *Brachionus patulus* (Rotifera). Revista de Biología Tropical 51: 399–408.

Pfnur, R. B., G. Burkhardt, A. Peither & J. P. Lay, 1991. Chronic ecotoxicity of 3,4-dichloroaniline to freshwater ecosystems. Toxicology and Environmental Chemistry 31/32: 367–373.

Polishchuk, L. K., 1978. Determination of birth and death rates in natural populations of planktonic organisms. Zhurnal Obshchej Biologii 39: 189–193.

Pontin, R. M., 1978. A key to the freshwater planktonic and semiplanktonic Rotifera of the British Isles. Freshwater Biological Association, Ambleside, Scientific Publications 38 pp.

Pourriot, R., 1983. Reproductive strategies in rotifers. Comptes rendus de l'Academie des Sciences 296: 1109–1111.

Pourriot, R. & T. W. Snell, 1983. Resting eggs in rotifers. Hydrobiologia 104: 213–224.

Razlutskij, V. I., 2000. Estimating cladoceran birth rate: use of the egg age distribution to estimate mortality of ovigerous females and eggs. Hydrobiologia 428: 135–144.

Reitan, K. I., J. R. Rainuzzo, G. Oie & Y. Olsen, 1993. Nutritional effects of algal addition in first-feeding of turbot (*Scophthalmus maximus* L.) larvae. Aquaculture 118: 257–275.

Ringelberg, J. & J. Steenvoorden, 1986. Diel variation in the egg ratio of rotifers throughout the season (preliminary report). Hydrobiological Bulletin 19: 153–158.

Rombaut, G., P. Dhert, J. Vandenberghe, L. Verschuere, P. Sorgeloos & W. Verstraete, 1999. Selection of bacteria enhancing the growth rate of axenically hatched rotifers (*Brachionus plicatilis*). Aquaculture 176: 195–207.

Ruttner-Kolisko, A., 1974. Plankton Rotifers: Biology and Taxonomy E. Schweizerbarťsche Verlagsbuchhandlung Stuttgart,. Die Binnengewässer 26: 1–146.

Sarma, S. S. S., 1987. Experimental studies on the ecology of *Brachionus patulus* (Müller) (Rotifera) in relation to food, temperature and predation. Ph.D. Thesis, University of Delhi, Delhi, India.

Sarma, S. S. S., 1996. Some relationships between size structure and fertility of rotifer populations. In Kaul, B. L. (ed.), Advances in Fish and Wildlife Ecology and Biology. vol. 1. Daya Publishing House, Tri Nagar, Delhi, India, pp. 37–50.

Sarma, S. S. S., 2000. The use of rotifers for ecotoxicological studies in Mexico. In Ríos-Jara, E. (ed.) Estudios sobre plancton en México y el Caribe. Sociedad Mexicana Planctología/Universidad de Guadalajara, México: 8–11 .

Sarma, S. S. S. & T. R. Rao, 1990. Population dynamics of *Brachionus patulus* Müller (Rotifera) in relation to food and temperature. Proceedings of the Indian Academy of Sciences (Animal Sciences) 99: 335–343.

Sarma, S. S. S. & T. R. Rao, 1991. The combined effects of food and temperature on the life history parameters of *Brachionus patulus* Muller (Rotifera). Internationale Revue der gesamten Hydrobiologie 76: 225–239.

Sarma, S. S. S., E. D. Fiogbe & P. Kestemont, 1999. Population growth of *Brachionus calyciflorus* Pallas (Rotifera) in relation to algal (Dictyosphaerium chlorelloides) density. In Kaul, B. L. (ed.), Advances in Fish and Wildlife Ecology and Biology. vol. 2. Daya Publishing House, Tri Nagar, Delhi, India, pp. 83–93.

Sarma, S. S. S., P. S. Larios-Jurado & S. Nandini, 2001. Effect of three food types on the population growth of *Brachionus calyciflorus* and *Brachionus patulus* (Rotifera: Brachionidae). Revista de Biología Tropical 49: 75–82.

Saunders, P. A., K. G. Porter & B. E. Taylor, 1999. Population dynamics of *Daphnia* spp. and implications for trophic interactions in small, monomictic lake. Journal of Plankton Research 21: 1823–1845.

Shiel, R. J. & W. Koste, 1986. Australian Rotifera. In De Deckker, P. & W. D. Williams (eds), Limnology in Australia. Dr W. Junk Publ. Dordrecht, CSIRO Melbourne, 141–150.

Snell, T. W. & C. R. Janssen, 1995. Rotifers in ecotoxicology. Hydrobiologia 313/314: 231–247.

Snell, T. W., M. J. Childress, E. M. Boyer & F. H. Hoff, 1987. Assessing the status of rotifer mass culture. Journal of World Aquaculture Society 18: 270–277.

Thorp, J. H. & A. Covich (eds), 2001. Ecology and Classification of North American Freshwater Invertebrates. (2nd edn.) Academic Press, San Diego.

Threlkeld, S. T., 1983. Empty loricas and the dynamics of *Kellicottia longispina* in a subalpine, oligotrophic lake. Hydrobiologia 104: 367–372.

Urabe, J., 1992. Midsummer succession of rotifer plankton in a shallow eutrophic pond. Journal of Plankton Research 14: 851–866.

Walz, N., S. S. S. Sarma & U. Benker, 1995. Egg size in relation to body size in rotifers: an indication of reproductive strategy?. Hydrobiologia 313/314: 165–170.

Weithoff, G. & N. Walz, 1995. Influence of the *Planktothrix agardhii* on population growth and reproductive pattern of the rotifer *Brachionus calyciflorus*. Hydrobiologia 313/314: 381–386.

Wolfinbarger, W. C., 1999. Influences of biotic and abiotic factors on seasonal succession of zooplankton in Hugo Reservoir, Oklahoma, USA. Hydrobiologia 400: 13–31.

Hydrobiologia (2005) 546:375–380

A. Herzig, R.D. Gulati, C.D. Jersabek & L. May (eds.) Rotifera X: Rotifer Research: Trends, New Tools and Recent Advances
DOI 10.1007/s10750-005-4279-y

Factors affecting swimming speed in the rotifer *Brachionus plicatilis*

Manuel Yúfera[1,*], E. Pascual[1] & J.M. Olivares[2]
[1]*Instituto de Ciencias Marinas de Andalucía (CSIC), Apartado Oficial, 11510, Puerto Real, Cádiz, Spain*
[2]*Instituto de Acuicultura de Torre de la Sal (CSIC), Ribera de Cabanes, 12595, Torre La Sal, Castellón, Spain*
*(*Author for correspondence: E-mail: manuel.yufera@icman.csic.es)*

Key words: swimming speed, Rotifera, *Brachionus plicatilis*, female status, temperature, feeding condition

Abstract

This study examines the swimming speed in amictic females of *Brachionus plicatilis* in laboratory cultures. Five different stages were examined: recently hatched females, juveniles, adult non-ovigerous females, ovigerous females with 1 attached egg and ovigerous females with 2 attached eggs. We tested the speed at two temperatures, 15 °C and 25 °C, and two feeding conditions, presence and absence of microalgal cells. An automated motion analysis system was used to measure speed which was then video recorded. Swimming speed ($\mu m\ s^{-1}$) increased with increasing body size. There was a slight decrease in the speed of adult females as the number of attached eggs increased. Swimming activity was higher at 25 °C than at 15 °C and in the absence of food than if microalgae were present. Average values under the different experimental conditions ranged between 500 $\mu m\ s^{-1}$ for the recently hatched and fed females and 1500 $\mu m\ s^{-1}$ for the adult non-ovigerous females in the absence of microalgae. Mass-specific swimming speed decreased with body mass increase.

Introduction

Swimming in rotifers is an activity that demands high energy (Epp & Lewis, 1984). It can, therefore, be affected by intrinsic and extrinsic factors that influence metabolic activity. In a previous study (Yúfera et al., unpublished) temperature was found to have a relatively low effect on the respiration rates of *Brachionus plicatilis*, but, in contrast, fecundity of the population had a strong effect; the population with higher eggs/female ratio had higher, specific oxygen consumption. Interestingly, ovigerous females exhibited higher oxygen consumption than expected from the sum of oxygen consumption of non-ovigerous females plus that of isolated eggs. This difference suggests a supplemental cost for reproduction and carrying the eggs. Clément (1977b) reported that the number of cilia and the filtering surface increased with the size in case of *Notommata* and *Trichocerca*. Light microscopical observations indicate a similar increase in *Brachionus*. During body growth of *Brachionus* females from hatching to maturity, the size of corona also increases, but mature females with small variations in body size, weight and shape increase in mass due to egg deposition. Although there are several studies on rotifer locomotion (Viaud 1940 in Clément, 1977a; Clément, 1977a; Wallace, 1980; Coulon et al., 1983; Luciani et al., 1983; Epp & Lewis, 1984; Stemberger & Gilbert, 1987; Snell et al., 1987; Charoy & Clément, 1993; Janssen et al., 1993; Charoy, 1995; Korstad et al., 1995; Hagiwara et al., 1998; Preston et al., 1999; Santos-Medrano et al., 2001), information on swimming ability at the different stages of development is scarce.

The aim of this study was to determine to what extent the swimming speed of *B. plicatilis* depends on female biomass and specifically to examine the swimming response of ovigerous females. In addition, the influence of temperature and the presence and absence of microalgal food was assessed for obtaining preliminary information about the influence of other environmental factors on swimming with potential involvement of metabolic rates of filter-feeding planktonic organisms.

Materials and methods

The morphological and physiological characteristics of the rotifer strain used in this study, *Brachionus plicatilis* (strain S1), have been described in detail in previous papers (Yúfera, 1982; Yúfera et al., 1997). The body length of *B. plicatilis* ranges between 140 and 280 μm, and the body dry mass, eggs included, between 0.25 and 1.00 μg. The swimming speed was measured using an automated motion analysis system, including video recording (Software: Motility – J.M. Olivares, Instituto de Acuicultura de Torre de la Sal, CSIC). For the swimming speed determinations, the rotifers were maintained in batch cultures in 3-l beakers containing *Nannochloropsis gaditana* as food, at a salinity of 33 psu and at two different temperatures (15 °C and 25 °C). They were acclimatized under these conditions for at least 1 month before the start of the experiments. Speed determinations were carried out in amictic females both fed ca. 10^7 cells ml^{-1} and unfed. In the latter case, the video recording was done between 30 and 60 min after the rotifers were transferred to cell-free seawater. Some microalgal remains could still be observed in the gut of these females. Determinations were made in a glass chamber (5 mm deep) with 3 ml of medium, at the experimental temperature. Five different stages were examined: recently hatched females, juveniles, adult non-ovigerous females, ovigerous females with 1 attached egg and ovigerous females with 2 attached eggs. The lorica sizes were: 140–160 μm, for recently hatched individuals; 190–220 μm, for juveniles; and >240 μm, for adults. For each stage and and food and temperature condition, 15 animals were used. The motion analysis system allows calculating the speed in different segments within the recorded trajectory of an animal. Thus, for each rotifer, we calculated the mean of the speed determined in the different segments. The system measures the distances of the curvilinear and helical paths. Only trajectories with continuous swimming (no stops and collisions) and in only one plane were considered. For each stage, the maximum episodic speed was recorded. To calculate the mass-specific speed (mm μg dry-mass^{-1} s^{-1}), the dry weight of the female at different stages (Table 1) was estimated using the formula described in Yúfera et al. (1997) for the average individual weight of populations with varying egg/female ratio. Calculations were made assuming an egg/female ratio of 0.05, 1 and 2 as the weight of a single adult female, a female with one egg and a female with two eggs, respectively. The weight of a newborn was considered the same as the egg and the weight of a juvenile as the geometric mean of newborn and adult female.

The effect of different factors on the swimming speed was examined by a three-way ANOVA over log-transformed speed data. Tukey multiple comparison tests were then performed on all statistically significant ANOVAs. Tests were considered significant at $p < 0.05$.

Results

The absolute swimming speed (μm s^{-1}) increased with increasing body mass in non-ovigerous females in all treatments. The females with one and two attached eggs showed a slightly decreasing speed, although the trend was non-significant

Table 1. Individual dry weight (ng) of the different stages considered in this study at 15 and 25 °C

	Newborn	Juvenile adult	Adult	Adult + 1Egg	+2 Eggs
15 °C	194	276	395	841	1042
25 °C	171	251	369	650	758

Values estimated from the formula described in Yúfera et al. (1997).

(Fig. 1). The swimming speed was always higher at 25 °C than at 15 °C for each given stage and treatment. In addition, the speed of the animals was always higher in the absence of algal cells. The ANOVA indicates that the three tested factors affect the swimming speed (Table 2). The maximum average speed values of adult females with microalgae in the medium were 941 μm s^{-1} and 1274 μm s^{-1} at 15 °C and 25 °C, respectively; in absence of algal cells the maximum speed values were 1029 μm s^{-1} and 1507 μm s^{-1} at 15 and 25 °C, respectively.

The difference between the maximum episodic speed and the mean speed recorded in each stage increased gradually with the body size in the fed animals. In absence of food this difference was higher but similar in all stages (Table 3).

When converted into mass specific speed, higher values were obtained with younger animals (2.85 and 4.53 mm μg^{-1} s^{-1} at 15 °C and 25 °C, respectively) and lower ones with females with 2 attached eggs (0.77 and 1.47 mm μg^{-1} s^{-1} at 15 and 25 °C respectively) (Fig 2).

Discussion

Values of absolute speed determined in the present study are similar to those described in other studies with *B. plicatilis* (Luciani et al., 1983; Epp & Lewis, 1984; Korstad et al., 1995). All the three factors examined in this study, i.e. female stage, temperature and presence of algal cells, had significant effect on swimming speeds. During size increase of females, primarily in the first 30–40 h after hatching, the swimming speed also increased, probably because of a corresponding increase of

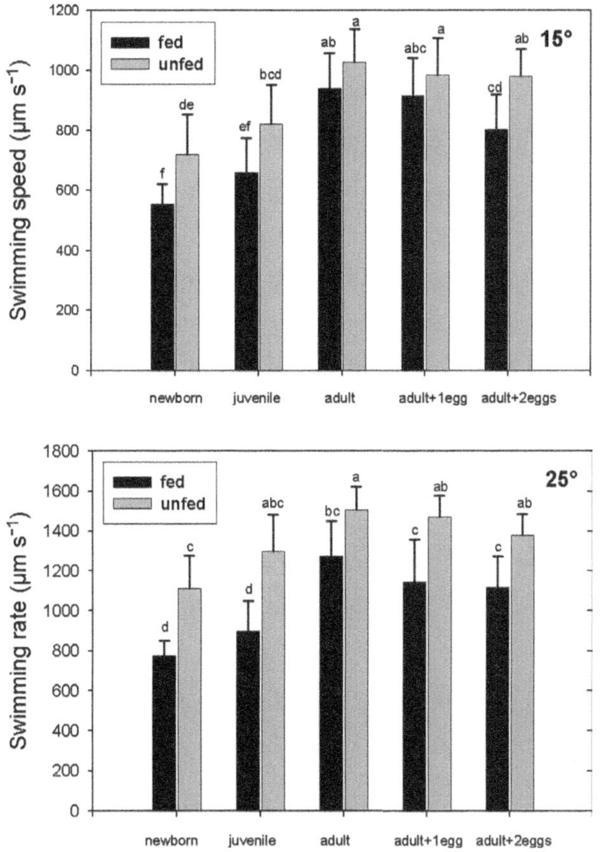

Figure 1. Swimming speed (mean ± SD) of different stages of *Brachionus plicatilis* females in presence and absence of microalgal cells at 15 °C (upper graph) and at 25 °C (lower graph). The letters above the bars (a to f) were used to indicate the significant differences obtained with the Tukey comparison test ($p > 0.05$). For each temperature, same letters indicate no statistically significant difference.

Table 2. Results of the three-way ANOVA of the effect of female stage, temperature and feeding condition on the swimming speed (μm s^{-1}) of *B. plicatilis*

Source	Mean squares	df	*p*-value	
Female stage	$1.303 \cdot 10^6$	4	<0.0001	N < J < AEE < AE < A
Temperature	$8.263 \cdot 10^6$	1	<0.0001	15 < 25
Feeding	$3.006 \cdot 10^6$	1	<0.0001	F < UF
Stage × temperature	$5.931 \cdot 10^3$	4	0.861	
Stage × feeding	$24.126 \cdot 10^3$	4	0.262	
Temperature × feeding	$447.160 \cdot 10^3$	1	<0.000	F15 < UF15 < F25 < UF25
Stage × temperature × feeding	$24.387 \cdot 10^3$	4		
Residuals	$18.233 \cdot 10^3$	247		

Results of the *a posteriori* Tukey test applied to factors detected to be significant are also included. For each test factors joined by underline were not significantly different at the 5 % level. N, newborn; J, juvenile; A adult; AE, adult with 1 egg; AEE, adult with 2 eggs; F, algae present; UF, algae absent; 15, temperature 15 °C; 25, temperature 25 °C.

Table 3. Maximum episodic swimming speed (Max: μm s^{-1}) recorded and percentage of increase over mean speed (%Diff) at the different stages considered in this study

	Newborn	Juvenile	Adult	Adult + 1 Egg	Adult + 2 Eggs
Fed rotifers					
15 °C Max	676	947	1333	1260	1012
%Diff	21.6	43.27	41.66	37.40	25.87
25 °C Max	878	1143	1708	1684	1525
%Diff	13.29	27.28	34.07	47.07	36.28
Unfed rotifers					
15 °C Max	1040	1148	1328	1350	1190
%Diff	44.44	39.66	29.06	37.06	21.43
25 °C Max	1660	1863	1915	1953	1642
%Diff	49.26	43.75	27.07	32.77	18.99

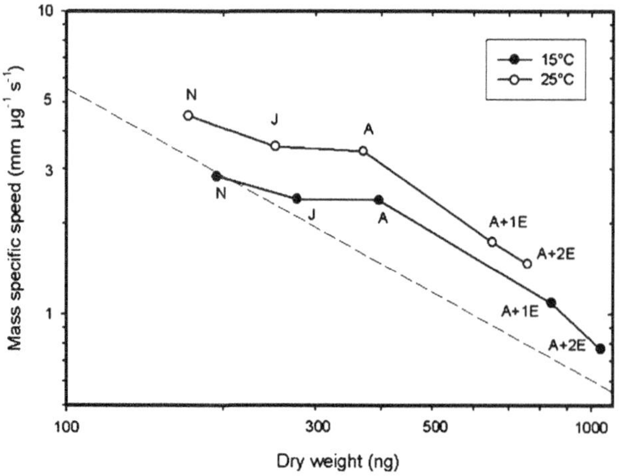

Figure 2. Mass specific swimming speed of different stages of *Brachionus plicatilis* females in presence of microalgae. The broken line indicates the regression calculated by Stemberger and Gilbert (1987) for different rotifer species.

the coronal area. Similar results were reported by Luciani et al. (1983) working with *B. plicatilis*. They found an increase in the swimming speed from young to mature females, but then a decrease in speed as animals became old. On the contrary, Beauvais & Enesco (1985) working with *Asplanchna brightwellii* and Hagiwara et al. (1998), also working with *B. plicatilis*, observed a decline in swimming speed with age. Nevertheless, a precise comparison with those studies is not possible because we did not examine the old individuals.

The mass-specific swimming speed which decreased as expected with increasing body mass, compares well with observations on different rotifer species (Stemberger & Gilbert, 1987). This decrease was relatively small during the body mass increase of immature females. Such a response would indicate a progressive improvement in swimming efficiency from hatching to maturity. In contrast, the supplemental weight of attached eggs involves a noticeable decrease of the mass specific speed.

The swimming speed of females without eggs and with one egg virtually did not differ. This indicates that the ovigerous females will be able to use the energy saved to maintain the swimming speed necessary for feeding.

We found an increase in swimming speed from 15 to 25 °C: 25–39% in fed rotifers and of 42–58% in unfed animals. Epp & Lewis (1984) found a plateau in the swimming speed in the temperature range from 20 to 32 °C, but the swimming speed almost doubled from 15 to 25 °C. On the other hand, Snell et al. (1987) found an increase in swimming activity for the temperatures in the range 10–30 °C, and increment of 60% in speed from 15 to 25 °C.

Finally, it seems that a quick removal of algal cells induces an immediate response in *B. plicatilis*, as found in *B. calyciflorus* (Charoy & Clément, 1993; Charoy, 1995). Thus, by increasing the swimming speed the rotifer effectively improves the encounter probability while searching for food. Such a response is more evident in the younger females. In fact, the highest episodic increases of speed were observed in the newborns. How long such a response can be maintained will depend on the internal reserves and on the swimming strategy that the animal exhibits under prolonged starvation conditions. Interestingly, Galkovskaya (1980; cited in Galkovskaya, 1985) found an increase of the oxygen consumption in rotifers after 2 h of starvation followed by a noticeable decline after 8 h of starvation. Such results probably reflect the swimming behaviour explained above.

Acknowledgements

This work was supported by the Ministry of Science and Technology, Spain (PN Project: AGL2000-0697-C02-01). We thank E. Ramos-García for helpful technical assistance.

References

Beauvais, J. E. & H. E. Enesco, 1985. Lifespan and age-related changes in activity level of the rotifer *Asplanchna brightwelli* influence of curare. Experimental Gerontology 20: 359–366.

Charoy, C., 1995. Modification of the swimming behaviour of *Brachionus calyciflorus* (Pallas) according to the food environment and individual nutritive status. Hydrobiologia 313/314: 197–204.

Charoy, C. & P. Clément, 1993. Foraging behaviour of *Brachionus calyciflorus* (Pallas): variation in the swimming path according to presence or absence of algal food (*Chlorella*). Hydrobiologia 225/256: 95–100.

Clément, P., 1977a. Phototaxis in rotifers (action spectra). Archiv für Hydrobiologie, Ergebnisse der Limnologie Beiheft 6: 47–49.

Clément P., (1977b). Ultrastructural research on rotifers Archiv für Hydrobiologie, Ergebnisse der Limnologie Beiheft 6: 270–297.

Coulon, P. Y., P. J. Charras, J. L. Chassé, P. Clément, A. Cornillac, A. Luciani & E. Wurdak, 1983. An experimental system for the automatic tracking and analysis of rotifer swimming behaviour. Hydrobiologia 104: 197–202.

Epp, R. W. & W. M. Lewis Jr., 1984. Cost and speed of locomotion for rotifers. Oecologia 61: 289–292.

Galkovskaya, G. A., 1985. Oxygen consumption rate in rotifers. Hydrobiologia 313/314: 147–156.

Hagiwara, A., N. Yamamiya & A. Belem & Araujo, 1998. Effect of water viscosity on the population growth of the rotifer *Brachionus plicatilis* Müller. Hydrobiologia 387/388: 489–494.

Janssen, C. R., M. D. Ferrando & G. Persone, 1993. Ecotoxicological studies with the freshwater rotifer Brachionus calyciflorus I. Conceptual framework and applications. Hydrobiologia 255/256: 21–32.

Korstad, J., A. Neyts, T. Danielsen, T. Overrein & Y. Olsen, 1995. Use of swimming speed and egg ratio as predictors of the status of rotifer cultures in aquaculture. Hydrobiologia 313/314: 395–398.

Luciani, A., J. L. Chassé & P. Clément, 1983. Aging in *Brachionus plicatilis*: the evolution of swimming as a function of age at two different calcium concentrations. Hydrobiologia 104: 141–146.

Preston, B. L., G. Cecchine & T. W. Snell, 1999. Effects of pentachlorophenol on predator avoidance behaviour of the rotifer *Brachionus calyciflorus*. Aquatic Toxicology 44: 201–212.

Santos-Medrano, G. E., R. Rico-Martínez & A. Velázquez-Rojas, 2001. Swimming speed and Reynolds numbers of eleven freshwater rotifer species. Hydrobiologia 446/447: 35–38.

Snell, T. W., M. J. Childress, E. M. Boyer & F. H. Hoff, 1987. Assessing the status of rotifer mass culture. Journal of the World Aquaculture Society 18: 270–277.

Stemberger, R. S. & J. J. Gilbert, 1987. Rotifer threshold food concentration and size-efficiency hypothesis. Ecology 68: 181–187.

Wallace, R. L., 1980. Ecology of sessile rotifers. Hydrobiologia 73: 181–193.

Yúfera, M., 1982. Morphometric characterisation of a small-sized strain of *Brachionus plicatilis* in culture. Aquaculture 27: 55–61.

Yúfera, M., G. Parra & E. Pascual, 1997. Energy content of rotifers (*Brachionus plicatilis* and *Brachionus rotundiformis*) in relation to temperature. Hydrobiologia 358: 83–87.

Hydrobiologia (2005) 546:381–386
A. Herzig, R.D. Gulati, C.D. Jersabek & L. May (eds.) Rotifera X: Rotifer Research: Trends, New Tools and Recent Advances
DOI 10.1007/s10750-005-4280-5

An evidence for vertical migrations of small rotifers – a case of rotifer community in a dystrophic lake

Andrzej Karabin & Jolanta Ejsmont-Karabin*
Centre for Ecological Research PAS, Hydrobiological Station, Leśna 13, 11-730, Mikołajki, Poland
*(*Author for correspondence: E-mail: jolanta@onet.pl)*

Key words: Rotifera, Crustacea, diel vertical migrations, dystrophic lake, Lake Kruczy Staw

Abstract

Diel vertical migrations of zooplankton were studied in a small, dystrophic Kruczy Staw Lake. Two rotifer species (*Synchaeta pectinata* Ehrenberg, *Trichocerca simonei* DeSmet) inhabiting the lake occurred near lake bottom (7–8 m depth) in the daytime. At night they were observed in surface waters (0–2 m). Both amplitude and speed of the rotifer migration were markedly higher than those of crustaceans. As invertebrate predators are scarce or altogether lacking in the lake, vertical stratification of rotifer and crustacean communities both seasonally and dielly may be caused by strong competition for very low food resources in the lake. This assumption is supported by the observed reverse changes in densities of zooplankton and their food (i.e. picoplankton) during a diel cycle.

Introduction

A two-years study of plankton in Kruczy Staw Lake showed that during the whole productive season, zooplankton species showed a distinct vertical stratification. However the stratification seemed to be ‚atypical'. As a rule, Rotifera occupied the near-bottom habitat where oxygen concentration is very low, whereas different crustacean species had their density maxima at different depths in upper water layers. Although in summer, rotifers occur commonly in deeper parts of hypolimnion, they are comprised of winter and spring species that avoid upper, warmer layers (Ruttner-Kolisko, 1975; Miracle, 1977). Pelagial rotifers in Kruczy Staw Lake, i.e. *Synchaeta pectinata* and *Trichocerca simonei*, do not belong to this community.

The aim of the present study is to examine if vertical segregation of certain species is observed also in their diel cycle and what causes this non-typical behaviour. In Lake Kruczy Staw, where food resources for zooplankton are probably limited but crustacean zooplankton is abundant, vertical segregation may be attributed to both exploitative competition, and mechanical interference competition (Gilbert, 1989a, b; MacIsaac & Gilbert, 1990, 1991).

Methods

Kruczy Staw Lake is situated in Piska Forest (Masurian Lake District, North-eastern Poland). Kruczy Staw is a small (area, ca. 1.5 ha) and deep (Z_{max}, 10 m) lake devoid of inflow and outflow. Lake is surrounded by a zone of moss vegetation of up to several dozen metres. The lake is acidic: the pH of water varies from 3.7 in spring to 5.4 in summer.

During the study period the lake water was very transparent (Secchi depth, 7.6 m) and stratified, both thermally and chemically (oxygen), with thermocline and chemocline at 5 m (Fig. 1).

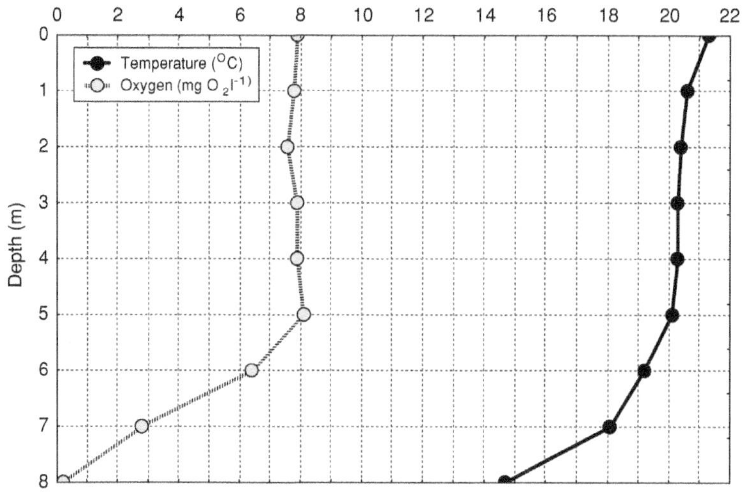

Figure 1. Profiles of temperature and oxygen concentration during the study.

A 24-hour sampling was carried out on August 2–3, 2000 in central part of the lake basin (depth ca. 8 m). Lake water was sampled for analyses of zooplankton and chlorophyll seven times every 4 h. Zooplankton was sampled with a 5 l Bernatowicz sampler, taking two samples (10 l) from each depth at 1 m intervals, from surface to bottom, concentrated using a plankton net of 30 μm mesh size and fixed with 4% formalin.

Sunset on August 2 was observed at 19.20 h, and sunrise on August 3 at 4.10 h.

Results

The lake is characterised by seasonal changes in phytoplankton structure. Large algae occur in the pelagic waters in early spring and late autumn. During the rest of the season, phytoplankton is comprised mainly of picoplankton genera belonging to Chlorophyceae (*Coccomyxa*) and Cyanophyceae (*Synechococcus*). Only qualitative composition of picoplankton was determined. The algal biomass was measured as chlorophyll *a*. The concentrations of chlorophyll *a* varied significantly with depth, ranging from 0.1 to >10 μg l^{-1}.

Despite poor food resources, as evident from the chlorophyll concentrations, both densities and species diversity of zooplankton were high. Rotifer community consisted mainly of *Synchaeta pectinata* and *Trichocerca simonei*, which occupied deeper layers during the day, and their densities ranged from 75 to 95 ind. l^{-1} and averaged 27 ind. l^{-1}. Crustacean community that was composed of *Eudiaptomus gracilis* Sars, *Holopedium gibberum* Zaddach, *Diaphanosoma brachyurum* (Lievin) and *Ceriodaphnia quadrangula* (O.F. Muller) had an average density in the water column of 290 ind. l^{-1}. Mean densities of Crustacea in other dystrophic lakes in Masurian Landscape Park, with pH 3.9–5.9, ranged from 45 to 59 ind. l^{-1} (A. Karabin, unpubl. data). Thus, crustacean densities in Lake Kruczy Staw are much higher. Pelagic, crustacean zooplankton was also poor in seven dystrophic lakes in the Wigry National Park, as it consisted of only 1–5 species. In most lakes, the zooplankton densities ranged from 46 to 96 ind. l^{-1}, with a maximum of about 225 ind. l^{-1} (Tunowski, 1992).

There are no invertebrate predators in the pelagial of Lake Kruczy Staw. Fish population that is low is restricted to predatory perch (*Perca fluviatilis*), which are stunted.

Amplitude and intensity of the diel migrations (Fig. 2) varied among the three species of Cladocera present. *Ceriodaphnia quadrangula* stayed for all 24 h close to the bottom where oxygen is low. *Holopedium gibberum* started upward migration in early afternoon and reached a dusk-time maximum at 1 m depth. *Diaphanosoma brachyurum* migrated upwards during sunset and stayed in surface waters (0–1 m) during the night, it started

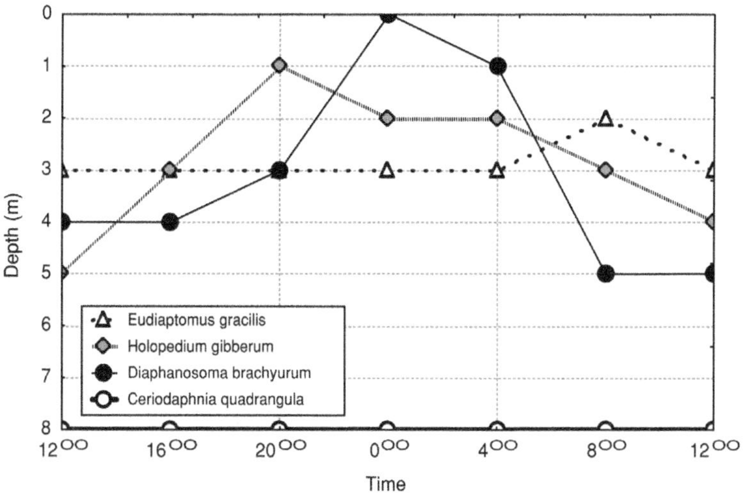

Figure 2. Depths of maximum densities (ind. l^{-1}) of four crustacean species during 24 h.

downward migration at sunrise, and its density increased in 5 m layer. *Eudiaptomus gracilis*, did not perform any diel migrations (Fig. 2).

From 12.00 to 16.00 h >95% of the *Synchaeta pectinata* population (20–32 ind. l^{-1}) was concentrated near the bottom (8 m). After sunset it moved intensively into surface waters and reached a high density there at midnight. *S. pectinata* was already absent in upper and middle water layers at 08.00 h and its population density was again high at 8 m at noon time (Fig. 3a). During the day time *Trichocerca simonei* occurred at 4–8m, 80% of its population (36–74 ind. l^{-1}) was concentrated in near-bottom layers (7–8 m). After sunset the species started migrating upward and at midnight its maximum was observed at 2 m. However, the animals starting descending already at night attaining at 04.00 h their maximum in 3–7 m stratum. At day time (8.00–12.00) the species was again concentrated in deep water layers (Fig. 3b).

Synchaeta pectinata needed about 8 h to ascend to surface water layers but only 4 h to descend back to bottom layers. Thus, the rate of upward migration was ca. 1.7 cm min^{-1} and of downward migration twice as high, 3.4 cm min^{-1}. *Trichocerca* had shorter amplitude of migration and its ascent and descent rate were almost same (1.2 cm min^{-1}). These rates of rotifer migration were markedly higher than those for crustaceans: *H. gibberum* – 0.5–0.8 cm min^{-1} and *D. brachyurum* – 0.8–1.1 cm min^{-1}.

Apart from avoiding invertebrate predation, another reason of vertical migration in zooplankton may be daily changes in food conditions. Comparison of vertical distribution of chlorophyll and zooplankton implies a strong grazing pressure of zooplankton on phytoplankton. The correlation analyses confirmed this (Table 1): chlorophyll concentrations and densities of rotifer and crustacean communities showed significant negative correlationships.

Discussion

Vertical migration of rotifers is thought to be less significant than other planktonic animals. The amplitude of migration is related to the animal body size (George & Fernando, 1970; Rey & Capblancq, 1975). Miracle (1977) observed the amplitude of daily migrations of several rotifer species ranging between 0.2 and 5.6 m. In Kruczy Staw Lake, the two rotifer species have high amplitudes of diel vertical migrations: *Synchaeta pectinata* – 8 m; *Trichocerca simonei* – 6 m. These amplitudes of migration were clearly larger than those of Cladocera species: *H. gibberum* – 3–4 m; *D. brachyurum* – 4–5 m. Thus, the rotifers dielly migrated into water layers with varying temperature, food and oxygen levels, whereas diurnal migrations of crustaceans were limited to same environmental conditions (Fig. 1). In addition, the rotifers migrated at faster rates

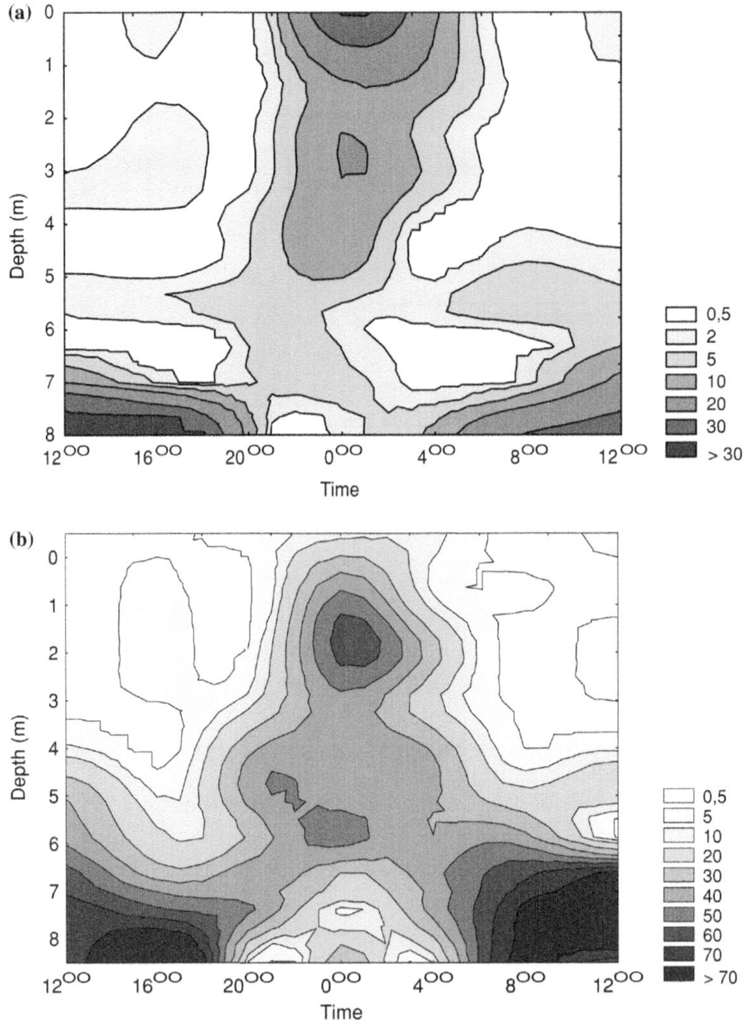

Figure 3. Vertical distribution of the densities (ind. l^{-1}) of *Synchaeta pectinata* (a) and *Trichocerca simonei* (b).

Table 1. Pearson correlation coefficients for the relationships between chlorophyll *a* concentration and taxonomic groups of zooplankton

	Taxonomic groups			
	Rotifera	Cladocera	Rotifera + Cladocera	*E. gracilis*
Chlorophyll *a*	−0.50 $p<0.0001$	−0.30 $p=0.023$	−0.41 $p<0.002$	0.05

than the crustaceans, apparently to cover greater amplitudes within the same time span.

Night-time ascent is considered to be advantageous to planktonic rotifers: (1) during dark, a period of cessation of primary production, the rotifers exploit the energy stored in algal biomass before it enters detritus food chain, and (2) the day-time descent facilitates predator avoidance (Miracle, 1977). However, invertebrate predators are absent in Kruczy Staw Lake and rotifers are not preyed upon by adult fish (perch) thus predator avoidance cannot be a factor that will induce such strong migrations whereas food availability probably is a factor that triggers migration.

In summer, zooplankton food resources of zooplankton are very poor, both quantitatively and qualitatively. Larger algae were absent and phytoplankton consisted of mainly picoplankton belonging to species of Chlorophycea and Cyanophyceae. In the lakes investigated by us picoplankton densities were, however, low as suggested by concentrations of chlorophyll and high Secchi-disc transparency (7.6 m).

Therefore, zooplankton grazing pressure on phytoplankton is probably strong. The inverse relationship between the vertical distribution of zooplankton densities and chlorophyll concentration confirms this. Concurrently with an increase in rotifer and crustacean densities chlorophyll concentration decreased rapidly at different depths and *vice-a-versa*. The conclusion is supported by correlation coefficients between concentration of chlorophyll *a* at different depths and densities of three taxonomic groups (Rotifera, Cladocera, Calanoida) of zooplankton in Lake Kruczy Staw (Table 1). Statistically significant negative correlation between chlorophyll and rotifer abundance suggests that, despite of high energetic costs, migrations of rotifers into upper, food-rich and warm water layers is generally profitable.

Competition for food resources may be a reason of vertical partition of rotifers and crustaceans both spatially (diel cycles) and temporally. According to 'classical' size-efficiency hypothesis, large-size species are stronger competitors at low food concentration than small ones as they can exploit a broader food size spectrum (Hall et al., 1976). It could explain why crustacean community of Kruczy Staw Lake is quantitatively and qualitatively richer than rotifer community and why rotifers are driven down to the bottom layers. However, we did not differentiate the size of available food particles (exclusively picoplankton of cell size $<2\ \mu$m). According to literature, Cladocera may feed on food particles $<2\ \mu$m but their filtering efficiency is then low (Gliwicz, 1977; Knoechel & Holtby, 1986; Yasuno & Zhang, 2002). Even though the grazing efficiency of Calanoida on such small-sized particles is expected to be even lower (Bogdan & Gilbert, 1984; Zankai, 1994), *E. gracilis* is the most abundant zooplankter in Kruczy Staw Lake. Both, the lack of an inverse correlation between chlorophyll a concentrations and *Eudiaptomus gracilis* densities, and less significant relationship between chlorophyll concentrations and Cladocera densities than between chlorophyll and Rotifera (Table 1) support the literature suggestions that picoplankton itself cannot satisfy food demands of Crustacea, and particularly *E. gracilis*. Moreover, the differences between the three zooplankton groups imply that picoplankton is relatively most efficiently exploited by rotifers. Crustaceans may, however, utilize alternative food sources but we did not analyse their gut contents.

Conclusion

- A lack of predators controlling rotifer abundance and, at the same time, significant correlation between chlorophyll concentrations and rotifer densities suggest that the main factor that triggers vertical migrations of Rotifera is food resource, both its quality and quantity.
- Energetic costs of migrations are high. Assuming that picoplankton itself does not satisfy food (and so energetic) demands of Cladocera, there should exist another factor inducing vertical migrations. It is well known (Lampert 1993), that diel vertical migration is prey's response to visual predator abundance. Perch population in Lake Kruczy Staw is small and stunted. Relatively high densities of large Cladocera indicate rather weak impact of the fish on zooplankton prey. However, predator presence and namely, their kairomone may induce vertical migrations (Loose & Dawidowicz, 1994).

Acknowledgements

We would like to thank Hendrik Segers and Willem H. DeSmet for their help with determination of *Trichocerca* species and two anonymous referees for constructive comments on an earlier version of this manuscript.

References

Bogdan, K. G. & J. J. Gilbert, 1984. Body size and food size in freshwater zooplankton. Proceedings of the National Academy of Sciences USA 81: 6427–6431.

George, M. G. & C. H. Fernando, 1970. Diurnal migration in three species of rotifers in Sunfish Lake, Ontario. Limnology & Oceanography 15: 218–223.

Gilbert, J. J., 1989a. The effect of Daphnia interference on a natural rotifer and ciliata community Short term bottle experiments. Limnology & Oceanography 34: 606–617.

Gilbert, J. J., 1989b. Competitive interactions between the rotifer *Synchaeta oblonga* and the cladoceran *Scapholeberis kingi* Sars. Hydrobiologia 186/187: 75–80.

Gliwicz, Z. M., 1977. Food size selection and seasonal succession of filter feeding zooplankton in an eutrophic lake. Ekologia Polska 25: 179–225.

Hall, D. J., S. T. Threlkeld, C. W. Burns & P. H. Crowley, 1976. The size-efficiency hypothesis and the size structure of zooplankton communities. Annual Review of Ecology and Systematics 7: 177–208.

Knoechel, R. & L. B. Holtby, 1986. Cladoceran filtering rate: body length relationship for bacterial and large algal particles. Limnology & Oceanography 31: 195–200.

Lampert, W., 1993. Ultimate causes of diel vertical migration of zooplankton: new evidence for the predator avoidance hypothesis. Archiv fur Hydrobiologie 39: 79–88.

Loose, C. J. & P. P Dawidowicz, 1994. Trade offs in diel vertical migration by zooplankton: the costs of predator avoidance. Ecology 75: 2255–2263.

MacIsaac, H. J. & J. J. Gilbert, 1990. Does exploitative or interference competition from *Daphnia* limit the abundance of *Keratella* in Loch Leven ? A reassessment of May and Jones (1989). Journal of Plankton Research 12: 1315–1322.

MacIsaac, H. J. & J. J. Gilbert, 1991. Discrimination between exploitative and interference competition between Cladocera and *Keratella cochlearis*. Ecology 72: 924–937.

Miracle, M. R., 1977. Migration, patchiness, and distribution in time and space of planktonic rotifers. Archiv fur Hydrobiologie 8: 19–37.

Rey, J. & J. Capblancq, 1975. Dynamique des populations et production du zooplancton du lac de Port-Bielh (Pyrenees Centrales). Annales Limnologie 11: 1–45.

Ruttner-Kolisko, A., 1975. The vertical distribution of planktonic rotifers in a small alpine lake with a sharp oxygen depletion. Internationale Vereinigung für Theotretische und Angewandte Limnologie, Verhandlungen 19: 1286–1294.

Tunowski, J., 1992. Zooplankton of dystrophic lakes of Wigry National Park. In B. Zdanowski (ed.), Lakes of Wigry National Park, Ossolineum, Wroitaw: 129–134.

Yasuno, M. & X. Zhang, 2002. Feeding experiments of *Moina macrocopa* (Cladocera) on picoplankton. Internationale Vereinigung für Theotretische und Angewandte Limnologie, Verhandlungen 28: 1199–1202.

Zankai, N. P., 1994. Feeding of copepodite and adult stage of *Eudiaptomus gracilis* (G.O. Sars, 1863) (Copepoda, Calanoida) on mixed plastic beads. Crustaceana 66: 90–109.

Hydrobiologia (2005) 546:387–395
A. Herzig, R.D. Gulati, C.D. Jersabek & L. May (eds.) Rotifera X: Rotifer Research: Trends, New Tools and Recent Advances
DOI 10.1007/s10750-005-4281-4

Structure distinctions of pelagic rotifer plankton in stratified lakes with different human impact

Galina A. Galkovskaya* & Inessa F. Mityanina
Institute of Zoology, National Academy of Sciences of Belarus, Academicheskaya st., 27, 220072 Minsk, Belarus
*(*Author for correspondence: E-mail: gal@biobel.bas-net.by)*

Key words: plankton rotifers, species diversity, spatial structure, stratified lakes, human impact

Abstract

Pelagic rotifer plankton was studied in four stratified lakes with different degrees of human impact from June to July 2001 and throughout 2002. Rotifer species diversity was closely correlated to temperature and oxygen concentration (correlation coefficients were 0.90 and 0.87, respectively) in the water column of the hypertrophic Lake Kruglik. In the mesotrophic lakes, the correlation coefficients were much lower and their reduction was related to decreasing human impact on the lakes. Species richness was similar in Lakes Kruglik and S. Volos, but the spatial structure of the community differed greatly. The maximum rotifer density was observed in the epilimnion of Lake Kruglik, with densities dropping sharply towards the hypolimnion. In the mesotrophic lakes, the highest rotifer density was recorded in the meta- and hypolimnion. A comparative analysis of the morphometric characteristics of *Keratella cochlearis* showed that (1) the lorica length of ovigerous females increased in all four lakes with decreasing temperature; (2) the shortest lorica length was in Lake Kruglik at the same temperature; (3) in the mesotrophic lakes a significant increase in lorica length occurred as the temperature decreased from 14.2 °C to 4.2 °C. There is the similar relationship in rotifers of the genus *Filinia*. Hypoxia in the clino- and hypolimnion of Lake Kruglik reduced the diversity of spatial niches created by thermal stratification. As a result, the number of non-overlapping niches for rotifers in Lake Kruglik is reduced by a factor of 2–5 compared to that in mesotrophic lakes, but the mean value of the overlapping index is significantly higher.

Introduction

Biodiversity in any ecosystem is sustained, to a great extent, by the structural diversity of the system. Under different levels and types of disturbance, the change in the system structure can optimize the functioning that preserves biological diversity. Unfortunately, when studying planktonic animals and, in particular, rotifers, little attention is usually paid to investigating of their community structure. However, knowledge of structural organization is important for obtaining a better understanding of the mechanisms of community self-sustention and of the principles that underlie the functioning of freshwater ecosystems. It is also important to understand the structural changes that occur in response to different types of disturbance, including variation in the temperature profile of the water column.

Since the 1970s, many authors have described differences in the vertical density distribution of individual species within the pelagic plankton, especially in relation to rotifers of the genus *Keratella* (Larsson, 1971; Hofmann, 1983a; Mikschi, 1989) We have substantiated, for the first time, the principle of layer functioning of *Keratella cochlearis* populations under stable and prolonged thermocline conditions (Galkovskaya & Mityanina, 1989). Later, this principle was verified in other rotifer species (*Filinia terminalis*, *Anuraeopsis fissa*,

Kellicottia longispina, *Polyarthra* spp.) in the stratified Lake Banyoles (Miracle & Alfonso, 1993). At present, layer functioning has been shown to be inherent in many rotifer species in lakes showing a stable thermocline during the summer months, with some species forming individual functional groups (local populations) within these layers while others are closely related to a certain layer (Geller et al., 1992; Galkovskaya et al., 2001).

The goal of this research was to make a comparative analysis of species diversity, the morphometric characteristics of some species, and the spatial structure of the rotifer communities in stratified lakes with different levels of human impact.

Materials and methods

The investigations were carried out in the following four lakes in the Vitebsk region of Belarus in June–July 2001 and throughout 2002: Lake South (S.) Volos (area 1.21 km^2, max. depth 40.4 m); Lake Dolgoe (area 2.6 km^2, max depth 53.6 m); Lake North (N.) Volos (area 4.21 km^2, max. depth 29.2 m); Lake Kruglik (area 0.4 km^2, max. depth 31.5 m). These lakes have a similar genesis type and different levels of human impact. The lakes showed steady thermal stratification from the end of May to September each year and differed in the oxygen content of the deeper waters. Changes in the temperature profile of the water column during the investigation are shown in Figure 1. The water column was separated into layers following Hutchinson (1957).

Some characteristics of the lakes investigated are given in Table 1. According to Carlson's classification (Carlson, 1977), Lake S. Volos is a mesotrophic lake with oligotrophic features and Lake Kruglik is a hypertrophic lake.

Zooplankton samples were collected at the deepest part of each lake. Samples were taken with a Ruttner sampler (1 litre) at several horizons in each layer: epilimnion – at 2 horizons, metalimnion – at 3–4 horizons, clinolimnion – at 2 horizons, hypolimnion – at 2 horizons. In total, 76 samples were taken. The samples were narcotized with CO_2 saturated water and fixed in 2%

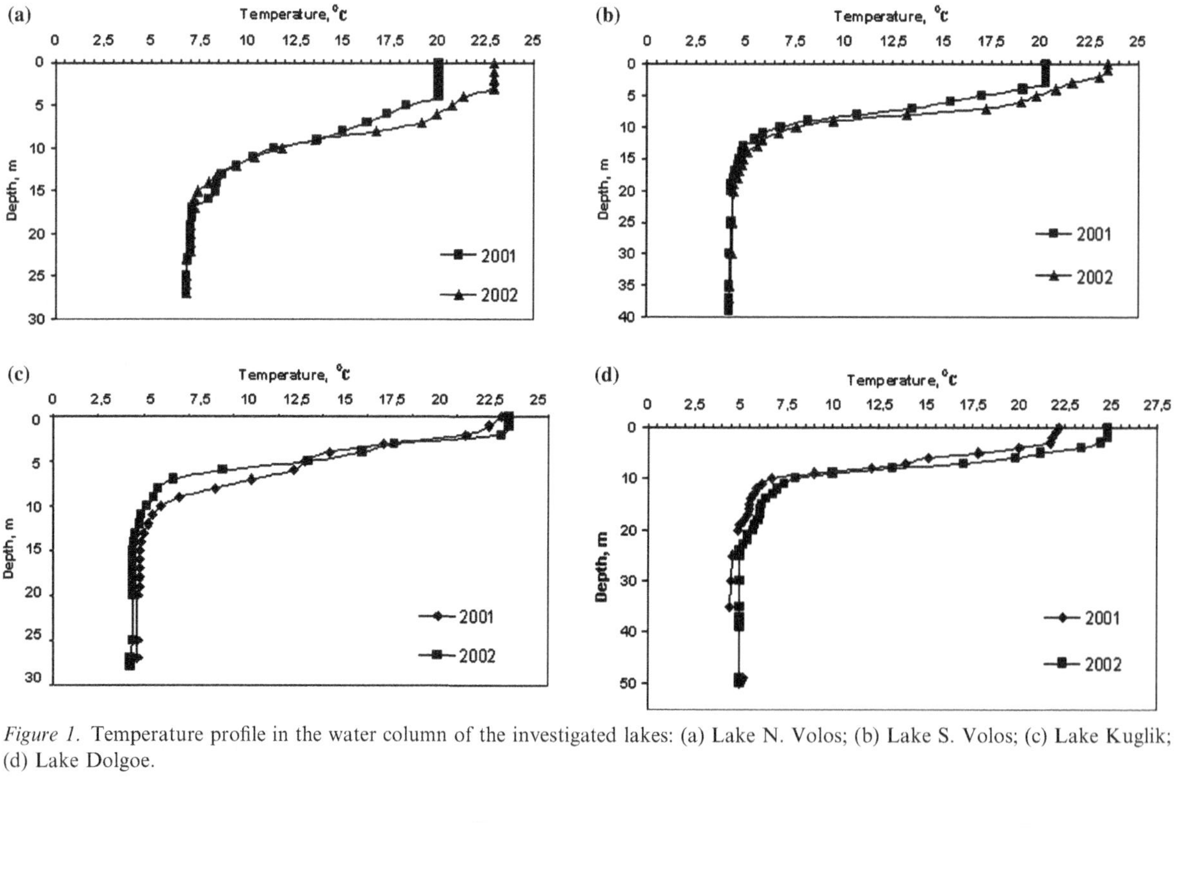

Figure 1. Temperature profile in the water column of the investigated lakes: (a) Lake N. Volos; (b) Lake S. Volos; (c) Lake Kuglik; (d) Lake Dolgoe.

Table 1. Some characteristics of the lakes investigated

Lake	S. Volos		Dolgoe		N. Volos		Kruglik	
Date	27.06.2001	12.07.2002	2.07.2001	16.07.2002	28.06.2001	13.07.2002	4.07.2001	17.07.2002
Oxygen saturation near bottom (%)	58.7	33.7	57.0	40.0	26.0	13.0	5.8	Anoxia
Secchi depth (m)	6.7	7.3	5.0	5.1	6.5	4.8	0.6	0.7
Epilimnion* (m)	3.5	2	4	3	4	3	2	2
Metalimnion (m)	12	11	11	10	12	13	10	7
Clinolimnion	19	15	13	23	17	16	13	14*

*Lower boundary is given for layers.

formalin. After settling for 3 days, the supernatant liquid was decanted until there was no more than 100 ml of the original sample volume left. All of the rotifers in the sample were counted. Lorica lengths were measured for the dominant species of loricate rotifers. Spatial niche overlapping in the stratified water column was calculated after Pianka (Pianka, 1974):

$$ON_{jk,kj} = \Sigma p_{ij}p_{ik}/\sqrt{\Sigma p_{ij}^2 \Sigma p_{ik}^2}$$

where:

$ON_{jk,kj}$ the overlapping index of species *j* and *k*;
p_{ij} – the ratio of the density of species *j* in the *i*th sample to the sum of densities of this species in all the samples of the vertical water column;
p_{ik} – the same ratio for the species *k*.
The values of ON can vary between 0 and 1.

Statistical processing of the material was carried out with the computer program STATISTICA (version 5.5) and EXCEL.

Results and discussion

Species diversity

Species richness ranged from 18 in 2001 to 25 in 2002. The corresponding number of species for each of these years in each lake was: Lake S. Volos – 20 and 18; Lake N. Volos – 18 and 21; Lake Dolgoe – 21 and 25; Lake Kruglik – 20 and 18. It should be noted that, because the data on species richness were obtained from samples with a volume of 1 l, the species density in each sample was never less than 1 ind.$^{-1}$. In total, 37 species and varieties of rotifer were recorded (Table 2). Three species were found only in Lake S. Volos: *Conochiloides natans*, *Lecane flexilis*, *L. luna*. Five species and varieties were observed only in Lake Dolgoe: *Keratella cochlearis hispida*, *Philodina* sp., *Pompholyx complanata*, *Synchaeta pectinata*, *Synchaeta tremula*. Four species and varieties were revealed only in Lake Kruglik: *Anuraeopsis fissa*, *Brachionus angularis bidens*, *Filinia longiseta*, *Trichocerca pusilla*.

As seen from Table 3, the highest values of the Chekanovsky–Sorensen similarity index were observed when comparing the species richness of rotifers in Lakes S. and N. Volos in both 2001 and 2002. The lowest coefficients were noted between Lake Kruglik and the other three lakes.

The number of species in individual layers of the stratified water column in Lake Kruglik was closely and significantly correlated with the temperature ($r = 0.90$, $n = 18$, $p < 0.001$) and dissolved oxygen ($r = 0.87$, $n = 18$, $p < 0.001$). In the other lakes, these correlations became weaker with decreasing human impact (see Tables 4 and 5). In Lake S. Volos, the number of species in each horizon does not depend on either temperature or oxygen concentration.

Spatial structure

A "layered" variation in total rotifer density was particularly pronounced in Lake Kruglik compared to the other three lakes. The mean values of the rotifer density in the epilimnion of this hypertrophic lake were approximately 10 times greater than in the epilimnion of the other lakes. In the metalimnion it was 2–5 times smaller than in the epilimnion. The smallest density was in the

Table 2. List of *Rotifera* found in the lakes investigated, June–July 2001 and 2002

		S. Volos	N. Volos	Kruglik	Dolgoe
1	*Anuraeopsis fissa* (Gosse, 1851)			*	
2	*Ascomorpha ecaudis* Perty, 1850	*	*	*	*
4	*Ascomorpha saltans* Bartsch, 1970	*		*	*
5	*Asplanchna priodonta* Gosse, 1850	*	*		*
6	*Bipalpus hudsoni* (Imhof, 1891)		*		*
7	*Brachionus angularis* Gosse, 1851		*		*
8	*Brachionus angularis bidens* Plate, 1886			*	
9	*Cephalodella catellina* (Müller, 1786)	*	*		
10	*Chromogaster ovalis* (Bergental, 1892)	*	*	*	*
11	*Collotheca pelagica (*Rousselet, 1893)	*	*	*	*
12	*Conochiloides natans* (Seligo, 1900)	*			
13	*Conochilus unicornis* Rousselet, 1892	*	*	*	*
14	*Filinia longiseta* (Ehrenberg, 1834)			*	
15	*Filinia terminalis* (Plate, 1886)	*	*		*
16	*Gastropus stylifer* Imhof, 1891	*	*	*	*
17	*Kellicottia longispina* (Kellicott, 1879)	*	*	*	*
18	*Keratella cochlearis* (Gosse, 1851)	*	*	*	*
19	*Keratella cochlearis hispida* (Lauterborn, 1898)				*
20	*Keratella cochlearis tecta* (Gosse, 1851)		*	*	
21	*Keratella quadrata* (Müller, 1786)	*	*	*	*
22	*Lecane flexilis* (Gosse, 1886)	*			
23	*Lecan luna* (Müller, 1776)	*			
24	*Philodina* sp.				*
25	*Polyarthra dolichoptera* Idelson, 1925	*	*		*
26	*Polyarthra major* Burckhardt, 1900	*	*	*	*
27	*Polyarthra remata* Skorikov, 1896	*	*		*
28	*Polyarthra vulgaris* Carlin, 1943	*	*	*	*
29	*Pompholyx complanata* Gosse, 1851				*
30	*Pompholyx sulcata* Hudson, 1885			*	*
31	*Synchaeta kitina* Rousselet, 1902	*	*	*	*
32	*Synchaeta pectinata* Ehrenberg, 1832				*
33	*Synchaeta tremula* (Müller, 1786)				*
34	*Trichocerca capucina* (Wierzejski et Zacharias, 1893)	*	*	*	*
35	*Trichocerca pusilla (*Lauterborn, 1898)			*	
36	*Trichocerca similis* (Wierzejski, 1893)	*	*	*	*
37	*Trichocerca stylata* (Gosse, 1851)			*	*
38	*Trichocerca rousseleti* (Voigt, 1902)	*	*	*	*
	Number of species	23	22	22	28

clino- and hypolimnion (zone of low temperature and hypoxia).

The water transparency and oxygen concentration near the bottom of the lake suggested that there was an increase in human impact in the three mesotrophic lakes in the following direction: S. Volos – Dolgoe – N. Volos. In these lakes, rotifer density in the meta- and clinolimnion was higher than in the epilimnion. At the same time, rotifer density, calculated as the weighted mean for the water column, proved to be similar for all four lakes; the interannual range of differences within each lake overlapped the differences among the lakes (Table 6).

Table 3. Similarity indices of rotifer species richness in the lakes studied

Rotifera				
2001				
	Kruglik	S. Volos	Dolgoe	N. Volos
Kruglik	1	0.56	0.56	0.63
S. Volos	0.67	1	0.75	0.77
Dolgoe	0.60	0.78	1	0.77
N. Volos	0.72	0.82	0.78	1
2002				

Table 4. Correlation coefficients (r) between species number and temperature (t, °C) in the stratified water column (n – sample number)

Lake	Δt, °C	n	r	p
Kruglik	20.7 (25.0 – 4.3)	18	0.90	<0.001
Dolgoe	19.8 (24.8 – 5.0)	18	0.66	0.004
S. Volos	18.8 (23.0 – 4.2)	18	0.18	0.46
N. Volos	16.2 (23.0 – 6.8)	17	0.70	0.002

Table 5. Correlation coefficients (r) between the number of species and dissolved oxygen concentration (O_2, ml l^{-1}) in the stratified water column (n – sample number)

Lake	ΔO_2 ml l^{-1}	n	r	p
Kruglik	8.80 (8.81 – 0.01)	18	0.87	<0.001
N. Volos	5.21 (6.37 – 1.16)	17	0.60	0.01
Dolgoe	4.19 (6.85 – 2.66)	18	0.61	0.009
S. Volos	3.46 (8.33 – 4.87)	18	0.35	0.016

Table 6. Mean rotifer density (ind. l^{-1}) in layers and in the whole water column, June–July 2001 and 2002

Lakes	S. Volos		N. Volos		Dolgoe		Kruglik	
Year of investigation	2001	2002	2001	2002	2001	2002	2001	2002
Epilimnion	180.7	29	211.5	291	171.5	213	1524.5	3086
Metalimnion	353	168	430.7	383.8	387.7	205.7	773	571
Clinolimnion	399	307	262	302	142	292	3	37.5
Hypolimnion	294	27.5	67	144.5	98	333	4	34
Weighted mean for water column	314.5	116	238.5	268.2	209.6	297.3	316.5	380.2

How is the pelagic rotifer community structured in the lakes studied? The ratio of species population density within the layers of the stratified water column in July is shown in Figure 2. *Polyarthra vulgaris* dominated in the epilimnion of all of the lakes. *Ascomorpha ecaudis, Ascomorpha saltans, Collotheca pelagica, Conochilus unicornis, Gastropus stylifer, Polyarthra remata* populations were concentrated in the metalimnion of the mesotrophic lakes. *Filinia. terminalis* was concentrated in the clinolimnion where the temperature was below 10 °C. This species is known as a cold stenotherm (Ruttner-Kolisko, 1980; Bērzinš & Pejler, 1989). Maximum densities of many eurythermic species

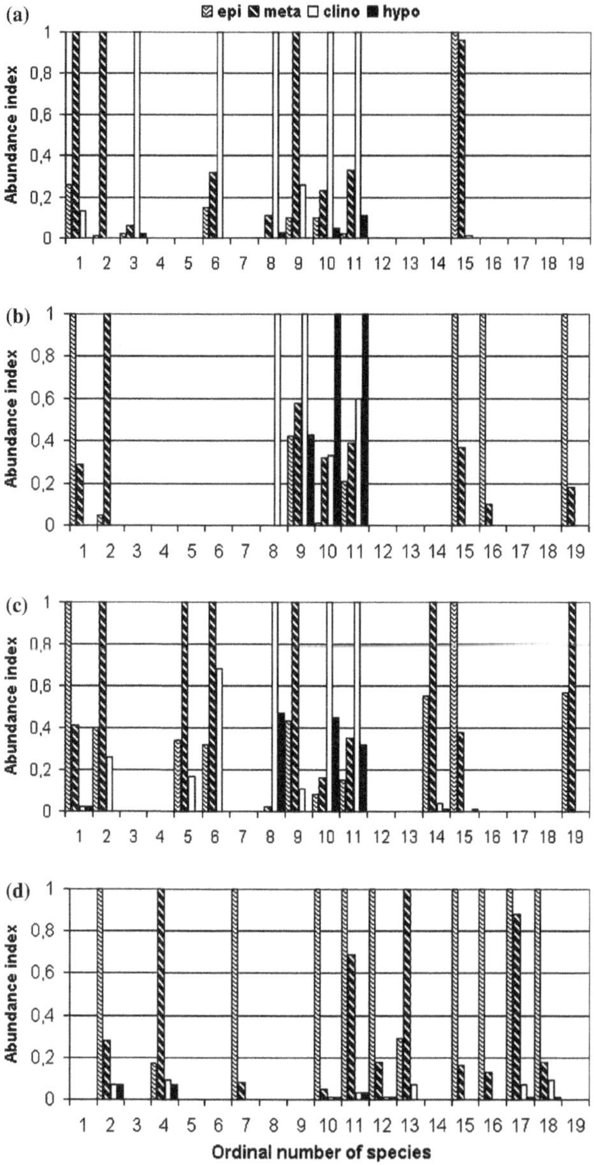

Figure 2. Distribution of rotifers by abundance index in the water layers of the lakes investigated: (a) Lake S. Volos, 12.07.2002; (b) Lake Dolgoe, 2.07.2002; (c) Lake N. Volos, 13.07.2002; (d) Lake Kruglik, 17.07.2002. The ratio between the given density of a species and its maximum value is used as an abundance index. The maximum density of the species is chosen as the maximum value occurring in the four layers. Species whose maximum density was above 10 ind l^{-1} are shown in this figure. Species are presented in the following order: 1 – *A. ecaudis*, 2 – *A. saltans*, 3 – *A. priodonta*, 4 – *B. angularis*, 5 – *C. pelagica*, 6 – *C. unicornis*, 7 – *F. longiseta*, 8 – *F. terminalis*, 9 – *G. stylifer*, 10 – *K. longispina*, 11 – *K. cochlearis*, 12 – *K. c. tecta*, 13 – *K. quadrata*, 14 –*P. remata*, 15 – *P. vulgaris*, 16 – *P. sulcata*, 17 – *S. kitina*, 18 – *T. pusilla*, 19 – *T. similis*.

were also observed in the clinolimnion, only *Asplanchna priodonta, C. unicornis* (S. Volos), *K. longispina*, *K. cochlearis* (S. Volos, N. Volos), *G. stylifer* (Dolgoe). Finally, *K. cochlearis* and *K. longispina* have shown maximum density in the hypolimnion of Lake Dolgoe (Fig. 2a–c).

The maximum density of any of the species was noted in the epilimnion of eutrophic Lake Kruglik (Fig. 2d). All of the species found here were typical of eutrophic lakes. Along with the typical form of *K. c. cochlearis, K. c. tecta* form was also abundant (accordingly, 300 and 935 ind. l^{-1}). The densities

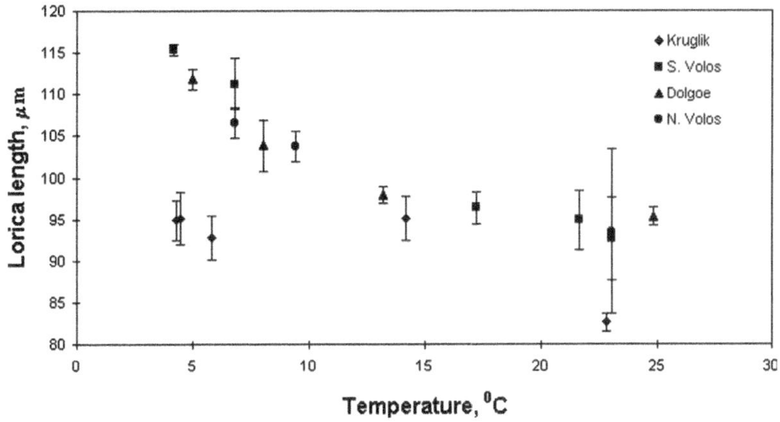

Figure 3. Lorica length of *Keratella cochlearis* ovigerous females in sampling horizons that differ in water temperature, July 2002.

of all species are very small in clino- and hypolimnion in comparison with epi- and metalimnion: hypoxia (oxygen saturation $<2.0\%$) is already noted at a depth of 7 m.

It follows from the above that "thermopreference", appearing in the form of maximum abundance in one of the layers, manifests itself in some eurythermic species in stratified mesotrophic lakes. A sharp increase in the density of crustacean plankton was observed in all of the lakes from the end of June to the beginning of July. Probably, stratification creates conditions for self-supporting rotifer populations at lower temperatures thereby reducing the risk of negative interactions with the crustacean plankton. Abundance of *A. priodonta* is usually associated with that of *K. cochlearis* (Pourriot, 1977; Hofmann, 1983b). Maximum densities of these species were observed in the clinolimnion of Lake S. Volos (temperature <10 °C). Just such conditions seem to be optimal for these species, which usually show a "predator-prey" type interaction within the general structure of summer zooplankton in stratified lakes.

In the species that do not display pronounced thermopreferences, morphofunctional groups may be formed under conditions of thermal stratification. These groups show a distinct difference in morphometric characteristics within the temperature range of the vertical water column. This was shown earlier by comparing the dimensional characteristics of the *K. cochlearis* lorica (Galkovskaya, 1998). Figure 3 shows variation in the *K. cochlearis* lorica length depending on the temperature of the sampling horizon (mean values for 10 measurements of females with eggs are given). One can easily see that (1) lorica length increases as temperature decreases, (2) under similar temperatures the individual size is smallest in Lake Kruglik, (3) in this group of mesotrophic lakes there are no significant differences in lorica size over a temperature interval of 17–23 °C, although significant increases in lorica length occur with temperatures decreasing from 14.2 °C to 4.2 °C. There is some eveidence that a similar tendency is also found in rotifers of the genus *Filinia* as, in Lake Kruglik, the mean body length of ovigerous females of *Filinia longiseta* was 148 ± 4.57 μm at 22.8 °C and 161.5 ± 4.17 μm at 4.5 °C. Unfortunately, it was not possible to obtain the required sample volume for studying the size variation in relation to temperature variation of the stratified water column for any other species.

The number of cases of non-overlapping niches (N_0) is lowest in Lake Kruglik. With similar species richness in Lakes Kruglik and S. Volos in 2001, the number of non-overlapping cases in Lake S. Volos was more than twice as many as in Lake Kruglik and, in 2002, the number of non-overlapping cases in Lake S. Volos was five times as many as in Lake Kruglik. With similar species richness in Lakes Dolgoe and N. Volos, the number cases of non-overlapping niches was much higher than in Lake Kruglik. The mean value of niche overlapping indices (ON $\pm$ SE) with similar species richness was considerably higher in the hypertrophic lake in comparison with the mesotrophic one (Table 7).

Table 7. Characteristics of rotifer niche overlapping in four stratified lakes, June–July 2001 and 2002

Number of cases (N)	Kruglik		Dolgoe		N. Volos		S. Volos	
	01	02	01	02	01	02	01	02
$N_{0 \to 1}$	190	153	190	300	153	210	190	153
N_0	18	6	37	86	38	23	39	31
$N_{0 \to 0.1}$	37	9	63	119	49	58	75	51
$N_{0 \to 0.1}/N_{0 \to 1}$	0.19	0.06	0.32	0.34	0.29	0.23	0.39	0.33
$N_0/N_{0 \to 1}$	0.09	0.04	0.20	0.29	0.22	0.11	0.20	0.20
Species number	20	18	21	25	18	21	20	18
ON ± SE	0.48 ± 0.025	0.62 ± 0.022	0.43 ± 0.025	0.28 ± 0.017	0.41 ± 0.027	0.39 ± 0.023	0.31 ± 0.023	0.35 ± 0.027

Formation of functional groups or local populations in stratified layers assumes the existence of a stable thermal stratification during a time comparable to the duration of at least some successive generations of the populations. Short-term sporadic disturbances of stratification within the water column occur in some stratified lakes. For instance, sporadic agitation due to methane discharge from the bottom takes place in the large eutrophic Lake Biva (Sakai et al., 2002). Also, epilimnion depth can change and turbulent flows can be formed in the metalimnion depending on wind strength. In this case, the peculiar morphometric characteristics of the lake, including its area, will result in disturbance of the plankton spatial structure (Saggio & Imberger, 2001).

Regular change in epilimnion depth under wind-induced mixing almost certainly occurs in the lakes that we studied; interannual differences are important too. For example, in June–July, 2000 the epilimnion depth in Lake S. Volos varied from 0–4 m to 0–6 m; at the end of June, 2001 it was 0–3.5 m and in July, 2002 – 0–2 m, with the lower boundary of the metalimnion being almost constant (11–12 m). This means that, during the stratification period, which lasts from the end of May to September, considerable perturbations occurred due to changes in the epi- and metalimnion boundary. A sharp gradient in water temperature and viscosity in the metalimnion decreases the rate of sedimentation of suspended organic matter that is formed in the epilimnion. This results in the accumulation of biogenic elements in the epilimnion, which promotes eutrophication. So, increases in the epilimnion depth favours the incorporation of a large amount of organic material into the food chain and sustaining zooplankton species richness in mesotrophic lakes. This supposition was verified by the investigations of zooplankton species richness in Lake S. Volos performed between May and September 2000: the epilimnion depth increased from May to July and, accordingly, species richness of both rotifers and zooplankton (Rotifera, Crustacea) as a whole increased.

The hypolimnion is practically anoxic in all of the series of eutrophic lakes to which Lake Kruglik belongs. Nevertheless, in the hypolimnion of Lake Kruglik we found some rotifer species living at densities of 8.4 ind. l^{-1} in 2001 and 34 ind. l^{-1} in 2002. Sulphur bacteria are known to develop in the hypolimnion of stratified lakes (Watanabe, 1992). Obtaining electron-donor from sulphide, they are capable of synthesizing organic material that favours the formation of specific consortia in the hypolimnion. More detailed investigations of biodiversity in stratified lakes, in particular within the hypolimnetic zone of eutrophic lakes, are required.

Conclusion

Thermal stratification of the water column creates conditions for structural heterogeneity of the pelagic zooplankton. This manifests itself in both changes in species richness and in differences in the spatial location of species populations. In eutrophic lakes with hypolimnetic hypoxia, species collect in the epilimnion; in mesotrophic ones disjunction of spatial niches of the species is observed within the whole vertical water column. Increased human impact promotes decreases in the individual sizes of the species population in keeping with temperature effects on sizes in the stratified vertical water column. The structural organization of the rotifer community, especially the structure of niche overlapping, may be a model in comparing trophic status of stratified lakes. The sustention of species diversity under high human impact, when the possibility of occupying spatial niches formed as a result of thermal stratification is reduced due to hypoxia, can be accounted for only by niche disjunction of species populations along other vectors of niches, in particular along the trophic vector.

Acknowledgements

The authors are indebted to the anonymous referees for valuable comments. Special thanks to Dr Linda May for correcting the English text.

References

Bērzinš, B. & B. Pejler, 1989. Rotifer occurrence in relation to temperature. Hydrobiologia 175: 223–231.

Carlson, R. E., 1977. A trophic state index for lakes. Limnology and Oceanography 22: 361–369.

Galkovskaya, G. A. & I. F. Mityanina, 1989. Morphological structure and functional patterns of *Keratella cochlearis* (Gosse) populations in stratified lakes. Hydrobiologia 182: 119–129.

Galkovskaya, G. A., 1998. Morphotypical diversity and morphometric characteristics of *Keratella cochlearis* (Gosse, 1851) in stratified lakes. Polish Journal of Ecology 46: 187–196.

Galkovskaya G. A., I. F. Mityanina & N. N. Voskobovich, 2001. Spacial structure peculiarities of the zooplankton species richness. In The Animal Biodiversity in Belarus: 8–9.

Geller, W., R. Pinto-Coelho & H.-R. Pauli, 1992. The vertical distribution of zooplankton (Crustacea, Rotatoria, Ciliata) and their grazing over the diurnal and seasonal cycles in lake Constance. Archiv für Hydrobiologie 35: 79–85.

Hofmann, W., 1983a. On temporal variation in the rotifer *Keratella cochlearis* (Gosse): the question of "Lauterborn-cycles". Hydrobiologia 101: 247–254.

Hofmann, W., 1983b. Interactions between *Asplanchna* and *Keratella cochlearis* in the Plußsee (north Germany). Hydrobiologia 104: 363–365.

Hutchinson, G. E., 1957. A Treatise on Limnology, Vol. 1. Wiley Sons, New York, 1015 pp.

Larsson, P., 1971. Vertical distribution of planktonic rotifers in a meromictic Lake; Blankvath near Oslo, Norway. Norwegian Journal of Zoology 19: 47–75.

Mikschi, E., 1989. Rotifer distribution in relation to temperature and oxygen content. Hydrobiologia 186/187: 209–214.

Miracle, M. R. & M. T. Alfonso, 1993. Rotifer vertical distributions in a meromictic basin of Lake Banyoles (Spain). Hydrobiologia 255/256: 371–380.

Pianka, E. R., 1974. Niche overlap and diffuse competition. Proceeding National Academy of Sciences USA 71: 2141–2145.

Pourriot, R., 1977. Food and feeding habits of Rotifera. Archiv für Hydrobiologie 8: 243–260.

Ruttner-Kolisko, A., 1980. The abundance and distribution of *Filinia terminalis* in various types of lakes as related to temperature, oxygen, and food. Hydrobiologia 73: 169–175.

Saggio, A. & J. Imberger, 2001. Mixing and turbulent fluxes in the metalimnion of a stratified lake. Limnology and Oceanography 46: 392–409.

Sakai, Y., J. Murase, A. Sugivoto, K. Okubo & E. Nakayama, 2002. Resuspension of bottom sediment by an internal wave in Lake Biwa. Lake Reservoirs Research and Management 7: 339–344.

Watanabe, J., 1992. Effect of thermal stratification on trophic linkages among plankton communities in eutrophic lakes. Archiv für Hydrobiologie 35: 1–12.

Hydrobiologia (2005) 546:397–404
A. Herzig, R.D. Gulati, C.D. Jersabek & L. May (eds.) Rotifera X: Rotifer Research: Trends, New Tools and Recent Advances
DOI 10.1007/s10750-005-4282-3

Changes in rotifer species composition and abundance along a trophic gradient in Loch Lomond, Scotland, UK

Linda May[1,]* & Matthew O'Hare[2]
[1]*Centre for Ecology and Hydrology, Bush Estate, Penicuik, Midlothian, EH26 0QB Scotland, UK*
[2]*Centre for Ecology and Hydrology, Winfrith Technology Centre, Dorchester, Dorset, DT2 8ZD, UK*
(**Author for correspondence: E-mail: lmay@ceh.ac.uk*)

Key words: Lakes, rotifers, indicators, trophic state, phosphorus

Abstract

Loch Lomond is the second largest body of freshwater in Great Britain. It is a long, narrow lake (36.4 km long, 8.8 km wide). The northern basin is fjord-like and surrounded by a mountainous, base-poor, rocky catchment. In contrast, the southern basin is much broader and shallower with a mainly lowland, calcareous, agricultural catchment. This causes a trophic gradient along the length of the loch that runs from the oligotrophic northern basin to the more mesotrophic southern basin. Rotifer samples were collected at monthly intervals between May and October 2002 at three locations along the length of the loch. More than 12 species were found, the commonest of which were *Keratella cochlearis* (Gosse) and *Trichocerca stylata* (Gosse). Although the species composition of the rotifer community varied very little among the sites, rotifer abundance increased markedly from north to south, apparently reflecting the trophic gradient along the length of the loch. The results suggest that rotifer abundance may be a more sensitive indicator of trophic state, and changes in trophic state, than species composition.

Introduction

Many studies have shown that rotifer species composition and abundance varies from lake to lake. This has led many authors to conclude that rotifers could be used as indicators of ecological status, especially in relation to lake trophy (e.g. Herzig, 1979; Blancher, 1984; Karabin, 1985; Berzins & Pejler, 1989). In general, inter-site comparisons have suggested that both species composition and abundance could be used as indicators of trophic state (Gannon & Stemberger, 1978; Berzins & Pejler, 1989; Matveeva, 1991, Duggan et al., 2001). However, within-site studies have indicated that abundance is probably the more sensitive indicator, especially during periods of change (e.g. Fuller et al., 1977; Karabin, 1985; Evans, 1986; Walz et al., 1987; Matveeva, 1991; Ejsmont-Karabin & Hillbricht-Ilkowsks, 1994). This is especially well illustrated by work on Lake Mikolajskie, Poland, where maximum total rotifer densities increased from 2000 ind l^{-1}, in 1963/64, to 8,000 ind l^{-1}, in 1989/90, as the lake became increasingly eutrophic (Ejsmont-Karabin & Hillbricht-Ilkowska, 1994), and on Loch Leven, Scotland, where maximum rotifer densities fell from 7,000 ind l^{-1} in 1978/1982 to 2,500 ind l^{-1} in 1991/1994 following a reduction of 50% in the phosphorus load (Gunn & May, 1997). In both cases, dramatic changes in rotifer abundance were recorded, while the species composition of the rotifer community changed very little.

This paper examines data from a single, predominantly oligotrophic, lake, Loch Lomond, Scotland, which has a subtle trophic gradient along its longitudinal axis. The aim of the study was to determine whether variations in rotifer communities could be used to highlight the very small differences in trophic state along the length of the lake better than the more traditional methods of determining chlorophyll *a* and total phosphorus concentrations.

Site description

Loch Lomond has the largest surface area of any freshwater body in Great Britain and is the second biggest, after Loch Ness, in terms of water volume. It is situated in Scotland between latitudes 56° N and 56° 19′ N and longitudes 4° 30′ W and 4° 43′ W (Fig. 1), in an area with a cool, wet and windy climate. The total catchment area of the loch is approximately 781 km^2 (Maitland, 1981). This includes a natural catchment area of 696 km^2 (Fig. 1), a lake surface area of 71 km^2, and three external areas that supplement the water supply for a hydro-electric power station.

The loch basin is of glacial origin and has a complex and varied shape consisting of two distinct parts that have merged together to form a single loch (Smith et al., 1981). The main, northern, basin is separated from the southern basin by numerous islands and outcrops of Old Red Sandstone associated with the geological Highland Boundary fault-line. Below water, a ridge separates the basins. This is thought to limit water from the south basin moving northwards into the mid and north basins.

The northern and southern basins differ in character because of the very different bedrock geology, topography, soil and land uses of their respective catchments (Maitland, 1981). The northern basin is fjord-like, being long (length 20 km), narrow (maximum width 1.5 km), deep (maximum depth 194 m) and oligotrophic, with a mountainous, base-poor, rocky catchment. In contrast, the southern basin is much broader (maximum width 8.8 km), shallower (typically between 5 m and 20 m deep at the southern end) and mesotrophic, with a mainly lowland, base-rich, agricultural catchment (Murray & Pullar, 1910; Tippett, 1994). The northern and southern catchments also differ markedly in relation to population density. The northern sub-catchment is sparsely populated (2.3 people km^{-2}: Maitland et al., 1981) with little urban development, while the southern part of the catchment is more densely populated (28

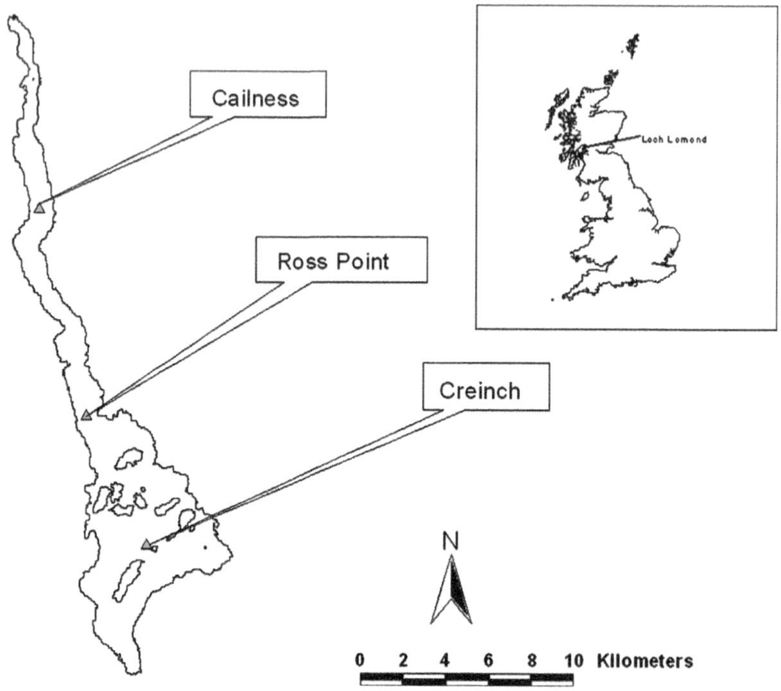

Figure 1. Location of sampling sites along the length of Loch Lomond; inset shows the location of Loch Lomond in Scotland.

people km^{-2}: Maitland et al., 1981) and has more urban development.

Within the lake, the north–south variation in catchment characteristics and, consequently, nutrient load, is reflected in a slight, but persistent trophic gradient along the longitudinal axis of the lake with the southern basin considered mesotrophic and the other basins oligotrophic (Best & Traill, 1994). This is illustrated by the mean total phosphorus concentrations for monthly data from 1995–1999, which were 8.4 μg l^{-1} for Cailness, in the northern basis, and 9.0 μg l^{-1} for Creinch, in the southern basin. Although the difference is subtle, it was thought sufficiently reliable to classify the northern and southern basins as being of different trophic status in a recent review of Scottish lochs by the Scottish Environment Protection Agency (SEPA).

Methods

Samples for chemical and biological analyses were collected at monthly intervals (i.e. 3 May, 24 June, 20 August, 24 September and 25 October) during 2002 at three sites on the mid-line that runs approximately north–south along the length of Loch Lomond (Fig. 1). These were integrated samples taken with a 10 m length of flexible tubing (10 mm internal diameter), which sampled the whole water column over a depth of 0–10 m. In total, two samples were taken at each site, one for chemical analyses and the other for the assessment of rotifer composition and abundance. Samples were always collected in late morning and the maximum time between starting to sample the first site and finishing sampling the last site was 90 min.

The first sample was analysed for chlorophyll *a* by the laboratories of the SEPA. Sub-samples of approximately 50 ml were taken for total phosphorus (TP) analysis. These samples were processed, unfiltered, by the Centre for Ecology and Hydrology (CEH) using a sulphuric acid-potassium persulphate digestion technique to convert all forms of phosphorus to orthophosphate, followed by the measurement of orthophosphate by the addition of ammonium molybdate and ascorbic acid (Murphy & Riley, 1962; American Public Health Association, 1975). Samples were stored frozen prior to analysis. The lower limit of detection of TP using this method is about 2 μg P l^{-1}, with a precision of about 3% (Mackereth et al., 1978).

The samples (1 l) taken for the determination of rotifer species composition and abundance were preserved in 4% formaldehyde. They were then concentrated by allowing the suspended matter to settle to the bottom of a measuring cylinder and siphoning off the overlying liquid. The rotifers were enumerated in a sedimentation chamber at ×100 magnification under an inverted microscope. Species were identified according to Koste (1976).

From May to October, thermistor chains (Aanderaa Instruments TR7) recorded temperature profiles within the three main basins. Going from north to south, the chains in each basin reached from the surface down to a depth of 50 m, 40 m and 20 m, respectively. Each thermistor chain recorded temperature at 11 equally spaced points along its length and at 30-minute intervals.

Results

The rotifer samples examined were collected from the upper 10 m of the water column at each site. The temperature profile data showed that this water layer was well mixed and almost isothermal throughout the sampling period, both within and across the sampling sites (Fig. 2). The northern and mid basins of the loch were stratified at a depth of around 15–20 m over the sampling period, so samples from the upper 10 m of the water column should have reflected the epilimnetic rotifer community well. It was assumed that rotifer numbers in the deep, cold and dark hypolimnion would be very low and these were ignored for present purposes. The temperature profiles also show that the south basin is well mixed, so samples from the upper 10 m should have provided representative samples from this part of the lake.

More than 12 rotifer species were recorded, with most occurring at relatively low densities (Table 1). The most common species, achieving population densities in excess of 50 ind. l^{-1}, were (in order of importance): *Keratella cochlearis*, *Trichocerca stylata*, *Conochilus hippocrepis*, *Polyarthra dolichoptera*, *Kellicottia longispina*, *Synchaeta* sp. Rotifer species diversity varied little across the sampling sites over the study period.

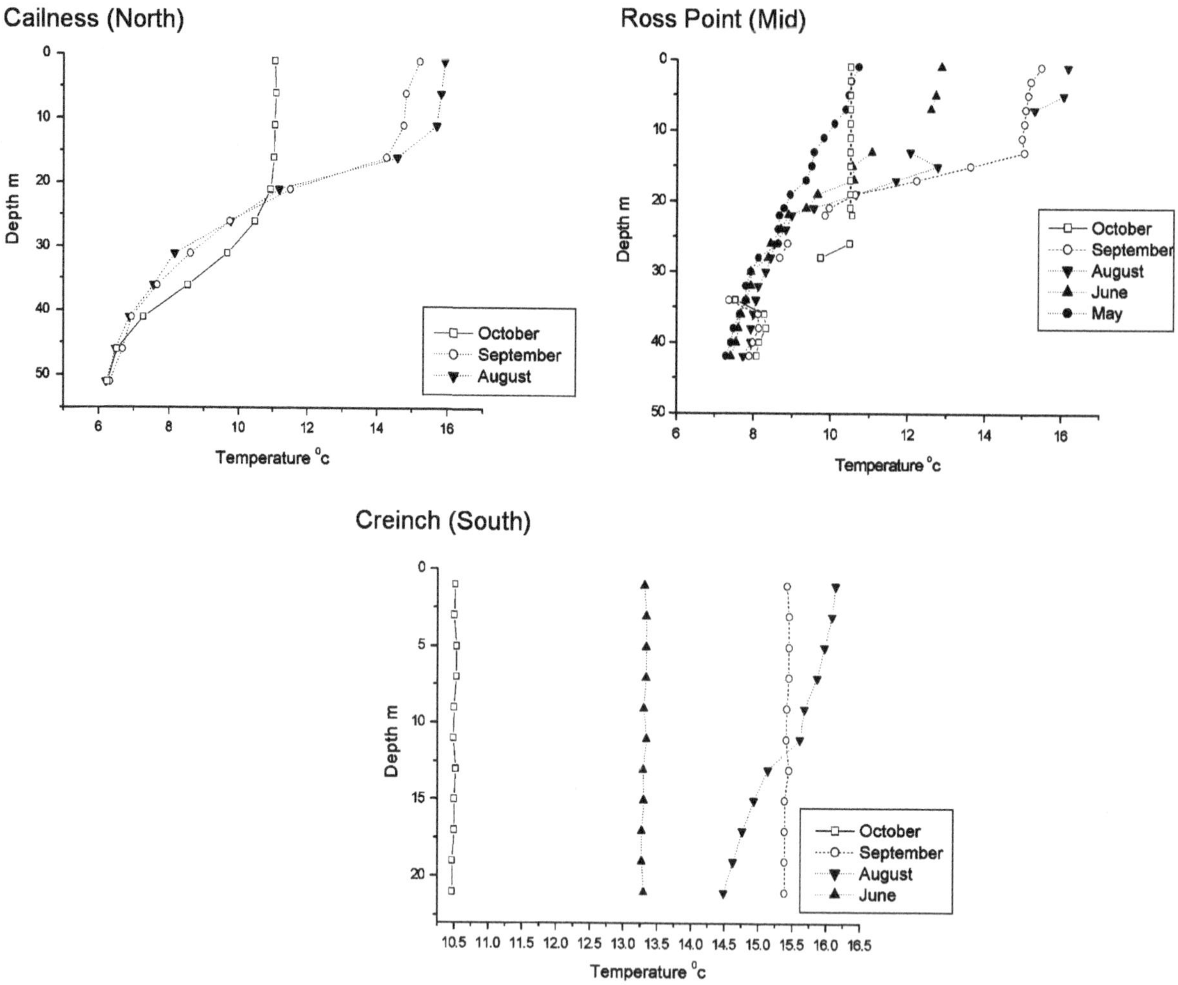

Figure 2. Temperature profiles in the northern, mid and southern basins of Loch Lomond, May–October 2002. Data collection at Cailness began August 2002.

Total rotifer abundance ranged from 0 ind. l^{-1} at the northern most site, Cailness, in October 2002 to 849 ind. l^{-1} at the southern most site, Creinch, in August 2002 (Fig. 3a–c). Of the more abundant species, only *K. cochlearis* and *T. stylata* showed any seasonal trend in abundance. *K. cochlearis* tended to be more abundant in the summer (Fig. 3e–g), while *T. stylata* was more common in spring (Fig. 3h–i).

A comparison of mean total rotifer abundance at each site over the study period showed a steady increase in density from north to south along the length of the loch. This pattern was shown even more clearly when the colonial species were ignored to reduce the effects of clumping. A similar pattern of increasing abundance from north to south was evident in both of the most numerous species, *K. cochlearis* and *T. stylata*. These trends seem to reflect the trophic gradient along the loch far more clearly than the chlorophyll *a* and TP concentrations that are more traditionally used for this purpose (Fig. 4).

Discussion

Although numerous during the spring and summer, the rotifers of Loch Lomond had received little attention until this study began in 2002. The available published information lists only two genera for the loch, *Asplanchna* spp. and *Notholca* spp. (Pomeroy, 1994), with no indication of their

Table 1. Rotifer species found in the northern, mid and southern basins of Loch Lomond between May and October 2002

Rotifer species	Sampling site		
	Cailness (Northern)	Ross Point (Mid)	Creinch (Southern)
Keratella quadrata (Müller)		*	
Keratella cochlearis (Gosse)	**	**	****
Kellicottia longispina Kellicott	*	*	*
Euchlanis dilatata Ehrb.	*	*	*
Trichocerca stylata (Gosse)	**	***	***
Trichocerca myersi (Hauer)		*	*
Synchaeta sp.	*	*	*
Polyarthra dolichoptera Idelson	*	**	*
Polyarthra vulgaris Carlin	*	*	*
Conochilus hippocrepis (Schrank)	*	***	*
Filinia longiseta (Ehrb.)		*	
Collotheca sp.	*	*	

Relative abundance of each species at each site, in terms of maximum density recorded, is indicated as follows: * < 100; ** > 100–200; *** > 200–400; **** > 400 ind. l^{-1}.

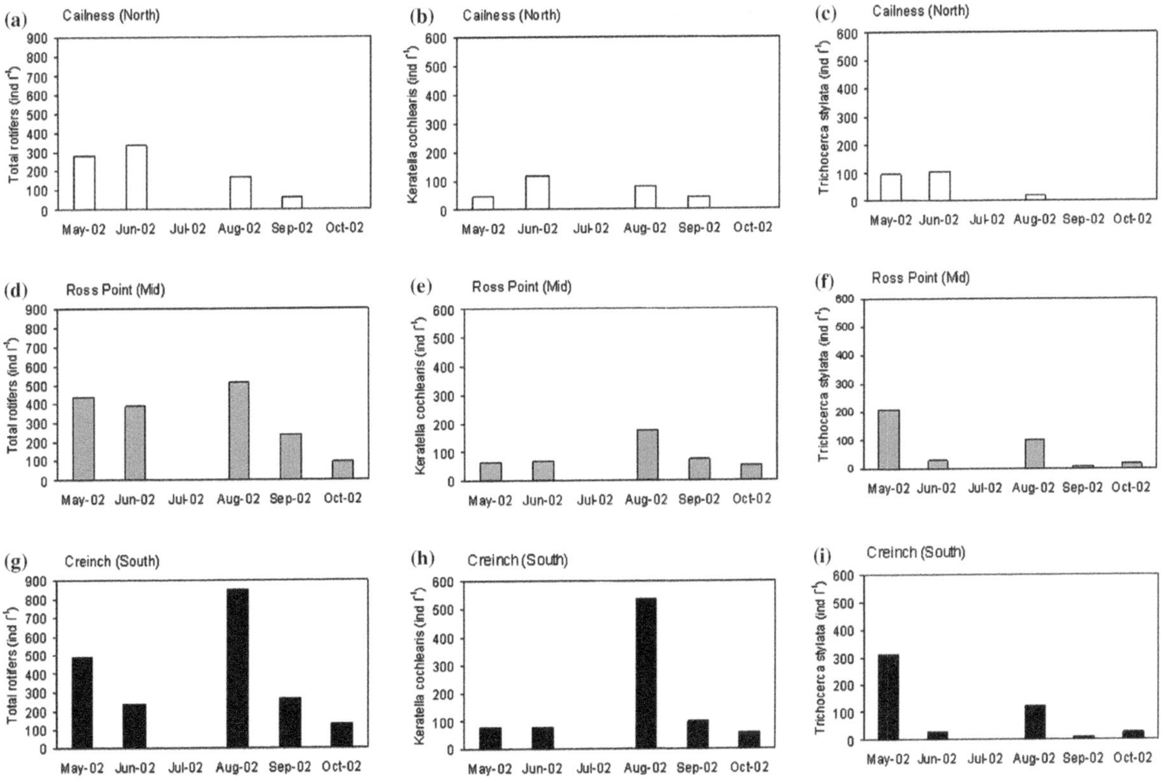

Figure 3. Temporal variation in abundance of total rotifers, *K. cochlearis*, and *T. stylata* in the northern (a–c), mid (d–f) and southern (g–I) basins of Loch Lomond, May–October 2002.

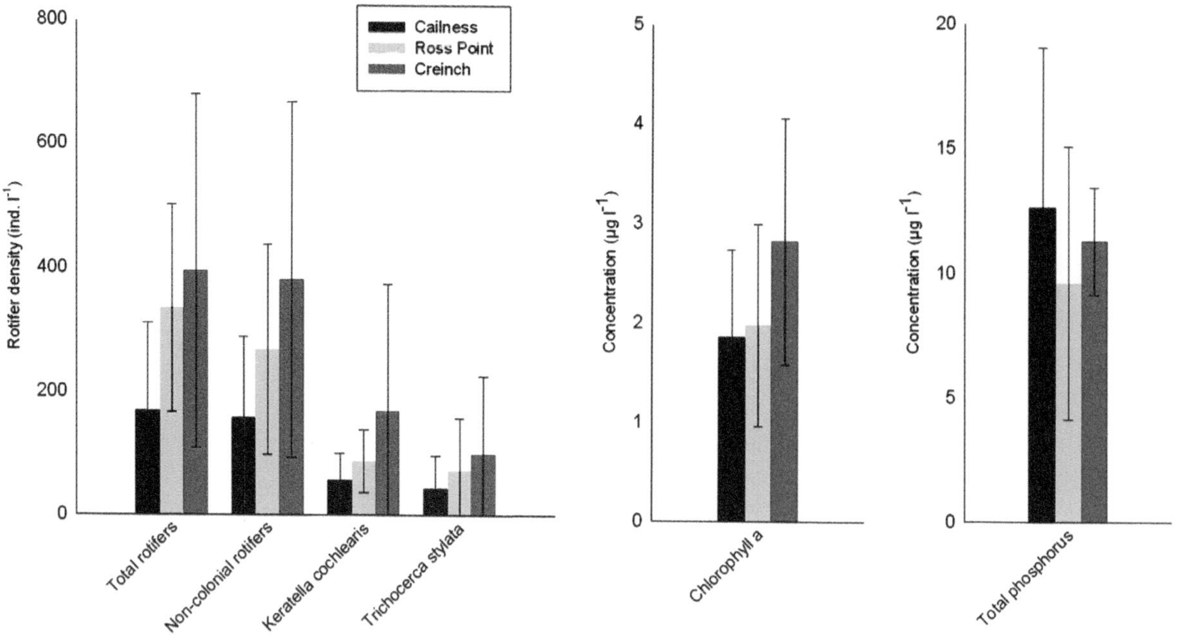

Figure 4. Variation in mean rotifer abundance and mean chlorophyll *a* and total phosphorus concentrations along the north–south trophic gradient in Loch Lomond, May–October 2002. Vertical bars show ± 1SE.

abundance. The present study found more than 12 rotifer species between May and October 2002, the most common being *Keratella cochlearis* and *Trichocerca stylata*. These 12 species did not include the two species previously recorded by Pomeroy (1994). The reason for this is unclear, as no information on sampling methods, site locations or seasonality is given in the original publication.

Rotifer abundances in the loch varied from a minimum of 0 ind. l^{-1} at Cailness in October 2002 to a maximum of 849 ind. l^{-1} at Creinch in August 2002. Values of more than 200 ind. l^{-1} were recorded across all of the sampling sites for most of the study period. In general, these relatively high levels of abundance are not unusual for an oligo-mesotrophic system with open water TP concentrations of around 10 $\mu g\ l^{-1}$. In Piburger See, Austria, (approx. 10 μg TP l^{-1}), for example, Schaber (1976) recorded rotifer densities ranging from 17 to 870 ind. l^{-1} during the mid-1970s. These values contrast with the higher densities usually recorded in meso-eutrophic systems (up to 2500 ind. l^{-1} – Radwan & Popiolek, 1989), eutrophic lakes such as Loch Leven, Scotland, and Lake Oglethorpe, Georgia, USA (up to 8000 ind. l^{-1} – Orcutt & Pace, 1984; Gunn & May, 1997) and hypereutrophic waterbodies, such as Priest Pot, England, (up to 40,000 ind. l^{-1} – Hewitt & George, 1987).

Rotifer abundance was found to increase markedly, in a southerly direction, along the mid-line of the loch. It is unlikely that this could have been caused by diurnal migration during the sampling process, because the sites were all sampled within a 90-minute period. It is also unlikely to have been caused by variations in water temperature among the sampling sites, because water temperature varied very little (< 1 °C) across the loch on any given sampling occasion. Although predation could have caused significant variations in rotifer abundance across the loch, this is unlikely because crustacean zooplankton numbers within this system are generally very low as a result of heavy predation from powan, *Coregonus lavaretus*, (Pomeroy 1994). The dominant crustacean zooplankton taxa in this loch attain maximum densities during the growing season of only 1.9 ind. l^{-1} (*Eudiaptomus gracilis* (Sars)), 1.5 ind. l^{-1} (*Daphnia hyalina* Sars) and 5 ind. l^{-1} (*Cyclops abyssorum* (Claus)) (Pomeroy 1994).

The most likely explanation of the observed increase in rotifer densities along the length of the loch is their response to the subtle trophic gradient that develops from north to south due to differential nutrient loadings and variations in topography. In Loch Lomond, trophic status in the different basins is not a straightforward function of nutrient concentration and predation, but also of other factors that affect productivity, such as mixing depth and light climate. The ratio of mixing depth to euphotic depth varies along the length of the loch, being much lower in the north basin (mean depth 170 m) than in the mid (mean depth 70 m) and south (mean depth 20 m) basins, especially during summer stratification. In addition, shading by the surrounding mountains limits the amount of light reaching the loch surface in the north and mid basins in comparison with the south basin, where the surrounding topography is lower (C. Adams, personal communication). Being higher up the food chain, rotifer abundance probably reflects the overall trophic conditions that develop within each basin of the loch more clearly than simple spot measurements of the more traditional indicators such as chlorophyll *a* and total phosphorus (TP) concentrations. These results also suggest that rotifer abundance is a more sensitive indicator of lake trophic status than species composition.

Acknowledgements

We thank the Scottish Environment Protection Agency for help with sample collection and analysis. This work is part-funded by the European Commission under the 5th Framework Programme (Project: EUROLAKES; Contract No: EVK1-CT1999-00004). Further information on this project can be found at: http://pcs0.hydromod.de/Eurolakes

References

American Public Health Association, 1975. Standard Methods for the Examination of Wastewater. Publ. American Public Health Association Washington 476: 481–482.

Berzins, B. & B. Pejler, 1989. Rotifer occurrence and trophic degree. Hydrobiologia 182: 171–180.

Best, G. A. & I. Traill, 1994. The physico-chemical limnology of Loch Lomond. Hydrobiologia 290: 29–37.

Blancher, E. C., 1984. Zooplankton-trophic state relationships in some north and central florida lakes. Hydrobiologia 109: 251–263.

Duggan, I. C., J. D. Green & R. J. Shiel, 2001. Distribution of rotifers in North Island, New Zealand, and their potential use as bioindicators of lake trophic state. Hydrobiologia 446/447: 155–164.

Ejsmont-Karabin, J. & A. Hillbricht-Ilkowsks, 1994. Illustration of the eutrophication process: comparison of rotifers from Mikolajskie Lake in the years 1989–1990 and 1963–1964. Polski Archiwum Hydrobiologii 41: 477–487.

Evans, M. S., 1986. Lake Huron rotifer and crustacean zooplankton, April–July, 1980. Great Lakes Research 12: 281–292.

Fuller, D. R., R. S. Stemberger & J. E. Gannon, 1977. Limnetic rotifers as indicators of trophic change. Journal of the Elisha Mitchell Science Society 93: 104–113.

Gannon, J. E. & R. S. Stemberger, 1978. Zooplankton (especially Crustacea and rotifers) as indicators of water quality. Transactions American Microscopical Society 97: 16–35.

Gunn, I. D. M. & L. May, 1997. Analysis of 1996 zooplankton samples – Loch Leven NNR. Report to Scottish Natural Heritage, 18 pp.

Herzig A., 1979. The zooplankton of the open lake. In Löffler, H. (ed.), Neusiedlersee, Limnology of a shallow lake in Central Europe. Dr W Junk Publishers, The Hague. Monographie Biologicae 37: 281–335.

Hewitt, D. P. & D. G. George, 1987. The population dynamics of *Keratella cochlearis* in a hypereutrophic tarn and the possible impact of predation by roach. Hydrobiologia 147: 221–227.

Karabin, A., 1985. Pelagic zooplankton variation in the process of lake eutrophication I. Structural and quantitative features. Ekologica Polska 33: 567–616.

Koste, W., 1976. Rotatoria: Die Rädertiere Mitteleuropas. Gebrüder Borntraeger, Berlin, Stuttgart, 673 pp.

Mackereth, F. J. H., J. Heron & J. F. Talling, 1978. Water analysis: some revised methods for limnologists. Freshwater Biological Association Scientific Publication No. 36, Titus Wilson & Son Ltd., Kendal.

Maitland, P. S., 1981. Introduction and catchment analysis. In Maitland, P. S. (ed.), Dr W Junk Publishers, The Hague, pp. 1–27.

Maitland, P. S., I. R. Smith, A. E. Bailey-Watts, B. D. Smith & A. A. Lyle, 1981. Comparison and synthesis. In Maitland, P. S. (ed.), Dr W Junk Publishers, The Hague, pp. 253–282.

Matveeva, L. K., 1991. Planktonic rotifers as indicators of trophic state. Bulletin of the Moscow Naturalist's Society, Biology Section 96: 54–62.

Murphy, J. & J. P. Riley, 1962. A modified single-solution method for the determination of phosphate in natural waters. Analytica Chimica Acta 27: 31–36.

Murray J. & L. Pullar, 1910. Bathymetrical survey of the freshwater lochs of Scotland. Challenger Office Publishers, Edinburgh.

Orcutt, J. D. & M. L. Pace, 1984. Seasonal dynamics of rotifer and crustacean zooplankton populations in a eutrophic, monomictic lake with a note on rotifer sampling techniques. Hydrobiologia 119: 73–80.

Pomeroy, P. P., 1994. Zooplankton in Loch Lomond: perspectives, predation and powan. Hydrobiologia 290: 75–90.

Radwan, S. & B. Popiolek, 1989. Percentage of rotifers in spring zooplankton in lakes of different trophy. Hydrobiologia 186/187: 235–238.

Schaber, P., 1976. Rotatorien und Crustaceen des Piburger See (1973–1975). Abteilung Limnologie/Institut für Zoologie, Universität Innsbruck, Jahresbericht 2: 78–94.
Smith, I. R., A. A. Lyle & A. J. Rosie, 1981. Comparative physical limnology. In Maitland, P.S. (ed.), Dr W Junk Publishers, The Hague, pp. 29–65.
Tippett, R., 1994. An introduction to Loch Lomond. Hydrobiologia 290: 11–15.
Walz, N., H.-J. Elster & M. Mezger, 1987. The development of the rotifer community structure in Lake Constance during its eutrophication. Archiv für Hydrobiologie, Supplement 74: 452–487.

Hydrobiologia (2005) 546:405–414
A. Herzig, R.D. Gulati, C.D. Jersabek & L. May (eds.) Rotifera X: Rotifer Research: Trends, New Tools and Recent Advances
DOI 10.1007/s10750-005-4283-2

Diversity and abundance of the planktonic rotifers in different environments of the Upper Paraná River floodplain (Paraná State – Mato Grosso do Sul State, Brazil)

Cláudia Costa Bonecker*, Christiane Luciana Da Costa, Luiz Felipe Machado Velho & Fábio Amodêo Lansac-Tôha
Nupélia, Postgraduate course in Ecology of Continental Aquatic Environments, State University of Maringá, Av. Colombo, 5790, 87020–900 Maringá-PR, Brazil
*(*Author for correspondence: E-mail: bonecker@nupelia.uem.br)*

Key words: rotifer, diversity, abundance, floodplain, Paraná River, Brazil

Abstract

This study proposes that diversity and abundance of rotifers show spatial and temporal variations in the Upper Paraná River floodplain due to heterogeneity of the environment and hydrological level fluctuations of the main river. The structure and dynamics of rotifer assemblages were investigated by samplings carried out during the rainy (February) and dry period (August) of the year 2000, in 36 environments (rivers, channels, backwaters, open and isolated floodplain lakes). The influence of phytoplankton biomass on rotifer diversity and abundance was also investigated. 104 taxa of rotifers were identified. The highest species richness was found in rivers and open floodplain lakes, the highest abundances in the isolated floodplain lakes, and the highest values of species diversity in the channels, especially during the rainy period. β_2-diversity values were higher in the channels, especially during the dry period. Flow differences and food availability were predominant factors influencing the structure and dynamics of the rotifer communities.

Introduction

Floodplains present a great environmental heterogeneity influenced by the variable flow regime of the main river. This fluctuation is responsible for the spatial and temporal distribution patterns of aquatic communities due to alterations of limnological characteristics of the environments during high and low water (Neiff, 1990). Dodson (2000) added that environmental heterogeneity is the result of patches of different kinds of habitat in the landscape and of processes occurring at different times.

Rotifers have ecological relevance in aquatic environments, filtering suspended material of different sizes (from bacteria to filamentous algae) and using different strategies to obtain food, which classifies them as generalists or specialists. Their high population renewal rates distinguish them as an important link in energy flow and nutrient cycling (Esteves, 1998). Another important characteristic is their high tolerance to changes in environmental conditions (Allan, 1976). All of these aspects probably explain the success of these organisms in aquatic environments.

Studies on the Upper Paraná River floodplain have shown that rotifers constitute a very diverse group, with about 230 taxa recorded (Bonecker et al., 1994; Lansac-Tôha et al., 1997; Serafim, 1997; Garcia et al., 1998), in addition to representing a large part of the zooplanktonic abundance in different environments of this floodplain (Lansac-Tôha et al., 1997).

The objective of this study is to analyze the rotifer richness and β_2-diversity, abundance and species diversity of the rotifer community in different environments of the Upper Paraná River floodplain, in addition to determining spatial and temporal changes in these variables and the influence of available food (phytoplankton biomass).

Materials and methods

This study was carried out in 36 environments including rivers, channels, open and isolated floodplain lakes, and backwaters. Backwaters are lentic environments connected to the river and formed by sedimentation of the island shore. These environments were located in the stretch of the Paraná River characterized by a braided channel, with low declivity and an extensive floodplain (Souza-Filho, 1993). The Environmental Protection Area of the Islands and Várzeas of the Paraná River (22° 40′–22° 50′ S and 53° 10′–53° 40′ W) is on this floodplain (Table 1 and Fig. 1).

Collections were made in February (rainy period) and in August (dry period) of the year 2000 (Fig. 2). These periods were differentiated according to daily data of the flow level of the Paraná River (Agência Nacional de Energia Elétrica – ANEEL), which demonstrated the absence of a prolonged period of flooding. Thomaz et al. (1997) suggested that the flood begins in this floodplain when the hydrological level is above 350 cm.

Rotifers were sampled below the surface of the pelagic region of each environment using a motorized pump and a plankton net of 70 μm mesh size by filtering 1000 l of water per sample. The concentrate was fixed in a final solution of 4% formaldehyde, buffered with calcium bicarbonate. Concurrently, water samples were collected using a Van Dorn bottle to analyze chlorophyll-*a* ($\mu g\ l^{-1}$) (Goltermann et al., 1978).

Rotifer richness was estimated according to the stabilization of the species increment curve per sample, using the basic literature (Koste, 1978; Koste & Robertson, 1983; José de Paggi, 1989; Segers, 1995) for identification.

Changes in the species composition of each environment and system (formed by different rivers – Paraná, Baía and Ivinheima), and in both hydrological periods, were evaluated using the β-diversity index (β_2) (Whittaker, 1960) from the equation $[(R/\alpha_{max})-1]/[n-1]$, where α_{max} is the maximum species richness value of all n samples analyzed and R is the sum of the number of species in n samples (Harrison et al., 1992). The species diversity (H') of the community was estimated using the Shannon–Wiener index (Pielou, 1975).

Abundance was determined using a Sedgwick–Rafter slide in an optical microscope at 100×

Table 1. List of the environments (sampling stations) studied on the Upper Paraná River floodplain in February and August 2000. Numbers of sampling stations as indicated in Figure 1

Rivers		Open floodplain lakes	System	Isolated floodplain lakes	System
Paraná – 25		Garças – 27	Paraná	Clara – 23	Paraná
Ivinheima – 8		Pombas – 14	Paraná	Pousada – 26	Paraná
Baía – 28		dos Patos – 6	Ivinheima	Genipapo – 22	Paraná
Channels	System	Finado Raimundo – 9	Ivinheima	Osmar – 17	Paraná
Ipoitã – 4	Ivinheima	Peroba – 1	Ivinheima	Capivara – 7	Ivinheima
Cortado – 13	Paraná	Sumida – 11	Ivinheima	Jacaré – 10	Ivinheima
Curutuba – 15	Baía	Boca do Ipoitã – 5	Ivinheima	Zé do Paco – 3	Ivinheima
Baía – 33	Baía	Porcos – 31	Baía	Cervo – 12	Ivinheima
Backwater		Maria Luiza – 34	Baía	Ventura –2	Ivinheima
Bilé – 20	Paraná	Onça – 36	Baía	Pousada das Garças – 30	Baía
Leopoldo – 21	Paraná	Guaraná – 19	Baía	Fechada – 29	Baía
Manezinho – 16	Paraná	Gavião – 35	Baía	Aurélio – 32	Baía
Pau-Véio – 24	Paraná			Traíra – 18	Baía

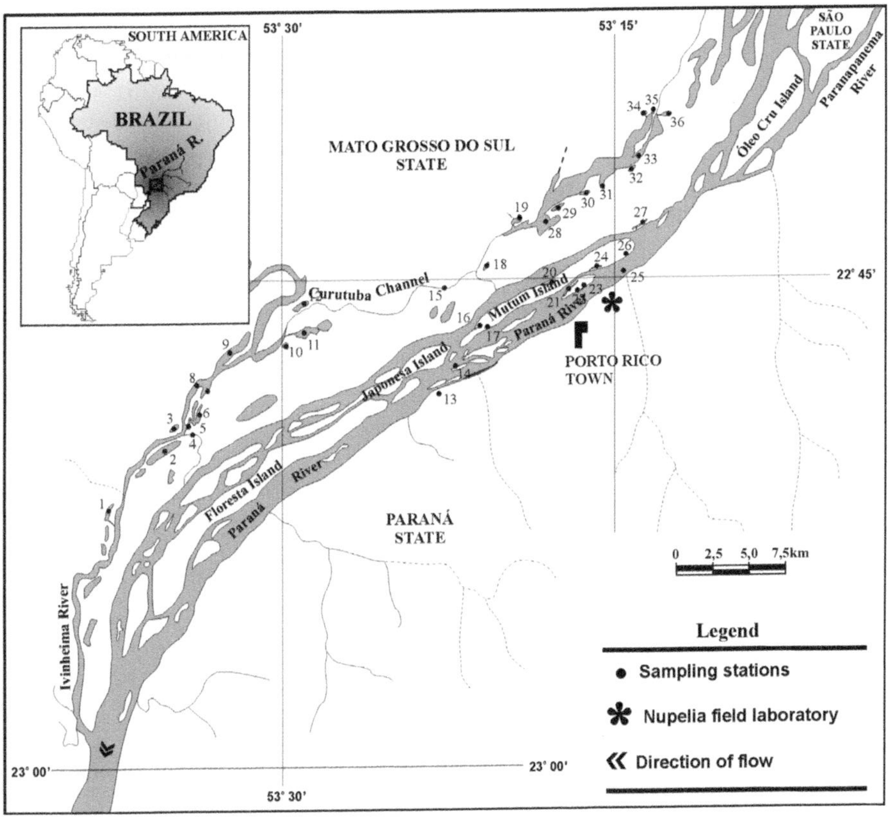

Figure 1. Location of the sampling area and collecting stations.

magnification, counting at least 50 individuals in three subsamples (1.7 ml), obtained with a Stempel pipette (Bottrell et al., 1976). In samples with low densities all individuals were counted. The final density was estimated in individuals m^{-3} and expressed as log-transformed data ($\log x + 1$).

Analysis of variance (ANOVA) was used to determine if the spatial variation in richness, abundance, and species diversity was related to the collection periods (Sokal & Rohlf, 1991), using these ecological attributes as the dependent variables and the two periods as the influencing factor. The abundance data were log-transformed.

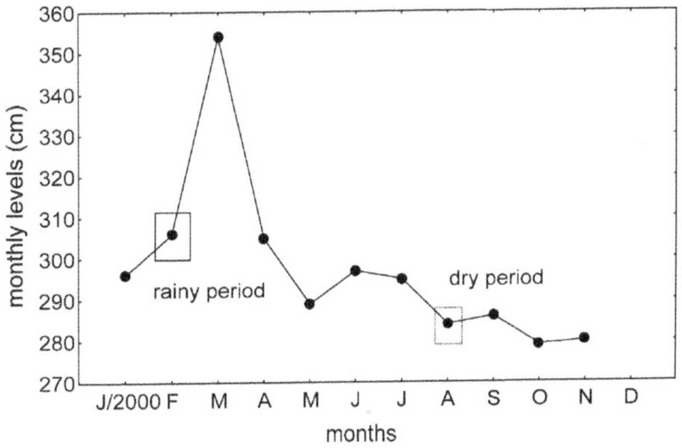

Figure 2. Monthly levels of the Paraná River recorded from January to November 2000. The collecting months are marked.

Differences between periods were considered significant at $p < 0.05$.

A Pearson Correlation Analysis was carried out (data log-transformed) in order to test if rotifer abundance was related to phytoplankton biomass, derived from concentrations of chlorophyll-*a*, for each environment and sampling period. Correlation was considered significant at $p < 0.05$. Data were analyzed using the statistical package STATISTICA version 5.0 (Statsoft Inc., 1996).

Results

Rotifer richness

One hundred and four rotifer taxa (19 families) were recorded, with the majority belonging to Brachionidae, Lecanidae and Trichocercidae (Table 2). Some of these taxa represent new floodplain occurrences: *Brachionus forficula*, *Kellicottia bostoniensis*, *Trichocerca dixonnuttalli* and *T. macera*. The highest species numbers were found in the rivers and open floodplain lakes, especially during the rainy period (Fig. 3).

β_2-diversity

The β_2-diversity results were not high, varying from 2.90% to 10.38%. The highest value was observed in the channels, followed by the one obtained in rivers (8.48%). The lowest values were recorded in the isolated (2.98%) and open (3.65%) floodplain lakes. Concerning the periods and systems studied, during the dry period, higher β_2-diversity values were observed in the Baía system (17.78%), although the systems did not differ markedly (Paraná: 17.48%, Ivinheima: 16.36%). In the rainy period, the spatial variation was similar to that of the dry period, i.e. higher values recorded in the Baía (17.53%) and Paraná (17.48%) systems and lower values in the Ivinheima (16.36%) system (Fig. 4).

Species diversity (Shannon–Wiener index)

The rotifer community had higher species diversity during the rainy period, especially in the channels, due to a higher evenness. In contrast, a lower diversity was found in the isolated floodplain lakes, probably due to lower evenness and lower richness values. During the dry period, the species diversity was similar in all environments, and so were evenness and species richness (Figs. 3 and 5).

Abundance

The rotifer community showed higher densities in the floodplain lakes during both sampling periods, particularly in the isolated floodplain lakes during the rainy period, when greater densities were also recorded in the rivers (Fig. 6). *Lecane proiecta* was abundant in the floodplain lakes, backwaters and rivers, showing wide distribution among the environments. *Brachionus falcatus* was abundant in the isolated floodplain lakes and *Filinia opoliensis* in the open floodplain lakes and backwaters. *Polyarthra* sp. was the most abundant species in the channels, and *Euchlanis dilatata* in the channels and rivers.

On the other hand, lowest abundances were found in the channels and rivers during the dry period (Fig. 6). In this period *Polyarthra* sp. was widely distributed among the environments with higher densities in the channels, rivers, open floodplain lakes and backwaters. *E. dilatata* was also dominant in the channels, while *Ploesoma truncatum* showed great abundance in the rivers. *Keratella cochlearis* was the most abundant species in the open floodplain lakes and backwaters, and *Synchaeta pectinata* in the open and isolated floodplain lakes and backwaters. *Keratella americana* and *Hexarthra intermedia* were dominant in the isolated floodplain lakes.

Integration of results

The ANOVA showed that only the abundance data varied between the environments in the two periods ($p = 0.0299$) (Fig. 6). The correlation between rotifer abundance and chlorophyll-*a* concentration in the two periods showed a positive relationship, mainly in the floodplain lakes (Fig. 7), where we found high phytoplankton densities. These results confirmed that phytoplankton, as measured by chlorophyll-*a* concentration, was an important food resource during the development of the rotifer communities during the rainy period.

Table 2. Checklist of the rotifers recorded in different environments of the Upper Paraná River floodplain in February and August 2000

Monogononta

Asplanchnidae
Asplanchna sieboldi (Leydig, 1854)
Asplanchna sp.

Brachionidae
Anuraeopsis fissa (Gosse, 1851)
Brachionus angularis Gosse, 1851
B. budapestinensis Daday, 1885
B. calyciflorus Pallas, 1766
B. caudatus Barrois & Daday, 1894
B. dolabratus Harring, 1914
B. forficula Wierzejski, 1891
B. mirus Daday, 1905
B. falcatus Zacharias, 1898
B. quadridentatus Hermann, 1783
B. q. mirabilis (Daday, 1897)
B. urceolaris (O. F. Müller, 1773)
Kellicottia bostoniensis (Rousselet, 1908)
Keratella americana Carlin, 1943
K. cochlearis (Gosse, 1851)
K. lenzi Hauer, 1953
K. tropica (Apstein, 1907)
Plationus macracanthus (Daday, 1905)
P. patulus (O.F. Müller, 1786)
Platyias q. quadricornis (Ehrenberg, 1832)
P. q. brevispinus (Daday, 1905)
P. leloupi (Gillard, 1957)

Conochilidae
Conochilus coenobasis (Skorikov, 1914)
C. dossuarius (Hudson, 1875)
C. natans (Seligo, 1900)
C. unicornis rousselet, 1892

Dicranophoridae
Dicranophorus claviger (Hauer, 1965)

Epiphanidae
Epiphanes macrourus (Barrois & Daday, 1894)
E. clavulata (Ehrenberg, 1832)

Euchlanidae
Beauchampiella eudactylota (Gosse, 1886)
Euchlanis dilatata Ehrenberg, 1832
E. incisa Carlin, 1939
Dipleuchlanis propatula (Gosse, 1886)

Gastropodidae
Ascomorpha ecaudis Perty, 1850
A. saltans (Bartsch, 1870)
Gastropus hyptopus (Ehrenberg, 1838)

Hexarthridae
Hexarthra intermedia (Wieszniewski, 1929)
H. mira (Hudson, 1871)

Flosculariidae
Floscularia sp.
Ptygura sp.

Lecanidae
Lecane aculeata (Jakubski, 1912)
L. amazonica (Murray, 1913)
L. bulla (Gosse, 1851)
L. closterocerca (Schmarda, 1859)
L. cornuta (O. F. Müller, 1786)
L. curvicornis (Murray, 1913)
L. elsa Hauer, 1931
L. hamata (Stokes, 1896)
L. leontina (Turner, 1892)
L. ludwigii (Eckstein, 1893)
L. luna (O. F. Müller, 1776)
L. lunaris Ehrenberg, 1832
L. monostyla (Daday, 1897)
L. papuana (Murray, 1913)
L. proiecta Hauer, 1956
L. quadridentata (Ehrenberg, 1832)
L. scutata (H. & M., 1926)
L. signifera (Jennings, 1896)
L. stichaea Harring, 1913
L. ungulata (Gosse, 1887)

Lepadellidae
Lepadella benjamini Harring, 1916
L. ovalis (O. F. Müller, 1786)

Mytilinidae
Mytilina macrocera (Jennings, 1894)
M. ventralis (Ehrenberg, 1832)

Notommatidae
Cephalodella mucronata Myers, 1924
Cephalodella sp.
Notommata sp.

Testudinellidae
Pompholyx sp.
Testudinella mucronata (Gosse, 1886)
T. ohlei Koste, 1972
T. patina (Hermann, 1783)

Trichocercidae
Elosa sp.
Trichocerca bicristata (Gosse, 1887)
T. bidens (Lucks, 1912)
T. capucina (Wierzejski & Zacharias, 1893)
T. cylindrica (Imhof, 1891)
T. chattoni (Beauchamp, 1907)
T. dixonnuttalli (Jennings, 1903)
T. elongata (Gosse, 1886)
T. iernis (Gosse, 1887)
T. inermis (Linder, 1904)
T. insignis (Herrick, 1885)
T. macera (Gosse, 1886)
T. pusilla (Lauterborn, 1898)
T. scipio (Gosse, 1886)
T. similis (Wierzejski, 1893)
T. stylata (Gosse, 1851)
Trichocerca sp.

Trichotriidae
Macrochaetus sericus (Thorpe, 1893)
M. collinsi (Gosse, 1867)
Trichotria tetractis (Ehrenberg, 1830)

Trochosphaeridae
Filinia longiseta (Ehrenberg, 1834)
F. opoliensis (Zacharias, 1898)
F. pejleri Hutchinson, 1964
Filinia sp.
Horaella thomassoni Koste, 1973

Synchaetidae
Ploesoma truncatum (Levander, 1894)
Polyarthra vulgaris Carlin, 1943
Polyarthra sp.
Synchaeta pectinata Ehrenberg, 1832
S. stylata Wierzejski, 1893

Bdelloidea

Philodinidae
Dissotrocha aculeata (Ehrenberg, 1832)

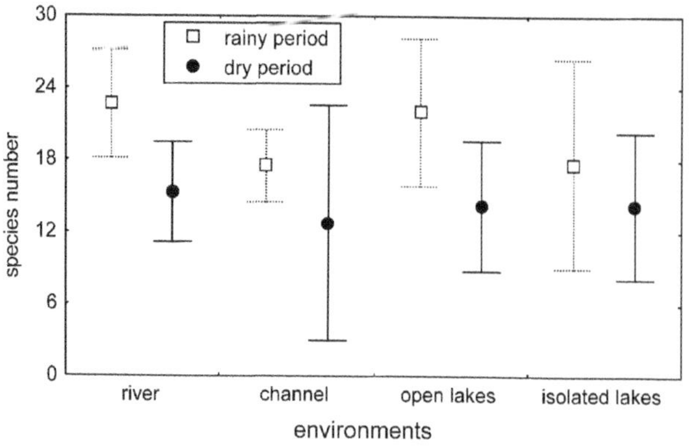

Figure 3. Rotifer richness recorded in the different environments (open lakes = floodplain lakes and backwater) during the rainy and dry periods (symbol = average, bar = standard deviation).

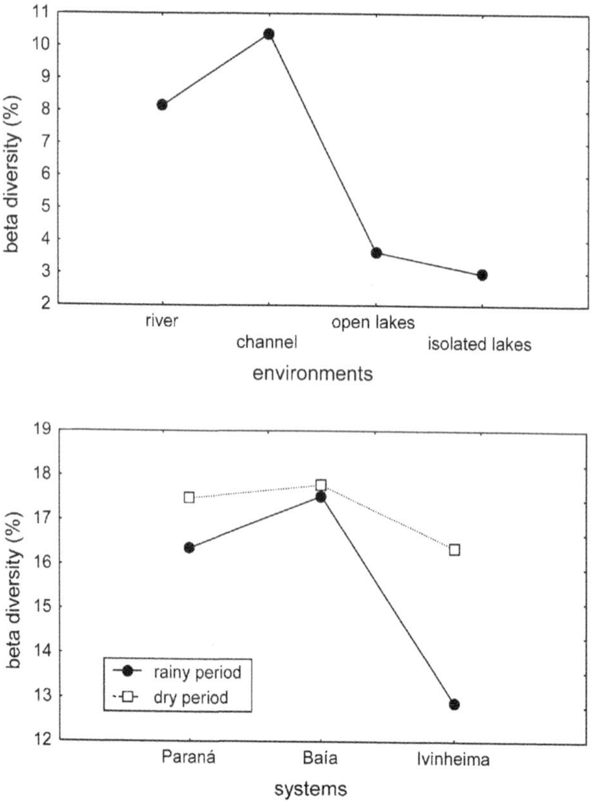

Figure 4. β-diversity of the rotifer community recorded in the different environments (open lakes = floodplain lakes and backwater) and systems formed by the main rivers (rainy and dry periods).

Discussion

Rotifer families with greater species richness (Brachionidae, Lecanidae and Trichocercidae) matched the typical species associations known from tropical floodplain environments. These families are commonly recorded in freshwater aquatic environments of Brazil (Bozelli, 1992; Bonecker et al., 1998; Lansac-Tôha et al., 1997; 1999).

The highest rotifer richness values, especially in rivers and open floodplain lakes during the

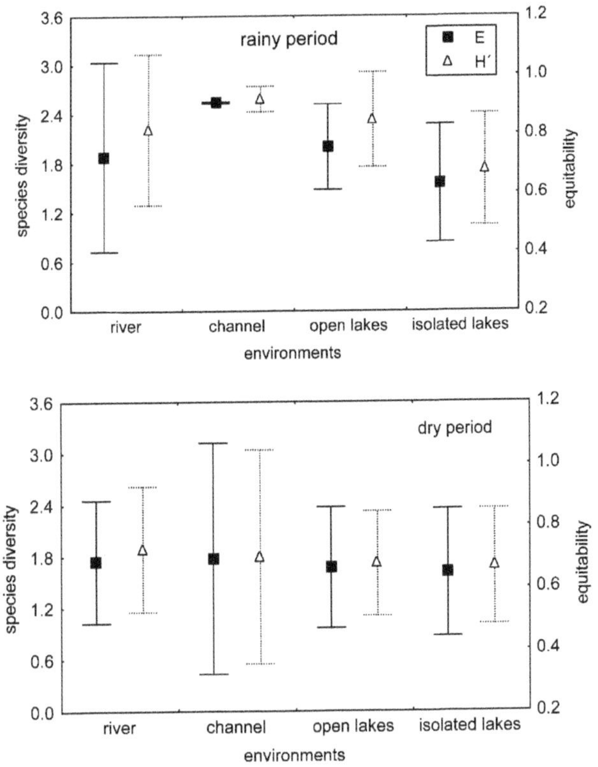

Figure 5. Species diversity of the rotifer communities recorded in the different environments (open lakes = floodplain lakes and backwater) during the rainy and dry periods (symbol = average, bar = standard deviation).

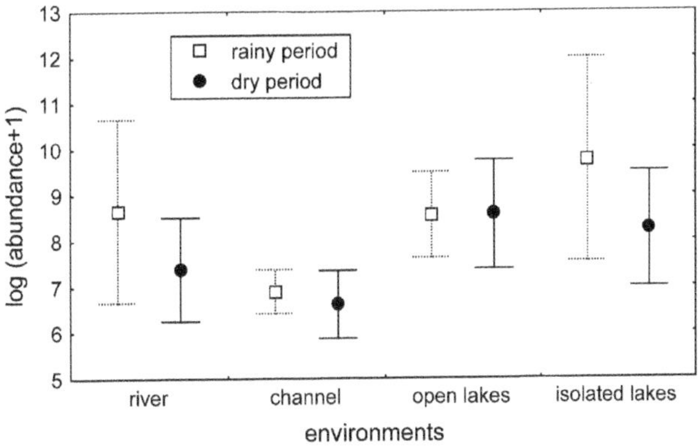

Figure 6. Rotifer abundance recorded in the different environments (open lakes = floodplain lakes and backwater) during the rainy and dry periods (symbol = average, bar = standard deviation).

rainy period, are probably contributed by species originating from lentic environments, not previously connected to the rivers. Koste & Robertson (1983) suggested that rotifer richness in lentic environments of flooded areas tends to increase during the period of higher flow levels, due to the incorporation of benthic and periphytic taxa associated with decomposing aquatic vegetation.

An increase in the lake area generally occurs during this hydrological period (rainy) due to marginal flooding. The exchange of water masses between the littoral and pelagic regions in these environments contributes to the increase in species

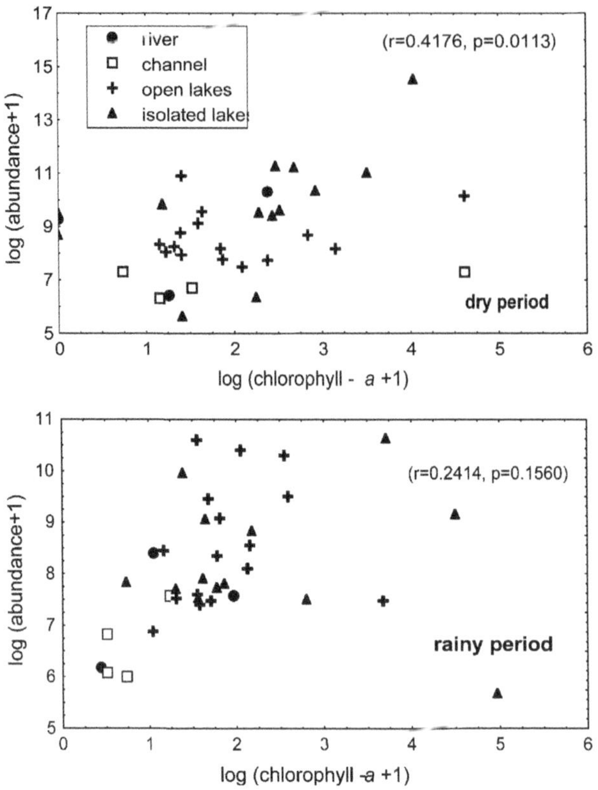

Figure 7. Relationship between rotifer abundance (ind. m^{-3}) and chlorophyll-*a* concentration ($\mu g\ l^{-1}$) recorded in the environments (open lakes = floodplain lakes and backwater) in the rainy and dry periods.

richness of the entire environment. This, along with the permanent connection to the river, favors the occurrence of a large number of taxa in the open floodplain lakes. This characteristic has been described in various studies on floodplains (Brandorff & Andrade, 1978; Bonecker et al., 1994; Lansac-Tôha et al., 1997). On the other hand, the lowest richness values observed during the dry period in all environments reinforces the importance of flows for the increase in diversity of floodplain environments.

Highest β-diversity values in channels, rather than in floodplain lakes, demonstrates that the change of rotifer community composition is more intense in lotic environments and less pronounced in the lentic ones. Just as for other biotic parameters analyzed, flow appears to be a predominant factor in this diversity estimate.

The lowest β-diversity values observed in the Ivinheima system may be related to the low Secchi disk values in previous studies (Thomaz et al., 1997), and so may be low diversity values found in some floodplain lakes in February. The reduced transparency appears to be due to a large quantity of suspended biogenic and/or abiogenic material. For the Ivinheima system, Thomaz et al. (1997) highlighted the large quantity of suspended inorganic material in the water column.

The higher species diversities during the rainy period is apparently influenced by the increase of species numbers in all environments and the absence of high species dominances, especially in the channels. According to Marzolf (1990), in environments with high current flow transport loss of organisms is higher than their reproductive rate, which prevents large populations from developing. On the other hand, Corrales (1979) and José de Paggi (1981) recorded higher zooplankton diversity in the Paraná River and secondary channels during the dry period.

The highest densities of rotifers in the isolated floodplain lakes during the rainy period reflect the large development of planktonic populations in lentic environments, probably due to the absence of pronounced flood and consequent overflow. In contrast, we recorded lowest abundance values in the channels and rivers during the dry period. This fact could be the result of the lower connectivity and faunal exchange between lotic and lentic environments during this period.

The variation of abundance between the environments in both the periods was especially due to differences in the number of individuals found in the isolated floodplain lakes and channels, as the other environments presented the same variation patterns. In an open floodplain lake and a river (Baía river and Guaraná lagoon) Bonecker et al. (2002) showed that densities were much more variable in the dry period than during the period of higher flow, which is also confirmed by us.

Phytoplankton seems to be an important food resource that influences the structure and dynamics of rotifer communities during the rainy period, mainly in the floodplain lakes where we recorded high phytoplankton densities. Bonecker & Lansac-Tôha (1996) and Train et al. (2001) described highest phytoplankton and rotifer abundances in floodplain lakes during the dry period and showed the importance of the algae community for the development of rotifer populations. Finally, we suggest that flow differences and food availability were predominant factors for the structure and dynamics of the rotifer communities.

Acknowledgements

We thank Dr. Luis Carlos Gomes for suggestions. The constructive criticism of the editor and one anonymous referee also is appreciated. Supported by PELD/CNPq and Nupélia/UEM.

References

Allan, J. D., 1976. Life history patterns in zooplankton. American Naturalist 110: 165–180.

Bonecker, C. C. & F. A. Lansac-Tôha, 1996. Community structure of rotifers in two environments of the high river Paraná floodplain (MS), Brazil. Hydrobiologia, 325: 137–150.

Bonecker, C. C., F. A. Lansac-Toha & A. Staub, 1994. Qualitative study of rotifers in different environments of the high Paraná river floodplain (MS), Brasil. Revista Unimar 6: 1–16.

Bonecker, C. C., F. A. Lansac-Tôha & D. C. Rossa, 1998. Planktonic and non planktonic rotifers in two environments of the upper Paraná river floodplain-MS, Brazil. Brazilian Archives of Biology and Technology 41: 447–456.

Bonecker, C. C., F. A. Lansac-Tôha, L. M. Bini & L. F. M. Velho, 2002. Daily fluctuation in rotifer population abundance in two environments of the upper Paraná River floodplain, Brazil. Amazoniana 17: 139–151.

Bottrell, H. H., A. Duncan, Z. Gliwicz, E. Grygierek, A. Herzig, A. Hillbricht-Illkowska, P. Larsson & T. Weglenska, 1976. A review of some problems in zooplankton production studies. Norwegian Journal of Zoology 24: 419–456.

Bozelli, R. L., 1992. Composition of the zooplankton of Batata and Mussurá lakes of the Trombeta River, State of Pará, Brazil. Amazoniana 2: 239–226.

Brandorff, G. O. & E. R. Andrade, 1978. The relationship between the water level of the Amazon river and the fate of the zooplankton population in Lago Jacaretinga, a varzea lake in the central Amazon. Studies on Neotropical Fauna and Environment 13: 63–70.

Corrales, M. A., 1979. Contribuición al conocimiento del zooplâncton de alto Paraná. Ecosur 6: 185–205.

Dodson, S. I., 2000. Effects of environmental heterogeneity in aquatic ecology. Internationale Vereinigung für Theoretische und Angewandte Limnologie, Verhandlungen 27: 3260–3263.

Esteves, F. A., 1998. Fundamentos de Limnologia (2nd edn.) Interciência/FINEP. Rio de Janeiro, 602 pp.

Garcia, A. P. P., F. A. Lansac-Tôha & C. C. Bonecker, 1998. Species composition and abundance of rotifers in different environments of the floodplain of the upper Paraná river, Brazil. Revista Brasileira de Zoologia 15: 327–343.

Golterman, H. L., R. S. Clymo & M. A. M. Ohmstad, 1978. Methods for Physical and Chemical Analysis of Fresh Waters. Blackwell Scientific Publications, Oxford, 214 pp.

Harrison, S., S. J. Ross & J. H. Lawton, 1992. Beta diversity on geographic gradients in Britain. Journal of Animal Ecology 61: 151–158.

José de Paggi, S., 1981. Variaciones temporales y distribuición horizontal del zooplancton en algunos cauces secundarios del rio Paraná Medio. Studies on Neotropical Fauna and Environment 6: 185–199.

José de Paggi, S., 1989. Rotíferos de algunas provincias del noroeste argentino. Revue ďHydrobiologie Tropicale 23: 297–311.

Koste, W., 1978. Rotatoria. Die Rädertiere Mitteleuropas, begründet von Max Voigt. Monogononta I. Gebrüder Borntraeger, Berlin, 673 pp.

Koste, W. & B. Robertson, 1983. Taxonomic studies of the Rotifera (Phylum Aschelminthes) from a Central Amazonian varzea lake, Lago Camaleão (Ilha de Marchantaria, rio Solimoes, Amazonas, Brazil). Amazoniana 8: 225–254.

Lansac-Tôha, F. A., C. C. Bonecker, L. F. M. Velho & A. F. Lima, 1997. Composição, distribuição e abundância da comunidade zooplanctônica. In Vazzoler, A. E. A. M., A. A. Agostinho, & N. S. Hahn (eds), Planície de Inundação do Alto Rio Paraná: Aspectos Físicos, Biológicos e Socioeconômicos. Maringá-PR, Eduem, 117–155.

Lansac-Tôha, F. A., L. F. M. Velho & C. C. Bonecker, 1999. Estrutura da comunidade zooplanctônica antes e após a formação do Reservatório de Corumbá-GO. In Henry, R. (ed.), Ecologia de reservatórios: estrutura, função e aspectos sociais. FAPESP/FUNDIBIO, Botucatu: 349–374.

Marzolf G. R., 1990. Reservoirs as environments for zooplankton. In Thornton K. W., B. L. Kimmel & F. E. Payne (eds), Reservoir Limnology: Ecological Perspectives. Wiley – Interscience Publication, New York, 7: 195–208.

Neiff, J. J., 1990. Ideas para la interpretación ecologica del Paraná. Interciencia 15: 424–441.

Pielou, E. C., 1975. Ecological Diversity. John Wiley, New York, 165 pp.

Segers, H., 1995. Rotifera 2: The Lecanidae (Monogononta). Guides to the identification of the microinvertebrates of the continental waters of the world 6. SPB Academic Publishing, The Hague, 226 pp.

Serafim, M. Jr., 1997. Heterogeneidade espacial e temporal da comunidade zooplanctônica do sistema rio Ivinhema-lagoa dos Patos, planície de inundação do alto rio Paraná (MS). Dissertação (Mestrado) – Ecologia de Ambientes Aquáticos Continentais. Universidade Estadual de Maringá, Paraná, 33 pp.

Sokal, R. R. & F. J. Rohlf, 1991. Biometry: The Principles and Practice of Statistics in Biological Research. W.H. Freeman and Company, New York, 859 pp.

Souza Filho, E. E., 1993. Aspectos da geologia e estratigrafia dos depósitos sedimentares do rio Paraná no segmento entre Porto Primavera (MS) e Guaíra (PR). Tese (Doutorado) – Universidade de São Paulo, São Paulo, 213 pp.

Statsoft Inc., 1996. Tulsa: Statistica. 3 v.

Thomaz, S.M., M.C. Roberto & L.M. Bini, 1997. Caracterização limnológica dos ambientes aquáticos e influência dos níveis fluviométricos. In Vazzoler, A. E. A. M., A. A. Agostinho & N. S. Hahn (eds), Planície de Inundação do alto rio Paraná: Aspectos Físicos, Biológicos e Socioeconômicos. EDUEM, Maringá, 74–102.

Train S., L. C. Rodrigues, P. F. Borges, A. Takeouyeda, M. M. Nacagava & E. M. Bovo, 2001. In Agostinho, A. A., S. M. Thomaz & K. Nakatani (eds), A planície de inundação do alto rio Paraná (Relatório PELD-CNPq/ Nupélia-UEM), Maringá, 87–100.

Whittaker, R. M., 1960. Vegetation of the Siskiyou Mountains. Oregon and California. Ecological Monographs 30: 279–338.

Hydrobiologia (2005) 546:415–421
A. Herzig, R.D. Gulati, C.D. Jersabek & L. May (eds.) Rotifera X: Rotifer Research: Trends, New Tools and Recent Advances
DOI 10.1007/s10750-005-4284-1

Relationships between rotifers, phytoplankton and bacterioplankton in the Corumbá reservoir, Goiás State, Brazil

Cláudia Costa Bonecker* & Anderson Setsuo Miyashiro Aoyagui
Nupélia, Posgraduate Course in Ecology of Continental Aquatic Environments, State University of Maringá, Av. Colombo, 5790, 87020-900, Maringá-PR, Brazil
*(*Author for correspondence: E-mail: bonecker@nupelia.uem.br)*

Key words: Brazil, bacterioplankton, phytoplankton, reservoir, rotifers

Abstract

This study evaluates the relationships between rotifers and phytoplankton and rotifers and bacterioplankton in a tropical reservoir. Fourteen stations in the reservoir were sampled, including in its arms and tributaries, in the dry and rainy seasons. The highest rotifer density was found in the dry season, mainly in the upper and intermediary stretches of the reservoir. *Brachionus calyciflorus*, *Polyarthra vulgaris*, *Keratella tropica*, *K. cochlearis*, *K. americana* and *Pompholyx complanata* were the most abundant species. Densities of *B. calyciflorus* and bacteria were significantly correlated. On the other hand, there was an inverse relationship between *P. vulgaris* and bacteria. Diatoms (Bacillariophyceae) were observed to be associated to *K. americana*. We suggest that the rotifer populations play a part in microbial and herbivory food webs.

Introduction

Rotifers are regarded opportunistic organisms, mainly because of their high tolerance to changes in environmental conditions and their high reproductive rates (Allan, 1976). Their feeding strategy consists of filtering of water and capturing the prey in the crown of cilia. The prey selection being dependent on particle size. This type of feeding may be classified as generalists or specialists (Gilbert & Bogdan, 1984) depending on extent of selection of food particles. The selection of particles is related, among other factors, to the great morphological diversification of the buccal apparatus.

Arndt (1993) considered that bacteria, heterotrophic flagellates and small ciliates constitute a large part of the food resources that may be used by rotifers. Branco & Senna (1996) and Caleffi (1998) have shown the importance of bacteria and phytoplankton as a food resource for rotifers in reservoirs. Also Ooms-Wilms (1997) emphasized the importance of bacteria as a food resource for rotifers, and highlighted the relationship between the size of the bacteria and the rotifer species that consume them. Thus, rotifers play an important role in the transfer of energy in the trophic webs and the nutrient recycling.

Spatial and seasonal variations in food supply can affect rotifer distribution. Sanders et al. (1989) who investigated rotifer seasonality in a temperate lake discussed the relationship between the abundance of some rotifer species that fed mainly on bacteria, and the density of bacterioplankton over a period of 8 months, which included both stratification and destratification periods. Mazumder et al. (1992), in a 2-year study of a lake, found that rotifers consumed mainly phytoplankton. These findings do not necessarily apply to lakes in different geographical regions, especially to tropical lakes, which are characterized by high temperatures and near-lack of seasonal variations, except for two extreme seasonal periods (dry and rainy seasons).

We evaluated the relationship between rotifers and phytoplankton and bacterioplankton in a tropical reservoir by studying their abundance and distribution. We hypothesized that this relationship is influenced by the spatial heterogeneity of the reservoir, and by the marked limnological differences between the rainy and dry seasons.

Materials and methods

The Corumbá reservoir (Goiás State, Brazil) is located in the Corumbá River, which is one of the tributaries of the Paranaíba River. It has an area of 65 km^2 and mean depth of 23 m. Average hydraulic residence time of water in the reservoir is 40 days. Its drainage area is 27,800 km^2 and main tributaries are the Santo Antônio, Peixe and Pirapetinga rivers, the last named rivers receives sewage from Caldas Novas and waste discharge from a slaughterhouse located in the flooded area.

Sampling stations and sampling

Fourteen sampling stations were selected in the central channel (CRB 00, CRB 05, CRB 10, CRB 17, CRB 20, CRB 25, CRB 27, CRB 30) and marginal channels (PRP 15, PRP 18, JQT 15, TAQ 15, BVT 15, STA 10) of the reservoir in the dry (September 1998) and rainy (March 1999) seasons (Fig. 1).

Rotifers were sampled using a motorized pump and a plankton net (70 μm mesh), in different depths at each station, depending on the maximum depth of the station. Six hundred liters of water were filtered and the concentrated plankton in the net was transferred to polyethylene bottles (400 ml) and the samples were fixed in a 4% formaldehyde solution, buffered with calcium carbonate. In the laboratory the samples were concentrated to 85 ml in glass jars. Bacterioplankton samples (50 ml) were collected from the surface water and preserved in 4% formaldehyde solution. Also phytoplankton samples were taken from the surface water and fixed in acetic-acid Lugol's solution. From these preserved samples aliquots of appropriate volume depending on the concentration of seston were taken for quantitative microscopical counting.

The identification of the rotifer using an optical microscope using the keys of Koste (1978) and Segers (1995). For this the animals from the concentrated samples were transferred to glass slides and covered with cover slips before the identifications.

Rotifer densities (ind. m^{-3}) were determined from quantitative sample counts in a Sedgwick-Rafter counting cell, under an optical microscope. At least 200 individuals per each sample were counted. The concentrated samples were subsampled using a Stempell pipette.

The data on bacterioplankton and phytoplankton densities were obtained from the Basic Limnology and Phytoplankton laboratories of Nupélia-UEM, respectively. In the laboratory, 1 ml aliquots from the bacterioplankton samples were filtered through 0.2 μm Nucleopore membranes in replicates. Membranes were stained for 5 min with DAPI (fluorocromo 4′6′ diamidino-2 phenylindole) (Porter & Feig, 1980). For control filters 0.2 μm filtered distilled water was used applying the following protocol. Bacterial density (cells ml^{-1}) was counted in 40 randomly chosen fields in a Zeiss epifluorescence microscope (resolution 1000×) (Hobbie et al., 1977; Booth, 1993). Phytoplankton density (ind. ml^{-1}) was estimated according to Uthermöhl (1958) using a Zeiss inverted microscope at a magnification of 400×. Individuals (cells, colonies, coenobiotes, and filaments) were counted in 150 fields, or in the case of samples with few algae, the minimal area method was used and as many fields were counted as were necessary to establish the number of taxa added per field.

Correlation analysis was carried out between the densities of rotifers and the bacterioplankton and the phytoplanktonic classes numerically predominant in the study area during the two hydrological periods.

Results

The rotifer maxima were found in the dry season (September 1998), mainly in the main channel in the intermediate segment of the reservoir (CRB 10, CRB17 and CRB20). In the rainy season (March

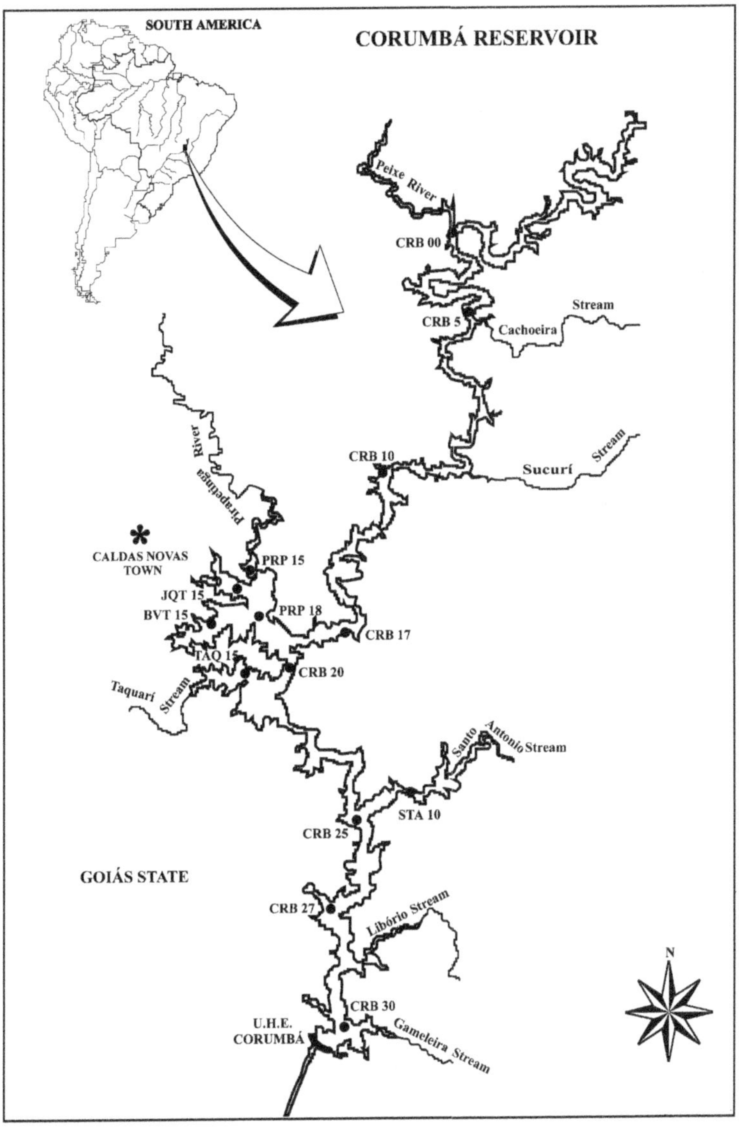

Figure 1. Corumbá Reservoir – GO (15° 79′ S and 48° 31′ W) and location of the sampling stations: CRB 00 (1 depth), CRB 05 (2 depths), CRB 10 (2 depths), CRB 17 (3 depths), CRB 20 (5 depths), CRB 25 (3 depths), CRB 27 (3 depths), CRB 30 (6 depths), PRP 15 (3 depths), PRP 18 (3 depths), JQT 15 (2 depths), TAQ 15 (2 depths), BVT 15 (2 depths), STA 10 (3 depths).

1999), the densities were higher in marginal channels of the intermediary stretch (PRP 15, JQT 15, TAQ 15 and BVT 15) (Fig. 2).

Rotifer average density was 28,083 ind. m^{-3}. The most commonly occurring rotifer taxa, which accounted for ca. 50% of the total densities, were *Brachionus calyciflorus* (mean density, 1733 ind. m^{-3}), *Polyarthra vulgaris* (mean density, 4577 ind. m^{-3}), *Keratella tropica* (mean density, 965 ind. m^{-3}), *K. cochlearis* (mean density, 912 ind. m^{-3}), *K. americana* (mean, 401 ind. m^{-3}), *Pompholyx complanata* (mean density, 4691 ind. m^{-3}). These species showed differential spatial distributions (Fig. 3).

In the dry season, *B. calyciflorus* and *K. tropica* had higher densities in the marginal channels (PRP 18), and were important in the upper (CRB 00 and CRB 10) and intermediate (CRB 20 and CRB 25) stretches of the reservoir. However, *K. cochlearis* was numerically more important in the central channel of the reservoir, especially in the upper

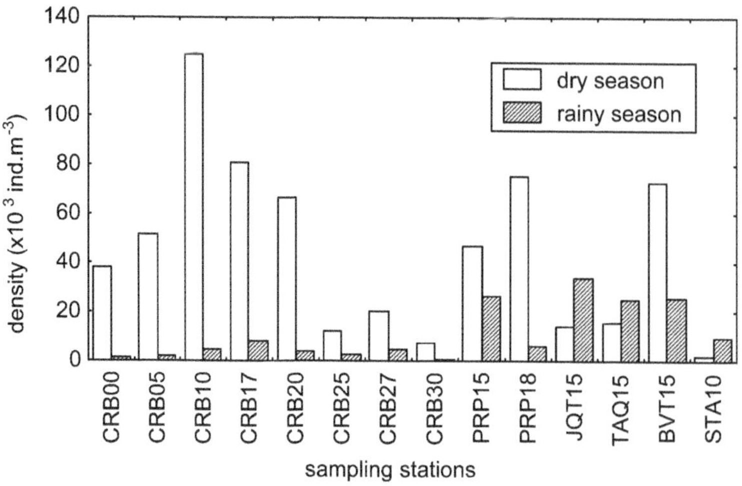

Figure 2. Average density of rotifers at the different sampling stations and hydrological periods.

(CRB 05) and intermediate (CRB 17) stretches. Densities of *P. complanata* were also high in the upper (CRB 10) and intermediate (CRB 20) stretches of the central channel of the reservoir. The species with high densities at almost all sampled sites were *P. vulgaris* and *K. americana*. They were the most common in the lower (CRB 27 and CRB 30) stretch of the central channel of the reservoir (Fig. 3).

On the other hand, in the rainy season *B. calyciflorus* was numerically important only in central channel, especially in the upper (CRB 00, CRB 05 and CRB 10) and intermediate (CRB 17) stretches. *K. tropica* was also more abundant in the central channel of the reservoir, mainly in the upper (CRB 00) stretch. However, *K. cochlearis* and *K. americana* were more abundant in the marginal channels (PRP 15 and PRP 18, and JQT 15 and TAQ 15, respectively). *P. vulgaris* maintained the same dry season pattern, i.e. it was important in the marginal channels (JQT 15 and TAQ 15). *Pompholyx complanata* was not observed in the rainy season (Fig. 3).

Correlations between rotifer density and bacterioplankton and rotifers and phytoplankton densities were significant for three rotifer taxa (*B. calyciflorus*, *P. vulgaris* and *K. Americana*) of the six taxa analysed (Table 1).

B. calyciflorus densities varied directly with those of bacteria. Both this rotifer and bacteria had relatively higher densities in the upper (CRB 00 and CRB 10) stretch of the central channel during the dry season. Densities of *P. vulgaris* and bacterioplankton were inversely related. Rotifer densities in the intermediate and lower stretches were higher in the longitudinal axis of the reservoir during the both hydrological periods (Fig. 3), whereas the bacterial densities were higher in the upper stretch. However, the rotifer densities and the main phytoplanktonic classes (Bacillariophyceae, Cryptophyceae, Cyanophyceae, and Chlorophyceae) were not correlated.

The association between diatoms and *K. americana* was observed mainly in the dry season. In the intermediate stretch of the reservoir the densities of both this rotifer and diatoms were high. According to Rodrigues (personal communication), diatoms were represented mainly by small-sized cells.

Discussion

The distribution of some rotifer species differed in the rainy and in the dry seasons, but some other species maintained their habitat preferences in both seasons. Generally, densities in the rainy season were dramatically lower than in the dry season, probably in relation to changes in the physical, chemical and biological conditions in the water column. During the rainy season a greater outflow and a lower residence time of the water in the res-

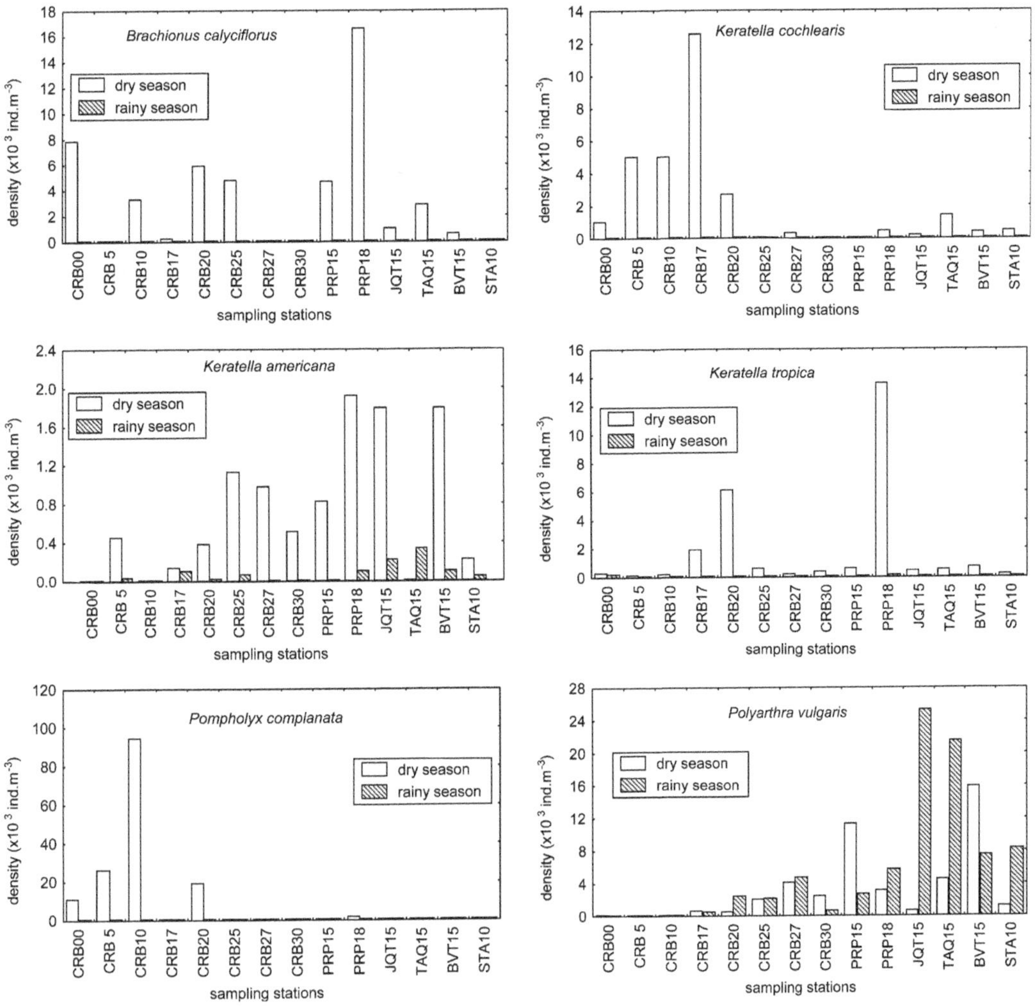

Figure 3. Average density of the most frequent rotifers at different sampling stations and hydrological periods.

Table 1. Pearson correlations between bacterioplankton (Bact), different phytoplankton families (Baci = Bacillariophyceae, Cyan = Cyanophyceae, Chlo = Chlorophyceae, Cryp = Cryptophyceae) and main rotifers species ($p < 0.05$)

Species/Groups	Bact		Baci		Cyan		Chlo		Cryp	
	r	*p*	*r*	*p*	*r*	*p*	*r*	*p*	*r*	*p*
B. calyciflorus Pallas, 1886	*0.55*	*0.02*	0.20	0.47	0.07	0.77	0.25	0.35	−0.16	0.55
K. cochlearis Gosse, 1851	0.39	0.14	0.15	0.57	0.30	0.26	0.43	0.10	−0.01	0.99
K. americana Carlin, 1943	−0.26	0.32	*0.59*	*0.01*	0.29	0.28	−0.04	0.88	0.06	0.83
K. tropica Apstein, 1907	0.46	0.07	0.35	0.19	0.27	0.32	−0.11	0.68	−0.01	0.96
P. complanata Gosse, 1851	0.33	0.21	0.34	0.20	0.41	0.12	0.49	0.05	0.06	0.833
P. vulgaris Carlin, 1943	*−0.70*	*0.02*	0.33	0.21	0.15	0.59	0.11	0.69	0.44	0.08

Italic values of *r* and *p* were significant.

ervoir appear to affect the rotifers reproduction rates (Schmid-Araya & Zuñiga, 1992). Also the increase of turbidity in the rainy season and consequently decrease in light penetration to the deeper water layers and changes in phytoplankton density appear to influence the development of rotifer populations (Nogueira et al., 1999). The decrease of pH which was also observed during the rainy season (Thomaz, personal communication) affected the density of these organisms as reported by Guimarães-Landa & Mourgués-Schurter (1999) in other tropical reservoir in Brazil (Minas Gerais State). Our results based on the correlation analysis between the density of some rotifer species and the bacterioplankton are consistent with the view that rotifers are important bacteria consumers. Starkweather et al. (1979) showed that even though *B. calyciflorus* is considered an herbivore, its diet depends on other available resources in the environment, such as bacteria, when their densities are high. They also reported that *Brachionus* may ingest large particles of detritus, on which bacteria may be present as aggregates. Arruda et al. (1983) considered that zooplanktonic organisms capable of ingesting particles enriched with organic material may benefited in such environments. This plasticity in the diet allowed the species belonging to this genus to be classified as generalists (Gilbert & Bogdan, 1984). On the other hand, Gilbert & Bogdan (1984) discussed that *P. vulgaris* feeds primarily on cells with flagella; and they classify it as a specialist. The preference of *Polyarthra* for flagellate cells was also observed by Bogdan & Gilbert (1982).

The significant correlation ($p < 0.05$) between diatoms (Bacillariophyceae) and *K. americana* densities suggests that algae are an important food resource for the rotifer species. This preference for diatoms indicates that *K. americana* is not necessarily a generalist in the sense as reported by Gilbert & Bogdan (1984). Also Infante (1978) and Cisneiros et al. (1991) consider that if abundant Bacillariophyceae may be used by zooplankton.

The study suggests that different rotifer populations participate in microbial and herbivore webs. This relationship is influenced by the hydrodynamics of the reservoir (spatial heterogeneity and differences in hydrological levels).

Acknowledgements

We thank Furnas Centrais Eléctricas S.A. and CNPq and DBI/Nupélia for financial and logistical support. Dr. Sidinei Magela Thomaz for suggestions. Dr. Barbara Robertson and Dr. Luis Carlos Gomes revised the English text. The constructive criticism of the editor and one anonymous referee also is appreciated.

References

Allan, J. D., 1976. Life history patterns in zooplankton. American Naturalist 110: 165–180.

Arndt, H., 1993. Rotifers as predators on components of the microbial web bacteria, heterotrophic flagellates, ciliates – a review. Hydrobiologia 255/256: 231–246.

Arruda, J. A., G. R. Marzolf & R. T. Faulk, 1983. The role of suspended sediments in the nutrition of zooplankton in turbid reservoir. Ecology 64: 1225–1235.

Bogdan, G. & J. J. Gilbert, 1982. Seasonal patterns of feeds by natural populations of *Keratella*, *Polyarthra* and *Bosmina*: clearance rates, selectivities, and contribution to community graze. Limnology and Oceanography 27: 918–934.

Booth, B. C., 1993. Estimating cell concentration and biomass of autotrophic plankton using microscopy. In Kemp, P. F. B. F. Sherr, & J. J. Cole (eds), Handbook of Methods in Aquatic Microbial Ecology. Boca Raton, Florida, 199–205.

Branco, C. W. C. & P. A. C. Senna, 1996. Relations among heterotrophic bacteria, chlorophyll-*a*, total phytoplankton, total zooplankton and physical and chemical features in the Paranoá Reservoir, Brasília, Brazil. Hydrobiologia 337: 171–181.

Caleffi, S., 1998. Guarapiranga Reservoir: study of the zooplankton community and aspects of its eutrophication. Internationale Vereinigung für Theoritische und Angewandte Limnologie, Verhandlungen 26: 1898–1903.

Cisneiros, R., E. Hooker & L. E. Velasquez, 1991. Natural diet of herbivorous zooplankton in lake Xolotlán (Managua). Hydrobiological Bulletin 25: 163–167.

Gilbert, J. J. & K. G. Bogdan, 1984. Rotifer grazing: *in situ* studies on selectivity and rates. In Meyers, D. G. & J. R. Strickler (eds), Trophic Interactions Within Aquatic Ecosystems. AAAS Selected Symposium 85: 97–133.

Guimarães-Landa, G. & L. R. Mourgués-Schurter, 1999. Composição e abundância do zooplâncton em um sistema artificial raso (Represa Pomar) no Campus da Universidade Federal de Lavras – Minas Gerais. Bios 7: 21–31.

Hobbie, J. E., R. Daley & S. Jasper, 1977. Use of nucleopore filters for counting bacteria by fluorescence microscopy. Applied and Environmental Microbiology 33: 1225–1228.

Infante, A., 1978. Untersuchungen über die Ausnutzbarkeit verschiedener Algen durch das Zooplankton. Archiv für Hydrobiologie 42: 340–405.

Koste, W., 1978. Rotatoria. Die Rädertiere Mitteleuropas begründet von Max Voight. Monogononta, 2 vols. Gebrüder Borntraeger, Berlin, 673 pp, 474 pp.

Mazumder A., D. R. S. Lean & W. D. Taylor, 1992. Dominance of small filter feeding zooplankton in the lake Ontario foodweb. Journal of Great Lakes Research 3: 456–466.

Nogueira, M. G., R. Henry & F. E. Maricatto, 1999. Spatial and temporal heterogeneity in the Jurumirim Reservoir, São Paulo, Brazil. Lakes and Reservoirs: Research and Management 4: 107–120.

Ooms-Wilms, A. L., 1997. Are bacteria na important food source for rotifers in eutrophic lake? Journal of Plankton Research 19: 1125–1141.

Porter, K. & Y. S. Feig, 1980. The use of DAPI for identifying and counting aquatic microflora. Limnology and Oceanography 25: 943–948.

Sanders, R. W., K. G. Porter, S. J. Bennet & A. E. De Biase, 1989. Seasonal patterns of bactivory by flagellates, ciliates, rotifers and cladocerans in a freshwater planktonic community. Limnology and Oceanography 34: 673–687.

Schmid-Araya, J. M. & R. L. Zuñiga, 1992. Zooplankton community structure in two Chilean reservoirs. Archiv für Hydrobiologie 123: 305–335.

Segers, H., 1995. Rotifera. The Lecanidae (Monogononta). Guides to the identification of the microinvertebrates of the continental waters of the world, 2. SPB Academic Publishing, The Hague 226 pp.

Starkweather, P. L., J. J. Gilbert & T. M. Frost, 1979. Bacterial feeding by the rotifer *Brachionus calyciflorus*: clearance and ingestion rates, behavior and population dynamics. Oecologia 44: 26–30.

Uthermöhl, H., 1958. Zur Vervollkommnung der quantitativen phytoplankton-methodic. Internationale Vereinigung fuer Theoretische und Angewandte Limnologie, Mitteilungen 9: 1–39.

Hydrobiologia (2005) 546:423–430
A. Herzig, R.D. Gulati, C.D. Jersabek & L. May (eds.) Rotifera X: Rotifer Research: Trends, New Tools and Recent Advances
DOI 10.1007/s10750-005-4285-0

Short time-response of psammic communities of Rotifera to abiotic changes in their habitat

Jolanta Ejsmont-Karabin
Centre for Ecological Research PAS, Hydrobiological Station, Leśna 13, 11-730, Mikołajki, Poland
E-mail: jolanta@onet.pl

Key words: Rotifera, psammon, community structure, community dynamics

Abstract

Sandy microhabitats and their fauna are usually characterised as extremely unstable and unpredictable. The aim of the paper is to evaluate the scale and the character of these phenomena. Rotifer composition and density and abiotic variables were investigated every half a day in the hygroarenal of the eutrophic Mikołajskie Lake. In total, 62 species were found with only two species occurring permanently throughout the study period. Densities in general were extremely variable and strong fluctuations were observed even at 12 h intervals. On average, psammophilic rotifers dominated the community (74%) while psammoxenes and psammobionts constituted only 18 and 8%, respectively. The most stable group were the psammophiles. The results support the presumption that fluctuating environments are ideal for organisms that reproduce quickly, may colonize vacant habitats rapidly and have wide niches, i.e. eurybionts.

Introduction

Interstitial microhabitats are usually characterised as extremely unstable and unpredictable and their fauna seems to be very variable and fluctuating (Pennak, 1951; Schmid-Araya, 1993; Pejler, 1995). This hypothesis has been confirmed with studies on seasonal dynamics of psammon rotifers in the mesotrophic Lake Kuc, where strong changes were observed in 1-week intervals in the community abundance and structure (Ejsmont-Karabin, 2003a).

Wiszniewski (1933, 1934a,b) distinguished three categories of psammon, i.e.: hydro-, hygro- and eupsammon. Hydropsammon inhabits the permanently submerged sands. Hygropsammon inhabits a zone periodically washed with waves and eupsammon the zone with constantly desiccated surface microlayer and a moist deeper one. From among the categories, hygropsammon communities are the most disturbed ones (Wiszniewski, 1933) as their habitat offers environmental conditions varying with waves and currents (Neel, 1948).

According to Wiszniewski (1934a, b; 1953) psammon rotifers may be separated into three categories: psammobiotic (species occurring only in sand), psammophilic (species occurring both in sand and open water) and psammoxenic (rotifers that are not able to survive in sands) ones.

There have been no quantitative studies at short-time intervals of rotifers in the hygropsammon. The purpose of the paper is to answer the following questions: can we observe large-scale fluctuations of psammon rotifers at 12 h' intervals? Does the short-time response vary among ecological groups in these communities?

Materials and methods

Rotifer composition, density and climatic conditions (temperature, wave range) were investigated with high sampling frequency (every half a day) in the hygroarenal of the eutrophic Mikołajskie Lake between 13 September and 9 October 2002. Mikołajskie Lake belongs to the system of the

Great Masurian Lakes (Northern Poland). The lake area is 460 ha, maximum depth –27.8 m, mean depth –11.0 m. Its littoral area covers 19% of the total lake surface.

Samples of sand and its interstitial water were taken from the hygroarenal at five sampling points along a 10 m transect in the zone wetted by lake waves. At low tide, 2 cm deep sample cores were taken with a 6 cm diameter sharp-edged cylinder. Samples were transferred to glass containers and rinsed six times with tap water. After the sand sedimented the upper layer of water was filtered through a 30 μm mesh-size plankton net. All rotifers were counted and identified in three subsamples each equal to 10% of the sample. The first subsample was analysed alive, while the others were fixed with 4% formalin.

Temperature in sand was determined at a depth of 1.0 cm with the thermometer. Wave range measurements were based upon observations of the width of the zone washed with lake waters.

The Shannon-Weaver diversity index (Margalef, 1957) for each sample was calculated.

The absolute and relative rates of community change were calculated for 12 h periods using the equations: $\delta_a = (N_{t+1} - N_t)$ and $\delta_r = 2(N_{t+1} - N_t)(N_t)^{-1}$, where N_{t+1} and N_t are densities at time $t + 1$ and t, respectively.

Statistical analyses were run with STATISTICA (Statsoft. Inc.) software. Probability levels of ≤ 0.05 were considered significant.

Results

September and October in Masurian Lakeland are usually months with strong winds and rapidly changing temperatures. However, despite of strongly changing air temperatures (very high differences were noted between day and night temperatures), temperatures of interstitial waters were markedly less fluctuating. This is because lake waters washing beach sands keep interstitial waters at temperatures close to those in the pelagic zone. Strong differences were observed in wave range due to wind action, especially at the beginning and the end of the study (Fig. 1).

Total rotifers density fluctuated strongly (Fig. 2a) with a 23-fold difference between the lowest (96 ind. 100 cm^{-2}) and the highest (2206 ind. 100 cm^{-2}) rotifer abundance. Rapid increases were followed by rapid decreases often

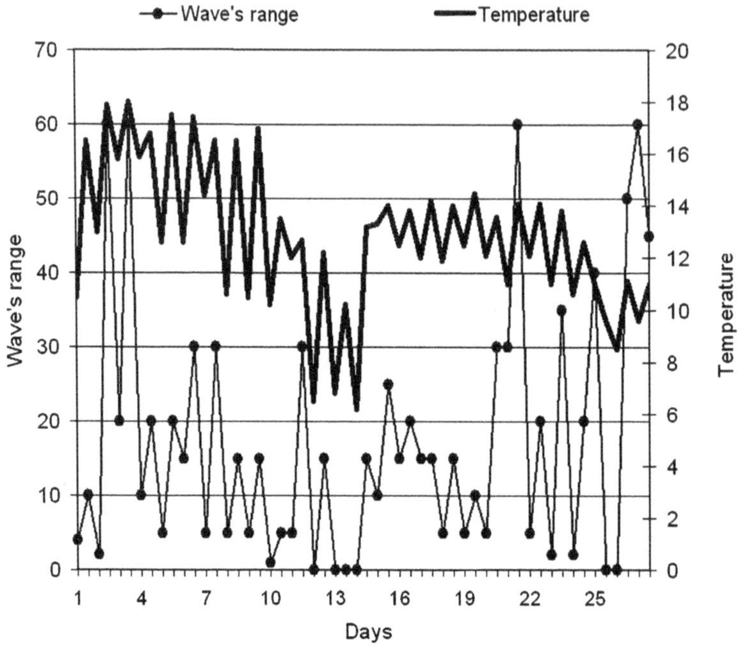

Figure 1. Changes in temperature of interstitial waters (°C) and wave range (cm) in a course of 27 days in Autumn 2002 in Mikolajskie Lake.

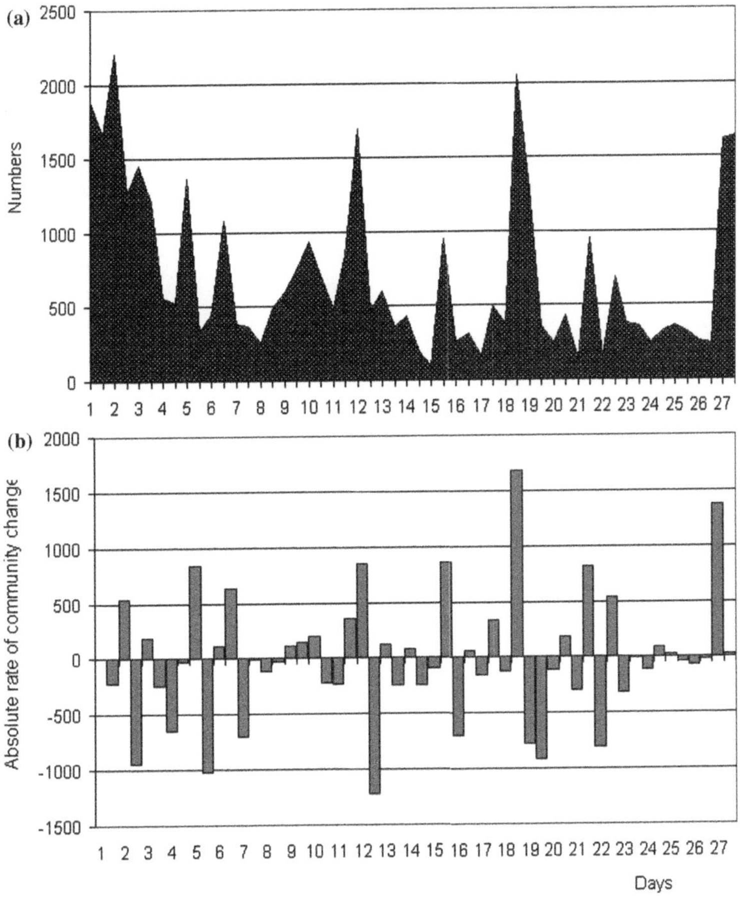

Figure 2. Changes in total abundance (ind. 100 cm^{-2}) (a) and rate of community change based on absolute density (ind. 100 cm^{-2} 12 h^{-1}) (b) in hygropsammon of Mikolajskie Lake in Autumn 2002.

within 12 h intervals (Fig. 2B). Nevertheless, there was no significant relationship between rotifer numbers and waves' reach neither as a direct effect nor as a 12 and 24 h time lag (Pearson correlation coefficients, $r = 0.03$, $p = 0.24$; $r = 0.01$, $p = 0.84$ and $r = 0.01$, $p = 0.94$, respectively). Rotifer densities were not influenced by temperatures directly ($r = 0.01$, $p = 0.47$), but weak correlation was found for 12 and 24 h time lag ($r = 0.31$, $p = 0.03$ – in both cases).

The total number of 62 rotifer psammon species found during the study is comparable to the rotifer species richness found in macrophytes. However, as many as 25 species occurred only sporadically. There were, among them, psammobionts like *Cephalodella compacta* Wiszniewski, *Encentrum diglandula* (Zavadovski), *E. sutor* Wiszniewski, *Trichocerca myersi* (Hauer), psammophiles (*Cephalodella exigua* (Gosse), *Encentrum marinum* (Dujardin), *Lecane arcuata* (Bryce), *L. clara* (Bryce), *L. stenroosi* (Meissner), *Lepadella quadricarinata* (Stenroos), *Trichocerca tenuior* (Gosse)) and psammoxenes (*Brachionus angularis* Gosse, *Cephalodella sterea* (Gosse), *C. tenuior* (Gosse), *C. ventripes* (Dixon-Nuttall), *Colurella geophila* Donner, *Dicranophorus forcipatus* (Muller), *Encentrum uncinatum* (Milne), *Euchlanis dapidula* Parise, *Gastropus stylifer* Imhof, *Lecane nana* (Murray), *Monommata longiseta* (Muller), *Paradicranophorus aculeatus* (Neiswestnowa-Shadina), *Synchaeta kitina* Rousselet, *Trichocerca capucina* (Wierzejski & Zacharias). Only two species – psammophilic *Lecane closterocerca* and *L. levistyla* – occurred at all sampling occasions throughout the study period (Table 1). The most frequent psammobionts were *Dicranophorus*

Table 1. List of rotifer species with frequency >5% of sampling points, their ecological status (psammophiles – PH, psammoxenes – PX, psammobionts – PB), mean abundance and frequency

Species	Group	Range of numbers	Mean (SE)	Frequency (in %)
Anuraeopsis fissa (Gosse)	PX	0–182	11 (4)	50
Ascomorpha ovalis (Bergendal)	PX	0–22	2 (1)	22
Ascomorpha saltans Bartsch	PX	0–80	9 (2)	46
Cephalodella auriculata (Muller)	PH	0–96	11 (3)	46
Cephalodella catellina (Muller)	PH	0–89	5 (2)	39
Cephalodella gibba (Ehrenberg)	PH	0–19	3 (1)	46
Cephalodella reimanni Donner	PH	0–436	29 (12)	35
Cephalodella tenuiseta (Burn)	PH	0–9	1 (0.3)	30
Collotheca mutabilis (Hudson)	PX	0–100	6 (2)	30
Collotheca pelagica (Rousselet)	PX	0–17	1 (0.4)	20
Collotheca wiszniewskii Varga	PB	0–12	1 (0.4)	28
Colurella adriatica Ehrenberg	PX	0–198	15 (4)	67
Colurella colurus (Ehrenberg)	PH	0–65	9 (2)	63
Colurella hindenburgi Steinecke	PH	0–146	45 (9)	94
Colurella obtusa (Gosse)	PH	0–60	8 (2)	50
Colurella uncinata (Muller)	PX	0–31	4 (1)	35
Dicranophorus grandis (Ehrenberg)	PX	0–38	1 (0.8)	9
Dicranophorus hercules Wiszniewski	PB	0–538	32 (11)	81
Dicranophorus luetkeni (Bergendal)	PH	0–24	3 (0.6)	44
Euchlanis contorta (Wulfert)	PH	0–389	26 (8)	57
Euchlanis dilatata Ehrenberg	PX	0–109	6 (2)	37
Keratella cochlearis (Gosse)	PX	0–574	28 (12)	65
Lecane bulla (Gosse)	PX	0–20	2 (0.5)	31
Lecane closterocerca (Schmarda)	PH	27–1906	327 (52)	100
Lecane levistyla (Olofsson)	PH	3–153	34 (4)	100
Lecane luna (Muller)	PX	0–42	9 (1)	78
Lecane lunaris (Ehrenberg)	PH	0–60	6 (1)	57
Lecane psammophila (Wiszniewski)	PB	0–57	11 (2)	63
Lepadella patella (Muller)	PH	0–68	12 (2)	85
Notommata cyrtopus Gosse	PH	0–20	1 (0.4)	11
Polyarthra vulgaris Carlin	PX	0–66	5 (2)	26
Trichocerca intermedia (Stenroos)	PB	0–10	1 (0.3)	13
Trichocerca pusilla (Lauterborn)	PX	0–23	2 (0.7)	26
Trichocerca similis (Wierzejski)	PX	0–15	1 (0.4)	19
Trichocerca taurocephala (Hauer)	PB	0–12	2 (0.4)	28
Wierzejskiella sabulosa (Wiszniewski)	PB	0–15	1 (0.4)	13
Wierzejskiella velox (Wiszniewski)	PB	0–43	5 (1)	52

hercules and *Lecane psammophila*. Strong fluctuations were observed in dominant and rare species.

Species diversity (Y) was also high (Fig. 3) and dependent on the number of species (X) as shown by regression equation, $Y = 0.84X^{0.42}$ ($R^2 = 0.24$; $p < 0.001$). Its values ranged from 0.97 to 4.16 with a mean value of 2.83 ± 0.75. A weak tendency was noted on the negative relationship between psammon rotifer densities (X) and their Shannon diversity index (regression equation, $Y = -0.0004X + 3.1$; $R^2 = 0.08$; $p < 0.04$).

In Mikołajskie Lake psammophiles dominated with their mean numbers of 509 ± 441 (SD) ind. 100 cm^{-2} despite the fact that there were rapid

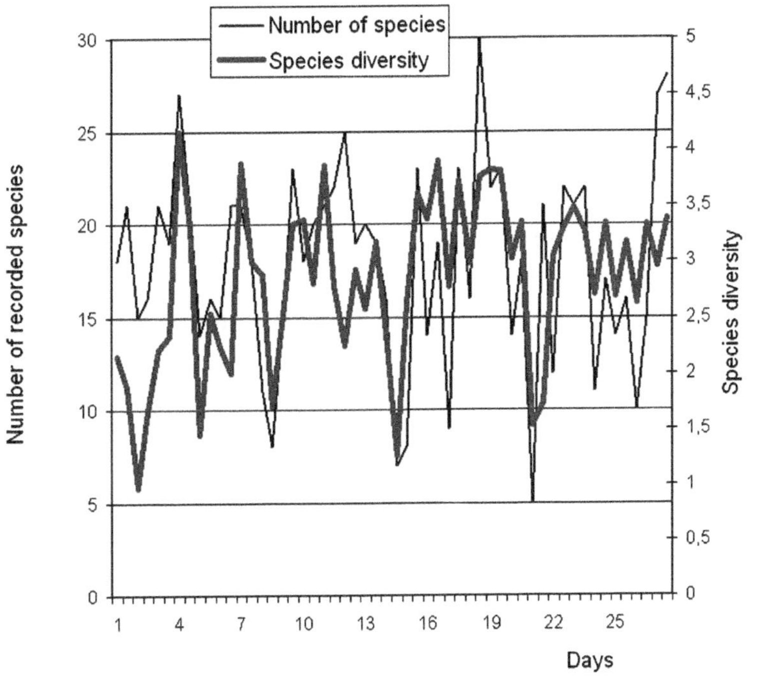

Figure 3. Comparison of short-time changes between species diversity and number of species of hygropsammon rotifers in Mikolajskie Lake in Autumn 2002.

fluctuations in their relative composition (Fig. 4). Less abundant were psammoxenes (mean 120 ± 179 ind.100 cm^{-2}) and psammobiotic species were only 8% of the total community (55 ± 101 ind. 100 cm^{-2}).

The relative rate of community increase for 12 h time intervals was the highest in psammoxenes (3.12 ± SD 7.84), lower in psammobionts (2.44 ± SD 9.98) and the lowest in psammophiles (0.90 ± SD 2.09). However, only the difference between psammoxenes and psammophiles was statistically significant at 95% ($p = 0.03$; *t*-test). The highest increase in numbers observed in psammophiles was that from 300 to 1256 ind. 100 cm^{-2}, thus ca. 3-fold within 12 h. The increase in psammobionts was stronger (6-fold) although at lower densities (from 67 to 469 ind. 100 cm^{-2}).

Discussion

The importance of the frequency of disturbances has been shown for phytoplankton succession in pelagial by Reynolds (1988). The author suggests that intermediate scale (20–200 h) disturbances tend to preserve high species diversity of communities. Short time and severe environmental disturbances may return the community succession to a more primitive stage, with *r*-selected species domination (Reynolds, 1988). Poor organization, high productivity, high losses, rapid cycling and weak competition characterize such communities. Psammon communities seem to have all above-mentioned features. As it is shown in this paper, the communities are relatively high in species diversity (Fig. 3). They are more abundant than pelagic communities (Ejsmont-Karabin, 2003a), but at the same their abundance fluctuates strongly (Figs. 2 and 4).

In this respect hygropsammon communities resemble benthic rotifer assemblages in gravel streams. Schmid-Araya (1998b) showed that rotifer inhabitants of hyporheic waters form a non-equilibrium community, which is non-predictable in time. Thus, the community exhibits a temporally near-random pattern.

The ephemeral nature of this microhabitat offers psammon rotifers little opportunity to increase their numbers by parthenogenetic reproduction (Neel, 1948). This type of reproduction

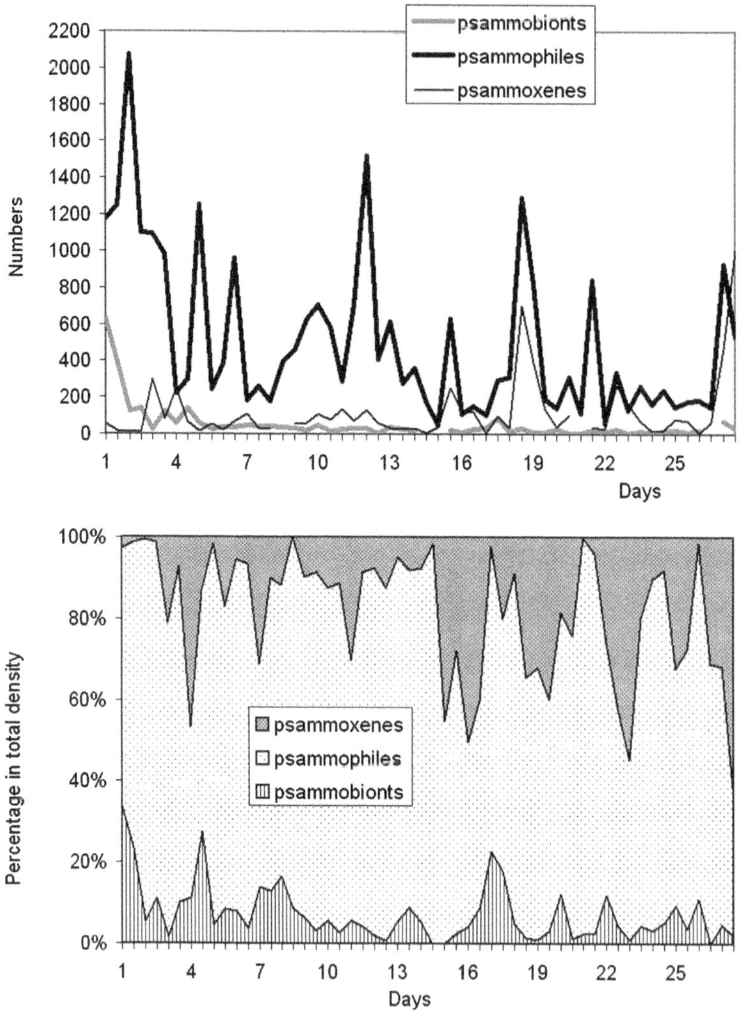

Figure 4. Abundance and relative percentage of total rotifer density of three ecological groups of psammon.

gives slow change in population densities as a female needs a few hours or even a day (Ejsmont-Karabin, et al., 1993) to mature and then its first parthenogenetic egg needs some time to hatch. Although parthenogenetic reproduction may play a substantial role during periods with more stable weather conditions it cannot explain 3- or 6-fold increase in rotifer densities. The rapid density changes of particular species in this study could be explained by the assumption that hatching from resting eggs is the important way to develop psammon rotifer populations after a rapid decrease in their abundance. Hatching from resting eggs may be very rapid, synchronous (if they receive a distinct environmental signal) and in great numbers (Pourriot & Snell, 1983). Wiszniewski (1932) observed frequently the occurrence of males of many species, some of them in high densities, which is not so common in pelagic waters.

Interstitial spaces of sand in the hygroarenal are completely filled with water due to wave action that by returning to the lake dilutes the interstitial water and transports materials (Neel, 1948; Pennak, 1951). This mechanical factor may determine the structure and functioning of psammon communities and may explain the high variability of the communities even in a short-time scale. Psammophiles inhabiting littoral waters may contribute to the rapid increases of psammon communities if their eggs attached to bottom substrates are transported in this way into the hygroarenal zone. This may also explain the important role played by psammophiles in psammon communities.

Psammophilic forms were the dominating ecological group of psammon rotifers in all 44 beaches of 18 lakes studied by Ejsmont-Karabin (2003b) in the Masurian Lake District (Poland). Communities inhabiting hygroarenal of a mesotrophic Lake Piaseczno were dominated by psammophiles or psammobionts (Radwan & Bielańska-Grajner, 2001), but in a eutrophic Lake Uściwierz and a hypertrophic Lake Głebokie Uścimowskie (Leczna-Wlodawa Lakeland, eastern Poland) psammophilous taxa predominated, reaching 67–100% and 58–72% of the total abundance of psammon, respectively (Radwan et al. 2001).

The great number of eurybionts with wide niches may lead to intense competitive interactions, whereas a great number of stenobionts may be constrained to live only in stable conditions. This is a reason why communities with high diversity should have some optimum proportion of stenobiotic and eurybiotic species (Protasov, 2002). The hygroarenal cannot be considered as a stable habitat and because of its scarce resources this habitat may be a place of strong competitive relationships. The proportion of eurybiotic to stenobiotic (i.e., psammophiles to psammobionts) rotifers is 9:1 in hygropsammon of Mikołajskie Lake in September–October. However, the proportion may differ among seasons and habitats. Mean proportion for lake beaches in Masurian Lakes was 2.5:1 in spring and 1.2:1 in summer (Ejsmont-Karabin, 2003b). The low proportion of eurybiotic to stenobiotic rotifers in summer may result from more stable weather conditions at that time (Ejsmont-Karabin, 2003b).

Thus, short time and severe environmental disturbances favour in many ways the dominance of psammophilic rotifers, mostly small, oportunistic and eurybiotic species occurring in different seasons and lake habitats. It also supports Schmid-Araya's (1998a) presumption that fluctuating environments are ideal for organisms that reproduce quickly, may colonize vacant habitats rapidly and have wide niches.

Acknowledgements

I would like to thank Jenny Schmid-Araya for constructive comments on an earlier version of the manuscript and for linguistic improvements. I am also grateful to Ramesh D. Gulati for his support and help in improving the text.

References

Ejsmont-Karabin, J., K. Siewertsen & R. D. Gulati, 1993. Changes in size, biomass and production of *Euchlanis dilatata lucksiana* Hauer during its lifespan. Hydrobiologia 255/256: 77–80.

Ejsmont-Karabin, J., 2003a. Is sandy beach of the lake an ecotone? Psammon Rotifera in a mesotrophic Lake Kuc (Masurian Lakeland, Northern Poland). Polish Journal of Ecology 51: 219–224.

Ejsmont-Karabin, J., 2003b. Rotifera of lake psammon: community structure *versus* trophic state of lake waters. Polish Journal of Ecology 1: 5–35.

Margalef, R., 1957. Information theory in ecology. General Systems 3: 36–71.

Neel, J. K., 1948. A limnological investigation of the psammon in Douglas Lake, Michigan, with especial reference to shoal and shoreline dynamics. Transactions of American Microscopical Society 67: 1–53.

Pejler, B., 1995. Relation to habitat in rotifers. Hydrobiologia 313/314: 267–278.

Pennak, R. W., 1951. Comparative ecology of the interstitial fauna of fresh-water and marine beaches. Annee Biologique 27: 449–480.

Pourriot, R. & T. W. Snell, 1983. Resting eggs in rotifers. Hydrobiologia 104: 213–224.

Protasov, A. A., 2002. Biodiversity and its Estimation. Conceptual Diversicology, Nat. Acad. Sci. Ukraine, Kiev, 106 pp. (in Russian).

Radwan, S. & I. Bielańska-Grajner, 2001. Ecological structure of psammic rotifers in the ecotonal zone of Lake Piaseczno (eastern Poland). Hydrobiologia 446/447: 221–228.

Radwan, S., T. Ozimek, I. Bielańska-Grajner & J. Sender, 2001. Structure of the water/land ecotones in trophically different Polish lakes. Internationale Vereinigung für Theoretische und Angewandte Limnologie, Verhandlungen 27: 3848–3851.

Reynolds, C. S., 1988. The concept of ecological succession applied to seasonal periodicity of freshwater phytoplankton. Internationale Vereinigung für Theoretische und Angewandte Limnologie, Verhandlungen 23: 683–691.

Schmid-Araya, J. M., 1993. Benthic Rotifera inhabiting the bed sediments of a mountain gravel Stream. Biologische Station Lunz, Jahresbericht 14: 75–101.

Schmid-Araya, J. M., 1998a. Rotifers in interstitial sediments. Hydrobiologia 387/388: 231–240.

Schmid-Araya, J. M., 1998b. Small-sized invertebrates in a gravel stream: community structure and variability of benthic rotifers. Freshwater Biology 39: 25–39.

Wiszniewski, J., 1932. Les rotiferes des rives sablonneuses du lac Wigry. Note preliminaire. Archiwum Hydrobiologiii i Rybactwa 6: 86–100.

Wiszniewski, J., 1933. Ożyciu w wilgotnych piaskach [On life in moist sands]. Wszechświat 1: 1–7(in Polish).

Wiszniewski, J., 1934a. Les Rotiferes psammiques. Annales musei zoologici Polonici 10: 339–396.

Wiszniewski, J., 1934b. Recherches ecologiques sur le psammon et specialement sur les Rotiferes psammiques. Archiwum Hydrobiologii i Rybactwa 8: 149–165.

Wiszniewski, J., 1953. Les Rotiferes de la faune polonaise et des regions avoisinantes. Polskie Archiwum Hydrobiologii 1: 317–490.

Hydrobiologia (2005) 546:431–440
A. Herzig, R.D. Gulati, C.D. Jersabek & L. May (eds.) Rotifera X: Rotifer Research: Trends, New Tools and Recent Advances
DOI 10.1007/s10750-005-4286-z

The influence of biotic and abiotic factors on psammic rotifers in artificial and natural lakes

Irena Bielańska-Grajner
Department of Ecology, University of Silesia, Bankowa 9, 40-007, Katowice, Poland
E-mail: igrajner@us.edu.pl

Key words: psammic rotifers, lakes, sand grain size, porosity, algae

Abstract

The effect of sand grain size, porosity and the abundance of algae on the community structure of psammic rotifers was tested in three anthropogenic lakes (Upper Silesia) and compared with three natural ones (West Pomerania). The structure of grain size in the studied beaches of artificial and natural lakes was similar but differences were found in the abundance of algae. The structure of the psammic rotifer community in Upper Silesian lakes was different from natural lakes in West Pomerania. The structure of psammic rotifer communities in anthropogenic reservoirs and lakes (West Pomerania) differed in numbers of species and numbers of psammobiont species, and also diversity index. The most altered structure of rotifer assemblages was observed in the psammon rotifers of the strongly contaminated Dzierżno Duże Lake.

Introduction

Although psammic rotifers have been investigated for many years (Wiszniewski, 1932, 1934a, b; Neiswestnowa-Shadina, 1935) more attention has been paid to rotifers inhabiting marine beaches (Turner, 1990, 1993; Tzschaschel, 1983; Saunders-Davies, 1998; Sørensen, 1998). Many papers have also been published on the sand and interstitial rotifers of rivers and streams (Zullini & Ricci, 1980; Pennak & Ward, 1986; Schmid-Araya, 1994, 1995, 1998 a, b; Turner, 1996; Turner & Distler, 1995). There is less information on psammic rotifers living in stagnant waters (Wiszniewski, 1932, 1934a, b; 1947; Neel, 1948; Segers, 1998; Ejsmont-Karabin, 2003; Bielańska-Grajner, 2001; Radwan & Bielańska-Grajner, 2001).

The relationship between the presence and abundance of rotifers and grain size has not been studied in detail (Wallace & Ricci, 2002). Giere (1993) described that the total pore volume of a sediment core determines the availiability of meiofauna to inhabit a sediment core. The size of the internal surfaces of sediment particles is an important determinant for meiobenthic life. It defines directly the area available for the establishment of biofilm (bacteria, fungi, diatoms) which forms a substantial biotic parameter for microscopic animal life.

One of the primary influences on meiofaunal distribution, especially in running waters, is the interaction between sediment particle size and advective exchange (Boulton et al., 2002).

The direct correlation between pore dimensions and body size of meiofauna could be experimentally proven (Williams, 1972). It has been observed that the largest particles of beach sand allowed the bigger organisms to colonise the beach (Jansson, 1967; Williams, 1972; Watling, 1991).

The objectives of this paper were to determine: (1) if the structure of psammic rotifer communities of natural and artificial lakes is similar and (2) if the rotifer community structure is related to grain size of sand and porosity and the presence and abundance of algae.

Study area

The study was conducted in three natural lakes situated in Northwest Poland and three artificial lakes located in the Upper Silesian region. Natural lakes are located in a relatively clean, non-industrial area of Poland. Siecino and Sarcze lakes are situated in the area of Ińsk Landscape Park and Jeleń lake near Bytów. Anthropogenic lakes are located in Upper Silesia which is the most degraded and contaminated part of the country. The general characteristics of the lakes are described in Table 1.

Materials and methods

Rotifers were collected during the spring (May), summer (July–August) and autumn (September–October) of 2002. Samples of psammon were collected at three sites located at the water edge and at 1 m above the non-submerged interstitial and below submerged interstitial. Samples were collected with a plastic corer of 3.5 cm diameter and 10 cm long. Only two fractions, 0–1 cm and 1–2 cm sediment depth, were analysed. Ten samples were collected at each site, five for qualitative analysis and the remaining five for quantitative analysis. Quantitative samples were preserved in a mixture of formaldehyde and glycerol (ratio 3:1). The qualitative analysis was conducted on live rotifer species. Rotifer density was expressed as the number of individuals per 1 dm^3 of sand. Samples of algae were collected similarly as rotifers and preserved with Lugol's and 4% formalin solutions. The investigated algae were identified and enumerated by the standard method in summer and autumn.

Interstitial water for chemical analysis were filtered through a bolting silk net and measured with Merck reagent.

The sand porosity of the tested lakes was evaluated by the alcohol method (Uggla, 1979). Soil samples were taken with a cylindrical core sampler of 18 cm^3 volume.

Estimation of bulk density of soil – C

Bulk density of soil (C) is the mass per unit volume of undisturbed soil, dried to a constant weight at 105 °C and calculated using the following equation:

$$C = [(b_1 - b)/v]\,\mathrm{g/cm^3},$$

where b_1 is the weight of a tube with soil after being dried at 105 °C, b the weight of an empty tube, v the volume of a tube.

Table 1. General characteristics of the study lakes and annual mean values of physico-chemical analysis

Lake	Siecino	Sarcze	Jeleń	Pławniowice D.	Dziećkowice	Dzierżno D.
Latitude (°N)	53° 37′	53° 38′	54° 12′	50° 23′	50° 08′	50° 21′
Longitude (°E)	16° 01′	16° 36′	17° 32′	18° 28′	19° 29′	18° 33′
Surface area (ha)	729.7	35.5	84	225	712	615
Max. depth [m]	44.3	4.9	33	18	12	20
Littoral water:						
Temperature (°C)	16	16.5	14.2	12	18.5	17.5
pH	8.0	8.5	7.1	8.5	7.8	8.2
Conductivity (uS cm^{-1})	220	240	50	490	180	531
Oxygen mgO_2/l	7.3	9.2	7.3	8.3	7.5	6.6
Phosphorus ($mgPO_4/l$)	0.34	0.2	0.2	0.2	0.25	2.5
Nitrogen (mg NO_3/l)	10	0.8	0	2.2	10.5	7.5
Interstitial water:						
Temperature (°C)	12.6	14.8	11.3	17	16.5	15.8
pH	7.3	7.5	7	7.8	7.3	7.5
Conductivity (uS cm^{-1})	220	240	120	590	130	950
Oxygen mgO_2/l	3.8	3.9	3.8	4.7	4.2	2.5
Phosphorus ($mgPO_4/l$)	1.7	0.42	0.67	1.5	0.65	4.3
Nitrogen (mg NO_3/l)	10	0	1.7	10	10	10.5

Table 2. Mean values of median sand grain size, minimum and maximum value of porosity and % of Bacillariophyceae density in psammon of reservoirs and lakes (eups – eupsammon, hygro – hygropsammon, hydro – hydropsammon)

Lakes	Seasons	Grain size	Porosity	% of Bacillariophyceae		
				Eups	Hygro	Hydro
Siecino	Spring	0.337	34.89–39.64	–	–	–
	Summer	0.337	35.87–41.24	90	40	85
	Autumn	0.337	40.11– 40.29	84	94	96
Sarcze	Spring	0.44	37.7–40.02	–	–	–
	Summer	0.45	40.83–43.39	24	21	24
	Autumn	0.337	36.22–37.24	98	96	97
Jeleń	Spring	0.375	35.32–39.62	–	–	–
	Summer	0.375	20.42–35.57	31	43	37
	Autumn	0.375	37.25–40.44	65	79	70
Pławniowice D	Spring	0.375	40.85–45.20	–	–	–
	Summer	0.412	44.09–46.42	50	50	69
	Autumn	0.3	45.6–49.43			
Dzierżno D	Spring	0.375	34.05–42.34	–	–	–
	Summer	0.337	40.02–40.69	91	79	94
	Autumn	0.3	42.56–42.56	48	69	98
Dziećkowice	Spring	0.337	39.81–42.34	–	–	–
	Summer	0.335	32.83–41.16			
	Autumn	0.375	36.98–41.16	91	75	86

Estimation of solid particle density of soil – C_1

Solid particle density is the weight of solid particles in a standard volume of those solid particles. It was determined using the alcohol (volumetric) method. A soil sample (10 g) was dried at 105 °C and then placed in a graduated measuring flask (50 ml volume). Denaturated alcohol was poured into the flask with the soil and shaken for 5 min. The flask then was filled with alcohol to a scale mark and shaken again for 2 min. If the volume of solution lowered below the scale mark, more alcohol was poured in.
C_1 was calculated as

$$C_1 = m/(50 - x)$$

where m is the soil sample dried at 105 °C (10 g), 50 the volume of a volumetric flask (in ml), x the amount of alcohol used to fill a flask (in ml).

Estimation of capillary soil porosity – P_k

Capillary soil porosity (P_k) is the sum of all capillary soil pores, which can retain water against gravity force. A tube with soil was placed in a dish with water to observe capillary conduction. The tube was removed from the dish when the water reached the upper surface of soil and filled all capillary pores, and the remaining water drained away and the tube was weighed. Afterwards, the tube with soil was dried at 105 °C for 2–7 days to evaporate all water from the soil. Desiccation was carried until a constant weight of tubes with soil was obtained. P_k was calculated from the equation:

$$P_k = [(a_1 - b_1)/v] \times 100\%$$

where: a_1 is the weight of a tube with soil after capillary conduction, b_1 the weight of a tube with soil after being dried at 105 °C, v the volume of a tube (in cm^3).

Determination of sand grain size (S)

Sand grain sizes was determined by sieving the sand through a mesh (Richling, 1993). A sample of deposits (100 g weight) was dried at the temperature of 105 °C, weighed and sieved. The mesh sizes were as follows: 1.0; 0.8; 0.5; 0.4; 0.25; 0.2 and 0.1 mm. Deposits remaining on the sieve were

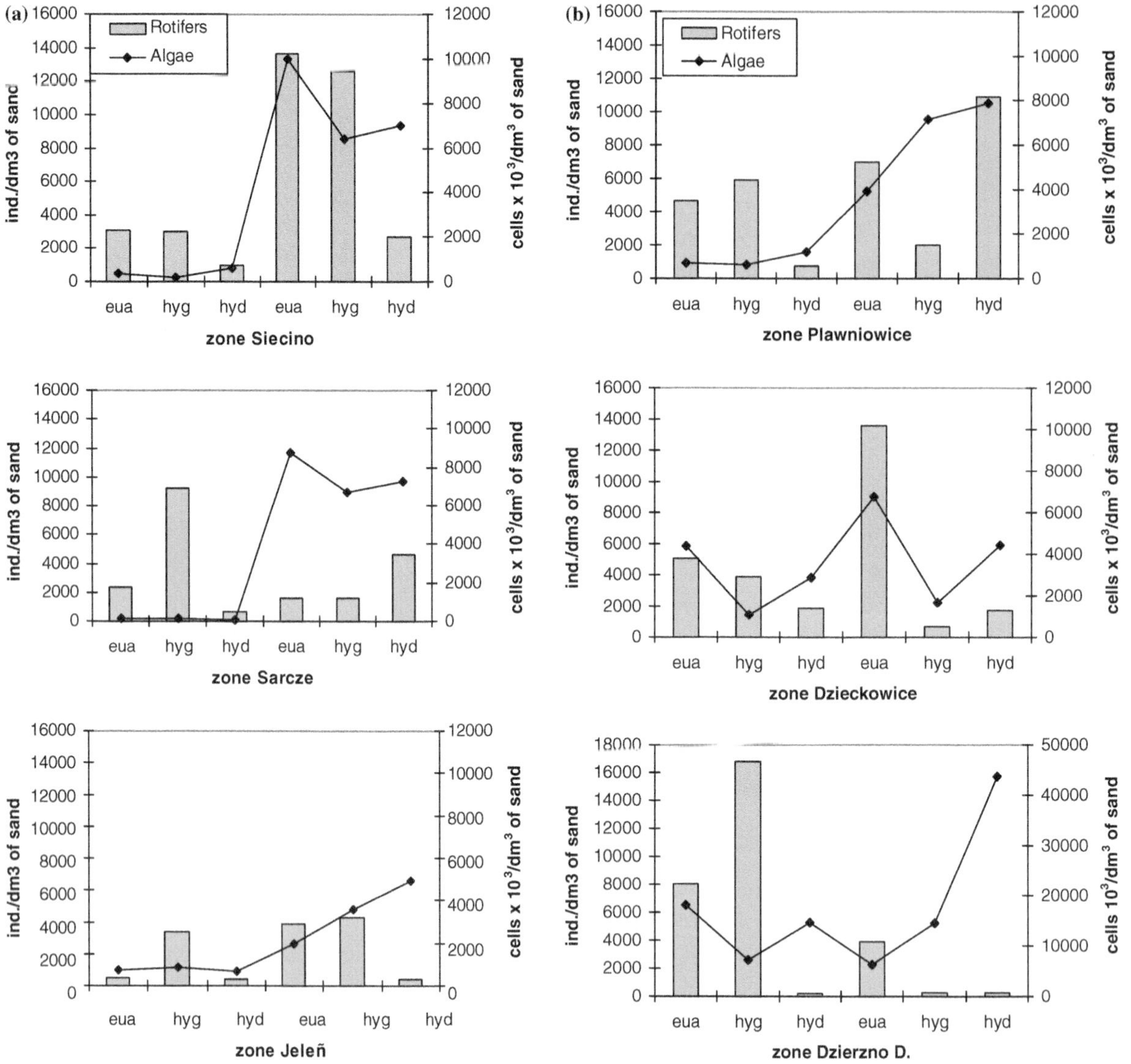

Figure 1. (a) Rotifer abundance (ind./dm^3 of sand) – bars, left-hand axis and algae (cells × 10^3/ dm^3 of sand) – solid line, right-hand axis, in summer and autumn in studied lakes. (b) Rotifer abundance (ind./dm^3 of sand) – bars, left-hand axis and algae (cells × 10^3/dm^3 of sand) – solid line, right-hand axis, in summer and autumn in studied reservoirs.

weighed and the proportion of soil particles calculated by the following formula:

$$S = (gs/Gs) \times 100\%$$

where *gs* is the mass of dry deposits, left on a sieve (in grams), *Gs* the total mass of dry deposits used for the analysis (in grams), *S* the proportion of a weighted fraction of deposits (in %).

Average grain size in the deposits was calculated for the statistical requirements. This is usually defined as a parameter of a positional cumulative curve and corresponds to a grain diameter for which there are as many observations larger as there are smaller, the median (*Md*) being:

$$Md = d\ 50\%$$

Cluster analysis of the similarity of the rotifer community and Shannon–Weaver index were calculated using computer software MVSP-3.

Table 3. Rotifer species in the psammon of studied lakes and reservoirs: diversity index; s – spring, su – summer, a – autumn; pb – psammobionts, p – psammophiles, px – psammoxenic

Lakes	Siecino	Sarcze	Jeleń	Pławniowice	Dziećkowice	Dzierżno	Ecol. Group
Diversity index	2.9	2.8	2.3	2.2	1.9	0.8	
Species							
Adineta gracilis Janson		a					pph
Brycella tenella (Bryce)	a	s, a					pph
Cephalodella auriculata (Mull.)	s, su, a	s		a	su, a	su, a	pph
C. catellina (Mull.)	s, su	s	s, su, a	s, su, a	su, a	s, su, a	pph
C. delicata Wulf.	s						px
C. eva (Gosse)	a						px
C. gibba (Ehrb.)	s, su, a	s, su, a	s, su a	s, su, a	su, a	a	pph
C. gracilis (Ehrb.)	s, su a	su, a	s, a	su, a	su, a	s, su, a	pph
C. hoodi (Gosse)	a						px
C. misgurnus Wulf.	su	su	su				px
C. paxi Wulf.	a						px
C. tachyphora Myers						s	pph
C. ventripes (Dix.-Nutt.)						s	pph
Colurella adriatica Ehrb.	s, su, a	s, su, a	s, su, a	su, a			px
C. colurus (Ehrb.)		su, a		s	su		pph
C. hindenburgi Stein.	s	s		su			px
C. obtusa (Gosse)		a		s			px
Dicranophorus forcipatus (Mull.)	a						px
D. hercules Wiszn.	s, a	s	a	a	a		pb
D. leptodon Wiszn.	a	a					pb
D. luetkeni (Bergen.)		s					pb
Dissotrocha aculeata (Ehrb.)			s				px
Elosa worallii f. *spinifera* Wiszn.	a	s, su, a	s		a		pb
Encentrum acrodon Wulf.		su					px
E. marinum (Dujar.)	su, a		su, a	a	su, a	s, su, a	px
E. saundersiae (Huds.)					su		px
E. tyrphos Wulf.	s, su, a		a	a		a	px
Itura aurita (Ehrb.)			s				px
Lecane bulla (Gosse)		su					px
L. clara (Bryce)	s						pph
L. closterocerca (Schm.)	s, su, a	s, su, a	s, su, a	s, su, a	su		pph
L. flexilis (Gosse)	s	s					px
L. hamata (Stokes)	s	s					px
L. levistyla (Olofs.)	s	s	s				pb
L. luna (Mull.)	s, a	s, su	s	s, su, a	su		px
L. lunaris (Ehrb.)	s, a	s	s, a	s, su, a	su		px
L. psammophila Wiszn.	s, a	s, su	s, su, a				pb
L. scutata (Harr. et Myers)	s, su, a	s, a	a	s, su, a	su, a		px
L. tenuiseta Harr.						s	px
Lepadella koniari Bartos			a				px
Lepadella patella (Mull.)	s, su, a	s, su	s, a	s, su, a	su, a		pph
Lindia janickii Wiszn.		s					pb
Lindia pallida Harr. et Myers	a		s, a			a	px

Continued on p. 436

Table 3. (Continued)

Lakes	Siecino	Sarcze	Jeleń	Pławniowice	Dziećkowice	Dzierżno	Ecol. Group
Diversity index	2.9	2.8	2.3	2.2	1.9	0.8	
Monommata astia Myers	s	a	a		su		pph
Mytilina mucronata (Mull.)			s, su				px
M. ventralis (Ehrb.)		s					px
Notholca labis (Gosse)		s					pph
Notommata cyrtopus (Gosse)	s, a	s, su, a	s, a	A	su		px
Philodina aculeata (Ehrb.)	s, su, a	s, a	a	s, a	su		px
Proales minima (Montet.)		a		a			pph
Rotaria rotatoria Pall.	s	s, sa		s, su			px
R. tardigrada (Ehrb.)				a		a	pph
Taphrocampa annulosa Gosse		s	s				px
Trichocerca intermedia Sten.	su, a	su, a	s, su	s, su, a	su, a		pph
T. rousseleti Voigt	a						px
T. taurocephala Hauer	su	s	a	s, su	su		pb
T. tenuior (Gosse)	a	s	s	a	a		pph
T. uncinata (Voigt)				s			px
T. weberi (Jenn.)		s		su	su		px
Tylotrocha monopus Jenn.	s						px
Wierzejskiella elongata (Wiszn.)		s	a				px
W. sabulosa (Wiszn.)	a						pb
W. velox (Wiszn.)	su, a		a	a	a		pb
Wigrella depressa Wiszn.	a						pb
Total species	41	40	30	27	22	11	

Results and discussion

The chemical parameters of the lake water and the interstitial water of the beaches varied considerably. Interstitial water contained less oxygen but a high concentration of phosphate, especially in the anthropogenic reservoirs, which indicates their high trophy (Table 1).

Grain sizes and porosity of all the investigated beach sand were very similar. Table 2 presents data on the median diameter and mean values of sand porosity for the beaches studied. In most lakes, algae were most abundant in autumn, Bacillariophyceae being the most abundant group in all the lakes (Table 2). There was no significant correlationship between the number of rotifers and sand grain size, nor between porosity and number of algae. An exeption to this was Pławniowice Duże reservoir where more rotifers and more algae was observed in autumn (Fig. 1).

Arov (1990) found that in Lake Bajkał larger species occurred in sand with a bigger median size of sand grain, and that sand with smaller grain size was inhabited by smaller species. In the studied lakes we did not observe such relationships. According to Ejsmont-Karabin (2004) there is positive significant correlation between the rotifer numbers and the share of grain size fraction 0.25–1.00 mm.

In total, 64 species of rotifers were found in the psammon of the investigated lakes. The rotifer community structure in lakes (West Pomerania) and artificial reservoirs (Upper Silesia) differed distinctly. Much more taxons of rotifers including psammobiont, occurred in the beaches of the natural lakes than in artificial lakes (Table 3). The highest number of species (41) was found in Siecino Lake, Sarcze Lake (40) and Jeleń Lake (30). The number of rotifer species found was 27 in Pławniowice Duże, 22 in Dziećkowice, and only 11 in Dzierżno Duże reservoir. In the natural lakes of West Pomerania a higher contribution of psammobionts and higher taxon of rotifers was observed. In Siecino Lake they constituted about 20% of the total number of psammobionts, in Sarcze and Jeleń Lake 13 and 16%, respectively. In the artificial reservoirs psammophiles exhibited the highest frequency. In Pławniowice Duże psammophiles *Cephalodella catellina* and *Lecane closterocerca* constituted 50%, *Cephalodella gibba*, *Cephalodella catellina* and *Lecane closterocerca* in Dziećkowice reservoir composed 49% of total rotifer numbers. In Dzierżno Duże *Cephalodella catellina* constituted as many as 80% of the rotifer community (Fig. 2).

In lakes Siecino, Sarcze and Jeleń the contribution of bdelloid species reached 4.88, 7.5, and 6.6%, respectively, whereas in lakes Pławniowice Duże, Dziećkowice and Dzierżno Duże Bdelloidea contributed 11.11, 4.5 and 9.1%, respectively.

Cluster analysis revealed that the lake Dzierżno Duże differed from the rest of lakes (Fig. 3). Diversity index of psammon rotifer communities was high in natural lakes and low in artificial reservoirs (Table 3).

According to Myers (1936), beaches of lakes with higher trophy should be rich in psammon rotifers and have psammophiles present more frequent when submerged macrophytes are abundant. Studies in Polish lakes do not support this hypothesis. Ejsmont-Karabin (2003) studying psammon rotifers in 44 beaches of 18 lakes with different trophy, found that psammobionts were more abundant in lakes with lower trophy and psammoxen more in lakes with higher trophy.

In four lakes of the ęczyńsko-Włodawskie Region differing in levels of trophy, psammon of a mesotrophic lake was dominated by psammobionts, and psammophiles were less numerous, whereas in two eutrophic lakes and one

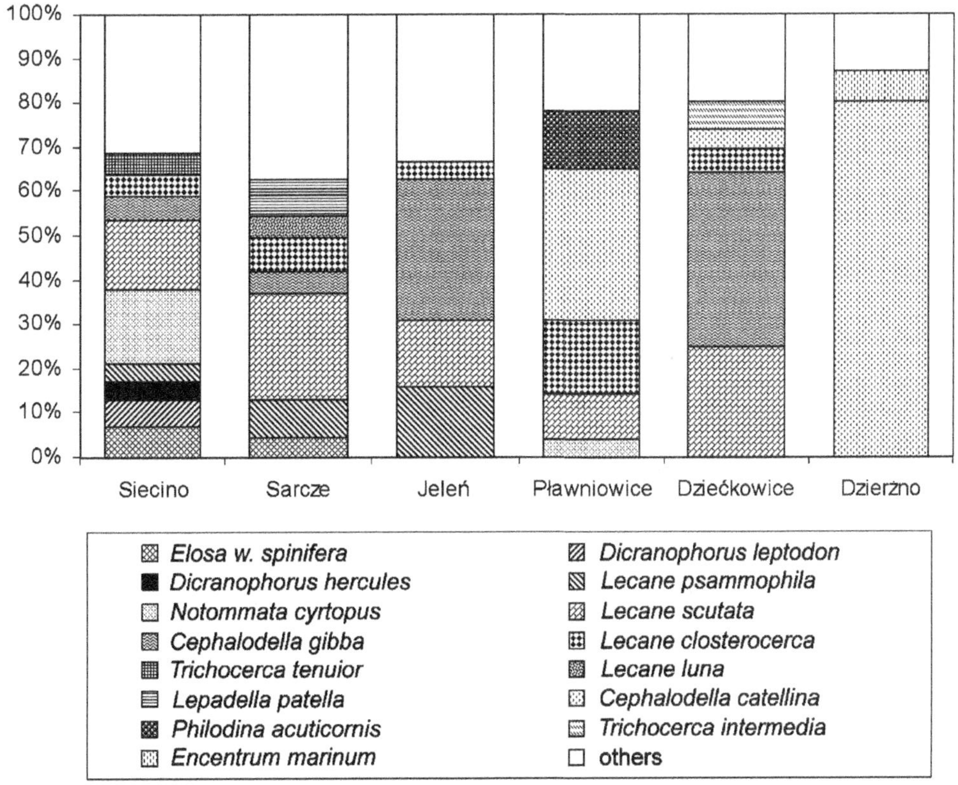

Figure 2. Relative contribution of dominant species to the total abundance.

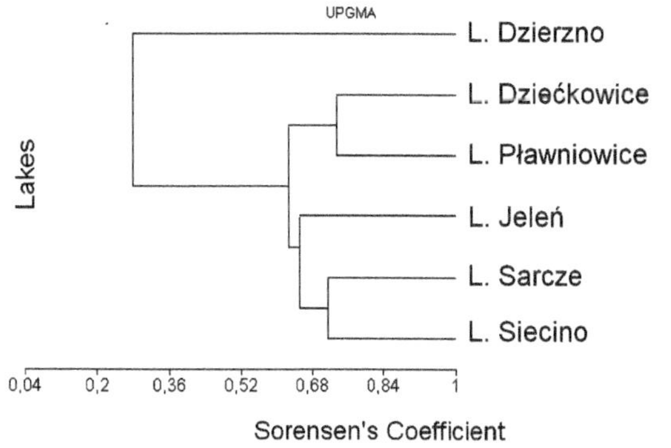

Figure 3. Cluster analysis of the investigated lakes and reservoirs on the basis of rotifer abundance.

hypertrophic lake psammophiles were most abundant (Radwan et al., 1998).

Also, our study does not confirm the observations of Myers (1936), because we observed more rotifer species in lakes with lower trophy, and their highest abundance in the autumn when macrophytes were less numerous or when they were not yet well developed. In addition, there were no submerged macrophytes close to the majority of investigated beaches.

In all the investigated lakes higher abundance of rotifers was observed in the hygroarenal and euarenal than in hydroarenal (Fig. 4), as also reported by many authors (Wiszniewski 1934a, b, 1947; Radwan et al., 1998; Bielańska-Grajner 2001; Radwan & Bielańska-Grajner 2001; Ejsmont-Karabin 2003).

In conclusion, the structure of psammic rotifers communities in anthropogenic reservoirs and lakes (West Pomerania) differed in their species number and number of psammobiont species, as well as diversity index. The number of rotifers and sand grain size and soil porosity and the number of algae were not significantly correlated.

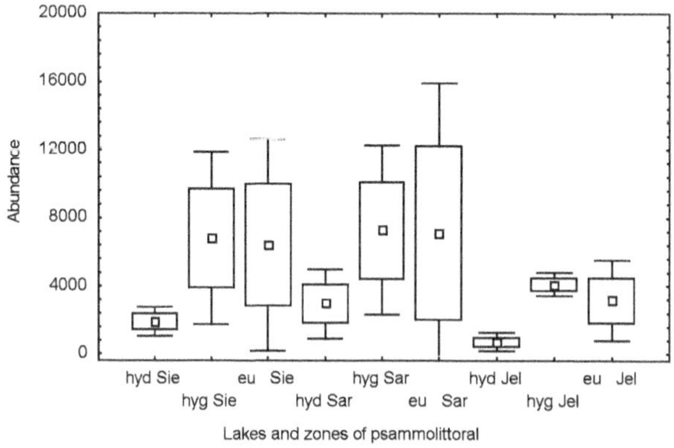

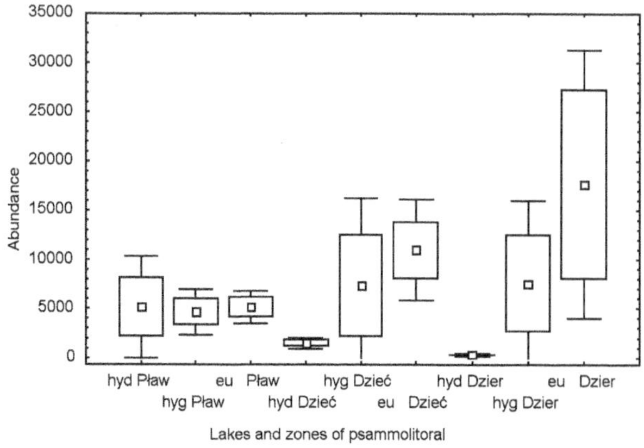

Figure 4. Mean and standard error of rotifer abundance, in lakes and artificial resfervoirs: hyd – hydropsammon, hyg – hygropsammon, eu – eupsammon, Sie – Siecino, Sar – Sarcze, Jel – Jeleń, Pław – Pławniowice Duże, Dzieć – Dziećkowice, Dzier – Dzierżno Duże.

Acknowledgements

I am grateful to Dr Teresa Skalska for the identification and analysis of algae, Dr Tadusz Molenda for the measurements of size of grains of sand and Dr Halina Lekacz and Małgorzata Kędziorska M.Sc for the determination of porosity of sand. I thank Mr Michael Tregenza for correcting the English. The research was supported by the State Council for Scientific Research, Project No. 3 PO4F 042 22.

References

Arov, I. V., 1990. Zhiznennye formy kolovratok v psammali ozera Bajkal. (Living forms of Rotifers of the Baikal Lake). In Skarlato, O. A (ed.), Kolovratki (Rotifers). Proceedings of the Third All-Union Rotifer Symposium. Zool. Inst. AN SSSR, Leningrad, 107–112 (in Russian).

Bielańska-Grajner, I., 2001. The psammic rotifer structure in three Lobelian Polish lakes differing in pH. Hydrobiologia 446/447: 149–153.

Boulton, A., C. Hakenkamp, M. Palmer & D. Stryer, 2002. Freshwater meiofauna and surface water-sediment linkages: a conceptual framework for cross-system comparisons. In Rundle, S. D., A. L. Robertson & J. M. Schmid-Araya (eds), Freshwater Meiofauna:Biology and Ecology. Backhuys Publishers, 241–259.

Ejsmont-Karabin, J., 2003. Rotifera of lake psammon: community structure versus trophic state of lake waters. Polish Journal of Ecology 51: 5–35.

Ejsmont-Karabin, J., 2004. Are community composition and abundance of psammon Rotifera related to grain-size structure of beach sand in lakes? Polish Journal of Ecology 52: 363–368.

Giere O., 1993. Meiobenthology. The Microscopic Fauna in Aquatic Sediments. Springer-Verlag, Berlin, 330 pp.

Jansson, B., 1967. The significance of grain size and pore water content for the interstitial fauna of sandy beaches. Oikos 18: 311–322.

Myers, F. J., 1936. Psammolitoral rotifers of Lenape and Union Lakes, New Jersey. American Museum Novitates 830: 1–22.

Neel, J. K., 1948. A limnological investigation of the psammon in Douglas Lake, Michigan, with especial reference to shoal and shoreline dynamics. Transactions of the American Microscopical Society 67: 1–54.

Neiswestnowa-Shadina, K., 1935. Zur Kenntnis des rheophilen Mikrobenthos. Archiv für Hydrobiologie 28: 555–582.

Pennak, R. W. & J. V. Ward, 1986. Interstitial faunal communities of the hyporheic and adjacent groundwater biotopes of a Colorado mountain stream. Archiv für Hydrobiologie/Supplement 74: 356–396.

Radwan, S. & I. Bielańska-Grajner, 2001. Ecological structure of psammic rotifers in the ecotonal zone of Lake Piaseczno (eastern Poland). Hydrobiologia 446/447: 221–228.

Radwan, S., I. Bielańska-Grajner & B. Popiołek, 1998. Rotifer communities in different littoral biotopes and in the pelagic zone of the Polesie lakes. In Radwan, S. (ed.), Freshwater Ecotones, Structure, Types and Function. UMCS Publishers, Lublin, 43–49 (in Polish).

Richling, A., 1993. Detailed Methods of Physical Geography. PWN Warszawa, 283 pp. (in Polish).

Saunders-Davies, A., 1998. Differences in rotifer population of the littoral and sub- littoral pools of a large marine lagoon. Hydrobiologia 387/388: 225–230.

Schmid-Araya, J. M., 1994. Spatial and temporal distribution of micro-meiofaunal groups in a alpine gravel stream. Internationale Vereinigung für Theoretische und Angewandte Limnologie, Verhandlungen 25: 1649–1655.

Schmid-Araya, J. M., 1995. Disturbance and population dynamics of rotifers in bed sediments. Hydrobiologia 313/314: 279–290.

Schmid-Araya, J. M., 1998a. Rotifers in interstitial sediments. Hydrobiologia 387/388: 231–240.

Schmid-Araya, J. M., 1998b. Small-sized invertebrates in a gravel stream: community structure and variability of benthic rotifers. Freshwater Biology 39: 25–39.

Segers, H., 1998. Notes on the taxonomy and distribution of the interstitial Rotifera from a Dune Pool. Belgian Journal of Zoology 128: 35–47.

Sørensen, M. V., 1998. Marine Rotifera from a sandy beach at Disko Island, West Greenland, with the description of *Encentrum porsildi* n. sp. and *Notholca angakkoq* n. sp. Hydrobiologia 386: 153–165.

Turner, P. N., 1990. Some interstitial Rotifera from a Florida, U.S.A., beach. Transaction of the American Microscopical Society 109: 417–421.

Turner, P. N., 1993. Distribution of rotifers in a Floridian saltwater beach, with a note on rotifer dispersal. Hydrobiologia 255/256: 435–439.

Turner, P. N., 1996. Preliminary data on rotifers in the interstitial of the Ninnescah river, Kansas, USA. Hydrobiologia 319: 179–184.

Turner, P. N. & D. A. Distler, 1995. Notes on the Hyportheic Rotifera of the Ninnescah river, Kansas, USA. Transactions of the Kansas Academy of Science 98: 92–101.

Tzschaschel, G., 1983. Seasonal abundance of psammon rotifers. Hydrobiologia 104: 275–278.

Uggla H., 1979. Soil Science. PWN, Warszawa, 120 pp. (in Polish).

Wallace, R. L. & Ricci, C., 2002. Rotifera. In Rundle, S. D., A. L. Robertson & J. M. Schmid-Araya (eds), Freshwater Meiofauna: Biology and Ecology. Backhuys Publishers: 15–44.

Watling, L., 1991. The sedimentary milieu and its consequences for resident organisms. Journal of American Zoology 31: 789–796.

Williams, R., 1972. The abundance and biomass of the interstitial fauna of a graded series of shell-gravels in relation to the available space. Journal of Animal Ecology 41: 623–646.

Wiszniewski, J., 1932. Les Rotifères des rives sablonneuses du lac Wigry. Archives d' Hydrobiologie et d'Ichthyologie 6: 86–100.

Wiszniewski, J., 1934a. Les Rotifères psammiques. Annales Musei Zoologici Polonici 10: 339–399.

Wiszniewski, J., 1934b. Recherches ecoloques sur le psammon et specialment sur les Rotiferes psammiques. Archives d' Hydrobiologie et d'Ichthyologie 8: 149–165.

Wiszniewski, J., 1947. Remarques relatives aux recherches recentes sur le psammon d'eaux douces. Archives d' Hydrobiologie et d'Ichthyologie 13: 7–36.

Zullini, A. & C. Ricci, 1980. Bdelloid rotifers and nematodes in a small Italian stream. Freshwater Biology 10: 67–72.

Part VII
Long-term Studies

Hydrobiologia (2005) 546:443–450

A. Herzig, R.D. Gulati, C.D. Jersabek & L. May (eds.) Rotifera X: Rotifer Research: Trends, New Tools and Recent Advances
DOI 10.1007/s10750-005-4287-y

Seasonal rotifer dynamics in the long-term (1969–2002) record from Lake Kinneret (Israel)

Moshe Gophen
Kinneret Limnological Laboratory (Emeritus), MIGAL-Galilee Technology Center, P.O. Box 831, 11016 Kiryat Shmone, Israel
E-mail: Gophen@Migal.Org.IL

Key words: Rotifera, long term record, Lake Kinneret, seasonal dynamics

Abstract

Long-term (1969–2002) data record of biomass distribution of rotifers in Lake Kinneret is combined with previously published information on their metabolic activity and newly calculated population dynamics parameters to synthesize a model of their seasonal dynamics in Lake Kinneret. Nineteen rotifer species were recorded in routine samples collected in Lake Kinneret (Israel) in 7 offshore (deeper than 5 m), stations, at 12 discrete depths during 1969–2002. Organisms were sorted and counted (including external egg carrying females), biomass was measured and calculated for the entire lake stock ($g_{w.w}$ m^{-2}; mg l^{-1}). Rates of grazing, respiration and production were measured experimentally at three different temperature ranges. Results were extrapolated to the lake community for months with similar temperatures. Rotifera comprised 7% of total zooplankton biomass in Lake Kinneret whilst Cladocera and Copepoda 58 and 35% respectively. Rotifers were found to be more abundant during December–June and decline in summer months. Monthly (1969–2001) means indicated total grazing capacity of rotifers as 11%, respiration as 9% and production as 3.7% of the total zooplankton metabolic activity. Positive relations were indicated between rotifer and small bodied cladoceran numerical concentrations. Population growth models suggest that rotifers are not food limited in Lake Kinneret but that fish predation plays an important role in regulating abundance in spring-summer and fall.

Introduction

Kinneret is the only natural freshwater lake in Israel. Water levels maintained between the upper and lower formulated red-lines of 208.90 m and 215.50 m below sea level respectively. Maximum depth is 44 m at water level of 209 m bsl with mean depth of 26 m and the surface area of 170 km^2. The lake is mixed in winter (mid-December through February) and thermal stratification sets up in spring, persisting for 7–8 months. Epilimnetic temperatures typically range from 15–17 °C in winter and 22–30 °C in summer (LKDB 1969–2001). Lake Kinneret supplies >50% of Israel's drinking water. Therefore intensive long term monitoring and water quality management of the lake is of a national significance. The monitoring programme has included routine sampling of all levels of the lake food web, including fish, zooplankton, protozoa, bacteria, phytoplankton, chemical, physical, meteorological and hydrological parameters.

Zooplankton data in Lake Kinneret was reported by Richard (1890), Gurney (1913), Bodenheimer (1935) , Komarovsky (1959), Yashouv & Alhunis (1961), and Gophen (1978). The number of recorded rotifer species varied between 10 and 35. This paper sythesize previous published rotifers data of metabolic parameters and recalculated values of population dynamics together with species and community biomass distribution in lake Kinneret. The physical conditions (temperature and hydrology) are cosiderably included as well.

The aim of the study is a comprehensive analysis of the rotifer seasonal dynamics in Lake Kinneret during 1969–2002. The data is utilysed for the analysis, of predation, food resources, hydrology and temperature, as drives of population dynamics. Here, published estimates of the metabolic rates are used to determine community-level contribution of the rotifers to whole-lake productivity and nutrient recycling for the study of their seasonal dynamics. Additionally, the Lake Kinneret rotifer dietary habits is summarized, and long-term zooplankton data are used to estimate population birth and death rates, such, that the importance of predation and food limitation may be infered across seasons.

There are a 24 of native and introduced fish species in Lake Kinneret and 8 are commercially harvested. Most Lake Kinneret fishes are planktivorous and the most common fish in the lake (*Acanthobrama* spp.) is zooplanktivore. Several are benthivorous, and the level of piscivory is low (Gophen et al., 1990). The most abundant, the endemic bleak, *Acanthobrama* spp. contributes ca. 80% of the total fish numbers and more than 50% of fish stock biomass in the lake (Walline et al., 2000). For 8 years commencing 1994–1995, the Israel Water Commission has subsidized removal of sub-commercial sized bleak from Lake Kinneret (ca. 400–750 tonnes per year) (Walline et al., 2000; Gophen, 2002). Their removal was a response to zoopplankton biomass sharp decline and shifts in the size structure of the bleak population and concerns about the consequent negative effect on water quality.

Methods

Zooplankton monitoring was carried out during 1969–2001 by a weekly and biweekly sampling program in 7 stations at 12 discrete depths. Sample analyses included the counting of sorted species, external egg carrying females, individual biomass measurements (Table 1), followed by routine computation of lake biomass concentration (Table 1) and population dynamic parameters. Metabolic parameters were measured experimentally at three temperatures with respect to the seasonal mean epilimnetic temperatures (Table 2) (Gophen & Azoulay, 2002). These results were extrapolated to the measured densities of the lake community for months with similar temperatures to calculate the activity of the entire natural lake assemblages. Methods for the calculation of carbon content are reported in Gophen & Azoulay (2002) and Serruya et al. (1980). Abiotic (physical and chemical) and biotic long term data (monthly means) were taken from the Lake Kinneret Data Base (LKDB 1969–2001) (for sampling procedures see Gophen et al., 1999; Serruya, 1978): chemistry – A. Nishri, phytoplankton – U. Pollingher and T. Zohary, primary production – Y. Z. Yacobi and T. Berman, zooplankton – M. Gophen. The parameters of population dynamics are correspond to species which carry external eggs which were counted routinely.

Table 1. Biomass (μg ww/Ind.) of rotifers in Lake Kinneret

Keratella spp.	0.1
Asplanchna brightwelli	36.0
Asplanchna priodonta	37.5
Synchaeta pectinata	4.5
Synchaeta oblonga	0.2
Hexarthra fennica	0.5
Brachionus angularis	0.1
Philodina sp.	1.0
Trichocerca sp.	1.0
Collotheca sp.	1.0
Brachionus calyciflorus	1.0
Anureopsis fissa	1.0
Lecane sp.	0.3
Filinia sp.	0.5
Conochiloides coenobasis	4.0
External eggs	0.03

Table 2. Metabolic parameters of lake Kinneret rotifers (*Keratella* spp., *Brachionus* spp., *and Synchaeta* spp.) measured at three temperature ranges (15–20 °C, 20–24 °C, and 24–28 °C) and implied to respective months with similar mean epilimnetic temperatures, given in mg C/mg C(body content)/day

Month	Jan–Apr	May; Nov–Dec	Jun–Oct
Temp.	15–20	20–24	24–28
Production (P)	0.047	0.080	0.123
Respiration (R)	0.153	0.190	0.143
Food intake (C)	0.750	2.375	4.000
Metabolic efficiency (%)(R + P/C)	27	11	13

The following parameters were used for the analysis of rotifer population dynamics (Edmondson, 1965):

$$B = E/D \quad \text{finite birth rate} \qquad (1)$$

where E = no. of eggs per female; D = embryonic development time (an average of 19 hours = 0.8 day was used) (Herzig, 1983);

$$T = 1/B \quad \text{population turnover time} \qquad (2)$$

$$b = \ln(E+1)/D \quad \text{instantaneous birth rate} \qquad (3)$$

$$r = (\ln N_t - \ln N_o)/t \quad \text{instantaneous rate of population change} \qquad (4)$$

where: t = sampling time interval (14 days), N_t = population at the end of 14 days, N_o = population at the beginning of 14 days;

$$d = b - r \quad \text{instantaneous death rate} \qquad (5)$$

These parameters were annually averaged and their temporal fluctuations are presented in Figure 3 and the impact of inflow discharges on those parameters in Figure 4. The monthly discharges values were sorted into three groups: low (50–60 mcm/month), medium (60–70 mcm month^{-1}) and high (70–100 mcm month^{-1}). ANOVA was tested ($p < 0.05$) against annual means of birth rate (b) and population turnover time (T) values. Temporal changes of rotifers biomass concentrations (ppm) were carried out by sorting the period of 1969–2002 into three year groups with respect to changes of total zooplankton biomass concentration: (1) 1969–1982 biomass decline; (2) 1983–1993 low level of biomass, and (3) 1994–2002 biomass increase. Monthly means of rotifers biomass concentration (ppm) of each year group were comparatively analysed by ANOVA). Seasonal changes of rotifer biomass density were analysed by sorting months in two groups (seasons): 1 = December–May and 2 = June–November. An ANOVA test was performed ($p < 0.05$) to learn if the rotifer biomass concentration (g m^{-3}; mg l^{-1}; ppm) differed significantly between the two periods.

Results

The zooplankton community in Lake Kinneret is dominated by cladocerans (58% of zooplankton biomass) throughout most of the year, most notably *Ceriodaphnia* spp. *Diaphanosoma* sp., and *Bosmina* spp. (Gophen, 1984). Another major zooplankton group is the copepods (35% of zooplankton biomass). The third major zooplankton group is rotifers, comprising 7% of the total zooplankton biomass (Fig. 1) (Gophen, 1978). A list of the species of the lacustrine/planktonic rotifers recorded by the author excluding organisms inhabiting inshore (<5 m depth) waters is given in

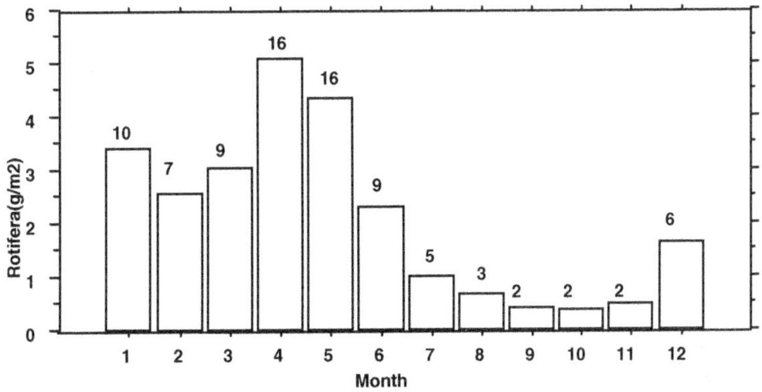

Figure 1. Monthly averages (1969–2001) of rotifers biomass (g $_{ww}$ m^{-2}) in Lake Kinneret. Averaged percentages of rotifer biomass from total zooplankton are given for each month.

Gophen (1978). In addition to species presented by Gophen (1978), Dr B. Azoulay, in collaboration with M. Gophen identified the organisms *Conochiloides coenobasis*, *Anureopsis fissa*, *Lecane* sp. and *Filodina* sp. The biomass of the most common species is given in Table 1. The metabolic parameters of the rotifers (Gophen & Azoulay, 2002) are given in Table 2. Food preferences by Lake Kinneret rotifers (Table 3), based on microscopical observation (Serruya et al., 1980) and $\Delta^{13}C$ analyses (Zohary et al., 1994) indicates seasonal dietary variations of zooplankton, with nanophytoplankton their predominant source, whilst chemosynthetic bacteria and their protist grazers constituted a supplemental food source in winter. Small dinoflagellates are ingested in spring.

Previous data of the Lake Kinneret food-web structure (Gophen & Azoulay, 2002; Walline et al., 1993) revealed the following multiannual averages: bacterial biomass – 7 $g_{w.w}$ m^{-2}; protozoa biomass – 4 $g_{w.w}$ m^{-2} ; herbivorous zooplankton (including cladocerans, copepods and rotifers) biomass – 27.4 $g_{w.w}$ m^{-2}; food requirements of herbivore zooplankton – 56 $g_{w.w}$ m^{-2}; food demands of protozoa – 7 $g_{w.w}$ m^{-2}; daily P/B ratio of bacteria – 0.58 (i.e., production of 4.1 $g_{w.w}$ m^{-2} d^{-1}); daily P/B ratio of protozoa – 0.82 (i.e., production of 3.3 $g_{w.w}$ m^{-2} d^{-1}); daily P/B ratio of herbivore zooplankton – 0.15 (i.e., production of 4.1 $g_{w.w}$ m^{-2} d^{-1}). Evaluation of documented data (Gophen, 1978, 1984; Gophen & Azoulay, 2002) indicated that the biomass of small herbivore zooplankters (copepod nauplii and small rotifers – all rotifers excluding *Asplanchna* spp.) is 1.5 $g_{w.w}$ m^{-2} and that of large herbivore zooplankters is 25.9 $g_{w.w}$ m^{-2} (94.5% of total grazing capacity).

Table 3. Food preferences of Lake Kinneret rotifers

Food Source	Preference Index
*Peridinium**	0.05
Microcystis	0.05
Chlorophyta	0.10
Detritus	0.75
Predation*	
Juvenile copepod stages	0.03
Cladocerans	0.01
Rotifers	0.01

*By *Asplanchna* spp.

Micrograzers consume 3.96 mg C m^{-2} h^{-1} (5.6 g C m^{-2} $month^{-1}$) and macrograzers ingest 73.2 g C m^{-2} $month^{-1}$. Consequently, the micrograzing (including small rotifers) capacity is 7.1% of the total grazing activity carried out by zooplankton in Lake Kinneret. Results in Figure 1 show the seasonality of rotifer densities and their percentage of total zooplankton in Lake Kinneret: higher biomass (absolute and %) during December–June and lower in July–November. Inverse relationship between the biomass density of rotifers and the epilimnetic mean temperature was documented: higher densities in low temperatures and vice versa (Fig. 2). Similar relations were also documented between several abiotic parameters and the two seasons: lake residence time was shorter in winter (i.e., water exchange was faster), inflow discharges are higher in winter and epilimnetic nutrient inventories are higher in winter as well.

The three groups of monthly discharge values (low, medium, high) (see methods) were ANOVA tested, against annual means of birth rate (b) and population turnover time ($1/B$) values. Results indicated strong relationships of higher birth – rate and lower turnover time ($p = 0.0484$ and 0.0284 respectively) when discharges were high. Rate of population changes (r) was negative (−0.014) when discharge was "low" and positive when discharge was "medium" and "high" ($p = 0.004$ and 0.0015 respectively) .On the other hand death rate (d) was higher (p = 0.04.7) when discharge was "high" in comparison with "medium". The two periods (winter and summer) (see methods) differed significantly ($p = 0.015$) with regard to rotifers biomass concentration: winter > summer.

Discussion

The analysis of zooplankton seasonal dynamics in lakes is including a wide variety of potential factors, biotic and abiotic, which might have an impact on it. These effects might cause significant fluctuations of the population density, species or demographic composition, and other characteristics of the assemblages (Wallace & Snell, 2001; Wetzel, 2001). The review and analysis in this paper are aimed at understanding the ecological status of rotifers in Lake Kinneret during 1969–2002 and their seasonal dynamics. Climatological,

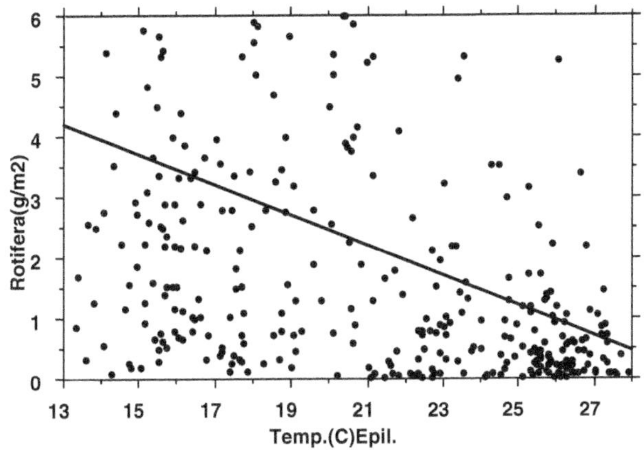

Figure 2. Interrelationship (simple regression: $R^2 = 0.128$; $p < 0.0001$) between monthly (1969–2001) values of epilimnetic temperature and rotifer biomass ($g_{ww}\ m^{-2}$) in Lake Kinneret.

hydrological, chemical and biological long-term data were incorporated into a population growth model of rotifers in Lake Kinneret.

Results in Table 2 indicate that the highest metabolic efficiency were measured under the lowest temperatures, which correspond with winter months in Lake Kinneret. The highest food demands and level of respiration are at high temperatures, prominently corresponded to the summer months. Results in Table 3 show a high preference for detritus and chlorophytes, and a low preference for diatoms. The high availability of detritus and bacteria in Lake Kinneret in winter–spring and early summer was documented by Serruya et al. (1980). Moreover, the suitable temperatures for the rotifers metabolism in winter–spring time is consequently concluded. The suitability of the winter–spring period for the rotifer growth is resulted in by the high inflow discharges accompanied by intensive nutrient inputs, and high level of algal production. The result is a seasonal (March–early June) algal bloom of *Peridinium* which abruptly crash during May–June accompanied by high concentration of detritial particles and dissolved organic matter in the epilimnion (Serruya et al., 1980). On the other hand during summer months the epilimnion of Lake Kinneret resemble a chemostat ecosystem: high temperatures, low nutrient inputs, and very high food demands of secondary consumers like fish and zooplankton. For the Kinneret fish community, the summer epilimnion is an ecosystem under stress. The temperatures regime dictate high metabolic demands and food availabilities are low. After the annual crash of the *Peridinium* bloom and the consequent organic matter recycling, nutrient supply for primary consumers are limited. Nevertheless, the densities of suspended detritial particles, bacteria, and nanophytoplankton are sufficient to cover the metabolic requirements of the rotifer community in summer. The outcome of the Kinneret stable stratification in summer is low level of mixing and low fluxes of nutrient from lower to upper layers which is accompanied by the low external inputs (Serruya et al., 1980). Therefore primary consumers are food (especially phosphorus) limited (Gophen et al., 1999; Gophen, 2002) and zooplankton is pressed by the high food requirements of fish (Serruya et al., 1980). The population dynamic parameters of rotifers in summer reflect these stressed condition (Fig. 3): negative values of rate of change, low values of instantaneous birth rate and high values of population turnover time.

The final result is decline of the rotifer population densities in summer. Rotifer predatory habits are due to *Asplanchna* spp. which mostly prey *Peridinium* during its blooming season and zooplankton anytime. On the other hand *Asplanchna* is highly vulnerable to fish predation and therefore its feeding habits is an additional pressure on other rotifers in summer. Consequently, no food limitation for rotifers in the Kinneret ecosystem throughout a full year cycle and a significant fish pressure on rotifer in summer are suggested. The metabolic structure of the

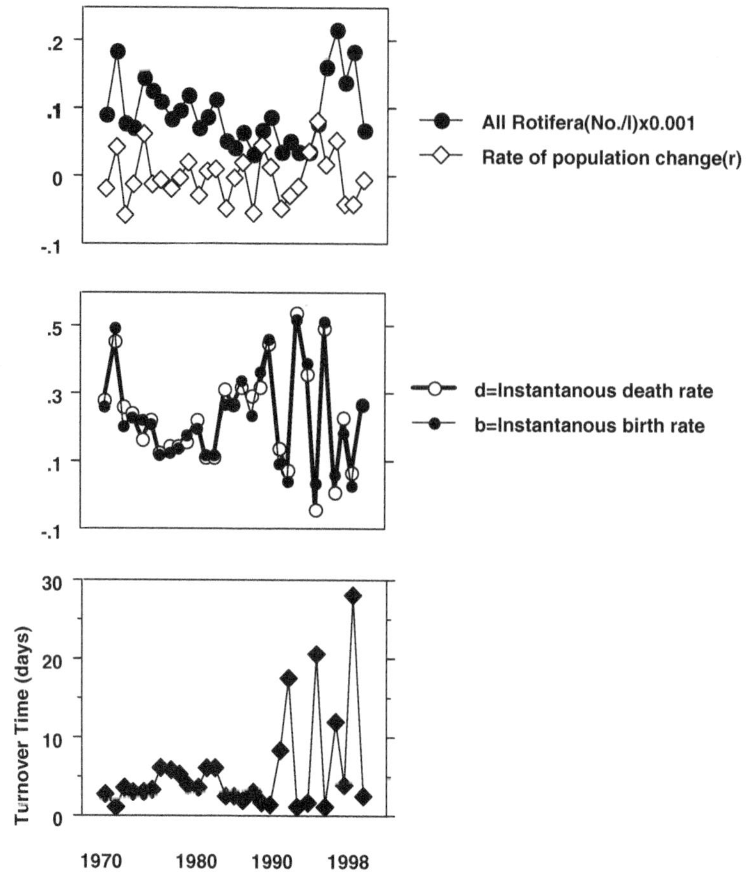

Figure 3. Annual averages (1969–1998) of rotifer population dynamics parameters in Lake Kinneret. Upper panel: No. L^{-1} of all rotifers and rate of population change (*r*); middle panel: instantanous birth (*b*) and death (*d*) rates; lower panel: turnover time (1/*B*) (in days) (see text).

Kinneret food web (Walline et al., 1993; Gophen & Azoulay, 2002) and the role of rotifers in it indicate the greater importance in nutrient recycling of herbivore zooplankton in comparison to that of protozoans. Only 7% of the food consumption by rotifers is channeled to production and the rest is recycled. Ciliates assimilate a much higher portion of ingested food and less is recycled. It is suggested that the role of rotifers in the energy flow within the Kinneret food web is minor and their contribution to recycled fluxes as well. These seasonal differences are a result of the summer–winter subtropical changes of the geographical region. Rotifer food resources, nanophytoplankton, detritus, bacteria and protozoa are less abundant in summer months. Nevertheless, fish predation pressure is stronger in summer months (Serruya et al., 1980; Gophen, 1984; Gophen et al., 1990 Easton & Gophen, 2001) and ecophysiological conditions in summer dictate lower efficiencies of food utilization and therefore lower production. Gophen (1980) has suggested that an intensive influx of rotifer biomass from the drainage basin through inflow river discharges significantly support the lake populations. The applicability of the "egg ratio" model for the analysis of natural rotifer population was thoroughly described by Dumont (1977). The rotifer biomass and numerical densities are seasonally fluctuated in Lake Kinneret: low in summer and high in winter. Results indicate that the respiration of rotifer assemblages comprised 6% of the total respiration capacity of the entire zooplankton community in the lake. The food intake of rotifers comprised 7.6% of the total grazing capacity of herbivore zooplankton and rotifers production – 3.7% of total production of

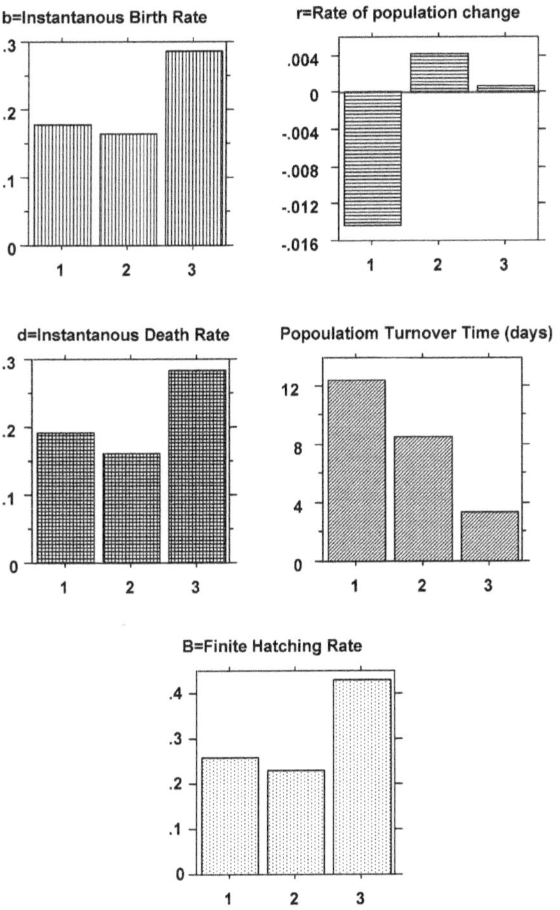

Figure 4. Relationships between three groups of monthly inflow discharge (1 = 50–60 mcm month^{-1}, 2 = 60–70 mcm month^{-1}, and 3 = 70–100 mcm month^{-1}) and parameters of population dynamics (see text).

zooplankton in Lake Kinneret. A weak relationship was indicated between monthly values of primary production and rotifers biomass (1969–1999). The long-term average of rotifers production percentage from primary production is 0.43%. A weak relationship was also indicated between monthly values of percentage of rotifers production from primary production and time (months). These values (%) in 1972, 1987, 1990, and 1991 were lower and in 1973, 1976, and 1982 significantly higher than the mean level but no long-term temporal trend was indicated. It is therefore suggested that rotifers are not food limited, their role in the energy flow within the Kinneret food web is minor and their contribution to recycled fluxes as well.

Although long term stability of biomass and numerical densities (No. L^{-1}) were documented, species composition was slightly modified (data are not shown here): densities of *Keratella* spp. and *Asplanchna* spp. declined. The numerical densities of poorly escaped zooplankters from fish predation, rotifers and small cladocerans, indicated strong positive relationship. Increasing values of the ratio of small/large body cladocerans (Gophen, 2002) in the lake assemblages is a good indicator of intensified pressure by fish visual predation (mostly by Lavnun) (Drenner et al., 1882): the large organisms preferably preyed upon than small zooplankters and their relative number in the population is increasing. Filter feeding fish remove preferably small cladocerans and most of the rotifer species (excluding "jumpers" like *Polyarthra* spp. and *Hexarthra* spp.) which has lower escapability than larger individuals in the lake community. Therefore, the strong relationship that was found between rotifers and small cladocerans in Lake Kinneret is reflecting a mutual suppressing factor, suggested, as visual and filter feeding fish predation. The high amplitudes of the calculated population dynamics parameters during the 1990s (Fig. 3) are probably due to the fluctuations of fish predation pressures resulted by fishery management.

References

Bodenheimer, F. S., 1935. Animal Life in Palestine, 506 pp.

Drenner, R. W., G. L. Vinyard, M. Gophen & S. R. McComas, 1882. Feeding behavior of the cichlid *Sarotherodon galilaeus*: selective predation on Lake Kinneret. Hydrobiologia 87: 17–20.

Dumont, H. J., 1977. Biotic factors in the population dynamics of rotifers. Archiv für Hydrobiologie, Ergebnisse der Limnologie 8: 98–122.

Easton, J. & M. Gophen., 2001. Trophic relations between zooplankton and bleaks (*Acanthobrama* spp.) in Lake Kinneret (Israel). Internationale Vereinigung für Theoretische und Angewandte Limnologie, Verhandlungen 28: 1258–1261.

Edmondson, W. T., 1965. Reproductive rates of rotifers in natural populations. Ecological Monographs 35: 61–111.

Gophen, M., 1978. Zooplankton in Lake Kinneret. In Serruya, C. (ed.), Lake Kinneret. Junk Publishers, The Hague, Monographiae Biologicae 32: 297–310.

Gophen, M., 1980. The influx of carbon into Lake Kinneret in 1971–1972 by Jordan zooplankton. Hydrobiologia 71: 47–50.

Gophen, M., 1984. The impact of zooplankton status on the management of Lake Kinneret (Israel). Hydrobiologia 113: 249–258.

Gophen, M. Val H. Smith, A. Nishri & S. T. Threlkeld 1999Nitrogen deficiency, phosphorus sufficiency, and the invasion of Lake Kinneret, Israel, by the N2-fixing cyanobacterium *Aphanizomenon ovalisporum* Aquatic Sciences61293306.

Gophen, M., 2002. The management of Lake Kinneret (Israel): Water supply, water quality and food web structure perspectives. Proceedings of The Water and Environmental Research ICWRER 2002 (G. H. Schmitz, ed.), Dresden, Germany. Vol. 2: 275–285.

Gophen, M., S. Serruya & S. Threlkeld, 1990. Long term patterns in nutrients, phytoplankton and zooplankton of Lake Kinneret and future predictions for ecosystem structure. Archiv für Hydrobiologie 118: 449–460.

Gophen, M. & B. Azoulay, 2002. The trophic status of zooplankton communities in Lake Kinneret (Israel). Internationale Vereinigung für Theoretische und Angewandte Limnologie, Verhandlungen 28: 836–839.

Gurney, R., 1913. Entomostraca from Lake Tiberias. Journal Asian Society Bengal IX: 231–232.

Herzig, A., 1983. Comparative studies on the relationship between temperature and duration of embryonic development of rotifers. Hydrobiologia 104: 237–246.

Komarovsky, B., 1959. The Plankton of Lake Tiberias. Bulletin Research Council Israel 8B, 65–69.

Richard, J., 1890. Copepods recueillis par. Mr. Le Dr. Barrois de Egypt, en Syrie et en Palestine. Revue boil. Nord. France 5: 400–405, 433–443, 458–475.

Serruya, C., M. Gophen & U. Pollingher, 1980. Lake Kinneret: Carbon flow patterns and ecosystem management. Archiv für Hydrobiologie 88: 265–302.

Serruya, C. (ed.), 1978. Lake Kinneret. Junk Publishers, The Hague, Monographiae Biologicae 32, 501 pp.

Yashouv, A. & M. Alhunis, 1961. The dynamics of biological processes in Lake Tiberias. Bulletin Research Council Israel 10B: 12–35.

Wallace R. L. & Snell T. W., 2001. Rotifera. In Thorpe J. H. & Covich A. P. (eds), Ecology and Classification of North American Freshwater Invertebrates. Academic Press (2nd edn), 195–254.

Walline, P. D., S. Pisanty, M. Gophen & T. Berman, 1993. The ecosystem of Lake Kinneret, Israel. In Christensen, V. & D. Pauley (eds) Trophic Models of Aquatic Ecosystems. ICLARM: 103–109.

Walline, P. D., J. A. Tyler, S. B. Brandt, I. Ostrovsky & J. M. Jech, 2000. Bleak abundance: how changes may affect consumption of Lake Kinneret zooplankton. Archiv für Hydrobiologie, Advances in Limnology 55: 493–511.

Wetzel, R. G., 2001. Planktonic communities: Zooplankton and their interaction with fish. In Wetzel, R. G. (ed.), Limnology. (3rd edn). Academic Press, 396–489.

Zohary, T., J. Erez, M. Gophen, I. Berman-Frank & M. Stiller, 1994. Seasonality of stable carbon isotopes within the pelagic food web of Lake Kinneret. Limnolology and Oceanography 39: 1030–1043.

Hydrobiologia (2005) 546:451–455
A. Herzig, R.D. Gulati, C.D. Jersabek & L. May (eds.) Rotifera X: Rotifer Research: Trends, New Tools and Recent Advances
DOI 10.1007/s10750-005-4288-x

Seasonality of rotifers and temperature in Lough Neagh, N. Ireland

Tony E. Andrew[1,*] & J.A.M. Andrew[2]
[1]*Biological and Environmental Sciences, University of Ulster, Coleraine, Co. Derry, BT52 1SA, N. Ireland*
[2]*Department of Earth & Ocean Sciences, University of Liverpool, L69 3GP, Liverpool*
*(*Author for correspondence: E-mail: TE.Andrew@ulster.ac.uk)*

Key words: long rotifer datasets, Fourier analysis

Abstract

Long runs of seasonal rotifer population data allow analysis of seasonal occurrence using mathematical tools. The application of Fourier analysis to a 15 year dataset describes seasonality in simple mathematical terms. This facilitates comparison of population expression with potential population driving variables and provides a basic modelling tool. Results show that annual patterns of occurrence and density have linkages with annual maximum and minimum environmental temperature, although the exact relationships are not clear.

Introduction

The majority of rotifer population studies, like other zooplankton, span a period of one to three years. These studies are usually concerned with attributes such as seasonality or production of a single species or taxon (Andrew & Fitzsimons, 1992); more rarely the whole zooplankton community is encompassed (e.g. Johnson & Walker, 1974). These short, in-depth studies, often treated with sophisticated analytical techniques, leave the impression that zooplankton communities change relatively little in species composition, abundance or the timing of seasonal occurrence from year to year, unless major environmental changes that affect the zooplankton community have occurred, e.g. perturbation created by the catastrophic introduction of a new predator (Richards et al., 1975). Despite logistical difficulties in maintaining lengthy limnological studies, a variety of long-term datasets exist (e.g. Carlin, 1943; Goldman et al., 1979; Herzig, 1979, 1987; Kratz et al., 1987). These studies employ a variety of descriptive, analytical and experimental methods often leading to annual estimates of species composition, abundance or production, which are useful as indicators of change. Despite reasoned pleas from practising limnologists (Herzig, 1987), disappointingly few such studies are published.

The objective of this study is to examine 14 years of rotifer population data, together with the likely driving variables, from a single lake, Lough Neagh, in order to discover what the relationships are between years and, what are the ‚normal' limits of change of community composition and abundance.

Study area

Lough Neagh is a large shallow eutrophic lake that has been the subject of a variety detailed limnological studies for more than a quarter of a century. With a surface area of 367 km^2, a mean depth of 8.6 m and extreme exposure to wind action, it is well mixed and exhibits only transient thermal or oxygen stratification. Detailed descriptions of the limnology, paleoecological structure and history of the lake may be found in Wood & Smith (1993). Algal crops are high, exceeding 50 mg m^{-3} chlorophyll *a*, exceptionally 150 mg m^{-3} chlorophyll *a*, during peak periods comprising mainly diatom and cyanophyte species. These chlorophyll values are indicative of high nutrient loading consistent with Battarbee's (1978) description of changing trophic status. Water control measures during the past 150 years, resulting in water level reductions, have affected the littoral regions and marginal development of the lake with consequent effects on the plankton communities (Wood et al., 1990).

Quantitative zooplankton sampling has been carried out consistently since 1968 (Graham & Logan, 1970; Fitzsimons & Andrew, 1993), predated by the early study of Dakin & Latarche (1913). Other relevant zooplankton work includes that of Andrew (1992), Andrew & Fitzsimons (1992), Edgar & Andrew (1990), Fitzsimons & Andrew (1992), and Andrew & Woodward (1993).

Eight species of rotifers have been recorded from the plankton during the course of this study, *Keratella cochlearis*, *Keratella quadrata*, *Kellicottia longispina*, *Filinia longiseta*, *Polyarthra dolichoptera*, *Brachionus angularis*, *Cephalodella gibba* and *Pompholyx sulcata* (Andrew & Fitzsimons, 1992). A detailed description of the seasonality of this limited rotiferan fauna is found in Fitzsimons & Andrew (1993), population dynamics and production are described in Fitzsimons & Andrew (1992). The contemporary crustacean zooplankton is dominated by the omnipresent *Cyclops abyssorum*. *Cyclops vicinus* occurs seasonally in some years at low densities. *Eudiaptomus gracilis* is found on most sampling occasions but at lower densities than *C. abyssorum*. Two species of *Daphnia*, *D. hyalina* and *D. longispina* occur in the plankton, together with three species of carnivorous Cladocera, *Leptodora kindti*, *Bythotrephes longimanus* and *Polyphemus pediculus*. The semi-planktonic *Mysis relicta*, a potential predator and competitor of the herbivorous zooplankton, is also present in the lake. Fish populations, particularly *Coregonus pollan*, *Perca fluviatilis*, *Scardinus erythrophthalmus*, and more recently *Rutilus rutilus*, predate on zooplankton either as adults or juvenile stages. Since the earlier study of Dakin & Latarche (1913), the zooplankton composition has changed from a bosminid dominated to a daphnid dominated zooplankton (Wood et al., 1990).

Methods

Sampling strategies and methodology for the long term sampling programme on Lough Neagh are described in detail in Andrew & Fitzsimons (1992) and Fitzsimons & Andrew (1993). Our analysis of data from 1968 to 1982 is based on zooplankton population samples collected using vertical net hauls from four stations on the Lough. These were subsequently treated as bulked samples considered to be representative of the lake as a whole. In order to facilitate the subsequent analysis, the data have been transformed to monthly mean values using a Hanning filtered, running average of individual sampling day results, typically used in Fourier analysis (Davis, 1986). Up to five sampling days may represent one month. Environmental temperature, algal standing crops and species composition, were taken, with permission, from the data of R.V. Smith and C. E. Gibson (Aquatic Sciences Research Division, Department of Agriculture and Rural Development N. Ireland) (Wood & Smith, 1993).

The analysis of population trends used Fourier analysis with standard Excel packages and Matlab. Fourier was chosen as an unprejudiced de-constructor of cyclical data allowing subsequent comparisons of species and potential drivers of the system.

Fourier analysis is frequently used for analysis of time series data with cyclic patterns such as seasonality (Davis, 1986) although this has not been applied to zooplankton population data. Any periodic function can be expressed as the sum of an infinite number of sinusoidal functions but relatively few contribute to any one dataset. The data is deconstructed, using regression methods, into a series of sinusoidal harmonics, which may be ranked by their power contribution to the dataset. The major harmonics may be then used to reconstruct a noise-free curve for modelling purposes or used for comparison with related datasets. Individual contributory harmonics may be compared between populations or other physical, chemical or biotic features of the environment. In addition, trends in a fluctuating population may be defined with more mathematical certainty. Fourier analysis has been used in a limited number of plankton population studies (e.g. Marn, 1997; Barciela et al., 1999).

This analysis is applied only to the rotifer and temperature relationships as the phytoplankton component is controlled partly by crustacean zooplankton not included here. In order to simplify the analysis of the patterns seen in this approach, the rotiferan plankton has been grouped as a whole, effectively considering these as, the population'. Some species are always present, *Keratella cochlearis*, *Keratella quadrata*, *Kellicottia longispina*, these are grouped with all other species which occur for limited seasons and not every year. This grouping masks the individual species expression, seasonal timing, individual population densities, etc., but enables the identification of major trends and potential controlling influences.

Results

Rotifer density shows pronounced annual variation (Fig. 1). Direct comparisons with potential food, algal biomass (Fig. 1c) and temperature (Fig. 1b), reveal similarities of timing and general trends of higher or lower population densities on a cyclic basis separate from the annual recurring cycles together with algal biomass. A general conclusion is that higher rotifer populations coincide with lower algal population expression. Temperature trends are not so easily related although there is a longer term cycle beyond the annual mode. The data show a general coincidence of population density of rotifers with higher annual temperatures, i.e. they are summer species. However, the peak densities rarely coincide with temperature maxima, there is a phase shift but not unidirectional.

Figure 1a, describing the rotiferan seasonality, is fitted with sinusoidal envelopes derived from the Fourier analysis which is a more representative descriptor of longer term trends than a simple trend approach. The characteristics of these are presented in Table 1. The upper line, representing the maximum densities achieved, reveals the annual variation in density, up to four times. The lower line, representing the mean annual population density, does not share the same amplitude i.e. annual production does not vary as much as individual extremes of population expression. This line has a different underlying cycle out of phase with the maxima fit. Although there are distinct

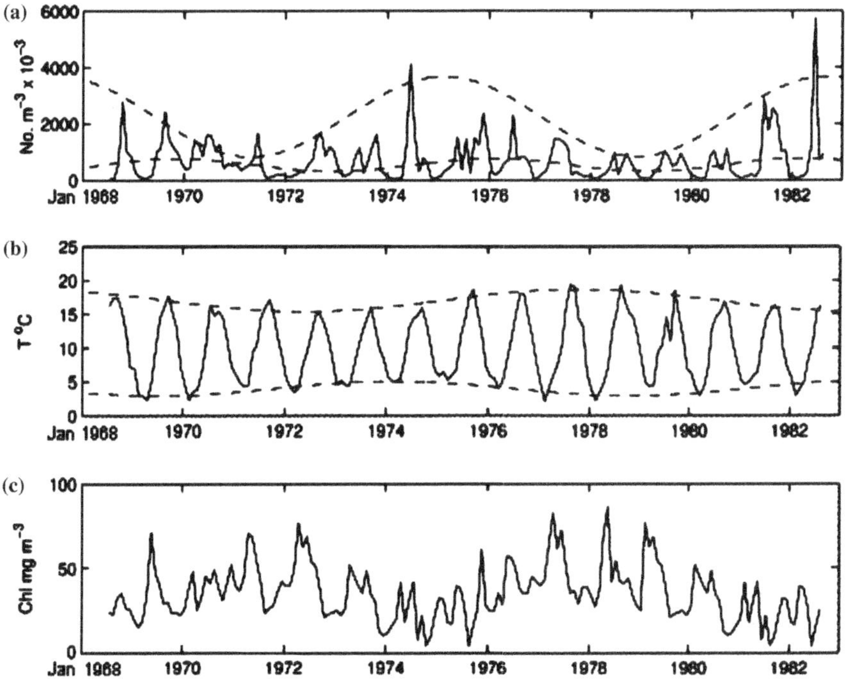

Figure 1. (a) The population density of all rotifer zooplankton 1968–1982 together with sinusoidal envelopes fitted to annual maxima and annual medians. (b) Environmental temperature 1968–1982 together with sinusoidal envelopes fitted to annual maxima and annual minima. (c) Chlorophyll-*a* concentration 1968–1982.

Table 1. Fourier derived characteristics of rotifer population and temperature characteristics of the period 1968–1982 in L. Neagh

	Max. sinusoidal period (years)	Max. annual mean (°C)	Max. amplitude variation (°C)	Min. sinusoidal period (years)	Min. annual mean (°C)	Max. amplitude variation (°C)
Rotifers	7.58			6.17		
Temperature	10.83	17.3	1.61	9.25	3.6	1.09

Figures relate to the sinusoidal envelopes described in the text.

annual changes in species composition and abundance the overall rotiferan numbers change relatively little during these years although there is an apparent downward trend until 1981. Rather than maximum population density being a good measure of population standing crop, the annual mean may be more informative.

The sinusoidal envelope approach is applied to the maximum and minimum temperature regime of the lough (Fig. 1b and Table 1). Fourier analysis reveals the maxima and minima of the wave do not entirely coincide, colder periods being phase shifted to the right. One interesting feature of this pattern is that warm summers tend to be followed by cold winters and vice versa.

Figure 1c shows the annual standing crop of phytoplankton represented as chlorophyll *a*. It can be seen that there is a longer term cyclical pattern which is inversely related to the rotiferan density pattern.

Discussion

The rotiferan species production ultimately depends on the primary production in the plankton or its derivatives including bacterio-plankton, protozoa and detritus. Factors controlling phytoplankton growth are well documented for L. Neagh (Wood & Smith, 1993). Part of this control is by zooplankton grazing but the results of this study suggest that primary production does not limit either rotifer or crustacean growth either. This is noted by Fitzsimons & Andrew (1993). Phytoplankton growth and succession are controlled by light and nutrient availability as well as zooplankton grazing. The phytoplankton biomass is more than sufficient in quality and quantity to maintain similar levels of production in all zooplankton species in most years.

The temperature environment may not appear to be very different from year to year. Closer examination shows subtle variations occurring on a longer term basis. This longer term variability has been considered by Gibson & Stewart (1973) and also by Wood et al. (2000) who conclude that the lake behaves like a colder, deeper lake. Periods of cold winters prevent over wintering of cold-intolerant zooplankton species delaying the onset of population growth of the rotifers. This relationship can be seen comparing the minimum temperature and mean rotifer density relationships. Warmer summers promote the growth of rotifer populations, maximum population densities occurring in warmer periods.

However, temperature does not exert a single simple influence on rotifers. For example, growth functions such as embryonic development (Herzig, 1983) or somatic growth and other physiological functions are temperature dependent (Doohan, 1973). That rotifer species have a preferred temperature range is well established (May, 1983). The data presented here reveal that rotifer densities do not show a simple, direct correlation with temperature outside the annual, seasonal, temperature cycle. These are revealed by the use of a time series analysis.

The use of time series analysis and its derivatives enables an unprejudiced examination and analysis indicating where changes occur and other explanations should be sought.

There are two other possibilities afforded by Fourier analysis that are not explored here. Firstly, the technique allows the direct comparison of multi-species and multi-factor cyclic datasets to be directly compared by multiple regression analysis and therefore to identify possible interactions and common driving variables. Secondly, the simplified mathematical expressions obtained may be used for other analytical or predictive models. All this is impossible without a sufficient run of data of sufficient quality to satisfy the technique.

Acknowledgements

Thanks are due to Tony Fitzsimons and Rory Quinn. Also Aquatic Sciences Research Division, Department of Agriculture and Rural Development, N. Ireland for access to much of the data.

References

Andrew, T. E., 1992. The occurrence of Bosmina longirostris (OFM) in Lough Neagh. Irish Naturalist's Journal 24: 76–77.

Andrew, T. E. & A. G. Fitzsimons, 1992. Seasonality, population dynamics and production of planktonic rotifers in Lough Neagh, Northern Ireland. Hydrobiologia 246: 147–164.

Andrew, T. E. & E. M. Woodward, 1993. Some observations on the populations of Mysis relicta (Loven). In Wood, R. B. & R. V. Smith (eds), Lough Neagh. Kluwer, The Hague, 327–338, 529 pp.

Barciela, R. M., E. Garca & E. Fernandez, 1999. Modelling primary production in a coastal embayment affected by upwelling using dynamic ecosystem models and artificial neural networks. Ecosystem Modelling 120: 199–211.

Batterbee, R. W., 1978. Observations on the recent history of Lough Neagh and its drainage basin. Philosophical Transactions of the Royal Society 28: 303–344.

Carlin, B., 1943. Die Planktonrotatorien des Motalaström. Meddelanden från Lunds Universitets Limnologiska Institution, Vol. 5, 256 pp.

Dakin, W. J. & M. Latarche, 1913. The plankton of Lough Neagh. Proceedings of the Royal Irish Academy 30: 20–96.

Davis, J. C., 1986. Statistics and Data Analysis in Geology (2nd edn). John Wiley & Sons, 646 pp.

Doohan, M., 1973. Energetics of planktonic rotifers applied to populations in reservoirs. Unpubl. Ph.D Thesis. University of London, 226 pp.

Edgar, A. J. & T. E. Andrew, 1990. A simple method for fitting Belehradek's equation to embryonic development data of zooplankton. Hydrobiologia 194: 177–181.

Fitzsimons, A. G. & T. E. Andrew, 1992. The co-existence of three species of carnivorous Cladocera in the zooplankton of Lough Neagh. Irish Naturalist's Journal 24: 114–115.

Fitzsimons, A. G. & T. E. Andrew, 1993. The zooplankton of Lough Neagh, 1968–1978; the seasonal succession of the zooplankton species. In Wood, R. B. & R. V. Smith (eds), Lough Neagh. Kluwer, The Hague, 281–326, 529 pp.

Gibson, C. E. & D. A. Stewart, 1973. The annual temperature cycle of Lough Neagh. Limnology and Oceanography 18: 791–793.

Goldman, C. R., M. D. Morgan, S. T. Threkeld & N. Angelli, 1979. A population dynamics analysis of the cladoceran disappearance in Lake Tahoe, California. Limnology and Oceanography 24: 289–297.

Graham, T. R. & K. G. Logan, 1970. Preliminary studies of Lough Neagh ostracods. Irish Naturalist's Journal 16: 326–328.

Herzig, A., 1979. The zooplankton of the open lake. In H. Löffler (ed.), Neusiedlersee: The Limnology of a Shallow Lake in Central Europe. Monographie Biologicae 37: 281–335.

Herzig, A., 1983. Comparative studies on the relationship between temperature and duration of embryonic development of rotifers. Hydrobiologia 104: 237–246.

Herzig, A., 1987. The analysis of planktonic rotifer populations: a plea for long term investigations. Hydrobiologia 147: 163–180.

Johnson, D. & A. F. Walker, 1974. The zooplankton of Loch Leven, Kinross. Proceedings of the Royal Society of Edinburgh (B)74: 285–294.

Kratz, T. K., T. M. Frost & J. J. Magnuson, 1987. Inferences from spatial and temporal variability in ecosystems – long term zooplankton data from lakes. American Naturalist 129: 830–846.

Marn, V. H., 1997. A simple-biology, stage structured population model of the spring dynamics of *Calanus chilensis* at Mejillones del Sur Bay. Ecological Modelling 105: 1, 65–87.

May, L., 1983. Rotifer occurrence in relation to water temperature in Loch Leven, Scotland. Hydrobiologia 104: 311–315.

Richards, R. C., C. R. Goldman, T. C. Frantz & R. Wickwire, 1975. Where have all the *Daphnia* gone? The decline of a major Cladoceran predator in Lake Tahoe, California-Neveada. Proceedings of the International Association of Theoretical and Applied Limnology 19: 835–842.

Wood, R. B., T. E. Andrew & J. M. Redfern, 1990. Cladoceran remains from the recent sediments of Lough Neagh, Northern Ireland. Proceedings of the International Association of Theoretical and Applied Limnology 24: 560–562.

Wood, R. B., T. E. Andrew & C. E. Carter, 2000. Lough Neagh – eutrophic, temperate, shallow – behaving as an oligotrophic, cold, deep lake?. Proceedings of the International Association of Theoretical and Applied Limnology 27: 1240–1242.

Wood, R. B. & R. V. Smith (eds), 1993. Lough Neagh. Kluwer, The Hague, 529 pp.

Hydrobiologia (2005) 546:457–472 © Springer 2005
A. Herzig, R.D. Gulati, C.D. Jersabek & L. May (eds.) Rotifera X: Rotifer Research: Trends, New Tools and Recent Advances
DOI 10.1007/s10750-005-4289-9

Abiotic vs. biotic factors: lessons drawn from rotifers in the Middle Loire, a meandering river monitored from 1995 to 2002, during low flow periods

Nicole Lair
Laboratoire de Géographie Physique, UMR 6042, Université Blaise Pascal, Maison de la recherche, 63560, Clermont-Ferrand Cedex, France
E-mail: Nicole.LAIR@univ-bpclermont.fr

Key words: large rivers, water discharge, river morphology, rotifer distribution

Abstract

The numbers of rotifers in large rivers never achieve the abundances observed in eutrophic lakes. The adverse conditions of the current have conduct biologists to establish links between particular geomorphologies and biological processes in the plankton of rivers. Recent attempts to examine specific adaptations of rotifers have shown that among several planktonic forms, the loricate species appeared to be better adapted to the current than soft-bodied or littoral-epibenthic species. The eutrophic Middle Loire provides rotifers with more edible biomass than that necessary for their production so, the aim of this study was to determine which other factors were responsible for the origins, growth and survival of rotifer populations in the river. Samples were taken bi-monthly in the current, from end-June to early-October during 8 years, in two sections of the River Loire situated at 550 and 640 km from the source. Flow rate, temperature, dissolved oxygen, suspended matter, biological oxygen demand and algal densities were examined in parallel, and among the 61 rotifer species collected at each site, the 30 dominant species were retained for the analysis. Planktonic loricate species were dominant in the Middle Loire, followed by epibenthic species, soft-bodied species being least abundant. The densities of rotifers and algae changed in parallel and in relation to temperature; flow was clearly unfavourable to algae, represented by the Chlorophytes and to rotifers, whatever the sites. Co-inertia analysis revealed that the assemblage of species was closely grouped at Dam where the immediate environment was dominated by numerous scattered islands. This analysis also illustrated that the consequence of eutrophication in the water quality was more marked downstream. Lessons drawn from this experience of the Middle Loire, which ranks among the richest rivers in terms of species, allowed to highlight that dominance of the Brachionidae is a rule in numerous rivers and may be explained by the capacity of several species to continue growing in a current of 0.2 m s^{-1}. Trichocercids may be relevant indicators of sandy rivers. The flexibility of rotifers in the face of hydraulic conditions, the question of the rotifers' origin, the respective roles of downstream transfer and processes, as well as the role of the rotifers in the river food-web are discussed.

Introduction

As in lacustrine environments, the metazoan potamoplankton that grows in lowland rivers, consists of rotifers and crustaceans, the former usually being dominant (Shiel et al., 1982; Walz, 1995). In streams (Ejsmont-Karabin & Kruk, 1999), but principally in large rivers, the abundance of rotifers can be greater than a few thousand individuals per litre, as has been illustrated for several European rivers, including the Danube (Reckendorfer et al., 1999), Elbe (Holst et al., 2002), Loire (Lair & Reyes-Marchant, 1997), Meuse (Gosselain et al., 1994), Rhine (De Ruyter et al., 1990), Oder

(Schröder, 2001), Po (Ferrari et al., 1989), Pripyat (Galkovskaya & Molotkov, 2001) and Thames (May & Bass, 1998). However, river densities never achieve those observed in eutrophic lakes and such a disparity raises questions about the origin, growth and survival of river zooplankton (i.e. metazooplankon sensu stricto).

In transient habitats, aquatic animals are challenged by a cyclic deterioration in environmental conditions and climate, as a regulator of water flow, exerts an important control over fluvial communities (cf. Admiraal et al., 1994). The 'Age of Water' (Kofoïd, 1903; Baranyi et al., 2002) is important in plankton production, while discharge often initiates changes in river geomorphology and thus plankton samples taken in the river are the result of these complex and unpredictable processes. The floodplains, dead zones and lentic areas were widely suspected to play an essential role, as places exporting planktonic crustaceans and rotifers into the river (Tan & Shiel, 1993; Lair & Reyes-Marchant, 1997; Ejsmont-Karabin & Kruk, 1999; Frisch, 2001) and the success of the latter within the river certainly rests on their shorter development times. Lately several attempts have been made by biologists to establish real links between particular local geomorphology and biological processes in plankton (Reckendorfer et al., 1999; Schiemer et al., 2001; Holst et al., 2002).

Certainly, apart from the typically planktonic rotifers, several species that are caught in the current, mainly inhabit the littoral zone and superficial sediments, where their foot may act as an anchor, preventing displacement, a useful adaptation in flowing water (Ricci & Balsamo, 2000; Schröder 2001). In addition, among several planktonic species the loricate forms appear to be well adapted to the current, in contrast to the soft-bodied forms which are not. Indeed, an analysis of the rotifer distributions in five habitat types on a cross-section of the Middle Loire, shows their affinities for distinctive areas: Brachionids are dominant in the 'current', a region that remains the least populated, while in places with characteristics of open water (in 'pools', above 'sands', behind 'islands', in 'submerged macrophytes'), soft-bodied aloricate species are dominant and Trichocercidae are concentrated above the sandy areas (Picard, 2003). Established species can adapt, and Schröder (2001) investigated how monogonont rotifers cope with alternating aquatic and terrestrial phases and showed that different patterns occurred in different species. However, specific adaptations of rotifers to some adverse conditions remained poorly documented. They have also to overcome the difficulties of feeding and mating, or of activating their responses to predators, in flowing water.

The small particles that dominate in such unstable environments usually provide a surplus of edible food. Indeed, rivers carry high densities of algae and bacteria, associated with numerous heterotrophic flagellates and ciliates which, taken together, also remain poorly studied (e.g., Lair et al., 1998, 1999; Servais et al., 2000; Picard, 2003). However, eating to be fruitful and multiply, remains a challenge for potamoplankton. In fact, although the feeding mechanisms of the rotifers are already well known, one of the questions that remains to be discussed is their ability to feed in lotic conditions, within the limits imposed by their respective feeding mechanisms. Activity measurements are usually carried out in confined conditions (Gosselain et al, 1996; Kobayashi et al., 1996; Viroux, 2000) and the results reflect the potential of entire communities, or of species, that can be far from the reality. The effects of turbulence have been tested solely by measuring grazing rates on *Brachionus calyciflorus* (Pallas) (Miquelis et al., 1998). These authors have shown that in an agitated medium at 0.18–0.22 m s^{-1}, the grazing activity of starved amictic females harvested from the River Seine was not reduced by agitation of the water. Modelling data are generally derived from measurements made in lentic conditions (Schöl et al., 2002), but certainly energy transfers cannot be so efficient in turbulent waters. Indeed, in the Middle Loire, modelling calculations with the bioenergetic model ECOPATH (Chistensen & Pauly, 1992), based on data including algae, bacteria and protozoans caught in the same water stream, have indicated that the high quantities of edible food could theoretically support up to 70 000 rotifers per litre (Picard, 2003), a value ten times as that found in reality.

Furthermore, several workers have concluded that some species reproduce in the main channel at low water (Rzoska, 1978; Saunders & Lewis, 1988; Vasquez & Rey, 1989; Holst et al., 2002), but this has rarely been adequately demonstrated, because estimation of plankton activity remains technically problematic in running water. Experiments have therefore been carried out with rotifers caught in the Middle Loire, using an experimental procedure that mimicked the water flow. The demographic results for the dominant rotifers species reared both in a current of 0.2 m s^{-1} velocity and in lentic conditions, showed large differences between them (Picard & Lair, 2003). Three distinctive groups of species emerged from these results: loricate species such as *Brachionus angularis* (Gosse) and *Keratella cochlearis v. tecta* (Gosse) whose growth rates were never negative, showed that they were capable of growing in the current; the soft-bodied *Asplanchna priodonta* (Gosse) and *Polyarthra dolichoptera* (Idelson) appeared to be unable to grow in the current, while epibenthic species such as *Epiphanes macrourus* (Barrois & Daday), with growth rates sometimes positive, sometimes negative, gave contrasting answers. This again underlines the importance of the physical characteristics of the river in the development of rotifer populations.

The seasonal succession of the potamoplankton in the Middle Loire, studied in 1995 during a period of low water flow, illustrated the importance of localised lentic areas in plankton abundance in the stream (Lair & Reyes-Marchant, 1997). Considering a series of monitoring data collected from 1995 to 2002, the present study therefore aims to show to what extent river rotifer communities are made up by the development of benthic species, limnoplankton inocula or by potamoplankton, distinguishable as discrete functional groups of river born rotifers and what are the factors that control their development. Then, on the basis of our actual knowledge of animal potamoplankton ecology, questions for the future, related to the geomorphological characteristics of the rivers, will be reviewed.

Description of the site, sampling procedure and methods

Principal characteristics of the Middle Loire and its living communities

The River Loire, rises in the Massif Central of France crosses the 'Val de Loire' and flows to the Atlantic coast (Fig. 1). With a length of 1012 km, this rain-fed river makes it the longest large river in the country. It is known as a rare, wild European river. Indeed, a few reservoirs have been built in the upper basin, but even the more downstream reservoir of Villerest is situated only at 250 km from the source. Extensive studies, including physical and chemical characteristics of the water quality, phytoplankton and zooplankton communities, have been conducted since 1977 within the framework of the surveillance programmes initiated by Electricity of France (EDF), in the region of their four nuclear power plants. Since 1995, at two sites: Dampierre-en-Burly (Dam) situated 550 km from the source and Saint-Laurent-des-Eaux (Slb), situated 90 km downstream of Dam, the records have been supplemented by animal plankton studies. This part of this nutrient-rich river (stream order 8) flows in meandering stretches, with continuous shifting sand banks that creates shallows in the channel and produces areas of low or standing waters among the numerous gravel islands. It is characterised by sharply contrasting flows, with severe low waters and can be very shallow during summer. At low discharge, the current was estimated at 0.2 m s^{-1} in

Figure 1. Situation of the power plants on the Middle Loire. The sampling sites are (a) Dampierre-en-Burly (Dam), 123 m elevation, 550 km from the source, the width of the river at this location oscillates around 300 m. (b) Saint-Laurent-des-Eaux (Slb), 85 m elevation, 640 km from the source, it extends approximately between 200 and 325 m width. (Belleville-sur-Loire (Bel) and Chinon (Ch) were studied extensively only in 1999).

the middle of the river. The immediate environment of Dam offers larger areas of lentic water and scattered islands than at Slb (Lair & Reyes-Marchant, 1997, Fig. 1).

The Middle Loire, in which chlorophyll *a* can reach values higher than 200 μg l^{-1} is among the richest European rivers. The algal and rotifer communities have been extensively described by Lair & Reyes-Marchant (1997). The light attenuation varies with the concentration of suspended material and, during low flow, this usually corresponds to the amount of phytoplankton. Briefly, during summer, the algae are classically dominated by small Chlorophytes, known to be an important food source for Brachionids (Pourriot, 1977), associated with numerous centric diatoms. Cyanobacteria, more numerous at Slb, are few and sporadic in occurrence. The microbial food web has also been studied and its importance as additional food for rotifers was determined (Lair et al., 1998; 1999; Picard, 2003). The average densities of the algae (including 77–81% of cells $< 20\ \mu$m) obtained in 1999 from the end of June to early-October, ranged from 20.8 $\pm$ 10.2 10^6 cells l^{-1} (Dam) to 20.9 $\pm$ 12.7 10^6 cells l^{-1} (Slb). Heterotrophic bacteria (including 75–80% of cells $< 1.3\ \mu$m) ranged from 8.6 $\pm$ 1.8 10^9 cells l^{-1} (Dam) to 10.0 $\pm$ 2.5 10^9 cells l^{-1} (Slb), those attached were rare. Heterotrophic flagellates (including 79–93% of cells $< 5\ \mu$m) ranged from 4.1 $\pm$ 2.0 10^6 cells l^{-1} (Dam) to 4.5 $\pm$ .6$10^6$ cells l^{-1} (Slb). Ciliates (including 81–84% of cells $< 50\ \mu$m) ranged from 45.4 $\pm$ 26.4 10^3 cells l^{-1} (Dam) to 50.6 $\pm$ 24.8 10^3 cells l^{-1} (Slb). With regard to the rotifer densities, the biomass of unicells of edible size, was always higher than the incipient limiting level (Walz, 1995) and, since the grazing activity of rotifers was usually saturated at 0.5–1.5 mg C l^{-1} (Rothhaupt, 1990), the edible biomass available was always greater than that necessary for production (Picard, 2003). As confirmed by the ECOPATH calculations noted in the introduction, food can be considered as a non-limiting factor.

Processes of sampling and analysis

Rotifers were sampled from the end of June to early-October, during the period of low discharge that usually lasts for about three to four months. At each site in the Middle Loire (Dam & Slb), three sampling stations were chosen in relation to the reactors. The first station is situated upstream of the plant, the other two are downstream of the cooling water discharge (from a few hundred metres to a few km). Samples were taken with a Van Dorn sampler, from floating platforms situated above the water flow, at approximately 0.50 m below the surface. Each sample consisted of three sub-samples (10 l $\times$ 3) mixed together in a large tank and used for chemical and biological analysis. In the approach taken in 1995, sampling stations were considered separately, but on account of displacements of sand banks that occur locally from year to year during flood periods, inducing inevitable morphological changes, the results obtained at each sampling station were grouped in order to have an overview of the situation at every site. Briefly, in the present study, each data set corresponds to the mean of a series of samples (3 samples at each site), taken bimonthly (8 dates), during 8 years (1995–2002) giving, at each site, 64 data extracted from the 192 samples analysed. The procedures for the examination and determination of algae and rotifers are classical and have been described in detail in Lair & Reyes-Marchant (1997).

Of the 29 physical and chemical variables recorded every year in situ with a WTW 196 apparatus or analysed at the Municipal Laboratory, registered by the Ministry of Health, we retained the following: flow (continuously monitored by Electricity de France), temperature (t °C), dissolved Oxygen (O_2), Suspended Matter (SM) and Biological Oxygen Demand (BOD_5), expected to be the more discriminating variables at this time of the year, in the rotifers' distribution. Pearson correlation coefficients were calculated to evaluate the relationships between algae and rotifer densities. In addition, at each site, a co-inertia analysis derived from a classical Principal Component Analysis, made with centred and normed tables of the species densities and environmental variables, was carried out, and the co-structure tested to determine which factors control the species distributions, using the ADE procedures (Chessel & Dolédec, 1992; Dolédec & Chessel, 1994).

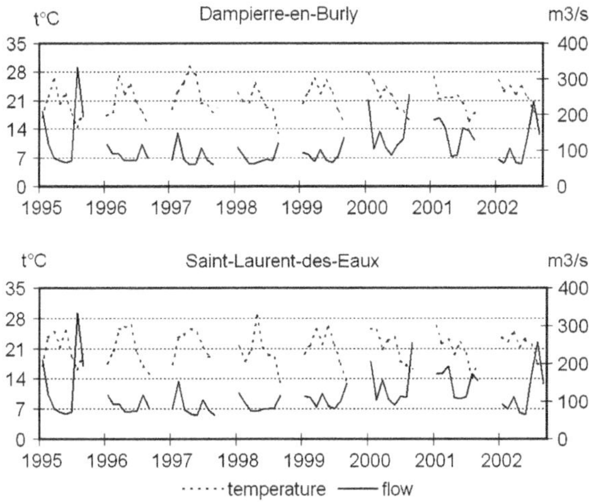

Figure 2. Time series of the variations in flow and temperature in the sampling areas.

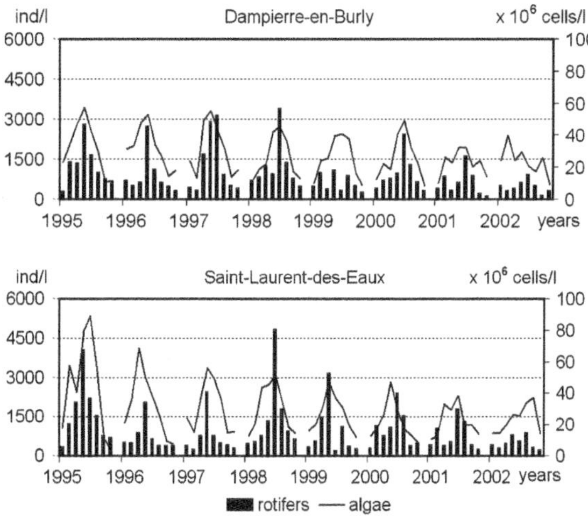

Figure 3. Time series of the variations in density of the algae and rotifers.

Results

During the eight sampling periods, the water temperature varied between 14.5 and 29.0 °C. The flow rates ranged between 60 and 334 $m^3 s^{-1}$ at Dam and between 63 and 340 $m^3 s^{-1}$ at Slb and were characteristic of the low discharges of the Val de Loire during summer. The highest flows in the period, generally occur from September onward and, except in 2000–2001, they stayed below 100 $m^3 s^{-1}$ in the middle of summer, the time of high temperatures (Fig. 2). Cross-correlation between the variables measured at these two sites revealed that they functioned in parallel, with regard to flow rates ($R^2 = 0.94$) and temperature ($R^2 = 0.85$). Every year the most common riverine plankton groups, rotifers and algae, peaked at low river discharges and high water temperatures. The changes in rotifer density were parallel to those of the algae and the distribution patterns revealed summer maxima, with large peaks in some years and several successive maxima in others (Fig. 3). The algal densities ranged on average from 26×10^6 cells l^{-1} (Dam) to 28×10^6 cells l^{-1} (Slb) and the rotifer densities ranged on average from 897 to 945 ind l^{-1}, confirming the up-downstream increase observed in 1995. The densities of algae ($R^2 = 0.61$), and among them Chlorophytes ($R^2 = 0.56$), total rotifers ($R^2 = 0.61$), and filter feeding rotifers ($R^2 = 0.70$) were also very closely related from Dam to Slb, but not those of the predaceous species ($R^2 = 0.35$) which peaked in some years, depending on the sites.

Rotifer species composition of the Middle Loire

In the course of the eight years of investigation, a total of 13 families, 25 genera and 61 rotifer species were collected at these sites. The variations in percentage representation of the families illustrate the marked dominance of the Brachionidae and Trichocercidae (Table 1). The principal species (30 retained in total) were *Anuraeopsis fissa* (Gosse), *Ascomorpha ovalis* (Bergendal), *Asplanchna priodonta*, *Brachionus angularis*, *B. bennini* (Lessling), *B. bidentata* (Anderson), *B. calyciflorus*, *B. leydigii* (Cohn), *B. quadridentatus* (Hermann), *B. urceolaris* (O.F. Muller), *Cephalodella gibba* (Ehrenberg), *Colurella adriatica* (Ehrenberg), *Epiphanes macroura*, *Euchlanis dilatata* (Ehrenberg), *Gastropus stylifer* (Imhof), *Keratella cochlearis* (Gosse) and its form *tecta*, *Lecane bulla* (Gosse), *L. closterocerca* (Schmarda), *L. luna* (O.F. Muller), *L. lunaris* (Ehrenberg), *Polyarthra dolichoptera*, *P. major* (Burckhardt), *P. vulgaris* (Carlin), *Rhinoglena frontalis* (Ehrenberg), *Sinantherina socialis* (Ehrenberg), *Trichocerca brachyura* (Gosse), *T. elongata* (Gosse), *T. pusilla* (Lauterborn) and *T. similis* (Wierzejski).

These species have been classified with reference to their affinities to the current, giving a view

Table 1. Percentages of the different rotifer families in samples from the Middle Loire for the period 1995–2002

	1995	1996	1997	1998	1999	2000	2001	2002
Dampierre-en-Burly								
Asplanchnidae	3	6	3	4	6	16	4	14
Brachionidae	49	40	35	60	51	25	41	45
Colurellidae	4	2	5	3	1	3	2	2
Epiphanidae	4	3	1	1	5	13	6	10
Euchlanidae	1	1	2	1	2	1	1	1
Flosculariidae	2	4	3	0	1	18	13	5
Gastropodidae	1	2	1	7	5	3	3	3
Hexarthridae	0	1	0	0	0	1	0	0
Lecanidae	7	7	5	5	5	3	9	2
Notommatidae	7	8	9	4	4	2	7	3
Synchaetidae	7	5	7	6	6	6	6	5
Testudinellidae	2	4	3	1	1	1	1	1
Trichocercidae	11	16	26	7	14	6	8	10
Saint-Laurent-des-Eaux								
Asplanchnidae	5	9	7	11	26	22	2	13
Brachionidae	50	33	32	49	25	23	26	39
Colurellidae	4	2	3	5	2	3	1	2
Epiphanidae	7	4	5	1	4	11	3	10
Euchlanidae	2	1	1	1	2	2	1	1
Flosculariidae	2	2	2	1	0	3	17	4
Gastropodidae	1	1	1	9	4	3	5	5
Hexarthridae	0	1	0	0	0	0	0	0
Lecanidae	6	5	4	6	4	4	8	3
Notommatidae	6	13	12	4	3	4	8	6
Synchaetidae	7	6	7	6	8	12	18	5
Testudinellidae	2	2	2	1	1	1	1	1
Trichocercidae	7	22	22	6	21	13	10	13

of the importance of their preferred habitats (Fig. 4). The first group (39% at Dam & 40% at Slb) corresponded to the loricate species, typically planktonic and better adapted than the rest to growing in the current. Depending on the years, they were less abundant at Dam than at Slb. The second group (37% at Dam & 35% at Slb) consisted of the littoral-epibenthic taxa and, depending on the years, they appeared more numerous at Dam. The third group (24% at Dam & 25% at Slb), composed of soft-bodied planktonic taxa, also occurred in significant quantities. Globally, 63% of the species at the downstream site Slb, were planktonic while at the upstream site Dam, there were fewer planktonic species (60%) and littoral-epibenthic taxa were more frequent.

Interactions between environmental variables, algae and rotifers

Among these living communities, the algae were most strongly correlated with flow at both sites (Table 2). Soft-bodied and loricate species of rotifers were more strongly correlated with flow at Dam than at Slb, in contrast to the epibenthic species. Temperature remained a discriminating factor for rotifers and algae at both sites, epibenthic species appearing reactive and the soft-bodied species not {correlations between algae, rotifers and the other variables (SM, O_2 and BOD_5) were not significant}. Correlations between rotifers and the three dominant groups of algae, were particularly significant with the green algae

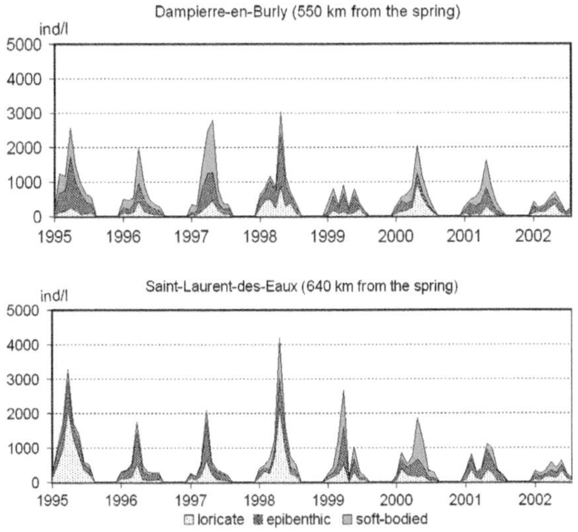

Figure 4. Time series of the variations in density of the epibenthic and planktonic ('loricate' and 'soft-bodied') rotifers at two sites in the Middle Loire.

(determination coefficients (50%). Total rotifers and their sub-categories were positively correlated with density and biomass of the green algae. Planktonic loricate rotifers were the most strongly correlated (determination coefficients (50%), epibenthic species to a lower extent, and soft-bodied forms least. To a lesser extent, significant correlations were observed with the other groups of algae, especially the Cyanobacteria.

In the river, the dominant predaceous species were *C. gibba* (a small epibenthic species) and *A. priodonta* (a large planktonic soft-bodied species). *Asplanchna* can occasionally reach densities >1000 ind l^{-1} (a maximum of 1490 ind l^{-1} was counted at a station at Slb in 1999). Correlations giving a view of the potential preys encountered showed that *Cephalodella gibba* appeared linked to algae ($r = 0.59$, $p = 0.0001$ at Dam and Slb) and particularly Chlorophytes ($r = 0.54$, $p = 0.0001$ at Dam and $r = 0.58$, $p = 0.0001$ at Slb) but not *Asplanchna*. In relation to potentially encountered species, the presence of *Asplanchna* was correlated with epibenthic or soft-bodied species, which are more diversified at Slb (Table 3).

In the multivariate analysis carried out with the 30 species and the abiotic variables previously listed, the density of Chlorophytes was added as an indicator of food. In the co-inertia analysis, the co-repartition between environmental and biotic variables was significant at both sites. The results illustrate that, in contrast to t °C, flow was clearly unfavourable to algae (represented by the Chlorophytes) and to rotifers, that were totally rejected in opposition, whatever the sites. SM were few significant giving the view that in this shallow river, light is not the essential limiting factor of algal growth and rotifers feeding. However, differences did appear between the sites.

At the upstream site of Dam, Chlorophytes (and SM) were more strongly linked to the F1 axis (that explained 86% of the inertia) than flow and also t °C, and the F2 axis (that explained 10% of the inertia) clearly discriminated BOD_5, (Fig. 5). Such a scheme illustrated the classical links between low water discharge, high temperature, high oxygen production and high density of algae. Most of the species were grouped together close to the origin, the 7 species more distant were those more numerous in the samples: the trio *B. angularis, B. calyciflorus* and *K. tecta* (three loricate species), situated in the upper part of the F1 axis, was linked to Chlorophytes, while the trio *T. brachyura, T. pusilla* and *C. gibba* (three epibenthic species) appeared more linked to temperature. *R. frontalis* (a soft-bodied species) isolated close to the F2 axis, may be more reactive to the water quality, as expressed by the BOD_5 and to a lesser extent by O_2.

At the downstream site of Slb, flow and temperature linked to the F1 axis (that explained 76% of the inertia) were the discriminant variables for living material. SM linked to the F2 axis (that explained only 12% of the inertia), was opposed to Chlorophytes that were linked to BOD_5 (Fig. 6). Such a pattern illustrated the answer to the river eutrophication and its consequences for the water quality. The species distribution was more spread out from the origin of the axis than at the upstream site, except *S. socialis* which peaked only in 2001 (cf. Table 1). *A. priodonta* that strongly peaked in this area in 1999 and 2000 remained close to *E. macrourus* and *P. major* (2 soft-bodied species that peaked in the same years) and to *L. luna* (an epibenthic species). Among species far away from the origin of the axis, the genus *Brachionus*, *Anuraeopsis* and *Keratella* were more distant from the F1 axis (strictly linked to t °C) than *T. pusilla, P. vulgaris, C. gibba, E. dilatata* and *P. dolichoptera.* These epibenthic species may have greater affinity to the temperature. On the F2 axis, SM data were opposed to O_2 and *L. closterocerca* was the least abundant species at the site.

Discussion

Lessons drawn from experience of the Middle Loire

Among European rivers, the Middle Loire remains a place of extremes. Summer is the time in which the slightest change in water flow induces increase or decrease of the lentic areas which are strongly swept by thermal winds. These areas offer a high diversity of habitats and are good for the growth of opportunistic species. During low water discharge, the river looks like a shallow lake, productive from surface to bottom, shallow depths resulting also in high ambient light levels and, as a consequence, high primary production. Indeed from 1995 to 2002, the absence of flood events has maintained numerous lentic areas and the main bed has remained isolated from lateral backwaters (the nearest reservoir is situated 290 km upstream from Dam). The results showed that the two sites function similarly with regard to flow rates (there is no tributary in between) and water temperature, and the densities of the classical components of the potamoplankton were very similar, increasing from up to downstream.

The cladocerans found in the digestive tracts of young fish caught in the open water of the margins

Table 2. Set of Pearson correlations fitted to evaluate the relationships between abiotic variables, available food and rotifer densities in the Middle Loire

Variables	Flow rate				Temperature			
	Dam	R^2	Slb	R^2	Dam	R^2	Slb	R^2
Total algae	−0.47***	0.29	−0.39***	0.23	0.43***	0.19	0.44***	0.22
Total rotifers	−0.38**	0.16	−0.29*	0.13	0.34**	0.16	0.27*	0.12
Soft-bodied species	−0.33**	0.15	−0.24*	0.14	0.14	0.08	0.18	0.07
Loricate species	−0.34**	0.16	−0.26*	0.08	0.29*	0.11	0.23	0.08
Epibenthic species	−0.26*	0.11	−0.29*	0.13	0.38**	0.14	0.31*	0.14
Asplanchna priodonta	−0.14	0.02	−0.14	0.02	0.12	0.02	0.18	0.03
	Total algae				Chlorophytes			
	Density		Biomass		Density		Biomass	
	Dam	Slb	Dam	Slb	Dam	Slb	Dam	Slb
Total rotifers	0.73***	0.66***	0.45***	0.40***	0.77***	0.78***	0.71***	0.63***
Soft-bodied species	0.38**	0.29*	0.39**	0.25*	0.38**	0.26*	0.39***	0.25*
Loricate species	0.71***	0.69***	0.39**	0.33**	0.72***	0.74***	0.68***	0.67***
Epibenthic species	0.66***	0.60***	0.40***	0.41***	0.65***	0.56***	0.66***	0.55***
	Diatoms				Cyanobacteria			
	Density		Biomass		Density		Biomass	
	Dam	Slb	Dam	Slb	Dam	Slb	Dam	Slb
Total rotifers	0.22	0.01	0.24*	0.14	0.35**	0.31**	0.05	0.00
Soft-bodied species	0.09	0.03	0.24*	0.18	0.24*	0.33**	0.23	0.05
Loricate species	0.21	0.05	0.17	0.08	0.31**	0.14	0.05	−0.09
Epibenthic species	0.24*	0.1	0.24*	0.15	0.36**	0.44***	0.01	0.04

Significance of statistical tests is indicated by either one ($p < 0.05$), two ($p < 0.01$) or three ($p < 0.001$) asterisks.

Table 3. Set of Pearson correlations fitted to evaluate the relationships between *Asplanchna priodonta* and other rotifer species in the Middle Loire

Potential prey items	*Asplanchna priodonta*	
	Dam	Slb
Colurella adriatica	0.45***	–
Epiphanes macrourus	0.59***	0.56***
Euchlanis dilatata	–	0.55***
Polyarthra dolichoptera	0.25*	0.57***
Polyarthra major	–	0.60***
Polyarthra vulgaris	–	0.54***
Trichocerca brachyura	–	0.72***
Trichocerca elongata	–	0.62***
Trichocerca similis	–	0.54***
Trichocerca pusilla	–	0.45***

and littoral vegetation, are solely littoral species (e.g., Steiner C. and Ducher C, university degree projects, unpublished), extremely rarely caught in the current. In addition, in the 64 data series obtained during the low water periods of 1995–2002, typically planktonic crustaceans were absent from the samples collected in the current. Obviously, the role of the 'Age of Water', underlined at the dawn of the XXe century by Kofoïd, and demonstrated one century later by Baranyi et al. (2002) for Danube zooplankton, cannot be doubt in the control of rotifer production.

In comparison with the species richness of rotifer populations among rivers of the world, listed by Kobayashi et al. (1998), the Middle Loire ranks among the highest. The changes in dominance from site to site, as well as from year to year, illustrated once again the opportunism of these organisms. The changes in rotifer density were parallel to those of the algae, a general scheme usually associated with increases in temperature and decreases in water discharge (Pace et al., 1992; Thorpe et al., 1994; Kobayashi et al. 1998; Schröder, 2001), since small Chlorophytes constitute a large part of their edible food. The progressive decrease in density of the algae observed during the course of these 8 years resulted from the reduction in phosphates discharges from wastewater treatment plants in the catchment area, while nitrates have also decreased every summer (Lair, 2002, synthesis paper published in Hydroécologie Appliquée, the national revue of EDF). In contrast to the algae, the changes in rotifer densities, linked in some years to several excavation works (documented in Lair et al., 1999), were not significant.

As for the algae (e.g., Bauer et al., 2002), the proportion of rotifer species that grow in storage zones illustrates the importance of habitats to the distribution of rotifers (Picard, 2003). Soft-bodied species remained the least numerous in the Middle Loire, and the main group continued to be loricate species, with the short-spined *B. angularis* (on average in the eight years survey, this species was 1st in density at Dam and 3rd at Slb) and *K. cochlearis* v. *tecta* (2nd at Dam, 4th at Slb), as well as *B. calyciflorus* (5th at both sites). The dominance of Brachionids is a general rule in numerous rivers, insofar as several species of this genus can continue to grow in a current experienced at 0.2 m s^{-1}, while the soft-bodied forms cannot. In addition, current may act as a refuge from predation by other plankton. Indeed, because of the passive transportation of plankton in such an unsettled and viscous environment, it appears evident that predators have difficulty in encountering, detecting, capturing, piercing, sucking out, and/or ingesting prey. Filtering, also remains problematic. Are the feeding mechanisms of rotifers, with a well-developed buccal rotatory organ at the apex of a protective lorica, more efficient in these circumstances?

The epibenthic species that can attach to substrates came second in abundance in the Middle Loire, with the dominance of *C. gibba* (4th in density at Dam, 2nd at Slb). Trichocercids (*T. brachyura* and *T. pusilla*) were among the more numerous and their abundance seemed to be particularly relevant as indicators of the river habitats. Indeed, as underlined in the introduction, in the Middle Loire, they were found concentrated above the sandy areas (Picard & Lair, 2004). This complements the results of Holst et al. (2002) who considered exceptional the striking dominance of *Trichocerca pusilla* in the potamal region of the Elbe, another 'sandy sediments, near-natural river'. In their small scale approach with a transverse section of the river, these authors did not find significant differences in the abundance of rotifers between the main stream and groyne fields. However, this section of the river was straightened, current velocity was not documented and the

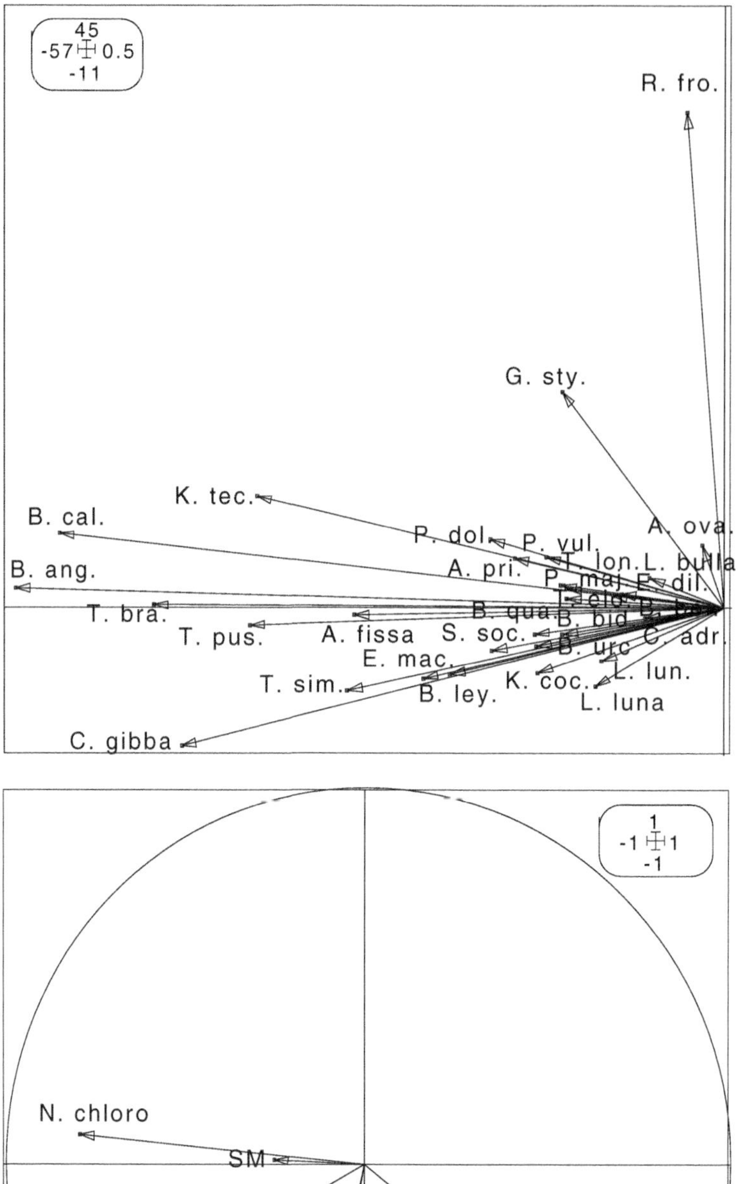

Figure 5. Co-inertia analysis between environment and species variables for the site at Dampierre-en-Burly.

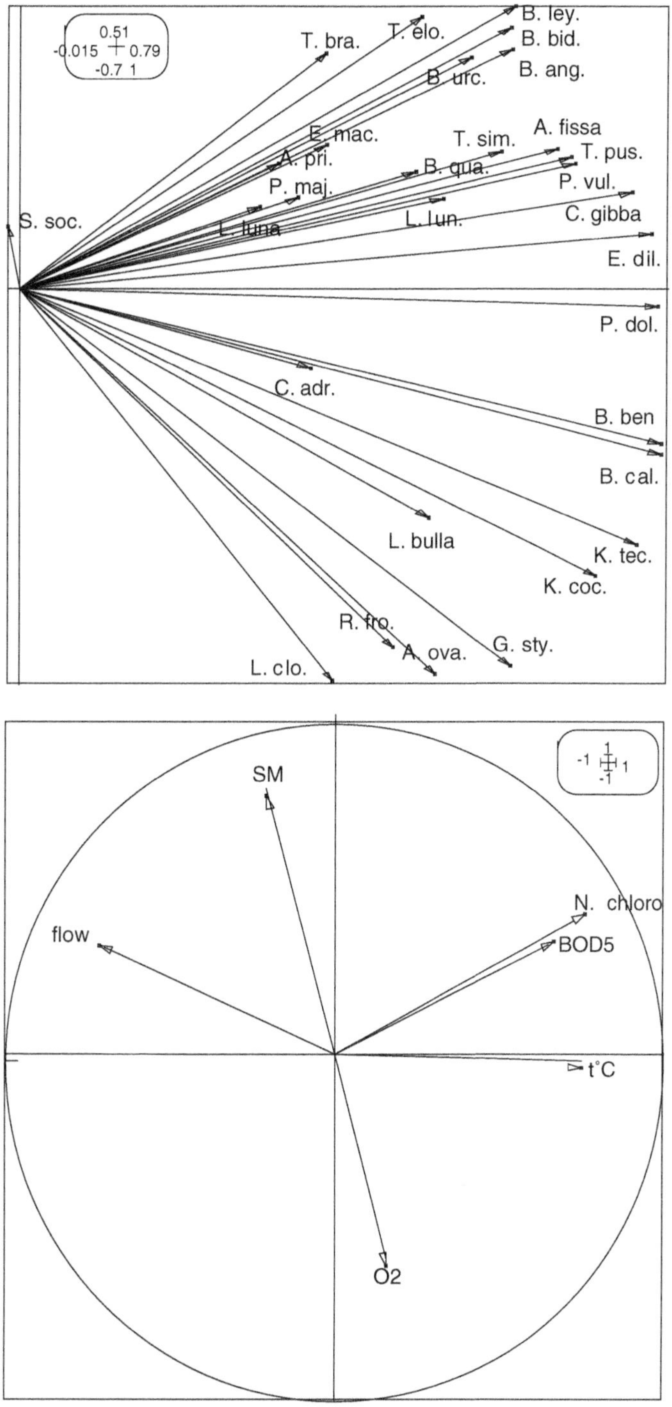

Figure 6. Co-inertia analysis between environment and species variables for the site at Saint-Laurent-des-Eaux.

supposition of a simple dilution effect was not excluded.

Top-down control by predaceous rotifers is also at the heart of the matter. *A. priodonta*, whose density achieved first position at Slb, reached high densities, depending on the years. This predator is known to be capable of adjusting its diet to its environment and to be more of a grazer than a predator (Kappes et al., 2000) and its success in the Middle Loire was certainly because the river provides more food resources than are required. The strong correlation between *Asplanchna* and other species, such as the soft-bodied *E. macrourus*, *P. major*, may be more indicative of their co-existence in the same environment, the comparison with the very low density of small rotifer prey being in contradiction with theoretical predator-prey relationships.

The results derived from the co-inertia analysis illustrate that flow was clearly unfavourable whatever the living material and the sites. This confirms that low water discharge, high temperature and food are essential to allow planktonic rotifers to multiply, and that Monogononts of the Order Ploima are incapable of withstanding high flow velocities (Linhart, 2002). While retention zones tended to expand in the region close to Dam, in this section the water flow weighted slightly less in the rotifer species distribution than at Slb. The aggregation close to the origin of most of the species at Dam contrasted with the more dispersed distribution observed at Slb, giving two distinct patterns between the sites. Dominant species, such as *B. angularis, B. calyciflorus* and *K. cochlearis* v. tecta which largely dominated the rotifer communities at both sites and discriminated at Dam, were diluted with the pool of other species at Slb. Moreover, several epibenthic species appeared more sensitive to the temperature at both sites. The morphological characteristics of the river (and as a consequence the current velocity) may act as the dominant force, lentic areas being more numerous in the section at Dam than at Slb. This pattern may underline more local processes at Dam, with more species swept from stagnant water, while longitudinal transfers would be more important at Slb. Several species that have appeared more linked to the temperature were exported at both sites.

The great flexibility of rotifers in responding to hydraulic conditions

Plankton communities are undoubtedly carried downstream at various speeds (Reynolds, 1988), and refuge habitats from the disturbance of flow remain numerous in rivers. Rivers such as the Middle Loire, in which typically planktonic crustaceans are absent from the current during low flow, emphasize that rotifers would be best adapted to running water conditions and particularly those species of Brachionids that proliferate also in other rivers. Their shorter turnover time than that of crustaceans is beneficial. However, during dispersal, their habitat is not chosen by the rotifers and may be unsuitable for their growth. Admitting that fast growing rotifers can proliferate in a place, the production of protective cover under the influence of sudden and stochastic increases in water flow seems to be problematic. It poses the question of how the installed populations persist locally, particularly in sections in which hydraulic backwaters remain sparse, as in regulated rivers in which the structural properties are degraded (Schiemer et al., 2001). Facilities for several species to feed in the current were underlined above, and the difficulties for males to come across females, is a constraint deleted by heterogony. In addition, the production of resting eggs can take place either when population growth reaches its maximum, or at low densities in some populations (Schröder, 2001). Moreover, this author showed that rotifers can produce resting eggs during inundation periods, but also in the permanent water during the rest of the year, adjacent storage zones being able to preserve potential colonisers producing resting eggs, and rebuild populations with few individuals. Such a situation is particularly relevant in unregulated rivers. William & Smith (1996) showed that individual rotifers may respond to hydraulic conditions in a species-specific way, and the results from the Middle Loire confirm existing differences among species. Moreover, Schröder (2001) emphasized the dominance of the ubiquitous *B. calyciflorus, K. cochlearis, Keratella quadrata* and *Polyarthra* spp. in the permanent waters of the floodplain, and they can stock the rivers. The example of the Middle Loire has confirmed that lentic areas are more densely populated than flowing reaches; but, these aquatic biota are also refuges for fish that reside there, particularly during droughts (Magoulick & Kobza, 2003). However, the constraint remains and the final chance for species to become dominant may be their capability of sustaining a minimum feeding activity and continuing to grow and reproduce at low rates in the current, and of hatching a few resting eggs at just the right time, while the current acts as a refuge from predation by predaceous plankters.

Are rivers unique and the plankton origin too?

Rivers provide dispersal routes for both animals and plants that are important for maintaining the diversity of the riparian zooplankton. It has been postulated that zooplankters may be effectively dispersed by waterfowl. If so, and assuming that dispersal can be provided by wind, rain or waterfowl, the experiments of Jenkins & Underwood (1999) do not support 'the premise of readily dispersed zooplankton'. The origin of plankton remains a complex issue, developed for example in the 'flood pulse concept' of Junk et al. (1989), and in the 'inshore retention concept' of Schiemer et al. (2001), but the importance of the physical environment has often been neglected by planktonologists.

The plankton communities are undoubtedly carried downstream at various speeds (Reynolds, 1988) and numerous authors have explained the downstream increase in zooplankton by the time needed to build significant populations. Downstream transfer certainly plays a key role, thanks to the provision of organic matter and algae provided to rotifer plankton and other communities and, if the food available can be limited in headwaters, it is certainly not in the lowland parts of rivers. However, the longitudinal changes in geomorphological features, responsible for retention time and water discharge, and the complexity of the hydraulics often make it difficult to evaluate accurately the extent of longitudinal transportation of plankton.

Most investigations have focused on large-scale longitudinal changes of abundance and species composition, based on samples taken from the main stream, but few have considered small-scale transverse variations (e.g. Holst et al., 2002). The recent increase in research at smaller scales has indicated an enhanced role of localised areas, and the results from the Middle Loire where flood pulses still occur, have contributed greatly to underlining the importance of local processes in which inoculation points are possibly located within distances as short as a hundred meters (Lair & Reyes-Marchant, 1997). The increases in density observed over short distances cannot be explained by the transit time, because it is not the 'Age of Water', *sensus* turnover time in which most of the organisms built their successive generations. Indeed, their origin, strictly linked to the morphology of the rivers, remains an important aspect and, in rivers characterised by a paucity of hydraulic backwaters in their catchment areas, dispersal of potamoplankton localised to the margins could be a more general rule as was suspected.

Examining the lateral distribution of the river plankton in the Meuse, Marneffe et al. (1996) observed the maxima in algae and zooplankton close to the banks. An attempt to illustrate this was also made by Reckendorfer et al. (1999) in the Austrian part of the river Danube, to which end they developed an index of lentic habitat availability, based on length of the shoreline and water velocity. In a similar way, the example of the Middle Loire emphasizes the importance of fine-scale patchiness and illustrates the role of lentic areas in the supply of rotifers to the main channel. For example, the observations made at the scale of the sites in the Middle Loire, showed that during summer, the rotifer density increased over short distances up and downstream of the plants, despite the fact that they need a minimum of 34 km to double their biomass (Lair & Reyes-Marchant, 1997). Sampling only in the channel might result in an underestimate of global rotifer densities, which occur in higher densities in the lentic areas of this river (Picard, 2003).

Abiotic vs. biotic factors

In the light of these considerations, the presence of plankton in rivers, classically linked to potamal environments, needs to be reassessed. Indeed, in headwater reaches characterised by gravel bedloads with riffle and pool morphology, diel and seasonal drift have been observed for several species such as *E. dilatata, K. cochlearis* etc. (Schram et al., 1998), and rivers with slow-flowing upper reaches can show much higher plankton densities upstream than downstream (Reckendorfer et al., 1999). These results underline once again the importance of the river morphology.

In the middle reaches, the example of rotifer distribution in the Middle Loire showed that downstream transport was linked to the availability of lentic areas situated in the immediate environment and the downstream increase in density observed from Dam to Slb was not the general rule in this river. Indeed, the site at Belleville-sur-Loire, situated 40 km upstream of Dam (Fig. 1), whose communities were studied again in 1999, was more densely populated than the site at Dam (Picard, 2003). In the section at Bel, the river frequently extends up to 500 m in width and more local sources of plankton may supply the main stream. In contrast, in regulated rivers in which the channel is deeply incised into its floodplain and few permanent waters occur, plankton increase downstream may be a more general rule. Drifting from adjacent water bodies connected to the river is inevitable, and this would happen particularly at those periods when large flood events occur. Rotifers originating from reservoirs, from which zooplankton species are flushed passively and then observed in downstream outlets after a long time (Chandler, 1937), are not questioned and there is a widespread literature on this matter. Zooplankton drift in rivers, such as the Rhine (Admiraal et al., 1994), also reproduced the seasonal succession of plankton classically observed in lakes, indicating its origin. Further, downstream the rotifer density may be limited by the presence of suspended matter (Miquelis et al., 1998), that also limits light penetration and primary production. Continuing down the course of the river, large changes in flow rates at the river-estuary interface induce not only large changes in animal plankton densities, but also in species composition.

Conclusion

Among the many interesting aspects of studying rotifers in rivers, the question that remains is their role in the food web. Several investigations have indirectly shown the modest to considerable influence of grazing by zooplankton (e.g., Schöl et al., 2002) and the role of rotifers in the regulation of the algae, which is open to debate, is linked to the problem of their feeding efficiency in rivers. The grazing effect of the numerous filter-feeders was suspected to cause the summer decline in the algae of the Middle Loire (Lair & Reyes-Marchant, 1997). However, experiments carried out in agitated waters have demonstrated the low efficiency of the species concerned. Furthermore, to our knowledge, studies including all the actors involved in the trophic network are lacking and, in addition, to quantify the top down control by fish remains a problem. Obviously, the observations during eight years, of the dynamics of algae and rotifers that have evolved in parallel in the Middle Loire cannot support the assertion 'the more phytoplankton, the more rotifers in' as much as the food available (including other protists and bacteria) is excessively greater in abundance than are the rotifers. In addition, their population development, enhanced during low discharge/high temperature periods, remains limited to a few weeks. We can reasonably admit that top-down control of the algae by river rotifers, preferentially exerted at low or null flow, may remain very limited over all, whereas predation by predaceous plankton may certainly be exerted in lentic areas. Vertebrate predation, whose intensity depends on the synchronisation of the respective timing of births that vary from year to year, may be limited to fish larvae during a short period, the young of the year being observed to go quickly from cladocerans to insect larvae (Reyes-Marchant et al., 1992, and unpublished data of Steiner and Duchet, noted above). The shift in the rotifers density observed every year at the end of the summer, is certainly due essentially to the increase in water discharge, decrease in temperature (and as a result, decrease in food), that largely act as forcing variables.

As a general rule, from headwater to middle reaches and estuaries, the origin of rotifers seems to be strictly linked to the morphological characteristics of the rivers (including man-made structures), the rotifer species caught in the course of the rivers acting as indicators of their various origins. Finally, the increase in density of rotifers may undoubtedly be limited by 'the Age of Water' and, if the role of the floodplain is fundamental during high flow, the importance of local processes illustrated by several authors may become the key factor in the functioning of river plankton.

For the future, programmes are needed that work towards a better knowledge of rotifers' strategies and feeding ecology at a species level, which remains poorly documented in rivers. Further, to obtain a holistic approach to the rotifers' ecology in rivers, multidisciplinary research is needed. A good example was that conducted from Vienna on the Danube (Reckendorfer et al., 1999) including river habitats and river inhabitants. In fact, we now have at our disposal a formidable arsenal of techniques including computer processing of aerial photographs and digitisation for field mapping of storage zones, including backwaters, which can be finalised with Geographical Information Systems. Calibration of storage zones, including submerged macrophytes, using a Global Positioning System can be associated with series sampling of living material, based on random sampling strategies (as was done in a section of the Middle Loire in 1999) in successive stretches, in order to obtain a better knowledge of the life in rivers.

Acknowledgements

Data used in this work comes from monitoring programmes supported by Electricité de France. We thank Patricia Reyes-Marchant, Jean-Luc Peiry and Aude Beauger for their help with data and work processing and anonymous reviewers for their valuable suggestions that improved the manuscript. Mary Burgis was the English reviewer.

References

Admiraal, W., L. Breebaart, G. M. J. Tubbing, B. Zanten Van, E. D. De Ruyter Van Steveninck & R. Bijkerk, 1994. Seasonal variation in composition and production of planktonic communities in the lower River Rhine. Freshwater Biology 32: 519–531.

Baranyi, C., T. Hein, C. Holarek, S. Keckeis & F. Schiemer, 2002. Zooplankton biomass and community structure in a Danube River floodplain system: effects of hydrology. Freshwater Biology 47: 473–482.

Bauer, D. E., M. E. Conde & N. Gomez, 2002. Phytoplankton of a small lowland stream related to water quality and hydraulic discontinuities. Archiv für Hydrobiologie 153: 421–442.

Chandler, D. C., 1937. Fate of typical lake plankton in streams. Ecological Monographs 7: 445–479.

Chessel, D. & S. Dolédec, 1992. ADE Software. Multivariate Analysis and Graphical Display for Environmental Data. Université Lyon 1.

Christensen, V. & D. Pauly, 1992. ECOPATH II – a software for balancing steady – state ecosystem models and calculating network characteristics. Ecological Modelling 61: 169–185.

De Ruyter, E., D. Van Steveninck, W. Admiraal & B. Van Zanten, 1990. Changes in plankton communities in regulated reaches of the lower Rhine river. Regulated Rivers 5: 67–75.

Dolédec, S. & D. Chessel, 1994. Co-inertia analysis: an alternative method for studying species-environment relationships. Freshwater Biology 31: 277–294.

Ejsmont-Karabin, J. & M. Kruk, 1999. Effects of contrasting land use on free-swimming rotifer communities of streams in Masurian Lake District, Poland. Hydrobiologia 387/388: 241–249.

Ferrari, I., A. Farabegoli & R. Mazzoni, 1989. Abundance and diversity of planktonic rotifers in the Po River. Hydrobiologia 186/187: 201–208.

Frisch, D., 2001. Life cycles of the two freshwater copepods *Cyclops strenuus* Fischer and *Cyclops insignis* Claus (Cyclopoida, Copepoda) in an amphibious floodplain habitat. Hydrobiologia 453/454: 285–293.

Galkovskaya, G. A. & D. V. Molotkov, 2001. Species diversity and dominance in the planktonic rotifer community of the Pripyat River in the Chernobyl region (1988–1996). Hydrobiologia 446/447: 179–185.

Gosselain, V., J. -P. Descy & E. Everbecq, 1994. The phytoplankton community of the river Meuse, Belgium: seasonal dynamics (year 1992) and the possible incidence of zooplankton grazing. Hydrobiologia 289: 179–191.

Gosselain, V., C. Joaquim-Justo, L. Viroux, M. Mena, A. J. -P. Metens Descy & J. -P. Thomé, 1996. Laboratory and in situ grazing rates of large river rotifers and their contribution to community grazing rates. Archiv für Hydrobiologie, Supplement 113: 353–361.

Holst, H., H. Zimmermann-Timm & H. Kausch, 2002. Longitudinal and transverse distribution of plankton rotifers in the potamal of the River Elbe (Germany) during late summer. International Review of Hydrobiology 87: 267–280.

Jenkins, D. G. & M. O. Underwood, 1999. Zooplankton may not disperse readily in wind, rain, or waterfowl. Hydrobiologia 387/388: 15–21.

Junk, W. J., P. B. Bayley & R. E. Sparks, 1989. The flood pulse concept in River-Floodplain Systems. Canadian Special Publication of Fisheries and Aquatic Sciences 106: 110–127.

Kappes, H., C. Mechenich & U. Sinsch, 2000. Long-term dynamics of *Asplanchna priodonta* in Lake Windsborn with comments on the diet. Hydrobiologia 432: 91–100.

Kobayashi, T., P. Gibbs, P. Dixon & R. J. Shiel, 1996. Grazing by a river zooplankton community: importance of microzooplankton. Marine and Freshwater Research 47: 1025–1036.

Kobayashi, T., R. J. Shiel, P. Gibbs & P. I. Dixon, 1998. Freshwater zooplankton in the Hawkesbury-Nepean River: Comparison of community structure with other rivers. Hydrobiologia 377: 133–145.

Kofoïd, C. A., 1903. The plankton of the Illinois River, 1894–1896, with introductory notes upon the hydrography of the

Illinois River and its basin. Part I. Quantitative investigations and general results Ill. State Laboratory of Natural History Bulletin 6: 95 629.
Lair, N. & P. Reyes-Marchant, 1997. The potamoplankton of the Middle Loire and the role of the moving littoral' in downstream transfer of algae and rotifers. Hydrobiologia 356: 33–52.
Lair, N, P. Reyes-Marchant & V. Jacquet, 1998. Développement du phytoplancton, des ciliés et des rotifères sur deux sites de la Loire moyenne (France), en période d'étiage. Annales de Limnologie 34: 35–48.
Lair, N., V. Jacquet & P. Reyes-Marchant, 1999. Factors related to autotrophic potamoplankton, heterotrophic protists and micrometazoans abundance at two sites in a lowland temperate river, during low water flow. Hydrobiologia 394: 13–28.
Linhart, J., 2002. Bdelloidea/Monogononta abundance ratio: a possible measure of the relation of stream rotifers to flow velocity ?. Biologica 39/40: 101–110.
Magoulick, D. & R. M. Kobza, 2003. The role of refugia for fish during drought: a review and synthesis. Freshwater Biology 40: 1186–1198.
Marneffe, Y., J. -P. Descy & J. -P. Thomé, 1996. The Zooplankton of the lower river Meuse, Belgium: seasonal changes and impact of industrial and municipal discharges. Hydrobiologia 319: 1–13.
May, L. & J. A. B. Bass, 1998. A study of rotifers in the River Thames, England, April-October, 1996. Hydrobiologia 387/388: 251–257.
Miquelis, A., C. Rougier & R. Pourriot, 1998. Impact of turbulence and turbidity on the grazing rate of the rotifer *Brachionus calyciflorus*: Pallas. Hydrobiologia 386: 203–211.
Pace, M., S. Findlay, E. G. Stuart & D. Links, 1992. Zooplankton in advective environments: the Hudson River community and a comparative analysis. Canadian Journal of Fisheries and Aquatic Sciences 49: 1060–1069.
Picard, V. 2003. Structure et dynamique du potamoplancton de la Loire moyenne (France) en période de basses eaux. Analyse spatio-temporelle de la distribution des algues, des bactéries hétérotrophes, des protozoaires flagellés et ciliés et des rotifères. Influence des variables hydrogéomorphologiques et modélisation écotrophique. Thèse de Doctorat, Université de Clermont-Ferrand (France), 145 pp.
Picard V. & N Lair, 2003. Laboratory approach of the growth of rotifers sampled in Middle Loire (France) under turbulence. Journal de Recherche Oceanographique 28: 196–199.
Pourriot, R., 1977. Food and feeding habits of Rotifera. Archiv für Hydrobiologie 8: 243–260.
Reyes-Marchant, P., A. Cravino & N. Lair, 1992. Food and feeding behaviour of roach (*Rutilus rutilus*, Linné 1758) juveniles in relation to morphological changes. Journal of Applied Ichthyology 8: 77–89.
Reckendorfer, W., H. Keckeis, G. Winkler & F. Schiemer, 1999. Zooplankton abundance in the River Danube, Austria: The significance of inshore retention. Freshwater Biology 41: 583–591.
Reynolds C. S, 1988. Potamoplankton: paradigms, paradoxes and prognoses. In Round FE (Ed), Algae and the aquatic environment. Biopress Ltd, Bristol, 285–311.
Ricci, C. & M. Balsamo, 2000. The biology and ecology of lotic rotifers and gastrotrichs. Freshwater Biology 44: 15–28.
Rothhaupt, K. O., 1990. Changes of the functional responses of the rotifers *Brachionus rubens* and *Brachionus calyciflorus* with particle size. Limnology and Oceanography 35: 24–32.
Rzoska, J., 1978. On the Nature of Rivers. Dr W. Junks. The Hague, 67 pp.
Saunders, J. F. & W. M. Lewis Jr., 1988. Zooplankton abundance in the Caura River, Venezuela. Biotropica 20: 206–214.
Schiemer, F., H. Keckeis, W. Reckendorfer & G. Winkler, 2001. The 'inshore retention concept' and its significance for large rivers. Large Rivers 12. Archiv für Hydrobiologie Supplement 135: 509–516.
Schöl, A., V. Kirchesch, T. Bergfeld, F. Schöll, J. Borcheding & D. Müller, 2002. Modelling the Chlorophyll a content of the River Rhine – Interrelation between riverine algal production and population biomass of grazers, rotifers and the Zebra Mussel, *Dreissena polymorpha*. International Review of Hydrobiology 87: 295–317.
Schram, M. D., M. Gunter & D. B. Engle, 1998. Diurnal vertical distribution and drift of zooplankton in an Ozark headwater stream pool. Journal of Freshwater Ecology 13: 47–54.
Schröder, T., 2001. Colonising stategies and diapause of planktonic rotifers (Monogononta, Rotifera) during aquatic and terrestrial phases in a floodplain (Lower Oder Valley, Germany). International Review of Hydrobiology 86: 635–660.
Servais, P., V. Gosselain, C. Joaquim-Justo, S. Becquevort J.-P. Thomé & J.-P. Descy, 2000. Trophic relationships between planktonic micro-organisms in the River Meuse (Belgium): a carbon budget. Archiv für Hydrobiologie, 149: 625–653.
Shiel, R. J., K. F. Walker & W. D. Williams, 1982. Plankton of the Lower River Murray, South Australia. Australian Journal of Marine and Freshwater Research 33: 301–327.
Tan, L.-W. & R. J. Shiel, 1993. Responses of billabong rotifer communities to inundation. Hydrobiologia 255/256: 361–369.
Thorpe, J. H, A. R. Black, K. H. Haag & J. D. Wehr, 1994. Zooplankton assemblages in the Ohio River: seasonal, tributary, and navigation dam effects. Canadian Journal of Fisheries and Aquatic Sciences 51: 1634–1643.
Vasquez, E. & J. Rey, 1989. A longitudinal study of zooplankton along the Lower Orinoco River and its delta (Venezuela). Annales de Limnologie 25: 107–120.
Viroux, L., 2000. An attempt to evaluate zooplankton longitudinal dynmics in a lowland river using a small still-standing 'potamocorral' enclosure. Internationale Vereinigung für Theoretische und Angewandte Limnologie, Verhandlungen 27: 2979–2982.
Walz, N., 1995. Rotifer populations in plankton communities: energetics and life history strategies. Experientia 51: 437–453.
Williams, D. D. & M. R. Smith, 1996. Colonization dynamics of river benthos in response to local changes in bed characteristics. Freshwater Biology 36: 237–248.

Part VIII
Trophic Interactions

Hydrobiologia (2005) 546:475–489
A. Herzig, R.D. Gulati, C.D. Jersabek & L. May (eds.) Rotifera X: Rotifer Research: Trends, New Tools and Recent Advances
DOI 10.1007/s10750-005-4290-3

Freshwater copepods and rotifers: predators and their prey

Zdenek Brandl
Faculty of Biological Sciences, University of South Bohemia, Branisovská 31, CZ-370 05 Ceské Budéjovice, Czech Republic, and Hydrobiological Institute, Academy of Sciences of the Czech Republic, Na Sádkách 7, 370 05 Ceské Budejovice, Czech Republic
E-mail: zdbrandl@bf.jcu.cz

Key words: Rotifers, cyclopoid copepods, calanoid copepods, feeding, predation

Abstract

Three main groups of planktonic animals inhabit the limnetic zone of inland waters and compete for common food resources: rotifers, cladocerans and copepods. In addition to competition, their mutual relationships are strongly influenced by the variable, herbivorous and carnivorous feeding modes of the copepods. Most copepod species, at least in their later developmental stages, are efficient predators. They exhibit various hunting and feeding techniques, which enable them to prey on a wide range of planktonic animals from protozoans to small cladocerans. The rotifers are often the most preferred prey. The scope of this paper is limited to predation of freshwater copepods on rotifer prey. Both cyclopoid and calanoid copepods (genera *Cyclops*, *Acanthocyclops*, *Mesocyclops*, *Diacyclops*, *Tropocyclops*, *Diaptomus*, *Eudiaptomus*, *Boeckella*, *Epischura* and others) as predators and several rotifer species (genera *Synchaeta*, *Polyarthra*, *Filinia*, *Conochilus*, *Conochiloides*, *Brachionus*, *Keratella*, *Asplanchna* and others) as prey are reported in various studies on the feeding relationships in limnetic communities. Generally, soft-bodied species are more vulnerable to predation than species possessing spines or external structures or loricate species. However, not only morphological but also behavioural characteristics, e.g., movements and escape reactions, and temporal and spatial distribution of rotifer species are important in regulating the impact of copepod predation. The reported predation rates are high enough to produce top-down control and often achieve or even exceed the reproductive rates of the rotifer populations. These findings are discussed and related to the differences between the life history strategies of limnetic rotifer species, with their ability to quickly utilize seasonally changing food resources, and adjust to the more complicated life strategies of copepods.

Introduction

Copepods are extensively represented in the seas where most of about 10,000 known species and a similar number of so far undescribed species (Hairston & Bohonak, 1998) are widely distributed. Huys & Boxshall (1991), evaluating their abundance, diversity and size, call them „the insects of the seas''. However, some groups of copepods successfully colonize inland waters and together with Cladocera and Rotifera are quite abundant in freshwater planktonic communities. Although herbivorous at least during some life stages, copepods also represent one of the main groups of invertebrate predators in both limnetic and littoral regions of inland waters. Rotifers, due to their usually much smaller size and restricted mobility, become often preferred food items in the diet of freshwater copepods. Williamson (1983a) reviewed and summarized information on invertebrate predation on planktonic rotifers. The present review updates this earlier work but is

Table 1. Copepod predators on planktonic rotifers (based on data in publications from 1980 on)

Copepod predator	Rotifer prey and predation rate (if given), prey.predator^{-1} day^{-1}	Data presented	Reference
Cyclopoida			
Acanthoc. robustus	*Ag, Ap, Bc, Bd, Fit, Kco, Kq, N, Pm, Poms*	C: R: Beh. + Prob.	Roche (1987)
Acanthoc. robustus	*Ap* 11.3, *Bc* 3.4, *Bd* 6.3, *Kco* 1–3.2, *Pm* 9.8, *Poms* 5.7, *Sk* 7.8, *Sp* 22.3	CR, PR, PS, FR	Roche (1990a)
Acanthoc. robustus	*Ag, Ap, Bc, Bd, Fit, Kco, Kq, N, Pm, Poms, Sk, Sp*	C: Beh., R: Beh. + Prob.	Roche (1990b)
Acanthoc. robustus	*A-spp, Bc, Cu, Kco, Kq, Pm, Pv, Poms,*	horizont. + vertical distribution	Roche (1990c)
Acanthoc. robustus	*A*-spp 0.3, *Bd* 0.3–1.3, *Bru* 1.0–2.6, *Pm* 0–0.3, non: *Ba, Bc, Fil, Kc, Kq*	PR, C: Beh.	Yang & Brandl (1996)
Acanthoc. robustus	*K*-spp, *P*-spp, *S*-spp	C: life hist. param.	Hopp et al. (1997)
Acanthoc. robustus	*Kco, Kq*	C: + R: life history parameters	Conde-Porcuna & Declerck (1998)
Acanthoc. robustus	*A*-sp, *Ba, Bru, Cu, Kco, Kq, Pd, Poms, S*-sp	PR, PI	Brandl & Prazakova (2002)
Cyclops abysorum	*K*-spp, *P*-spp, *S*-spp	C: life hist. param.	Hopp et al. (1997)
Cyclops vicinus	*Ch, Noc, Pd, Pm*		Brandl & Fernando (1981)
Cyclops vicinus	*Ba, Kellg,Pv, So* all sp. 1.2 – 14, non: *Kco, Kq, Kt*	Prey selectivity	Zánkai (1984)
Cyclops vicinus	*Brach. rubens* 6–13 (*C.v.* males), 16–43 (fem.),	PR, C: egg production	Santer (1993)
Cyclops vicinus	*Brach. rubens*	C: life history parameters	Hansen & Santer (1995)
Cyclops vicinus	*K*-spp, *P*-spp., *S*-spp.	C: life hist. param.	Hopp et al. (1997)
Cyclops vicinus	*P*-spp 1.6–13.6, *S*-spp 0.2–36.6, non: *Ap, Asc*-sp, *C*-sp, *Kellg, Kco* 0–0.5, *Kq* 0–0.3	PR, FR, PS	Plassmann et al. (1997)
Cyclops vicinus	*A*-sp, *C-spp, Fil, Kellg, Kco, Kq, Pd, Pm*	PR	Brandl (1998b)
Cyclops vicinus	*A*-sp, *Ba, Bru, Cu, Kco, Kq, Pd, Poms, S*-sp	PR, PI	Brandl & Prazakova (2002)
Diac. b. odessanus	*Bp, Bi, Bt*	C: Beh. + Prob., R: Beh.	Lapesa et al. (2002)
Diac. bicuspidatus thomasi	*Bc*, non: *A*-sp.	PR, C: + R: Beh. + Prob.	Williamson & Gilbert (1980)
Diac. b. thomasi	*S*-spp, *Pm, Pr*, non: *Kco, Kellg, Pv, C*-sp	seasonal changes, R: + C: life hist. param.	Stemberger & Evans (1984)
Diac. b. thomasi	*Sp, Pm, Pr*, non: *Ap, Ase, Kco, Pv*	PS, CR	Stemberger (1985)
Diac. b. thomasi	*Sp* 1.54 – 24, *Kco* 1.2 – 2.4	PR, CR, FR	Stemberger (1986)
Diac. b. thomasi	*Kco, P*-spp	C: life history parameters	Adrian & Frost (1993)
Diac.b. thomasi	*Kco* eggs	CR	LeBlanc et al. (1997)
Mesocyclops edax	*Ag*	C: + R: Beh. + Prob.	Williamson (1980)
Mesocyclops edax	*Ag, Bc, Pv, Kco*	PR, C: + R: Beh. + Prob.	Williamson & Gilbert (1980)
Mesocyclops edax	*Pm, Poms*, non: *Kco*	PI	Brandl and Fernando (1981)
Mesocyclops edax	*Cd, Kco, Kelbo, P*-spp	gut content	Williamson & Magnien (1982)

Mesocyclops edax	*Aspl. girodi* 1.6–3.2	PR, C: Beh. + Prob.	Williamson (1983b)
Mesocyclops edax	*Ker. cochlearis* spine 3.50, unsp. 3.95	PR, PS,	Stemberger & Gilbert (1984)
Mesocyclops edax	*Ag* male 0.8, female 0.3, *Pd* 47.0, *Sp* 0.3	PR, PS, PI, FR	Williamson (1984)
Mesocyclops edax	*Ap, Ba, Fil, Kco, Pe, Poms*	PR, C: Beh.	Brandl & Fernando (1986)
Mesocyclops edax	*Ap, Bc*	FR, C: + R: Beh.	Williamson (1986)
Mesocyclops edax	*Ag* 0.4–1.3, *Aso, Cu* 1.6–1.7, *Ga, Ka, Kco, Kcr, Pe* 0.1–0.9, *Pr* + *Pv* 1.9–4.9	CR, PR, PS	Schoeneck et al (1990)
Mesocyclops edax	*Kp, K*-spp	C: life history parameters	Adrian & Frost (1993)
Mesocyclops kieferi	*Brach. calyciflorus* 28.8–65.8	PR	Matsumura-Tundisi et al. (1990)
Mesoc. leuckarti	*Brach. rubens* 22.0	PR, C: life history parameters	Hansen & Santer (1995)
Mesoc. leuckarti	*K*-spp, *P*-spp, *S*-spp	C: life hist. param.	Hopp et al. (1997)
Mesoc. longisetus	*Brach. calyciflorus* 35.8–79.9	PR	Matsumura-Tundisi et al. (1990)
Mesoc. thermocyclopoides	*Ba, Bc, Bru*	C: life hist. param.	Kumar & Rao (1999a)
Mesoc. thermocyclopoides	*Ba, Bc*	C: life hist. param.	Kumar & Rao (1999b)
Mesoc. thermocyclopoides	*Ai, Ba, Bc*	R: life hist. param.	Kumar & Rao (2001)
Thermoc. crassus	*K*-spp, *P*-spp, *S*-spp	C: life hist. param.	Hopp et al. (1997)
Tropoc. extensus (= *T. pras. mexicanus* of earlier authors)	*Polyrthra remata* 2–8	PR, R: Beh. + growth rate	Diéguez & Gilbert (2002)
Tropoc. prasinus mexicanus	*Ker. cochlearis* – spine 1.05, unsp. 4.10	PR, PS	Stemberger & Gilbert (1984)
Tropoc. prasinus mexicanus	*P*-spp (0.1–0.6), *S*-spp (1.2–1.4), *G*-spp (0.03), *Kco* (0.04), non: *Ktau, T*-spp.	PR, mass ingestion rate	Adrian & Frost (1992)
Tropoc. prasinus mexicanus	*Kco, Kp, P*-spp	C: life history parameters	Adrian & Frost (1993)
Calanoida			
Acanthodiaptomus denticornis	*Kco*	C: Beh.	Lair & Hilal (1992)
Boeckella hamata	*Anu* 0.20	PR, C: life history parameters	Couch et al. (1999)
Boeckella major	*Kp, Cn, E*-sp, *T*-spp, *N*-sp, *Lec*-spp, *Lep*-spp, *T*-spp, *Tes*-sp, *Tri*-sp	C: gut and mouthpart analysis	Green & Shiel (1999)
Boeckella major	*Kp, Cn, E*-sp, *T*-spp, *N*-sp, *Lec*-spp, *Lep*-spp, *T*-spp, *Tes*-sp, *Tri*-sp	gut content, PS	Green et al. (1999)
B. pseudochelae	*T*-spp, *Lec*-sp	C: gut and mouthpart analysis	Green & Shiel (1999)
B. pseudochelae	*T*-spp, *Lec*-sp	gut content, PS	Green et al. (1999)
B. triarticulata	*Anu* 0.47–5.10, *Ktec* 0.48, *Pd* 0.8–3.77,	PR, C: life history parameters	Couch et al. (1999)
Diaptomus pallidus	*So* 8.2, *Pr* 42.0, non: *Kco* 0.0	PR, CR, PS	Williamson & Butler (1986)

Continued on p. 478

Table 1. (Continued)

Copepod predator	Rotifer prey and predation rate (if given), prey.predator^{-1} day^{-1}	Data presented	Reference
Diaptomus pallidus	*Ba, Fit, Ka, Kco, Pm, So*	C: + R: Beh.	Williamson (1987)
Eudiapt. gracilis	*K*-spp, *P*-spp, *Pom*-sp	C: life history parameters	Berger & Maier (2001)
Epischura lacustris	*Euchlanis dilatata*	C: life history parameters	Schulze & Folt (1990)
Eurytemora affinis	*Fil, Kco, Pv, Tr*	R: life hist. param., seasonal changes	Yoshida et al. (2000)
Hemiboeckella searli	*T*-spp, *Lep*-sp, *Lec*-spp, *E*-sp, *Tri*-sp	C: gut and mouthpart analysis	Green & Shiel (1999)
Hemiboeckella searli	*T*-spp, *Lep*-sp, *Lec*-spp, *E*-sp, *Tri*-sp	gut content, PS	Green et al. (1999)
Diaptomus (Hesperod.) arcticus	*Pd, Kq*	PI	Paul & Schindler (1994)
Limnocalanus macrurus	*K*-spp	CR	Warren (1985)
Limnoc. macrurus	*K*-sp, *Kel*-sp, *P*-sp	CR, (PI)	Nero & Sprules (1986)
Parabroteas sarsi	*Ch*	PS, C: + R: PM	Diéguez & Balseiro (1998)
Senecella calanoides	*K*-sp, *Kel*-sp, *P*-sp	CR, (PI)	Nero & Sprules (1986)

Abbreviations: PR, predation rate, prey.predator^{-1} day^{-1}; CR, clearance rate, ml.predator^{-1} day^{-1}; PS, prey selectivity; FR, functional response of PR to prey density; PI. predatory impact, loss in %. day^{-1}; PM, predator or prey morphology; C, copepod; R, rotifer; Beh., feeding behavior; Prob., probability of attack, capture, and ingestion after encounter..Rotifer species : -sp, -spp, one or more undetermined species of the genus, non: nonpreferred species, A, *Asplanchna*; Ap, *A. priodonta*; Ag, *A. girodi*; Anu, *Anuraeopsis fissa*; Ase, *Ascomorpha ecaudis*; Aso, *A. ovalis*; Ba, *Brachionus angularis*; Bc, *B. calyciflorus*; Bd, *B. diversicornis*; Bi, *B. ibericus*; Bp, *B. plicatilis*; Bru, *B. rubens*; Brot, *B. rotundiformis*; C, *Conochilus*; Ch, *C. hippocrepis*; Cu, *C. unicornis*; Cd, *C. dossuarius*; Cn, *C. natans*; E, *Euchlanis*; Fil, *Filinia longiseta*; Fit, *F. terminalis*; G, *Gastropus*; Ga, *Gastropus stylifer*; K, *Keratella*; Ka, *Keratella americana*; Kco, *K. cochlearis*; Kq, *K. quadrata*; Kcr, *K. crassa*; Kp, *K. procurva*; Ks, *K. slacki*; Ktau, *K. taurocephala*; Ktec, *K. tecta*; Kel, *Kellicottia*; Kelbo, *Kellicottia bostoniensis*; Kellg, *K. longispina*; Lec, *Lecana*; Lep, *Lepadella*; N, *Notommata*; P, *Polyarthra*; Pe, *P. euryptera*; Pd, *P. dolichoptera*; Pm, *P. major*; Pr, *P. remata*; Pv, *P. vulgaris*; Pom, *Pompholyx*; Poms, *P. sulcata*; Sk, *Synchaeta kitina*; So, *S. oblonga*; Sp, *S. pectinata*; T, *Trichocerca*; Tr, *T. rousseleti*; Tes, *Testudinella*; Tri, *Trichotria*.

limited to predation on rotifers by mostly freshwater, limnetic, copepods. Greene (1983, 1985) also published a detailed study on selective predation and on prey selection in zooplankton. Stemberger & Gilbert (1987a) reviewed the works on rotifer defenses against predators, and Havel (1987) discussed morphological, chemical and behavioural defenses induced by predators. Recently, also Walz (1993, 1995) reviewed strategies of rotifer life history and Snell (1998) summarized chemical ecology of rotifers, including chemical signals in feeding and induction of defenses against predators. An extensive body of literature is now available on the feeding modes of marine calanoid copepods on various types of prey other than rotifers and some of these are quite helpful in further evaluating feeding on rotifers by freshwater calanoid copepods.

Materials and methods

Both the ISI Web of Science Database and the author's own reference database, revealed more than 200 publications containing some information relevant to predatory feeding by freshwater copepods on rotifers. About 70% of these papers were published during the last two decades. Each five-year period showed about twice as many publications as the previous one.

Results and discussion

Copepods as predators of rotifers: the copepod species

Most likely, all of about 600 known species of cyclopoid copepods (Family Cyclopidae) feed on some species of rotifers. However, specific data about feeding on rotifers exists for only less than 30 species, 20 of which are included in the review by Williamson (1983a). The literature published subsequently added an additional seven species, and supplementary information on the ones studied earlier, and included copepod species inhabiting the limnetic region of lakes and reservoirs. Feeding by the cyclopoid copepod species dwelling among macrophytes in the littoral region has not been studied since the pioneering works by Geoffrey Fryer on the feeding mechanism (Fryer, 1957 a) and food composition (Fryer, 1957 b) of freshwater cyclopoid copepods.

The last two decades have produced new insight into the opportunistic and omnivorous character of cyclopoids' feeding – see e.g., Adrian (1991), Adrian & Frost (1993), Hansen & Jeppesen (1992), Hansen & Santer (1995), Santer (1993) or Santer & van den Bosch (1994) as representatives of investigators studying the algal component of the diet of cyclopoid copepods (more references in Brandl, 1998 a). Generally, the smaller the cyclopoid copepod species and/or the younger its developmental stage, the more important is the algal component in its diet (Adrian & Frost, 1993). However, even small species of limnetic cyclopoid copepods (e.g., species of the genera *Tropocyclops* and *Thermocyclops*) prey on rotifers (Table 1). Nevertheless, the size itself does not determine the extent to which a species is a raptorial feeder. Hansen & Santer (1995) compared the development, growth, survivorship and egg production of two common European species, *Cyclops vicinus* Uljanin and the slightly smaller *Mesocyclops leuckarti* Claus, reared on an algal diet with or without *Brachionus rubens* Ehrenberg as supplemental prey. The large *C. vicinus* developed to the adult stage on a purely algal diet, whereas the smaller *M. leuckarti* required the rotifer prey. Even the younger copepodid stages of *M. leuckarti* were carnivorous. Survival and egg production were higher on the *Brachionus* diet. The predation rate of adult females of *Mesocyclops* on *Brachionus* was unaffected by the availability of algal food. On the other hand, Santer (1993), studying *C. vicinus*, found the predation rate reduced by one-third in the presence of edible algae. Thus, the smaller adult females of *M. leuckarti* seem to be more carnivorous than that of *C. vicinus*.

Hopp et al. (1997) compared adult longevity and reproductive parameters of five cyclopoid copepod species fed algal and animal food including rotifers of the genera *Polyarthra*, *Synchaeta* and *Keratella*. The copepod species were the large *C. vicinus*, the medium-sized *C. abyssorum* Sars and *Acanthocyclops robustus* (Sars), and the relatively small *M. leuckarti* and *Thermocyclops crassus* (Fischer). The greatest effect of diets with and without rotifers was found in the medium-sized *A. robustus*. This species produced larger egg clutches for a longer period, with about three times higher lifetime egg production, and also survived more than twice as long when fed with animal food alone. The smaller *M. leuckarti* did not produce more than one egg clutch when cultivated on non-animal diet. The effects of different food regimes on reproductive parameters and longevity were the slightest in the relatively large *C. abyssorum* as well as in the smallest *T. crassus* (Hopp et al., 1997).

Summarizing, the extent to which a cyclopoid copepod is a raptorial and carnivorous predator is rather a species-specific characteristic than size-related. Even congeneric species may differ in their needs for animal food (e.g., *C. vicinus* vs. *C. abyssorum*). Most natural populations of cyclopoid copepods consist mainly of younger developmental stages. The extent to which a particular species needs animal food depends on whether early copepodid stages feed on animal prey (like in *M. leuckarti* – Hansen & Santer, 1995) or not. This is valid for a continuously reproducing population; on the other hand, a population that has just emerged from diapause and contains only 4th and 5th copepodid stages consists of solely subadult and adults with a high level of carnivory.

An additional caution is warranted here. There are some questions about the species identity of palearctic and nearctic populations of some taxa believed to have a holarctic distribution (*Diacyclops*, *Tropocyclops*, and *Acanthocyclops*): some of them may represent different species with more limited distributions. Therefore, information about species-specific, raptorial feeding should not be simply generalized.

Knowledge on feeding of calanoid copepods on rotifers covers even fewer freshwater species than in cyclopoid copepods. Williamson's (1983a) lists 11 species and only about 5 more have been added subsequently (Table 1). Higher species diversity and a wider range of body sizes in calanoid copepods draw attention to the need for more information about the animal component of their diets. Predatory feeding by some freshwater calanoid copepods has been described before, e.g., by Elster (1936) for *Heterocope borealis* (Fischer), or for various nearctic diaptomids, e.g., *Diaptomus shoshone* Forbes – by Anderson (1967), *D.* arcticus Marsh, and *D.* nevadensis Light - by Anderson (1970), or *D. arcticus* – by Paul & Schindler (1994). More information exists about predatory feeding by freshwater calanoid copepods on other crustaceans, e.g., by Wong & Sprules (1985), Warren (1985), Kerfoot (1987, 1988), Folt & Byron (1989), Schulze & Folt (1989) or O'Brien (2001) – or on protozoans (Burns & Gilbert, 1993). Although the selective nature of calanoid feeding is well known from old studies (Lowndes, 1935; Fryer, 1954), the common, European limnetic species of Diaptomidae were generally regarded as herbivorous. There were occasional observations of rotifers in the gut of *Eudiaptomus* (Fryer, 1954). More recent work of Berger & Maier (2001) revealed the importance of rotifers *Polyarthra*, *Pompholyx* and *Keratella*, in the diet of the common small European species *Eudiaptomus gracilis* (Sars). Adult females live more than twice as long, carry larger egg clutches and produce five times more eggs during their lifespan if fed a mixed algal and animal diet. Apparently, we need to reappraise the diet of small diaptomid species to include an animal component.

The omnivorous feeding of medium and large sized species is better documented. Williamson & Butler (1986) who analyzed the animal component of the diet of the *Diaptomus pallidus* Herrick found it to feed on rotifers with clearance and ingestion rates 5–6 times greater than on algae in comparable concentration. Feeding on rotifers positively influenced both the survival and reproduction of this copepod. The ingestion rate on *Polyarthra remata* (Skorikov) was much higher than on *Synchaeta oblonga* Ehrenberg, while *Keratella cochlearis* (Gosse) was not consumed, although its eggs were. Also, for cyclopoid copepod *Diacyclops thomasi* predation on the eggs of *K. cochlearis*, but not on the rotifer itself, was reported by LeBlanc et al. (1997). Williamson (1987) examined predation by *D. pallidus* on seven rotifer species, *S. oblonga*, *Polyarthra major*, *Filinia terminalis*, *K. cochlearis*, *K. americana*, *Brachionus angularis* and *Rotaria* sp. *Diaptomus* was able to capture all seven species, but only five were ingested. *Synchaeta* was the most vulnerable species with no effective defense mechanism. Three of the seven species exhibited an escape response. *Keratella*'s swimming escape response and the rapid appendage elevation responses of *Filinia* and *Polyarthra* were frequently used when these rotifers were entrained in the feeding currents of the copepod, before the rotifers came into contact with the predator. Williamson & Vanderploeg (1988) analyzed the predatory suspension-feeding behaviour of *D. pallidus* using high-speed film and naupliar larvae and three rotifers, *S. oblonga*, *Polyarthra vulgaris* and *K. cochlearis* as prey. *Keratella* was the least reactive species, *Synchaeta* responded by rolling up at distances of about 0.07 mm and *Polyarthra* showed a highly effective escape response at distances of about 0.40 mm, by using a series of sequential elevation and return of its four triplets of lateral appendages. Also *Diaptomus* exhibited distinctly different levels of response to the prey species examined.

Paul & Schindler (1994) compared the development of rotifer populations in large *in situ* enclosures with and without *Hesperodiaptomus arcticus* in Snowflake Lake, Alberta, for two years. They found densities of *Polyarthra dolichoptera* significantly reduced in the enclosures with *Hesperodiaptomus* in both years, whereas the densities of *Keratella quadrata* were reduced significantly in one year but not in the other.

Couch et al. (1999) studied the feeding of the southern hemisphere calanoid copepods *Boeckella triarticulata* Sars and *B. hamata* Brehm. Both these calanoids to some extent ingested *Anuraeopsis fissa* (Gosse) (up to 4% of daily carbon intake), but *B. triarticulata* also consumed *P. dolichoptera* Idelson in higher amounts (6–30% of daily carbon intake). *Keratella cochlearis tecta* (Gosse) was consumed only in very low amounts. Green & Shiel (1999) and Green et al. (1999) found *Boeckella major*, *B. pseudochelae* and *Hemiboeckella searli* inhabiting Australian ephemeral pools to feed on both rotifer and crustacean species, and the proportion of rotifers in the diet increased with decreasing size of the predatory copepod.

Rotifers as prey of copepods: the rotifer species

Practically all of the limnetic rotifer species co-occurring with predatory copepods have been reported as copepod prey. Even predatory rotifers of the genus *Asplanchna* or colonial species of *Conochilus* form prey for some copepod species (Table 1). The ability of copepods to feed on practically any limnetic rotifers does involve both selection preference for certain prey species: in addition to size limitation and behaviour of the predators, escape and/or protection characteristics of rotifer species determine the degree to which a given species is vulnerable to copepod predation. Soft-bodied rotifers, *Synchaeta* sp., are often most vulnerable. Stemberger (1985) examined prey selection by *Diacyclops thomasi* (Forbes) (= *D. bicuspidatus t.*) offering 8 rotifer and 2 cladoceran prey species. The copepods consistently selected for the soft-bodied rotifers *Synchaeta pectinata* Ehrenberg, *Polyarthra major* Burckhardt and *P. remata*, but did not feed on other small, soft-bodied (*P. vulgaris* Carlin, *Ascomorpha ecaudis* Perty) and loricate (*Keratella cochlearis*, *K. crassa* Ahlstrom) rotifers. The vulnerability of *Polyarthra* species to predation by *Diacyclops* was species-specific and perhaps related to the speed of the escape response: the larger and slowly moving *P. major* was easily captured. Stemberger & Evans (1984) studied seasonal succession of the rotifer species in Lake Michigan in relation to predation by *D. bicuspidatus thomasi* (Forbes). Before the copepod population began to develop, the spring community was dominated by the vulnerable soft-bodied species of the genus *Synchaeta*. The *Synchaeta* species declined after the population increase of *Diacyclops* in late spring and were replaced by a summer assemblage of rotifers dominated by spiny (genera *Kellicottia* and *Keratella*) and colonial (*Conochilus*) species. Later, in September, the assemblage was dominated by soft-bodied species of the genus *Polyarthra*, which exhibit rapid and efficient evasive reaction.

How to get the prey: copepod abilities to catch rotifers

Cyclopoid and calanoid copepods differ in their methods of obtaining food. The former are relatively slow, cruising hunters that swim in the hop-and-sink manner. They depend largely on the proper recognition of signals produced by objects moving in the water around them, which may be potential prey or large, sometimes dangerous predators of the cyclopoids themselves. To distinguish between the prey and predator they are equipped with sensory setae on the widely extended first antennae. Potential prey size could match the size of the predator itself. Bigger prey items means more food but the need to tear the large prey apart also means significant losses of prey biomass, estimated to be about two-thirds of the biomass of large prey (Brandl & Fernando, 1975). On the other hand, small prey can be easily seized and completely ingested. In fact, most of the data on cyclopoid copepod prey selection show a preference for small-sized prey (see Brandl, 1998a) for review). Whereas, large prey, like larger cladocerans, can often successfully escape the attack of a cyclopoid copepod, the result of the encounter with small prey depends on predator's behaviour, prey protective morphology and escape abilities. The most complete descriptions of predator behaviour are probably those by Kerfoot (1978) and Williamson (1980). Using a high-speed motion camera, Kerfoot analyzed swimming, attack and handling of prey by *D. bicuspidatus thomasi*, *Acanthocyclops vernalis* (Fischer) and *Epischura nevadensis* Lilljeborg. While during the sinking phase of the hop-and-sink movement, the cyclopoid copepod reaches a velocity of about 1 mm s^{-1}, which during the few milliseconds of the power stroke of the thoracic legs can reach a speed about 28 mm s^{-1}. The result is a forward thrust of the body by about one body length. However, the animal can repetitively apply such strokes. When such an attack movement is properly oriented it must result in contact with a target prey, if the prey is unable to produce an unexpected escape maneuver, which is exactly what some rotifers do.

The movements of calanoid copepods entirely differ and their hunting methods are more species specific. Kerfoot (1978) observed *Epischura* exhibit smooth, almost continuous motion by rotating the second antennae during the power strokes, lasting about 80 ms, with only a short pause between strokes. Acceleration can give an animal a maximum speed of about 34 mm s^{-1}. When contacted, the prey is grasped by a basket-like structure consisting of the first and second maxillae, the maxillipeds and the first pair of swimming legs.

Extensive data on feeding of marine calanoid copepods have been published since 1980 (e.g., Landry, 1980, 1981; Koehl & Strickler, 1981; Cowles & Strickler, 1983; Price et al., 1983; Greene & Landry, 1985, 1988;Yen, 1985, 1988; Greene, 1986, 1988; Gifford & Dagg, 1988; Landry & Fagerness, 1988; Paffenhöffer, 1988; Price et al., 1988; Wong, 1988; Costello et al., 1990; Marrasé et al., 1990; Hwang et al., 1994; Strickler & Costello, 1996; Hwang & Strickler, 2001; Bundy & Vanderploeg, 2002; Doall et al., 2002; Jiang et al., 2002 b) and even on some freshwater calanoid copepods feeding on algal food (Vanderploeg et al., 1988; DeMott & Watson, 1991). Wong (1984) analyzed the gut contents and examined the mouthpart morphology of five freshwater calanoid copepods and found *Senecella calanoides* and *Epischura lacustris* feeding on rotifers. DeMott & Watson (1991) examined the ability of *Tropocyclops prasinus* and four freshwater calanoid copepods (*Diaptomus birgei*, *D. oregonensis*, *D. minutus* and *E. lacustris*) to detect and capture food particles and found mechano-reception as the primary mechanism for the remote detection of large particles by diaptomid copepods. Jiang et al. (2002 a) described swimming behaviour of calanoid copepods in the laboratory and analyzed its components.

Calculations of Rothschild & Osborn (1988), corrected by Evans (1989), emphasized the relative speed of swimming predator and prey and the effect of turbulence. Price (1988), in his review on feeding, emphasized on the wide variety of mechanisms in different taxa of suspension-feeders capable of responding to individual food particles. Obviously, calanoids are able to perceive particles at distance by mechano-reception and to evaluate their quality from either mechanic or chemical signals.

The chances and abilities of rotifers to escape copepod predation

Williamson (1983a) summarized morphological properties or behavioural abilities protecting individual species of rotifers, based on published works. Morphological properties include presence of a lorica, its hardness, and the presence or absence of spines, gelatinous sheath in colonial *Conochilus* or, within a species, simply the size of an individual. Whereas some populations of rotifers exhibit quite a uniform size of amictic females, others have individuals of a wide range of sizes. In *Brachionus angularis* Gosse, some summer populations consist of two otherwise similar morphs of quite different sizes (Bartos, 1959) without any intermediate individuals. Brandl & Fernando (1978) reported a more than three times higher predation rate of *Cyclops vicinus* on the smaller of the two morphs (possibly different cryptic species).

Many loricate rotifers produce progeny equipped with protective structures of their lorica when exposed to chemical signals of potential predators (Gilbert, 1966: and numerous later authors – for reviews see, e.g., Havel, 1987; Stemberger & Gilbert, 1987a or Snell, 1998). This strategy of population protection can be useful against copepod predation, too. However, the result depends on the predator, e.g., its size. Thus, rotifer species may differ in their morphological response to a water-soluble factor (presumably metabolic wastes) released by cyclopoid copepods. *Keratella slacki* examined by Gilbert & Stemberger (1984) developed longer spines induced by a *Asplanchna*-conditioned medium but not in *Tropocyclops prasinus* (Fischer) medium. In contrast, spineless *Keratella cochlearis* (forma *tecta*) cultivated in the medium conditioned by *Mesocyclops edax* (Forbes) or *T. prasinus* produced a significant proportion of spined progeny (*K.c.* forms *tuberculata*, *micracantha* and *typica*) as well as in the medium conditioned by *Asplanchna*. In predation experiments, *T. prasinus*, but not *M. edax*, selected positively for the unspined phenotypes of *K. cochlearis* and consumed almost four times more of them per day than of the spined ones. Similarly, spineless females of *Keratella testudo* (Ehrenberg) produced offspring with posterior spines induced in the medium conditioned by various planktonic animals including the copepods *T. prasinus*, *Diaptomus minutus* Lilljeborg and *Epischura lacustris* Forbes (Stemberger & Gilbert (1987 a, b). Stemberger (1988) evaluated the costs of this armament development for the population growth parameters in various life conditions of *K. testudo*.

Any protection resulting from the spine armament becomes inefficient when the predator size is relatively great compared with the size of the prey and its defensive structures. This is the case in the nonselective feeding by large *M. edax* on spined and unspined morphs of *K. cochlearis* (Stemberger & Gilbert, 1984). On the other hand, predation by large copepods on large prey species may change the predatory relations in the plankton community, e.g., in case of copepod predation on the predatory rotifer *Asplanchna*. Williamson & Gilbert (1980) documented how predation by *M. edax* on *Asplanchna girodi* de Guerne may greatly reduce the predation by *Asplanchna* on *K. cochlearis*. Similarly, Kumar & Rao (2001) found that feeding by *Mesocyclops thermocyclopoides* Harada on *Asplanchna intermedia* Hudson reduced its predation pressure on *Brachionus calyciflorus* Pallas, thus giving this species a better chance to successfully compete with *B. angularis*.

Formation of colonies is another protective, morphological response in the rotifer genus *Conochilus*. Diéguez & Balseiro (1998) analyzed the seasonal changes of colony size in *C. hippocrepis* (Schrank) and found it to be related to the size of the maxilliped of the predatory calanoid copepod *Parabroteas sarsi* (Daday) during the season in which this copepod preyed upon *Conochilus*.

Instead of structures that might complicate the successful attack and efficient seizure of prey by predatory copepods, some rotifers develop structures that greatly increase their escape abilities. Species of *Polyarthra* can escape predation by rapid movement of their antero-lateral appendages (Gilbert & Williamson, 1978). The appendages are arranged in four triplets, with one more pair of small paddles in some species, and are capable of quick movement. This allows the animal to jump far from its original position and, from the predator's point of view, in a random orientation: Gilbert (1985) used a high-speed cinematographic analysis (from 120 to 300 frames s^{-1}) of the movement of *Polyarthra* under the microscope to study the details of the paddles' function and to measure the speed of an animal during its escape reaction. While the normal swimming velocity of *P. vulgaris* was about 0.35 mm s^{-1} (equivalent to 2.6 body lengths s^{-1}), it increased about 100-fold during the escape response for about 0.05 s, displacing the animal by about 15 body lengths. Thus, such escape reactions are very efficient and explain how *Polyarthra* often survives in habitats with a high density of planktonic predators. Also, Williamson (1987) has reported similar escape response for *Polyarthra* from *Diaptomus pallidus*. Stemberger (1985) evaluated prey selection by *Diacyclops bicuspidatus thomasi* and found *P. vulgaris* to be the least vulnerable species of the genus. Kirk & Gilbert (1988) describe details of escape behaviour of *P. remata* in a flow field produced by a siphon and simulating the flow field produced either by a large cladoceran competitor or by a predator.

Two other rotifer genera, *Filinia* and *Hexarthra*, are also equipped with specialized extensions of the body wall, which are moved by muscles and provide an active defense response. When *Filinia terminalis* (Plate) approaches the feeding current of the calanoid *D. pallidus*, it stretches its movable lateral bristles into an anterior direction, roughly parallel to the caudal bristle, thus stretching itself lengthwise. This response does not move the rotifer substantially but confronts *Diaptomus* with a long object that triggers an avoidance response that leads to rejection of the *Filinia* prey (Williamson, 1987).

Even loricate species can increase their movement to escape predation. This was noted first by Williamson (1987) for *Keratella* which actively escaped from being predated by *D. pallidus*. When entrained in the feeding currents of *D. pallidus*, *K. cochlearis* rapidly increased its swimming speed briefly (<1 s) to avoid being captured by the copepod. Later, Gilbert & Kirk (1988) studied in detail the reactions of *K. cochlearis* and *K. testudo* to stimuli caused either by an approaching predator or by the feeding currents of large suspension-feeders. These two species differ slightly in their swimming speed, which is about 0.5 mm s^{-1}. Their maximum velocity during the immediate escape responses was about 1.8 mm s^{-1}, but it gradually declined. The response that lasted for about 2 s resulted in active displacement of 2 mm, which is about 12 and 18 body lengths for *K. testudo* and *K. cochlearis*, respectively. During the escape, the rotifers continued to rotate on their longitudinal axes and moved away in a constant direction.

Recently, Lapesa et al. (2002) used video recording to examine the swimming abilities and vulnerability of three *Brachionus* species (*B. rotundiformis* Tschugunoff, *B. plicatilis* Müller and *B. ibericus* Ciros-Pérez, Gómez & Serra) when exposed to the predatory attacks of *Diacyclops bicuspidatus odessanus* (Schmankewitch). The rotifers did not exhibit any noticeable escape responses but, after contact, were attacked by the copepod to varying degrees: the smallest species, *B. rotundiformis*, with the highest frequency, the largest species, *B. plicatilis*, with the lowest frequency. Again, the most vulnerable species was the competitively superior one so that differential predation by a copepod acted against the competitive abilities of these sibling congeners.

Another strategy to reduce or avoid predation is to avoid the space where the predators occur. Since some populations of predatory copepods exhibit regular diurnal vertical migrations to escape their predators (visually orienting fish or insects) the endangered rotifers may do the same but in the opposite direction. Gilbert & Hampton (2001) determined vertical distributions of various species in a shallow (max. 2 m) pond inhabited by the predatory notonectid *Buenoa macrotibialis* Hungerford. Whereas *Tropocyclops extensus* (Kiefer) (= *T. prasinus mexicanus* of other authors) showed a pronounced diel vertical migration, avoiding the surface and staying near the bottom during the day, its main prey, *P. remata*, concentrated near the surface during daylight hours and avoided the bottom. At night, both of these two species were uniformly distributed across the water column.

Predation rates and predatory impact of copepods on rotifer populations

The above discussion has shown that there are various factors determining the predation rate of copepods preying on rotifers. Many of them are species-specific for either the predator or the prey species and some are specific for a given combination of prey and predator. Others may result from the food satiation level of the predator, which can change its preference if hungry. Stemberger's study (1985) on the selective feeding of *Diacyclops thomasi* on various combinations of rotifer prey is a good example of how this factor can change: while hungry *Diacyclops* consumed proportionally larger *Synchaeta pectinata*, satiated copepods preferred a smaller prey, *Polyarthra major*.

Prey density is yet another feature that strongly influences the result of the hunting effort and thus predation rate. However, data concerning the functional response of the predators to changing densities of prey, i.e., the relation between prey density and a predator's daily ration, are not often readily available. Some published works report great variability in predation rates of a predator on given prey species (Table 1). I found the same when summarizing the values I measured for *Cyclops vicinus* in various Czech water bodies. Predation rates depend also on the sex and developmental stage of copepods: Diéguez & Gilbert (2002) found significantly higher predation rates for adult females than adult males or 5th copepodids of *Tropocyclops extensus* feeding on *Polyarthra remata*. Although rotifers are able to achieve high reproductive rates (Walz, 1993, 1995), the impact of copepod predation on rotifer population density can be high (Brandl & Fernando, 1981; Walz et al., 1987; Fussmann, 1996; Plassmann et al., 1997; Couch et al., 1999; Yoshida et al., 2000; Diéguez & Gilbert, 2002) and in some cases responsible for the decline of a rotifer population and seasonal extermination of a species from a community. Plassmann et al. (1997) studied the causes of the decline of *Synchaeta oblonga* and *S. pectinata* in Lake Constance in May and found that predation by *C. vicinus* controlled the spring abundance of *Synchaeta* spp. *Synchaeta* always declined about one month earlier than other dominant species. Between 30 and 90% of the *Synchaeta* population was cropped daily by adult *C. vicinus*. However, predation rates of *C. vicinus* on *Polyarthra vulgaris* and *P. dolichoptera* were much lower (6–36%) than on *Synchaeta*. Brandl & Fernando (1981) reported daily consumption rates between 13 and 24% of rotifer individuals present in three Canadian lakes by *Mesocyclops edax*. For a Czech reservoir, the assemblage of three copepod species, *C. vicinus*, *Mesocyclops leuckarti* and *Acanthocyclops robustus* consumed daily about 9% of the rotifer individuals on a yearly average, with a somewhat higher mean value of 12.5% for the May–August period. Couch et al. (1999) estimated that at its maximum density *Boeckella hamata* would daily clear one-third the existing population of *Anuraeopsis fissa* from the reservoir. This implies that the normal *Boeckella* densities can still significantly reduce the ambient prey populations. Diéguez & Gilbert (2002) demonstrated how *T. extensus* could deplete a natural population of the susceptible prey species *P. remata* and thus shift the species structure of the rotifer assemblage in favour of predation-resistant species. Thus, the coexistence of rotifer and copepod populations in plankton communities is the result of their interspecific relationships in which cyclopoid and calanoid copepods act as important predators.

In stable but seasonally changing ecosystems of temperate lakes and ponds, these relationships manifest in a form of interconnected seasonal events, which are determined by life histories of participating species. Contrasting life strategies of some rotifers together with their different vulnerability to predation set limits to their occurrence. *Keratella cochlearis* and *S. pectinata* represent typical cases of the K_S and the r_{max} strategies (Walz, 1993) (*Keratella*: $r = 0.17$ day^{-1} – Fussmann, 1996; *Synchaeta*: $r = 0.80$ day^{-1} – Stemberger & Gilbert, 1985; $r = 0.61$ day^{-1} – Fussmann, 1996). Moreover, *Keratella* has much lower food requirements than *Synchaeta* (Stemberger & Gilbert, 1985) and possesses an efficient armament against predation in its lorica and spines. In North American (Stemberger & Evans, 1984) and European lakes (Plassmann et al., 1997) and ponds (Brandl & Prazakova, 2002), spring blooms of small algae are utilized by soft-bodied fast reproducing species (e.g., *Synchaeta* spp.). Their dense populations form food resource for cyclopoid copepods (*Diacyclops* in N. America, *Cyclops* in Europe). The bulk of spring *Cyclops* populations emerges from the over wintering diapausing 4th and 5th copepodid stages, which quickly mature and reproduce. The new generation of naupliar larvae finds a still high level of algal food. Later, soft-bodied rotifers are replaced by loricate species (e.g., *Keratella*) with lesser food requirements, able to sustain predation by *Cyclops* and competition by cladocerans. Most of the *Cyclops* copepodid stages enter diapause later during summer and disappear from plankton (Brandl, 1994). In Central European carp ponds, with a two-year cycle of carp cultivation, such an event takes place only in alternating years if carp overwinters in the pond. In the years without fish in the ponds during early spring, large cladocerans of the genus *Daphnia* dominate and consume most algae and outcompete smaller herbivores.

In less stable fishless temporary waters, copepod predation may be an important force structuring the communities. Green et al. (1999) found species diversity significantly reduced and community size structure shifted to larger-bodied zooplankton in Australian ephemeral pools inhabited by the predatory copepod *Boeckella major*, which fed on as many as 13 rotifer species.

Conclusions

(1) Cyclopoid and calanoid copepods are efficient predators of rotifers. They exhibit widely different feeding techniques to which many rotifer species respond using a variety of defense mechanisms or escape reactions.
(2) Predation by copepods may have a significant impact on rotifers causing often a seasonal decline of the rotifer populations and replacement of species that are more susceptible to predation by the less susceptible ones.
(3) Both the predation rates and the predatory impact of this predation may be species-specific. Some soft-bodied rotifer species are especially vulnerable but their size and different escape abilities can considerably influence the resulting predation rates.
(4) Behavioural and morphological adaptations, together, may serve as important antipredator defenses for some rotifer species against copepod predation.

Acknowledgements

This work was supported by the grants of the Ministry of Education, Youth and Sports of the Czech Republic No. MSMT-CEZ 123100004 and of the Grant Agency of the Academy of Sciences, Czech Republic, No. K 6005114. Dr K. Edwards greatly improved the English of the paper. The comments and recommendations made by two anonymous referees were extremely useful to the author and greatly helped to improve the comprehensibility and the language of this review.

References

Adrian, R., 1991. Filtering and feeding rates of cyclopoid copepods feeding on phytoplankton. Hydrobiologia 210: 217–223.

Adrian, R. & T. M. Frost, 1992. Comparative feeding ecology of *Tropocyclops prasinus mexicanus* (Copepoda, Cyclopoida). Journal of Plankton Research 14: 1369–1382.

Adrian, R. & T. M. Frost, 1993. Omnivory in cyclopoid copepods: comparison of algae and invertebrates as food for three, differently sized species. Journal of Plankton Research 15: 643–658.

Anderson, R. S., 1967. Diaptomid copepods from two mountain ponds in Alberta. Canadian Journal of Zoology 45: 1043–1047.

Anderson, R. S., 1970. Predator-prey relationships and predation rates for crustacean zooplankters from some lakes in western Canada. Canadian Journal of Zoology 48: 1229–1240.

Bartos, E., 1959. Virnici – Rotatoria. Fauna CSR, Vol. 15. Czechoslovak Academy of Sciences, Praha: 969.

Berger, I. & G. Maier, 2001. The mating and reproductive biology of the freshwater planktonic calanoid copepod *Eudiaptomus gracilis*. Freshwater Biology 46: 787–794.

Brandl, Z., 1994. The seasonal dynamics of zooplankton biomass in two Czech reservoirs: a long-term study. Archiv für Hydrobiologie, Beihefte Ergebnisse der Limnologie 40: 122–135.

Brandl, Z., 1998a. Feeding strategies of planktonic cyclopoids in lacustrine ecosystems. Journal of Marine Systems 15: 87–95.

Brandl, Z., 1998b. Life strategy and feeding relations of *Cyclops vicinus* in two reservoirs. International Revue of Hydrobiology 83: 381–388.

Brandl, Z. & C. H. Fernando, 1975. Food consumption and utilization in two freshwater cyclopoid copepods. Internationale Revue der Gesamten Hydrobiologie 60: 471–494.

Brandl, Z. & C. H. Fernando, 1978. Prey selection by the cyclopoid copepods *Mesocyclops edax* and *Cyclops vicinus*. Internationale Vereinigung für Theoretische und Angewandte Limnologie, Verhandlungen 20: 2505–2510.

Brandl, Z. & C. H. Fernando, 1981. The impact of predation by cyclopoid copepods on zooplankton. Internationale Vereinigung für Theoretische und Angewandte Limnologie, Verhandlungen 21: 1573–1577.

Brandl, Z. & C. H. Fernando, 1986. Feeding and food consumption by *Mesocyclops edax*. In: Schriever G., H. K.Schminke & C.-T.Shih (eds), Proceedings of the Second International Conference on Copepoda, Ottawa, Canada, August 13–17, (1984). Syllogeus, N. Museum of Canada, Ottawa, 58: 254–258.

Brandl, Z. & M. Prazakova, 2002. Impact of predation by cyclopoid copepods (Copepoda: Cyclopoida) on zooplankton in a carp pond in Czech Republic. Acta Societatis Zoologicae Bohemicae 66: 169–175.

Bundy, M. H. & H. A. Vanderploeg, 2002. Detection and capture of inert particles by calanoid copepods: the role of the feeding current. Journal of Plankton Research 24: 215–223.

Burns, C. W. & J. J. Gilbert, 1993. Predation on ciliates by freshwater calanoid copepods: rates of predation and relative vulnerabilities of prey. Freshwater Biology 30: 377–393.

Conde-Porcuna, J. M. & S. Declerck, 1998. Regulation of rotifer species by invertebrate predators in a hypertrophic lake: selective predation on egg-bearing females and induction of morphological defences. Journal of Plankton Research 20: 605–618.

Costello, J. H., J. R. Strickler, C. Marrasé, G. Trager, R. Zeller & A. J. Freise, 1990. Grazing in a turbulent environment: behavioral response of a calanoid copepod, *Centropages hamatus*. Proceedings of the National Academy of Sciences USA 87: 1648–1652.

Couch, K. M., C. W. Burns & J. J. Gilbert, 1999. Contribution of rotifers to the diet and fitness of *Boeckella* (Copepoda: Calanoida) Freshwater Biology 107–118.

Cowles, T. J. & J. R. Strickler, 1983. Characterization of feeding activity patterns in the planktonic copepod *Centropages typicus* Kroyer under various food conditions. Limnology and Oceanography 28: 106–115.

DeMott, W. R. & M. D. Watson, 1991. Remote detection of algae by copepods: responses to algal size, odors and motility. Journal of Plankton Research 13: 1203–1222.

Diéguez, M. & E. Balseiro, 1998. Colony size in *Conochilus hippocrepis*: defensive adaptation to predator size. Hydrobiologia 387/388: 421–425.

Diéguez, M. C. & J. J. Gilbert, 2002. Suppression of the rotifer *Polyarthra remata* by the omnivorous copepod *Tropocyclops extensus*: predation or competition. Journal of Plankton Research 24: 359–369.

Doall, M. H., J. R. Strickler, D. M. Fields & J. Yen, 2002. Mapping the free-swimming attack volume of a planktonic copepod, *Euchaeta rimana*. Marine Biology 140: 871–879.

Elster, H. J., 1936. Einige biologische Beobachtungen an *Heterocope borealis*. Internationale Revue der Gesamten Hydrobiologie und Hydrographie 33: 357–433.

Evans, G. T., 1989. The encounter speed of moving predator and prey. Journal of Plankton Research 11: 415–417.

Folt, C. L. & E. R. Byron, 1989. A comparison of the effects of prey and non-prey neighbours on foraging rates of *Epischura nevadensis* (Copepoda: Calanoida). Freshwater Biology 21: 283–293.

Fryer, G., 1954. Contributions to our knowledge of the biology and systematics of the freshwater Copepoda. Schweizerische Zeitschrift für Hydrologie 16: 64–77.

Fryer, G., 1957a. The feeding mechanism of some freshwater cyclopoid copepods. Proceedings of the Zoological Society of London 129: 1–25.

Fryer, G., 1957b. The food of some freshwater cyclopoid copepods and its ecological significance. Journal of Animal Ecology 26: 263–286.

Fussmann, G., 1996. The importance of crustacean zooplankton in structuring rotifer and phytoplankton communities: an enclosure study. Journal of Plankton Research 18: 1897–1915.

Gifford, D. J. & M. J. Dagg, 1988. Feeding of the estuarine copepod *Acartia tonsa* Dana: Carnivory vs. herbivory in natural microplankton assemblages. Bulletin of Marine Science 43: 458–468.

Gilbert, J. J., 1966. Rotifer ecology and embryological induction. Science 151: 1234–1237.

Gilbert, J. J., 1985. Escape response of the rotifer *Polyarthra*: a high-speed cinematographic analysis. Oecologia 66: 322–331.

Gilbert, J. J. & S. E. Hampton, 2001. Diel vertical migrations of zooplankton in a shallow, fishless pond: a possible avoidance-response cascade induced by notonectids. Freshwater Biology 46: 611–621.

Gilbert, J. J. & K. L. Kirk, 1988. Escape response of the rotifer *Keratella*: description, stimulation, fluid dynamics, and ecological significance. Limnology and Oceanography 33: 1440–1450.

Gilbert, J. J. & R. S. Stemberger, 1984. *Asplanchna*-induced polymorphism in the rotifer *Keratella slacki*. Limnology and Oceanography 29: 1309–1316.

Gilbert, J. J. & C. E. Williamson, 1978. Predator-prey behavior and its effect on rotifer survival in associations of *Mesocyclops edax*, *Asplanchna girodi*, *Polyarthra vulgaris*, and *Keratella cochlearis*. Oecologia 37: 13–22.

Green, J. D. & R. J. Shiel, 1999. Mouthpart morphology of three calanoid copepods from Australian temporary pools: evidence for carnivory. New Zealand Journal of Marine and Freshwater Research 33: 385–398.

Green, J. D., R. J. Shiel & R. A. Littler, 1999. *Boeckella major* (Copepoda: Calanoida): a predator in Australian ephemeral pools. Archiv für Hydrobiologie 145: 181–196.

Greene, C. H., 1983. Selective predation in freshwater zooplankton communities. Internationale Revue der Gesamten Hydrobiologie 68: 297–315.

Greene, C. H., 1985. Planktivore functional groups and patterns of prey selection in pelagic communities. Journal of Plankton Research 7: 35–40.

Greene, C. H., 1986. Patterns of prey selection: implications of predator foraging tactics. American Naturalist 128: 824–839.

Greene, C. H., 1988. Foraging tactics and prey-selection patterns of omnivorous and carnivorous calanoid copepods. Hydrobiologia 167/168: 295–302.

Greene, C. H. & M. R. Landry, 1985. Patterns of prey selection in the cruising calanoid predator *Euchaeta elongata*. Ecology 66: 1408–1416.

Greene, C. H. & M. R. Landry, 1988. Carnivorous suspension feeding by the subarctic calanoid copepod *Neocalanus cristatus*. Canadian Journal of Fisheries and Aquatic Sciences 45: 1069–1074.

Hairston, N. G. Jr. & A. J. Bohonak, 1998. Copepod reproductive strategies: life-history theory, phylogenetic pattern and invasion of inland waters. Journal of Marine Systems 15: 23–34.

Hansen, A. -M. & E. Jeppesen, 1992. Life cycle of *Cyclops vicinus* in relation to food availability, predation, diapause and temperature. Journal of Plankton Research 14: 591–605.

Hansen, A. -M. & B. Santer, 1995. The influence of food resources on the development, survival and reproduction of the two cyclopoid copepods: *Cyclops vicinus* and *Mesocyclops leuckarti*. Journal of Plankton Research 17: 631–646.

Havel, J. E., 1987. Predator-induced defenses: a review. In Kerfoot, W. C. & A. Sih (eds), Predation. Direct and Indirect Impacts on Aquatic Communities. University Press of New England, Hanover and London: 263–278.

Hopp, U., G. Maier & R. Bleher, 1997. Reproduction and adult longevity of five species of planktonic cyclopoid copepods reared on different diets: a comparative study. Freshwater Biology 38: 289–300.

Huys, R. & G. A. Boxshall, 1991. Copepod Evolution. The Ray Society, London, 468 pp.

Hwang, J. S., J. H. Costello & J. R. Strickler, 1994. Copepod grazing in turbulent flow: elevated foraging behavior and habituation of escape responses. Journal of Plankton Research 16: 421–431.

Hwang, J. S. & J. R. Strickler, 2001. Can copepods differentiate prey from predator hydromechanically?. Zoological Studies 40: 1–6.

Jiang, H., T. R. Osborn & C. Meneveau, 2002a. The flow field around a freely swimming copepod in steady motion. Part I: theoretical analysis.: Journal of Plankton Research 24: 167–189.

Jiang, H., T. R. Osborn & C. Meneveau, 2002b. Hydrodynamic interaction between two copepods: a numerical study. Journal of Plankton Research 24: 235–253.

Kerfoot, W. C., 1978. Combat between predatory copepods and their prey: *Cyclops*, *Epischura*, and *Bosmina*. Limnology and Oceanography 23: 1089–1102.

Kerfoot, W. C., 1987. Translocation experiments: *Bosmina* responses to copepod predation. Ecology 68: 596–610.

Kerfoot, W. C., 1988. Defensive spines: inverse relationship between coefficients of variation and size. Limnology and Oceanography 33: 1412–1429.

Kirk, K. L. & J. J. Gilbert, 1988. Escape behavior of *Polyarthra* in response to artificial flow stimuli. Bulletin of Marine Science 43: 551–560.

Koehl, M. A. R. & J. R. Strickler, 1981. Copepod feeding currents: food capture at low Reynolds number. Limnology and Oceanography 26: 1062–1073.

Kumar, R. & T. R. Rao, 1999a. Effect of algal food on animal prey consumption rates in the omnivorous copepod, *Mesocyclops thermocyclopoides*. International Revue of Hydrobiology 84: 419–426.

Kumar, R. & T. R. Rao, 1999b. Demographic responses of adult *Mesocyclops thermocyclopoides* (Copepoda, Cyclopoida) to different plant and animal diets. Freshwater Biology, 42: 487–501.

Kumar, R. & T. R. Rao, 2001. Effect of the cyclopoid copepod *Mesocyclops thermocyclopoides* on the interactions between the predatory rotifer *Asplanchna intermedia* and its prey *Brachionus calyciflorus* and *B angularis*. . Hydrobiologia 453/454: 261–268.

Lair, N. & M. Hilal, 1992. *Acanthodiaptomus denticornis*, another omnivorous copepod: description of its mouth appendages and feeding experiments on animal prey. Hydrobiologia 248: 137–142.

Landry, M. R., 1980. Detection of prey by *Calanus pacificus*: implications of the first antennae. Limnology and Oceanography 25: 545–549.

Landry, M. R., 1981. Switching between herbivory and carnivory by the planktonic marine copepod *Calanus pacificus*. Marine Biology 65: 77–82.

Landry, M. R. & V. L. Fagerness, 1988. Behavioral and morphological influences on predatory interactions among marine copepods. Bulletin of Marine Science 43: 509–529.

Lapesa, S., T. W. Snell, D. M. Fields & M. Serra, 2002. Predatory interactions between a cyclopoid copepod and three sibling rotifer species. Freshwater Biology 47: 1685–1695.

LeBlanc, J. S., W. D. Taylor & O. E. Johannsson, 1997. The feeding ecology of the cyclopoid copepod *Diacyclops thomasi* in Lake Ontario. Journal of Great Lakes Research 23: 369–381.

Lowndes, A. G., 1935. The swimming and feeding of certain calanoid copepods. Proceedings of the Zoological Society of London : 687–715.

Marrasé, C., J. H. Costello, T. Granata & J. R. Strickler, 1990. Grazing in a turbulent environment: energy dissipation, encounter rates, and efficacy of feeding currents in *Centropages hamatus*. Proceedings of the National Academy of Sciences USA 87: 1653–1657.

Matsumura-Tundisi, T., A. C. Rietzler, E. L. G. Espindola, J. G. Tundisi & O. Rocha, 1990. Predation on *Ceriodaphnia cornuta* and *Brachionus calyciflorus* by two *Mesocyclops* species coexisting in Barra Bonita reservoir (SP, Brazil). Hydrobiologia 198: 141–151.

Nero, R. W. & W. G. Sprules, 1986. Predation by three glacial opportunists on natural zooplankton communities. Canadian Journal of Zoology 64: 57–64.

O'Brien, W. J., 2001. Long-term impact of an invertebrate predator, *Heterocope septentrionalis*, on an arctic pond zooplankton community. Freshwater Biology 46: 39–45.

Paffenhöfer, G.-A., 1988. Feeding rates and behavior of zooplankton. Bulletin of Marine Science 43: 430–445.

Paul, A. J. & D. W. Schindler, 1994. Regulation of rotifers by predatory calanoid copepods (subgenus *Hesperodiaptomus*) in lakes of the Canadian Rocky Mountains. Canadian Journal of Fisheries and Aquatic Sciences 51: 2520–2528.

Plassmann, T., G. Maier & H. B. Stich, 1997. Predation impact of *Cyclops vicinus* on the rotifer community in Lake Constance in spring. Journal of Plankton Research 19: 1069–1079.

Price, H. J., 1988. Feeding mechanisms in marine and freshwater zooplankton. Bulletin of Marine Science 43: 327–343.

Price, H. J., G.-A. Paffenhöfer & J. R. Strickler, 1983. Modes of cell capture in calanoid copepods. Limnology and Oceanography 28: 116–123.

Price, H. J., G.-A. Paffenhöfer, C. M. Boyd, T. J. Cowles, P. L. Donaghay, W. M. Hammer, W. Lampert, L. B. Quetin, R. M. Ross, J. R. Strickler & M. J. Youngbluth, 1988. Future studies of zooplankton behavior: questions and technological developments. Bulletin of Marine Science 43: 853–872.

Roche, K. F., 1987. Post-encounter vulnerability of some rotifer prey types to predation by the copepod *Acanthocyclops robustus*. Hydrobiologia 147: 229–233.

Roche, K., 1990a. Prey features affecting ingestion rates by *Acanthocyclops robustus* (Copepoda: Cyclopoida) on zooplankton. Oecologia 83: 76–82.

Roche, K., 1990b. Some aspects of vulnerability to cyclopoid predation of zooplankton prey individuals. Hydrobiologia 198: 153–162.

Roche, K., 1990c. Spatial overlap of a predatory copepod, *Acanthocyclops robustus*, and its prey in a shallow eutrophic lake. Hydrobiologia 198: 163–183.

Rothschild, B. J. & T. R. Osborn, 1988. Small-scale turbulence and plankton contact rates. Journal of Plankton Research 10: 465–474.

Santer, B., 1993. Potential importance of algae in the diet of adult *Cyclops vicinus*. Freshwater Biology 30: 269–278.

Santer, B. & F. Bosch, 1994. Herbivorous nutrition of *Cyclops vicinus*: the effect of a pure algal diet on feeding, development, reproduction and life cycle. Journal of Plankton Research 16: 171–195.

Schoeneck, L. J., C. E. Williamson & M. E. Stoeckel, 1990. Diel periodicity and selectivity in the feeding rate of the predatory copepod *Mesocyclops edax* . Journal of Plankton Research 12: 29–40.

Schulze, P. C. & C. L. Folt, 1989. Effects of conspecifics and phytoplankton on predation rates of the omnivorous copepods *Epischura lacustris* and *Epischura nordenskioldi*. Limnology and Oceanography 34: 444–450.

Schulze, P. C. & C. L. Folt, 1990. Food resources, survivorship, and reproduction of the omnivorous calanoid copepod *Epischura lacustris*. Ecology 71: 2224–2240.

Snell, T. W., 1998. Chemical ecology of rotifers. Hydrobiologia 387/388: 267–276.

Stemberger, R. S., 1985. Prey selection by the copepod *Diacyclops thomasi*. Oecologia 65: 492–497.

Stemberger, R. S., 1986. The effects of food deprivation, prey density and volume on clearance rates and ingestion rates of *Diacyclops thomasi*. Journal of Plankton Research 8: 243–251.

Stemberger, R. S., 1988. Reproductive costs and hydrodynamic benefits of chemically induced defenses in *Keratella testudo*. Limnology and Oceanography 33: 593–606.

Stemberger, R. S. & M. S. Evans, 1984. Rotifer seasonal succession and copepod predation in Lake Michigan. Journal of Great Lakes Research 10: 417–428.

Stemberger, R. S. & J. J. Gilbert, 1984. Spine development in the rotifer *Keratella cochlearis*: induction by cyclopoid copepods and *Asplanchna*. Freshwater Biology 14: 639–647.

Stemberger, R. S. & J. J. Gilbert, 1985. Body size, food concentration, and population growth in planktonic rotifers. Ecology 66: 1151–1159.

Stemberger, R. S. & J. J. Gilbert, 1987a. Defenses of planktonic rotifers against predators. In: Kerfoot, W. C. & A. Sih (eds), Predation. Direct and Indirect Impacts on Aquatic Communities. University Press of New England, Hanover and London, 227–239.

Stemberger, R. S. & J. J. Gilbert, 1987b. Multiple-species induction of morphological defenses in the rotifer *Keratella testudo*. Ecology 68: 370–378.

Strickler, J. R. & J. H. Costello, 1996. Calanoid copepod behavior in turbulent flows. Marine Ecology Progress Series 139: 307–309.

Vanderploeg, H. A., G.-A. Paffenhöfer & J. R. Liebig, 1988. *Diaptomus* vs. net phytoplankton: effects of algal size and morphology on selectivity of a behaviorally flexible, omnivorous copepod. Bulletin of Marine Science 43: 377–394.

Walz, N., 1993. Life history strategies of rotifers. In N. Walz (ed.), Plankton Regulation Dynamics. Springer-Verlag, Berlin and Heidelberg, Germany, 193–214.

Walz, N., 1995. Rotifer populations in plankton communities: energetics and life history strategies. Experientia 51: 437–453.

Walz, N., H. J. Elster & M. Mezger, 1987. The development of the rotifer community structure in Lake Constance during its eutrophication. Archiv für Hydrobiologie/Supplement (Monographische Beiträge) 74: 452–487.

Warren, G. J., 1985. Predaceous feeding habits of *Limnocalanus, macrurus*. Journal of Plankton Research 7: 537–552.

Williamson, C. E., 1980. The predatory behavior of *Mesocyclops edax*: predator preferences, prey defenses, and starvation-induced changes. Limnology and Oceanography 25: 903–909.

Williamson, C. E., 1983a. Invertebrate predation on planktonic rotifers. Hydrobiologia 104: 385–396.

Williamson, C. E., 1983b. Behavioral interactions between a cyclopoid copepod predator and its prey. Journal of Plankton Research 5: 701–711.

Williamson, C. E., 1984. Laboratory and field experiments on the feeding ecology of the cyclopoid copepod *Mesocyclops edax*. Freshwater Biology 14: 575–585.

Williamson, C. E., 1986. The swimming and feeding behavior of *Mesocyclops*. Hydrobiologia 134: 11–19.

Williamson, C. E., 1987. Predator-prey interactions between omnivorous diaptomid copepods and rotifers: the role of prey morphology and behavior. Limnology and Oceanography 32: 17–177.

Williamson, C. E. & N. M. Butler, 1986. Predation on rotifers by the suspension-feeding calanoid copepod *Diaptomus pallidus*. Limnology and Oceanography 31: 393–402.

Williamson, C. E. & J. J. Gilbert, 1980. Variation among zooplankton predators: the potential of *Asplanchna*, *Mesocyclops*, and *Cyclops* to attack, capture, and eat various rotifer prey. In W. C. Kerfoot (ed.), Evolution and Ecology of Zooplankton Communities. University Press of New England, Hanover, NH, USA, 509–517.

Williamson, C. E. & R. E. Magnien, 1982. Diel vertical migration in *Mesocyclops edax*: implications for predation rate estimates. Journal of Plankton Research 4: 329–339.

Williamson, C. E. & H. A. Vanderploeg, 1988. Predatory suspension-feeding in *Diaptomus*: prey defenses and the avoidance of cannibalism. Bulletin of Marine Science 43: 561–572.

Wong, C. K., 1984. A study of the relationships between the mouthparts and food habits in several species of freshwater calanoid copepods. Canadian Journal of Zoology 62: 1588–1595.

Wong, C. K., 1988. Effects of competitors, predators, and prey on the grazing behavior of herbivorous calanoid copepods. Bulletin of Marine Science 43: 573–582.

Wong, C. K. & W. G. Sprules, 1985. Size-selective feeding by the predatory copepod *Epischura lacustris* Forbes. Canadian Journal of Fisheries and Aquatic Sciences 42: 189–193.

Yang, Y. -F. & Z. Brandl, 1996. Feeding of *Acanthocyclops robustus* on zooplankton. Chinese Journal of Oceanology and Limnology 14: 17–26.

Yen, J., 1985. Selective predation by the carnivorous marine copepod *Euchaeta elongata*: laboratory measurements of predation rates verified by field observations of temporal and spatial feeding patterns. Limnology and Oceanography 30: 577–597.

Yen, J., 1988. Directionality and swimming speeds in predator-prey and male-female interactions of *Euchaeta rimana*, a subtropical marine copepod. Bulletin of Marine Science 43: 395–403.

Yoshida, T., S. Ban, T. Takenouchi, T. Aono, Y. Ishikawa, H. Mikami, K. Takano, K. Imada, R. Yasutomi & K. Takeuchi, 2000. Top-down control of population dynamics of the dominant rotifers in two mesotrophic lakes in Hokkaido, Japan. Archiv für Hydrobiologie 148: 481–498.

Zánkai, N. P., 1984. Predation of *Cyclops vicinus* (Copepoda: Cyclopoida) on small zooplankton animals in Lake Balaton. Archiv für Hydrobiologie 99: 360–378.

Hydrobiologia (2005) 546:491–501
A. Herzig, R.D. Gulati, C.D. Jersabek & L. May (eds.) Rotifera X: Rotifer Research: Trends, New Tools and Recent Advances
DOI 10.1007/s10750-005-4291-2

Life history characteristics of *Asplanchnopus multiceps* (Rotifera) fed rotifer and cladoceran prey

S. Nandini[1,]* & S.S.S. Sarma[2]
[1]*Division of Research and Postgraduate Studies, National Autonomous University of Mexico, Campus Iztacala, Av. de Los Barrios No. 1, Los Reyes, AP 314, 54090 CP, Tlalnepantla, State of Mexico, Mexico; E-mail: sarma@servidor.unam.mx*
[2]*Laboratory of Aquatic Zoology, Division of Research and Postgraduate Studies, National Autonomous University of Mexico, Campus Iztacala, Av. de Los Barrios No. 1, Los Reyes, AP 314, 54090 CP, Tlalnepantla, State of Mexico, Mexico*
*(*Author for correspondence: Fax: +52-555-623-1256; E-mail: nandini@servidor.unam.mx)*

Key words: *Asplanchnopus multiceps*, population dynamics, life table, cladocera, rotifera

Abstract

Members of rotifer family Asplanchnidae are important invertebrate predators in freshwater communities. Although a considerable amount of information exists on species of *Asplanchna*, relatively less is known about *Asplanchnopus*. We isolated *Asplanchnopus multiceps* from the littoral of a small river in the State of Hidalgo in Central Mexico and separated a clone in our cultures. The gut content analysis of some animals collected from the field revealed the presence of cladocerans and rotifers, and therefore we cultured *A. multiceps* on a food mixture comprising littoral rotifers and cladocerans. We conducted population growth experiments of *A. multiceps* using six prey types (cladocerans: *Macrothrix triserialis, Alona rectangula* and *Pleuroxus aduncus;* rotifers, *Brachionus patulus, B. macracanthus* and *B. urceolaris*). The prey species (*A. rectangula* and *B. patulus*) on which the highest growth rates were observed were used to test the life-table demographic patterns in *A. multiceps*. All experiments were conducted in 50 ml containers with 25 ml of the medium and at three food levels (0.5, 1.0 and 2.0 ind. ml^{-1} for the cladocerans, and 2.0, 4.0 and 8.0 ind. ml^{-1} for the rotifers) with four replicates at each treatment. The spines of *M. triserialis* and *B. macracanthus* were apparently effective deterrents against *Asplanchnopus* predation since both these diets resulted in low, and sometimes negative, growth rates of the predator. The average lifespan and net reproductive rate of *A. multiceps* ranged from 3.8 to 8.4 days and 2.6 to 12.2 ind. female^{-1}, respectively, on *A. rectangula*; and from 5.0 to 9.4 days and 1.6–18.4 ind. female^{-1}, respectively, on *B. patulus*. The rate of population increase of *A. multiceps* ranged from 0.1 to 0.8 d^{-1}, depending on the prey type and density. The role of *A. multiceps* in structuring littoral rotifer and cladoceran communities is discussed.

Introduction

Among predatory rotifers, members of family Asplanchnidae are important, being commonly found in many freshwater bodies. They feed on a wide spectrum of prey ranging from protozoans to micro-crustaceans (Stemberger & Gilbert, 1987). The genus *Asplanchna* has been well studied, particularly with respect to population growth and life history characteristics (Gilbert, 1976; Iyer & Rao, 1996), prey selection (Sarma, 1993; Hampton & Starkweather, 1998), functional response (Nandini & Sarma, 1999) and its impact on prey morphology (Gilbert, 1999). However, other genera such as *Asplanchnopus* are predominantly littoral dwelling in ponds and lakes (Koste, 1978). Information on *Asplanchnopus* is scarce, as most studies concentrate on the planktonic taxa and because this genus relatively rare compared with *Asplanchna*.

Asplanchnopus feeds on smaller cladocerans, rotifers and ciliates (Koste, 1978). Under natural conditions competition between *Asplanchnopus* and *Asplanchna* is limited since the latter is typically planktonic (Gilbert, 1980). The trophic state of the water body determines the distribution of these genera, the former is usually encountered in eutrophic and the latter in mesotrophic water bodies. These two genera, due to the differences in their habit, influence the diversity and abundance of different prey types. *Asplanchna* has been well studied using rotifer (e.g., *Brachionus*, *Anuraeopsis*, *Keratella*) and protozoan (*Paramecium*, *Chlamydomonas*) prey (Gilbert, 1980; Stemberger & Gilbert, 1987; Dumont et al., 1995). Although we observed (authors' unpublished data) that *Asplanchna* could grow on an exclusive diet of micro-crustaceans under laboratory conditions, this is not often their natural diet (Salt, 1977; Kappes et al., 2000). *Asplanchnopus*, on the other hand, inhabits littoral areas, where some chydorid cladocerans are common and abundant. These chydorids are often found in the stomach content analysis of *Asplanchnopus* from field samples (Salt et al., 1978). However, the effects of rotifers or micro-crustaceans as prey on the life history variables of *Asplanchnopus* may differ.

Predators tend to stabilize their prey populations through different mechanisms. In general, the functional and aggregative responses are considered as having a stronger stabilizing impact than the numerical and developmental responses primarily due to long time lags between predator and prey populations (Murdoch & Bence, 1987). Predatory rotifers, having life cycles similar to or shorter than their prey, would have a stronger impact on their prey populations. This signifies the importance of studying the life history variables of various predatory rotifers in order to extrapolate their relative importance in field situations. In general, predators, which are capable of reproducing at an early age, may cause more rapid depletions of the prey than those that reproduce later (Murdoch & Bence, 1987). In order to quantify the age-dependent responses of a predator to a given prey population, life table demographic studies are useful. Both population growth and life table studies are complementary and necessary to understand the life history variables of a given species (Krebs, 1985).

Considering that *Asplanchnopus* commonly occurs in the littoral region of mesotrophic lakes and rivers in Mexico and the paucity of information on this genus, we studied the life history variables of *Asplanchnopus multiceps* on different littoral rotifers and cladocerans using life table demography and population growth experiments.

Materials and methods

The predator (*Asplanchnopus multiceps*) was isolated from Rio Venados in the State of Hidalgo, Mexico and maintained in the laboratory for two months prior to use in the experiments. We isolated *Brachionus patulus* from Lake Santa Elena (Mexico State), *B. macracanthus* from a pond in Morelia (Michoacan State), *B. urceolaris* from a pond in Cozumel (Yucatan State), *Macrothrix triserialis* from a pond in Veracruz (Veracruz State), *Alona rectangula* from Lake Chapultepec in Mexico City and *Pleuroxus aduncus* from Lake Xochimilco in Mexico City. All the prey species were maintained in cultures for more than two years under laboratory conditions using moderately hardwater (EPA medium, prepared daily by dissolving 96 mg $NaHCO_3$, 60 mg $CaSO_4$, 60 mg $MgSO_4$ and 4 mg KCl in one litre of distilled water) (Anonymous, 1985). Prey rotifers and cladocerans were cultured separately in 1-l beakers and were fed daily the green alga *Chlorella vulgaris* at a density of 0.5×10^6 cells ml^{-1}. *C. vulgaris* was mass cultured in 2 l transparent bottles using Bold's basal medium. Log phase alga was harvested, centrifuged at 4000 rpm for 5 min, and rinsed and resuspended in distilled water. The density of the alga was estimated using a haemocytometer.

A. multiceps was cultured in 2 l glass containers and fed daily with a mixture of prey zooplankton mentioned earlier. The general conditions that favoured healthy cultures were: pH 7.0–7.5, temperature 22–25 °C, continuous but diffused fluorescent illumination, and change of the culture medium every alternate day with a prey density of about 20 ind. ml^{-1} (rotifers) or 2 ind. ml^{-1} (cladocerans). Both population growth and life table demography experiments were conducted separately but simultaneously using 50 ml transparent jars containing 25 ml medium and one of the specified prey types at a chosen density.

For measuring the population growth we used mixed population of *A. multiceps*. We introduced five 2–4-day-old individuals of *A. multiceps* into each of the 72 test jars (6 prey species × 4 replicates × 3 prey concentrations) containing one of the prey species at a specified density (2, 4 and 8 ind. ml^{-1} for rotifers and 0.5, 1 and 2 ind. ml^{-1} for cladocerans). These prey densities were based on those used in previous studies on other members of Asplanchnidae (e.g., Sarma et al., 1996).

During the experiments we daily quantified the number of living individuals of *A. multiceps* which were then transferred to test jars containing fresh prey of the appropriate type and at specified density. Dead predators and uneaten prey were discarded. The experiment was terminated after 12–15 days by which time *Asplanchnopus* populations in most replicates began to decline. We derived the rate of population increase during the exponential phase of population growth, following the formula: $r = (\ln N_t - \ln N_0)/t$, where, N_0 and N_t are the initial and final population densities respectively, and t is time in days (Krebs, 1985). Data on the population maxima and rates of population growth were analysed using analysis of variance (ANOVA) (Sokal & Rohlf, 2000).

Based on the population growth studies, the life table experiments were performed using two zooplankton species one each from Rotifera and Cladocera (*B. patulus* and *A. rectangula*), which supported the predator's growth. We used the same prey densities as employed in the population growth studies. For life table demography we used neonate (<3 h following hatching) *A. multiceps*. This was easily achieved since *A. multiceps* laid eggs on the walls of the beakers; therefore we transferred the stock culture into a fresh container, filled up the original beaker with EPA medium and then isolated the newly hatched individuals after 2–3 hours. Into each of the 24 test jars (2 prey species × 3 densities × 4 replicates) containing one of the two specified prey at chosen density, we introduced 5 neonates of *A. multiceps*. On initiation of life table experiments, we counted from each replicate, every 12 h the number, the individuals of live *A. multiceps* of the original cohort. Dead individuals of the original cohort and the offspring produced, if any, were counted and discarded. The surviving adults were transferred to fresh jars containing appropriate prey type at desired density. In both, the population dynamics and life table demography study the eggs laid by *A. multiceps* on the walls of the beaker were difficult to separate without damaging them. Hence, while changing the medium daily, these vessels were filled with EPA medium with appropriate prey at the desired density, to allow hatching of the individuals, which were then included in the final analyses (life table experiments) or returned to the original population (population growth studies). The experiments were terminated when all individuals died. We analysed the survivorship and reproduction related parameters following standard life table formulae (Krebs, 1985):

$$\text{Gross reproductive rate} = \sum_{0}^{\infty} m_x$$

$$\text{Net reproductive rate } (R_o) = \sum_{0}^{\infty} l_x m_x$$

$$\text{Generation time } (T) = \frac{\sum l_x m_x \cdot x}{R_0}$$

$$\text{Rate of population increase}(r) = \sum_{x=0}^{n} e^{-rx} l_x m_x = 1$$

The statistical significance of the differences in the survival and reproductive parameters of *A. multiceps* fed rotifers and cladocerans were tested using ANOVA (Sokal & Rohlf, 2000).

Results

The population growth curves of *Asplanchnopus multiceps* fed 3 rotifer prey species at 3 different densities are presented in Figure 1. Regardless of the prey species, in general, an increase in the availability of their density in the medium resulted in higher population densities of *A. multiceps*. However, comparing all prey species used, *B. patulus* supported higher population density of

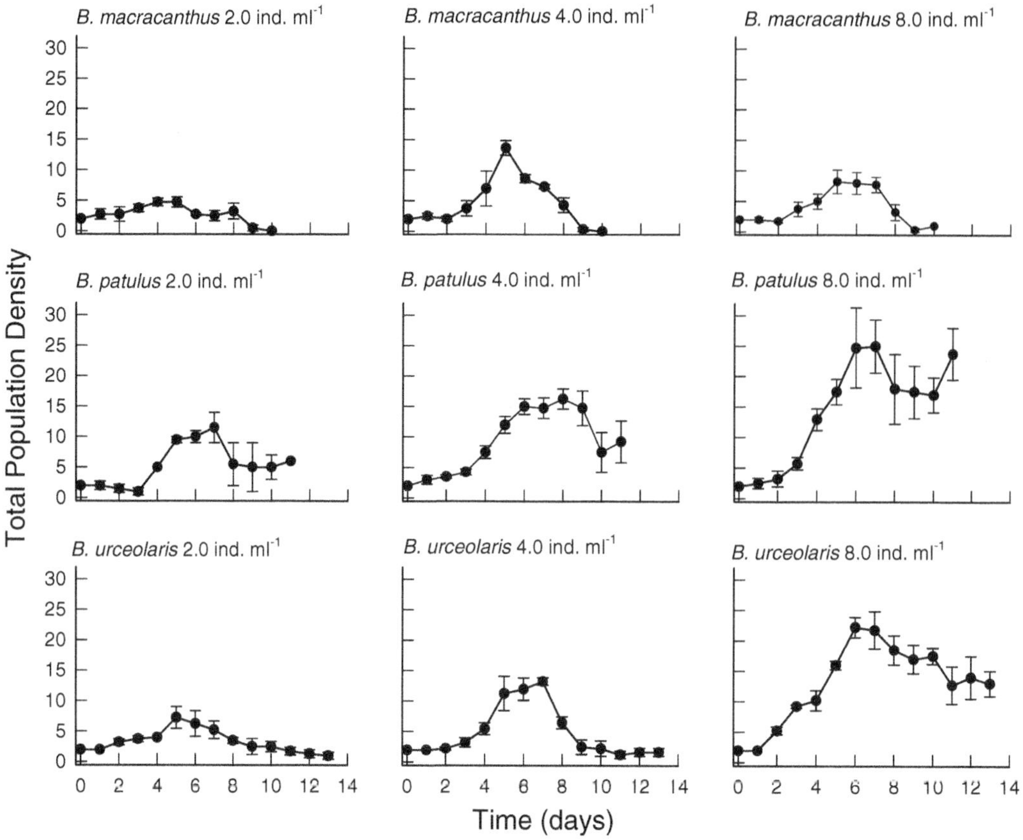

Figure 1. Population dynamics of *Asplanchnopus multiceps* offered *Brachionus macracanthus*, *B. patulus* and *B. urceolaris* at densities of 2.0, 4.0 and 8.0 ind. ml^{-1}. Shown are densities (ind.), mean ± standard error based on four replicates.

the predator. The highest peak population abundance of *A. multiceps* was observed on a diet of *B. patulus* at the highest prey density offered. The predator population declined before the end of the experimental duration on a diet of *B. macracanthus* even at the highest density of the latter. *A. multiceps* showed a positive population growth when offered the cladocerans *Alona rectangula* and *Pleuroxus aduncus* but not *Macrothrix triserialis* (Fig. 2). Of these three cladocerans, the first-named consistently supported higher population abundances of the predator. Regardless of *A. rectangula* or *P. aduncus*, an increase in the prey density resulted in increase of population of *A. multiceps*. The rate of population increase from population growth studies of *A. multiceps* varied from negative (when fed *M. triserialis* at even the highest test density) to positive (for the remaining prey species) (Fig. 3). The highest r (0.4 d^{-1}) of *A. multiceps* was recorded when fed *B. patulus*. Statistically, prey density significantly influenced both peak population densities and growth rates ($p < 0.05$, F-test, ANOVA Table 1).

Data on the age-specific survivorship curves of *A. multiceps* fed *B. patulus* (Fig. 4a) and *A. rectangula* (Fig. 4b) revealed improved survival with increasing prey density in the medium. Thus, nearly rectangular survivorship patterns were observed at a prey density of 4 or 8 ind. ml^{-1} for *B. patulus* or 1 and 2 ind. ml^{-1} for *A. rectangula*. The trends in survivorship are also reflected in the age-specific life expectancy curves of the predator (Fig. 4c, d). The age-specific fecundity curves of *A. multiceps* fed *B. patulus* and *A. rectangula* (Fig. 4e, f) showed a nearly normal distribution pattern of offspring production with a peak around 6 days. Regardless of the prey type, lower food levels contributed to a single peak in the offspring production while the highest prey density caused two peaks. The maximum number of

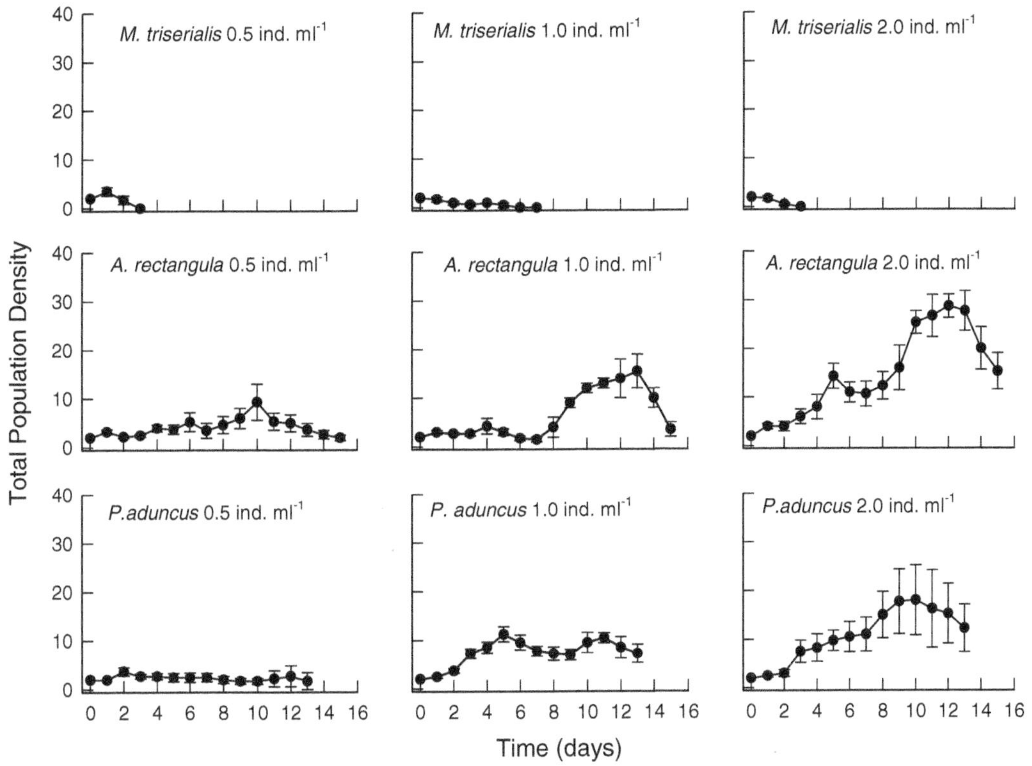

Figure 2. Population dynamics of *Asplanchnopus multiceps* offered *Macrothrix triserialis*, *Alona rectangula* and *Pleuroxus aduncus* at densities of 0.5, 1.0 and 2.0 ind. ml^{-1}. Shown are densities (ind.), mean ± standard error based on four replicates.

offspring produced (11 ind. $female^{-1}$ d^{-1}) was observed on a diet of *B. patulus* at a density of 8 ind. ml^{-1}.

Selected life history variables of *A. multiceps* fed *B. patulus* and *A. rectangula* at different densities are presented in Table 2. Regardless of prey type and density, *A. multiceps* had an average lifespan about 6 days, while life expectancy at birth was about 5 days. Regardless of prey species, both gross and net reproductive rates were lower at lower prey densities. The highest net reproductive rate (about 11 offspring $female^{-1}$ $lifespan^{-1}$) was recorded when *A. multiceps* was fed *B. patulus* at a density of 8 ind. ml^{-1}. The generation time for animals fed both *B. patulus* or *A. rectangula* was about 4 days. The rate of population increase derived from life table demographic studies varied from 0.21 to 0.74 d^{-1} depending on prey density. In general, higher prey densities resulted in elevated r (Table 3).

Survivorship related parameters were less affected by prey density than those related to reproduction. The average lifespan, life expectancy at birth and generation time were statistically not influenced by prey density. However, both gross and net reproductive rates and the rate of population increase were significantly influenced by the prey density (Table 1).

Discussion

The food and feeding habits of members of Asplanchnidae are diverse. While *A. priodonta* is primarily herbivorous (Kappes et al., 2000), all other species of *Asplanchna* are carnivorous (Salt, 1977). Very little is known about the feeding habits of species belonging to the genus *Asplanchnopus*. Koste (1978) has recorded all of them as carnivorous, feeding on small species of littoral rotifers and cladocerans. We too found *Chydorus* in a preliminary analysis of the stomach contents of freshly collected *A. multiceps*. Thus, the use of rotifer and cladoceran prey in this study is within the wide spectrum of food items generally selected

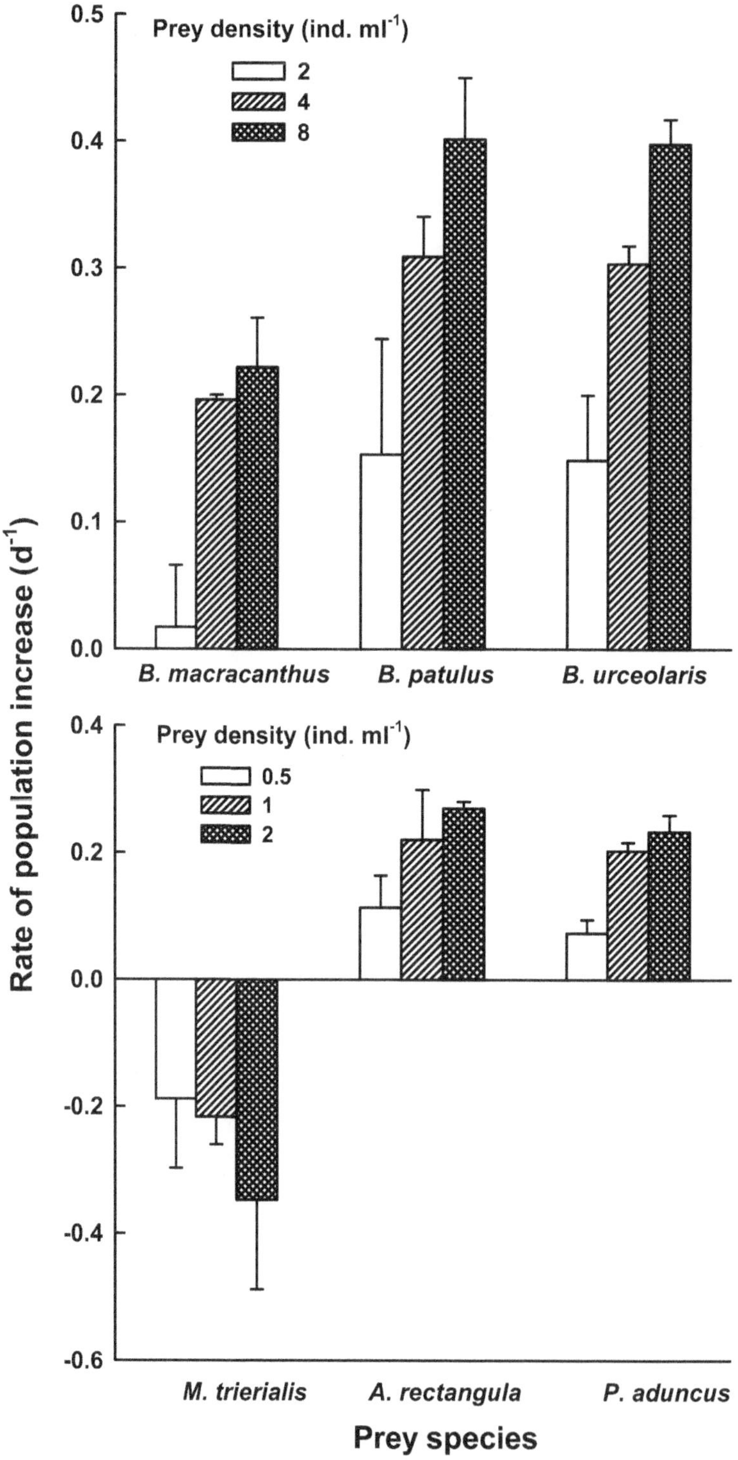

Figure 3. Population growth rates of *Asplanchnopus multiceps* offered rotifers (*Brachionus macracanthus*, *B. patulus* and *B. urceolaris* at densities of 2.0, 4.0 and 8.0 ind. ml^{-1}) and cladocerans (*Macrothrix triserialis*, *Alona rectangula* and *Pleuroxus aduncus* at densities of 0.5, 1.0 and 2.0 ind. ml^{-1}). Shown are mean ± standard error based on four replicates.

Table 1. Life history variables of *Asplanchnopus multiceps* offered *Brachionus patulus* and *Alona rectangula* at three prey densities

Prey density (ind. ml^{-1})	Life history variable					
	Lifespan	Expectancy at birth	Gross Reprod. rate	Net Reprod. Rate	Gen. time	Rate of pop. increase
B. patulus						
2.0	5.50 ± 0.19	5.00 ± 0.19	4.5 ± 0.74	2.3 ± 0.35	4.03 ± 0.16	0.21 ± 0.05
4.0	6.10 ± 0.21	5.60 ± 0.21	14.98 ± 5.11	6.10 ± 0.99	4.96 ± 0.65	0.41 ± 0.03
8.0	6.65 ± 0.92	6.15 ± 0.92	16.88 ± 5.80	10.85 ± 2.64	3.94 ± 0.53	0.74 ± 0.03
A. rectangula						
0.5	5.40 ± 0.56	4.90 ± 0.56	7.88 ± 1.56	2.95 ± 0.35	5.21 ± 0.19	0.22 ± 0.02
1.0	7.30 ± 0.17	6.80 ± 0.17	6.55 ± 0.69	4.45 ± 0.45	4.53 ± 0.18	0.38 ± 0.03
2.0	7.70 ± 0.29	7.20 ± 0.29	17.72 ± 4.42	10.50 ± 0.81	4.47 ± 0.15	0.68 ± 0.05

Shown are means ± standard error based on four replicate observations.

by *Asplanchnopus* under natural conditions. All the prey species as well as the predator used in our study are generally found near vegetation. Whereas both *B. patulus* and *B. macracanthus* are considered as tychoplanktic, *B. urceolaris* is often found attached to aquatic vegetation. *Alona rectangula, Macrothrix triserialis* and *Pleuroxus aduncus* are benthic-littoral cladocerans (Smirnov, 1992; Dodson & Frey, 2001).

Although de Paggi (2002) has reported that *A. multiceps* is ovoviviparous, we observed only ovipary. In our experiments as well as under mass

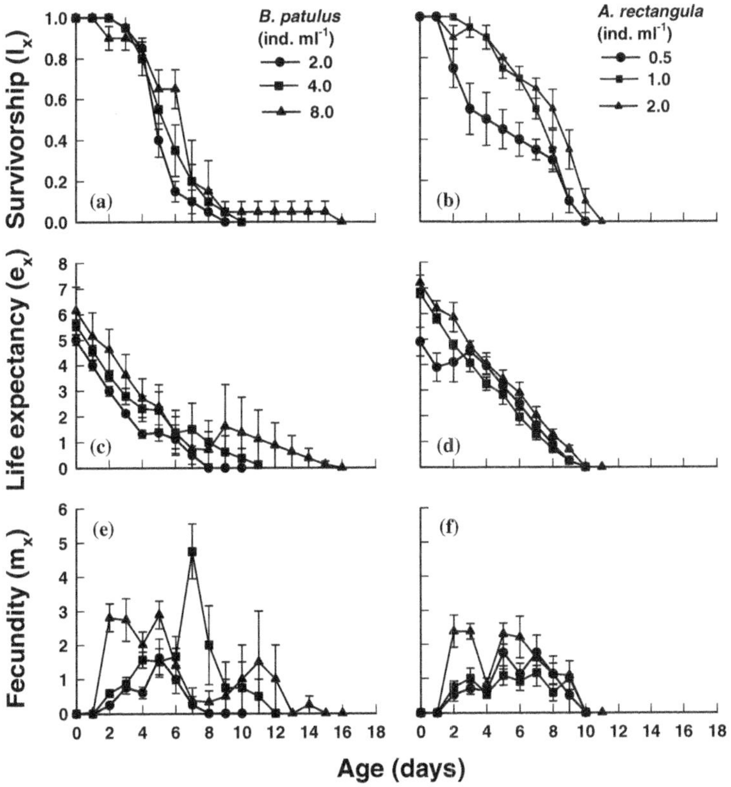

Figure 4. Patterns of survivorship (a and b), life expectancy (c and d) and fecundity (e and f) of *Asplanchnopus multiceps* offered *Brachionus patulus* (at densities of 2, 4 and 8 ind. ml^{-1}) and *Alona rectangula* (at densities of 0.5, 1.0 and 2.0 ind. ml^{-1}). Shown are mean ± standard error based on four replicates.

Table 2. Analysis of variance (ANOVA) performed for selected life history variables of *Asplanchnopus multiceps* offered *Alona rectangula* and *Brachionus patulus* as prey at three food levels

Source	DF	SS	MS	*F*-ratio
A. rectangula				
Average life span				
Between prey density	2	12.08	6.04	10.62***
Error	9	5.12	0.569	
Life expectancy at birth				
Between food level	2	12.08	6.04	10.12***
Error	9	5.12	0.567	
Gross reproductive rate				
Between food level	2	298.02	149.01	4.96*
Error	9	268.98	29.89	
Net reproductive rate				
Between food level	2	127.81	63.93	49.16***
Error	9	11.7	1.30	
Generation time				
Between food level	2	1.36	0.681	5.33*
Error	9	1.15	0.128	
Population growth rate				
Between food level	2	0.437	0.219	42.85***
Error	9	0.046	0.005	
B. patulus				
Average life span				
Between prey density	2	2.65	1.32	1.06^{ns}
Error	9	11.15	1.24	
Life expectancy at birth				
Between food level	2	2.65	1.32	1.07^{ns}
Error	9	11.15	1.24	
Gross reproductive rate				
Between food level	2	509.12	254.56	4.61*
Error	9	497.04	55.23	
Net reproductive rate				
Between food level	2	146.81	73.40	6.82*
Error	9	96.75	10.75	
Generation time				
Between food level	2	2.55	1.28	1.29^{ns}
Error	9	8.90	0.99	
Population growth rate				
Between food level	2	0.577	0.288	51.05***
Error	9	0.051	0.006	

DF = degrees of freedom, SS = sum of squares; MS = mean square, F = F-ratio, $p < 0.01$, *** $p < 0.001$, ns = non-significant ($p > 0.05$).

culture conditions, this species reproduced by depositing its parthenogenetic eggs at the bottom or sides of the culture vessels. The eggs were so strongly attached to the vessels that even a powerful suction from Pasteur pipette did not easily pull them out. While some members of Asplanchnidae such as *A. girodi* are amphoteric (a single female capable of both parthenogenetic

Table 3. Results of analysis of variance (ANOVA) performed on population growth rate of *A. multiceps* in relation to the prey species and prey density tested

Source of variation	df	SS	MS	*F*
Rotifers				
Prey Type (A)	2	0.158	0.08	9.11***
Prey Density (B)	2	0.345	0.17	19.88***
Interaction of A × B	4	0.006	0.001	0.16^{ns}
Error	27	0.226	0.01	
Cladocerans				
Prey Type (A)	2	1.526	0.76	37.60***
Prey Density (B)	2	0.031	0.02	0.77^{ns}
Interaction of A × B	4	0.134	0.03	1.53^{ns}
Error	27	0.528	0.02	

DF = degrees of freedom, SS = sum of squares; MS = mean square, *F* = *F*-ratio, *** $p < 0.001$, ns = non-significant ($p > 0.05$).

and sexual reproduction) we did not observe this for *A. multiceps*. However, the males produced by a mictic female attempted to mate the same female and in some cases we observed the same female to simultaneously produce resting eggs and males. This is in contrast to *Asplanchna intermedia* or *A. brightwelli* where, once a mictic female starts producing resting eggs, male production ceases (Nogrady et al., 1993). We observed that both unfertilised mictic eggs (which upon birth become males) and resting eggs (mictic eggs once fertilized become resting eggs or cysts) were observed in a single female of *A. multiceps* similar to that reported in a few other non-planktonic rotifers such as *Collotheca tenuilobata* (Sarma & Rao, 1986).

The population growth curves of *A. multiceps* observed by us are similar to those for *A. intermedia* (Iyer and Rao, 1996), *A. sieboldi* (Nandini & Sarma, 1999) and *A. brightwelli* (Sarma et al., 1998). All of them usually reach peak abundances in less than a week and thereafter either fluctuate around the carrying capacity or crash rapidly, depending on prey type and density. Nevertheless, while *Asplanchna* reaches its peak population density in 2–4 days, under similar conditions, *Asplanchnopus* took 5–7 days.

Spines on the lorica of rotifers offer protection against *Asplanchna* predation (Gilbert, 1999; Nandini et al., 2003). Among the rotifers that we used, *B. urceolaris* lacks both posterior and posterolateral spines and has six reduced anterior spines (usually not more than 20 μm in length). On the other hand, both *B. patulus* and *B. macracanthus* have not only much longer posterior and posterolateral spines but also a long (about 40 μm) antero-median pair on the dorsal plate. For the three brachionid prey species, *A. multiceps* peak population abundances and growth rates were lowest on *B. macracanthus*. This is to be expected considering that the length of the spines is 50% that of the lorica. For *B. patulus* this length is about 40% while for *B. urceolaris* it is only 15%. However, population growth of *A. multiceps* is comparable on these two prey species, rather than higher on *B. urceolaris*, which usually strongly attaches to the substratum (the walls of the test jars). *A. multiceps* probably had to spend more time to capture it (Sarma, 1993). Among the cladoceran prey, *Macrothrix* is morphologically characterized by the presence of a large number of spine-like structures throughout the carapace, which offers it considerable protection against invertebrate predation (Sarma et al., 2003). The other two cladocerans did not possess spines on their carapace and hence *Asplanchnopus* could reach densities as high as 1.0–1.5 ind. ml^{-1} on them. When prey is provided with spines or other related structures, the predators like *A. multiceps* may be also damaged, especially because it is soft-bodied and non-loricate. Thus, the low or negative growth rate of *A. multiceps* on *B. macracanthus* and *M. triserialis* may be not only due to the handling difficulties but also to a damage of the predator by the spines from these prey species. Feeding behaviour of *A. multiceps* using the different prey species was not examined in our study. However, there is some evidence suggesting that *B. macracanthus* is difficult to handle by *Asplanchna*. For example, Nandini et al. (2003) who studied the feeding behaviour of *A. sieboldi* offered five *Brachionus* species (*B. calyciflorus*, *B. havanaensis*, *B. macracanthus*, *B. rubens* and *B. patulus*) found that *A. sieboldi* had the lowest number of captures and ingestions on *B. macracanthus*. We did not quantify the prey-inflicted damage to the predator. However, results of Threlkeld (1987) on *Asplanchna* fed spined prey suggest this. *Asplanchnopus* induced morphological changes have not been recorded either in the field or under laboratory conditions (Gilbert, 1999). We did not also evaluate this aspect in our study.

Data on the life history variables measured from demographic studies also reveal some resemblance to *Asplanchna*. In general, both *Asplanchna* and *Asplanchnopus* have a short average lifespan, life expectancy at birth and generation time compared with brachionid rotifers (Dumont & Sarma, 1995; Nandini, 2000). There is some indication that *r* values, 0.2–0.7 d^{-1} derived for *Asplanchnopus* from population growth rates are lower than those obtained from demographic studies because in the former situation, the intraspecific competition for food exists between neonates and adults, while in the latter situation this rarely occurs since the neonates are continuously removed from the population. Regardless of the method of deriving of *r*, an increase in the growth rate with increasing prey density, as observed here for *A. multiceps*, is in agreement with many other studies on rotifers, both herbivores and carnivores (Gilbert, 1994; 1996; Dumont et al., 1995; Iyer & Rao, 1996).

In conclusion, both the population growth and life-table demography studies on *A. multiceps* indicate that this species is capable of growing and reproducing on brachionid prey rotifers and smaller chydorid cladocerans. It appears that *Asplanchna* relies less on a diet of small cladocerans than does *A. multiceps*. Further studies on the natural dietary preferences of *Asplanchnopus* would shed light on its role in regulating populations of littoral rotifers and cladocerans.

Acknowledgements

We thank the Division of Postgraduate Studies, FES Iztacala, for a grant given to SNI members. We also thank The National System of Investigators (CONACyT) (20520 and 18723 respectively) and PAPIIT IN234602 for financial assistance.

References

Anonymous, 1985. Methods of measuring the acute toxicity of effluents to freshwater and marine organisms. US Environment Protection Agency EPA/600/4-85/013.

Dodson, S. I. & D. G. Frey, 2001. Cladocera and other Branchiopoda. In Thorp, J. H. & A. P. Covich (eds), Ecology and Classification of North American Freshwater Invertebrates. Academic Press, London, 850–914.

Dumont, H. J. & S. S. S. Sarma, 1995. Demography and population growth of *Asplanchna sieboldi* (Rotifera) as a function of prey (*Anuraeopsis fissa*) density. Hydrobiologia 306: 97–107.

Dumont, H. J., S. S. S. Sarma & A. J. Ali, 1995. Laboratory studies on the population dynamics of *Anuraeopsis fissa* (Rotifera) in relation to food density. Freshwater Biology 33: 39–46.

Gilbert, J. J., 1976. Polymorphism in the rotifer *Asplanchna sieboldi*: Biomass, growth and reproductive rate of the saccate and campanulate morphotypes. Ecology 57: 542–551.

Gilbert, J. J., 1980. Observations on the susceptibility of some protists and rotifers to predation by *Asplanchna girodi*. Hydrobiologia 73: 87–91.

Gilbert, J. J., 1994. Susceptibility of planktonic rotifers to a toxic strain of *Anabaena flos-aquae*. Limnology and Oceanography 39: 1286–1297.

Gilbert, J. J., 1996. Effect of temperature on the response of planktonic rotifers to a toxic cyanobacterium. Ecology 77: 1174–1180.

Gilbert, J. J., 1999. Kairomone-induced morphological defenses in rotifers. In Tollrian, R. & C. D. Harvell (eds), The Ecology and Evolution of Inducible Defenses. Princeton Univ. Press, NJ, 127–141.

Hampton, S. E. & P. L. Starkweather, 1998. Differences in predation among morphotypes of the rotifer *Asplanchna silvestrii*. Freshwater Biology 40: 595–605.

Iyer, N. & T. R. Rao, 1996. Responses of the predatory rotifer *Asplanchna intermedia* to prey species differing in vulnerability: Laboratory and field studies. Freshwater Biology 36: 521–533.

Kappes, H., C. Mechenich & U. Sinsch, 2000. Long-term dynamics of *Asplanchna priodonta* in Lake Windsborn with comments on the diet. Hydrobiologia 432: 91–100.

Koste, W., 1978. Rotatoria. Die Rädertiere Mitteleuropas. Ein Bestimmungswerk begründet von Max Voigt. Bornträger, Stuttgart. Vol. 1, Textband 673 pp. Vol. 2, Tafelband 234 pp.

Krebs, C. J., 1985. Ecology. The Experimental Analysis of Distribution and Abundance (3rd edn.). Harper & Row, New York 800.

Murdoch, W. W. & J. Bence, 1987. General predators and unstable prey populations. In Kerfoot, W. C. & A. Sih (eds), Predation: direct and Indirect impacts on Zooplankton Communities. University Press of New England, NH, 17–30.

Nandini, S. & S. S. S. Sarma, 1999. Effect of hunger level on the prey capture behaviour, functional response and population growth of *Asplanchna sieboldi* (Rotifera). Freshwater Biology 42: 121–130.

Nandini, S., 2000. Responses of rotifers and cladocerans to *Microcystis aeruginosa* (Cyanophyceae): A demographic study. Aquatic Ecology 34: 227–242.

Nandini, S., R. Pérez-Chávez & S. S. S. Sarma, 2003. The effect of prey morphology on the feeding behaviour and population growth of the predatory rotifer *Asplanchna sieboldi*: a case study using five species of *Brachionus* (Rotifera). Freshwater Biology 48: 2131–2140.

Nogrady, T., R. L. Wallace & T. W. Snell, 1993. Rotifera. 1. Biology, Ecology and Systematics. Guides to the Identification of the Microinvertebrates of the Continental Waters of the World Vol. 4SBP Academic Publishers, The Hague 142 pp.

Salt, G. W., 1977. Components of feeding behaviour in rotifers. Hydrobiologia 147: 271–281.

Salt, G. W., G. F. Sabbadini & M. Commins, 1978. Trophi morphology relative to food habits in six species of rotifers (Asplanchnidae). Transactions of the American Microscopical Society 97: 469–485.

Sarma, S. S. S., 1993. Feeding responses of *Asplanchna brightwelli* (rotifera): laboratory and field studies. Hydrobiologia 255/256: 275–282.

Sarma, S. S. S., N. Iyer & H. J. Dumont, 1998. Feeding preference and population growth of *Asplanchna brightwelli* (Rotifera) offered two non-evasive prey rotifers. Hydrobiologia 361: 77–87.

Sarma, S. S. S. & T. R. Rao, 1986. Observations on the egg types and males of *Collotheca tenuilobata* Anderson (Rotifera: Collothecidae). Proceedings of Indian National Science Academy B52: 729–731.

Sarma S. S. S., H. J. Dumont & S. Nandini, 2004. Interactions between the anomopod cladocerans *Ceriodaphnia dubia*, *C. cornuta*, *Simocephalus vetulus* and *S. serrulatus*, and the worm *Aeolosoma* sp.: neither predation nor competition, or both? Hydrobiologia 526: 147–156.

Smirnov, N. N., 1992. The Macrothricidae of the World. Guides to the Identification of the Microinvertebrates of the Continental Waters of the World. SPB Acad. Publ., The Netherlands 143 pp.

Sokal, R. R. & F. J. Rohlf, 2000. Biometry. WH Freeman and Company, San Francisco 887.

Stemberger, R. S. & J. J. Gilbert, 1987. Defenses of planktonic rotifers against predators. In Kerfoot, W. C. & A. Sih (eds), Predation: Direct and Indirect Impacts on Zooplankton Communities. University Press of New England, NH, 227–239.

Threlkeld, S. T. & E. Choinski, 1987. Rotifers, cladocerans and planktivorous fish: What are the major interactions? Hydrobiologia 147: 239–243.

Hydrobiologia (2005) 546:503–508
A. Herzig, R.D. Gulati, C.D. Jersabek & L. May (eds.) Rotifera X: Rotifer Research: Trends, New Tools and Recent Advances
DOI 10.1007/s10750-005-4292-1

Susceptibility of ephemeral pool *Hexarthra* to predation by the fairy shrimp *Branchinecta mackini*: can predation drive local extinction?

Peter L. Starkweather
Department of Biological Sciences, University of Nevada, Las Vegas, Nevada, 89154-4004, USA
E-mail: strkwthr@ccmail.nevada.edu

Key words: Anostraca, *Branchinecta mackini*, *Hexarthra*, rotifer, predation

Abstract

Isolated rock pools in the US desert southwest often develop dense populations of a lithophilic species of the rotifer *Hexarthra*. I hypothesized that rotifers persist in these isolated ponds due to the absence of either competition with or predation by potentially sympatric crustaceans, especially anostracans. I tested the latter idea with laboratory and field experiments, in each case exposing rotifers to adult fairy shrimp (*Branchinecta mackini*, the dominant anostracan in this region in winter ponds) in 200 ml microcosms. In most cases individual female fairy shrimp had distinct negative effects on rotifer suspensions due to direct predatory consumption of the smaller animals. Estimated effective water processing rates ranged from ca. 50 to over 300 ml ind^{-1} h^{-1} and rotifer consumption rates by female *B. mackini* were between 280 and >600 *Hexarthra* consumed per individual predator per hour. Male *B. mackini* never significantly reduced rotifer numbers in either laboratory or field microcosms. The results indicate that, while perhaps not the sole determinant of rotifer distribution in these ephemeral pools, fairy shrimp predation can have a strong negative influence on natural *Hexarthra* populations.

Introduction

It is widely recognized that rotifers and crustaceans are the dominant metazoan zooplankton of lentic freshwaters. These taxa are directly linked through ecological interactions including competition and predator–prey relationships, and indirectly via their differential exploitation by visual, tactile and suspension-feeding predators. This observation has been confirmed in a wide variety of ecological settings, from large permanent lakes through small ephemeral pools.

In the many temporary waters of the US desert southwest, rotifers and crustaceans are brought together in very simple physical and trophic systems, which may accentuate the interactions of these taxa. Such systems may create situations in which prey refuges are rare and food webs are of low complexity, and where predator–prey relationships are thus very important influences in structuring communities (Brendonck et al., 2002).

Anostracans usually are regarded as suspension-feeders (see Brendonck, 1993a, b) except for the notorious raptorial predator *Branchinecta gigas* (Fryer, 1966; White et al., 1969; Brown & Carpelan, 1971; Boudrias & Pires, 2002). There is, however, some strong evidence for other anostracans having predatory effects on smaller invertebrates, including rotifers, at least in laboratory settings. Dumont et al. (1994) measured feeding behavior of *Streptocephalus proboscideus* on the rotifer *Anuraeopsis fissa*, noting effective predation complete with characteristic predator–prey functional responses. This work (and that of others, Dierckens et al., 1995; Ali et al., 1996) also indicated that consumption of rotifers by female fairy shrimp was up to 90% greater than that of males. A more recent *in vitro* study using female *Chirocephalus diaphanus* (Sarma & Nandini, 2002) showed a differential susceptibility of small zooplankton, including certain rotifers, to fairy shrimp predation. However, these laboratory studies used rotifer species not well known for their predator evasion abilities (Stemberger & Gilbert, 1987) and Ali et al. (1996) noted that *S. proboscideus*, at least, did not eat 'jumping' rotifers. *Hexarthra* is noted for its ability to produce avoidance jumps with its conspicuously setose 'arms' (Stemberger & Gilbert, 1987); the species used here showed active and frequent saltatory movements in both fresh collections and in the laboratory.

In this work I pose four simple questions. First, does the fairy shrimp *Branchinecta mackini* consume individuals of this species of *Hexarthra* in the laboratory, and, if so, are there any differences between the shrimp genders in this feeding activity? Second, is comparable behavior observed in a field setting? Third, how much variability is there among individual predators in this activity? And fourth, what are the rates at which observed feeding occurs, and are these rates adequate to suggest that *B. mackini* has a significant influence on *Hexarthra* in the systems in which they may co-occur? That is, might the fairy shrimp be able to drive these rotifers to transient local extinction if *B. mackini* were to invade the rotifers' resident ponds?

Materials and methods

The field site for this work, and the source of animals for the laboratory experiment, was Brownstone Canyon, an anthropological and ecological reserve near Las Vegas, NV (Clark County) administered by the US Bureau of Land Management. GPS coordinates for the elevated, high-density rotifer pond are: N 36 10.983, W 115 25.860 (elev. 1465 m); the source pond for the *B. mackini* was one of many inter-connected pools in a large catchment area ca. 350 m to the southwest.

For each of these experiments, whether run in the lab or field, I established 200 ml microcosms in 250 ml glass beakers, each containing the approximate *in situ* concentration of rotifers on the date of the experiment. I collected the rotifers as whole water samples, processed through 200 μm mesh nylon screening to exclude any larger organisms and/or coarse detrital material, retaining the concentrated *Hexarthra* in a 64 μm plankton bucket until diluted back to sampled densities with 20 μm-filtered pond water. I collected the *Branchinecta* with a dip net, re-suspended them in their source pond water and incubated them in the experimental setting for at least 1 h before exposing them to the rotifers. I attempted no systematic pre-feeding or starvation treatment of the fairy shrimp. Each experiment was started by adding a single *B. mackini* to each of several individual 200 ml microcosms (n = 3–9); a pilot observation had indicated that using multiple fairy shrimp per vessel resulted in frequent collisions and at least transient changes in swimming (and therefore possibly feeding) behavior. In the lab I incubated the microcosms at 12 °C in continuous dim illumination; field incubations were in sediment-filled enamel trays holding the beakers under ambient conditions; these are noted below when discussing experimental results. At each experimental time point I removed a 5 ml aliquot with an automatic pipette from each gently-stirred microcosm, fixing samples with equal volumes of 100% ethanol. At the end of each incubation I similarly fixed individual *B. mackini* in 50% ethanol and disposed of all residual experimental water in the lab or in unconsolidated alluvium far removed from the rotifer source ponds, this to avoid introduction of branchiopods, their embryos or cysts to the rotifer-dominated systems.

For rotifer counts I made no attempt to distinguish among different age or size classes nor did I perform differential counts of ovigerous vs. non-ovigerous females; I did not consider male *Hexarthra* since my collection procedures likely would not have sampled them with consistent efficiency. I measured the ethanol-fixed fairy shrimp with a stereomicroscope to the nearest 0.1 mm and subsequently dried them at 65 °C for at least 2 days (following 2 DI-water rinses), weighing them on a Cahn electrobalance to the nearest μg.

The species identification of *Hexarthra* from these unique systems is still unresolved, but, from examination of the trophi, it appears to be a member of the *Hexarthra mira-intermedia* group (Ruttner-Kolisko, 1974).

I calculated fairy shrimp effective water processing (clearance) and feeding rates based on negative exponential changes in rotifer densities over time intervals (as in Dierckens et al., 1995), correcting these values for progressive reduction in microcosm volume due to sampling and changes rotifer density due to consumption for each incubation interval. I evaluated the statistical significance of those changes for each treatment with one-way Analysis of Variance, or, in the case of non-normal rotifer count distributions, Kruskal-Wallis ranks comparisons, of *Hexarthra* numbers per unit volume vs. time.

Results

Figure 1 shows a 3 h time course for a 12 °C laboratory incubation of *Hexarthra* with individual male and female *B. mackini* compared to a control microcosm containing only rotifers. The control showed no significant differences in rotifer numbers per ml (mean ~12.5 ml^{-1}) over the 3 h incubation (ANOVA, $F = 0.87$, $p = 0.53$). Changes in rotifer numbers over time were highly significant for the female fairy shrimp (K-W, $H = 27.03$, $p < 0.001$) but were not so for the male *Branchinecta* (ANOVA, $F = 1.44$, $p = 0.46$). Two points are apparent: one, *B. mackini* can be a potent predator on this species of *Hexarthra* in a laboratory setting and two, female *B. mackini* would appear to be dramatically more effective at removing rotifers from suspension than are males. Since the number of rotifers per ml drops progressively over time in this type of experiment, it is possible to estimate the feeding activity of the (female) fairy shrimp when dynamically exposed to different prey densities. Estimates of effective water processing by the individual female *B. mackini* at 12 °C varied from ~80 ml processed (or 'cleared') per hour at 10–12.5 rotifers per ml, to over 300 ml h^{-1} as rotifer numbers fell below 2 per ml. These estimates result from impressive rotifer consumption values of hundreds of *Hexarthra* consumed per fairy shrimp per hour.

In the field I found a similar situation, as shown in Figure 2. Once again there was a significant decline in the numbers of *Hexarthra* in the treatments with female fairy shrimp (ANOVA, $F = 4.44$, $p = 0.004$). However, the control and male feeding treatments showed significant variation over time (ANOVA, $F = 3.23$, $p = 0.019$ and $F = 4.62$, $p = 0.003$ for control and male treatment, respectively), likely reflecting the heterogeneity of rotifer distributions within the microcosms despite attempts at establishing uniformity during field set-up and sampling. Over the 2.5 h time course of the experiment, water temperature rose from 7.0 to just over 9 °C, oxygen

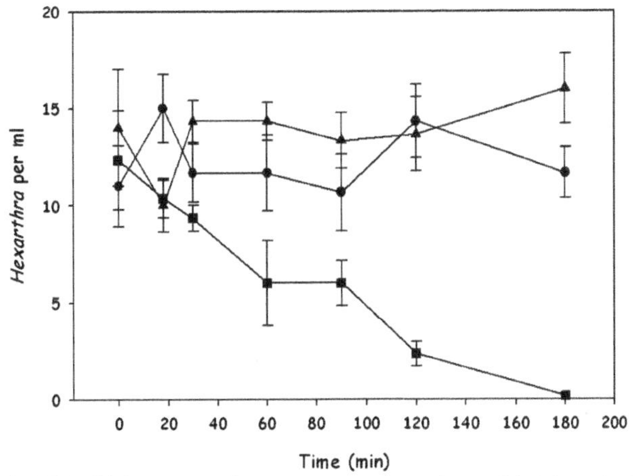

Figure 1. Numbers of *Hexarthra* in 200 ml laboratory microcosms over time when subject to potential predation by the fairy shrimp *Branchinecta mackini*; 12 C, continuous dim illumination. Circles, control with no fairy shrimp present; triangles, male *B. mackini*; squares, female *B. mackini*. Values shown are means ($n = 6$) ± 1 SE.

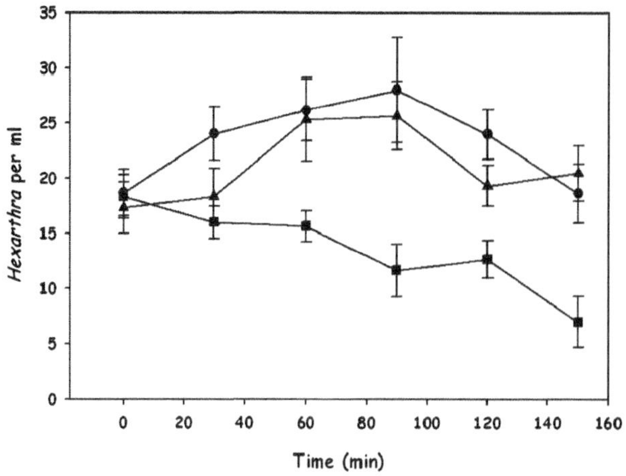

Figure 2. Numbers of *Hexarthra* in 200 ml field-incubated microcosms over time when subject to potential predation by the fairy shrimp *B. mackini*; physical conditions described in the text. Circles, control with no fairy shrimp present; triangles, male *B. mackini*; squares, female *B. mackini*. Values shown are means ($n = 6$) ± 1 SE.

tension increased from 5.8 to 6.6 mg l^{-1} and conductivity dropped from 150 to 144 μS; none of these changes would likely explain the control variation or the differential feeding patterns of female and male fairy shrimp. Taken together, the fundamental result appears the same as in the first (laboratory) experiment, *vis.*, that female *B. mackini* can have a distinct negative affect on these *Hexarthra* populations while males of the crustacean species are ineffective rotifer predators.

In an attempt more accurately to assess the quantitative impact of female *B. mackini* on desert pool *Hexarthra* populations, I ran an additional experiment in the field using replicate microcosms with only female fairy shrimp. On this day, field water temperatures ranged from 3 to 7.5 °C, oxygen tension from 7.4 to 8.2 mg l^{-1} and conductivity from 225 to 234 μS. Of seven animals evaluated, five caused significant declines (ANOVA, $5.71 < F < 13.58$, p always < 0.038) in rotifer numbers in microcosms over the 2 h incubation period (the control showed no significant difference in estimated rotifer numbers for this interval, ANOVA, $F = 0.89, p = 0.53$). Fig. 3 shows these trajectories. Combining the results for all the female *B. mackini*, I calculated an overall effective water processing rate of 46.2 ± 19.5 (SE) ml per animal per hour integrated over the 2 h incubation. Using an estimate of 6.1 rotifers per ml (the mean for all rotifer counts in this experiment) gives an individual feeding rate for female *B. mackini* of 281.6 ± 116.3 rotifers consumed per hour. These values are lower than those for the laboratory experiment shown in Fig. 1, most likely reflecting the ~6–9 °C lower temperature of this run.

I compared the hourly rates of both effective water processing (clearance rate) and rotifer consumption by individual *B. mackini* females of differing body length and dry weight. While there was a suggestion of an upward trend in effective water processing and rotifer consumption with both *B. mackini* body length and dry weight, in neither case was there a significant correlation between female fairy shrimp size and feeding activity (Spearman Rank Coefficients, 0.548 ($p = 0.139$), 0.395 ($p = 0.290$) for length and dry weight, respectively vs. clearance rate and −0.048 ($p = 0.885$), 0.012 ($p = 0.931$) for length and dry weight, respectively vs. *B. mackini* feeding rate).

Discussion

Previous work in the laboratory has shown that at least three species of anostracan can be effective predators on cultured non-evasive rotifers (Dumont et al., 1994; Dierckens et al., 1995; Ali et al., 1996; Sarma & Nandini, 2002). In addition, my own field observations of ephemeral desert pools in the US Southwest, and comparable survey results from North Africa (H. Dumont, personal communication), suggest that anostracans and

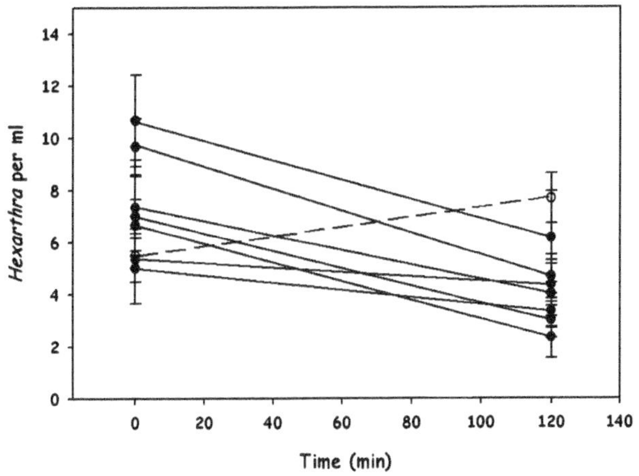

Figure 3. Numbers of *Hexarthra* in 200 ml field-incubated microcosms during a 2 h incubation when subject to predation by female *B. mackini*; physical conditions described in the text. Open circles with dashed line, control with no fairy shrimp present; filled circles with solid lines, with female *B. mackini*. Values shown are means ($n = 6$) ± 1 SE.

rotifers are often mutually exclusive inhabitants of these ephemeral sites. This work is the first to evaluate whether or not fairy shrimp can have a direct negative affect on rotifer populations in a field setting via predation, and the first to determine if the quantitative impact of this interaction might be sufficient to drive this observed habitat segregation. While I did not test younger developmental stages of the fairy shrimp, it is appropriate to add *B. mackini* (at least adult females) to the list of taxa which are effective predators on rotifers. This is particularly impressive since the rotifer genus *Hexarthra* is generally regarded as an effective evader of crustacean predators (Stemberger & Gilbert, 1987; Kak & Rao, 1998); clearly, in this case, the predatory capacity of female *Branchinecta* is adequate to counter this normally-robust behavioral defense. Male *B. mackini* are not significant predators on this *Hexarthra* species, having shown no capacity to remove the rotifers from suspension. While male and female fairy shrimp have shown little or no gender differentiation in small particle suspension-feeding (Beladjal et al., 1997; Brendonck, 1993a, b), the current results are consistent with those for *Streptocephalus* species when preying on rotifers; that is, females are many times more effective as rotifer predators than are males. In the field setting for this work, the sex ratio of *B. mackini* is strongly biased toward females (PLS, unpublished), accentuating the likely influence of fairy shrimp predation in structuring the local zooplankton communities.

In terms of quantitative impacts, female *B. mackini* predation potential is high. Maximum effective water processing rates range from approximately 50 to 300 ml per individual per hour, increasing with decreasing rotifer density. This equates to hundreds of rotifers consumed per hour or potentially thousands of rotifers per day for a single female fairy shrimp. Since in catchment drainage ponds in Brownstone Canyon total fairy shrimp populations can reach peak densities of 1–3 per liter (but are usually less abundant, Starkweather et al., 2001), the impact of anostracans on any incipient rotifer population may be substantial. As noted in Fig. 3, there was variation among females in observed predation on *Hexarthra*. This could be due to a number of factors including individual (and different) nutritional histories, adult instar, phase of the molt cycle during the trials or reproductive condition. These same factors might be responsible for the lack of observed significant positive relationships between fairy shrimp size and water processing or feeding rate estimates.

Conclusions

1. Female *B. mackini* are effective predators of the rotifer *Hexarthra* when tested in either laboratory or field settings. Males of this species of fairy shrimp do not appear to prey on *Hexarthra*.

2. Fairy shrimp predation rates are high given the ecological conditions of the source field sites and *B. mackini* predation is a credible driver for the local extinction of rotifers in ponds dominated by fairy shrimp.
3. Ephemeral ponds which do contain large rotifer populations (those geographically elevated above the more-common catchment ponds) represent a refuge for *Hexarthra* as long as *B. mackini* do not invade those spatially and hydrologically isolated systems.

Acknowledgements

Many thanks to the UNLV intramural faculty research grants committee for financial support during the experimental phase of this work and with appreciation for the hospitality of the CNR – Istituto per lo Studio delgi Ecosistemi (Verbania-Pallanza, Italy) during manuscript preparation.

References

Ali, A. J., S. S. S. Sarma, G. Murugan & H. J. Dumont, 1996. Effect of zooplankton type and abundance on prey consumption by the fairy shrimp *Streptocephalus proboscideus* (Anostraca: Crustacea). Hydrobiologia 319: 191–202.

Beladjal, L., N. Peiren, K. R. Dierckens & J. Mertens, 1997. Feeding strategy of two sympatric anostracan species (Crustacea). Hydrobiologia 359: 207–212.

Boudrias, M. A. & J. Pires, 2002. Unusual sensory setae of the raptorial *Branchinecta gigas* (Branchiopoda:Anostraca). Hydrobiologia 486: 19–27.

Brendonck, L., 1993a. Feeding in the fairy shrimp *Streptocephalus proboscideus* (Frauenfeld) (Branchiopoda:Anostraca) I. Aspects of the feeding biology. Journal of Crustacean Biology 13: 235–244.

Brendonck, L., 1993b. Feeding in the fairy shrimp *Streptocephalus proboscideus* (Frauenfeld) (Branchiopoda:Anostraca) II. Influence of environmental conditions on feeding rate. Journal of Crustacean Biology 13: 245–255.

Brendonck, L., E. Michels, L. DeMeester & B. Riddoch, 2002. Temporary pools are not 'enemy-free'. Hydrobiologia 486: 147–159.

Brown, L. R. & L.H. Carpelan, 1971. Egg hatching and life history of a fairy shrimp *Branchinecta mackini* Dexter (Crustacea: Anostraca) in a Mohave Desert Playa. Ecology 52: 41–54.

Dierckens, K. R., S. S. S. Sarma, J. Mertens & H. J. Dumont, 1995. Feeding the fairy shrimp *Streptocephalus* (Anostraca, Crustacea) with the rotifer *Anuraeopsis*. Hydrobiologia 308: 29–33.

Dumont, H. J., A. Jawahar Ali, S. S. S. Sarma & J. Mertens, 1994. Predatory filter-feeding in fairy shrimps: functional response of *Streptocephalus proboscideus* (Crustacea:Anostraca) fed *Anuraeopsis fissa* (Rotifera). Internationale Revue der gesamten Hydrobiologie 79: 511–519.

Fryer, G., 1966. *Branchinecta gigas* Lynch, a non-filter-feeding raptorial anostracan, with notes on the feeding habits of certain other anostracans. Proceedings of the Linnean Society London 177: 19–35.

Kak, A. & R. Rao, 1998. Does the evasive behavior of *Hexarthra* influence its competition with cladocerans? Hydrobiologia 387/388: 409–419.

Ruttner-Kolisko A., 1974. Plankton Rotifers, Biology and Taxonomy. Schweizerbart'sche Verlagsbuchhandlung, Stuttgart. Die Binnengewässer Supplement 26(1), 146 pp.

Sarma, S. S. S. & S. Nandini, 2002. Studies on the functional response and prey selection using zooplankton in the anostracan *Chirocephalus diaphanus* Prevost, 1803. Hydrobiologia 486: 169–174.

Starkweather, P. L., N. Vaskov & T. C. Ng, 2001. Population dynamics and life history patterns of branchiopod crustaceans in Mojave Desert ephemeral rock pools. Internationale Vereinigung für Theoretische und Angewandte Limnologie, Verhandlungen 27: 3770–3773.

Stemberger R. S. & J. J. Gilbert, 1987. Defenses of planktonic rotifers against predators. In: Kerfoot W.C. & A. Sih (eds.), Predation: Direct and Indirect Impacts on Aquatic Communities, University Press of New England, pp. 227–239.

White, G., G. Fabris & R. Hartland-Rowe, 1969. The method of prey capture by *Branchinecta gigas* Lynch, 1937 (Anostraca). Crustaceana 16: 158–160.

Hydrobiologia (2005) 546:509–518
A. Herzig, R.D. Gulati, C.D. Jersabek & L. May (eds.) Rotifera X: Rotifer Research: Trends, New Tools and Recent Advances
DOI 10.1007/s10750-005-4294-z

Decline of clear-water rotifer populations in a reservoir: the role of resource limitation

Miloslav Devetter[1,]* & Jaromír Sed´a[2]
[1]*Institute of Soil Biology, Czech Academy of Sciences, Na sádkách 7, 370 05, Ceské Budejovice, Czech Republic*
[2]*Hydrobiological Institute, Czech Academy of Sciences, Na sádkách 7, 370 05, Ceské Budejovice, Czech Republic*
*(*Author for correspondence: E-mail: devetter@upb.cas.cz)*

Key words: resource limitation, rotifers, birth rate, clear water phase, zooplankton, population decline

Abstract

The relative importance of and changes in resource limitation of herbivorous rotifers were assessed during the clear-water phase in the Rímov Reservoir, Czech Republic, using *in situ* manipulative experiments. Resource limitation was tested experimentally as the difference in population growth rate (Δr) among various experimental treatments on four occasions. The reservoir community of rotifers was exposed to three treatments: (i) control, (ii) diluted and (iii) diluted and fertilized. Significant responses to these experimental manipulations were shown by *Synchaeta* spp., *Polyarthra* spp. and *Keratella cochlearis*. Growth rate was usually highest during the spring rotifer maximum and decreased during the clear water phase. The highest intensity of food limitation (expressed as 'Chlorophyll-*a*' limitation) was found in *Synchaeta* spp. *K. cochlearis* had low food limitation during the spring peak, high food limitation during the second experiment and low food limitation, again, during the later experiment. In contrast, *Polyarthra* spp. had the same Chlorophyll-*a* limitation throughout the whole experimental period. Linear regression was used to estimate the relative proportion of Δr variability explained by Chlorophyll-*a* concentration and rotifer density in all of the experiments. Chlorophyll-*a* concentration explained 89, 97 and 92% of the resource limitation in *Synchaeta* spp., *Polyarthra* spp. and *K. cochlearis*, respectively. The proportion of variability explained by rotifer density-dependent factors was lower: 60% for *Synchaeta* spp. and 68% for *Polyarthra* spp.

Introduction

A clear water phase is typical of the spring algal succession in many mesotrophic and eutrophic lakes in temperate zones (Lampert et al., 1986; Sommer et al., 1986; Arndt et al., 1993; Scheffer et al.,1997). This occurs when the spring bloom, consisting of small, rapidly growing algae, is followed by a period of clearer water characterised by low concentrations of food particles and high Secchi-disk transparency. The clear water phase usually coincides with a spring peak of large, filter-feeding zooplankton (Sommer et al. 1986; Brandl et al., 1989; Komárková, 1989; Sed´a, 1989; Vyhnálek et al., 1991; Fussmann, 1996; Deneke & Nixdorf, 1999). This, through competition, is often considered to be one of the main factors causing the decline of rotifers during the clear water phase (Gilbert, 1985, 1988; Lampert et al., 1986; May & Jones, 1989; MacIsaac & Gilbert, 1991; Conde-Porcuna et al., 1994). It has been suggested that the rotifer community is highly food limited during the clear water phase (Lampert et al., 1986; Devetter, 1998; Devetter & Sed´a, 2003), and that this resource limitation has an important impact on the dynamics of the zooplankton communities (Rothhaupt 1990; Müller-Navarra & Lampert, 1996; Ciros-Pérez et al., 2001; Kirk, 2002).

To assess the importance of resource limitation, we require a measure of the intensity of limitation expressed as an increase in population growth rate caused by the experimental increase in resource availability (Merriman & Kirk, 2000; Cordova et al., 2001). Knowledge of this parameter will allow us to look for differences in how resource limitation varies through time and among species. Several approaches can be used. The effects of food limitation can be tested by experiments in which the level of food available to the zooplankton community is enhanced. Several studies have been conducted on food limitation in planktonic rotifers using this method (González & Frost, 1992; Merriman & Kirk, 2000; Cordova et al., 2001; Kirk, 2002). However, there are some limitations to using this approach, because the addition of cultured algae to the experimental media may change the structure of the available food. Fertilization experiments can also be used, and may be preferable, as these tend to support the development of the native phytoplankton communities. However, it also is possible that different species may respond differently to the additional nutrient supply (González, 2000). Food level manipulation is based on the assumption that density-dependent factors relating to the consumer are not important. This might not be always true, especially under the high population densities found during seasonal peaks. Diluting the zooplankton community at any given food level supports the density-dependent requirements of species and also changes the ratio between the abundance of food and abundance of consumers. By manipulating the consumer density but not food level the part of variability explained by density-dependent factors can be obtained. The results of the experiment, therefore, will reflect the combined effects of density-dependent limitation and food limitation forces.

In the present paper we examine the relative importance of and changes in resource limitation of herbivorous rotifer species through the clear-water phase caused by food-dependent and consumer density-dependent factors in a reservoir.

Methods

All investigations were done in the meso-eutrophic Rímov Reservoir in South Bohemia, Czech Republic, which is described in detail by Brandl et al. (1989) and Sed'a & Kubecka (1997). The reservoir has a diverse assemblage of zooplankton, but is dominated by only a small number of taxa. *Keratella cochlearis*, *Synchaeta lakowitziana*, *Polyarthra dolichoptera*, *Polyarthra vulgaris*, *Kellicottia longispina* and *Conochilus hippocrepis* are the most important rotifers (Devetter, 1998), *Daphnia galeata*, *Bosmina longirostris*, *Diaphanosoma brachyurum* and *Ceriodaphnia quadrangula* dominate the cladoceran community (Sed'a, 1989) and *Cyclops vicinus*, *Mesocyclops leuckarti*, *Thermocyclops crassus* and *Eudiaptomus gracilis* dominate the copepods (Brandl et al., 1989). A high frequency sampling programme was carried out at 2–5 day intervals during spring 2000. The sampling and experimental sites were situated at the deepest point of the reservoir profile, 250 m from the dam.

Rotifers were sampled between 9 and 11 am on each sampling day with a wide mouth flexible plastic industrial hose, using the modified approach of Pennak (1962). The hose was used for integrated sampling of the top 8 m of the water column. The sampled water (effective volume of *ca* 60 l) was filtered through a 40 μm mesh nylon net. The rotifers were anesthetized with carbonated water to prevent contraction of soft-bodied species, preserved with a 4% formaldehyde solution and counted under the microscope. At least three subsamples and a minimum of 400 animals were counted per sample. The possible loss of rotifer eggs due to filtration (Likens & Gilbert, 1970; Pace & Orcutt, 1981) was assessed repeatedly, but no significant number of eggs was found in the filtered water.

Instantaneous birth rate was calculated according to Paloheimo (1974):

$$b = \ln(E + 1)/D$$

where b is birth rate, E is egg ratio (number of eggs per animal) and D is embryonic development time (days). Embryonic development time for each species was calculated according to Herzig (1983). As no embryonic development time for *S. lakowitziana* is given, the value of D for a very similar species,

S. oblonga, was used in the calculations for this species (Zoufal, 1989). Similarly, the parameters for *K. cochlearis* were used for *K. longispina* (Herzig, 1983). Egg ratio was calculated using both attached and free eggs (*Synchaeta* does not carry its eggs). Free eggs were identified to species (*K. cochlearis, K. quadrata, K. longispina*) or genus (*Synchaeta, Polyarthra*) level.

Samples for determining Chlorophyll-*a* concentration were taken as a mixed sample from the upper 4 m of the water column, with a plastic tube. Samples were filtered through glassfibre GF/C filters, extracted by acetone and their absorbance measured (Lorenzen, 1967). Separate samples were collected for determining the algal fraction theoretically available to rotifers as food (i.e. <20 μm). Water transparency was estimated with a Secchi-disk, 20 cm in diameter.

Resource limitation experiments

Experiments were started during the spring phytoplankton and rotifer maxima and continued throughout the clear water phase. They comprised four *in situ* experiments in transparent 19-l circular PET (polyethylenetereftalate) containers suspended at a depth of 1 m at the sampling site. Experiments were carried out on 20 April, 11 and 24 May, and 6 June 2000, and the exposure time was 12 days in April and 6–7 days in May and June. Exposure times were selected to represent about three times the expected rotifer embryonic development time at the relevant reservoir temperature.

Water samples for the experimental enclosures were collected using a flexible hose. Some was filtered through a 275 μm net to remove crustaceans and provide a sample of the natural rotifer community. The remainder was filtered through a 40 μm mesh net to remove crustaceans and rotifers and provide experimental dilution water.

The experiments consisted of three treatments: C – control – the natural rotifer community without crustaceans; D – diluted – experimental dilution water with a 1.5 L inoculum of the natural rotifer community; DF – diluted and fertilized – as D with the addition of 200 μg PO_4^{3-} P per litre. Three replicates were incubated in all treatments. After incubation, the whole volume of the container was filtered and counted. A mixed sample of the three replicates was taken to measure the concentration of Chlorophyll-*a* from each treatment and processed as described above. It was assumed that the quality of the algal food did not change drastically during the course of the experiments. In the experimental analysis, only Chlorophyll-*a* concentrations representing the 'edible' fraction of algae (i.e., <20 μm) was considered.

The instantaneous rate of population increase in all treatments was calculated according to Edmondson (1960):

$$r = (\ln Nt - \ln N_0)/t$$

where N_0 is the population size at the start of the experiment, Nt is the population size after exposure and t is duration of experiment in days. Two way ANOVA (fixed effects) was used for comparison of r among treatments and among experiments.

The combined effects of food and density-dependent limitation is always to inhibit the growth of rotifer populations during their development. However, the relative importance of these two factors may change in the different developmental phases. Experimental manipulation in enclosures can discriminate between them, and their relative importance, under the prevailing conditions.

The different responses among the control and treatments were used to describe the resource limitation

$$\Delta r_D = r_D - r_C,$$

$$\Delta r_{DF} = r_{DF} - r_C,$$

where r_D and r_{DF} are population growth rate in the diluted and the diluted and fertilised enclosures, respectively, and r_C is population growth rate in the control enclosures.

The relative increase in Chlorophyll-*a* concentrations (log-transformed) between the D and DF treatments, weighted by initial concentration, was determined as follows:

$$CHL_D = \log((\mathrm{Chl} - a_D - \mathrm{Chl} - a_C)/\mathrm{Chl} - a_C)$$

$$CHL_{DF} = \log((\text{Chl} - a_{DF} - \text{Chl} - a_C)/\text{Chl} - a_C)$$

where Chl-a_D, Chl-a_{DF} and Chl-a_C are the mean Chlorophyll-*a* concentrations in the two treatments and the control, respectively. Linear plots of the above parameters were used to assess food limitation and consumer density dependence, as shown in Figure 1. The method makes the assumption that the response of the rotifer community to the experimental manipulations is not density independent, especially over periods when rotifer densities are changing rapidly. The final response of Δr in relation to an increase of Chlorophyll-*a* after fertilization is expected to be a hyperbolic function. However, under the limited range of Chlorophyll-*a* concentrations used in the experiment, we have assumed a more or less linear response.

The parameters of the following equation

$$\Delta r_{D,DF} = y0 + a * \text{CHL}_{D,DF}$$

were derived from the experiment, and used to characterise the density-dependent limitation (intercept – $y0$) and Chlorophyll-*a* dependent limitation (slope – a).

The proportion of Δr variability explained by parameters $y0$ and a was tested using a linear regression. All of the data sets for correlations were tested for normality and variance homogeneity. They passed in all cases.

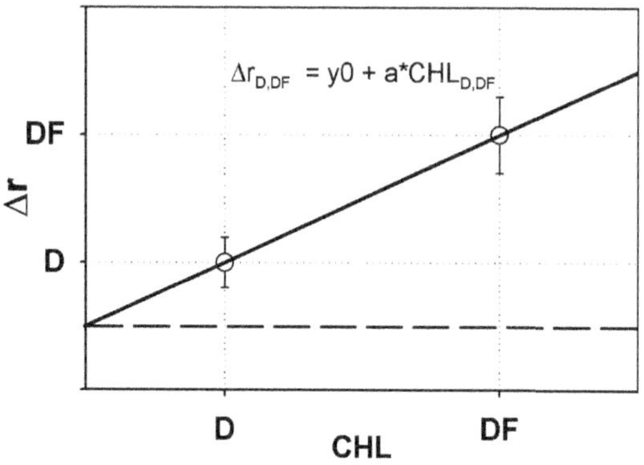

Figure 1. Linear plot of resource limitation in relation to increase in Chlorophyll-*a* concentration. Dashed line indicates the level of $y0$.

Results

Species dynamics

The presence of rotifer species in the Římov Reservoir can be divided into two phases in time. The first (spring) phase is dominated by *Synchaeta lakowitziana, Polyarthra dolichoptera* and *Keratella cochlearis.* The later phase is dominated by *K. cochlearis, Polyarthra vulgaris* and *Trichocerca similis.* These phases are separated by the clear water phase. All species reached high abundances during the spring peak in April (Fig. 2). This was followed by a marked decrease in abundance during the clear water phase in May (*Synchaeta* became extinct). By June, some of the species (*Polyarthra* spp., *K. cochlearis*) had increased again. The timing of the clear water decline differed according to species. *Synchaeta lakowitziana* was the first to decline, while *K. cochlearis* was the last. In all species, the birth rate declined from high values during the spring peak to lower values during the clear water phase. The birth rate of *Polyarthra* spp. increased again in mid May, but the increase in birth rate of the other species occurred later (Fig. 2).

Clear water phase

Temperature changes, Secchi-disk transparency and Chlorophyll-*a* concentration before and during the clear water phase are shown in Figure 3. The clear water phase began with a rapid increase in Secchi-disk transparency from about 2 m at the end of April to about 5 m in May. It ended with a slow decrease in the middle of June. Increased transparency coincided with a decline in phytoplankton, as characterized by Chlorophyll-*a* concentrations decreasing to $<1\ \mu g\ l^{-1}$ during the maximum transparency period. The proportion of large phytoplankton cells ($>20\ \mu m$) in the phytoplankton increased during the clear water phase. The epilimnetic mean temperature increased slowly over the investigation period.

Resource limitation experiments

Two way ANOVA (fixed effects) was used to analyse the overall species response to the experimental manipulations. Table 1 shows that the

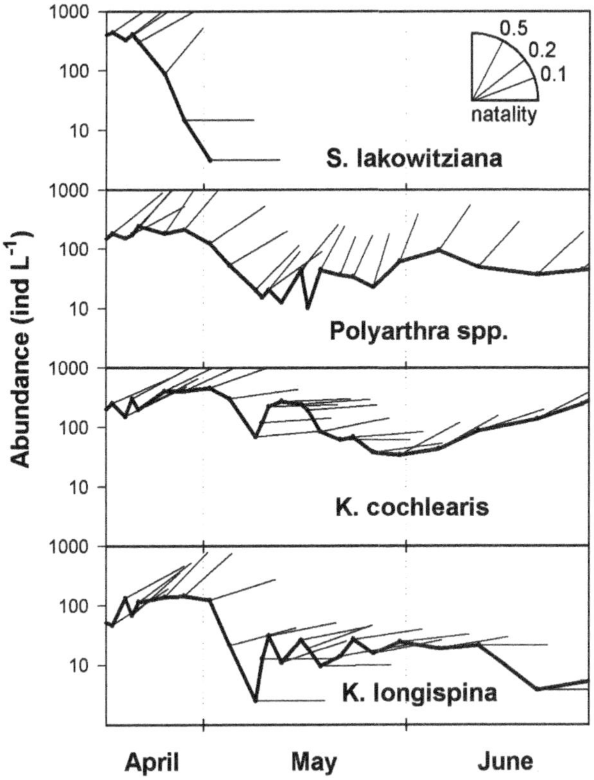

Figure 2. Rotifer abundance dynamics and instantaneous birth rates in Římov Reservoir in spring. The thick line shows abundance; the thin abscissae show actual natality (instantaneous birth rate). Mortality can be deduced from the difference between the slope of instaneous birth rate and the slope of the thick line, which reflects instantaneous growth rate (*r*).

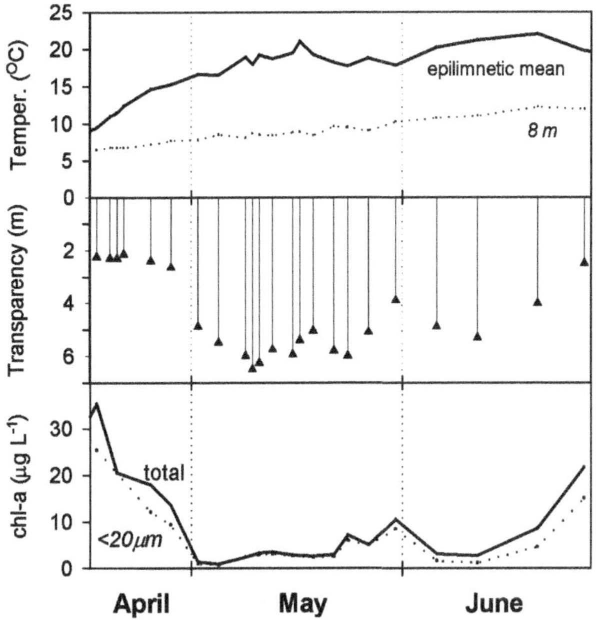

Figure 3. Changes in epilimnetic temperature, Secchi-disk transparency and the concentration of two size fractions of Chlorophyll-*a* in Římov Reservoir in spring.

Table 1. Two way ANOVA (fixed effects) on the population growth rate of rotifer populations in the experiments

Species	Experiment (df = 3)		Treatment (df = 2)		Experiment × Treatment (df = 6)		Error (df = 23)
	SS	F	SS	F	SS	F	SS
Synchaeta spp.	1.401	10.69***	1.884	21.59***	1.092	4.18**	1.012
Polyarthra spp.	0.123	12.59***	0.496	76.32***	0.150	7.71***	0.069
K. cochlearis	0.825	227.1***	0.112	46.22***	0.018	2.49	0.023
K. longispina	0.294	5.40**	0.102	2.80	0.276	2.56*	0.414

'Treatment' shows the sensitivity of species to experimental manipulations and 'Experiment' shows changes in the populations over the period of investigation.
$^*p<0.05$, $^{**}p<0.01$, $^{***}p<0.001$.

population growth rate of all rotifer species varied significantly among the experiments. Similarly, experimental manipulations among treatments showed a highly significant response in *Synchaeta* spp., *Polyarthra* spp. and *K. cochlearis*. However, *K. longispina* showed no significant response to the experimental manipulations. It is important to note that the interaction between "experiment" and "treatment" is significant in *Synchaeta* spp. and *Polyarthra* spp. This means that sensitivity to the experimental manipulations in these taxa changed over time. Otherwise, the effect of these two factors would be multiplicative. In contrast, the importance of this interaction in *K. cochlearis* was not significant.

Figure 4 shows the concentrations of Chlorophyll-*a* corresponding to 'edible' algae (<20 μm) at the start and end of the manipulation experiments. It is evident that both manipulations

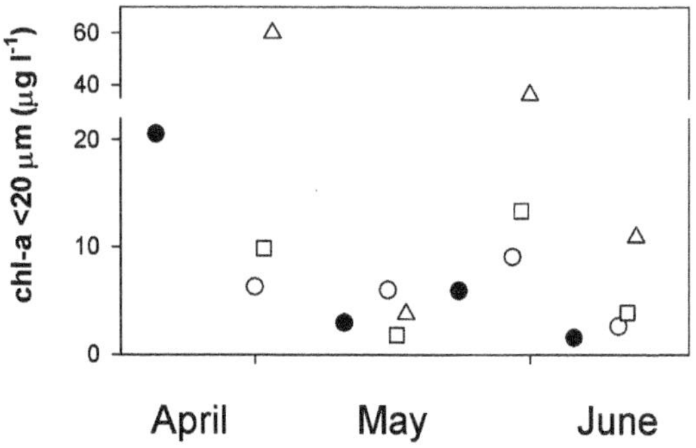

Figure 4. Changes in the fraction of Chlorophyll-*a* concentration representing algae <20 μm during the limitation experiments, shown on a real-time scale. The shaded circles mark the begining of each experiment. The end of the experiments are marked as follows: open circle – control, open square – diluted (D) and triangle – diluted and fertilized (DF).

promoted an increase in Chlorophyll-*a* concentration in all of the experiments, except the second one. In this case, the increase in Chlorophyll-*a* in treatment D was assigned zero and the increase in Chlorophyll-*a* in treatment DF was related to D instead of C.

The temporal changes of intensity of limitation (Δr) throughout the experimental period, computed for treatments D and DF, are shown in Table 2. Variability in response among the replicates was generally low. The most important changes were those associated with the parameters of the equation derived from this limitation. Parameter a (equivalent to slope), which is a measure of the importance of Chlorophyll-*a* limitation, increased in *Synchaeta* spp. from the spring peak through to the clear water phase. In contrast, in *Polyarthra* spp., this parameter changed very little throughout the experimental period. In *K. cochlearis*, Chlorophyll-*a* limitation was very low during the spring peak, relatively high during the start of the clear water phase and lower again, later. Parameter $y0$ (intercept) reflects the intensity of density-dependent limitation. In *Synchaeta* spp., the value of $y0$ was very high in the first (spring peak maximum) and last experiments. Similar patterns of variation between the first and last experiments were also found in *Polyarthra* spp., whereas $y0$ in *K. cochlearis* varied less dramatically (Table 2).

The method of linear regression was used to estimate the relative proportion of Δr variability explained by Chlorophyll-*a* and species density in all of the experiments (Tables 3 and 4, respectively). Chlorophyll-*a* concentration explained 89, 97 and 92% of the resource limitation in *Synchaeta* spp., *Polyarthra* spp. and *K. cochlearis*,

Table 2. Changes in resource limitation (Δr) in all of the experiments for each species and parameters of the equation $\Delta r = y0 + a$ CHL derived from the experiments

Species	Experiment	Δr_D		Δr_{DF}		Intercept	Slope
		Mean	SD	Mean	SD	$y0$	a
Synchaeta spp.	1	0.45	0.03	0.58	0.02	0.423	0.070
	2	0	0	0.06	0.39	0.000	0.204
	3	0.17	0.30	0.47	0.09	0.060	0.293
	4	0.78	0.16	1.20	0.12	0.624	0.406
Polyarthra spp.	1	0.31	0.02	0.41	0.01	0.295	0.052
	2	0.10	0.10	0.11	0.08	0.100	0.051
	3	0.17	0.05	0.13	0.07	–	–
	4	0.19	0.05	0.45	0.06	0.355	0.067
K. cochlearis	1	0.17	0.01	0.18	0.01	0.164	0.007
	2	0.02	0.02	0.06	0.03	0.022	0.140
	3	0.12	0.08	0.14	0.03	0.116	0.017
	4	0.12	0.10	0.16	0.06	0.097	0.048
K. longispina	1	0.07	0.01	−0.24	0.23	–	–
	2	0.02	0.10	0.08	0.10	0.018	0.221
	3	−0.03	0.22	0.06	0.03	−0.061	0.088
	4	0.01	0.30	−0.31	0.01	–	–

respectively. The proportion of variability explained by density-dependent factors was lower: 60% for *Synchaeta* spp. and 68% for *Polyarthra* spp. The proportion of variability explained by density-dependent factors in *K. cochlearis* was low and not significant. Surprisingly, Δr did not correlate with Chlorophyll-*a* concentration for any of the species in either the initial values, the whole data set or the experiments in the clear water phase (experiments 2–4).

Table 3. Linear regression using intensity of limitation Δr as the dependent variable and a*CHL explained as $a\text{*CHL} = \Delta r / y0$ as the independent variable

Species	df	R	R^2	F
Synchaeta spp.	7	0.774	0.60	8.96*
Polyarthra spp.	5	0.823	0.68	12.62*
K. cochlearis	7	0.235	0.06	0.35
K. longispina	3	0.389	0.15	0.36

R^2 shows the proportion of species Δr variability explained by density dependent factors over the experimental period.
*$p < 0.05$.

Discussion

The rapid decline of rotifers during the clear water phase is a well-known phenomenon as it is frequently observed in many lakes and reservoirs (Conde-Porcuna et al., 1994; Fussmann, 1996). The authors commonly refer to the rotifers as being outcompeted by large cladocerans, whose grazing causes a strong depression of Chlorophyll-*a* concentrations (Gilbert, 1988; May & Jones, 1989;

Table 4. Linear regression using intensity of limitation Δr as the dependent variable and $y0$ explained as $y0 = \Delta r / a\text{*CHL}$ as the independent variable

Species	Df	r	R^2	F
Synchaeta spp.	7	0.889	0.79	22.53**
Polyarthra spp.	5	0.967	0.94	58.92**
K. cochlearis	7	0.918	0.84	32.17**
K. longispina	3	0.575	0.33	0.99

R^2 shows the proportion of species Δr variability explained by Chlorophyll-*a* over the experimental period.
**$p < 0.01$.

MacIsaac & Gilbert, 1991). In our study, an important increase in food limitation, as indicated by Chlorophyll-*a* concentrations, was found between the first and later experiments in *Synchaeta* spp. and *K. cochlearis*. Food limitation increased throughout the experiments, especially in *Synchaeta* spp. This could explain the exclusion of this taxon from the plankton by the end of April. The highest impact of food limitation in *Synchaeta* supports the conclusion of Stemberger & Gilbert (1985) that this genus has a relatively high food threshold level and requires high food concentrations to saturate population growth rate. The experiments showed that the early spring population of *S. lakowitziana* had strongly decreased by the clear water phase, but persisted in undetectable densities. When the limiting conditions were experimentally removed by fertilization + population dilution, *Synchaeta* spp. was induced to grow. However, in the last experiment, the spring species *Synchaeta lakowitziana* was partially replaced by the summer *S. pectinata*. In addition, the effect of food limitation, an alternative explanation for the decrease of *S. lakowitziana* could be high susceptibility to interference from *Daphnia* (Gilbert, 1988), as well as the negative effect of predation by Cyclops vicinus (Devetter & Sed'a, in prep; Plasmann, 1997).

In contrast, *Polyarthra* spp. remained almost unaffected by Chlorophyll-*a* limitation throughout the experiments. This is in contrast to the observations of Gonzalez & Frost (1992), Merriman & Kirk (2000) and Cordova et al. (2001) who recorded considerable temporal variation in the intensity of food limitation in *Polyarthra*. This discrepancy can be explained by the complex feeding habits of *Polyarthra* (raptorial feeder), i.e., it does not feed exclusively on algae (Pourriot, 1977). However, in the longer-term, the availability of algal food seems to significantly affect the population growth of this species (Devetter & Sed'a, 2003). The relatively low food limitation of K. cochlearis can be explained by its ability to survive long periods of starvation, its non-selective feeding mode and its low threshold food concentration (Stemberger & Gilbert, 1985; Kirk, 1997; Cordova et al, 2001).

The results of our clear water limitation experiments showed resource limitation in three of the four rotifer taxa investigated. *Polyarthra* spp. *K. cochlearis* and *Synchaeta* spp. responded positively to dilution and fertilization in all of the experiments. *Kellicottia longispina* did not respond well to the experimental manipulations, probably due to its sensitivity to enclosures.

It is surprising that low food conditions persisted in the second experiment after dilution of the rotifer populations in the D treatment and fertilization in DF treatment. We suspect some imbalances in the nutrient cycling and probably high interference by bacteria in competition for phosphorus (Vrba, personal communication). As a result of this, the rotifers in the second experiment were definitely not saturated by the food and their Δr was primarily food limited.

The intensity of limitation of the rotifer taxa varied greatly among the experiments. Our Δr values for *Synchaeta* and *K. cochlearis* were similar to those found by Merriman & Kirk (2000). It is evident that the intensity of competition for algal food in the spring peak is lower than during the clear water phase (or stays unchanged for *Polyarthra* spp.). González & Frost (1992) conducted 15 food supplement experiments throughout the whole vegetative period and found *K. cochlearis* to be food limited in 47% of them in oligotrophic conditions, compared with 68% found by Merriman & Kirk (2000) in similar experiments with *K. cochlearis*.

We conclude that: (1) resource limitation is an important factor in structuring rotifer population dynamics, and (2) the importance of limitation by density-dependent factors and food varies with time and species. Resource limitation from food is more important than from consumer density-dependent factors.

Acknowledgements

This research was funded by a research grant AVOZ60660521. J. Macháček and anonymous referees are acknowledged for constructive comments on a previous version of the manuscript. Special thanks go to Viera Straskrabová for universal support and Keith Edwards for correcting the English.

References

Arndt, H., M. Krocker, B. Nixdorf & A. Kohler, 1993. Long-term annual and seasonal changes of metazooplankton and protozooplankton in Lake Müggelsee (Berlin) – effects of eutrophication, grazing activities, and the impact of predation. Internationale Revue der gesamten Hydrobiologie 78: 379–402.

Brandl, Z., B. Desortová, J. Hrbácek, J. Komárková, V. Vyhnálek, J. Sed'a & M. Straskraba, 1989. Seasonal changes of zooplankton and phytoplankton and their mutual relations in some Czechoslovak reservoirs. Archiv für Hydrobiologie, Beiheft Ergebnisse der Limnologie 33: 597–604.

Ciros-Perez, J., M. J. Carmona & M. Serra, 2001. Resource competition between sympatric sibling rotifer species. Limnology and Oceanography 46: 1511–1523.

Conde-Porcuna, J. M., R. Morales-Baquero & L. Cruz-Pizarro, 1994. Effects of *Daphnia longispina* on rotifer populations in a natural environment – relative importance of food limitation and interference competition. Journal of Plankton Research 16: 691–706.

Cordova, S. E., J. Giffin & K. L. Kirk, 2001. Food limitation of planktonic rotifers: field experiments in two mountain ponds. Freshwater Biology 46: 1519–1527.

Deneke, R. & B. Nixdorf, 1999. On the occurrence of clear-water phases in relation to shallowness and trophic state: a comparative study. Hydrobiologia 409: 251–262.

Devetter, M., 1998. Influence of environmental factors on the rotifer assemblage in an artificial lake. Hydrobiologia 387: 171–178.

Devetter, M. & J. Sed'a, 2003. Rotifer fecundity in relation to components of microbial food web in a eutrophic reservoir. Hydrobiologia 504: 167–175.

Edmondson, W. T., 1960. Reproductive rates of rotifers in natural populations. Memories Istituto Italiano Idrobiologia 12: 21–77.

Fussmann, G., 1996. The importance of crustacean zooplankton in structuring rotifer and phytoplankton communities – an enclosure study. Journal of Plankton Research 18: 1897–1915.

Gilbert, J. J., 1985. Competition between rotifers and *Daphnia*. Ecology 66: 1943–1950.

Gilbert, J. J., 1988. Suppression of rotifer populations by *Daphnia*: A review of the evidence, the mechanisms, and the effects on zooplankton community structure. Limnology and Oceanography 33: 1286–1303.

González, E. J., 2000. Nutrient enrichment and zooplankton effects on the phytoplankton community in microcosms from El Andino reservoir (Venezuela). Hydrobiologia 434: 81–96.

González, M. J. & T. M. Frost, 1992. Food limitation and seasonal population declines of rotifers. Oecologia 89: 560–566.

Herzig, A., 1983. Comparative studies on the relationship between temperature and duration of embryonic development of rotifers. Hydrobiologia 104: 237–246.

Kirk, K. L., 1997. Life-history responses to variable environments: starvation and reproduction in planktonic rotifers. Ecology 78: 434–441.

Kirk, K. L., 2002. Competition in variable environments: experiments with planktonic rotifers. Freshwater Biology 47: 1089–1096.

Komárková, J., 1989. Changes of phytoplankton assemblage during the spring period in the moderately eutrophic Rímov reservoir (South Bohemia). Archiv für Hydrobiologie, Beiheft Ergebnisse der Limnologie 33: 419–433.

Lampert, W., W. Fleckner, H. Rai & B. E. Taylor, 1986. Phytoplankton control by grazing zooplankton – A study on the spring clear-water phase. Limnology and Oceanography 31: 478–490.

Likens, G. E. & J. J. Gilbert, 1970. Notes on quantitative sampling of natural populations of planktonic rotifers. Limnology and Oceanography 15: 816–820.

Lorenzen, C. J., 1967. Determination of Chlorophyll and pheopigments: spectrophotometric equations. Limnology and Oceanography 12: 343–346.

MacIsaac, H. J. & J. J. Gilbert, 1991. Discrimination between exploitative and interference competition between Cladocera and *Keratella cochlearis*. Ecology 72: 924–937.

May, L. & D. H. Jones, 1989. Does interference competition from *Daphnia* affect populations of *Keratella cochlearis* in Loch Leven, Scotland. Journal of Plankton Research 11: 445–461.

Merriman, J. L. & K. L. Kirk, 2000. Temporal patterns of resource limitation in natural populations of rotifers. Ecology 8: 141–149.

Müller-Navarra, D. & W. Lampert, 1996. Seasonal patterns of food limitation in *Daphnia galeata*: Separating food quantity and food quality effects. Journal of Plankton Research 18: 1137–1157.

Pace, M. L. & J. D. Orcutt, 1981. The relative importance of protozoans, rotifers, and crustaceans in a freshwater zooplankton community. Limnology and Oceanography 26: 822–830.

Paloheimo, J. E., 1974. Calculation of instantaneous birth rate. Limnology and Oceanography 19: 692–694.

Pennak, R. W., 1962. Quantitative zooplankton sampling in littoral vegetation areas. Limnology and Oceanography 7: 487–489.

Plassmann, T., G. Maier & H. B. Stich, 1997. Predation impact of *Cyclops vicinus* on the rotifer community in the Lake Constance in spring. Journal of Plankton Research 19: 1069–1079.

Pourriot, R., 1977. Food and feeding habits of Rotifera. Archiv für Hydrobiologie, Beiheft Ergebnisse der Limnologie 8: 243–260.

Rothhaupt, K. O., 1990. Resource competition of herbivorous zooplankton – A review of approaches and perspectives. Archiv für Hydrobiologie 118: 1–29.

Scheffer, M., S. Rinaldi, Y. A. Kuznetsov & E. H. Vannes, 1997. Seasonal dynamics of *Daphnia* and algae explained as a periodically forced predator-prey system. Oikos 80: 519–532.

Sed'a, J., 1989. Main factors affecting spring development of herbivorous Cladocera in the Rímov Reservoir (Czechoslovakia). Archiv für Hydrobiologie, Beiheft Ergebnisse der Limnologie 33: 619–630.

Sed'a, J. & J. Kubecka, 1997. Long-term biomanipulation of Rímov Reservoir (Czech-Republic). Hydrobiologia 345: 95–108.

Sommer, U., A. Duncan, Z. M. Gliwicz & W. Lampert, 1986. The PEG-model of seasonal succession of planktonic events in freshwaters Archiv für Hydrobiologie 106: 433–471.

Stemberger, R. S. & J. J. Gilbert, 1985. Body size, food concentration and population growth in planktonic rotifers. Ecology 66: 1151–1159.

Vyhnálek, V., J. Komárková, J. Sed'a, Z. Brandl, K. Simek & N. Johanisová, 1991. Clear-water phase in the Rímov Reservoir (South Bohemia): controlling factors. Internationale Vereinigung für Theoretische und Angewandte Limnologie, Verhandlungen 24: 1336–1339.

Zoufal, W, 1989. Developmental times of *Synchaeta oblonga* eggs from the Danube (Austria). Hydrobiologia 186/187: 163–165.

Hydrobiologia (2005) 546:519–534
A. Herzig, R.D. Gulati, C.D. Jersabek & L. May (eds.) Rotifera X: Rotifer Research: Trends, New Tools and Recent Advances
DOI 10.1007/s10750-005-4295-y

Combined effects of food concentration and temperature on competition among four species of *Brachionus* (Rotifera)

Mario A. Fernández-Araiza[1,*], S.S.S. Sarma[2] & S. Nandini[3]
[1]*Doctoral Programme, Autonomous Metropolitan University, Campus Iztapalapa, Av. Michoacán y la Purísima, Col. Vicentina, Mexico City, 09340, Mexico*
[2]*Laboratory of Aquatic Zoology, Division of Research and Postgraduate Studies, National Autonomous University of Mexico, Campus Iztacala, Av. de Los Barrios, No. 1, AP 314, CP 54090, Tlalnepantla, State of Mexico, Mexico*
[3]*UIICSE, Division of Research and Postgraduate Studies, National Autonomous University of Mexico, Campus Iztacala, Av. de Los Barrios, No. 1, AP 314, CP 54090, Tlalnepantla, State of Mexico, Mexico*
*(*Author for correspondence: E-mail: mafa@campus.iztacala.unam.mx)*

Key words: multispecies competition, *Brachionus* Rotifers, temperature, population density

Abstract

We evaluated the effect of algal food density (1.5×10^6, 3.0×10^6 and 4.5×10^6 cells ml^{-1} of *Chlorella*) and temperature (22° and 28 °C) on competition among the rotifers *Brachionus calyciflorus*, *Brachionus havanaensis*, *Brachionus patulus* and *Brachionus rubens*, based on population growth experiments for 24 days. The growth experiments were conducted seperately for each individual rotifer species (i.e., controls), and in mixtures of all four species in equal initial proportions (i.e., under competition). The population growth of *B. calyciflorus*, *B. havanaensis*, *B. patulus* and *B. rubens* grown separately at two temperatures and at three algal food densities showed typical patterns of lag, exponential and retardation phases in the controls. This pattern differed considerably under competition. In general, we observed that in all of the test species, the highest growth rates were observed at higher food levels and in the absence of congenerics. At 22 °C, under the lowest food level, the differences in the population abundances of *B. havanaensis*, *B. patulus* and *B. rubens* grown alone and in the presence of competition were large. However, these differences reduced as food density was increased from 0.5×10^6 to 4.5×10^6 cells ml^{-1}. At 28 °C and at the lowest food level, all of the other rotifer species eliminated *B. havanaensis* in mixed cultures. Each brachionid species had a higher rate when grown alone than when cultured with other species. The highest *r* (mean ± standard error: 0.54 ± 0.01 day^{-1}) was recorded for *B. havanaensis* at 28 °C under 4.5×10^6 cells ml^{-1} of algal food density. At 28 °C at low algal food density, the presence of competitors resulted in negative population growth rates for three of the four rotifer species tested.

Introduction

Competition and predation are the two strongest natural forces structuring freshwater zooplankton communities (Chase et al., 2002). Within zooplankton communities, natural competitive processes occur over several centuries (Dumont, 1994). While competitive interactions under field conditions can be analyzed using zooplankton collections over several seasons, laboratory studies yield quantitative information over relatively short periods, which can be used for developing predictive models. Basically competition among zooplankton can be analyzed using experimental population growth where two or more species are maintained over a period of time and changes in the relative abundances of each species are quantified.

Several competition experiments have been conducted for rotifers using the population growth method. These include those of Gilbert (1988) who used this approach to study exploitative and interference competition between cladocerans and rotifers. In contrast, Rothhaupt (1988) took a more mechanistic approach to understanding competition between two *Brachionus* species (*B. calycflorus* and *B. rubens*). He found that mechanical interference among zooplankton species of similar size is nearly absent.

Exploitative competition depends mainly on the capacity to reproduce at minimal food levels (Stemberger & Gilbert, 1985) thus driving competitions dynamics (Grover, 1997). A low food threshold can be achieved through high food gathering efficiency (Gilbert, 1988; Schneider, 1990), resource storage capability, (Kirk, 1997) high assimilation efficiency, and so on. In turn, some of these factors can be improved through optimal resource allocation in reproduction and survival (Glazier & Calow, 1992). These aspects form a basis for understanding the role of competition in zooplankton community structure; nevertheless, there is some concern that these are mostly based on two species interactions and are thus simplified (Sarma et al., 2003a). Therefore they may not be helpful for understanding competition among many species, which occurs in natural waterbodies with variable environmental conditions such as diel changes in algal production (Hutchinson, 1967). In tropical waterbodies, for example, this is largely evident where some times up to 10 *Brachionus* species coexist (Nogrady et al., 1993). This does not suggest the absence of competition, but that the intensity of competition may not be high enough to lead to exclusion of a particular species.

Members of *Brachionus* have very similar dietary preferences often feeding on bacteria and smaller non-colonial and non-filamentous algae (Dumont, 1977; Pourriot, 1977). So, competition among members of *Brachionus* could be dependent on environmental factors. In nature two of the most important factors affecting the survival and reproduction of rotifers are food concentration and temperature (Edmondson, 1946). These factors may also have some influence on the competitive outcome. For example, Sarma et al. (1996) have shown that depending on food level, *B. calyciflorus* was able to replace or was out-competed by the other rotifer *Anuraeopsis fissa*. Another important factor is the initial density of competitors, which has been shown to affect the outcome of competition in both rotifers and cladocerans (Hurtado-Bocanegra et al., 2002).

In several water bodies in Mexico we find that different species of brachionid rotifers coexist. For example, in the Lake Xochimilco, more than 10 *Brachionus* species have been found simultaneously in a single locality, including the species chosen for this study (Flores-Burgos et al., 2003). Therefore, we decided to test the effect of two levels of temperature (22 and 28 °C) and three food levels (1.5×10^6, 3.0×10^6 and 4.5×10^6 cells ml^{-1} of *Chlorella vulgaris*) on the outcome of competition between four species of *Brachionus* (*B. calyciflorus*, *B. havanaensis*, *B. patulus* and *B. rubens*).

Materials and methods

The experimental strains of *Brachionus* species were isolated from the following water bodies: *Brachionus calyciflorus* from Lake Chapultepec, Mexico City; *Brachionus rubens* from Lake Aragon, Mexico City and *Brachionus havanaensis* from Lake Xochimilco, Mexico City and *Brachionus patulus* from the Santa Elena Reservoir, State of Mexico. All of the species were maintained under laboratory conditions for more than a year prior to experimentation. All of the rotifer species were maintained in EPA medium (Anonymous, 1985), and fed on the single-celled green alga, *Chlorella vulgaris*. The EPA medium was prepared by dissolving 96 mg $NaHCO_3$, 60 mg $CaSO_4$, 60 mg $MgSO_4$ and 4 mg KCL in 1 l of distilled water. *Chlorella* was mass cultured using Bold's basal medium (Borowitzka & Borowitzka, 1988). Algae in the logarithmic phase of growth were harvested and centrifuged at 4000 rpm for 5 min and later resuspended in distilled water. Alga density was estimated using haemocytometer.

The experiments were conducted at 22 °C and 28 °C, using three different food levels, i.e., 1.5×10^6, 3.0×10^6 and 4.5×10^6 cells ml^{-1} of *Chlorella*. The experiments were performed in 60 ml capacity jars containing 20 ml EPA medium and *C. vulgaris* at the appropriate concentration.

The experiments were initiated with a density of 1 ind. ml^{-1} of each rotifer species, The following treatments were set up for each food level and temperature combination: *B. calycilorus* (B. c.) alone, *B. rubens* (B. r.) alone, *B. havanaensis* (B. h.) alone, *B. patulus* (B. p.) alone and a 1:1:1:1 mixture (Five individuals from each species) of all four species, Four replicates were set up for each treatment, Thus, in total, there were 120 test jars.

Every day, the rotifer density was counted, initially taking total counts and later using three or four 1-ml aliquots. After counting, the individuals were transferred to fresh culture medium. The experiments were terminated after 20 days by which time the populations had begun to decline, Population growth rates were derived for the exponential phase of growth using the formula $r = (\ln Nt - \ln N_o)/t$ where N_o and N_t are the initial and final population densities and t is the time in days (Krebs, 1985). In treatments where a clear peak of population abundance was not evident, the growth rate was calculated from the slope of ln N_t against time. The differences among treatments for peak population density and growth rates were statistically evaluated using ANOVA following Sokal & Rohlf (2000) and the statistical package Statistica version 6.0.

Results

The population densities of *B. calyciflorus*, *B. havanaensis*, *B. patulus* and *B. rubens* cultured separately and together at 22 °C and at three algal food densities (1.5×10^6, 3.0×10^6 and 4.5×10^6 cells ml^{-1} of *Chlorella*) are presented in Figures 1–3. Whichever rotifer species was involved, an increase in food density resulted in an increase in population density. Similarly, regardless of the food level, all the tested rotifer species showed higher population abundances when grown alone than under competition. In most cases *B. calyciflocus* reached peak abundances earlier than the rest of the species. *Branchionus havanaensis* and *B. patulus* had a longer initial lag phase (i.e., about 3–5 days) than the other species. At the lowest food level, the differences in the population abundances of *B. havanaensis*, *B. patulus* and *B. rubens* grown alone and in the presence of competition were large. However, these differences narrowed as food density increased from 1.5×10^6 to 4.5×10^6 cells ml^{-1}.Population growth of all of the brachionid species grown at 28 °C differed to some extent from that recorded under similar conditions at 22 °C (Figs. 4–6). At higher food levels, the presence of competitors had a greater effect on the population growth of *B. calyciflonus*. In contrast, at this temperature and at low food level, the rest of the competing species eliminated *B. havanaensis*. In general, regardless of the food or temperature conditions, the presence of other species caused a considerable reduction in the abundance of any of the four rotifer species.

The trends in the population growth were also reflected in the rate of population increase (r day^{-1}) (Fig. 7). In general, regardless of temperature, an increase in food level enhanced r for in the majority of cases. When grown alone, each brachionid species had a higher growth rate than when cultured together with others. The highest r value (0.54 ± 0.01 day^{-1}) was recorded for *B. havanaensis* at 28 °C under 4.5×10^6 cells ml^{-1} of algal food density. At 28 °C and at low algal food density, the presence of competitors caused negative population growth rates for three of the four tested rotifer species.

Statistically, food density, temperature and the presence of competitors had a significant influence on the peak population abundance and the rate of population increase of all the tested species ($p < 0.001$, 3-way ANOVA, Table 1). The interaction of the three factors (food × temperature × competition) was significant ($p < 0.001$) for the rate of population increase for all the four *Brachionus* species but not for peak population density ($p < 0.05$).The effect of food × temperature on the peak population density was significant for *B. havanaensis* and *B. rubens* and for the rate of population increase for *B. calyciflorus* and *B. rubens*; other cases were non-significant. In contrast, food × competition had a significant influence on r in all of the species and on peak population density in three of the four species.

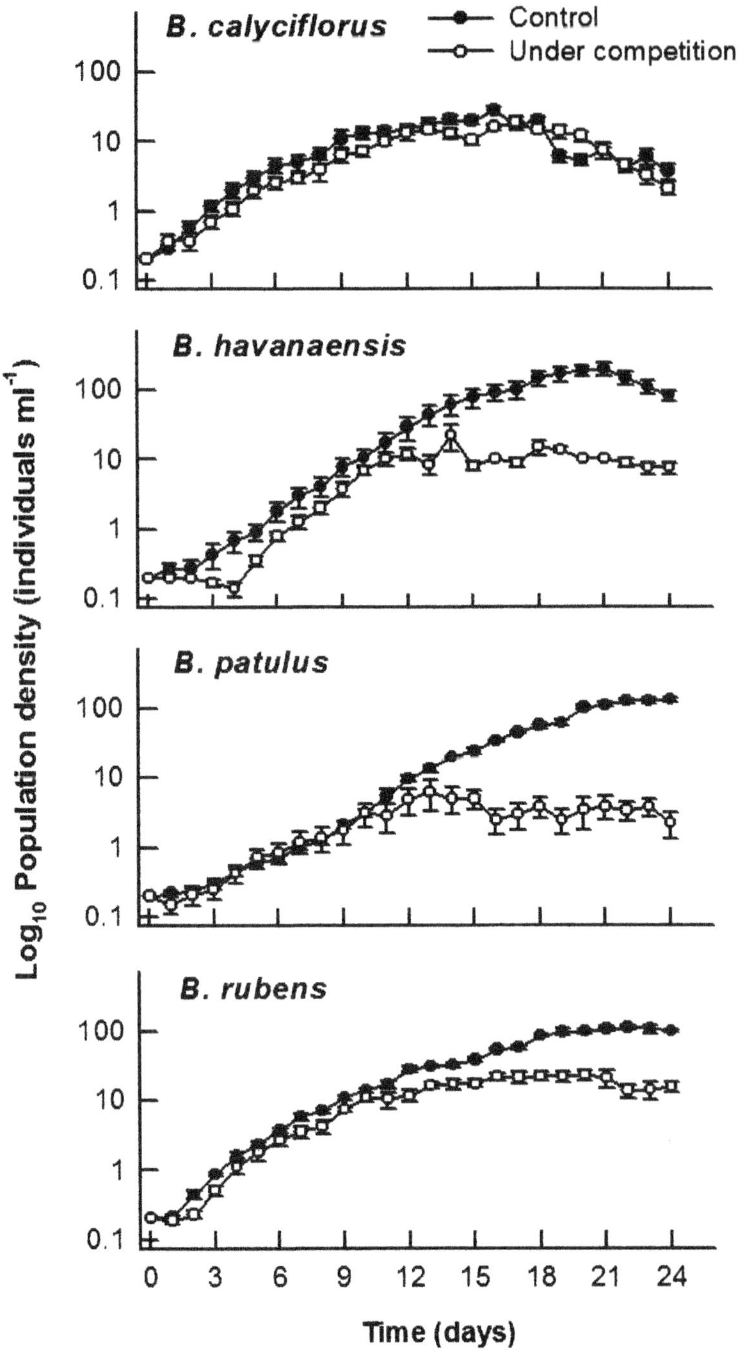

Figure 1. Population growth curves of *B. calyciflorus*, *B. havanaensis*, *B. patulus* and *B. rubens* grown separately and together at 22 °C under 1.5×10^6 cells ml^{-1} of *Chlorella* density. Shown are the mean ± standard error based on four replicates.

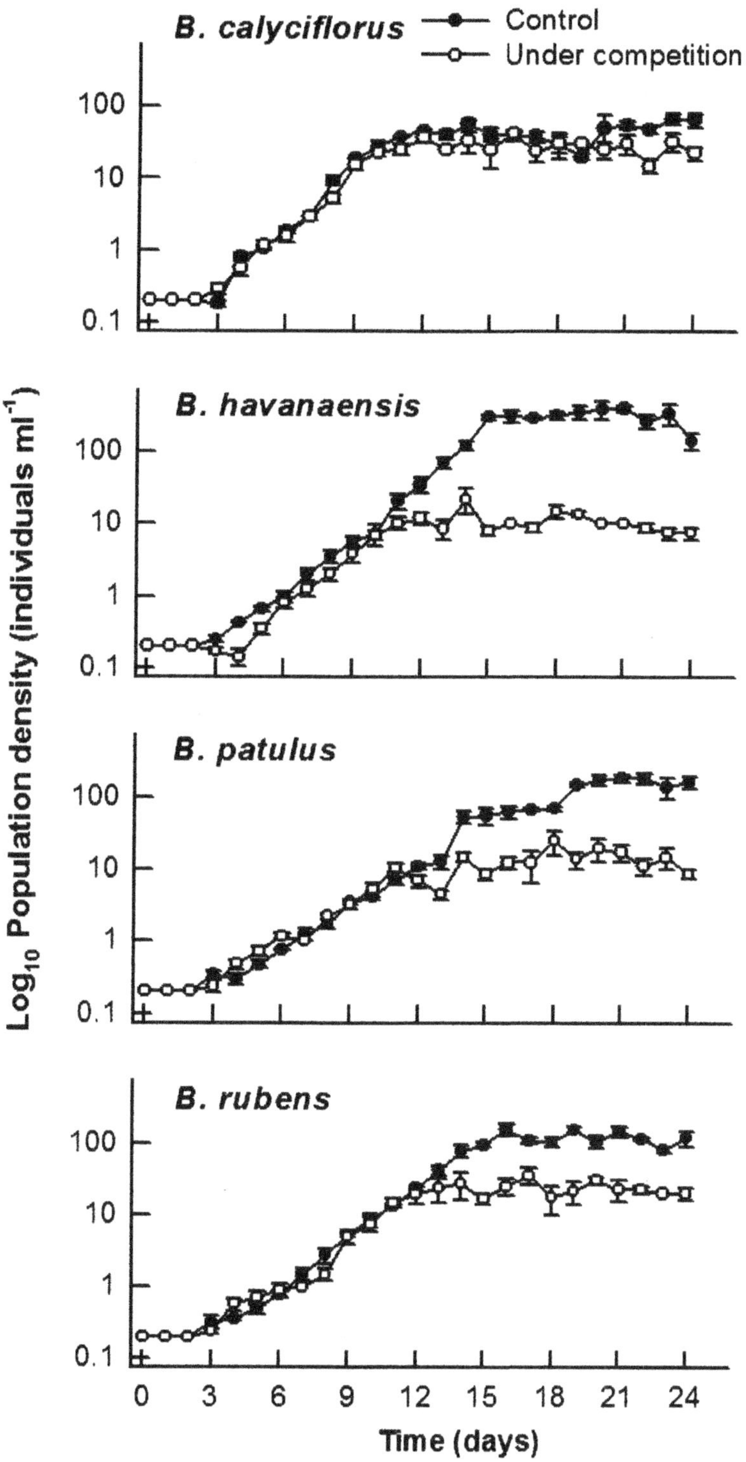

Figure 2. Population growth curves of *B. calyciflorus*, *B. havanaensis*, *B. patulus* and *B. rubens* grown separately and together at 22 °C under 3.0×10^6 cells ml^{-1} of *Chlorella* density. Shown are the mean ± standard error based on four replicates.

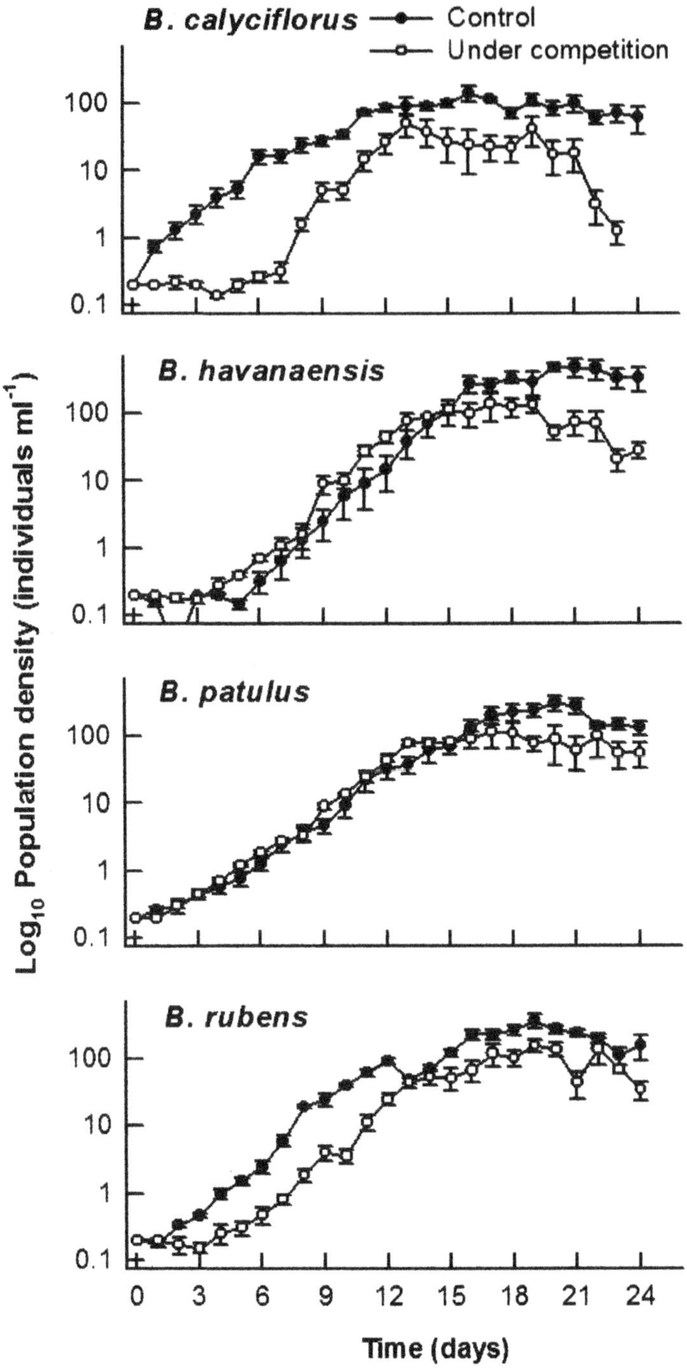

Figure 3. Population growth curves of *B. calyciflorus*, *B. havanaensis*, *B. patulus* and *B. rubens* grown separately and together at 22 °C under 4.5×10^6 cells ml^{-1} of *Chlorella* density. Shown are the mean ± standard error based on four replicates.

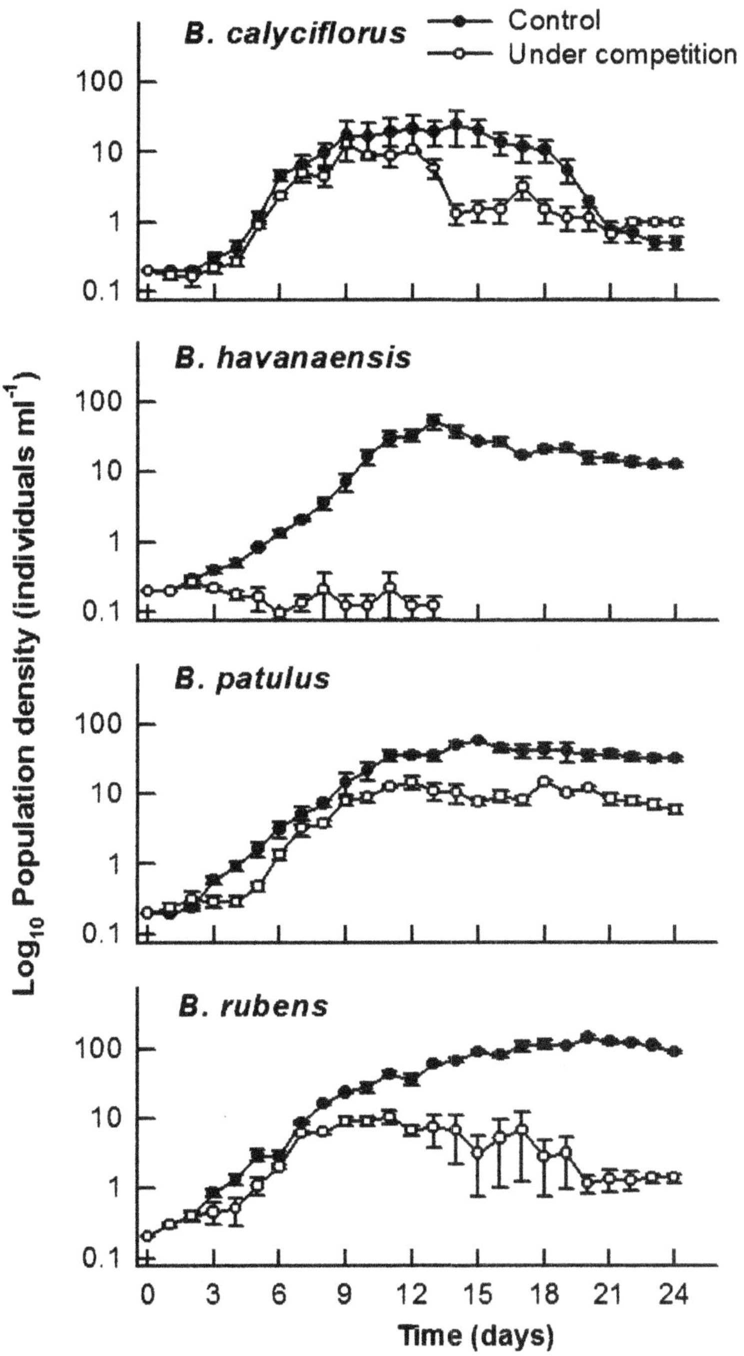

Figure 4. Population growth curves of *B. calyciflorus*, *B. havanaensis*, *B. patulus* and *B. rubens* grown separately and together at 28 °C under 1.5×10^6 cells ml^{-1} of *Chlorella* density. Shown are the mean ± standard error based on four replicates.

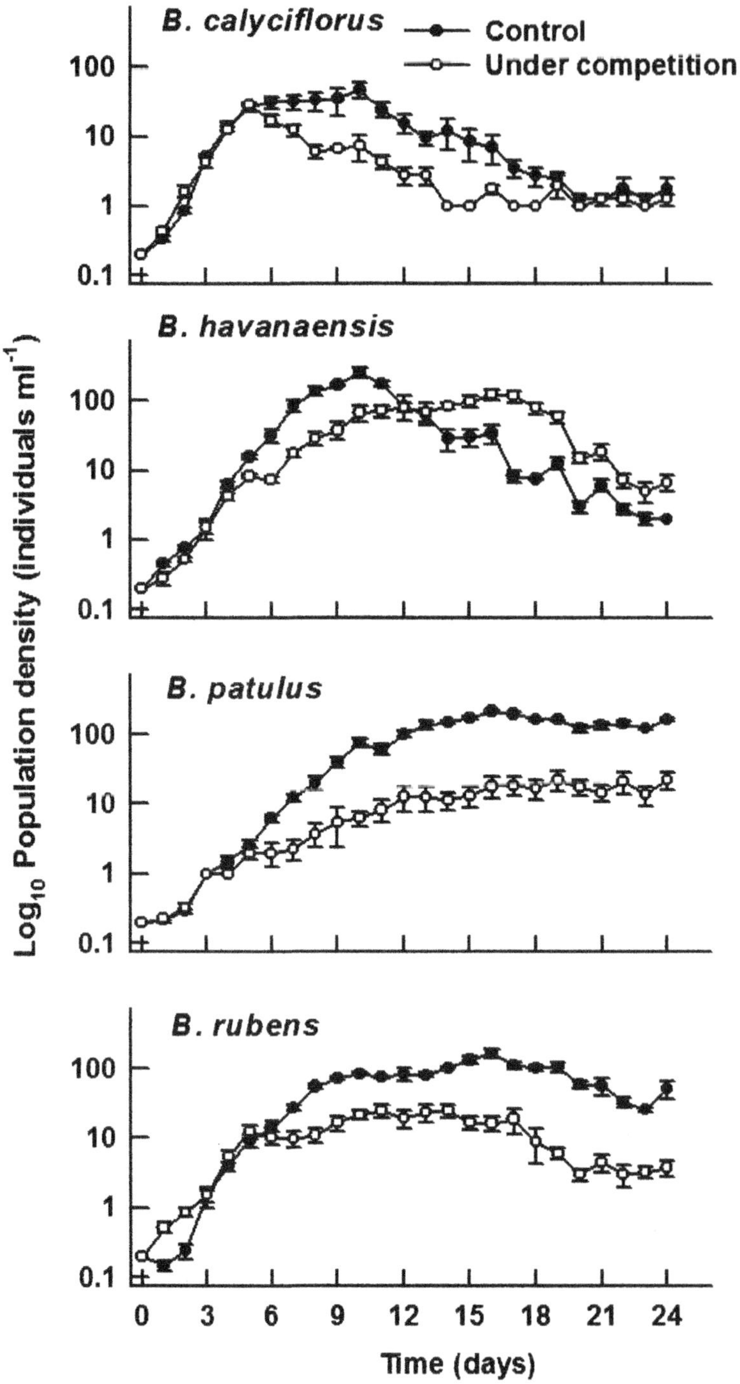

Figure 5. Population growth curves of *B. calyciflocus*, *B. havanaensis*, *B. patulus* and *B. rubens* grown separately and together at 28 °C under 3.0×10^6 cells ml^{-1} of *Chlorella* density. Shown are the mean ± standard error based on four replicates.

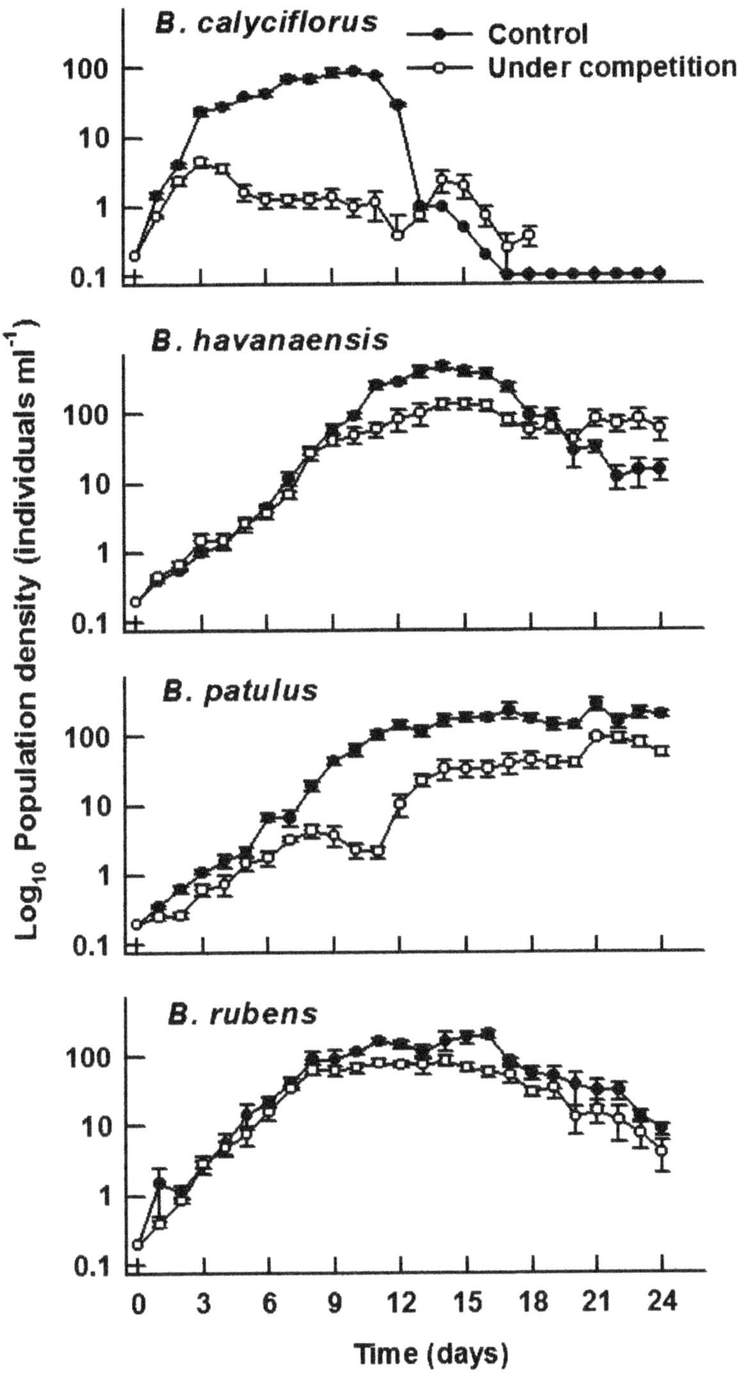

Figure 6. Population growth curves of *B. calyciflocus*, *B. havanaensis*, *B. patulus* and *B. rubens* grown separately and together at 28 °C under 4.5×10^6 cells ml^{-1} of *Chlorella* density. Shown are the mean ± standard error based on four replicates.

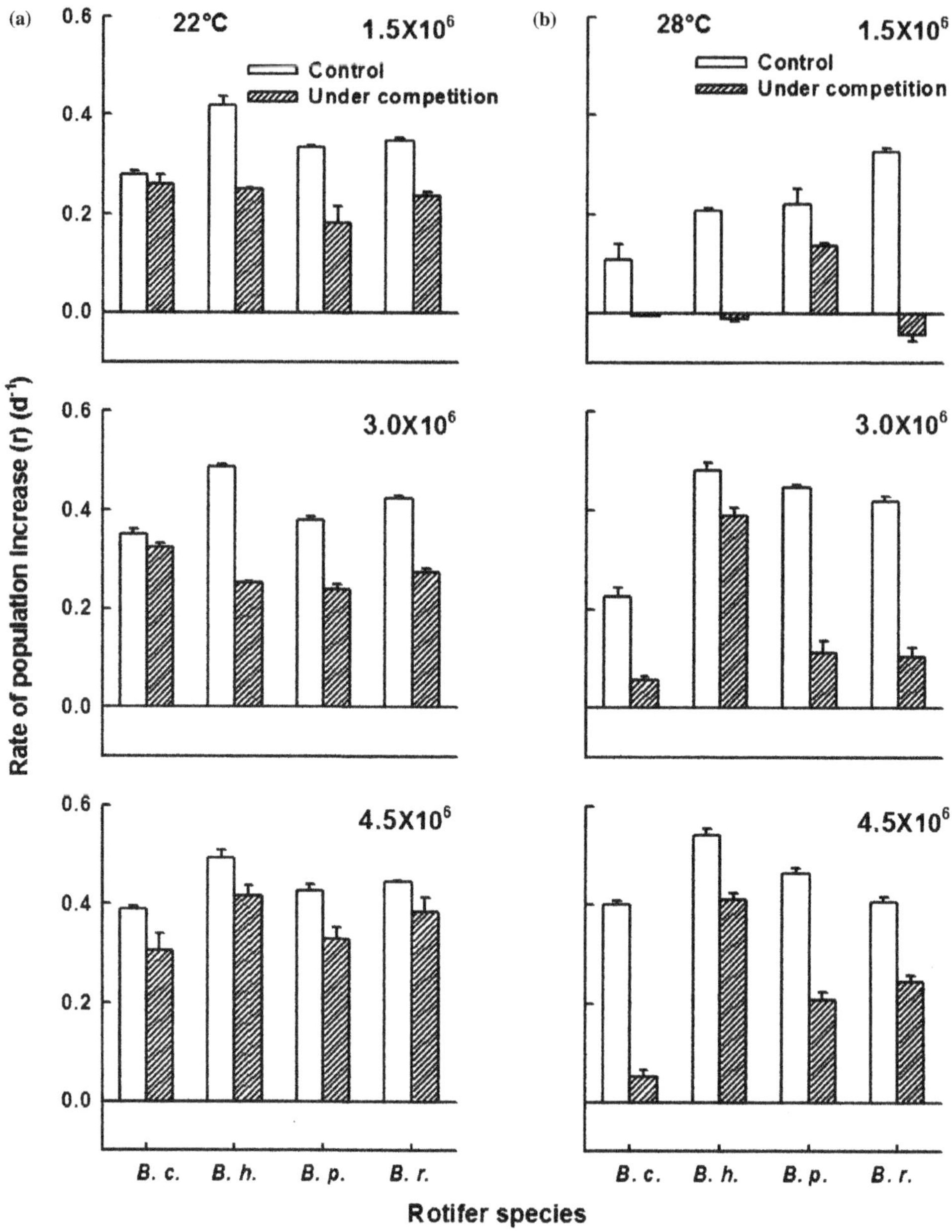

Figure 7. Rate of population increase r per day of *B. calyciflocus* (B. c.), *B. havanaensis* (B. h.), *B. patulus* (B. p.) and *B. rubens* (B. r.) grown separately and together at 22° (Column A) and 28 °C (Column B) under 1.5×10^6, 3.0×10^6 and 4.5×10^6 cells ml^{-1} of *Chlorella* density. Shown are the mean ± standard error based on four replicates.

Temperature × competition had a significant effect on peak population density only in case of *B. havanaensis*; and on r for *B. patulus* and *B. rubens*.

Discussion

Competition is best studied using specialised devices such as chemostats and turbidostats, where

Table 1. Results of the 3-way analysis of variance performed on the peak population density and rate of population increase per day of *B. calyciflorus*, *B. havanaensis*, *B. patulus* and *B. rubens* grown at three temperatures, two food levels and with and without the presence of competitors

Variable/Source	DF	SS	MS	F-ratio
Peak population abundance				
B. calyciflorus				
Food density (A)	2	25 642.38	12 821.19	19.02***
Temperature (B)	1	14 334.79	14 334.79	21.26***
Competition (C)	1	21 189.01	21 189.01	31.43***
Interaction (A × B)	2	7426.98	3713.49	5.51**
Interaction (A × C)	2	13 108.57	6554.29	9.72***
Interaction (B × C)	1	263.68	263.68	0.39ns
Interaction (A × B × C)	2	324.02	162.01	0.24ns
Error	36	24 272.31	674.23	
B. havanaensis				
Food density (A)	2	632 365.25	316 182.63	38.52***
Temperature (B)	1	57 073.00	57 073.00	6.95*
Competition (C)	1	809 937.00	809 937.00	98.68***
Interaction (A × B)	2	2172.50	1086.25	0.13ns
Interaction (A × C)	2	133 305.00	66 652.50	8.12**
Interaction (B × C)	1	98 541.75	98 541.75	12.01**
Interaction (A × B × C)	2	36 839.50	8207.38	2.24 ns
Error	36	295 465.50	8207.38	
B. patulus				
Food density (A)	2	264 577.31	132 288.66	67.42***
Temperature (B)	1	12 080.88	12 080.88	6.16*
Competition (C)	1	354 062.56	354 062.56	180.43***
Interaction (A × B)	2	5541.75	2770.88	1.41 ns
Interaction (A × C)	2	3282.31	21 641.16	11.03**
Interaction (B × C)	1	6475.13	6475.13	3.30 ns
Interaction (A × B × C)	2	4433.63	2216.81	1.13 ns
Error	36	70 642.75	1962.30	
B. rubens				
Food density (A)	2	222 719.75	111 359.88	46.34***
Temperature (B)	1	30 250.50	30 250.50	1239**
Competition (C)	1	227 838.50	227 838.50	94.80***
Interaction (A × B)	2	51 044.88	25 522.44	10.62**
Interaction (A × C)	2	3024.38	1512.19	0.63 ns
Interaction (B × C)	1	105.00	105.00	0.04 ns
Interaction (A × B × C)	2	5037.75	2518.88	1.05 ns
Error	36	86 517.25	2403.26	
Rate of population increase				
B. calyciflorus				
Food density (A)	2	0.16	0.08	60.07***
Temperature (B)	1	0.35	0.35	69.33***
Competition (C)	1	0.22	0.22	166.21***
Interaction (A × B)	2	0.01	0.01	5.38**
Interaction (A × C)	2	0.04	0.02	15.72***

Continued on p. 530

Table 1. (Continued)

Variable/Source	DF	SS	MS	F-ratio
Interaction (B × C)	1	0.07	0.07	2.66 ns
Interaction (A × B × C)	2	0.02	0.01	9.05**
Error	36	0.05	0.001	
B. havanaensis				
Food density (A)	2	0.54	0.27	434.45***
Temperature (B)	1	0.03	0.03	47.66***
Competition (C)	1	0.28	0.28	453.79***
Interaction (A × B)	2	0.22	0.11	175.10***
Interaction (A × C)	2	0.02	0.01	14.21***
Interaction (B × C)	1	0.001	0.001	0.81 ns
Interaction (A × B × C)	2	0.03	0.01	20.71***
Error	36	0.02	0.001	
B. patulus				
Food density (A)	2	0.15	0.08	56.85***
Temperature (B)	1	0.03	0.03	22.09***
Competition (C)	1	0.38	0.38	279.11***
Interaction (A × B)	2	0.01	0.005	2.02 ns
Interaction (A × C)	2	0.03	0.01	10.10***
Interaction (B × C)	1	0.03	0.03	20.28***
Interaction (A × B × C)	2	0.04	0.02	15.03***
Error	36	0.05	0.001	
B. rubens				
Food density (A)		0.19	0.09	132.84***
Temperature (B)	1	0.14	0.14	200.04***
Competition (C)	1	0.46	0.46	653.81***
Interaction (A × B)	2	0.01	0.01	7.77**
Interaction (A × C)	2	0.04	0.02	30.15***
Interaction (B × C)	1	0.09	0.09	132.65***
Interaction (A × B × C)	2	0.01	0.01	9.10***
Error	36	0.03	0.001	

DF = degrees of freedom, SS = sum of squares, MS = mean square, F-ratio = Fisher), *** = $p < 0.001$, ** = $p < 0.01$, * = $p < 0.05$, ns = non-significant ($p < 0.05$).

the competing species receive a constant and continuous supply of food during the entire experimental period (Rothhaupt, 1988). In static renewal systems, such as those used in this study, there is a continuous decline of food concentration as a result of rotifer feeding. This happens more drastically as the zooplankton population density increases. So, the test individuals receive a fixed quantity of food per day, the availability of which could depend on the next change in the population density. Rothhaupt (1990) determined the filtration rate of *B. calyciflorus* to be about 2.5 μl ind.$^{-1}$ h^{-1}. If this filtration rate is applied to the total density (about 500 ind. ml^{-1} of *B. havanaensis* (a species much smaller than *B. calyciflorus*) in these experiments, one might conclude that no food would be left in the test jars shortly after the food and culture medium had been replaced. However, such as an extrapolation is not reasonable. Firstly, in our study, *B. calyciflorus* never reached a density of 500 ind. ml^{-1}. Secondly, Rothhaupt's (1990) data are based on a single species population (i.e., filtration rates not measured under competition); under competition, filtration rates may be much lower. Though not quantified, our observations of the test jars every day revealed that certain algal cells always remained in the water column. Moreover, populations of all species increased since the initiation of the experiment. The use of chemostats and turbidostats is restricted because of the limitations posed during the standardization, especially when a large number of treatments are involved (Nandini & Sarma, 2003), Nevertheless, the use of static renewal systems to study competition has also often been used and is relevant to natural systems since it has been shown that zooplankton experience periodic starvation over a 24 h diel cycle (Zaret, 1980).

One possibility for studying competition in static renewal system is by transferring a certain part of the original test population daily but by maintaining the same age-structure of the test individuals. In this way, the problems of food depletion could be reduced (Ciros-Pérez et al., 2001). However, this eliminates the possibility of measuring the relative peak abundances of the competing species, which are also the objectives in our work. Peak population density has been shown earlier to be sensitive to competition (Sarma et al., 1999). We intended to verify this in the present experiment too. Life table method in competition also maintains relatively stable algal densities in daily renewal experimental designs (Nandini & Sarma, 2002; Milbrink et al., 2003).However, such an approach is more complicated (e.g., identification of amictic eggs) when several brachionid species are used simultaneously.

Despite the difficulties, competition among time or more zooplankton species has been studied more often under field conditions (Ortells et al., 2003) than under controlled laboratory situations; most experiments with zooplankton focus on only two species (Martìnez & Medel, 2002; Sarma et al., 2003a). In tropical waterbodies, the most common genus is Brachionus with about 40 species, many of which occur simultaneously in the same habitat even though they have similar food and feeding habits (Koste, 1978; Koreneva, 1989), This is apparently in contrast to the competition exclusion principle, considering the degree of niche overlap (Krebs, 1985). However, small differences in the relative tolerances to certain abiotic and biotic factors permit them to coexist during many months of the year (DeMott, 1989; Walz, 1995).

It has been well documented that threshold food level largely determines the competitive superiority of a species; species with lower rather than higher threshold food levels on a particular diet are favoured (Gliwicz, 1990), Nevertheless, other factors such as the initial density of competitors are just as important; in temporary ponds for instance, a species with a higher threshold food level may have left more resting eggs. Sarma et al., (1996) have shown that, depending on the initial density, algal food concentration has a decisive role in the competitive outcome between *B. calyciflorus* and *Anuraeopsis fissa*. Regardless of the initial density, at high food levels, *B. calyciflorus* out competed *A. fissa* and *vice versa* at lower algal levels. In another study on competition between two *Brachionus* species, Sarma et al. (1999) have shown that *B. calyciflorus* was not a superior competitor to *B. patulus*, especially at lower food levels. These studies are based on two species interactions. When three zooplankton species are grown together under similar test conditions, all of them could show reduced population abundances and growth rates as has been documented in some cladoceran species (Sarma et al., 2003a). In the present study, we cultured each of the four rotifer species separately and together. Our results also showed similar trends in that the presence of competitors resulted in reduced peak population abundances as well as population growth rate than when grown alone.

Food and temperature interact within a certain range causing changes in egg production and, hence, rate of population increase for most zooplankton species (Sarma & Rao, 1990; Weetman & Atkinson, 2002). When food concentration was increased from 1.5×10^6 to 4.5×10^6 cells ml^{-1}, the r value was enhanced considerably, but at high temperature (and especially at low algal density), growth rates were lower. The interaction of food level and temperature was statistically significant. This was further influenced by the presence of competitors.

The densities reached by rotifer species are dependent on their body size. In general an inverse relationship exists between the zooplankton body size and the peak population abundances when cultured under similar conditions (Nandini & Sarma, 2003). Of the four rotifer species used in this study, *B. calyciflorus* was the largest. *B. rubens* and *B. patulus* were almost the same size and *B. havanaensis* was the smallest rotifer. These differences in the body sizes were reflected in their abundances in the control experiments. For example, *B. calyciflorus* (the largest species) reached densities of no more than 120 ind. ml^{-1}. On the other hand, *B. havanaensis* reached peak abundances as high as 600 ind. ml^{-1}. Regardless of the test condition, the peak population densities observed here for all of the rotifer species were within the range reported in the literature (*B. calyciflorus* (600 ind. ml^{-1}: Sarma et al., 1996; *B. patulus* (300 ind. ml^{-1}: Sarma et al., 1999; *B. rubens* (250 ind. ml^{-1}; Sarma et al., 2003b; *B. havanaensis* (700 ind. ml^{-1}: Pavón-Meta et al., 2004).

Brachionus is a pantropical genus with a capacity to successfully reproduce at temperatures higher than 25 °C. In fact *B. patulus* has been grown at 35 °C (Sarma & Rao, 1990). They are also most often found in eutrophic ecosystems (Erikson, 1998). In the present study the algal concentrations i.e., 1.5×10^6, 3.0×10^6 and 4.5×10^6 cells ml^{-1} of *Chlorella* in terms of carbon are equivalent to 8.73, 17.46 and 26.18 μg ml^{-1}, respectively (Nandini & Sarma, 2003). Thus both the temperature and food levels chosen here represent the natural conditions under which the tested rotifer species are frequently encountered. The coexistence of at least three rotifer species at 28 °C especially under higher food levels observed in this study is supported by the field observations where the simultaneous occurrence of more than 10 *Brachionus* species, has been documented (e.g., Flores-Burgos et al., 2003). However, the competitive abilities of the tested species are reflected in their relative abundances rather than their complete absence under mixed conditions. Complete elimination of one rotifer species by the other though not rare (Kirk, 2002), more commonly the partial suppression on the growth rates has been documented (Sarma et al., 1999). Since all the rotifer species grew well on *Chlorella* when cultured separately, the differences in the relative abundances when grown together must be construed as the effect of competition.

The duration of experiments for competition studies could be one factor to be considered for extending the results to the field conditions. In this study, we terminated our experiments at day 24 by which time a clear trend was visible. Moreover, the duration chosen here is based on the fact of more than one generation time for *Brachionus* (Sarma & Rao, 1991) and that in the literature the majority of rotifer competition studies were limited to two to three week duration (Sarma et al., 1996; Kirk, 2002). Field observations have revealed that strong seasonal fluctuations in the abundances of zooplankton occur within this period (DeMott, 1989). Fluctuations in the environmental factors such as food concentration and temperature may permit several zooplankton taxa coexist since different species are competitively superior in different conditions (Grover, 1997) and that competitive exclusion might occur after a long time (Dumont, 1994), Thus, from the Figures 1–6, it is evident that coexistence was possible in many cases. Further experiments on multispecies competition possibly on field-collected seston can throw light on the mechanisms of coexistence of zooplankton species in nature.

Acknowledgements

Two anonymous reviewers and Dr. Linda May have greatly improved our manuscript. We thank MC Ramiro Jesús Sandoval, CONACyT (Ref. 188488, SNI-18723 and SNI- 20520) and PAPIIT-IN219405 for support.

References

Anonymous, 1985. Methods of Measuring the Acute Toxicity of Effluents to Freshwater and Marine Organisms. US Environment Protection Agency EPA/60014–85/013, Washington DC.

Borowitzka, M. A. & L. J. Borowitzka, 1988. Micro-Algal Biotechnology. Cambridge University Press, United Kingdom.

Ciros-Pérez, J., M. J. Carmona & M. Serra, 2001. Resource competition between sympatric sibling rotifer species. Limnology and Oceanography 46: 1511–1523.

Chase, J. M., P. A. Abrams, J. P. Grover, S. Diehl, P. Chesson, R. D. Holt, S. A. Richards, R. M. Nisbet & T. J. Case, 2002. Interaction between predation and competition: a review and synthesis. Ecology Letters 5: 302–315.

DeMott, W. R., 1989. The role of competition in zooplankton succession. In Sommer, U. (ed.) Plankton Ecology: Succession in Plankton Communities. Springer, New York, 195–252.

Dumont, H. J., 1977. Biotic factors in the population dynamics of rotifers. Archiv für Hydrobiologie, Beihefte 8: 98–122.

Dumont, H. J., 1994. Ancient lakes have simplified pelagic food webs. Archiv für Hydrobiologie, Beihefte 44: 223–234.

Edmondson, W. T., 1946. Factors in the dynamics of rotifer populations. Ecological Monographs 16: 357–362.

Erikson, R., E. Hooker, M. Mejia, A. Zelaya & K. Vammen, 1998. Optimal conditions for primary production in a polymictic tropical lake (Lake Xolotlán, Nicaragua). Hydrobiologia 382: 1–16.

Grover, J. P., 1997. Resource Competition. Chapman Hall, New York.

Juan-Burgos F., S. S. S. Sarma & S. Nandini, 2003. Estudio preliminar sobre la fauna de rotiferos de Xochimilco (México), El agua de cuenca de México. Sus problemas históricos y perspectivas de solución. Proceedings of the International Conference on Xochimilco, Ecological Park of Xochimilco, U. A. M. Xochimilco, Mexico City, Mexico Vol. 1: 163–171.

Gilbert, J. J., 1988. Suppression of rotifer populations by *Daphnia*: a review of the evidence, the mechanisms, and the effects on zooplankton community structure. Limnology and Oceanography 33: 1286–1303.

Glazier, D. S. & P. Calow, 1992. Energy allocation rules in *Daphnia magna*: clonal and age differences in the effects of food limitation. Oecologia 90: 540–549.

Gliwicz, Z. M., 1990. Food thresholds and body size in cladocerans. Nature 343: 638–640.

Hurtado-Bocanegra, M. D., S. Nandini & S. S. S. Sarma, 2002. Combined effects of food level and inoculation density on competition between *Brachionus patulus* (Rotifera) and the cladocerans *Ceriodaphnia dubia* and *Moina macrocopa*. Hydrobiologia 468: 13–22.

Hutchinson, G. E., 1967. A Treatise on Limnology. 2. Introduction to Lake Biology and Limnoplankton. John Wiley, New York, 1115 pp.

Kirk, K. L., 1997. Life-history responses to variable environments: starvation and reproduction in planktonic rotifers. Ecology 78: 434–441.

Kirk, K. L., 2002. Competition in variable environments: experiments with planktonic rotifers. Freshwater Biology 47: 1089–1096.

Koreneva, E. A., 1989. Fine structure of trophi of rotifers of the genus *Platyias* (Rotifera, Brachionidae). Trudy Institut Biologii Vnutrennikh vod Akad. Nauk SSSR 56: 83–94 (In Russian).

Koste, W., 1978. Rotatoria. Die Râdertiere Mitteleuropas. Ein Bestimmungswerk begründet von Max Voigt. Borntrisäger, Stuttgart, 2 vols.

Krebs, C. J., 1985. Ecology: The Experimental Analysis of Distribution and Abun2dance (3rd edn). Harper, Row, New York.

Martinez, G. & R. Medel, 2002. Indirect interactions in a microcosm-assembled cladoceran community: implications for apparent competition. Oikos 97: 111–115.

Milbrink, G., M. L. Kruse & J. Bengtsson, 2003. Competitive ability and life history strategies in four species of *Daphnia: D. obtusa, D. magna, D. pulex*, and *D. longispina*. Archiv für Hydrobiologie 157: 433–453.

Nandini, S. & S. S. S. Sarma, 2002. Competition between *Moina macrocopa* and *Ceriodaphnia juba*: a life table demography study. International Review of Hydrobiology 87: 85–95.

Nandini, S. & S. S. S. Sarma, 2003. Population growth of some genera of cladocerans (Cladocera) in relation to algal food (*Chlorella vulgaris*) levels. Hydrobiologia 491: 211–219.

Nogrady, T., R. L. Wallace & T. W. Snell, 1993. Rotifera. Vol. 1. Biology, Ecology and Systematics. SBP Academic Publishing, The Hague, 142 pp.

Ortells, R., A. Gomez & M. Serra, 2003. Coexistence of cryptic rotifer species: ecological and genetic characterisation of Brachionus plicatilis. Freshwater Biology 48: 2194–2202.

Pavón-Meta, E. L., S. S. S. Sarma & S. Nandini, 2004. Combined effect of food (Chlorella vulgaris) concentration and temperature on the population growth of *Brachionus havanaensis* (Rotifera: Brachionidae). Journal of Freshwater Ecology 19: 521–530.

Pourriot, R., 1977. Food and feeding habits of the Rotifera. Archiv Mr Hydrobiologie, Beihefte 8: 243–260.

Rothhaupt, K. O., 1988. Mechanistic resource competition theory applied to laboratory experiments with zooplankton. Nature 333: 660–662.

Rothhaupt, K. O., 1990. Changes of the functional responses of the rotifers *Brachionus rubens* and *Brachionus calyciflorus* with particle sizes. Limnology and Oceanography 35: 24–32.

Sarma, S. S. S. & T. R. Rao, 1990. Population dynamics of *Brachionus patulus* Müller (Rotifera) in relation to food and temperature. Proceedings of the Indian Academy of Sciences. (Animal Sciences) 99: 335–343.

Sarma, S. S. S. & T. R. Rao, 1991. The combined effects of food and temperature on the life history parameters of *Brachionus patulus* Muller (Rotifera). Internationale Revue der gesamten Hydrobiologie 76: 225–239.

Sarma, S. S. S., N. Iyer & H. J. Dumont, 1996. Competitive interactions between herbivorous rotifers; importance of food concentration and initial population density. Hydrobiologia 331: 1–7.

Sarma, S. S. S., M. A. Fernández-Araiza & S. Nandini, 1999. Competition between *Brachionus calyciflorus* Pallas and *Brachionus patulus* (Müller) (Rotifera) in relation to algal food concentration and initial population density. Aquatic Ecology 33: 339–345.

Sarma, S. S. S., E. Mangas-Ramírez & S. Nandini, 2003a. Effect of ammonia toxicity on the competition among three species of cladocerans (Crustacea: Cladocera). Ecotoxicology and Environmental Safety 55: 227–235.

Sarma, S. S. S., H. E. Trujillo-Hernández & S. Nandini, 2003b. Population growth of herbivorous rotifers and their predator (Asplanchna) on urban wastewaters. Aquatic Ecology 37: 243–250.

Schneider, D. W., 1990. Direct assessment of the independent effects of exploitative and interference competition between Daphnia and rotifers. Limnology and Oceanography 35: 916–922.

Sokol, R. R. & F. J. Rohlf, 2000. Biometry. W.H. Freeman and Company, San Francisco.

Stemberger, R. S. & J. J. Gilbert, 1985. Assessment of threshold food levels and population growth in planktonic rotifers. Archiv für Hydrobiologie, Beihefte 21: 269–275.

Walz, N., 1995. Rotifer populations in plankton communities: Energetics and life history strategies. Experientia 51: 437–453.

Weetman, D. & D. Atkinson, 2002. Antipredator reaction norms for life history traits in *Daphnia pulex*: dependence on temperature and food. Oikos 98: 299–307.

Zaret, T. M., 1980. Predation and Freshwater Communities. Yale University Press, New Haven, 180 pp.

Hydrobiologia (2005) 546:535–549
A. Herzig, R.D. Gulati, C.D. Jersabek & L. May (eds.) Rotifera X: Rotifer Research: Trends, New Tools and Recent Advances
DOI 10.1007/s10750-005-4296-x

Application of stable isotope tracers to studies of zooplankton feeding, using the rotifer *Brachionus calyciflorus* as an example

Antonie M. Verschoor*, Harry Boonstra & Thijs Meijer
Netherlands Institute of Ecology (NIOO-KNAW), Centre for Limnology, Rijksstraatweg 6, 3631 AC Nieuwersluis, The Netherlands
*(*Author for correspondence: E-mail: a.verschoor@nioo.knaw.nl)*

Key words: ^{13}C, ^{15}N, isotope ratio mass spectrometer (IRMS), grazing, ingestion, assimilation

Abstract

We present a protocol and calculation methods for the determination of zooplankton ingestion and assimilation rates with stable isotope tracers. These methods have been developed from experiments with the rotifer *Brachionus calyciflorus* that had been fed ^{13}C-labelled *Scenedesmus obliquus*. Stable isotope tracers offer the same advantages as radioisotopes. These include the possibility for direct and accurate quantification of ingestion and assimilation rates, short sample analysis times and low animal densities requirements. However, the use of stable isotope tracers requires relatively long sample preparation times and specialist equipment and is, thus, relatively costly for most laboratories. The application of stable isotope tracers in zooplankton feeding studies offers several advantages in comparison with radioisotopes. Firstly, they do not emit harmful radiation and can therefore be applied safely both in the laboratory and in the field. Secondly, the samples can be dried for safe storage and easy transportation. Thirdly, no aggressive chemicals are required for sample analysis.

Introduction

Throughout the history of ecology, new measurement techniques have usually stimulated rapid development in related areas of science when they have become widely available. For example, the availability of purified radioisotopes, in combination with sensitive laboratory-scale equipment to measure radioactive decay, made it possible to follow the fate of these isotopes within organisms. So, it became possible to measure ingestion, assimilation and respiration rates of zooplankton by direct rather than indirect methods. Peters (1984) reviewed and discussed many of the techniques available at that time. Since then, only a few really new techniques to study zooplankton feeding have been published. These include the use of gut fullness methods (Penry & Frost, 1990), chlorophyll fluorescence spectra (Lürling & Verschoor, 2003), biologically digestible tracer particles (Lindemann & Kleinow, 2000; Hammer et al., 2001), and stable isotope tracers.

Stable isotopes are increasingly being used in ecological research. With isotope ratio mass spectrometers (IRMS) becoming more and more available to ecological laboratories, it has become possible to accurately measure the isotopic composition of organisms, and to use this isotopic composition in ecological studies (Peterson & Fry, 1987). Because different sources of primary production have different isotopic signatures, and different isotopes of the same element are incorporated at different rates in biological tissue (isotopic fractionation), the natural isotope ratios of an element can be used to infer trophic linkages (e.g., Bearhop et al., 1999, Vander Zanden et al., 1999). Apart from using their natural abundance, stable isotopes are increasingly being used as biological tracers. Nowadays, an increasing number of different labelled compounds is becoming available with the heavy isotope making up most of the atoms of the tracer element (e.g., $NaHCO_3$, 99 at% ^{13}C, or glycine, 98 at% ^{15}N). This, combined with the high sensitivity of IRMS systems (changes in relative ^{13}C content of 0.001% are still detectable: Boschker & Middelburg, 2002), yields promising opportunities for the use of these tracers in the study of trophic interactions, both in the laboratory and in the field (e.g., Boschker & Middelburg, 2002). In zooplankton studies, stable isotopes have been used for the creation of nitrogen budgets for marine zooplankton species (Hino et al., 1997; Hasegawa et al., 2001), and to identify the role of zooplankton in the creation and disappearance of deep water chlorophyll maxima (Pilati & Wurtsbaugh, 2003; Lampert & Grey, 2003). There is, however, only very little published information about zooplankton feeding using stable isotope tracers. This may be due to the fact that there are no well-described protocols and calculation examples on how to use stable isotopes as tracers in feeding studies.

This study aims to make the stable isotope tracer method more available for zooplankton feeding studies. A detailed protocol for labelling and feeding experiments and examples of mass balance-based calculation methods are presented, and the advantages and disadvantages of the stable isotope tracer method for zooplankton feeding studies are discussed.

Materials and methods

General calculations with stable isotopes

Isotope ratio mass spectrometers (IRMS) generate two important types of data: data on the mass of the sample, given as peak height or peak area of the measured voltage, and data on the isotopic ratio of the sample, expressed as delta values (δ). Peak height and area correlate linearly with the mass of the specific element, which is shown by regression of different weights of reference material with a known, fixed, content of the specific element (dry rye powder: Fig. 2a). Delta values (δ) give the relative degree of isotopic enrichment compared to a standard reference material, per mille (‰). The delta value of a sample containing a specific isotope n of element E is calculated by subtracting the isotopic ratio (ratio of the heavy, rare isotope to the most abundant isotope) of a reference material (R_R) from the isotopic ratio of this sample (R_S), dividing by R_R, and multiplying by 1000‰:

$$\delta^n E = ((R_S - R_R)/R_R) \cdot 1000‰ = ((R_S/R_R) - 1) \cdot 1000‰ \quad (1a)$$

Similarly, isotopic ratios are calculated from the isotopic ratio of the reference material. For example, for ^{13}C the most commonly used reference material, Vienna PeeDee Belemnite limestone (VPDB), has an isotopic ratio of 0.0112372. The isotopic ratio of a sample containing ^{12}C and ^{13}C will then be:

$$R_S = ((\delta^{13}C/1000) + 1) \cdot 0.0112372 \quad (1b)$$

and the isotopic fraction:

$$F_s = R_s/(R_s + 1) \quad (1c)$$

Mass balances form the basis of all of the calculations presented. To set up elemental and isotopic balances, fractions are multiplied by mass to form the total mass of a specific isotope in a certain sample. As long as ratios or fractions are lower than a few percent, which should always the case for feeding studies, no corrections for the atomic mass difference between the stable and the common isotope are required. A sample with a certain mass signal, M_t, and a certain isotopic fraction, F_t, is decomposed into the two elements that make up the sample: the mass signal and isotopic fraction of the background (M_b, F_b), and the mass and signal of the sample itself (M_s, F_s). Thus, an elemental mass balance and an isotopic mass balance are made:

Overall elemental mass balance:

$$M_t = M_b + M_s \quad (2a)$$

Isotopic mass balance:

$$M_t \cdot F_t = M_b \cdot F_b + M_s \cdot F_s \quad (2b)$$

The background contribution is calculated by choosing an appropriate background reference; these are usually empty or medium-filled measurement cups. Estimates of background mass (M_b) and background isotopic fractions (F_b) are made by

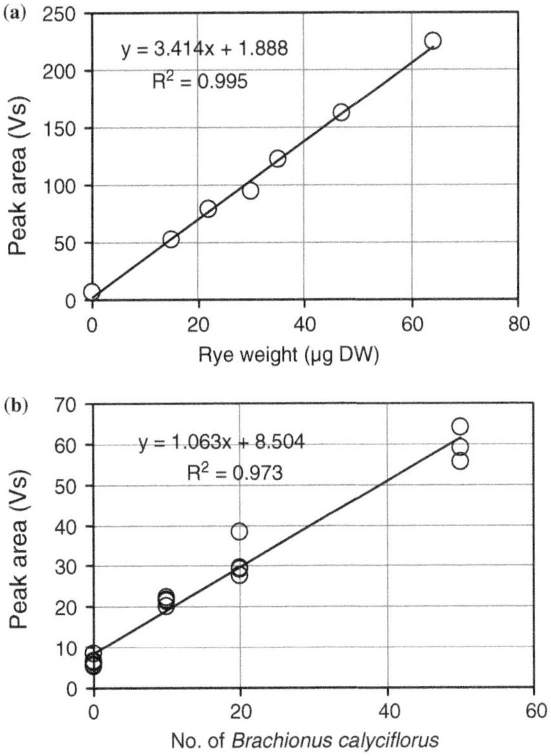

Figure 1. Regressions of known quantities of sample materials against mass, as measured by the IRMS (in this case, peak area). (a, Upper panel): Regression of dry weight of the elemental standard (rye powder) against sample peak area; a series of empty tin cups were used as zero values. (b, Lower panel): Regression of samples containing a fixed number of the rotifer *Brachionus calyciflorus* against sample peak area; a series of tin cups containing only medium was used for zero values.

averaging several background samples. Rearranging Equation (2a) yields the estimated real mass of the sample:

$$M_s = M_t - M_b \tag{2c}$$

and rearrangement of Equation 2b gives the estimated isotopic fraction of this sample:

$$F_s = (M_t \cdot F_t - M_b \cdot F_b)/(M_s) \tag{2d}$$

Food labelling

Prior to the zooplankton feeding experiment, the food cells should be labelled with the stable isotope tracer. Autotrophic food cells (algae and cyanobacteria) are labelled with heavy-isotope containing inorganic nutrients (e.g., $NaH^{13}CO_3$, $Na^{15}NO_3$), whereas heterotrophic food cells (yeast and most bacteria) are labelled with heavy isotope containing organic compounds (e.g., ^{13}C-labelled fatty acids, or ^{15}N-labelled amino acids). The rate of tracer accumulation in the food cells depends strongly on the physiological characteristics of the food cells (affinity with nutrient, maximum growth rate) and the environmental characteristics (temperature, tracer medium used). Therefore, differences in label uptake in different food fractions (e.g., structural and soluble compounds) may occur, but this does not necessarily cause differences in measured feeding rates (Lampert, 1977a). Differential labelling of different size classes and over- or underestimation of feeding rates may occur due to selective feeding (Baars & Oosterhuis, 1985; Gulati, 1985), but this effect is limited when using monocultures. Longer labelling times may prevent differential labelling problems.

Once a dynamic equilibrium is reached after labelling, all elements in the system will acquire the same, heavy-isotope enriched, δ value. In a closed

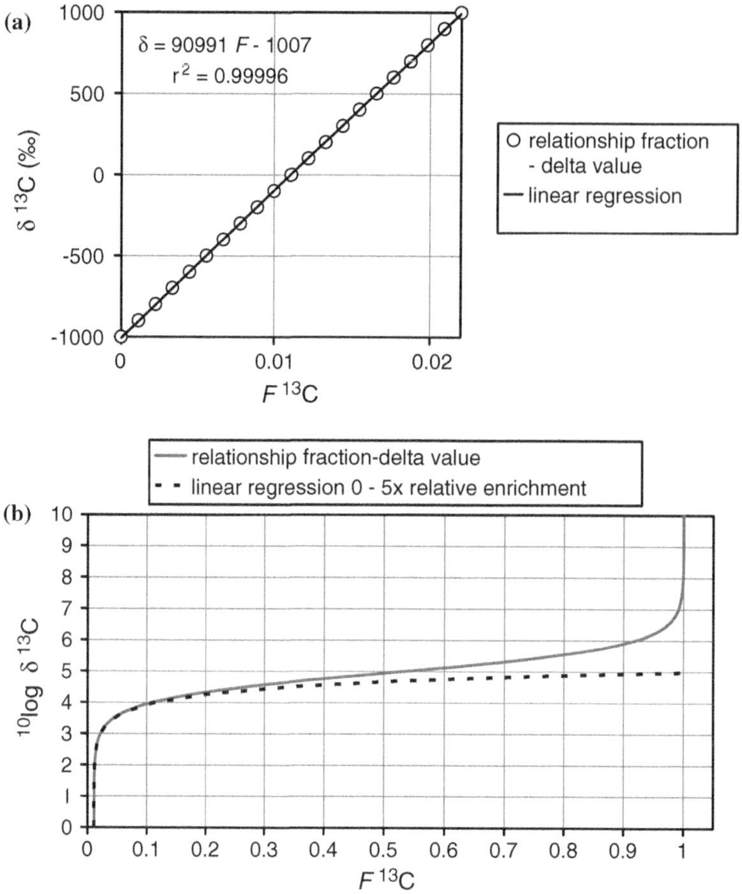

Figure 2. Relationships between isotopic fraction (F) and δ values for carbon. (a, Upper panel): The relationship between F and δ values for the range 0–2 times relative enrichment compared to the Vienna PeeDee Belemnite limestone (VPDB) standard. (b, Lower panel): As (a), but with the F–δ relationship for the range $1–1\cdot10^7 \times$ relative enrichment compared to the VPDB standard, together with a linear regression based on the range $0–5 \times F_{\text{reference}}$.

system, the following mass balances apply. These are applied to obtain the desired isotopic enrichment:

Overall mass balance:

$$M_{f,i} + M_{r,i} = M_{f,e} + M_{r,e} \tag{3a}$$

Isotopic mass balance:

$$M_{f,i} \cdot F_{f,i} + M_{r,i} \cdot F_{r,i} = M_{f,e} \cdot F_{f,e} + M_{r,e} \cdot F_{r,e} \tag{3b}$$

where subscripts 'f' and 'r' denote food and free dissolved resource material (= tracer + non-tracer), respectively, and subscripts 'i' and 'e' stand for the times of start of incubation and equilibrium, respectively. Furthermore, under equilibrium conditions, the isotopic fractions of food and free dissolved resource will be similar, so the equation can be simplified to:

$$F_{f,e} = F_{r,e} = F_e \tag{3c}$$

$$M_{f,i} \cdot F_{f,i} + M_{r,i} \cdot F_{r,i} = (M_{f,e} + M_{r,e}) \cdot F_e \tag{3d}$$

The equilibrium isotopic fraction of the food is then calculated using Equation 3d, regardless of the equilibrium mass of the food or remaining free dissolved mass:

$$F_e = (M_{f,i} \cdot F_{f,i} + M_{r,i} \cdot F_{r,i})/(M_{f,i} + M_{r,i}) \tag{3e}$$

Feeding experiments

For the determination of ingestion rates using stable isotope tracers, animals should first be acclimatised to the experimental food type and concentration for at least 1 h. Then, either labelled ('hot') food is added, or animals are transferred into the labelled food suspension. The duration of the experiment should be long enough to overcome the possible disturbance of ingestion rates caused by adding food or transferring animals (Peters, 1984). On the other hand, the experimental duration should be shorter than the gut passage time of the zooplankton, otherwise feeding rates will be underestimated (Starkweather & Gilbert, 1977a; Peters, 1984) due to egestion of the tracer (Lampert 1977a). Peters (1984) gives published gut passage times for various zooplankton taxa, and also supplies methods of estimating these when they are unknown. After the feeding period the animals are rinsed to remove any adhered food (Baars & Oosterhuis, 1985; Gulati, 1985), inactivated with carbonated or hot water, and preserved (Gulati, 1985). The time between the end of the feeding experiment and the picking of the animals should be minimised to prevent losses of tracer (Peters, 1984; Gulati, 1985).

Assimilation rates are determined in a similar way, but with tracer incorporation corrected for egestion of the tracer. Lampert (1977b) already reported that direct measurement of assimilation (= absorption of material from the gut) is only possible by using tracers. A commonly used method is to feed animals with tracer for a certain period of time, and then place them into clean medium or into the same food type without tracer. Assimilation studies should be longer than the gut passage time but short enough to prevent respiratory losses of tracer when ^{13}C is used.

Calculation of feeding rates

For further calculations on feeding rates, the measurement series should contain reference samples of zooplankton that have been fed on unlabelled food of the same kind (z), samples of the pure labelled food (f) and appropriate background samples. Zooplankton mass (M_z) is estimated from linear regression between number of animals and sample mass (e.g., Fig. 1b), in which the intercept represents the background contribution of cups plus medium (cf. (2c)). Average isotopic fractions (F_z, F_f) are calculated using equation 2d. The mass and isotope fractions of labelled food-fed zooplankton in a sample are then decomposed into a contribution from the labelled food (M_f,F_f) and from the zooplankton (M_z,F_z):

$$M_s = M_z + M_f \qquad (4a)$$

$$M_s \cdot F_s = M_f F_f + M_z \cdot F_z \qquad (4b)$$

so that the ingested food mass becomes:

$$M_f = (M_s \cdot F_s - M_z \cdot F_z)/F_f \qquad (4c)$$

Knowing the duration of the experiment (T) and the number of animals (N), we can calculate the ingestion rate (IR) or assimilation rate (AR):

$$\mathrm{IR,AR}(\text{mass mass}^{-1} t^{-1}) = M_f/(N \cdot T) = (M_s \cdot F_s - M_z \cdot F_z)/(N \cdot F_f \cdot T) \qquad (4d)$$

For calculation of elemental mass flows (e.g., energy budgets, modelling), the interest is usually more focussed on mass-specific feeding rates. Furthermore, mass-specific rates offer the benefit that IRMS mass signals can be used directly, without having to relate these to real masses (Fig. 1a). For these rates, the number of animals (N) is replaced by the zooplankton mass,

$$\mathrm{SIR,SAR}(\text{mass mass}^{-1} t^{-1}) = M_f/(M_z T) = (M_s \cdot F_s - M_z \cdot F_z)/(M_z \cdot F_f \cdot T) \qquad (4e)$$

The assimilation efficiency (AE) can then be calculated by dividing AR by IR:

$$AE(\%) = 100\% \cdot \mathrm{(S)AR/(S)IR} = 100\% \cdot (M_{s,a} \cdot F_{s,a} - M_{z,a} \cdot F_z)/(M_{s,i} \cdot F_{s,i} - M_{z,i} \cdot F_z)$$

with i or a denoting variables derived from ingestion or assimilation experiments, respectively.

As an alternative to these calculations, we give a preferred approach here, namely the case for which F is replaced by δ values and M_z is approximated. As δ values represent relative enrichment compared to a certain standard, they have the advantage that no background corrections are required. Furthermore, direct IRMS output (δ values) can be used and δ values increase method sensitivity since F values are typically very small. The δ value approach is possible because of the highly linear correlation between δ value and isotopic fraction in the range $0\text{–}2 \times F_{\text{reference}}$ (Fig. 2a). Also for higher enrichments, this linearity still holds: e.g., for carbon samples $0\text{–}5 \times F_{\text{reference}}$ ($\delta = -1000 - 5000$), $r^2 = 0.9998$, and for samples $0\text{–}20 \times F_{\text{reference}}$, we still have $r^2 = 0.9969$. Only at very high relative enrichments (Fig. 2b: above $F \approx 0.1$, $\delta^{13}C \approx 10000‰$), does the true relationship between F and δ values deviate seriously from linearity. Furthermore, we choose to approximate M_z because it cannot be determined accurately when animals cannot be counted directly, or when there is large variation in number – mass regressions (cf. Fig. 1b). In short-term feeding experiments, the mass of ingested or assimilated food is usually small with respect to the mass of the zooplankton. If we disregard the contribution of the food to the sample, and we replace F by δ values, we obtain:

$$M_f = M_s \cdot (\delta_s - \delta_z)/\delta_f \tag{5a}$$

Similar to (4d), we calculate the ingestion (IR) and assimilation rates (AR):

$$\text{IR,AR}(\text{mass ind}^{-1} t^{-1}) = M_s \cdot (\delta_s - \delta_z)/(\delta_f N \cdot T) \tag{5b}$$

Since M_z now equals M_s, only isotopic fractions and the feeding period are required to calculate mass-specific ingestion rates (SIR) and mass-specific assimilation rates (SAR):

$$\text{SIR, SAR}(\text{mass ind}^{-1} t^{-1}) = (\delta_s - \delta_z)/(\delta_f T) \tag{5c}$$

To calculate assimilation efficiencies (AE) i.e. the ratio of assimilation and ingestion rates, we use Equations (5b) or (5c), which leaves only the averaged δ values of the samples used for assimilation ($\delta_{s,a}$), ingestion ($\delta_{s,i}$) and zooplankton background (δ_z).

$$\begin{aligned} AE(\%) &= 100\% \cdot \text{(S)AR(S)IR} \\ &= 100\% \cdot (\delta_{s,a} - \delta_z)/(\delta_{s,i} - \delta_z) \end{aligned} \tag{5d}$$

This implies that, when the interest is in the assimilation efficiency only, it is sufficient to measure δ_s and δ_z values (compare with (4f)), and it is not necessary to put too much effort into measuring δ_f precisely. This may be particularly useful with very low specific ingestion and assimilation rates or very short feeding times, when high food enrichment is required for significant discrimination, but δ value of undiluted food samples may fall outside the reliable detection range of the IRMS.

Preparation and processing of samples for analyses

All samples were measured into 5 × 8 mm pressed tin cups (Elemental Microanalysis Ltd, Okehampton, UK). Tin cups were prepared by washing them in a 50/50 (vol/vol) methanol/chloroform solution to remove all possible (carbon-containing) contamination. Tin cups and sample material were handled with forceps and needles that were heated until red-hot over a gas flame, again to remove any contamination. A typical run first contained a number of empty samples, and then samples increasing in expected $\delta^{13}C$ value, with additional blanks after high δ values because of possible memory effects. One isotope standard was added to every ~10 samples. Isotope standards were prepared by weighing dry rye powder ($\delta^{13}C = -30‰$) into tin cups on an Ohaus AS120 microbalance to the nearest μg. Zooplankton samples were taken by putting individuals into tin cups, algal samples were taken by taking a known volume of concentrated algae, and additional background samples were taken to correct for the media that the algae and zooplankton were in. Samples were put into series of 50 into 96-well microtiterplates and oven dried overnight at 60 °C. Dried samples were folded and stored in a desiccator until measurement. Samples were measured in a Carlo Erba 1106 Elemental Analyser (EA) coupled online with a Finnigan Delta-S IRMS.

^{13}C tracer incorporation and loss in Scenedesmus obliquus

We investigated the $\delta^{13}C$ enrichment of a strain of the chlorophyte *Scenedesmus obliquus* Turpin (Kützing) that originated from the Max-Planck-Institute of Limnology, Plön, Germany. Label uptake was studied in two experiments. In the first experiment, we studied algal ^{13}C enrichment in an open system, using 1l Erlenmeyer bottles that were closed with cellulose plugs, still allowing exchange of $^{12}CO_2$ and $^{13}CO_2$ with the surrounding air. Samples were taken after 0, 1, 3, 6, 9, 12, 18, 24, 30, 36, and 48 h. For the second experiment, we studied algal ^{13}C enrichment in a closed system, using 1l screw cap bottles that were closed with rubber septa and as little air headspace as possible. Samples were taken after 0, 12, 24, 36, 48, 60, 72, 96 and 144 h. Algae for both experiments were grown in modified WC medium (Guillard, 1975) with an additional 5.0 mg $NaH^{13}CO_3$ l^{-1}, and approximately 10 mg C algae l^{-1}. Bottles for each experiment were put into an incubator in triplicate at 120 μmol quanta (PAR) m^{-2} s^{-1} (continuously), 20 °C and 100 revolutions per minute (RPM). At each sampling time, 1 ml of sample was taken from each bottle, rinsed on pre-combusted Whatman GF/F filters and resuspended in WC medium without $NaHCO^3$ or carbon-containing buffers. From each sampling time and each replicate, one 100 μl sample of resuspended algae was put into tin cups. Samples were further dried and processed as described above.

Ingestion and assimilation of ^{13}C-labelled Scenedesmus obliquus *by* Brachionus calyciflorus

We studied feeding in the rotifer *Brachionus calyciflorus* Pallas over a range of concentrations of the ^{13}C-labelled food alga *Scenedesmus obliquus*. Animals were obtained from previously hatched cysts (Microbiotest Inc., Nazareth, Belgium). Before the actual feeding experiment, animals were acclimatised to the different concentrations of unlabelled algae (i.e., 0.1, 0.2, 0.5, 1, 2, 5, and 10 mg C l^{-1}). Approximately 1000 animals were put into 250 ml of food suspension in 500 ml Erlenmeyer bottles. These bottles were placed on a slowly rotating table (40 RPM) to keep the food algae in suspension and incubated at 20°C for at least 1 h.

As food algae, we used a strain of *Scenedesmus obliquus* Turpin (Kützing), from the culture collection of the University of Texas at Austin (UTEX 2630). After pre-culturing, about 10 mg C algae l^{-1} were incubated overnight in modified WC medium containing an additional 5.0 mg $NaH^{13}CO_3$ l^{-1}. The algae were incubated in four 1-l bottles sealed with rubber septa, in an incubator set at 20 °C, 120 μmol quanta (PAR) m^{-2} s^{-1} and at 100 RPM. After incubation, algae were centrifuged for 10 min at 3000 RPM, and the resulting pellet was resuspended in clean WC medium (which did not contain additional $NaH^{13}CO_3$). After being centrifuged for a second time, the algae were first resuspended in a small volume of clean WC medium to enable samples of the concentrated algae to be taken for later analyses in the elemental analysers. On the basis of previously made calibrations of optical densities at 750 nm (Unicam helios δ spectrophotometer) and algal carbon content (UNICARB carbon analyser), different suspensions of the labelled algae were made. Algal suspensions were prepared in quadruplicate in 100 ml Erlenmeyer bottles, taking into account that, after addition of the animals (5 ml), concentrations of 0.1, 0.2, 0.5, 1, 2, 5, and 10 mg C l^{-1} would be obtained in a total volume of 50 ml.

The feeding experiment was done under the same conditions as the acclimatisation. Animals were gently filtered out of the acclimatisation bottles and resuspended in 25 ml of clean medium. Per concentration and per replicate, 5 ml of animal suspension was pipetted into the labelled algae suspension. After addition of the animals, they were allowed to feed for 10 min, which is half of the gut passage time for *B. calyciflorus* (Starkweather & Gilbert, 1977a). After feeding, all animals were filtered out of the suspension immediately and rinsed gently in clean WC medium. For the determination of ingestion, approximately half of the animals were washed out of the filter into small Petri dishes with 30 g l^{-1} NaCl solution. This solution was used as killing agent and caused immediate inactivation of the animals, thus preventing tracer loss by egestion. The remaining half of the animals was gently washed out of the filter again into clean WC medium and left for egestion to take place. After 1 h, these animals were filtered, rinsed and washed out into Petri dishes, as described above. From the animals collected in the Petri dishes, a sufficient number (at least 50) were picked out with a modified Pasteur pipette (with a very fine tip, attached to a silicon rubber tube with a mouthpiece) and as little as possible adhered NaCl solution, and put into tin cups.

Results

Relative errors when calculating with δ values and approximating zooplankton mass

Relative errors in estimated feeding parameters can be calculated and hence corrected for, exactly, using the formulae given in the Appendix. The relative errors of ingestion and assimilation rates are very much dependent on *m*, the specific ingestion/assimilation (M_f/M_z, which is usually small in zooplankton feeding experiments), and δ_f (Appendix, Fig 3a). The relative error of assimilation efficiencies is directly dependent on m_i/m_a (=1/assimilation efficiency) and decreases with increasing food enrichment (Appendix, Fig. 3b). The relative size of the possible overestimation does not exceed 3%, showing that corrections are often unnecessary.

Labelling of the algae

The enrichment of *Scenedesmus obliquus* in time is shown in Fig. 4. The labelling process was so rapid

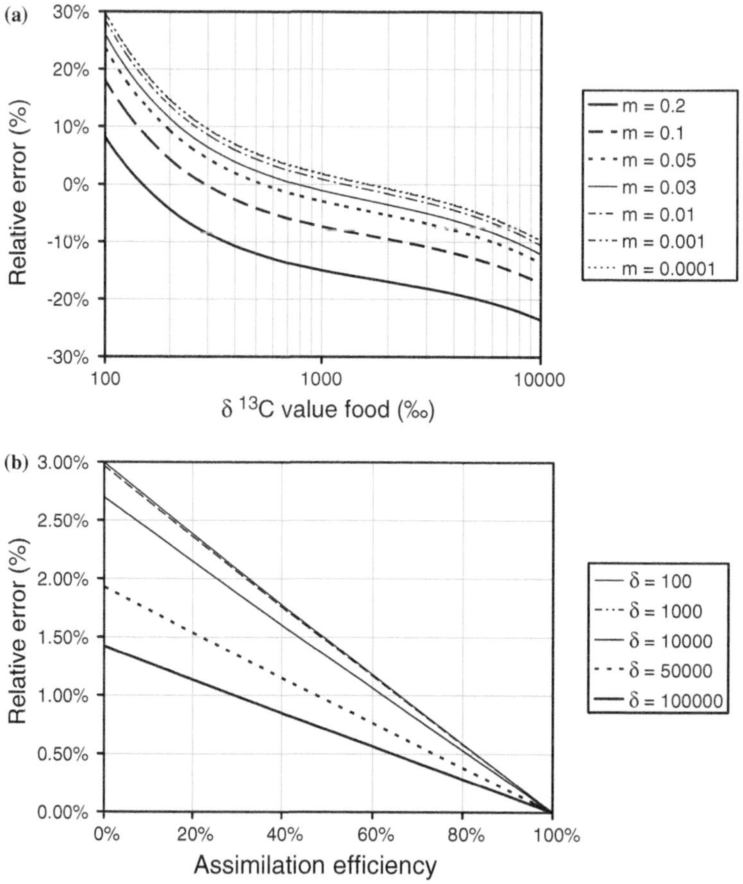

Figure 3. Relative errors of ingestion/assimilation rates and assimilation efficiencies, using the δ value approximation when exact zooplankton biomass (M_z) is unknown. Errors are given for calculations for ^{13}C over a range of δ values of the food for different values of specific ingestion rate (m_i) and assimilation (m_a), and $\delta_z = -30‰$. Typical maximum values of *m* for *Brachionus calyciflorus* are 0.03. (a, Upper panel): Relative error of SIR or SAR; note that errors overlap at low values of $m < 0.01$). (b, Lower panel): Relative error of assimilation efficiencies for different assimilation efficiencies and different δ values of the food.

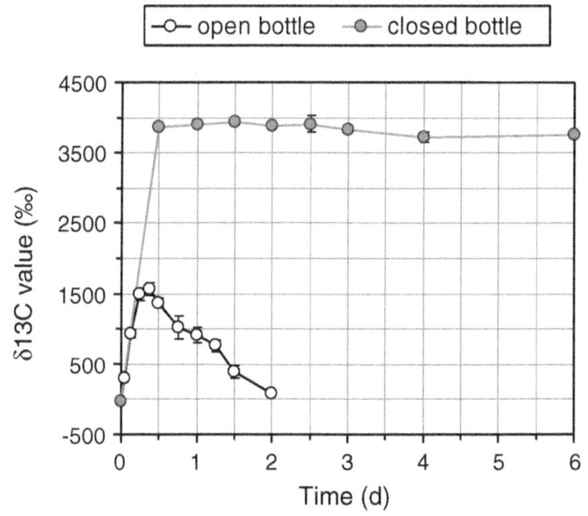

Figure 4. $\delta^{13}C$ values of *Scenedesmus obliquus* following labelling with $NaH^{13}CO_3$ (t = 0) in an incubator at 120 μE quanta (PAR) m^{-2} s^{-1}, 20 °C and 100 RPM. Average values (n = 3) are given for two different experiments: one using cellulose-stoppered Erlenmeyer flasks ('open bottle') the other using rubber septum-sealed bottles ('closed bottle'). Error bars are ± 1 SD.

that complete labelling was attained within 12 h. The initial slope of the lines is similar, showing the maximum uptake rates of $NaHCO_3$ (mainly $NaH^{13}CO_3$) by the green algae being similar (~7750‰ $\delta^{13}C$ d^{-1}). However, in the open bottles, respiratory losses were already visible after 6 h, when the enrichment slope started to decline. After 9 h, tracer uptake and respiratory loss of assimilated tracer were in equilibrium, and after that net loss of tracer started to occur, until all signs of enrichment had disappeared after 2 days. In the closed bottles, on the other hand, equilibrium was reached within 24 h, and δ values did not significantly differ up to the end of the experimental period (6 d, one-way ANOVA, $F = 0.8053$, $p = 0.596$). This illustrates the isotopic equilibrium principle (3e): as long as the system is closed and in equilibrium, tracer will remain present in the same quantities in all compartments of the system.

Ingestion and assimilation rates, and assimilation efficiency

Figure 5a shows the uncorrected estimates of mass-specific ingestion and assimilation rates based on δ values and unknown zooplankton. The $\delta^{13}C$ value for the algae was around 5900‰, for which the relative error is 8% underestimation (Fig. 3a). This can be used to correct the estimated rates. Figure 5a also shows the high sensitivity of the stable isotope method: mass-specific ingestion and assimilation rates as low as 0.025 mg C mg C^{-1} h^{-1} (i.e., 2.5 ng C ind^{-1} h^{-1}) can be measured with fairly high reproducibility. Also the maximum ingestion rates compare favourably with reported maximum ingestion rates for this species with radioisotope methods (e.g., Starkweather & Gilbert, 1977b, Rothhaupt, 1990). The (δ value) based assimilation efficiencies of *B. calyciflorus* (Fig. 5b) show a distinct pattern: assimilation efficiencies decrease with increasing food concentrations.

Discussion

Calculations

The commonly used method for calculations when zooplankton mass is unknown in feeding studies using stable isotopes as tracers is based on fractions. With this method, the excess fraction in the sample is calculated after background subtraction ($F_{excess} = F_{sample} - F_{background}$). Although the fraction approach is suitable for mass balances, we can prove that neglecting the mass difference between fed and unfed zooplankton may lead to serious errors in estimated feeding rates (not shown). We recommend using δ value approximations because these (a) can be applied directly on IRMS data instead of first having to calculate fractions (1b–c), therefore (b) do not require laborious background corrections, and (c) require less parameters for calculation. We offer methods to perform exact calculations on feeding rates and assimilation efficiencies, which may be preferable to more laborious and error-sensitive calculation methods based on fractions.

Determination of assimilation rates and efficiencies

We showed that ingestion rates can be determined and corrected accurately. However, determination of assimilation rates and assimilation efficiencies brings several potential complications. Firstly, the decreasing assimilation efficiencies in Figure 5b

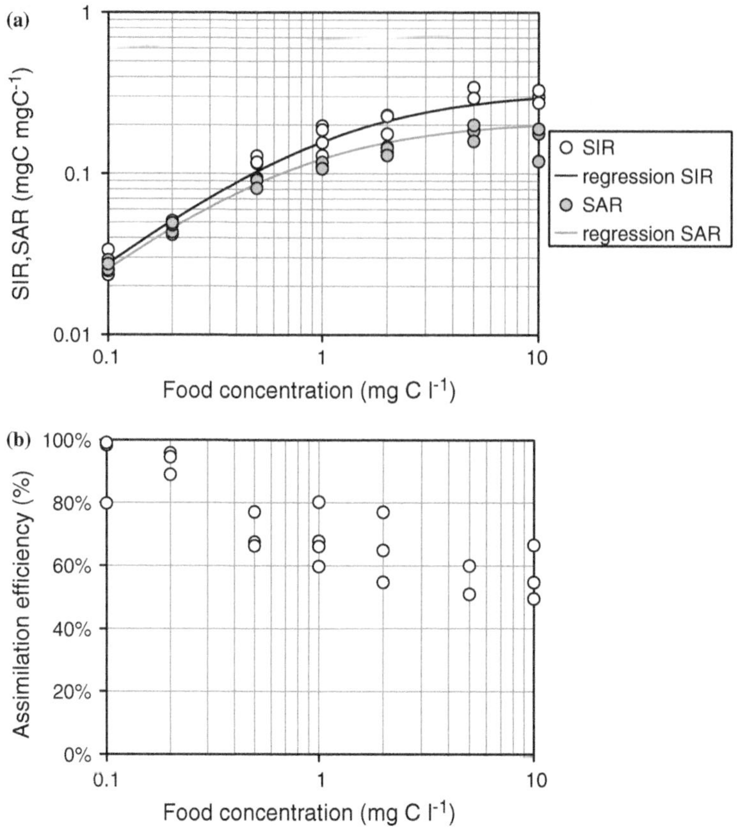

Figure 5. Ingestion and assimilation rates, and assimilation efficiencies, calculated by approximation of zooplankton biomass and using the δ value method. (a, Upper panel): Mass-specific ingestion (SIR) and assimilation rates (SAR) of *Brachionus calyciflorus* that have been feeding on different concentrations of ^{13}C-labelled *Scenedesmus obliquus*; the lines show two type II functional responses that were fitted through these data, based on the least squares of the log-transformed ingestion rates; note the overlap between SIR and SAR at the lower concentrations. (b, Lower panel): Assimilation efficiencies of *Brachionus calyciflorus* feeding on *Scenedesmus obliquus* derived from the same experiment.

with increasing food concentration indicate a methodological problem. Animals that have been feeding on very low food concentrations have very long gut passage times (Peters, 1984). When the gut residence times become larger than the experimental egestion time, the assimilation efficiency may approach 100%, because: (1) no tracer is lost through egestion, and (2) hardly any losses occur due to excretion or respiration. Because all samples are combusted, it is impossible to measure the same individuals for both ingestion and assimilation. This, by chance, may lead to calculated assimilation efficiencies higher than the theoretical maximum value of 100%. So, we recommend that samples from different replicates are pooled in these calculations. To avoid gut residence time problems, Rigler (1971, cited in Lampert, 1977a), gives an alternative method in which the tracer is measured at two different times after the gut has been filled with labelled food. This method also prevents other potential problems such as: (1) nutritional adaptation and handling stress, (2) different tracer respiration losses of empty and full animals, (3) loss of previously assimilated substances through the gut wall, and (4) incomplete replacement of labelled with unlabelled food (Lampert, 1977a).

Secondly, respiratory carbon tracer losses may be another problem when measuring assimilation over longer periods of time (see Fig. 4). Although it is possible to measure $^{13}CO_2$ and to add this carbon to the assimilated material to estimate real assimilation, animals probably consist of at least two compartments that affect the kinetics of any kind of tracer (Lampert & Gabriel, 1984). This may lead to disproportionately high losses of $^{13}CO_2$ when animals are exposed to enriched food, hence leading to overestimation of respiration and (corrected) assimilation. Lampert and Gabriel (1984) offer a method to correct for this problem, based on the linear relationship between tracer activity and respiration, which is also applicable to the presented calculations with stable isotopes.

Thirdly, isotopic fractionation is a problem that may occur when working with stable isotopes in assimilation studies. Fractionation is a phenomenon whereby organisms tend to favour the incorporation of the rare, heavy isotope rather than the common, lighter isotope. The latter is caused by the fact that, in each chemical transformation, heavy isotopes concentrate in the molecules where bond strengths are greatest (Peterson & Fry, 1987). We neglected fractionation effects in all calculations, because we can prove that these effects are insignificant. We give a numerical example for carbon, with a feeding time of 1 h, $m_a = 0.1$, $\delta_z = -30‰$, and $\delta_f = 1000‰$. Typical fractionation values are 1‰ for $\delta^{13}C$ (DeNiro & Epstein, 1978), so when assuming that all ingested carbon is built into animal structure, δ_f will become $1000‰ + 1‰ = 1001‰$, and δ_s will become $m_a \cdot 1001‰ + -30‰ = 70.1‰$. The mass-specific assimilation rate using Equation 5d will then be $(70.1‰$ to $-30‰)/1000‰ = 0.1001$ mg C mg C^{-1} h^{-1}. Compared to the 'real' assimilation rate of $m_a \cdot T = 0.1$, this is a 0.1% overestimation, equalling the relative fractionation of the algae in the animal. Compared to the relative error that is made due to the two-compartment tracer distribution (Lampert & Gabriel, 1984), fractionation is indeed negligible in most cases.

Optimisation of feeding experiments

An increase of method sensitivity can then only be achieved by optimisation of the signal to noise ratio in the IRMS. Large improvements can often be made by critically following each step in the feeding experiment, from labelling to measurement. With many of these steps, there is the risk of contamination from the laboratory environment. Apart from cleaning tin cups, using special needles and forceps, sometimes the use of gloves may increase precision. Our algal labelling results (Fig. 4) showed that it is advisable to use airtight vessels to avoid respiratory losses of tracer. The zooplankton biomass itself can be optimised to the most sensitive measurement range of the IRMS, based on regressions between known quantities of zooplankton and mass signal (see Fig. 1b). In the case of large zooplankton, even single individuals may give too high mass peaks. In that case, it is necessary to dry the zooplankton first, homogenise the material and weigh this very accurately into the measurement cups (cf. Lampert & Grey, 2003). Another option is adjusting the dilution of the gas flow entering the IRMS by changing the helium flow. By reducing the helium flow, Carman & Fry (2002) could use sample material masses as low as 1 μg N for ^{15}N and 2 μg C for ^{13}C. However, these low quantities are more sensitive to background effects, so when applying this method, again the risk of contamination should be minimised (Carman & Fry, 2002).

Dual labelling for differential uptake

To study selectivity of zooplankton on different food types, the different food types of interest can be labelled with different stable isotopes. Feeding studies employing different types of radioisotopes have, for example, been carried out on mixtures of differently labelled bacteria and algae (Gophen et al., 1973; Lampert, 1974), on mixtures of algae and yeast (Starkweather & Gilbert, 1977b) and on mixtures of differently labelled algae species (Lampert & Taylor, 1985). The same method can be applied to stable isotopes, where the limiting factors are the number and type of isotopes that can be handled by the IRMS, and the number of suitable stable isotopes.

Stable isotopes and biomarkers

Stable-isotope labelled biomarkers are very well applicable for studies of metabolic processes in zooplankton. Boschker & Middelburg (2002) have given an extensive review of these possibilities for microbial ecology, and more and more biomarkers are becoming commercially available. An interesting example of this has been the use of ^{13}C-labelled amino acids to investigate amino acid metabolism

Table 1. Attributes of some commonly used methods for the study of zooplankton feeding in the laboratory

Method	Stable isotope tracer	Radiotracer	Gut fluorescence	Tracer particles	Microscopic cell counts	Optical particle counts	Electronic particle counts
Quantification of ingestion	(Semi-) direct	(Semi-)direct	Direct	Semi-direct	Indirect	Indirect	Indirect
Quantification of assimilation	Direct	Direct	No	No	Indirect	No	Indirect
Requirement of special equipment	Yes	Yes	Yes	No	No	Yes	Yes
Costs per sample analysis	High	High	Low	High	Low	Low	Low
Accuracy of method	High	High	Low	Low	Low	High	Low
Preparation time	Long	Long	Low	Short	Short	Short	Short
Feeding time	Short	Short	Short	Short	Long	Long	Long
Sample analysis time	Short	Short	Short	Long	Long	Short	Short
Animal density required	Low	Low	Low	Low	High	High	High
Mixed food items possible?	Limited	Limited	No	No	Unlimited	Limited	No
Additional variates	Element type	element type	Pigment type	Size (+pigment type)	Algal species (+morphology)	Size (+pigment type)	Size
Safety precautions necessary?	No	Yes	No	Yes	No	No	No
Sample conservation method	Desiccator	Fixative	Fixative	Fixative	Fixative	Fixative	Fixative

The term semi-direct is used when only part of all ingested particles can be measured.

and amino acid essentiality *in vivo* in chickens (Berthold et al., 1991). Since there is much debate about what specific compounds limit zooplankton growth (e.g., Hessen, 1992, 1993; Urabe & Watanabe, 1992, 1993; Brett, 1993; Urabe et al., 1997; Müller-Navarra et al., 2000, 2004), and because certain essential compounds can be transformed into others (Von Elert, 2002), stable-isotope labelled biomarkers can be applied to give more insight into the mechanisms behind food quality limitation of zooplankton population growth.

There are certain stable isotopes of elements lacking radioisotopes (e.g., nitrogen) that offer options to make elemental budgets for zooplankton. There are also elements that do not have stable isotopes, but only radioisotopes (e.g., phosphorus). Various combinations of stable isotopes, radioisotopes and/or isotope-labelled biomarkers offer promising and almost unlimited possibilities for studies of physiological pathways and elemental budgets.

Which method should be used?

Commonly applied methods in quantitative zooplankton feeding studies are summarized in Table 1. The costs of the tracer particle method (e.g., beads) are determined by the price of the tracer particles, which is generally high. The accuracy of optical particle counting depends on the type of equipment: e.g., fluorescence-based methods allow measurements of very low signals and/or deconvolution of different wavelengths. The possibilities of measuring feeding on mixed food items simultaneously is limited by the type of equipment used, and the additional variables that can be measured with the method, e.g., the number of available isotopes in isotope studies. With tracer particles and radiotracers, sometimes aggressive chemicals have to be used in order to dissolve and store animals, and furthermore radioisotopes require laboratories that are designed for the use of these isotopes.

The main disadvantages of stable isotopes are: requirement of special equipment, the requirement of a clean room for sample handling, the (still) relatively high costs of sample analysis and the long time required for sample preparation (i.e., cleaning of tin cups, weighing of reference material, picking of animals, drying of samples, folding of tin cups) (Table 1). On the other hand, the sample analysis times are short, the methodological accuracy is high and can be optimised, required feeding times are short, and ingestion and assimilation rates can be measured directly.

Compared to radioisotopes, the application of stable isotope tracers in zooplankton feeding studies offers several advantages:

1. Compared to direct disintegration counts, no self-absorption will occur (Peters, 1984; Gulati, 1985), and compared to scintillation counts, no potentially harmful tissue solubilisers or scintillation cocktails are required.
2. No potentially harmful radiation is emitted, so no separate laboratories are required for feeding experiments, nor is handling or storage of stable isotopes and samples subject to legal restrictions.
3. Dried samples can be stored in a desiccator for a long period of time without loss of label, which is advantageous when handling large numbers of samples.
4. Elements that do not have radioisotopes sometimes have stable isotopes (e.g., ^{15}N), and very specific stable-isotope labelled biomarkers are increasingly available.

The first three points allow safe and uncomplicated application of stable isotopes in any laboratory or field experiment, as long as sample drying and dry storage is feasible. Although we are aware that there is not a single 'best method' for measuring zooplankton feeding, we think stable isotopes offer an attractive alternative or addition to available traditional methods for measurements on zooplankton feeding.

Acknowledgements

We would like to thank Virgilio Floris (Department of Microbial Ecology, NIOO-CL) for EA-IRMS assistance, Yegor Zadereev (Institute of Biophysics, Siberian Branch of the Russian Academy of Science, Krasnoyarsk, Russia) for cooperation on feeding experiments, and Koos (J.) Vijverberg and two anonymous reviewers for valuable suggestions for improving the manuscript.

References

Baars, M. A. & S. S. Oosterhuis, 1985. Zooplankton grazing in natural water with high concentration of ^{14}C bicarbonate: variable controls and gut passage time. Hydrobiological Bulletin 19: 71–80.

Bearhop, S., D. R. Thompson, S. Waldron, I. S. Russell, G. Alexander & R. W. Furness, 1999. Stable isotopes indicate the extent of freshwater feeding by cormorants *Phalacrocorax carbo* shot at inland fisheries in England. Journal of Applied Ecology 36: 75–84.

Berthold, H. K., D. L. Hachey, P. J. Reeds, O. P Thomas & S. P. D. Hoeksema Klein, 1991. Uniformly ^{13}C-labeled algal protein used to determine amino acid essentiality *in vivo*. Proceedings of the National Academy of Sciences (USA) 88: 8091–8095.

Boschker, H. T. S. & J. Middelburg, 2002. Minireview: Stable isotopes and biomarkers in microbial ecology. FEMS Microbiology Ecology 40: 85–95.

Brett, M. T., 1993. Comment on "Possibility of N or P limitation for planktonic cladocerans: and experimental test" (Urabe and Watanabe) and "Nutrient limitation of zooplankton production." (Hessen). Limnology and Oceanography 38: 1333–1337.

Carman, K. R. & B. Fry, 2002. Small-sample methods for delta C-13 and delta N-15 analysis of the diets of marsh meiofaunal species using natural-abundance and tracer-addition isotope techniques. Marine Ecology Progress Series 240: 85–92.

DeNiro, M. J. & S. Epstein, 1978. Influence of the diet on the distribution of carbon isotopes in animals. Geochimica et Cosmochimica Acta 42: 495–506.

Gophen, M., B. Z. Cavari & T. Berman, 1973. Zooplankton feeding on differentially labelled algae and bacteria. Nature 247: 393–394.

Guillard, 1975. Cultures of phytoplankton for feeding of marine invertebrates. In Smith, (ed.) Culture of Marine Invertebrate Animals. Plenum, New York, 29–60.

Gulati, R. D., 1985. Zooplankton grazing methods using radioactive tracers: technical problems. Hydrobiological Bulletin 19: 61–69.

Hammer, A., C. Grüttner & R. Schumann, 2001. New biocompatible tracer particles: use for estimation of microzooplankton grazing, digestion, and growth rates. Aquatic Microbial Ecology 24: 153–161.

Hasegawa, T., I. Koike & H. Mukai, 2001. Fate of food nitrogen in marine copepods. Marine Ecology Progress Series 210: 167–174.

Hessen, D. O., 1992. Nutrient element limitation of zooplankton production. The American Naturalist 140: 799–814.

Hessen, D. O., 1993. The role of mineral nutrients for zooplankton nutrition: Reply to the comment by Brett. Limnology and Oceanography 38: 1340–1343.

Hino, A., S. Aoki & M. Ushiro, 1997. Nitrogen-flow in the rotifer *Brachionus rotundiformis* and its significance in mass cultures. Hydrobiologia 358: 77–82.

Lampert, W., 1974. A method for determining food selection by zooplankton. Limnology and Oceanography 19: 995–998.

Lampert, W., 1977a. Studies on the carbon balance of *Daphnia pulex* as related to environmental conditions. I. Methodological problems of the use of 14C for the measurement of carbon assimilation. Archiv für Hydrobiologie 48: 287–309.

Lampert, W., 1977b. Studies on the carbon balance of *Daphnia pulex* as related to environmental conditions II. The dependence of carbon assimilation on animal size, temperature, food concentration and diet species. Archiv für Hydrobiologie 48: 310–335.

Lampert, W. & W. Gabriel, 1984. Tracer kinetics in *Daphnia*: an improved two-compartment model and experimental test. Archiv für Hydrobiologie 100: 1–20.

Lampert, W. & J. Grey, 2003. Exploitation of a deep-water algal maximum by *Daphnia*: a stable-isotope tracer study. Hydrobiologia 500: 95–101.

Lampert, W. & B. E. Taylor, 1985. Zooplankton grazing in a eutrophic lake: implications of diel vertical migration. Ecology 66: 68–82.

Lindemann, N. & W. Kleinow, 2000. A study of rotifer feeding and digestive processes using erythrocytes as microparticulate markers. Hydrobiologia 435: 27–41.

Lürling, M. & A. M. Verschoor, 2003. F_0-spectra of chlorophyll fluorescence for the determination of zooplankton grazing. Hydrobiologia 491: 145–157.

Müller-Navarra, D. C., M. T. Brett & A. M. Liston, 2000. A highly unsaturated fatty acid predicts carbon transfer between primary producers and consumers. Nature 403: 74–77.

Müller-Navarra, D. C., M. T. Brett, S. Park, S. Chandra, A. P. Ballantyne, E. Zorita & C. R. Goldman, 2004. Unsaturated fatty acid content in seston and tropho-dynamic coupling in lakes. Nature 427: 69–72.

Penry, D. L. & B. W. Frost, 1990. Re-evaluation of the gut-fullness (gut fluorescence) method for inferring ingestion rates of suspension-feeding copepods. Limnology and Oceanography 35: 1207–1214.

Peters, R. H, 1984. Methods for the study of feeding, grazing and assimilation by zooplankton. In Downing, J. A. & F. H. Rigler (eds) A manual for the assessment of secondary production in fresh waters. IBP Handbook 17 (2nd edn., pp. 336–412). Blackwell, Oxford, 336–412.

Peterson, B. J. & B. Fry, 1987. Stable isotopes in ecosystem studies. Annual Reviews of Ecology and Systematics 18: 293–320.

Pilati, A. & W. A. Wurtsbaugh, 2003. Importance of zooplankton for the persistence of a deep chlorophyll layer: A Limnocorral experiment. Limnology and Oceanography 48: 249–260.

Rigler, F. H., 1971. Methods for the measurement of assimilation of food by zooplankton. In Edmondson, W. T. (ed.) A manual on methods for the assessment of secondary production. IBP Handbook No. 17. Blackwell Scientific Publications, Oxford/Edinburgh, 264–269.

Rothhaupt, K. O., 1990. Changes of the functional responses of the rotifers *Brachionus rubens* and *Brachionus calyciflorus* with particle sizes. Limnology and Oceanography 35: 24–32.

Starkweather, P. L. & J. J. Gilbert, 1977a. Radiotracer determination of feeding in *Brachionus calyciflorus*: The importance of gut passage times. Archiv für Hydrobiologie, Ergebnisse der Limnologie 8: 261–263.

Starkweather, P. L. & J. J. Gilbert, 1977b. Feeding in the rotifer *Brachionus calyciflorus* 2. Effect of food density on feeding rates using *Euglena gracilis* and Rhodotorula glutinis. Oecologia 28: 133–139.

Urabe, J. & Y. Watanabe, 1992. Possibility of N or P limitation for planktonic cladocerans: an experimental test. Limnology and Oceanography 37: 244–251.

Urabe, J. & Y. Watanabe, 1993. Implications of sestonic elemental ratio in zooplankton ecology: Reply to the comment by Brett. Limnology and Oceanography 38: 1337–1340.

Urabe, J., J. Clasen & R.W. Sterner, 1997. Phosphorus limitation of *Daphnia* growth: Is it real? Limnology and Oceanography 42: 1436–1443.

Van der Zanden, M. J., B. J. Shuter, N. Lester & J. B. Rasmussen, 1999. Patterns of food chain lengths in lakes: A stable isotope study. The American Naturalist 154: 406–416.

Von Elert, E., 2002. Determination of limiting polyunsaturated fatty acids in *Daphnia galeata* using a new method to enrich algae with single fatty acids. Limnology and Oceanography 47: 1764–1773.

Appendix: Error calculation

We quantified the relative errors that occur when approximating zooplankton biomass and using δ values by comparing this method with the 'fraction method'. These formulae can be used directly to correct calculated feeding and assimilation rates. We have chosen not to show the intermediate steps of calculation, but these can be supplied upon request. Briefly, we introduced an additional variable $m(=M_f/M_z)$, representing the relative ingestion (m_i) or assimilation (m_a), and expressed all masses in terms of m. For implementation of these sometimes lengthy equations, it should be taken in mind that, apart from the chance of typing errors, one very commonly used spreadsheet programme even requires extra brackets, e.g., for multiplications within denominators.

The relative error (RE) of ingestion rates (5b,c), compared to the original method (4d,e), is given in Eq. [1].

We also calculated the relative error (RE) of this approach for assimilation efficiencies, using specific ingestion (m_i) and assimilation (m_a). We introduced an additional variable to relate the different masses during both experiments, but this variable fell out of the final result, leaving Eq. [2]:

$$\mathrm{RE} = \left(\frac{\left(\dfrac{m * \dfrac{\frac{\delta_f}{1000}+1}{\frac{\delta_f}{1000}+1+\frac{1}{R_R}} + \dfrac{\frac{\delta_z}{1000}+1}{\frac{\delta_z}{1000}+1+\frac{1}{R_R}}}{R_R * \left((m+1) - \left(m * \dfrac{\frac{\delta_f}{1000}+1}{\frac{\delta_f}{1000}+1+\frac{1}{R_R}} + \dfrac{\frac{\delta_z}{1000}+1}{\frac{\delta_z}{1000}+1+\frac{1}{R_R}} \right) \right)} - 1 \right) * 1000 - \delta_z}{m * \delta_f} - 1 \right) * 100‰ \qquad [1]$$

$$\mathrm{RE} = \left(\frac{\left(\left(\dfrac{m_a * \dfrac{\frac{\delta_f}{1000}+1}{\frac{\delta_f}{1000}+1+\frac{1}{R_R}} + \dfrac{\frac{\delta_z}{1000}+1}{\frac{\delta_z}{1000}+1+\frac{1}{R_R}}}{R_R * \left((m_a+1) - \left(m_a * \dfrac{\frac{\delta_f}{1000}+1}{\frac{\delta_f}{1000}+1+\frac{1}{R_R}} + \dfrac{\frac{\delta_z}{1000}+1}{\frac{\delta_z}{1000}+1+\frac{1}{R_R}} \right) \right)} - 1 \right) * 1000 - \delta_z \right) * m_i}{\left(\left(\dfrac{m_i * \dfrac{\frac{\delta_f}{1000}+1}{\frac{\delta_f}{1000}+1+\frac{1}{R_R}} + \dfrac{\frac{\delta_z}{1000}+1}{\frac{\delta_z}{1000}+1+\frac{1}{R_R}}}{R_R * \left((m_i+1) - \left(m_i * \dfrac{\frac{\delta_f}{1000}+1}{\frac{\delta_f}{1000}+1+\frac{1}{R_R}} + \dfrac{\frac{\delta_z}{1000}+1}{\frac{\delta_z}{1000}+1+\frac{1}{R_R}} \right) \right)} - 1 \right) * 1000 - \delta_z \right) * m_a} - 1 \right) * 100‰ \qquad [2]$$

Part IX
Aquaculture and Ecotoxicology

Hydrobiologia (2005) 546:553–558
A. Herzig, R.D. Gulati, C.D. Jersabek & L. May (eds.) Rotifera X: Rotifer Research: Trends, New Tools and Recent Advances
DOI 10.1007/s10750-005-4297-9

Screening methods for improving rotifer culture quality

Adriana Araujo & Atsushi Hagiwara*
Graduate School of Science and Technology, Faculty of Fisheries, Nagasaki University, Bunkyo, 852-8521, Nagasaki, Japan
*(*Author for correspondence: E-mail: hagiwara@net.nagasaki-u.ac.jp)*

Key words: strain selection, GABA, *Brachionus plicatilis*

Abstract

We studied whether the selection of rotifer *B. plicatilis* strain (Japanese, Russian or Australian), as well as the addition of gamma-aminobutyric acid (GABA) to the culture water, are useful in stabilising rotifer cultures. We examined the effect of a combination of the following stressors: unionized ammonia (2.4 mg l^{-1}), contamination by protozoa *Euplotes* sp. (10 cells ml^{-1}) and addition of methyl cellulose to increase the culture water viscosity at 15 cp. Rotifer reproductive tests and enzyme activity measurements (glucosidase) were conducted to determine the effect of the treatments. All tests were conducted at 25 °C and rotifers were fed *Nannochloropsis oculata* at 7×10^6 cells ml^{-1}. The combined effects of the stressors caused a significant decrease in lifespan, fecundity and glucosidase activity. The effect of the stressors on reproductive characteristics and glucosidase activity could be neutralized if rotifers were treated with GABA.

Introduction

Mass cultures of rotifers are prone to collapse, so it is important to develop techniques to stabilize them. Previous reports have identified the optimal environmental conditions for rotifer population growth in relation to temperature (Hirayama & Kusano, 1972; Rumengan & Hirayama, 1990), salinity (Hagiwara et al., 1988; Hoff & Snell, 1989), and the species and density of their dietary phytoplankton (Chotiyaputta & Hirayama, 1978; Snell & Boyer, 1988; Hirayama et al., 1989). Placing a filter in the culture tank has also been found useful for reducing suspended solids, bacterial biomass and protozoan contamination of rotifer culture medium (Balompapueng et al., 1997; Yoshimura et al., 1997). Nevertheless, collapses of rotifer cultures still occur. In hatcheries, the selection of rotifer strains for culture is important as strains can vary in reproductive rate, individual size, optimum culture conditions, and frequency of mixis (Meragelman et al., 1985; Lubzens, 1987; Fushimi, 1989; Fukusho, 1989; James & Abu-Rezeq, 1989; Lubzens et al., 1989). Several other factors are also important when selecting a strain for culture, including its tolerance of environmental perturbations. Some studies have demonstrated considerable variation between the responses of the rotifers *Brachionus plicatilis* and *Brachionus rotundiformis* to single stressors, such as unionized ammonia, viscosity and protozoan contamination, in the culture environment (Araujo et al., 2000, 2001). Others have reported variation among strains within the same species in relation to biometric variation as a direct response to abiotic variables (Serra & Miracle, 1987), genetic variation and function of mate recognition (Kotani et al., 1997), and mate recognition at different temperatures (Kotani & Hagiwara, 1999).

Gallardo et al. (1997) reported that the neurotransmitter gamma-aminobutyric acid (GABA) at

50 mg l^{-1} was effective in increasing the population growth of rotifers. Later, Gallardo et al. (1999) treated batch rotifer cultures with GABA at low food and high ammonia levels and suggested that this was an effective method of enhancing rotifer reproduction when culture conditions were not optimal. The authors also observed that GABA did not have the same effect in individual cultures, possibly because the environment was optimal. They suggested that, in batch cultures, where the environment deteriorates, GABA may compensate for the deterioration in environmental conditions by increasing the efficiency of assimilation and utilization of nutrients from the limited available food. Morse (1984) stated that some hormones have this effect, rather than acting by increasing feeding activity and ingestion.

The purpose of the current study is to develop techniques that could be used for maintaining the stability of mass rotifer cultures. GABA was applied to determine whether it was effective in enhancing the growth of cultured rotifers exposed to a combination of stressors, as is often found in mass culture environments. Furthermore, a screening was conducted to select rotifer strains that are the most resistant to culture stress and, therefore, more appropriate for use in mass culture practices.

Materials and methods

Responses of rotifer strains to stressors

Individual Culture

Individual culture followed the protocol described in Araujo et al. (2000). Neonates hatched from resting eggs of three strains of *Brachionus plicatilis* (i.e., Tokyo clone NH1L, Russian strain and Australian strain) were exposed to adverse environmental culture conditions. For the Tokyo strain, resting eggs were mass-produced (Hagiwara et al., 1993) and preserved in cans (Balompapueng et al., 1997). To produce resting eggs, the Russian and Australian strains were collected from stock cultures, cultured in the laboratory in 1-l beakers with 22 ppt seawater at 25 °C, and fed *Nannochloropsis oculata* (Droop) at 7×10^6 cells ml^{-1}. These cultures were maintained until the rotifers produced resting eggs, which were then collected and stimulated to hatch under light at 25 °C for about 48 h. *N. oculata* was supplied in the hatching medium to prevent starvation of the hatched animals. Newly hatched rotifers were individually introduced into each well of 96-well culture plates containing 100 μl of *N. oculata* suspension at a density of 7×10^6 cells ml^{-1}. Maternal females were transferred into fresh medium every 24 h. Fecundity and lifespan were determined by daily monitoring at 10× magnification under a stereomicroscope. The cultures were maintained at 25 °C in darkness, except during feeding and observation. Eight replicates were used for each treatment.

Hatched neonates were exposed to a treatment of combined stressors, which were unionized ammonia, *Euplotes* sp. and viscosity. Test concentrations were chosen based on the fecundity and lifespan LOECs (lowest observed effect concentration) reported by Araujo et al. (2001). The ciliate *Euplotes* sp. was collected from rotifer mass culture tanks in Kamiura Station, Japan Sea Farming Association, Japan in 1995 and maintained in our laboratory at 25 °C (Jung et al., 1997). *Euplotes* sp. was added to the designated treatment at a density of 10 cells ml^{-1}. Unionized ammonia concentrations were prepared at 2.4 mg l^{-1}, according to the method described by Yu & Hirayama (1986). Viscosity of the experimental seawater was regulated by dissolving methyl cellulose (15cP – WAKO Pure Chemical Industries, LTD., Tokyo, Japan) to a concentration of 0.05%. When compared to 22 ppt control seawater without methyl cellulose, the relative viscosity of the treated seawater was 1.078 (Hagiwara et al., 1998). Unionized ammonia levels were regulated in the seawater after adjusting the designated viscosity level. *Euplotes* sp. were simply pipetted into the seawater after the unionized ammonia and viscosity levels had been regulated. Treated samples were compared to a control (i.e., no addition of stressors).

Enzyme activity measurement

Enzyme testing also followed the protocol described in Araujo et al. (2000). One hundred neonate rotifers were inoculated into 24-well plates containing 1 ml of medium and incubated for 2 h within the designated treatments. At the end of the incubation period, rotifers transferred to fresh seawater at 22 ppt with a pipette. This procedure was repeated twice. Neonates were not fed during the test.

Glucosidase enzyme was chosen for this test because, compared to other enzymes studied, glucosidase activity showed the highest correlation with reproductive tests (Araujo et al., 2000, 2001). Rotifers were exposed to the fluorogenic substrate fluorescein di-β-D-glucopyranoside (FDGlu – Molecular Probes Inc), which is cleaved by the enzyme glucosidase. The reaction medium was prepared according to Burbank & Snell (1994): 5 mg of FDGlu was dissolved in 500 μl of DMSO making a 15.2 mM stock. Glucosidase substrate was diluted to 1.9 mM and then divided into several 50 μl aliquots and stored at −80 °C until use.

A 1.3 μl aliquot of FDGlu solution was added to the respective rotifer samples, replicated three times, and incubated at 25 °C in darkness for 15 min. At the end of the 15 min incubation period, 20 μl of sodium dodecyl sulfate (SDS) was added to the reaction tubes. They were then vortexed for 15 s, and centrifuged at 9000 rpm for 5 min. Approximately 450 μl of the supernatant was transferred to a 6 × 50 mm borosilicate microcuvette cylinder, which was placed into a fluorometer (Turner TD-700) to measure the enzyme activity. Fluorescence emission was read at 515 nm with an excitation wavelength of 490 nm (Moffat & Snell, 1995).

A two-way analysis of variance (ANOVA) was performed to determine any interaction between treatments and rotifer strains. Multiple comparison by Tukey's method was also conducted to determine significant differences between mean values of fecundity, lifespan and glucosidase activity among the rotifer strains (Systat version 8.0, SPSS inc. 1998).

Effect of GABA on the rotifer response to stressors

Individual culture and enzyme testing followed the protocol described above. Neonates of *B. plicatilis* (Tokyo strain) were exposed to (1) a non-stressing medium (control; no addition of stressors), (2) a treatment with stressors (unionized ammonia + viscosity + *Euplotes* sp.) at the levels described above), and (3) a treatment with stressors plus 50 mg l^{-1} of γ-amino-butyric acid (GABA), as described in Gallardo et al. (1999). GABA was purchased from Sigma Chemical Co.

A one-way ANOVA was applied to identify significant differences among the treatments. Pairwise comparison (Dunnett test) was used to detect differences between treatments and controls in fecundity, lifespan and glucosidase activity data (Systat version 8.0, SPSS inc. 1998).

Results

Responses of rotifer strains to stressors

Table 1 shows the results of the test to determine the response of different rotifer strains to the stressors outlined above. The effect of treatment on fecundity, lifespan and glucosidase activity differed among the strains studied (Two-way ANOVA, $p < 0.05$). In both non-stressed and stressed strains, a progressive decline was observed in fecundity of the Tokyo, Russian and Australian strains, in respective order. Responses of lifespan and glucosidase activity also showed a similar trend.

Under stress, the fecundity of the Tokyo strain decreased by 43.7% in comparison with those grown under control conditions, while that of the Russian and Australian strains decreased by 67.5 and 65.1%, respectively. In a non-stressing medium, the fecundities of the Russian and Australian strains did not differ, but both strains had significantly lower fecundities than the Tokyo strain. In a stressing medium, similar results were found (Tukey test, $p < 0.01$).

The lifespan of the Tokyo, Russian and Australian strains when exposed to stress decreased by 39.1, 65.7 and 74.1%, respectively, in comparison with the controls and they were significantly lower ($p < 0.01$). In a non-stressing medium, lifespan among strains did not differ. The lifespans of the Russian and Australian strains when exposed to stress did not differ, but they had significantly lower lifespans than the Tokyo strain.

Glucosidase activity declined by 30.9% of its control for the Tokyo strain, 19.9% for the Russian and 17.3% for the Australian strain ($p < 0.01$). In a non-stressing medium, glucosidase

Table 1. Mean (±se) fecundity, lifespan and glucosidase activity of the three *B. plicatilis* strains under extreme environmental conditions (n = replicates per treatment)

Strain	Treatment	Fecundity (eggs/female) $p = 0.000$; $n = 8$		Lifespan (days) $p = 0.000$; $n = 8$		Glucosidase activity (fluorescence) $n = 3$	
Tokyo	No stress	22.9 (2.7)		15.0 (0.5) a		420.7 (2.6)	
	Stress	12.9 (2.8)		9.13 (1.7)		290.5 (32.8) a	
Russia	No stress	16.1 (2.1) a		14.6 (1.1) a		357.6 (9.9) b	
	Stress	5.3 (1.8) b		5.0 (0.9) b		286.4 (24.3) a	
Australia	No stress	16.1 (3.7) a		14.0 (0.5) a		348.6 (7.1) b	
	Stress	5.6 (1.9) b		3.6 (0.9) b		288.2 (48.0) a	
Two-way ANOVA Source of variation		F	p	F	p	F	p
Strain		33.539	0.000	25.152	0.000	7.333	0.008
Treatment		171.29	0.000	239.89	0.000	126.26	0.000
Strain × treatment		3.461	0.041	15.926	0.000	5.426	0.021

activities in the Russian and Australian strains did not differ, but both strains had significantly lower glucosidase activities than the Tokyo strain. No significant difference in glucosidase activity was found among strains exposed to stress.

Effect of GABA on the rotifer response to stressors

Table 2 illustrates the results of the test to determine the effect of GABA on rotifer cultures. ANOVA showed statistically significant differences among treatment means in relation to lifespan, fecundity and glucosidase activity tests ($p < 0.01$). The combined effects of stressors caused a significant reduction in lifespan, fecundity and glucosidase activity compared to the control ($p < 0.01$). With the addition of GABA (3), reproductive characteristics and glucosidase activity of the rotifers were significantly higher ($p < 0.01$, Dunnett test) than those with stressors and without GABA (2), but were not significantly different than those in the control (1).

Table 2. Mean (±se) fecundity, lifespan and glucosidase activity of *B. plicatilis* (Tokyo strain) under different environmental conditions

Treatment	Fecundity (eggs/female) $p = 0.000$; $n = 8$	Lifespan (days) $p = 0.000$; $n = 8$	Glucosidase activity (fluorescence) $p = 0.002$; $n = 3$
No stress	20 (2.77)	15.6 (0.66)	682.13 (14.61)
Stress	11.29 (1.48)**	13.9 (0.6)**	529.7 (6.2)**
Stress + GABA	19.13 (2.64)	15.4 (0.46)	634.23 (33.42)

Stress = unionized ammonia, viscosity and *Euplotes* sp. One-way ANOVA p-values are indicated for each variable. Level of significant difference in relation to the control (Dunnett test) is indicated as $^{**}p < 0.01$.

Discussion

Our statistical analyses showed that the effects of each treatment depend on the rotifer strain (Table 1). The different responses of rotifer strains to the stressors indicate that they were not physiologically uniform. The results of the reproductive tests revealed that the Tokyo strain (clone NH1L) was the most stress-resistant strain. The highest fecundity was observed for the Tokyo strain at optimal culture conditions. Furthermore, the decline observed in fecundity of the Tokyo strain under stress, compared with its control, was considerably lower (43.7%) than that of the Russian and Australian strains.

Lifespans were not different among strains in the non-stressing environment, but the lifespans of the Russian and Australian strains under stress were significantly lower than that of the Tokyo strain under stress. The longest lifespan among the rotifer strains studied was that of the Tokyo strain. Furthermore, the amount of decline in lifespan of the Tokyo strain under stress was the lowest (39.1%) among the rotifer strains.

The decline observed in glucosidase activity in stressed Tokyo strain compared with its control was higher (30.9%) than that of Russian and Australian strains. However, the highest mean value of glucosidase activity was seen in the Tokyo strain in the non-stressing environment. It would be helpful to conduct additional tests on other *B. plicatilis* strains to elucidate the stress resistance variation among them. Further tests should also be conducted in rotifers raised in continuous culture systems, where contaminants are more abundant.

Table 2 confirms that fecundities, lifespans and glucosidase activities of rotifer strains under stress were lower than those of their controls. It was found that the addition of GABA significantly increased reproductive characteristics and glucosidase activity. No significant differences were observed between a medium with GABA and stressor, and the control. This implies that GABA enhances the physiological condition of the stressed rotifers, confirming previous research reported by Gallardo et al. (1999) in which GABA was found to be effective for enhancement of rotifer reproduction when culture conditions were not optimal. This also suggests that GABA is useful for improving rotifer culture quality. The improvement with GABA treatment can be beneficial, especially when large volumes of rotifers are required for feeding to fish larvae. Considering that GABA costs about two dollars (US) per gram (Sigma Chemical Co.), and the dosage is low (50 mg l^{-1}), large-scale use of GABA would not be very expensive. Aside from the beneficial effect of GABA on rotifer reproduction in enrichment cultures (Gallardo et al., 2001), the use of GABA-treated rotifers as food for fish larvae may also improve larval health. Considering that GABA is an amino acid derivative that has positive effects on abalone larval settlement and metamorphosis (Morse et al., 1979; Morse, 1984), it may also have a positive effect on the fish larvae. However, the effect of GABA on fish larvae still awaits investigation.

Acknowledgments

We gratefully acknowledge the Ministry of Education, Culture, Sports, Science and Technology (MEXT), Japan (Monbukagaku-sho), the Nagasaki Industrial Technology Foundation and The Japan Science Society (Sasakawa Foundation). We also would like to thank Dr. Tomonari Kotani, Dr. Mavit Assavaaree, Dr. Wenresti Gallardo and the staff of Kamiura Station at Japan Sea Farm Association.

References

Araujo, A. B., A. Hagiwara & T. W. Snell, 2000. Effect of unionized ammonia, viscosity and protozoan contamination on the enzyme activity of the rotifer *Brachionus plicatilis*. Aquaculture Research 31: 359–365.

Araujo, A. B., A. Hagiwara & T. W. Snell, 2001. Effect of unionized ammonia, viscosity and protozoan contamination on reproduction and enzyme activity of the rotifer *Brachionus rotundiformis*. Hydrobiologia 446/447: 363–368.

Balompapueng, M. D., A. Hagiwara, Y. Nozakim & K. Hirayama, 1997. Preservation of the euryhaline rotifer (*Brachionus plicatilis*) resting eggs by canning method. Hydrobiologia 358: 163–166.

Burbank, S. E. & T. W. Snell, 1994. Rapid Toxicity Assessment using esterase biomarkers in *Brachionus calyciflorus* (Rotifera). Environmental Toxicology and Water Quality 9: 171–178.

Chotiyaputta, C. & K. Hirayama, 1978. Food selectivity of the rotifer *Brachionus plicatilis* feeding on phytoplankton. Marine Biology 45: 105–111.

Fukusho, K., 1989. Biology and mass production of the rotifer *Brachionus plicatilis*. International Journal of Aquaculture and Fishery Technology 1: 232–240.

Fushimi, T., 1989. Systematizing large-scale culture methods. In Fukusho, K. & K. Hirayama (eds), A Live Feed – The Rotifer, *Brachionus plicatilis*. Koseisha-Koseikaku, Japan, 118–134.

Gallardo, W. G., A. Hagiwara, Y. Tomita, K. Soyano & T. W. Snell, 1997. Effect of some vertebrate and invertebrate hormones on the population growth, mictic female production, and body size of the marine rotifer *Brachionus plicatilis* Müller. Hydrobiologia 358: 113–120.

Gallardo, W. G., A. Hagiwara, Y. Tomita & T. W. Snell, 1999. Effect of growth hormone and gamma-aminobutyric acid on *Brachionus plicatilis* (Rotifera) reproduction at low food or high ammonia levels. Journal of Experimental Marine Biololy and Ecology 24: 179–191.

Gallardo, W. G., A. Hagiwara & T. W. Snell, 2001. Use of GABA to enhance rotifer reproduction in enrichment culture. Aquaculture Research 32(3): 243–246.

Hagiwara, A., K. Hamada, A. Nishi, K. Imaizumi & K. Hirayama, 1993. Mass production of rotifer *Brachionus plicatilis* resting eggs in 50m3 tanks. Nippon Suisan Gakkaishi 59: 93–98.

Hagiwara, A., A. Hino & R. Hirano, 1988. Effects of temperature and chlorinity on resting egg formation in the rotifer *Brachionus plicatilis*. Nippon Suisan Gakkaishi 54: 569–575.

Hagiwara, A., N. Yamamiya & A. B. Araujo, 1998. Effect of water viscosity on the population growth of the rotifer *Brachionus plicatilis* Müller. Hydrobiologia 386/387: 489–494.

Hirayama, K. & T. Kusano, 1972. Fundamental studies on physiology of rotifer for its mass culture. II. Influence of water temperature on population growth of rotifer. Nippon Suisan Gakkaishi 38: 1357–1363.

Hirayama, K., I. Maruyama & T. Maeda, 1989. Nutritional effect of freshwater *Chorella* on growth of the rotifer *Brachionus plicatilis*. Hydrobiologia 186/187: 39–42.

Hoff, H. & T. W. Snell, 1989. Plankton Culture Manual (2nd ed.). Florida Aqua Farms, Florida 126.

James, C. M. & T. Abu-Rezeq, 1989. Production and nutritional quality of two small-sized strains of the rotifer *Brachionus plicatilis*. World Aquaculture Society 20: 261–267.

Jung, M.-M., A. Hagiwara & K. Hirayama, 1997. Interspecific interactions in the marine rotifer microcosm. Hydrobiologia 358: 121–126.

Kotani, T. & A. Hagiwara, 1999. Mate recognition of the rotifer *Brachionus plicatilis* Muller at different temperatures. Bulletin of the Faculty of Fisheries of Nagasaki University 80: 55–59.

Kotani, T., A. Hagiwara & T. W. Snell, 1997. Genetic variation among marine *Brachionus* strains and function of mate recognition pheromone (MRP). Hydrobiologia 358: 105–112.

Lubzens, E., 1987. Raising rotifers for use in aquaculture. Hydrobiologia 147: 245–255.

Lubzens, E., A. Tandler & G. Minkoff, 1989. Rotifers as food in aquaculture. Hydrobiologia 186/187: 387–400.

Meragelman, E., E. Lubzens & G. Minkoff, 1985. A modular system for small-scale mass production of the rotifers *Brachionus plicatilis*. Israel Journal of Zoology 33: 186–194.

Moffat, B. D. & T. W. Snell, 1995. Rapid toxicity assessment using an *in vivo* enzyme test for *Brachionus plicatilis* (Rotifera). Ecotoxicology and Environmental Safety 30: 47–53.

Morse, D. E., 1984. Biochemical and genetic engineering for improved production of abalones and other valuable molluscs. Aquaculture 39: 263–282.

Morse, D. E., N. Hooker, H. Duncan & L. Jensen, 1979. γ-Aminobutyric acid, a neurotransmitter, induces planktonic abalone larvae to settle and begin metamorphosis. Science 204: 407–410.

Rumengan, I. F. M. & K. Hirayama, 1990. Growth responses genetically distinct S and L type rotifer (Brachionus plicatilis) strains to different temperatures. The Second Asian Fisheries Forum, 33–36.

Serra, M. & M. Miracle, 1987. Biometric variation in three strains of *Brachionus plicatilis* as a direct response to abiotic variables. Hydrobiologia 147: 83–89.

Snell, T. W. & E. M. Boyer, 1988. Thresholds for mictic female production in the rotifer *Brachionus plicatilis* (Müller). Journal of Experimental Marine Biology and Ecology 124: 73–85.

SPSS inc., 1998. SYSTAT 8.0: New Statistics. SPSS inc., USA.

Yoshimura, K., K. Usuki, T. Yoshimatsu, C. Kitajima & A. Hagiwara, 1997. Recent development of a high density mass culture system for the rotifer *Brachionus rotundiformis* Tschugunoff. Hydrobiologia 358: 139–144.

Yu, J.-P. & K. Hirayama, 1986. The effect of un-ionized ammonia on the population growth of the rotifer in mass culture. Nippon Suisan Gakkaishi 52: 1509–1513.

Hydrobiologia (2005) 546:559–568
A. Herzig, R.D. Gulati, C.D. Jersabek & L. May (eds.) Rotifera X: Rotifer Research: Trends, New Tools and Recent Advances
DOI 10.1007/s10750-005-4300-5

Interaction among copper toxicity, temperature and salinity on the population dynamics of *Brachionus rotundiformis* (Rotifera)

José Luis Gama-Flores[1,2,*], S.S.S. Sarma[2] & S. Nandini[3]
[1]*Doctoral Programme, Autonomous Metropolitan University. Campus Xochimileo, Calzada de Hueso, No. 1100, Quietud, C.P. 04960, Mexico City, Mexico*
[2]*Laboratory of Aquatic Zoology, Division of Research and Postgraduate Studies, National Autonomous University of Mexico, Campus Iztacala, Los Reyes, A.P. 314, C.P. 54090, Tlalnepantla, State of Mexico, Mexico*
[3]*UIICSE, Division of Research and Postgraduate Studies, National Autonomous University of Mexico, Campus Iztacala, Reyes, A.P. 314, C.P. 54090, Tlalnepantla, State of Mexico, Mexico*
*(*Author for correspondence: E-mail: joluga@servidor.unam.mx)*

Key words: rotifers, temperature, salinity, copper toxicity

Abstract

Heavy metals may interact with ecological factors such as temperature, food level and salinity, causing both mortality and reduced reproduction in organisms. Among different heavy metals, copper compounds are commonly used for eliminating algal blooms in aquaculture tanks. At certain concentrations, copper is toxic to rotifers. In the present work, we evaluated the combined effects of salt concentrations (2.5 and 5.0 g l^{-1} NaCl), copper levels (0, 0.03125, 0.0625, 0.125 and 0.25 mg l^{-1} as $CuCl_2$) and two temperatures (20 and 25 °C) on the population growth of *B. rotundiformis* using *Chlorella* as the algal food (at 0.5×10^6 cells ml^{-1} for every 24 h). Regardless of salinity and temperature, copper at concentrations as low as 0.03 mg l^{-1} had an adverse effect on the population growth of rotifers and above 0.125 mg l^{-1}, the populations did not grow. The effect of the toxicant on *B. rotundiformis* was more severe at 25° than at 20 °C at lower salinity. In general, we observed peak densities of rotifers around day 12 at 20 °C but 6–8 days earlier at 25 °C. Peak population densities of *B. rotundiformis* in the controls at the salinity of 2.5 g l^{-1} ranged from 90 to 180 ind. ml^{-1}, depending on temperature; at a salinity of 5.0 g l^{-1}, these were lower. The population growth rates, *r*, in our study varied from +0.31 to –0.12 depending on the test conditions. There was a significant impact of temperature, salinity and toxicity level on the population growth rate of *B. rotundiformis*. Our results suggested that even narrow changes in salinity could negatively influence the toxicity of heavy metal on the population growth rates of *B. rotundiformis*.

Introduction

Changes in environmental conditions may influence the sensitivity of organisms to various toxicants such as pesticides and heavy metals. Abiotic factors such as temperature and salinity and biotic factors such as food influence both survival and reproduction of several species of saline, marine and freshwater rotifer species (Hutchinson, 1967). Food shortage, sub-optimal temperature, or unfavourable salinity levels in the ambient waters interact with toxicants thereby causing changes in the tolerance capacity of zooplankton (Heugens et al., 2001). In freshwater bodies, rotifers provide a link in the pelagic food chain between the phytoplankton and the bacterioplankton and higher trophic levels (Nogrady et al., 1993). Seasonal changes in the chemical constituents of water, mainly an increase or decrease of salt concentrations as a result of evaporation, could be one of the cyclical stresses to zooplankton, particularly in inland saline lakes. These together with changes in the abundance dynamics of phytoplankton and different temperature regimes usually regulate the rotifers populations, which are generally less salt resistant (Green, 1993).

Among the heavy metals, copper is one of the few with both positive and negative influences on the dynamics of plankton (Rico-Martinez et al., 1998). At low concentrations, this is an essential metal for some phytoplankton species. At high concentrations, compounds of copper are used as algicides. Therefore, depending on the concentration, copper compounds can be lethal to zooplankton via dietary ingestion (e.g. algal food) or directly through medium i.e., aqueous uptake (Yu & Wang, 2002). Natural copper concentrations vary from 0.0005 to 0.005 mg l^{-1}. In Mexico, however, certain petrochemical wastewaters contain copper compounds at concentrations ranging from 0.07 to 6.40 mg l^{-1} (Cervantes & Moreno-Sánchez, 1999). The impact of different environmental factors on the copper toxicity to freshwater rotifers has not been well investigated globally and Mexico is not an exception to this. Thus, seasonal changes in temperature, and dissolved salts concentrations together with the exposure to toxic metals can further influence the population dynamics of rotifers. This leaves a gap in predicting oscillations in zooplankton populations in metal contaminated waterbodies, which is essential for developing water quality guidelines for industries.

Among members of *Brachionus* spp., only a few of them are capable of tolerating salinities higher than 5 g l^{-1}. *B. plicatilis* and *B. rotundiformis* are normally found in waterbodies with different concentrations of salt (Koste, 1978; Sarma et al., 2002). In aquaculture practices, these species are widely used for live food for larval crustaceans and fishes of commercial value (Lubzens et al., 1997). In aquaculture ponds too, sometimes nuisance algal blooms are controlled using copper compounds (Hawkins & Griffiths, 1987). Hence it is of considerable importance to understand the extent to which copper influences the growth of rotifers.

Chronic toxicity tests with rotifers have been conducted using several approaches (Snell & Janssen, 1998). Some of these include swimming speed (Charoy et al., 1995), food consumption (Ferrando & Andreu, 1993), life-table demography (Rao & Sarma, 1986) and population growth (Gama-Flores et al., 1999). However, most commonly population growth studies are employed because of their relevance to aquacultural operations (Luna-Andrade et al., 2002). For example, while population growth rates can be obtained using both the demographic method and the population growth data, the density at which a species can be harvested, cannot be obtained using the former (Krebs, 1985). The information on peak population density is important for feeding larval shrimps and fishes so as to meet the critical prey densities for effective feed management (Lubzens et al., 1997).

In this study we analysed the interactions among $CuSO_4$ toxicity, temperature and salinity on the population growth of *B. rotundiformis*.

Materials and methods

The test rotifer *Brachionus rotundiformis* was isolated from a saline tropical waterbody in Cozumel, Yucatan (20.5° N 86.9° W). Salinity of the waterbody at the time of collection was 6 g l^{-1}. A clonal culture of this species was established using the single-celled green alga *Chlorella vulgaris* as the exclusive food and maintained in EPA medium (Anonymous, 1985) supplemented with 6 g l^{-1} NaCl. The EPA medium was prepared by dissolving 96 mg $NaHCO_3$, 60 mg $CaSO_4$, 60 mg $MgSO_4$ and 4 mg KCl in one litre of distilled water. *C. vulgaris* (CL-V-3, CICESE, Ensenada, Baja California Sur, Mexico) was mass cultured using Bold's basal medium (Borowitzka & Borowitzka, 1988). Log phase alga was harvested, centrifuged at 3000 rpm and resuspended in distilled water. The density of *Chlorella* was estimated using haemocytometer. The desired algal density was prepared daily using EPA medium supplemented with chosen salt level. For routine feeding of rotifer mass cultures, algal densities of 0.5×10^6 to 1.0×10^6 cells ml^{-1} were found appropriate without inhibitory effects on the population growth. Both mass cultures and experimental jars were maintained under continuous but diffused fluorescent illumination. Depending on algal density and temperature we could maintain *B. rotundiformis* at densities of 100–200 ind. ml^{-1}.

Based on a preliminary experiment, we selected 5 (i.e., 0 (=control), 0.03125, 0.0625, 0.125, 0.250, mg l^{-1}) concentrations of copper in the form of sulphate, for quantifying the heavy metal effect on the population growth of *B. rotundiformis*. We used two temperatures, 20 and 25 °C, and two salinity levels, 2.5 and 5.0 g l^{-1}. Before starting the experiments, we acclimatized the rotifer populations to the desired temperature and salinity levels for 6 months. For each salinity level, heavy metal concentration and temperature combination, we used three replicates. Population growth experiments were conducted using 50 ml transparent jars containing 25 ml medium at the chosen heavy metal, temperature and salinity levels. In all we used 60 test jars (5 metal concentrations × 2 temperatures × 2 salinities × 3 replicates). Into each test jar, we introduced *B. rotundiformis* at an initial density of 2 ind. ml^{-1} using a Pasteur pipette under a stereomicroscope. Following the start of the experiment, we counted the number of living individuals of *B. rotundiformis* in each container using either total count or 2 to 3 aliquots of 1–5 ml each. Following the estimation of density, the rotifer population was daily transferred to fresh jars containing chosen food level, temperature and toxicant combinations. The experiments were terminated after 16 days by which time most populations began to decline.

The rate of population increase per day (r) was derived using the following formula (Krebs, 1985):

$$r = (\ln N_t - \ln N_0)/t$$

where N_0 = initial population density, N_t = final population density and, t = time in days. We treated data on the peak population density and the rate of population growth using 3-way analysis of variance (ANOVA) following the statistical package Statistica ver. 6.

Results and discussion

Data on the population growth of *B. rotundiformis* exposed to different levels of copper under two salinity levels at 20 and 25 °C, are shown in Figures 1 and 2. Regardless of salinity and temperature, copper at concentrations as low as 0.03 mg l^{-1} had an adverse effect on the population growth of rotifers, as compared to the controls (toxicant free treatment). The effect of the toxicant on *B. rotundiformis* was more severe at 25 °C than at 20 °C at lower salinity. In general, we observed peak densities of rotifers around day 12 at 20 °C but 6–8 days earlier at 25 °C. Peak population densities of *B. rotundiformis* in controls at a salinity of 2.5 g l^{-1} ranged from 90 to 180 ind. ml^{-1}, depending on temperature; at a salinity of 5.0 g l^{-1}, these were lower (Fig. 3).

Irrespective of the salinity and temperature conditions, at toxicant levels above 0.125 mg l^{-1}, the populations did not grow. The population growth rates in our study varied from +0.31 to −0.12 depending on the test conditions (Fig. 4). There was a significant impact of temperature, salinity and toxicity level on the population growth rate of *B. rotundiformis* ($p < 0.001$, 3-way ANOVA, Table 1). However, there was no significant interaction among the three factors (temperature, salinity and copper concentration) on the population growth rate ($p > 0.05$, 3-way ANOVA, Table 1). In order to understand the joint influence of salinity and temperature on the lowest copper concentration that had a significant effect on the population growth rate, we subjected our data first for Tukeýs test and then for 3-way ANOVA. The results are presented in Table 2. Thus, copper at concentration as low as 0.0625 mg l^{-1}, had significant effect on r, especially at higher salinity.

Heavy metals in water tend to form complexes with organic and inorganic chemicals and this may influence their bioavailability to the exposed zooplankton organisms (Salomón et al., 1995). In our study, the medium did not contain dissolved organic substances, except those released due to rotifer metabolism and these were daily removed. The contribution of detritus from the death of *Chlorella* due to salinity levels used here is unlikely. For example, Sarma et al. (2002) have used the same strain of *Chlorella* at salinities from 0 to 12 g l^{-1}; and did not observe the death of algal cells. In general *Chlorella* is capable of tolerating salinities higher than those used in our study (He et al., 1993). The inorganic substances used in the EPA medium are necessary for physiological balance of rotifers. The importance of defined culture medium for toxicity testing to zooplankton has been documented. For example, Janssen et al. (1994) have shown that toxic substances simply

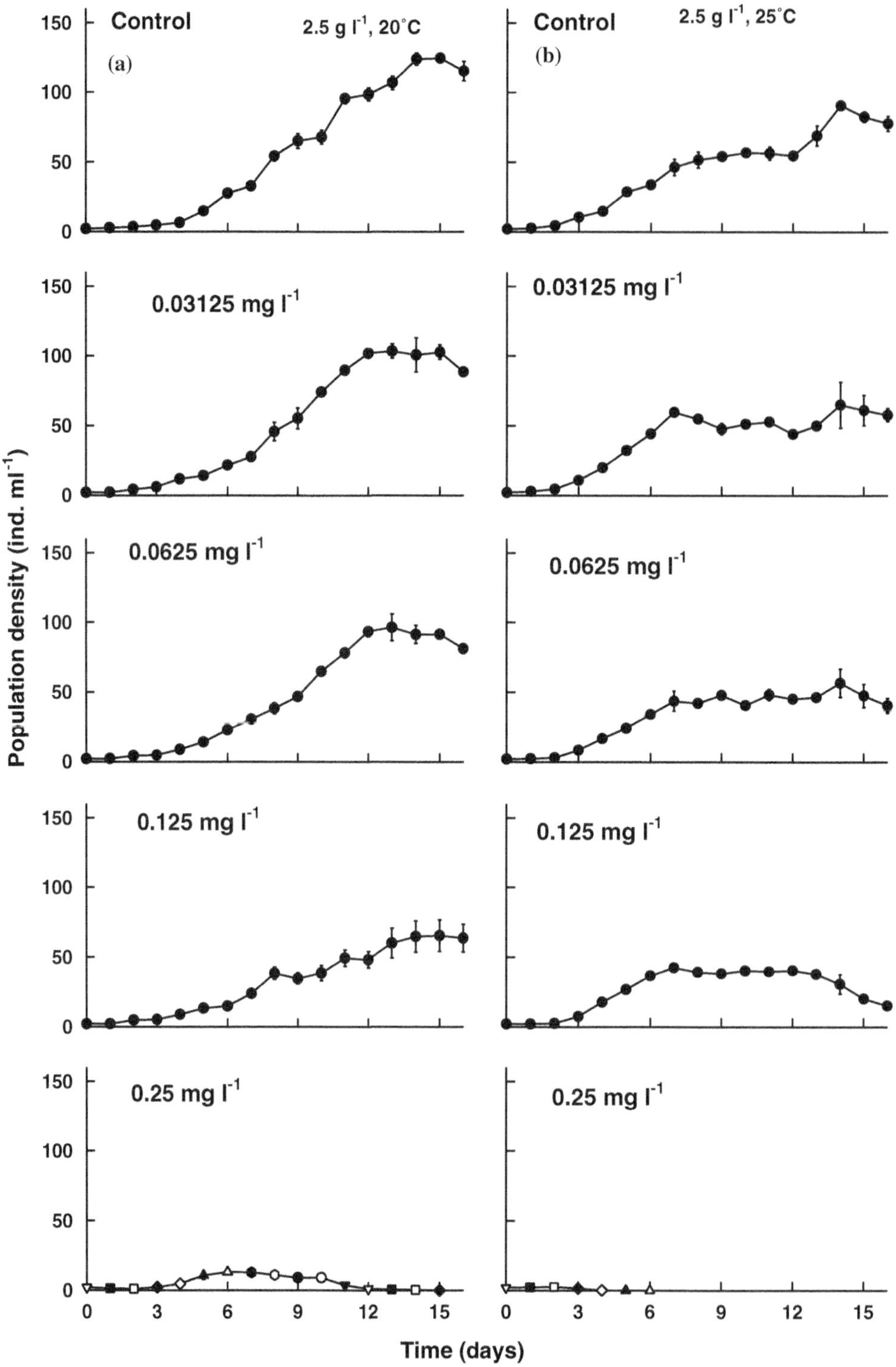

Figure 1. Population growth curves of *B. rotundiformis* cultured under different concentrations of copper sulphate at 20 and 25 °C and at the salinity of 2.5 g l^{-1} of NaCl. Values represent mean ± standard error based on three replicates.

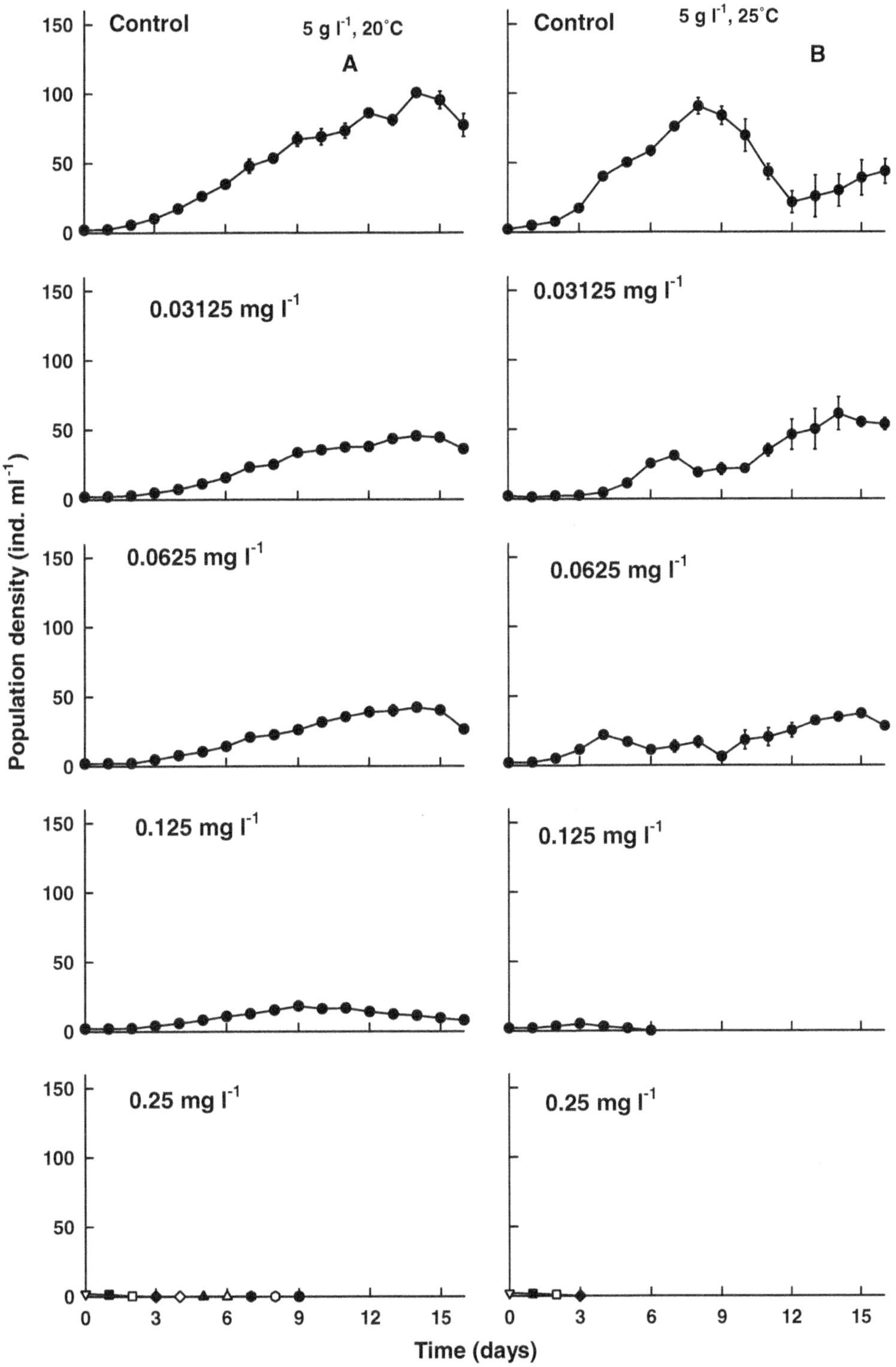

Figure 2. Population growth curves of *B. rotundiformis* cultured under different concentrations of copper sulphate at 20 and 25 °C and at the salinity of 5 g l^{-1} of NaCl. Values represent mean ± standard error based on three replicates.

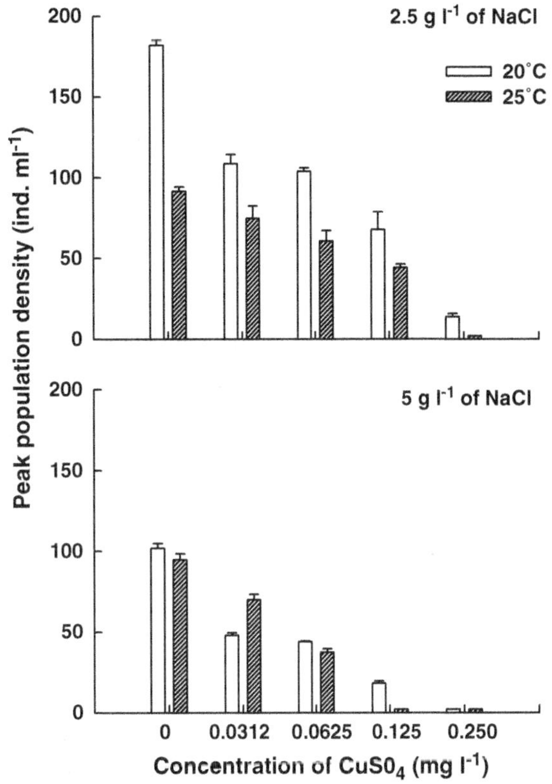

Figure 3. Peak population densities of *B. rotundiformis* grown in relation to different concentrations of copper sulphate at 20 and 25 °C and under two salinities (2.5 and 5 g l^{-1} of NaCl.

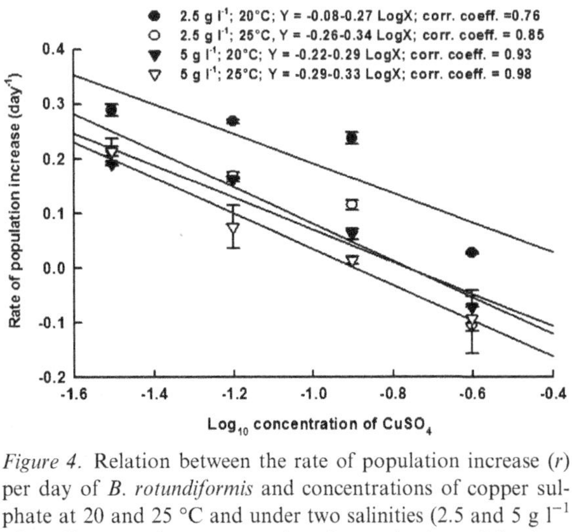

Figure 4. Relation between the rate of population increase (*r*) per day of *B. rotundiformis* and concentrations of copper sulphate at 20 and 25 °C and under two salinities (2.5 and 5 g l^{-1} of NaCl. For each treatment, plotted were the data of the three replicates.

exposed to aquatic organisms in distilled water, not only caused elevated mortalities in concentrations lower than those in the physiological medium, but also the pattern of survival was unpredictable. When using rotifers for bioassays, the Pourriot–Gilbert medium and the EPA medium are often preferred. Since the EPA medium is widely used not only for rotifers but also for other zooplankton such as cladocerans (Soucek et al., 2001), we preferred this in our studies.

While many members of *Brachionus* are pantropical, *B. rotundiformis* can be found at temperatures lower than 20° to higher than 30°. This taxon once thought to be a part of species complex of *B. plicatilis* is adapted to a wide range of salinity levels from freshwater to marine conditions (Ortells et al., 2003). Though it is possible to culture this species at salinities higher than 18 g l^{-1} (Sarma et al., 2002), the clone established was adapted to a salinity of 4–6 g l^{-1}. Toxicants affect biological processes such as respiration and feeding, the combined effects of which are ultimately reflected in the population abundance and growth rates (Halbach, 1984). The interaction of toxicants with both biotic and abiotic factors on zooplankton can thus be easily evaluated using the population growth approach. For rotifer species inhabiting freshwater, changes in salinity levels may cause severe osmoregulatory problems leading to an extinction of population even if other conditions (such as food density and temperature) are optimal. The presence of toxicants may further aggravate this situation. For example, Buikema & Cairns (1974) have shown that the toxicity of heavy metals to rotifers is severe at higher temperatures. Similarly, McLusky et al. (1986) have shown that the toxicity of heavy metals to estuarine organisms is negatively influenced by temperature. This was evident in the present study. Rotifer strains adapted to lower salinities may experience heavier mortalities when toxicant concentrations, salt and temperature levels increase. This was evident in the growth rates of rotifers at the minimum copper concentration at which significant effect was detected (see Table 2). Though certain metals in water may tend to form complexes with other inorganic substances in the medium and thereby reducing their bioavailability to the exposed organisms (Salomón et al., 1995), changes in salinity itself may offset this advantage.

Table 1. Statistical evaluation using 3-way analysis of variance (ANOVA) on the peak population density and the rate of population increase of *B. rotundiformis* subjected to two temperatures, two salinity levels and 5 heavy metal (copper) concentrations

Source	DF	SS	MS	F
Peak population density				
Temperature (A)	1	6651.8	6651.80	131.02***
Error	40	2030.70	50.76	
Salinity (B)	1	6227.93	6227.93	319.65***
Error	40	2030.70	50.76	
Heavy metal (C)	4	87400.56	21850.14	430.39***
Error	40	2030.70	50.76	
Interaction of A × B	1	5718.38	5718.38	112.63***
Error	40	2030.70	50.76	
Interaction of A × C	4	3736.28	934.07	18.39***
Error	40	2030.70	50.76	
Interaction of B × C	4	3008.60	752.15	14.81***
Error	40	2030.70	50.76	
Interaction of A × B × C	4	3006.04	751.51	14.80***
Error	40	2030.70	50.76	
Rate of population increase				
Temperature (A)	1	0.051	0.051	98.48***
Error	40	0.020	0.001	
Salinity (B)	1	0.050	0.050	96.76***
Error	40	0.020	0.001	
Heavy metal (C)	4	0.870	0.217	417.74***
Error	40	0.020	0.001	
Interaction of A × B	1	0.030	0.030	59.45***
Error	40	0.020	0.001	
Interaction of A × C	4	0.019	0.004	9.26***
Error	40	0.020	0.001	
Interaction of B × C	4	0.039	0.009	18.86***
Error	40	0.020	0.001	
Interaction of A × B × C	4	0.004	0.001	2.18ns
Error	40	0.020	0.001	

DF = degrees of freedom, SS = sum of squares, MS = mean square, F = F-ratio. Levels of significance: ***$p < 0.001$; ns = non-significant ($p > 0.05$).

Peak population density and growth rates are two variables sensitive to changes in the ambient medium. Many species of *Brachionus* can be grown to densities higher than 100 ind. ml^{-1} at algal (*Chlorella* or *Scenedesmus*) densities of about 0.5×10^6 cells ml^{-1}. At this food density, larger species such as *Brachionus calyciflorus* (body length about 200 μm) may reach peak abundances lower than 100 ind. ml^{-1}, while smaller taxa such as *Anuraeopsis fissa* (body length about 70 μm) may reach higher densities (about 400 ind. ml^{-1}) (Sarma et al., 1996). *B. rotundiformis* used in this study had a length of 160 ± 18 μm; its peak population densities and growth rates were within the range observed for similar sized rotifers (Moreno-Garrido et al., 1999). Sarma et al. (2001) have reviewed the population growth rates (r) of brachionid rotifers grown under non-stressful conditions. The r-values generally varied from 0.2 to 2.0 d^{-1} depending on the rotifer species, culture conditions and the source of data (e.g., life table or population growth study). In the present work, for controls r varied from 0.24 to 0.30 d^{-1} depending on temperature and salt concentration. When the

Table 2. Results of three-way analysis of variance on the minimum copper concentration that had significant effect on the rate of population increase of *B. rotundiformis*; other details as of Table 1

Source	DF	SS	MS	F
Test conditions: copper 0.0625 mg l^{-1}; 20 °C; salt 2.5 and 5 g l^{-1}; 25 °C; 5 g l^{-1}				
Temperature (A)	1	0.01	0.01	22.39***
Error	24	0.02	0.001	
Salinity (B)	1	0.02	0.02	28.95***
Error	24	0.02	0.001	
Heavy metal (C)	2	0.07	0.04	55.01***
Error	24	0.02	0.001	
Interaction of A × B	1	0.01	0.01	18.92***
Error	24	0.02	0.001	
Interaction of A × C	2	0.01	0.01	9.34***
Error	24	0.02	0.001	
Interaction of B × C	2	0.02	0.01	12.21***
Error	24	0.02	0.001	
Interaction of A × B × C	2	0.01	0.003	3.54*
Error	24	0.02	0.001	
Test conditions: copper 0.0125 mg l^{-1}; 25 °C; salt 5 g l^{-1}				
Temperature (A)	1	0.03	0.03	56.97***
Error	32	0.02	0.001	
Salinity (B)	1	0.06	0.06	102.12***
Error	32	0.02	0.001	
Heavy metal (C)	3	0.19	0.06	115.11***
Error	32	0.02	0.001	
Interaction of A × B	1	0.02	0.02	31.43***
Error	32	0.02	0.001	
Interaction of A × C	3	0.02	0.01	9.95***
Error	32	0.02	0.001	
Interaction of B × C	3	0.04	0.01	20.93***
Error	32	0.02	0.001	
Interaction of A × B × C	3	0.003	0.001	2.80ns
Error	32	0.02	0.001	

copper concentration in the medium was 0.250 mg l^{-1}, r became negative irrespective of salinity. We also observed significant interactions of factors (e.g., temperature and copper concentration) on the population growth rates of *B. rotundiformis*. While the growth rate in control at salinity of 2.5 g l^{-1} was higher at 20 °C than at 25 °C, the reverse was the case at 5 g l^{-1} of salt concentration. It is, therefore, important to conduct chronic toxicity studies under specific conditions, especially for saline or marine water rotifers.

Population growth of closely related taxon *B. plicatilis* in relation to copper stress has been previously studied. For example, Moreno-Garrido et al. (1999) have compared the population growth curves of *B. plicatilis* fed three different species of microalgae (*Dunaliella salina*, *Nannochloropsis gaditana* and *Isochrysis galbana*) using copper and cadmium as toxicants. They found that depending on the food combination, a copper concentration of 0.1 mg l^{-1} did not support rotifer growth in several treatments and whenever growth occurred, the density of *B. plicatilis* was always much lower than that at the corresponding toxicant free medium. Luna-Andrade et al. (2002) have also showed that the rate of population increase of *B. plicatilis* under copper stress decreased more or less linearly with increasing metal concentration in the medium. The levels of copper chosen here were lower than those in Moreno-Garrido et al. (1999) and

Luna-Andrade et al. (2002). The latter workers have observed that copper concentrations at or above 0.125 mg l^{-1} caused significant reduction of *r* values, while in the present study, copper level lower than half of this had negative influence on the growth rates of *B. rotundiformis* suggesting that this species was more sensitive to copper toxicity than *B. plicatilis*. The combined influence of salinity and copper toxicity was also evident on *r*. Our results suggest that even narrow changes in salinity could negatively influence the toxicity of heavy metal on the population growth rates of *B. rotundiformis*.

Acknowledgements

Two anonymous reviewers have greatly improved our manuscript for which we are grateful. We thank CONACyT (Ref. 185928, SNI-18723, SNI-20520 and Ref. 41786) and PAPIIT-IN234602 for financial support. Constructive criticism from Maria Elena Castellanos Paez and Ma. de Jesús Ferrara Guerrero is gratefully acknowledged.

References

Anonymous, 1985. Methods of Measuring the Acute Toxicity of Effluents to Freshwater and Marine Organisms. US Environment Protection Agency EPA/600/4–85/013, Washington DC.

Borowitzka, M. A. & L. J. Borowitzka, 1988. Micro-algal biotechnology. Cambridge University Press, United Kingdom.

Buikema, A. I. & J. Cairns Jr., 1974. Evaluation of *Philodina* (Rotifera) as a bioassay organism for heavy metals. Water Resources Bulletin 10: 648–661.

Cervantes, C. & R. Moreno-Sánchez, 1999. Contaminación ambiental por metales pesados. AGT Editorial, Mexico.

Charoy, C. P., C. R. Janssen, G. Persoone & P. Clement, 1995. The swimming behaviour of *Brachionus calyciflorus* (rotifer) under toxic stress. 1. The use of automated trajectometry for determining sublethal effects of chemicals. Aquatic Toxicology 32: 271–282.

Ferrando, M. D. & E. Andreu, 1993. Feeding behavior as an index of copper stress in *Daphnia magna* and *Brachionus calyciflorus* . Comparative Biochemistry and Physiology, C. Comparative Pharmacology and Toxicology 106: 327–331.

Gama-Flores, J. L., S. S. S. Sarma & M. A. F. Araiza, 1999. Combined effects of *Chlorella* density and methyl parathion concentration on the population growth of *Brachionus calyciflorus* (Rotifera). Bulletin of Environmental Contamination and Toxicology 62: 769–755.

Green, J., 1993. Zooplankton associations in East African lakes spanning a wide salinity range. Hydrobiologia 267: 249–256.

Halbach, U., 1984. Population dynamics of rotifers and its consequences for ecotoxicology. Hydrobiologia 109: 79–96.

Hawkins, P. R. & D. J. Griffiths, 1987. Copper as an algicide in a tropical reservoir. Water Research 21: 475–480.

He, Z., K. Qin, Y. Wang & W. Zhao, 1993. Biological resources in inland saline waters from southern Shanxi, China. Part 1. Lake Xiaochi. Journal of Dalian Fisheries College 8: 1–15.

Heugens, E. H. W., A. J. Hendriks, T. Dekker, N. M. van Stralen & W. Admiraal, 2001. A review of the effects of multiple stressors on aquatic organisms and analysis of uncertainty factors for use in risk assessment. Critical Reviews in Toxicology 31: 247–284.

Hutchinson, G. E., 1967. A Treatise on Limnology. 2. Introduction to Lake Biology and the Limnoplankton. John Wiley, New York, 1115 pp.

Janssen, C. R., M. D. Ferrando & G. Persoone, 1994. Ecotoxicological studies with the freshwater rotifer *Brachionus calyciflorus*. Ecotoxicology and Environmental Safety 28: 244–255.

Koste, W. 1978. Rotatoria. Die Rädertiere Mitteleuropas. Ein Bestimmungswerk begründet von Max Voigt. Bornträger, Stuttgart. Vol. 1, Textband 673 pp., Vol. 2, Tafelband, 234 pp.

Krebs, C. J., 1985. Ecology. The Experimental Analysis of Distribution and Abundance (3rd ed.). Harper & Row, New York.

Lubzens, E., G. Minkoff, Y. Barr & O. Zmora, 1997. Mariculture in Israel: Past achievements and future directions in raising rotifers as food for marine fish larvae. Hydrobiologia 358: 13–20.

Luna-Andrade, A., R. Aguilar-Duran, S. Nandini & S. S. S. Sarma, 2002. Combined effects of copper and microalgal (*Tetraselmis suecica*) concentrations on the population growth of *Brachionus plicatilis* Müller (Rotifera). Water, Air and Soil Pollution 141: 143–153.

McLusky, D. S., V. Bryant & R. Campbell, 1986. The effects of temperature and salinity on the toxicity of heavy metals to marine and estuarine invertebrates. Oceanography and Marine Biology: An Annual Review 24: 481–520.

Moreno-Garrido, I., L. M. Lubián & A. M. V. M. Soares, 1999. Growth differences in cultured populations of *Brachionus plicatilis* Müller caused by heavy metal stress as function of microalgal diet. Bulletin of Environmental Contamination and Toxicology 63: 392–398.

Ortells, R., A. Gomez & M. Serra, 2003. Coexistence of cryptic rotifer species: ecological and genetic characterisation of *Brachionus plicatilis*. Freshwater Biology 48: 2194–2202.

Nogrady, T., R. L. Wallace & T. W. Snell, 1993. Rotifera. Vol. 1. Biology, Ecology and Systematics. SBP Academic Publishers, The Hague 142.

Rao T. R. & Sarma, S. S. S., 1986. Demographic parameters of *Brachionus patulus* Müller (Rotifera) exposed to sublethal DDT concentrations at low and high food levels. Hydrobiologia 139: 193–200.

Rico-Martínez, R., I. A. Pérez-Legaspi, G. E. Quintero-Díaz, M. G. Rodríguez-Martínez, M. A. Hernández-Rodríguez & J. E. Zaragoza-Almaraz, 1998. Effects of copper addition to laboratory maintained microcosms of Presidente Calles Reservoir organisms (Aguascalientes, Mexico). Aquatic Ecosystem Health & Management 1: 323–332.

Salomón, W., U. Förstner & P. Mader (eds) 1995. Heavy metals. Problems and solutions. Springer-Verlag, Berlin.

Sarma, S. S. S., N. Iyer & H. J. Dumont, 1996. Competitive interactions between herbivorous rotifers: importance of food concentration and initial population density. Hydrobiologia 331: 1–7.

Sarma, S. S. S., P. S. Larios-Jurado & S. Nandini, 2001. Effect of three food types on the population growth of *Brachionus calyciflorus* and *Brachionus patulus* (Rotifera: Brachionidae). Revista de Biología Tropical 49: 75–82.

Sarma, S.S.S., B. Elguea-Sánchez & S. Nandini, 2002. Effect of salinity on competition between the rotifers *Brachionus rotundiformis* Tschugunoff and *Hexarthra jenkinae* (De Beauchamp) (Rotifera). Hydrobiologia 474: 183–188.

Snell, T. W. & C. R. Janssen, 1998. Microscale toxicity testing with rotifers. In Wells, P.P. K. Lee, K. & C. Blaise C.(eds.) Microscale Testing in Aquatic Toxicology: Advances, Techniques, and Practice. CRC Press, Florida, USA: 409–422.

Soucek, D. J., D. S. Cherry & C. E. Zipper, 2001. Aluminum-dominated acute toxicity to the cladoceran *Ceriodaphnia dubia* in neutral waters downstream of an acid mine drainage discharge. Canadian Journal of Fisheries and Aquatic Sciences 58: 2396–2404.

Yu, R. Q. & W X. Wang, 2002. Trace metal assimilation and release budget in *Daphnia magna*. Limnology and Oceanography 47: 495–504.

Hydrobiologia (2005) 546:569–575
A. Herzig, R.D. Gulati, C.D. Jersabek & L. May (eds.) Rotifera X: Rotifer Research: Trends, New Tools and Recent Advances
DOI 10.1007/s10750-005-4302-3

Effect of some pesticides on reproduction of rotifer *Brachionus plicatilis* Müller

Helen S. Marcial[1,*], Atsushi Hagiwara[1] & Terry W. Snell[2]
[1]*Graduate School of Science and Technology, Nagasaki University, 1-14 Bunkyo Machi, 852-8521, Nagasaki, Japan*
[2]*School of Biology, Georgia Institute of Technology, Atlanta, Georgia, 30332-0230, USA*
(**Author for correspondence: E-mail: d702125z@stcc.nagasaki-u.ac.jp*)

Key words: rotifera, *Brachionus plicatilis*, ecotoxicology, pesticides, reproduction, resting egg hatchability

Abstract

Pesticides have been major contributors to environmental pollution and they are now widely distributed in aquatic environments. Zooplankters are frequently used as test animals to detect aquatic contaminants because of their sensitivity and ecological importance. We investigated the effect of a 7-day exposure to four commonly used pesticides (diazinon, fenitrothion, methoprene and isoprothiolane) on reproduction of the rotifer *Brachionus plicatilis*. Pesticide concentrations of 3–7 times lower than the 24-h 50% lethal concentration (24-h LC_{50}) were tested to determine the 'no observed effect' concentration (NOEC), 'lowest observed effect' concentration (LOEC), and the '50% effective' concentration (EC_{50}) on specific growth rate (r), sexual reproduction, fertilization, resting egg production, and hatchability of resting eggs. Results showed that the lowest EC_{50} value of r, mixis, fertilization, and resting egg production of 1.4 μM for diazinon was 63 times lower than its 24-h LC_{50} of 88.4 μM, while for fenitrothion it was 66 times (3.5 and 229.8 μM, respectively). For isoprothiolane, the lowest EC_{50} value of r, mixis, fertilization, and resting egg production of 8.9 μM was 25 times lower than its 24-h LC_{50} of 220.7 μM, while for methoprene was 37 times (2.7 and 100.8 μM, respectively). In all pesticides, hatching rate of resting eggs consistently gave the lowest EC_{50} values which is about 2–40 times lower than the lowest EC_{50} of r, mixis, fertilization, and resting egg production. Hatchability of resting eggs therefore is the most sensitive parameter in detecting effects of pesticides exposure in rotifer *B. plicatilis*.

Introduction

The aquatic environment is often the final depository of most chemical contaminants. Among them, pesticides pose some of the most serious ecological problems because of their toxicity to both target and non-target organisms and their wide distribution. Some pesticides (e.g. atrazine, chlordane) are regarded as endocrine disruptors whereas others are persistent in the environment and bioaccumulate in food webs (e.g. organochlorine, carbamate). Several studies have shown that pesticides such as diazinon and fenitrothion have a severe impact in aquatic organisms (Ferrando et al., 1996; Sánchez et al., 1999; Marcial et al., 2002).

Zooplankton are frequently used to detect anthropogenic contamination because of their sensitivity to various toxicants and their important role in the ecosystem. Among the zooplankton species, rotifers are favored test animals because of their short generation time, ease of culture, and the commercial availability of their resting eggs (Snell & Janssen, 1995). In ecotoxicological studies using rotifers, end-points such as 24- or 48-h LC_{50}, swimming speed, filtration rate and enzyme activity have been used (reviewed by Snell & Janssen (1995)). Because only a portion of the rotifer life cycle is investigated in these studies, the true vulnerability of rotifer life cycles to toxicants is often underestimated (Preston & Snell, 2001). Hence, several studies assessed the effects of pesticides, heavy metals, and endocrine disrupting chemicals on the entire life cycle of rotifers (Snell & Carmona, 1995; Preston et al., 2000; Preston & Snell, 2001; Radix et al., 2002). Results showed that sexual reproduction and resting egg production are among the most sensitive endpoints. All of the above-mentioned studies, however, investigate only up to the production of resting eggs. No study has examined the effects of toxicants on the viability of resting eggs produced by rotifers exposed to chemical contaminants. This study aimed to (1) determine the acute and chronic effects of some commonly used pesticides on the life history of the rotifer *Brachionus plicatilis*, and (2) determine the most susceptible endpoint of pesticide exposure.

Materials and methods

Test animals

B. plicatilis is a monogonont rotifer that reproduces *via* cyclical parthenogenesis, incorporating both asexual (amictic) and sexual (mictic) reproduction (Snell & Carmona, 1995). Most of the time, amictic females produce amictic eggs mitotically. However, upon receiving certain environmental stimuli, amictic females may produce both amictic and mictic daughters. The mictic females then produce haploid eggs that are smaller than the amictic eggs and, if fertilized, develop into resting eggs. If these eggs remain unfertilized, they develop into males.

B. plicatilis NH1L strain (Hagiwara et al., 1993), which originated from the University of Tokyo and has been cultured in our laboratory for 15 years, was used in this study. This strain has the highest mixis rate among the *B. plicatilis* stocks maintained in our laboratory. Stock cultures were maintained in diluted seawater at 22%, stored in 25 ± 1 °C, and fed *Nannochloropsis oculata*. To obtain rotifers of similar age for the acute toxicity test, test animals were hatched from resting eggs. For the chronic toxicity test, amictic females were obtained by shaking egg-bearing females in a screw-capped bottle, and hatching the eggs in a 15 ml petri dish containing 10 ml of food solution. Amictic females were selected based on the morphology of their eggs (Hagiwara et al., 1988).

Test chemicals

Four pesticides were investigated: diazinon ($C_{12}H_{21}N_2O_3PS$), isoprothiolane ($C_{12}H_{18}O_4S_2$), fenitrothion ($C_9H_{12}NO_5PS$) and methoprene ($C_{19}H_{34}O_3$). Diazinon and fenitrothion are organophosphate insecticides known to inhibit acetylcholinesterase activity (Ecobichon & Joy, 1994), while methoprene is a juvenile hormone analogue known to mimic the action of juvenile hormones by disrupting the developmental processes of insects (Sehnal, 1983). These pesticides were purchased from WAKO Pure Chemical Industries Ltd., Japan. They were dissolved first in 100% dimethyl sulfoxide (DSMO), then in ultra-pure grade distilled water with the final test solution containing not more than 0.01% DMSO. Stock solutions and 10% DMSO were stored at 4 °C. Test solutions were prepared by the addition of appropriate aliquots of aqueous stock solution to filtered (Millipore 0.45 μm) and autoclave-sterilized seawater diluted to 22%.

Acute toxicity test

Using 24-h old rotifers hatched from the resting eggs, standard acute toxicity tests (ASTM, 1998) were carried out for each pesticide with seven concentrations each. Twenty rotifers were transferred to individual wells of a 6-well polystyrene plate, each containing 10 ml of a test solution, or a control solution (filter-sterilized seawater). The plates were incubated at 25 °C in darkness. Because of the short duration of the tests, the animals were not fed during the experiment. After 24 h, the rotifers were observed under the stereomicroscope. Rotifers were considered dead if no movement of the cilia and mastax was observed over a period of 30 s. The dead and alive rotifers were counted.

Chronic toxicity test

Range-finding tests consisting of five concentrations were conducted for 48 h. The concentration which did not result in mortality, was chosen to be the highest concentration in the definitive test. Final test concentrations were: diazinon (0.03, 0.3, 3.3, 16.4 and 32.8 μM), fenitrothion (0.04, 0.4, 3.6, 9.0 and 18.0 μM), isoprothiolane (0.3, 3.4, 17.2, 34.4 and 68.8 μM), and methoprene (0.03, 0.3, 3.0, 8.0

and 16.0 μM). The lowest concentration of each chemical corresponds to 0.01 mg l^{-1}. Three replicates were carried out for each treatment. A solvent control containing 0.01% DMSO was also tested. In each treatment, 10 amictic females bearing one egg were added in a 20 ml screw-capped bottles containing 10 ml of 7×10^6 cells ml^{-1} *N. oculata* and toxicant solution. After the addition of rotifers, the bottles were incubated at 25 °C in darkness for 6 days. From days 2–6, the bottles were emptied everyday into a glass petri dish and the number of non-ovigerous females, ovigerous amictic females, ovigerous mictic females, and fertilized mictic females were counted (Hagiwara et al., 1988). After counting, the rotifers were returned to the bottles. About half of the culture water was changed every day using pre-conditioned culture media (water from a culture where rotifers had been raised) with 5×10^6 cells ml^{-1} *N. oculata* and test pesticides. Preconditioned culture media was used because it is known to induce mixis (Carmona et al., 1993). On day 7, all resting egg-bearing females and resting eggs were collected and transferred to a new bottle containing diluted sterilized-seawater (22%) and stored at 25° C in darkness. After 3 weeks, the period of obligated diapause (Hagiwara & Hino, 1989), the resting eggs were induced to hatch by placing them in a glass petri dish with 10 ml of diluted sterilized seawater (22%) and exposed to continuous light (4000–5000 lux) at 25 °C. After 48 h, the hatched and unhatched resting eggs were counted. The observation of possible hatching was continued until day 4.

Statistical analysis

The 24-h LC_{50} was calculated using Probit Analysis (Minitab, Ver. 13). For the chronic toxicity test, specific growth rate (r) was calculated for each bottle according to: $r = (\ln N_t - \ln N_0)/t$, where N_t = number of female rotifers in the bottle on t, N_o = initial number of rotifers in each bottle (10), and t = 2, 3, 4, 5, or 6 days. One-way analysis of variance (ANOVA), with concentration as the independent variable and r, percent mixis, or percent fertilization as the dependent variable, followed by Dunnett's test was conducted for pairwise comparisons of each pesticide concentration relative to the control (Zar, 1999). From these results, NOEC, and LOEC were determined. In addition, the concentration of the toxicant that reduces the test parameter to 50% (EC_{50}) was calculated using regression analysis (Stephan & Rogers, 1985). Regression lines were calculated for mean r, percent mixis, percent fertilization, total number of resting eggs produced, and percent hatching against log toxicant concentration per treatment. The total number of resting eggs produced at different pesticide concentrations were compared to the control using Dunnett's test and the percent hatching using Chi-square contingency test. A p value of 0.05 or less was regarded as being significant for all tests.

Results

Results from the 24-h acute toxicity tests showed that diazinon was the most toxic among the pesticides tested, having the lowest 24-h LC_{50} value. The second most toxic was followed by methoprene (Table 1). In contrast, DMSO concentrations as high as 640 μM were not toxic to *B. plicatilis* (Table 1). In the chronic toxicity test, 0.01% DMSO (solvent control) was tested in addition to the negative control (seawater only) only in the diazinon experiment. This was because results showed that in all parameters it was not significantly different from the negative control (Fig. 3).

Daily NOEC, LOEC and EC_{50} values for r, mixis, and fertilization endpoints for fenitrothion are shown in Figure 1. In 7 out of 12 cases, EC_{50} values lay between NOEC and LOEC. The LOEC and NOEC of r, mixis, and fertilization, tend to increase with exposure time, while EC_{50} is less dependent on the duration of exposure. A similar trend was observed with other pesticides.

Table 1. Twenty-four-hour medial lethal concentrations (LC_{50}s) of test pesticides to the rotifer *B. plicatilis* at 25 °C

	LC_{50} (μM)
Diazinon	88.39 (74.58–102.51)
Fenitrothion	229.76 (219.30–240.22)
Isoprothiolane	220.74 (215.92–225.56)
Methoprene	100.81 (88.89–112.40)
DSMO	> 640

Values in parenthesis are 95% confidence intervals.

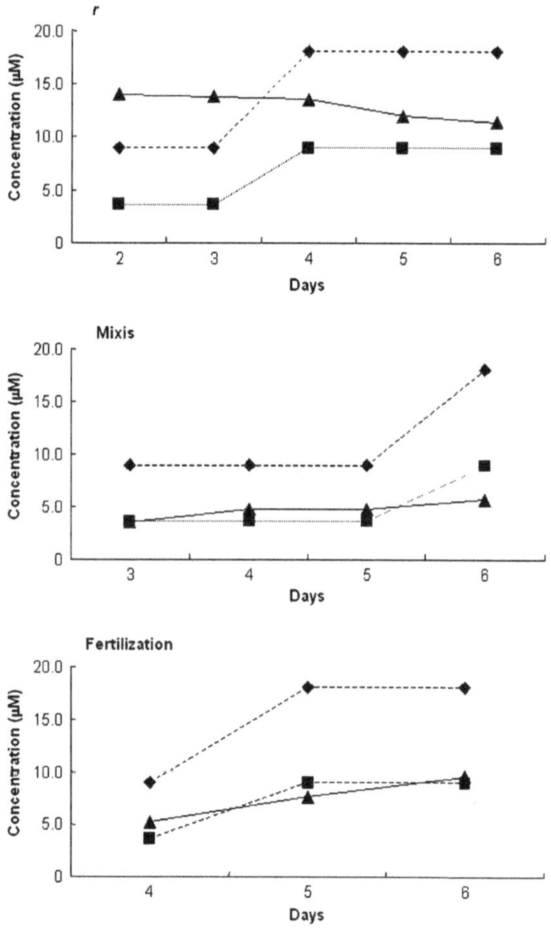

Figure 1. NOEC (■), LOEC (♦), and EC_{50} (▲) daily values of *r*, mixis, and fertilization of *B. plicatilis* exposed to fenitrothion.

EC_{50} values of *r*, mixis, fertilization, resting egg production (day 7) as well as percent hatching of resting egg of *B. plicatilis* exposed to the four pesticides are presented in Fig 2. In diazinon, EC_{50} values of mixis were not calculated because the correlation coefficient was less than 0.5. EC_{50} value of *r* was highest on day 2 and became stable from day 3 and onwards. Hatching rate of resting eggs gave the lowest EC_{50} value (0.2 μM) of all the endpoints. In fenitrothion, mixis consistently gave low EC_{50} values, however resting egg hatching rate gave the lowest EC_{50} value (1.7 μM). In methoprene exposures, the endpoints *r*, mixis, and fertilization, gave similar EC_{50}s for days 2–6. The EC_{50} value of resting egg production was highest (7.2 μM), however once again the hatching rate EC_{50} was the lowest endpoint (0.06 μM). In isoprothiolane, correlation coefficient of EC_{50} values of mixis were lower than 0.5, therefore it was not included in the graph. Hatching rate of resting eggs again gave the lowest EC_{50} value (2.7 μM).

The mean number of resting eggs produced in each treatment and their hatching rates are presented in Figure 3. Rotifers exposed to 16.4 μM and higher diazinon, 9.0 μM and higher fenitrothion, 8.0 μM and higher methoprene, and 68.8 μM isoprothiolane, produced significantly fewer resting eggs compared to the control.

The hatching rates of resting eggs in unexposed rotifers ranged from 56.7 to 67.9%. The hatching rate of the resting eggs from all exposed treatments was significantly lower than the control except for 0.3 and 17.2 μM isoprothiolane.

Discussion

Results from our chronic toxicity tests indicated that hatching rate of resting eggs produced by parents exposed to pesticides is the most sensitive endpoint. The resting egg hatchability EC_{50} (0.2 μM) of diazinon exposed rotifers was seven times lower than the lowest EC_{50} values of fertilization (1.4 μM), while that of fenitrothion was two times (1.7 and 3.5 μM, respectively), and more than three times in that of isoprothiolane (2.7 and 8.9 μM, respectively). Methoprene exposed rotifers gave a highest resting egg hatchability EC_{50} ratio of more than 40 times lower than mixis EC_{50} (0.06 and 2.7 μM). The EC_{50} values for resting egg hatchability were more than 80 times lower than their 24-h acute toxic concentration. Among the reproductive endpoints (*r*, mixis, and fertilization), none of these was consistently the most sensitive for detecting toxicity in *B. plicatilis* among the pesticides tested. The study of Ferrando et al. (1996) on the freshwater rotifer *Brachionus calyciflorus* exposed to fenitrothion showed that net reproductive rate and *r* were more sensitive endpoints than generation time and life expectancy. However, in the study of Preston & Snell (2001), using the same species but exposed to pentachlorophenol and copper, fertilization and resting egg production were the most sensitive. In addition, Preston et al. (2000) and Radix et al. (2002) have shown that mixis and fertilization were sensitive parameters in *B. calyciflorus* exposed to

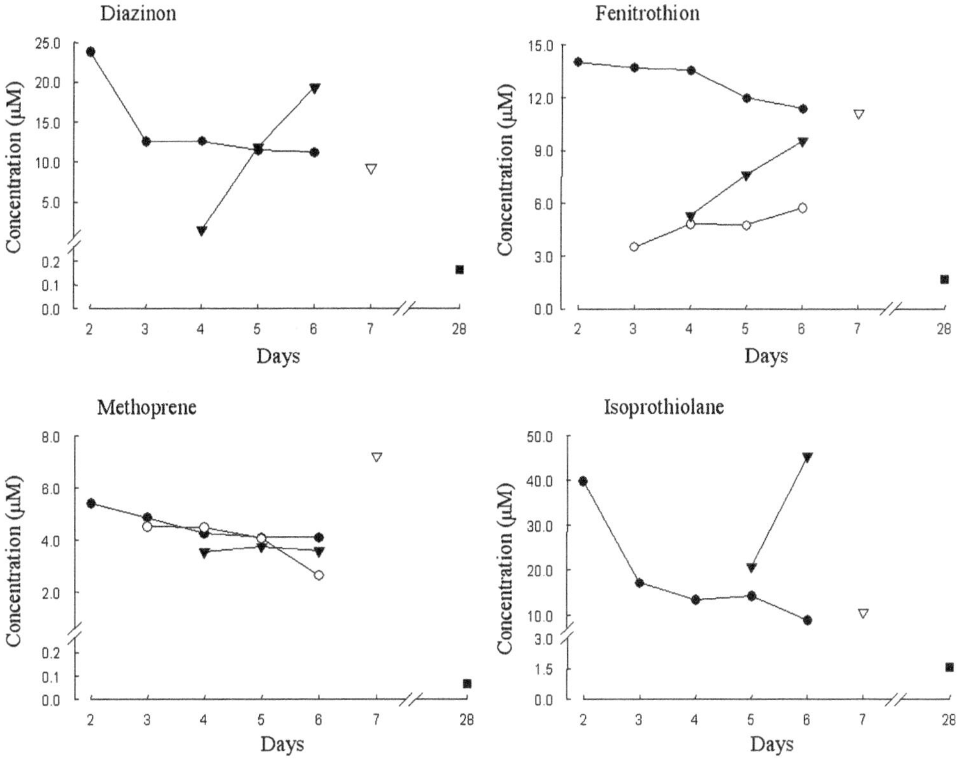

Figure 2. EC_{50} daily values of *r* (●), mixis (○), and fertilization (▼) of *B. plicatilis* exposed to diazinon, fenitrothion, methoprene and isoprothiolane. Resting egg production (▽) at day 7, and hatching rate of resting eggs (■) after 3 weeks are also indicated.

known and suspected endocrine disrupting chemicals. All of the above-mentioned studies however, did not investigate the hatchability of the resting eggs produced by the rotifers exposed to the test chemicals. The results of our study showed that rotifers exposed to 0.03–32.8 μM diazinon, 0.04–9.0 μM fenitrothion, 0.3–68.8 μM isoprothiolane, and 0.03–3.0 μM methoprene successfully produced resting eggs, however, their hatching rates were significantly lower than the control. Diazinon, fenitrothion, methoprene, and isoprothiolane affected the hatchability of the resting eggs at concentrations 2–40 times lower than *r*, mixis and fertilization EC_{50}s. The hatching rate of resting eggs therefore, is the most sensitive endpoint thus far described in rotifers for detecting the effects of pesticides.

The sensitivity of the hatching rate of resting eggs can be attributed to several factors. Resting eggs are the product of sexual reproduction, and sexual reproduction is induced by high asexual reproduction and under moderate environmental conditions (Snell, 1986; Hagiwara et al., 1988). Resting egg production and hatching therefore encompasses the complete life-cycle of rotifers. In our experiments the parents (females and males) were continuously exposed to the toxicants and resting eggs were formed in the presence of the pesticides. Pesticides might affect the oogenesis of females, spermatogenesis of males, and developmental stages of the resting eggs.

Several factors influence the hatchability of the resting eggs of rotifers. Among them are light, temperature, salinity, and maternal diet (Hagiwara & Hino, 1989, 1990; Hagiwara et al., 1995). In the present study, rotifers were reared under optimal culture conditions for resting egg formation, including provision of optimal food and hatching conditions described in Hagiwara et al. (1995). Therefore, the reduced hatchability could not be due to the culture conditions. Temporary exposure to two organochlorine compounds reduced the hatchability of eggs of the calanoid copepods *Eurytemora affinis* and *Acartia bifilosa* (Lindley et al.,

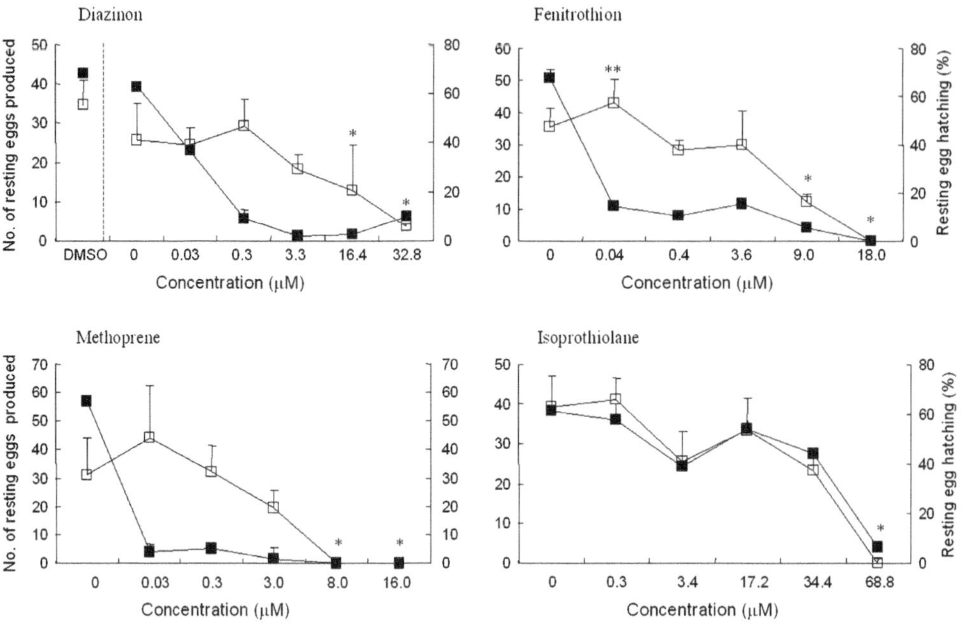

Figure 3. Mean number of resting eggs produced by *B. plicatilis* (□) and their hatching rates (■) exposed to different concentrations of diazinon, fenitrothion, methoprene, and isoprothiolane. *Significantly lower than the control ($p < 0.01$); **significantly higher than the control ($p < 0.05$) detected by ANOVA and Dunnett's test. The percent hatching of resting eggs of the exposed treatments were all significantly lower than the unexposed control except for 0.3 and 17.2 μM isoprothiolane. Vertical bars are standard error of the mean.

1999), and eggs laid in aqueous solutions of the two compounds did not yield viable nauplii. The same effects were observed by Naess (1991) in calanoid eggs exposed to the pesticide rotenone. This is the first report on the effect of toxicants on the hatchability of resting eggs in rotifers.

The effects of anthropogenic chemicals on the viability of resting eggs is an ecologically important parameter, because resting eggs enable rotifers to maintain their population during unsuitable environmental conditions. Modeling studies have indicated that a small decrease in r will reduce the number of resting eggs produced, resulting in loss of heterozygozity, depletion of the resting egg pool, and can lead to local population extinction (Snell et al., 1999; Snell & Serra, 2000).

Diazinon, fenitrothion and isoprothiolane have been detected in several rivers in Japan at 0.02–20 $\mu g\ l^{-1}$ (Fukushima et al., 1995; Sudo et al., 2002). Following field application of Altosid Liquid Larvicide (with 5% (S) methoprene) the expected environmental concentration of methoprene is 0.03 μM (Ross et al., 1994). The concentrations that affected r, mixis rate and fertilization rate are much higher than the concentrations found in the aquatic environments. However, the pesticides concentrations affecting the hatchability of rotifer resting eggs are within the levels reported in some aquatic environments. Therefore, rotifer resting egg hatching rate could be useful as endpoints in ecotoxicology to detect the effects of pesticides.

Acknowledgements

H. S. M. is grateful to the Ministry of Education, Culture, Sports, Science and Technology of Japan for the Ph.D. scholarship. This study was supported by a grant from Integrated Research Program for Effects of Endocrine Disrupters on Agriculture, Forestry and Fisheries and their Action Mechanisms on Domestic Animals as well as Prefectural Collaboration of Regional Entities for the Advancement of Technological Excellence to A. H. We acknowledge J.N. Nocillado for the critical review of the manuscript.

References

American Society of Testing and Materials (ASTM), 1998. Standard guide for acute toxicity test with the rotifer

Brachionus. In Allen, R. F., (ed.), Annual Book of ASTM Standards, Section II-Water and Environmental Technology. American Society of Testing Materials, West Conshohocken, PA, 837–843.

Ecobichon, D. J. & R. M. Joy, 1994. Organophosphorus esters insecticides. In Ecobichon, D. J., R. M. Joy & Z. B. Alfassi (eds) Pesticides and Neurological Diseases. CRC Press, Florida 171–249.

Ferrando, M. D., E. Sancho & E. Andreu-Moliner, 1996. Chronic toxicity of fenitrothion to an algae (*Nannochloropsis oculata*), a rotifer (*Brachionus calyciflorus*), and the cladoceran (*Daphnia magna*). Ecotoxicology and Environmental Safety 35: 112–120.

Fukushima M., Y. Yamaguchi, A. Yamada, 1995. Temporal trend of pesticide pollution in river water as a potable water. In IWSA Specialized Conference on Advance Treatment and Integrated Water System Management into the 21st Century, 158–163.

Hagiwara, A., A. Hino & R. Hirano, 1988. Effects of temperature and chlorinity on resting egg formation in the rotifer *Brachionus plicatilis*. Nippon Suisan Gakkaishi 54: 569–575.

Hagiwara, A. & A. Hino, 1989. Effect of incubation and preservation on resting egg hatching and mixis in the derived clones of the rotifer *Brachionus plicatilis*. Hydrobiologia 186/187: 415–421.

Hagiwara, A. & A. Hino, 1990. Feeding history and hatching of resting eggs in the marine rotifer *Brachionus plicatilis*. Nippon Suisan Gakkaishi 56: 1965–1971.

Hagiwara, A., A. Hino, K. Nishi, A. Imaizumi & K. Hirayama, 1993. Mass production of rotifer *Brachionus plicatilis* resting eggs in 50 m^3 tanks. Nippon Suisan Gakkaishi 59: 93–98.

Hagiwara, A., H. Hoshi, F. Kawahara, K. Tominaga & K. Hirayama, 1995. Resting eggs of the marine rotifer *Brachionus plicatilis* Müller: development, and effect of irradiation on hatching. Hydrobiologia 313/314: 223–229.

Lindley, J. A., P. Donkin, S. V. Evans, C. L. George & K. F. Uil, 1999. Effects of two organochlorine compounds on hatching and viability of calanoid copepod eggs. Journal of Experimental Marine Biology and Ecology 242: 59–74.

Marcial, H. S., A. Hagiwara & T. W. Snell, 2002. Effect of known and suspected endocrine disrupting chemicals on the demographic parameters of the copepod *Tigriopus japonicus*. Fisheries Science 68(Suppl. I) 863–866.

Naess, T., 1991. Tolerance of marine calanoid resting eggs: effects of freezing, dessication and Rotenone exposure – a field and laboratory study. Marine Biology 111: 455–459.

Preston, B. L., T. W. Snell, T. L. Robertson & B. J. Dingmann, 2000. Use of freshwater rotifer *Brachionus calyciflorus* in screening assay for potential endocrine disruptors. Environmental Toxicology and Chemistry 19: 2923–2928.

Preston, B. L. & T. W. Snell, 2001. Full life-cycle toxicity assessment using rotifer resting egg production: implications for ecological risk assessment. Environmental Pollution 114: 399–406.

Radix, P., G. Severin, K. W. Schramm & A. Kettrup, 2002. Reproduction of *Brachionus calyciflorus* (rotifer) for the screening of environmental endocrine disrupters. Chemosphere 47: 1097–1101.

Ross, D. H., D. Judy & B. R. Jacobson Howell, 1994. Methoprene concentrations in freshwater microcosm treated with sustained-release Altosid formulations. Journal of the American Mosquito Control Association 10: 202–210.

Sánchez, M., M. D. Ferrando, E. Sancho & E. Andreu, 1999. Assessment of the toxicity of a pesticide with a two-generation reproduction test using *Daphnia magna*. Comparative Biochemistry and Physiology Part C 124: 247–252.

Sehnal, F., 1983. Juvenile hormone analogues. In Downer, R. G. H. & H. Laufer (eds) Endocrinology of Insects. Alan R. Liss, New York, 657–672.

Snell, T. W., 1986. Effect of temperature, salinity and food level on sexual and asexual reproduction in *Brachionus plicatilis* (Rotifera). Marine Biology 92: 157–162.

Snell, T. W. & M. J. Carmona, 1995. Comparative toxicant sensitivity of sexual and asexual reproduction in the rotifer *Brachionus calyciflorus*. Environmental Toxicology and Chemistry 14: 415–420.

Snell, T. W. & C. R. Janssen, 1995. Rotifers in ecotoxicology: a review. Hydrobiologia 313/314: 231–247.

Snell, T. W., M. Serra & M. J. Carmona, 1999. Toxicity and sexual reproduction in rotifers: Reduced resting egg production and heterozygosity loss. In Forbes, V. E. (ed.) Genetics and Ecotoxicology. Taylor and Francis: 169–185.

Snell, T. W. & M. Serra, 2000. Using probability of extinction to evaluate the ecological significance of toxicant effects. Environmental Toxicology and Chemistry 19: 2357–2363.

Stephan, C. E. & J. R. Rogers, 1985. Advantages of using regression analysis to calculate results of chronic toxicity test. In Bahner, R. C. & D. J. H. Hansen (eds) Aquatic Toxicology and Hazard Assessment: Eight Symposium. American Society for Testing and Materials, Philadelphia, 328–339.

Sudo, M., T. Kunimatsu & T. Okubo, 2002. Concentration and loading of pesticide residues in Lake Biwa basin (Japan). Water Research 36: 315–329.

Zar J. H., 1999. Biostatistical Analysis, Vol. 4. Prentice Hall, New Jersey, 663 pp.

Hydrobiologia (2005) 546:577–585
A. Herzig, R.D. Gulati, C.D. Jersabek & L. May (eds.) Rotifera X: Rotifer Research: Trends, New Tools and Recent Advances
DOI 10.1007/s10750-005-4308-x

Heat shock protein 60 (HSP60) response of *Plationus patulus* (Rotifera: Monogononta) to combined exposures of arsenic and heavy metals

Judith V. Rios-Arana[1,*], J.L. Gardea-Torresdey[2], R. Webb[3] & Elizabeth J. Walsh[3]
[1]*University of Texas at El Paso, Environmental Science and Engineering Ph.D. Program, El Paso, TX, 79968-0519, USA*
[2]*University of Texas at El Paso, Department of Chemistry and Environmental Science and Engineering Ph.D. Program, El Paso, TX, 79968-0519, USA*
[3]*University of Texas at El Paso, Department of Biological Sciences and Environmental Science and Engineering Ph.D. Program, El Paso, TX, 79968-0519, USA*
(**Author for correspondence: E-mail: jvrios@utep.edu*)

Key words: stress protein, metal toxicology, aquatic toxicology, *Plationus patulus*, Rio Grande

Abstract

Organisms produce stress proteins as a response to natural and anthropogenic environmental changes. Induction of stress proteins has been reported in a variety of aquatic organisms, including rotifers, exposed to pollutants. Past studies on stress protein responses of rotifers have focused on exposure to single toxicants. In this study the rotifer *Plationus patulus* was exposed singly and in combination to various concentrations of As, Cr, Cu, Ni, Pb, and Zn. Following exposure, total protein was quantified (Bradford method) and stress protein 60 (HSP60) was identified using Western blotting. *P. patulus* induced HSP60 as a response to single exposures to Cr, Cu, Ni, Pb and Zn. HSP60 expression was increased (2 fold) in rotifers exposed to these single elements at both low and high concentrations as compared to unexposed rotifers. Arsenic exposure resulted in a 2 fold decrease in HSP induction. In rotifers exposed to metal mixtures, HSP60 was induced by the presence of As–Zn, As–Cr–Cu–Pb, As–Cr–Cu, As–Cr–Cu–Ni and As–Cr–Cu–Ni–Pb combinations in the media. HSP60 response to As and heavy metals toxicity depends on the type and number of elements present in the media as well as their concentrations and length of the exposure time.

Introduction

A pool of heat shock proteins (HSPs) is present in organisms under normal conditions (Cochrane et al., 1991). These proteins prevent severe damage at a cellular level when stressful conditions appear. Abrupt changes in the environment related to temperature or other physical factors may induce the synthesis of HSPs in stressed organisms (Morimoto et al., 1994; Lewis et al., 1999; Tedengren et al., 1999; Choresh et al., 2001). HSP60 synthesis is stimulated by conditions that lead to protein denaturation (Hightower, 1993). Because some pollutants stimulate their production, HSPs have been studied to determine their usefulness as biomarkers (Eckwert et al., 1997; Karouna-Renier & Zehr, 1999; Lewis et al., 1999; Chen et al., 1999). For example, when *Daphnia pulex* are stressed with different concentrations of arsenic (100–3000 $\mu g\ l^{-1}$ as arsenite or arsenate) synthesis of HSP83 mRNA increases and this induction is paralleled by changes in specific demographic responses (Chen et al., 1999). Thus HSP induction may be useful as a biomarker for environmental arsenic stress in populations of this cladoceran. Cochrane et al. (1991) found that heat shock protein 58 (HSP58) was induced in the marine rotifer *Brachionus plicatilis* in response to Cu and tributyltin stress. HSP58 synthesis was increased about 4 and 8 fold by Cu and tributyltin exposure, respectively. In polluted environments more than one element or toxicant is often present. The effect produced by exposures to single toxicants may be different from those observed in mixed exposures due to synergistic effects in rotifers (or any organism) (Folt et al., 1999; Fernandez & Beiras, 2001). The present study was performed to determine if exposure to heavy metals, singly or in combination, affects the production of HSP60 in the freshwater rotifer *Plationus patulus*.

Materials and methods

Rotifers (*P. patulus*) collected from the Rio Grande (El Paso Co., TX; 31° 53.166 N, 106° 35.915 W) using a 64 μm plankton net were cultured in filtered river water (Whatman; Qualitative No. 5) at 25 °C with a 12 h light:dark cycle. Rotifers were fed in excess with a mixture of *Chlamydomonas reinhardtii* (UTEX 90) and *Ankistrodesmus falcatus* (UTEX 749). Cultures were maintained until an adequate population size was reached for experiments. Growth media (fresh food, filtered river water) was replaced every other day.

Preliminary tests revealed that levels of metals found in river water used to culture the rotifers influenced the results obtained during metal stress experiments. Thus, 1 day prior to Cu exposure, rotifers were placed in synthetic media (Gilbert, 1975) to prevent the induction of HSP60 by any dissolved metal ions in the culture media. As an additional control, we confirmed the induction of HSP60 by heat using a modification of the procedure described by Wheelock et al. (1999).

Induction of HSP60 by copper sulfate exposure

Rotifers (100/treatment) were exposed to four concentrations of $CuSO_4$ (20, 35, 63, and 100 μg Cu l^{-1}) for 5, 10, 20, 30 or 60 min. Concentrations used for exposure were selected after considering the 24-h LC_{50} range found for *B. plicatilis* (Snell & Persoone, 1989) and previous studies by on HSP60 induction in this species (Cochrane et al., 1991; Wheelock et al., 1999). Rotifers were collected after exposure and concentrated in 200 μl of synthetic medium. Laemmli buffer (100 μl; Sigma) was added to each sample. Rotifers were homogenized for 30 s, placed on ice and boiled for 4 min. Total protein concentrations were determined by Bradford assay (Kruger, 2002) using a spectrophotometer (Optima 600) set at 595 nm. A series of dilutions (1, $^1/_2$, $^1/_4$, $^1/_8$, $^1/_{16}$, and $^1/_{32}$×) was prepared to qualitatively determine HSP60 induction using Western blotting. A 1× dilution corresponded to 48 μg of total protein. Samples (50–60 μl) were loaded on nitrocellulose membranes (NCM, Bio-Rad) using a vacuum system. After the NCM dried, non-specific sites were blocked with 2% gelatin (Bio-Rad) for 1 h. Nitrocellulose membranes were washed to remove excess gelatin with 1× Trizma Base-Tween 20 buffer (1× TBST) for 1 h, changing the buffer every 15 min. Anti-HSP60 rabbit polyclonal antibody (StressGen Biotechnologies; 1:4000 dilution) was reacted with the NCM overnight and washed as mentioned above. A secondary anti-rabbit goat antibody labeled with alkaline phosphatase (StressGen Biotechnologies; 1:8000 dilution) was reacted with the NCM for 1 h. Alkaline phosphatase substrate (1-Step NBT/BCIP; Pierce) was added to the membrane after washing to determine the presence of HSP60. The reaction was stopped with a 5% acetic acid solution. HSP60 standard (recombinant rat HSP60; Stressgen Biotechnologies) dilutions (1, $^1/_2$, $^1/_4$, $^1/_8$, $^1/_{16}$, and $^1/_{32}$×) were included as positive controls. The amount of HSP60 in standard dilutions corresponded to 480, 240, 120, 60, 30, and 15 ng, respectively.

Effect of As and heavy metal exposure on HSP60 induction

To evaluate induction of HSP60 in rotifers exposed to metals and arsenic, metal stock solutions (all at least 99.99% purity, unless noted) of 100 mg l^{-1} were prepared by dissolving appropriate amounts of arsenic(V) oxide hydrate (Aldrich), chromium(III) nitrate non-hydrate (Aldrich), nickel(II) nitrate hexahydrate (Aldrich), cupric sulfate pentahydrate (99.1%; Fischer Scientific), lead nitrate (Aldrich), and zinc sulfate heptahydrate (99.5%; EM Science) in distilled water. Stock solutions were diluted (10 mg l^{-1}) to prepare exposure concentrations. Single metals (two concentrations) were added to the synthetic medium containing 200 rotifers each. Rotifers were exposed for 1 h to low single metal concentration (10 μg l^{-1}, except for Zn: 20 μg l^{-1}) and 20 min to high single metal concentration (50 $\mu g\ l^{-1}$ of As, Cr, Cu and Ni; 100 $\mu g\ l^{-1}$ of Pb; 50 $\mu g\ l^{-1}$ of Zn) (all concentrations are given as nominal values). Exposure times were determined based on maximum HSP60 induction observed during $CuSO_4$ exposure experiments. Rotifers were homogenized and proteins were processed as previously described.

Effect of As and heavy metal mixtures on HSP60 induction

Three experiments were conducted to determine the effect of combined (two or more) elements on HSP60 production in rotifers:

(1) Rotifers (200 per treatment) were exposed to mixtures of As (50 $\mu g\ l^{-1}$), Cr (50 $\mu g\ l^{-1}$) and Zn (150 $\mu g\ l^{-1}$). Five treatments (As – Cu, As–Zn, Cu–Zn, and As–Cu–Zn) including a control (unexposed rotifers) were tested. Rotifers were exposed to the combined elements for 20 min. at 25 °C.

(2) Sets of 50 rotifers were exposed to combinations of four elements (As, Cr, Cu, Pb) at different concentrations. Low (L) concentrations were 10 $\mu g\ l^{-1}$ for all elements; high (H) concentrations varied depending on the element (40 $\mu g\ l^{-1}$ for As, 50 $\mu g\ l^{-1}$ for Cr and Cu, and 100 $\mu g\ l^{-1}$ for Pb). Exposure treatments were combinations of AsL–CrL–CuL–PbL, AsL–CrH–CuH–PbL, AsH–CrL–CuL–PbH, AsH–CrH–CuH–PbH, and an unexposed control. Rotifers were exposed for 20 min. at 25 °C.

(3) To evaluate HSP60 production as the number of elements is increased sequentially, 200 rotifers per treatment (6 treatments) were stressed by combinations of 2, 3, 4, 5 and 6 elements present at low concentrations (10 $\mu g\ l^{-1}$, except for Zn: 20 $\mu g\ l^{-1}$) and all elements at high concentrations: 50 $\mu g\ l^{-1}$ (As, Cr, Cu and Ni), 100 $\mu g\ l^{-1}$ (Pb) and 150 $\mu g\ l^{-1}$ (Zn). A control was carried out simultaneously with metal exposures. Rotifers were exposed for 1 h, except for those exposed to the mixtures of elements at high concentrations (20 min). After exposure, rotifers and NCM were handled as previously described.

Results

Copper sulfate exposure

Protein quantification revealed differences between treatments in total protein content. In rotifers exposed to the highest concentration of $CuSO_4$ (100 μg Cu l^{-1}), the amount of total protein increased (0.55–2.23 $\mu g\ l^{-1}$) with increasing exposure time (5–60 min.), while in treatments of rotifers exposed to 20 $\mu g\ l^{-1}$ Cu, total protein content ranged from 1.05 to 1.7 $\mu g\ l^{-1}$, which was similar to the protein content in the control (1.5 $\mu g\ \mu l^{-1}$).

Induction of HSP60 in rotifers exposed to $CuSO_4$ is related to metal concentration and exposure time. Rotifers exposed to 100 $\mu g\ l^{-1}$ apparently did not show differences in HSP60 expression ($^1/_4$ dilution) as compared with the control treatment (Fig. 1). Induction of HSP60 was not observed at this concentration (except at the 10 min exposure). Synthesis of HSP60 was decreased and reached a steady state after 30 min of $CuSO_4$ (63 $\mu g\ l^{-1}$) exposure (Fig. 1). At low concentration (20 $\mu g\ l^{-1}$), HSP60 increased 2–4 fold compared with control treatments. HSP60 response to $CuSO_4$ exposure differed at each concentration tested. Induction of HSP60 with varying exposure times was observed only in rotifers exposed to 20 and 35 $\mu g\ l^{-1}$ of Cu.

Arsenic and heavy metal exposure

The lowest levels of total protein were found in rotifers exposed to 20 $\mu g\ l^{-1}$ Zn; total protein content in these animals was 6 × lower than unexposed rotifers with 1.28 $\mu g\ l^{-1}$. High Zn concentration resulted in a 3-fold reduction in the amount of total protein. There is no apparent difference in protein levels between control and low concentration treatments of the other elements tested. HSP60 was induced in rotifers exposed to all elements (except As) at both low and high concentrations. Compared with control treatments, HSP60 synthesis was increased 2 fold in rotifers exposed to most elements (Fig. 2). The amount of HSP60 in rotifers exposed to low levels of As was approximately two times lower than the control treatment. The highest concentrations of HSP60

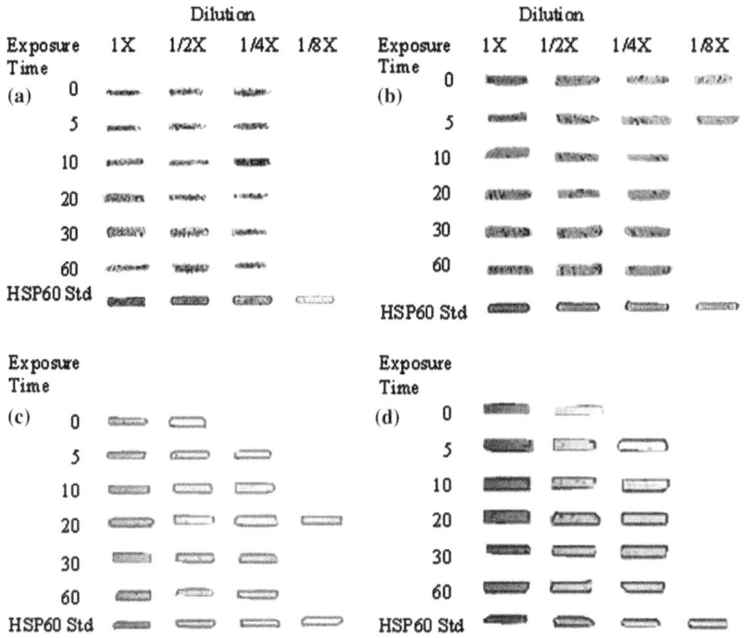

Figure 1. Induction response of HSP60 in *Plationus patulus* exposed to different concentrations of $CuSO_4$ for varying exposure times. Rotifers were exposed to (a) 100, (b) 63, (c) 35, and (d) 20 $\mu g\ l^{-1}$ of Cu. Exposure times were 5, 10, 20, 30, and 60 min. Dilutions were made based on total protein content: 48 μg (1×), 24 μg (1/2×), 12 μg (1/4×), 6 μg (1/8×). HSP60 standard (recombinant rat HSP60) dilutions were included as positive controls. HSP60 standard dilutions corresponded to 480, 240, 120, 60, 30, and 15 ng, respectively.

were noted in response to exposure to high Cu concentrations (Fig. 2).

Effects of combined elements on total protein content

In rotifers exposed to a combination of four elements, total protein content increased independent of their concentrations. When the effects of single and combined elements are compared, we observe that some elements such as As and Cr interacted, resulting in a similar total protein content as that produced by As alone. Similar effects are caused by As–Cu or As–Zn interactions. The opposite effect was observed when Cu interacted with Zn; the amount of protein increased most likely due to the effect of Cu predominating over the effect of Zn.

Effects of combined elements on HSP60 induction

The induction of HSP60 by As at low concentration was only half of that in the control treatment, while Cr at low concentration induced HSP60 (by 4 fold) as compared with the control (Fig. 2). When rotifers were exposed to an As–Cu solution, both at low concentrations, the amount of HSP60 observed was apparently the same as that for unexposed rotifers (Fig. 3). These results may indicate that the effect of As is greater than that of Cr and that As predominates when Cr and As are found together. The amount of HSP60 observed in controls and high As treatments were similar. Rotifers exposed to high level of Cu (50 $\mu g\ l^{-1}$) induced 2 fold more HSP60 than did the control treatment (Fig. 2). When As and Cu were placed together in the medium, the As effect again predominated over Cu since no induction of HSP60 was observed in the double exposure (Fig. 3). Zn by itself, at high concentration (150 $\mu g\ l^{-1}$), doubles the amount of HSP60 observed compared to the control treatment (Figure 2). The presence of Zn and As in the same exposure medium also induced HSP60 in rotifers (Fig. 3). The Zn effect predominated, increasing by 2 fold the amount of HSP60 in exposed rotifers. Cu and Zn may interact to decrease their effect on HSP60 induction by 4 and 2 fold, respectively. No difference in induc-

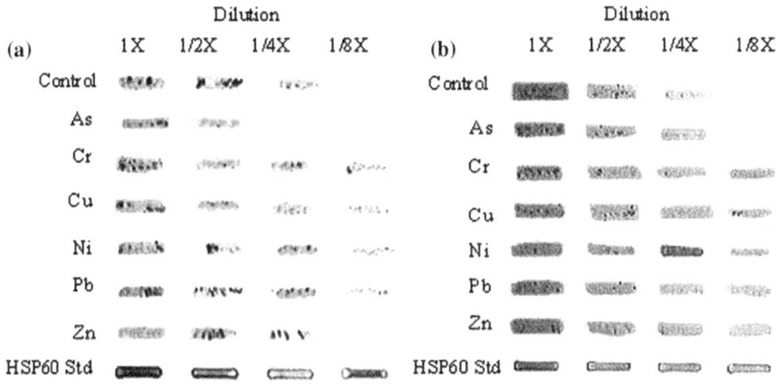

Figure 2. Induction response of HSP60 in *Plationus patulus* exposed to: (A) Single elements at low concentrations (10 $\mu g\ l^{-1}$ for all elements except for Zn [20 $\mu g\ l^{-1}$]), and (B) Single elements at high concentrations (50 $\mu g\ l^{-1}$ for As, Cr, Cu and Ni; 100 $\mu g\ l^{-1}$ for Pb; and 150 $\mu g\ l^{-1}$ for Zn). Samples were diluted in order to discern differences in HSP60 induction. Dilutions were made based on total protein amount: 48 μg (1×), 24 μg (1/2×), 12 μg (1/4×), and 6 μg (1/8×). Control samples were prepared with unexposed rotifers. The amount of HSP60 standard (recombinant rat HSP60) in dilutions corresponded to 480, 240, 120, and 60 ng.

tion was observed among Cu–Zn and control treatments. Figure 3 shows that mixtures containing the same elements at different concentrations produced different responses in HSP60 expression. AsL–CrH–CuH–PbL had the same HSP60 response as the control. Levels of HSP60 were 2 fold lower than the control in all other As–Cr–Cu–Pb treatments.

Some interesting interactions occurred among elements as their number in the media increased (Fig. 3). The inclusion of Cu in presence of As and Cr induced (2 fold) the synthesis of HSP60, however, when Ni was included in the media, the amount of HSP60 remained constant. The interaction of As–Cr–Cu–Ni–Pb increased HSP60 by 2 fold, while the HSP60 response was inhibited when exposure to six elements at low or high concentration (Fig. 3).

Discussion

Since an increase in HSP protein was observed with high temperature exposure, this confirms the characteristic of induction of HSP60 by heat stress (Cochrane et al., 1991). Our findings also agree with those of Wheelock et al. (1999) for *B. plicatilis* in which HSP60 response increased up to 3–4 fold when heat exposure occurred; HSP60 was induced 2–4 fold in *P. patulus* exposed to heat (data not shown).

Once we confirmed heat induced HSP60 in our system, we determined whether induction was related to the type of element, the number of elements in combination, their concentrations or exposure time. HSP60 expression was observed in samples exposed to 20 or 35 $\mu g\ l^{-1}$ of Cu. These results agree with those found by Cochrane et al. (1991) and Wheelock et al. (1999) in Cu exposure studies on *B. plicatilis*, where rotifers were exposed to lower (5 $\mu g\ l^{-1}$) or similar (25–50 $\mu g\ l^{-1}$) $CuSO_4$ concentrations to those tested here. In our case, the HSP60 response differed for all concentrations when the length of exposure time varied. An increase in HSP60 induction was observed in *P. patulus* exposed to low Cu concentrations over time, while HSP60 levels decreased or remained constant over time when high levels of Cu were present in the treatment. Exposure to high $CuSO_4$ concentrations (63 and 100 $\mu g\ l^{-1}$) may alter the physiological function of rotifers resulting in a lowering of HSP60 synthesis and therefore decreased cellular levels of HSP60 (Fig. 1). Although dead rotifers were not observed in samples before homogenization, the physiological effects of high $CuSO_4$ concentrations may be leading to death during longer exposure times. This assumptionn is supported by the observation of Cochrane et al. (1991) who found increased mortality when *B. plicatilis* was exposed to copper for 24 h.

After determining the effects of Cu alone, other elements were tested to determine if HSP60 was also induced. Low Cr, Cu, Ni, Pb (10 $\mu g\ l^{-1}$) and Zn (20 $\mu g\ l^{-1}$) concentrations induced HSP60 production in *P. patulus* (Fig. 2). However, arsenic did not induce HSP60 in rotifers exposed to high

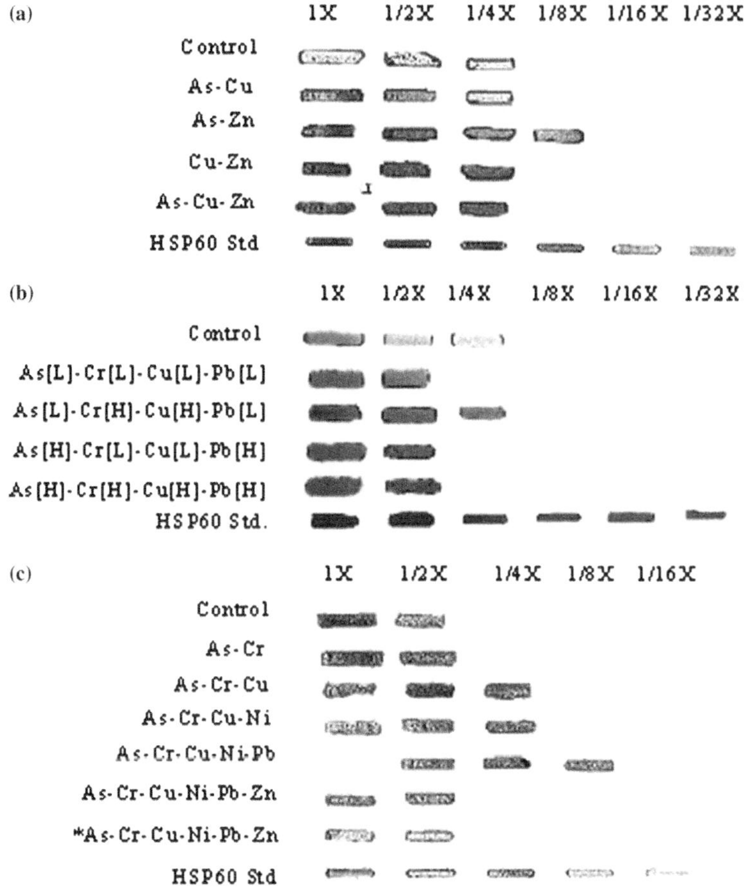

Figure 3. Induction response of HSP60 in *Plationus patulus* exposed to: (a) a combination of two and three elements at high concentrations: As (50 $\mu g\ l^{-1}$), Cu (50 $\mu g\ l^{-1}$), Zn (150 $\mu g\ l^{-1}$); (b) combinations of four elements at different concentrations: low (L) 10 $\mu g\ l^{-1}$ and high (H) to 50 $\mu g\ l^{-1}$ concentrations; and (c) combinations of more than two elements at low (10 $\mu g\ l^{-1}$ for all elements except for Zn [20 $\mu g\ l^{-1}$]) and high (50 $\mu g\ l^{-1}$ for As, Cr, Cu and Ni; 100 $\mu g\ l^{-1}$ for Pb; and 150 $\mu g\ l^{-1}$ for Zn) concentrations. Samples were diluted in order to discern differences in HSP60 induction. Dilutions were made based on total protein amount: 48 μg (1×), 24 μg (1/2×), 12 μg (1/4×), and 6 μg (1/8×). Control samples were prepared with unexposed rotifers. The amount of HSP60 standard (recombinant rat HSP60) in dilutions corresponded to 480, 240, 120, 60, 30, and 15 ng.

concentrations, even though rotifer mortality was not observed. This may be a similar response as seen in the high Cu exposures described above. Low As concentrations yielded less (2 fold) HSP60 induction, while HSP60 expression in high As concentrations was similar to the control.

An interesting finding of our work and others is that under stressful conditions, total protein levels decrease while HSP60 production increases. Wiegant et al. (1998) evaluated the effect of NaA_sO_2 and $CdCl_2$ on various HSPs' synthesis and induction in rat Reuber H35 hepatoma cells, and found they caused a reduction of 20–30% in total protein synthesis. In our case, total protein from rotifers exposed to high As_2O_5 decreased ca. 30% as compared with controls.

HSP60 responses of rotifers exposed to multiple element mixtures depended on type, number and concentration of elements present in the media and exposure time. Induction of HSP60 was observed in six treatments containing two, three, four or five elements. We assumed that expression of HSP60 would increase if two or more elements that singly induce HSP60 were combined, but the presence of additional toxic elements in the environment does not necessarily result in higher levels of HSP60 induction (Fig. 3).

As illustrated in Figure 3, exposure of *P. patulus* to diverse combinations of elements may induce or inhibit HSP60 response. Both effects could be different responses to the toxicity produced by the interactions of elements and their concentrations. According to Wah Chu et al. (2002), effects of combined metals are a result of synergism or neutralization. They describe a synergistic effect as a greater than 25% change in lethality response than the expected single metal response and a neutralizing effect as a change in endpoint response 25% lower than the expected. We observed both synergistic and neutral effects in rotifers, depending on which metal combinations were tested. The effect of binary combinations of As, Cu, and Zn on HSP60 showed that the interaction of As and Cu neutralize the single effect of both elements, while As-Zn act synergistically and increase the amount of HSP60 expressed. Heinz &Eckwert (1997) and Eckwert et al. (1997) studied the HSP60 and 70 responses of *Oniscus asellus* (Isopoda) to binary combinations of metals. They observed that HSP70 response to Zn, Cd and Pb exposure was synergistic when present as binary combinations. Eckwert et al. (1997) found that HSP70 expression was increased with increasing concentration until a maximum induction level was reached (216 mg kg^{-1} Cd and 2518 mg kg^{-1} Zn). Above these concentrations, HSP70 induction response started to decline possibly as a consequence of cellular damage. In our case, HSP60 induction was increased as we increased the number of elements in exposure mixtures without changing their respective concentrations (Fig. 3). Maximum HSP60 induction was produced by the As–Cr–Cu–Ni–Pb combination while the interaction of Zn with the other five elements reduced HSP60 expression. The introduction of a new element may shift HSP60 response due to synergism. In the case of HSP60 expression in *O. asellus*, Eckwert et al. (1997) observed a slight increase in induction as metal concentrations increased, independent of which elements were present in the mixtures. In contrast, an increase in metal concentrations may not increase HSP60 induction in *P. patulus* exposed to various combinations of As–Cr–Cu–Pb (Fig. 3). It is important to mention that the exposure route used by Eckwert et al. (1997) was different than ours. Metal exposure in *O. asellus* was through leaf litter material previously soaked in several metal solutions, while *P. patulus* may directly adsorb heavy metals from the environment. Our findings agree with Cochrane et al. (1991) in that not all toxicants necessarily induce stress protein expression. The results of our study demonstrate that even though total amounts of protein were increased in more than half of the treatments, HSP60 synthesis was not increased in these treatments. This may indicate that other defense mechanisms have been activated, confirming that not all heavy metals will switch on this particular defense mechanism (Locke, 2002). Based on the characteristics of the elements (such as oxidation state) used in the present study, it is possible to separate them in two groups according to their effects on HSP60 induction: (1) those that may compete for active sites in enzymes (Cu, Ni, Pb, Zn) due to their +2 oxidation states, and (2) those related to protein denaturation (As and Cr). The inhibitory effect of As in HSP60 induction and in total amount of protein may be due to protein oxidation since As_2O_5 could be attached to HSP60 and denature available proteins. It may also oxidize unfolded proteins by reacting to HS-protein groups reducing the amount of total protein in the organisms, and consequently, affecting several physiological functions (Friedman, 1973). The effects produced by Zn and Cu interactions may be related to induction of other proteins such as metallothioneins. It is well known that Cu, Ni, and Zn induce metallothionein gene transcription and protein synthesis while Pb is sequestrated by metallothioneins (Prasad, 1993). Finally, a faster response to HSP60 was observed in rotifers exposed to high levels of Cr, Cu, Ni, Pb, and Zn (Fig. 2). HSP60 response to multiple element exposure at low concentrations was higher than in rotifers exposed to high concentrations. It is possible that due to interactions of metals at high concentrations the toxicity increased, inhibiting HSP60 induction. Inhibition or degradation of HSP60 may indicate that the presence of combined elements at certain concentrations produce significant changes at the cellular level. These changes may switch on other defense or survival mechanisms. The effects that As and heavy metals produced on HSP60 and other cellular defense mechanisms may be vital for the survival and development of *P. patulus* populations in contaminated environments.

Conclusions

Physiological responses of *P. patulus* through HSP60 induction to As and heavy metal exposure are related to type and number of elements, concentration, and exposure time. In rotifers exposed to Cu, as metal concentration increases, HSP60 induction decreases. Exposure of rotifers to low and high Cr, Cu, Ni, Pb, and Zn concentrations increased HSP60 induction. Induction of HSP60 synthesis was not observed in rotifers exposed to low or high arsenic concentrations, as well as to some of the element combinations tested in the present study. This may imply that other defense mechanisms have been activated, or that the cellular damage has been irreversible. These mechanisms could also account for the decrease in total protein observed in most high, and some low metal exposure treatments. Future studies will be conducted to integrate these findings to those observed at a population level due to As and heavy metals exposure.

Acknowledgements

This research was supported by CONACYT, National Institutes of Health (NIH)/(NCRR) 5G12RR08124-08 and S06GM8012-33), State of Texas Tobacco Settlement Fund, Sigma Xi Grant-In-Aid of Research, and the UTEP Graduate School. We would like to thank Luisa Arroyo, Teresa de la Mora, Juan Remeriz, and Edith Soto for their invaluable help. Comments by the editor and two anonymous reviewers greatly improved the manuscript.

References

Chen, C. Y., K. B Sillett, C. L Folt, S. L. Whittemore & A. Barchowsky, 1999. Molecular and demographic measurements of arsenic stress in *Daphnia pulex*. Hydrobiologia 401: 229–238.

Choresh, O., E. Ron & Y. Loya, 2001. The 60-kDa heat shock protein (HSP60) of the sea anemone *Anemonia virdis*: a potential early warning system for environmental changes. Marine Biotechnology 3: 501–508.

Cochrane, B. J., R. B. Irby & T. W. Snell, 1991. Effects of copper and tributyltin on stress protein abundance in the rotifer *Brachionus plicatilis*. Comparative Biochemistry and Physiology 98C: 385–390.

Eckwert, H., G. Alberti & H. R. Köhler, 1997. The induction of stress proteins (hsp) in *Oniscus asellus* (Isopoda) as molecular marker of multiple heavy metal exposure: I. Principles and toxicological assessment. Ecotoxicology 6: 249–262.

Fernandez, N. & R. Beiras, 2001. Combined toxicity of dissolved mercury with copper, lead and cadmium on embryogenesis and early larval growth of the *Paracentrotus lividus* sea-urchin. Ecotoxicology 10: 263–271.

Folt, C. L., C. Y. Chen, M. V. Moore & J. Burnaford, 1999. Synergism and antagonism among multiple stressors. Limnology and Oceanography 44: 864–877.

Friedman, M., 1973. Oxidation reactions. The Chemistry and Biochemistry of the Sulfhydryl Group in Aminoacids, Peptides and Proteins. Pergamon Press Ltd. Headington Hill Hall, Oxford: 85.

Gilbert, J. J., 1975. Polymorphism and sexuality in the rotifer *Asplanchna*, with special reference to the effects of prey-size and clonal variation. Archiv für Hydrobiologie 75: 442–483.

Heinz-R. K. & H. Eckwert, 1997. The induction of stress proteins (HSP) in *Oniscus asellus* (Isopoda) as molecular marker of multiple heavy metal exposure: II. Joint toxicity and transfer to field situations. Ecotoxicology 6: 263–274.

Hightower, L. E., 1993. A brief perspective on the heat-shock response and stress proteins. Marine Environmental Research 35: 79–83.

Karouna-Renier, N. K. & J. P. Zehr, 1999. Ecological implications of molecular biomarkers: assaying sub-lethal stress in the midge *Chironomus tentans* using heat shock protein 70 (HSP-70) expression. Hydrobiologia 401: 255–264.

Kruger, N. J., 2002. The Bradford method for protein quantification. In Walker J. M., (ed.), The Protein Protocols Handbook. Humana Press Inc.: 15–21.

Lewis, S., R. D. Handy, B. Cordi, Z. Billinghurst & M. H. Depledge, 1999. Stress proteins (HSPs): methods of detection and their use as an environmental biomarker. Ecotoxicology 8: 351–368.

Locke, M., 2002. Overview of the stress response. In Locke M. & Noble E. G., (ed.), Exercise and Stress Response: The Role of Stress Proteins. CRC Press LLC, 1–11.

Morimoto, R., A. Tissières & C. Georgopoulus, 1994. Progress and perspectives on the biology of heat shock proteins and molecular chaperones. In Morimoto, R., A Tissières, & C. Georgopoulus (eds), The Biology of Heat Shock Proteins and Molecular Chaperones. Cold Springs Harbor Laboratory Press: 1–29.

Prasad, A. S., 1993. Biochemistry of metallothionein. In Frieden, Biochemistry of Zinc. Plenium Press New York: 77–91.

Snell T. W. & G. Persoone, 1989. Acute toxicity bioassays using rotifers. I. A test for brackish and marine environments with *Brachionus plicatilis*. Aquatic Toxicology 14: 65–80.

Tedengren, M., B. Olson, B. Bradley & L. Zhou, 1999. Heavy metal uptake, physiological response and survival of the blue mussel (*Mytilus edulis*) from marine and brackish waters in relation to the induction of heat-shock protein 70. Hydrobiologia 393: 261–269.

Wah Chu, K. & K. L. Chow, 2002. Synergistic toxicity of multiple heavy metals is revealed by a biological assay using a nematode and its transgenic derivative. Aquatic Toxicology 61: 53–64.

Wheelock, C. E., M. F. Wolfe, H. Olsen, M. L. Tjeerdema & M. L. Sowby, 1999. Hsp60-induced tolerance in the rotifer *Brachionus plicatilis* exposed to multiple environmental contaminants. Archives of Environmental Contamination and Toxicology 36: 281–287.

Wiegant, F. A. C., N. Spieker & R. van Wijk, 1998. Stressor-specific enhancement of hsp induction by low doses of stressors in conditions of self- and cross-sensitization. Toxicology 127: 107–119.

Hydrobiologia (2005) 546:587–595
A. Herzig, R.D. Gulati, C.D. Jersabek & L. May (eds.) Rotifera X: Rotifer Research: Trends, New Tools and Recent Advances
DOI 10.1007/s10750-005-4087-4

Subject Index

A

B

C

D

E

F

G

H

I

J

K

L

M

N

O

P

Q

R

S

T

U

V

W

Y

Z

Hydrobiologia (2005) 546:597–601
A. Herzig, R.D. Gulati, C.D. Jersabek & L. May (eds.) Rotifera X: Rotifer Research: Trends, New Tools and Recent Advances
DOI 10.1007/s10750-005-0662-y

Rotifer species index

(infrasubspecific categories not specified)

A

B

C

D

E

F

G

H

I

K

L

M

N

O

W

Z